THIRD EDITION

Laboratory Exercises in ANATOMY & PHYSIOLOGY with Cat Dissections

Gerard J. Tortora

Bergen Community College
Paramus, New Jersey

Robert B. Tallitsch

Augustana College
Rock Island, Illinois

Nicholas P. Anagnostakos

late of Bergen Community College
Paramus, New Jersey

MACMILLAN PUBLISHING COMPANY
New York
COLLIER MACMILLAN PUBLISHERS
London

Printed in the United States of America

Macmillan Publishing Company
866 Third Avenue, New York, New York 10022

Collier Macmillan Canada, Inc.

Printing: 4 5 6 7 8 Year: 1 2 3 4 5 6 7 8

CONTENTS

PREFACE

Laboratory Exercises in Anatomy and Physiology with Cat Dissections, Third Edition, has been written to guide students in the laboratory study of introductory anatomy and physiology. The manual was written to accompany most of the leading anatomy and physiology textbooks.

COMPREHENSIVENESS

This manual examines virtually every structure and function of the human body that is typically studied in an introductory anatomy and physiology course. Because of its detail, the need for supplemental handouts is minimized; the manual is a strong teaching device in itself.

USE OF THE SCIENTIFIC METHOD

Anatomy (the science of structure) and physiology (the science of function) cannot be understood without the practical experience of laboratory work. The exercises in this manual challenge students to understand the way scientists work by asking them to make microscopic examinations and evaluations of cells and tissues, to observe and interpret chemical reactions, to record data, to make gross examinations of organs and systems, to dissect, and to conduct physiological laboratory work and interpret and apply the results of this work.

ILLUSTRATIONS

The manual contains a large number and variety of illustrations. The illustrations of the body systems of the human have been carefully drawn to exhibit structures that are essential to students' understanding of anatomy and physiology. Numerous photographs, photomicrographs, and scanning electron micrographs are presented to show students how the structures of the body actually look. We feel that this laboratory manual has better and more complete illustrations than any other anatomy and physiology manual.

IMPORTANT FEATURES

Among the key features of this manual are (1) dissection of the white rat, cat, and selected mammalian organs; (2) numerous physiological experiments; (3) emphasis on the study of anatomy through histology; (4) lists of appropriate terms accompanying drawings and photographs to be labeled; (5) inclusion of numerous scanning electron micrographs and specimen photos; (6) phonetic pronunciations for many anatomical and physiological terms; (7) diagrams of commonly used laboratory equipment; (8) laboratory report questions and reports at the end of each exercise that can be filled in, torn out, and turned in for grading if the instructor so desires; and (9) three appendixes dealing with units of measurement, a periodic table of elements, and eponyms used in the laboratory manual.

NEW TO THIRD EDITION

Numerous changes have been made in the third edition of this manual in response to suggestions from instructors and students. We have added many new physiology experiments, numerous new line drawings, including ones that accompany the photomicrographs of cell division and epithelial and connective tissues, and additional phonetic pronunciations. The principal additions and revisions in each exercise are described below.

Exercise 1, Microscopy, now contains a section on microtomography. Exercise 2, Introduction to the Human Body, has new line art and a detailed procedure for dissection of the white rat. Exercise 3, Cells, is expanded to include a new section on facilitated diffusion. In Exercise 4, Tissues, new line art is added adjacent to the photomicrographs of epithelial and connective tissues. A revised discussion of the histology of the epidermis has been added to Exercise 5, Integumentary System.

Exercise 6, Osseous Tissue, contains a new experiment on the chemistry of bone and a table of bone markings. Exercise 7, Bones, includes a new table describing foramina of the skull. Exercise 8, Articulations, has an expanded section on types of synovial joints.

Exercise 9, Muscle Tissue, contains a new section on electromyography. Exercise 10, Skeletal Muscles, has been greatly expanded. New are sections on how skeletal muscles produce movement. Each muscle table also contains an overview of the muscles under consideration. In addition, there are new tables on muscles of the floor of the oral cavity, hand, and foot, and several new pieces of line art.

A major change in this third edition is the placement of Exercise 11, Surface Anatomy, following skeletal muscles. This exercise can be studied before, in conjunction with, or after the study of various body systems as a review.

Exercise 12, Nervous Tissue and Physiology, contains a new section on the reflex arc. Exercise 13, Nervous System, now contains new sections on nerve plexuses, spinal cord tracts, the flow of cerebrospinal fluid (CSF), sensory and motor areas of the cerebral cortex, the electroencephalogram (EEG), and the autonomic nervous system (ANS). In Exercise 14, Sensations and Sensory Pathways, there are new sections on adaptation of thermoreceptors, referred pain, astigmatism, binocular vision, afterimages, colorblindness, otoscopy, sound localization, audiometry, and sensory and motor pathways. Exercise 15, Endocrine System, contains a new section on hyperinsulinemia in fish.

Exercise 16, Blood, has a new section on the chemistry of plasma and oxyhemoglobin saturation. Exercise 17, Heart, now contains a new section on cardiac (coronary) circulation. In Exercise 18, Blood Vessels, there are new exercises on the scheme of circulation. Exercise 19, Cardiovascular Physiology, includes a new series of experiments on the cardiac cycle, microcirculation, mean electrical axis, and peripheral blood flow. Exercise 20, Lymphatic System, has been expanded.

Exercise 21, Respiratory System, has new sections on observing living lung tissue, mechanism of pulmonary ventilation, and the chemical regulation of respiration. Deglutition and movements of the gastrointestinal tract are new topics included in Exercise 22, Digestive System. The Urinary System, Exercise 23, has new experiments on chloride and sodium chloride in urine and renal blood flow.

In Exercise 24, Reproductive Systems, several new illustrations have been added. Exercise 25, Development, contains expanded discussions of spermatogenesis and oogenesis.

Exercise 26, Genetics, contains new sections dealing with Barr Bodies and observing phenotypes.

CHANGES IN TERMINOLOGY

In recent years, the use of eponyms for anatomical terms has been minimized or eliminated. Anatomical eponyms are terms named after various individuals. Examples include Fallopian tube (after Gabriello Fallopio) and Eustachian tube (after Bartolommeo Eustachio).

Anatomical eponyms are often vague and nondescriptive and do not necessarily mean that the person whose name is applied contributed anything very orginal. For these reasons, we have also decided to minimize their use. However, because some still prevail, we have provided eponyms, in parentheses, after the first reference in each chapter to the more acceptable synonym. Thus, you will expect to see terms such as ***uterine (Fallopian) tube*** or ***auditory (Eustachian) tube.*** See Appendix C.

INSTRUCTOR AIDS

A complimentary instructor's guide to accompany the manual is available from the publisher. This comprehensive guide contains: (1) a listing of materials needed to complete each exercise, (2) suggested audiovisual materials, (3) answers to illustrations and questions within the exercises, and (4) answers to laboratory report questions.

ACKNOWLEDGMENTS

We wish to thank the following reviewers for their outstanding contributions to the third edition of *Laboratory Exercises in Anatomy and Physiology with Cat Dissections*. They have shared their extensive knowledge, teaching experience, and concern for students with us, and their feedback has been invaluable: Wayne W. Carley, Lamar University; Elden W. Martin, Bowling Green State University; and Richard E. Welton, Southern Oregon State College. We also wish to thank Mary Dersch, whose outstanding medical illustrations have greatly improved the pedagogy and visual appeal of the manual, and Kristin Watts at Macmillan, who located several photographs and obtained the permissions.

As the acknowledgements indicate, the participation of many individuals of diverse talent and expertise is required in the production of a laboratory manual of this scope and complexity. For this reason, readers and users are invited to send their reactions and suggestions to me so that plans can be formulated for subsequent editions.

Gerard J. Tortora
Natural Sciences and Mathematics, S229
Bergen Community College
400 Paramus Road
Paramus, NJ 07652

LABORATORY SAFETY*

On August 21, 1987, the Centers for Disease Control (CDC) published a supplement to the *Morbidity and Mortality Weekly Report* (Vol. 36, no. 2S) entitled "Recommendations for Prevention of HIV Transmission in Health-Care Settings." The CDC guidelines recommend precautions to protect health-care workers from exposure to human immunodeficiency virus (HIV), the causative agent of acquired immune deficiency syndrome (AIDS). These guidelines are presented to reaffirm the basic principles involved in the transmission of not only the AIDS virus, but also any disease-producing organism.

Based on the CDC guidelines for health-care workers, as well as standard additional laboratory precautions and procedures, the following list has been developed for your safety in the laboratory. Although specific cautions and warnings concerning laboratory safety are indicated throughout the manual, you should read the following ***before*** performing any experiments.

A. General Safety Precautions and Procedures

1. Please arrive on time, since laboratory directions and procedures are given at the beginning of the laboratory period.

2. Read all experiments before you come to class to be sure that you understand all the procedures and safety precautions. Ask your instructor if you do not understand exactly what has to be done. Do not improvise any procedure.

3. Protective eyewear and laboratory coats or aprons must be worn by all students performing or observing experiments.

4. Do not perform any unauthorized experiments.

5. Do not bring any unnecessary items to the laboratory and do not place any personal items (pocketbooks, bookbags, coats, umbrellas, etc.) on the laboratory table or at your feet.

6. Make sure all apparatus is supported and squarely on the table.

7. Tie back long hair in order to prevent it from becoming a laboratory fire hazard.

8. Do not wear contact lenses when working with reagents.

9. Never remove equipment, chemicals, biological materials, or any other materials from the laboratory.

10. Do not operate any equipment until you are instructed in its proper use. If you are unsure of the procedures, ask the instructor.

11. Dispose of chemicals, biological materials, used apparatus, and waste materials according to the directions of your instructor. Not all liquids are to be disposed of in the sink.

12. Some exercises in the laboratory manual are designed to induce some degree of cardiovascular stress. Students should not participate in these exercises if they are pregnant or have hypertension or any other known or suspected condition that might compromise health. Before you perform any of these exercises, check with your physician.

13. Do not put anything in your mouth while in the laboratory. Never eat, drink, taste chemicals, lick labels, smoke, or store food in the laboratory.

14. Your instructor will show you the location of emergency equipment such as fire extinguishers, fire blankets, eye wash stations, and first-aid kits. Memorize their locations and know how to use them.

15. Wash your hands before leaving the laboratory and remove protective clothing, such as laboratory coats or aprons and protective eyewear.

B. Precautions Related to Working with Blood, Blood Products, or Other Body Fluids

1. Only single-use, disposable lancets and needles should be used. Never recap, bend, or break the lancet once it has been used. Place used lancets, needles or other sharp instruments in a *fresh* 1:10 dilution of household bleach (sodium hypochlorite) or commercially available disinfectant with tuberculocidal activity and then dispose of the instruments in a puncture-proof container. The fresh household bleach solution should be prepared for *each* laboratory session.

*The authors and publisher urge consultation with each instructor's institutional policies concerning laboratory safety and first aid procedures.

2. A laboratory disinfectant should be used to clean laboratory surfaces *before* and *after* procedures, and should be available for quick cleanup of any blood spills. A *fresh* 1:10 dilution of household bleach or commercially prepared disinfectant with tuberculocidal activity can be used. The fresh household bleach solution should be prepared for *each* session.

3. All reusable instruments, such as hemocytometers, well slides, and reusable pipettes should be disinfected with a *fresh* 1:10 solution of household bleach or a commercially prepared disinfectant with tuberculocidal activity and thoroughly washed with soap and hot water. The fresh household bleach solution should be prepared for *each* laboratory session.

4. Mouth pipetting should never be done. Mechanical pipetting devices should be used for the manipulation of all liquids in the laboratory.

5. All procedures and manipulations that have a high potential for creating aerosols or infectious droplets (such as centrifuging, sonicating, and blending) should be performed carefully. In such instances, a biological safety cabinet or other primary containment device is required.

C. Precautions Related to Working with Reagents

1. Use extreme care when working with reagents. Should any reagents make contact with your eyes, flush with water for fifteen minutes; or, if they make contact with your skin, flush with water for five minutes. Notify your instructor immediately should a reagent make contact with your eyes or skin and seek immediate medical attention.

2. Report all accidents to your instructor, no matter how minor they may appear.

3. The room in which you are working should be well-ventilated when you are working with chemicals or preserved specimens. Avoid breathing fumes for any extended period of time.

4. Never point the opening of a test tube containing a reacting mixture (especially when heating) toward yourself or another person.

5. Exercise care in noting the odor of fumes. Use "wafting" if you are directed to note an odor.

6. Do not force glass tubing or a thermometer into rubber stoppers. Lubricate the tubing and introduce it gradually and gently into the stopper. Protect your hands with toweling when inserting the tubing or thermometer into the stopper.

7. Never heat a flammable liquid over or near an open flame.

8. Use only glassware marked Pyrex or Kimex. Other glassware may shatter when heated. Handle hot glassware with test tube holders.

9. If you have to dilute an acid, always add acid (AAA) to water.

10. Don't use your fingers as a stopper for a test tube or bottle when shaking the tube to mix its contents.

11. Read the label on any chemical twice before using it.

12. Replace caps or stoppers on bottles immediately after use. Return spatulas to their correct place immediately after use and do not mix them up.

13. Mouth pipetting should never be done. Mechanical pipetting devices should be used for the manipulation of all liquids in the laboratory.

D. Precautions Related to Dissection

1. The room in which you are working should be well-ventilated when you are working with chemicals or preserved specimens. Avoid breathing fumes for any extended period of time.

2. In order to reduce the irritating effects of chemical preservatives to your skin, eyes, and nose, soak or wrap your specimen in a substance such as "Biostat." If this is not available, hold your specimen under running water for several minutes to wash away excess preservative and dilute the remaining preservative.

3. When dissecting, there is always the possibility of skin cuts or punctures from dissecting equipment or the specimens themselves, such as the teeth or claws of animals. Should you sustain a cut or puncture in this manner, wash your hands with a disinfectant soap, notify your instructor, and seek immediate medical attention to decrease the possibility of infection.

4. When cleaning dissecting instruments, always hold the sharp edges away from you.

PRONUNCIATION KEY

One of the unique features of this revised manual is the addition of phonetic pronunciations for many anatomical and physiological terms. The pronunciations are given in parentheses immediately after the particular term is introduced. The following key explains the essential features of the pronunciations.

1. The syllable with the strongest accent appears in capital letters, for example, bilateral (bī-LAT-er-al) and diagnosis (dī-ag-NŌ-sis).

2. A secondary accent is denoted by a single quote mark ('), for example, constitution (kon'-sti-TOO-shun) and physiology (fiz'-ē-OL-ō-jē). Additional secondary accents are also noted by a single quote mark, for example, decarboxylation (dē'-kar-bok'-si-LĀ-shun).

3. Vowels marked with a line above the letter are pronounced with the long sound as in the following common words:

ā as in *māke*	*ī* as in *ivy*
ē as in *bē*	*ō* as in *pole*

4. Unmarked vowels are pronounced with the short sound, as in the following words:

e as in *bet*	*o* as in *not*
i as in *sip*	*u* as in *bud*

5. Other phonetic symbols are used to indicate the following sounds:

a as in *above*	*yoo* as in *cute*
oo as in *soo*	*oy* as in *oil*

COMMONLY USED LABORATORY EQUIPMENT

Beaker

Erlenmeyer Flask

Florence Flask

Funnel

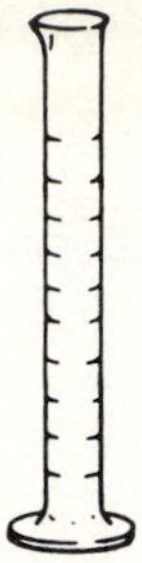
Graduated Cylinder

Crucible Cover

Crucible

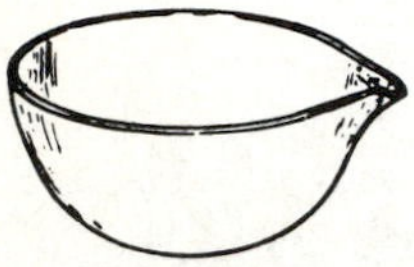
Evaporating Dish

Mortar and Pestle

Watch Glass

Stirring Rod

Test Tube

Test Tube Brush

Test Tube Holder

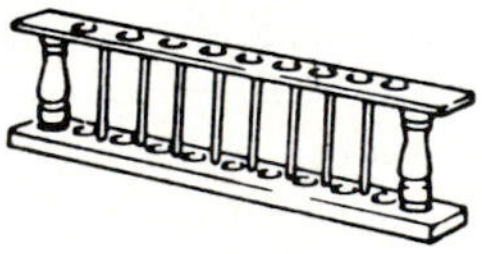
Test Tube Rack

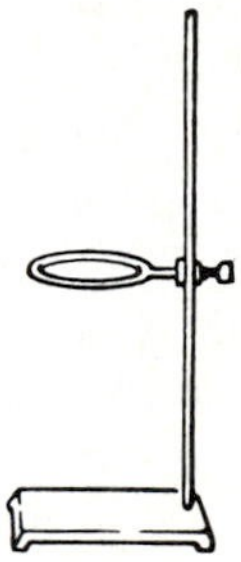
Ring Stand and Ring

Pinch Clamp

Utility Clamp

Tripod

Clay Triangle

Wire Gauze

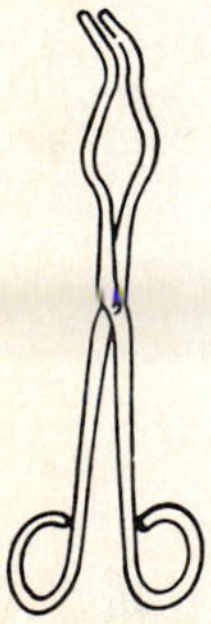
Crucible Tongs

Beaker Tongs

Forceps

Medicine Dropper

Nichrome Wire

Spatula

MICROSCOPY 1

NOTE: Before you begin any laboratory exercises in this manual, please read the section on LABORATORY SAFETY on page xi.

One of the most important instruments that you will use in your anatomy and physiology course is a compound light microscope. In this instrument, the lenses are arranged so that images of objects too small to be seen with the naked eye can become highly magnified, that is, apparent size can be increased, and their minute details revealed. Before you actually learn the parts of a compound light microscope and how to use it properly, discussion of some of the principles employed in light microscopy (mī-KROS-kō-pē) will be helpful. Later in this exercise, some of the principles used in electron microscopy will be discussed.

A. COMPOUND LIGHT MICROSCOPE

A ***compound light microscope*** uses two sets of lenses, ocular and objective, and employs light as its source of illumination. Magnification is achieved as follows: Light rays from an illuminator are passed through a condenser, which directs the light rays through the specimen under observation; from here, light rays pass into the objective lens, the magnifying lens that is closest to the specimen; the image of the specimen then forms on a mirror and is magnified again by the ocular lens.

A general principle of microscopy is that the shorter the wavelength of light used in the instrument, the greater the resolution. ***Resolution***, or ***resolving power***, is the ability of the lenses to distinguish fine detail and structure, that is, to distinguish between two points as separate objects. As an example, a microscope with a resolving power of 0.3 micrometers (mī-KROM-e-ters), symbolized μm[1], is capable of distinguishing two points as separate objects if they are at least 0.3 μm apart. The light used in a compound light microscope has a relatively long wavelength and cannot resolve structures smaller than 0.3 μm. This fact, as well as practical considerations, means that even the best compound light microscopes can magnify images only about 2000 times.

A ***photomicrograph*** (fō-tō-MĪ-krō′-graf), a photograph of a specimen taken through a compound light microscope, is shown in Figure 4.1. In later exercises you will be asked to examine photomicrographs of various specimens of the body before you actually view them yourself through the microscope.

1. Parts of the Microscope

Carefully carry the microscope from the cabinet to your desk by placing one hand around the arm and the other hand firmly under the base. Gently place it on your desk, directly in front of you, with the arm facing you. Locate the following parts of the microscope and, as you read about each part, label Figure 1.1 by placing the correct numbers in the spaces next to the list of terms that accompanies the figure.

Base—The bottom portion on which the microscope rests.

Body tube—The portion that receives the ocular.

[1] 1 μm = 0.000001 or 10^{-6} m. See Appendix A.

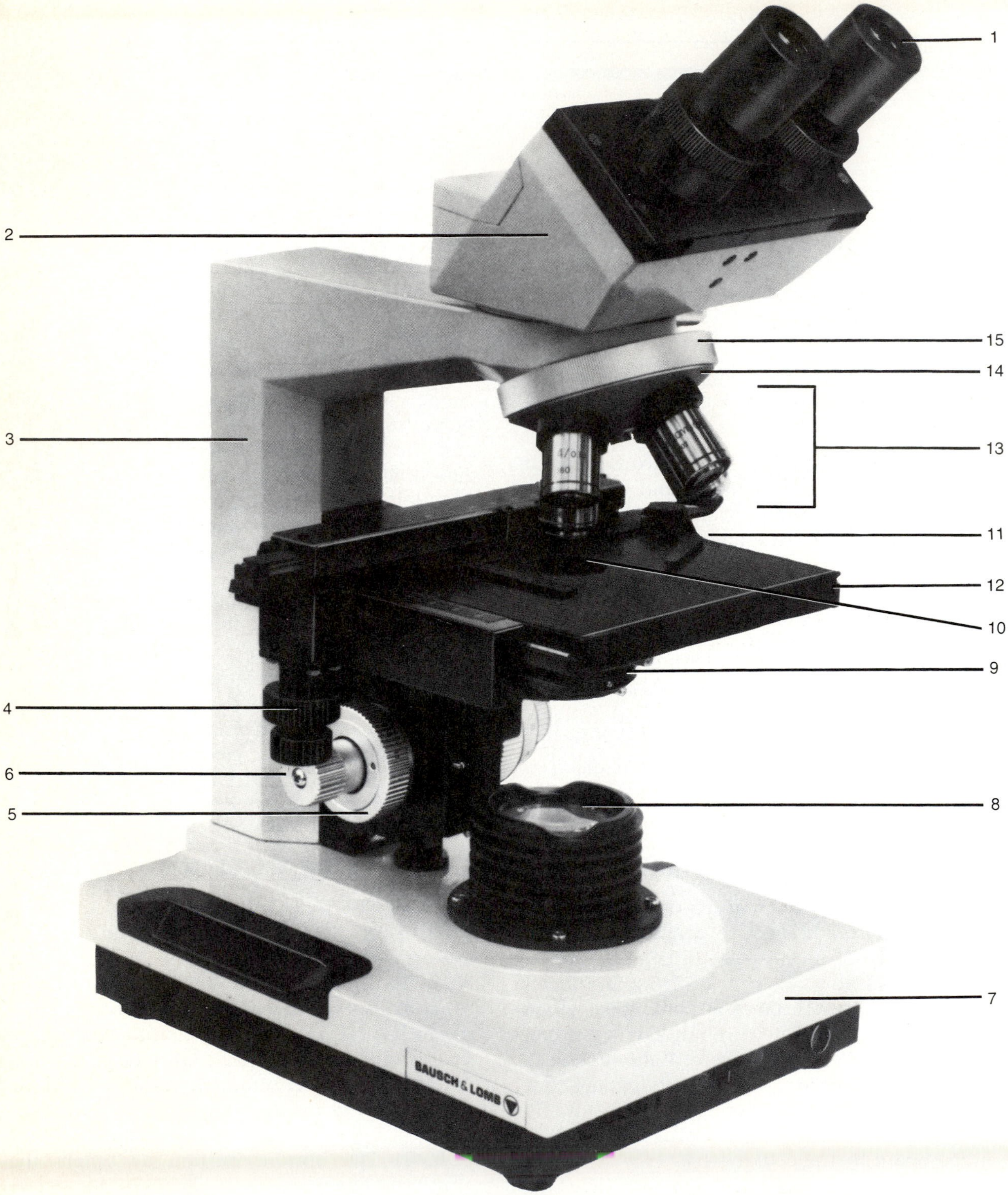

___ Arm
___ Base
___ Body tube
___ Coarse adjustment knob
___ Condenser
___ Diaphragm
___ Fine adjustment knob
___ Substage lamp
___ Mechanical stage knob
___ Nosepiece
___ Objectives
___ Ocular
___ Revolving nosepiece
___ Stage
___ Stage clip of mechanical stage

FIGURE 1.1 Galen II microscope.

Arm—Angular or curved part of the frame.

Inclination joint—A movable hinge in some microscopes that allows the instrument to be tilted to a comfortable viewing position.

Stage—A platform on which slides or other objects to be studied are placed. The opening in the center, called the ***stage opening***, allows light to pass from below through the object being examined. Some microscopes have a ***mechanical stage***. An adjustor knob below the stage moves the stage forward and backward and from side to side. With a mechanical stage, the slide and the stage move simultaneously. A mechanical stage permits a smooth, precise movement of a slide. Sometimes a mechanical stage is fitted with calibrations that permit the numerical "mapping" of a slide.

Stage (spring) clips—Two clips mounted on the stage that hold slides securely in place.

Substage lamp—The source of illumination for some light microscopes with a built-in lamp.

Mirror—A feature found in some microscopes below the stage. The mirror directs light from its source through the stage opening and through the lenses. If the light source is built in, a mirror is not necessary.

Diaphragm (DĪ-a-fram)—A device located between the condenser (a lens below the stage opening) and stage that regulates light intensity passing through the condenser and lenses to the observer's eyes. Such regulation is needed because transparent or very thin specimens cannot be seen in bright light. One of two types of diaphragms is usually used. An ***iris diaphragm***, as found in cameras, is a series of sliding leaves that vary the size of the opening and thus the amount of light entering the lenses. The leaves are moved by a ***diaphragm lever*** to regulate the diameter of a central opening. A ***disc diaphragm*** consists of a plate with a graded series of holes, any of which can be rotated into position.

Condenser—A lens located beneath the stage opening that concentrates the light beam on the specimen.

Condenser adjustment knob—A knob that functions to raise and lower the condenser. In its highest position, it allows full illumination and thus can be used to adjust illumination.

Coarse adjustment knob—A usually larger knob that raises and lowers the body tube (or stage) to bring a specimen into view.

Fine adjustment knob—A usually smaller knob found below or external to the coarse adjustment knob and used for fine or final focusing. Some microscopes have both coarse and fine adjustment knobs combined into one.

Nosepiece—A plate, usually circular, at the bottom of the body tube.

Revolving nosepiece—The lower, movable part of the nosepiece that contains the various objective lenses.

Scanning objective—A lens, marked 5× on most microscopes (× means the same as times); it is the shortest objective and is not present on all microscopes.

Low-power objective—A lens, marked 10× on most microscopes; it is the next shortest objective.

High-power objective—A lens, marked 43× or 45× on most microscopes; also called a ***high-dry objective***; it is the second longest objective.

Oil-immersion objective—A lens, marked 100× on most microscopes and distinguished by an etched colored circle (special instructions for this objective are discussed later); it is the longest objective.

Ocular (eyepiece)—A removable lens at the top of the body tube, marked 10× on most microscopes. An ocular is sometimes fitted with a pointer or measuring scale.

2. Rules of Microscopy

You must observe certain basic rules at all times to obtain maximum efficiency and provide proper care for your microscope.

1. Keep all parts of the microscope clean, especially the lenses of the ocular, objectives, condenser, and mirror. *You should use the special lens paper that is provided and never use paper towels or cloths, because these tend to scratch the delicate glass surfaces*. When using lens paper, use the same area on the paper only once. As

you wipe the lens, change the position of the paper as you go.

2. Do not permit the objectives to get wet, especially when observing a ***wet mount***. You must use a ***cover slip*** when you examine a wet mount or the image becomes distorted.

3. Consult your instructor if any mechanical or optical difficulties rise. *Do not try to solve these problems yourself.*

4. Keep *both* eyes open at all times while observing objects through the microscope. This is difficult at first, but with practice becomes natural. This important technique will help you to draw and observe microscopic specimens without moving your head. Only your eyes will move.

5. Always use either the scanning or low-power objective first to locate an object; then, if necessary, switch to a higher power.

6. If you are using the high-power or oil-immersion objectives, *never focus using the coarse adjustment knob*. The distance between these objectives and the slide, called ***working distance***, is very small and you may break the cover slip and the slide and scratch the lens.

7. When looking through the microscope, *never focus downward*. By observing from one side you can see that the objectives do not make contact with the cover slip or slide.

8. Make sure that you raise the body tube before placing a slide on the stage or before removing a slide.

3. Setting up the Microscope

PROCEDURE

1. Place the microscope on the table with the arm toward you and with the back of the base at least 1 inch (in.) from the edge of the table.

2. Position yourself and the microscope so that you can look into the ocular comfortably.

3. Wipe the objectives, the top lens of the eyepiece, the condenser, and the mirror with lens paper. Clean the most delicate and the least dirty lens first. Apply xylol or ethanol to the lens paper only to remove grease and oil from the lenses and microscope slides.

4. Position the low-power objective in line with the body tube. When it is in its proper place, it will click. Lower the body tube using the coarse adjustment knob until the bottom of the lens is approximately 1/4 in. from the stage.

5. Admit the maximum amount of light by opening the diaphragm, if it is an iris, or turning the disc to its largest opening.

6. Place your eye to the eyepiece (ocular), and adjust the light. When a uniform circle (the ***microscopic field***) appears without any shadows, the microscope is ready for use.

4. Using the Microscope

PROCEDURE

1. Using the coarse adjustment knob, raise the body tube to its highest fixed position.

2. Make a temporary mount using a single letter of newsprint, or use a slide that has been specially prepared with a letter, usually the letter "e." If you prepare such a slide, cut a single letter—"a," "b," or "e"—from the smallest print available and place this letter in the correct position to be read with the naked eye. Your instructor will provide directions for preparing the slide.

3. Place the slide on the stage, making sure that the letter is centered over the opening in the stage, directly over the condenser. Secure the slide in place with the stage clips.

4. Align the low-power objective with the body tube.

5. Lower the body tube as far as it will go *while you watch it from the side*, taking care not to touch the slide. The tube should reach an automatic stop that prevents the low-power objective from hitting the slide.

6. While looking through the eyepiece, turn the coarse adjustment knob counterclockwise, raising the body tube. Watch for the object to suddenly appear in the microscopic field. If it is in proper focus, the low-power objective is about 1/2 in. above the slide. When focusing, always *raise* the body tube.

7. Use the fine adjustment knob to complete the focusing; you will usually use a counterclockwise motion once again.

8. Compare the orientation of the letter as originally seen with its appearance under the microscope.

Has the orientation of the letter been changed?

9. While looking at the slide through the ocular, change the position of the slide by using your thumbs, or, if the microscope is equipped with them, the mechanical stage knobs. This exercise teaches you to move your material in various directions quickly and efficiently.

In which direction does the letter move when you move the slide to the left? ______

This procedure, "scanning" a slide, will be useful for examining living objects and for centering objects so you can observe them easily.

Make a drawing of the letter as it appears under low power in the space below on the left side.

Drawing of letter as seen under low power | *Drawing of letter as seen under high power*

10. Change your magnification from low to high power by carrying out the following steps:

a. Place the letter in the center of the field under low power. Centering is important because you are now focusing on a smaller area of the microscopic field. As you will see, microscopic field size decreases with higher magnification.

b. Make sure the illumination is at its maximum. Illumination must be increased at higher magnifications because the amount of light entering the lens decreases as the size of the objective lens increases.

c. The letter should be in focus, and if the microscope is ***parfocal*** (meaning that when clear focus has been attained using any objective at random, revolving the nosepiece results in a change in magnification but leaves the specimen still in focus), the high-power objective can be switched into line with the body tube without changing focus. If it is not completely in focus after switching the lens, a slight turn of the fine adjustment knob will focus it.

d. If your microscope is not parfocal, observe the stage from one side and carefully switch the high-power objective in line with the body tube.

e. While still observing from the side and using the coarse adjustment knob, *carefully* lower the objective until it almost touches the slide.

f. Look through the ocular and focus up slowly. Finish focusing by turning the fine adjustment knob.

g. If your microscope has an oil-immersion objective, you must follow special procedures. Place a drop of special ***immersion oil*** directly over the letter on the microscope slide, and lower the oil-immersion objective until it just contacts the oil. If your microscope is parfocal, you do not have to raise or lower the objectives. For example, if you are using the high-power objective and the specimen is in focus, just switch the high-power objective out of line with the body tube. Then add the oil and switch the oil-immersion objective into position; the specimen should be in focus. The same holds true when you switch from low power to high power. The special light-transmitting properties of the oil are such that light is refracted (bent) toward the specimen, permitting the use of powerful objectives in a relatively narrow field of vision. This objective is extremely close to the slide being examined, so when it is in position take precautions *never to focus downward* while you are looking through the eyepiece. Whenever you finish using immersion oil, be sure to saturate a piece of lens paper with xylol or alcohol and clean the oil-immersion objective and the slide if it is to be used again.

Is as much of the letter visible under high power as under low power? Explain. ______

Make a drawing of the letter as it appears under high power in the space provided on the right side following step 9.

11. Now select a prepared slide of three different colored threads. Examination will show that a specimen mounted on a slide has depth as well as length and width. At lower magnification the amount of depth of the specimen that is clearly in focus, the depth of field, is greater than that at higher magnification. You must focus at different depths to determine the position (depth) of each thread.

After you make your observation under low power and high power, answer the following questions about the location of the different threads:

What color is at the bottom, closest to the slide? ____________________

On top, closest to the cover slip? ____________

In the middle? ____________________

12. Your instructor might want you to prepare a wet mount as part of your introduction to microscopy. If so, the directions are given in Exercise 4, A.6, on page 51.

13. When you are finished using the microscope,

a. Remove the slide from the stage.

b. Clean all lenses with lens paper.

c. Align the mechanical stage so that it does not protrude.

d. Leave the scanning or low-power objective in place.

e. Lower the body tube as far as it will go.

f. Wrap the cord according to your instructor's directions.

g. Replace the dust cover or place the microscope in a cabinet.

5. Magnification

The total magnification of your microscope is calculated by multiplying the magnification of the ocular by the magnification of the objective used. Example: An ocular of 10× used with an objective of 5× gives a total magnification of 50×. Calculate the total magnification of each of the objectives on your microscope:

1. Ocular 10 × Objective 15 = 250
2. Ocular ____ × Objective ____ = ____
3. Ocular ____ × Objective ____ = ____
4. Ocular ____ × Objective ____ = ____

B. ELECTRON MICROSCOPE

Examination of specimens smaller than 0.3 μm requires an ***electron microscope***, an instrument with a much greater resolving power than a compound light microscope. An electron microscope uses a beam of electrons instead of light. Electrons travel in waves just as light does. Instead of glass lenses, magnets are used in an electron microscope to focus a beam of electrons through a vacuum tube onto a specimen. Since the wavelength of electrons is about 1/100,000 that of visible light, the resolving power of very sophisticated transmission electron microscopes (described shortly) is close to 10 nanometers (nm). 1 nm = 0.000,000,001 m or 10^{-9} m. See Appendix A. Most electron microscopes have a working resolving power of about 100 nm and can magnify images up to 200,000×.

Two types of electron microscopes are available. In a ***transmission electron microscope***, a finely focused beam of electrons passes through a specimen, usually ultrathin sections of material. The beam is then refocused, and the image reflects what the specimen has done to the transmitted electron beam. The image may be used to produce what is called a ***transmission electron micrograph*** (see Figure 6.2a). The method is extremely valuable in providing details of the interior of specimens at different layers but does not give a three-dimensional effect. Such an effect can be obtained with a ***scanning electron microscope***. With this instrument, a finely focused beam of electrons is directed over the specimen and then reflected from the surface of the specimen onto a television-like screen or photographic plate. The photograph produced from an image generated in this manner is called a ***scanning electron micrograph*** (see Figure 6.2b). Scanning electron micrographs commonly magnify specimens up to 10,000× and are especially useful in studying surface features of specimens.

The one drawback of the electron microscope is the fact that specimens must be killed because they are placed in a vacuum. With a process called ***microtomography***, three-dimensional images of *living* cells can be produced. Through microtomography, scientists hope to study how normal and abnormal cells move, reproduce, and grow. The process will also be used to follow developmental sequences in growing embryos. Other very important applications are observing the effects of drugs and cancer-causing substances on living cells.

Note: A section of ***LABORATORY REPORT QUESTIONS*** for each exercise is located at the end of each exercise. These questions can be answered by the student and handed in for grading at the direction of the instructor. Even if the instructor does not require you to answer these questions, we recommend that you do so anyway to check your understanding.

Some exercises have a section of ***LABORATORY REPORT RESULTS***, in which students can record results of laboratory exercises, in addition to laboratory report questions. As with the laboratory report questions, the laboratory report results are located at the end of selected exercises and can be handed in as the instructor directs. Instructions in the manual tell students when and where to record laboratory results.

ANSWER THE LABORATORY REPORT QUESTIONS AT THE END OF THE EXERCISE

110
15/50
100/

LABORATORY REPORT QUESTIONS

MICROSCOPY 1

Student ______________________ Date ______________________

Laboratory Section ______________________ Score/Grade ______________________

PART I. MULTIPLE CHOICE

b 1. The amount of light entering a microscope may be adjusted by regulating the (a) ocular (b) diaphragm (c) fine adjustment knob (d) nosepiece

c 2. If the ocular on a microscope is marked 10× and the low-power objective is marked 15×, the total magnification is (a) 50× (b) 25× (c) 150× (d) 1500×

d 3. The size of the light beam that passes through a microscope is regulated by the (a) revolving nosepiece (b) coarse adjustment knob (c) ocular (d) condenser

e 4. Parfocal means that (a) the microscope employs only one lens (b) final focusing can only be done with the fine adjustment knob (c) changing objectives by revolving the nosepiece will still keep the specimen in focus (d) the highest magnification attainable is 1000×

d 5. Which of these is *not* true when changing magnification from low power to high power? (a) the specimen should be centered (b) illumination should be decreased (c) the specimen should be in clear focus (d) the high power objective should be in line with the body tube

b 6. A microscope with which of the following resolving powers could distinguish the finest detail? (a) 0.01 μm (b) 10 nm (c) 0.001 μm (d) 1 nm

PART II. COMPLETION

7. The advantage of using immersion oil is that it has special ______________ properties that permit the use of a powerful objective in a narrow field of vision.

8. The uniform circle of light that appears when one looks into the ocular is called the ______________ .

9. In determining the position (depth) of the colored threads, the ______________ (red, blue, yellow) colored thread was in the middle.

10. If you move your slide to the right, the specimen moves to the ______________ as you are viewing it microscopically.

11. After switching from low power to high power, ______________ (more or less) of the specimen will be visible.

12. The ability of lenses to distinguish between two points at a specified distance apart is called ______________ .

13. The wavelength of electrons is about ______________ (what proportion?) that of visible white light.

14. A photograph of a specimen taken through a compound light microscope is called a(n) ____________________ .

15. The type of microscope that provides three-dimensional images of nonliving specimens is the ____________________ microscope.

PART III. MATCHING

____ 16. Ocular
____ 17. Stage
____ 18. Arm
____ 19. Condenser
____ 20. Revolving nosepiece
____ 21. Low-power objective
____ 22. Fine adjustment knob
____ 23. Diaphragm
____ 24. Coarse adjustment knob
____ 25. High-power objective

A. Platform on which slide is placed
B. Mounting for objectives
C. Lens below stage opening
D. Brings specimen into sharp focus
E. Eyepiece
F. An objective usually marked 43× or 45×
G. An objective usually marked 10×
H. Angular or curved part of frame
I. Brings specimen into general focus
J. Regulates light intensity

INTRODUCTION TO THE HUMAN BODY 2

In this exercise, you will be introduced to the organization of the human body through a study of its levels of structural organization, principal body systems, anatomical position, regional names, directional terms, planes of the body, body cavities, abdominopelvic regions, and abdominopelvic quadrants.

A. LEVELS OF STRUCTURAL ORGANIZATION

The human body is composed of several levels of structural organization associated with each other in various ways:

1. ***Chemical level***—Composed of all atoms and molecules necessary to maintain life.
2. ***Cellular level***—Consists of cells, the structural and functional units of the body.
3. ***Tissue level***—Formed by tissues, groups of similarly specialized cells and their intercellular material.
4. ***Organ level***—Consists of organs, structures of definite form and function composed of two or more different tissues.
5. ***System level***—Formed by systems, associations of organs that have a common function. The systems together constitute an ***organism.***

B. SYSTEMS OF THE BODY

Using an anatomy and physiology textbook, torso, wall chart, and any other materials that might be available to you, identify the principal organs that compose the following body systems.[1] In the spaces that follow, indicate the components and functions of the systems.

Integumentary

Components ________________________________

____

____

____

Functions ________________________________

____

____

____

Skeletal

Components ________________________________

____

____

____

Functions ________________________________

____

____

____

[1]You will probably need other sources, plus any aids the instructor might provide, to label many of the figures and answer some questions in this manual. You are encouraged to use other sources as you find necessary.

Muscular

Components

Functions

Lymphatic

Components

Functions

Nervous

Components

Functions

Respiratory

Components

Functions

Endocrine

Components

Functions

Digestive

Components

Functions

Cardiovascular

Components

Functions

Urinary

Components

Functions

Reproductive

Components ______________________

____

____

____

Functions ______________________

____

____

____

C. ANATOMICAL POSITION AND REGIONAL NAMES

Figure 2.1 shows anterior and posterior views of a subject in the ***anatomical position***. The subject is standing erect (upright positicn) and facing the observer when in anterior view, the upper extremities (limbs) are at the sides, the palms of the hands are facing forward, and the feet are flat on the floor. The figure also shows the common names for various regions of the body. When you as the observer make reference to the left and right sides of the subject you are studying, this refers to the *subject's* left and right sides. In the spaces next to the list of terms in Figure 2.1 write the number of each common term next to each corresponding anatomical term. For example, the skull (31) is cranial, so write the number *31* next to the term *Cranial*.

D. EXTERNAL FEATURES OF THE BODY

Referring to your textbook and human models, identify the following external features of the body:

1. ***Head (cephalic region or caput)***—This is divided into the ***cranium*** (brain case) and ***face***.
2. ***Neck (collum)***—This region consists of an anterior ***cervix***, two lateral surfaces, and a posterior ***nucha*** (NOO-ka).
3. ***Trunk***—This region is divided into the ***back*** (dorsum), ***chest*** (thorax), ***abdomen*** (venter), and ***pelvis***.
4. ***Upper extremity***—This consists of the ***armpit*** (axilla), ***shoulder*** (acromial region or omos), ***arm*** (brachium), ***elbow*** (cubitus), ***forearm*** (antebrachium), and ***hand*** (manus). The hand, in turn, consists of the ***wrist*** (carpus), ***palm*** (metacarpals), and ***fingers*** (digits). Individual bones of a digit (finger or toe) are called ***phalanges***.
5. ***Lower extremity***—This consists of the ***buttocks*** (gluteal region), ***thigh*** (femoral region), ***knee*** (genu), ***leg*** (crus), and ***foot*** (pes). The foot includes the ***ankle*** (tarsus), ***sole*** (metatarsus), and ***toes*** (digits).

E. DIRECTIONAL TERMS

To explain exactly where a structure of the body is located, it is a standard procedure to use ***directional terms***. Such terms are very precise and avoid the use of unnecessary words. Commonly used directional terms for humans are as follows:

1. ***Superior (cephalic*** or ***cranial)***—Toward the head or the upper part of a structure; generally refers to structures in the trunk.
2. ***Inferior (caudad)***—Away from the head or toward the lower part of a structure; generally refers to structures in the trunk.
3. ***Anterior (ventral)***—Nearer to or at the front or belly surface of the body. In the ***prone position***, the body lies anterior side down; in the ***supine position***, the body lies anterior side up.
4. ***Posterior (dorsal)***—Nearer to or at the back or backbone surface of the body.
5. ***Medial***—Nearer the midline of the body or a structure. The ***midline*** is an imaginary vertical line that divides the body into equal left and right sides.
6. ***Lateral***—Farther from the midline of the body or a structure.
7. ***Intermediate***—Between two structures.
8. ***Ipsilateral***—On the same side of the body.
9. ***Contralateral***—On the opposite side of the body.
10. ***Proximal***—Nearer the attachment of an extremity to the trunk; nearer to the point of origin.
11. ***Distal***—Farther from the attachment of an extremity to the trunk; farther from the point of origin.

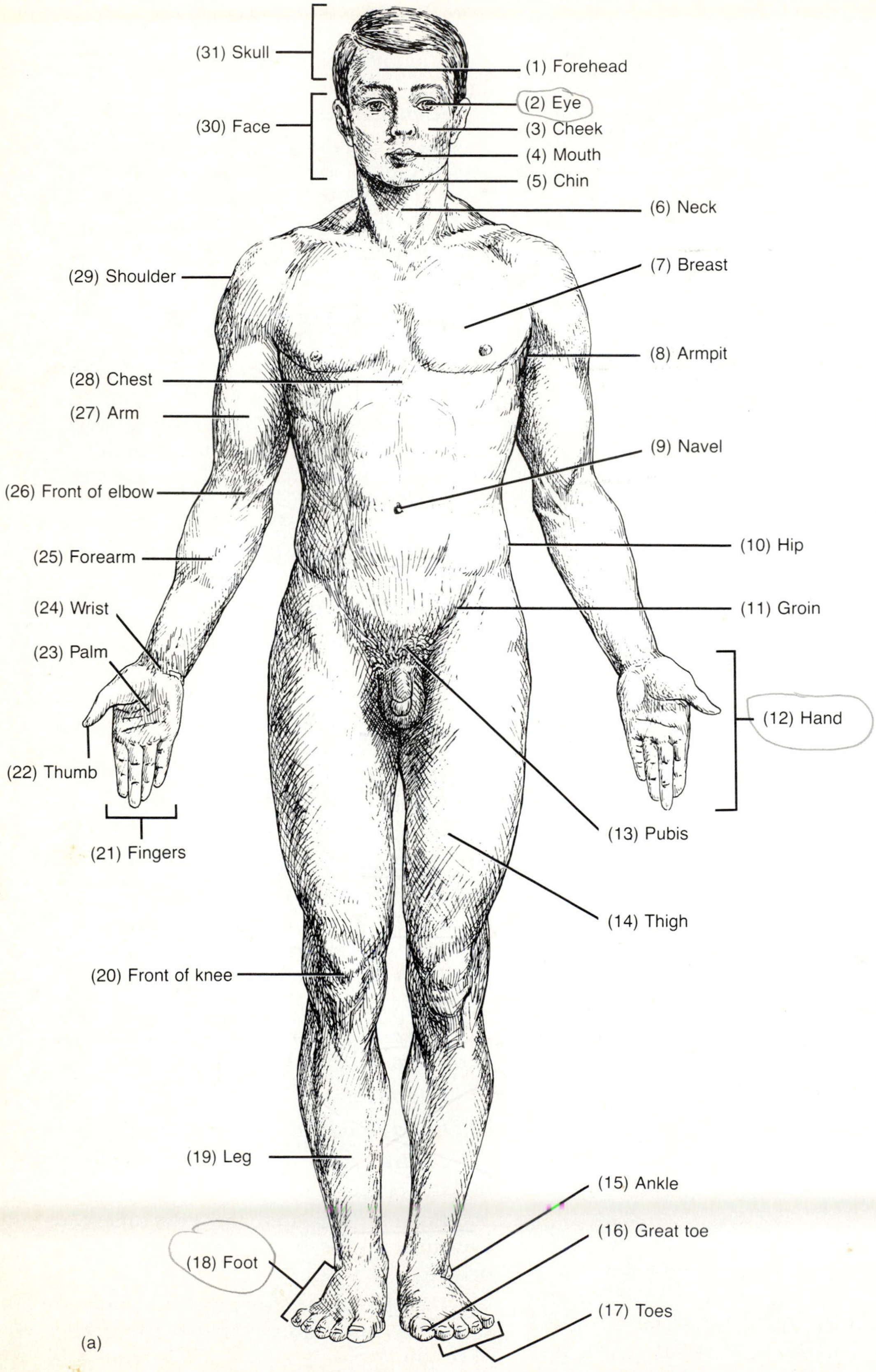

29 Acromial
25 Antebrachial
26 Antecubital
8 Axillary
27 Brachial
3 Buccal
24 Carpal
6 Cervical
10 Coxal
31 Cranial
19 Crural
21 Digital (hand)
17 Digital (foot)
30 Facial
14 Femoral
1 Frontal
16 Hallux
11 Inguinal
7 Mammary
12 Manual
5 Mental
23 Metacarpal
4 Oral
2 Orbital
20 Patellar
18 Pedal
22 Pollex
13 Pubic
15 Tarsal
28 Thoracic
9 Umbilical

FIGURE 2.1 The anatomical position. (a) Anterior view.

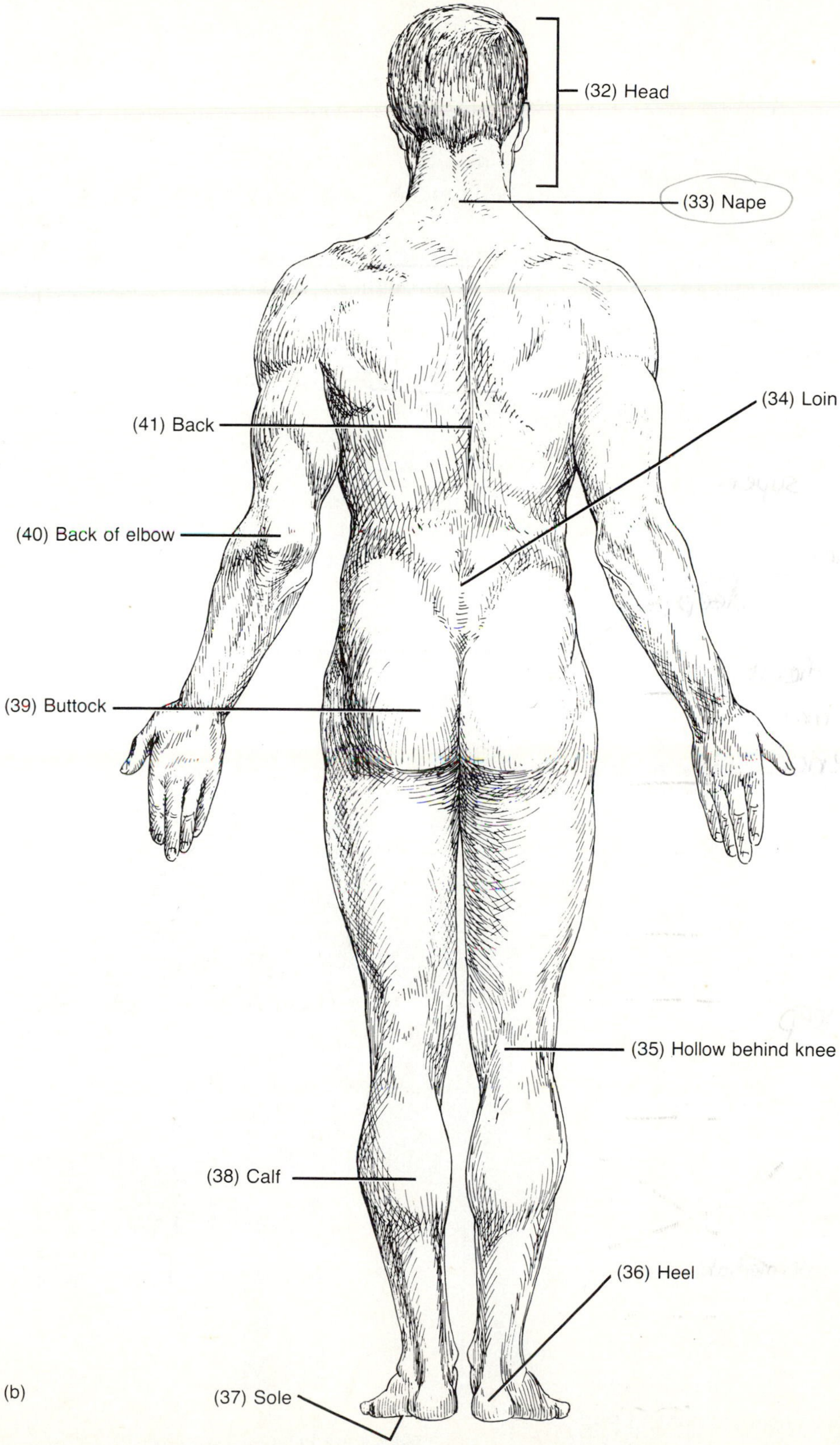

36 Calcaneal
32 Cephalic
41 Dorsal
39 Gluteal
34 Lumbar
33 Nuchal
40 Olecranal
37 Plantar
35 Popliteal
38 Sural

FIGURE 2.1 ***(Continued)*** The anatomical position. (b) Posterior view.

12. ***Superficial* (external)**—Toward or on the surface of the body.

13. ***Deep* (internal)**—Away from the surface of the body.

14. ***Parietal***—Pertaining to the outer wall of a body cavity.

15. ***Visceral***—Pertaining to the covering of an organ (viscus).

Using a torso and an articulated skeleton, and consulting with your instructor as necessary, describe the location of the following by inserting the proper directional term.

1. The ulna is on the ______ side of the forearm.

2. The ascending colon is superi______ to the urinary bladder.

3. The heart is superior to the liver.

4. The muscles of the arm are deep______ to the skin of the arm.

5. The sternum is superficial to the heart.

6. The humerus is Proximal to the radius.

7. The stomach is inferior to the lungs.

8. The muscles of the thoracic wall are ______ to the viscera in the thoracic cavity.

9. The ______ pleura forms the outer layer of the pleural sacs that surround the lungs.

10. The esophagus is deep______ to the trachea.

11. The phalanges are ______ to the carpals.

12. The ______ pleura forms the inner layer of the pleural sacs and covers the external surface of the lungs.

13. The ring finger is intermediate between the little (medial) and middle (lateral) fingers.

14. The ascending colon of the large intestine and the gallbladder are ______ .

15. The ascending and descending colons of the large intestine are ______ .

F. PLANES OF THE BODY

The structural plan of the human body may be described with respect to ***planes*** (imaginary flat surfaces) passing through it. Planes are frequently used to show the anatomical relationship of several structures in a region to one another.

Commonly used planes are:

1. ***Midsagittal (median)***—A vertical plane that passes through the midline of the body and divides the body or an organ into *equal* right and left sides.

2. ***Parasagittal***—A vertical plane that does not pass through the midline of the body and divides the body or an organ into *unequal* right and left sides.

3. ***Frontal* (coronal;** kō-RŌ-nal**)**—A vertical plane at a right angle to a midsagittal (or parasagittal) plane that divides the body or an organ into anterior and posterior portions.

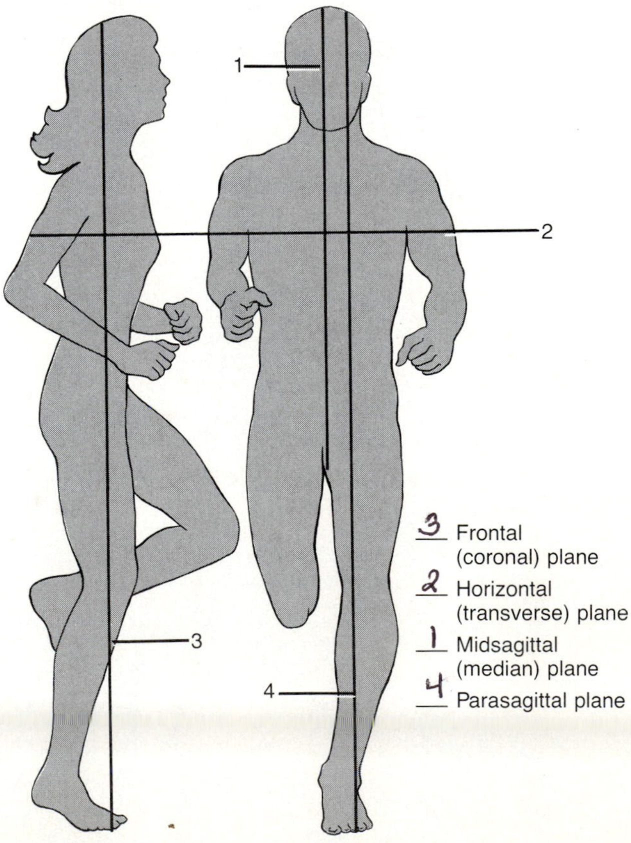

FIGURE 2.2 Planes of the body.

4. ***Horizontal (transverse)***—A plane that is parallel to the ground (at a right angle to midsagittal, parasagittal, and frontal planes) and divides the body or an organ into superior and inferior portions.

Refer to Figure 2.2 and label the planes shown.

G. BODY CAVITIES

Spaces within the body that contain various internal organs are called ***body cavities***. One way of organizing the principal body cavities follows:

Dorsal body cavity
  Cranial cavity
  Vertebral (spinal) cavity
Ventral body cavity
  Thoracic cavity
    Right pleural
    Left pleural
    Pericardial
  Abdominopelvic cavity
    Abdominal
    Pelvic

The space (actually a mass of tissue) between the pleurae of the lungs and extending from the sternum (breast bone) to the backbone is called the ***mediastinum***. It contains all structures in the thoracic cavity, except the lungs themselves. Included are the heart, thymus gland, esophagus, trachea, and many large blood and lymphatic vessels.

Label the body cavities shown in Figure 2.3. Then examine a torso or wall chart, or both, and determine which organs lie within each cavity.

H. ABDOMINOPELVIC REGIONS

To describe the location of viscera more easily, the abdominopelvic cavity may be divided into ***nine regions*** by using four imaginary lines: (1) an upper horizontal ***subcostal line*** that passes just below the bottom of the rib cage through the pylorus (lower portion) of the stomach, (2) a lower horizontal line, the ***transtubercular line***, that joins the iliac crests (top surfaces of the hipbones), (3) a ***right midclavicular line*** drawn through the midpoint of the right clavicle slightly medial to the right nipple, and (4) a ***left***

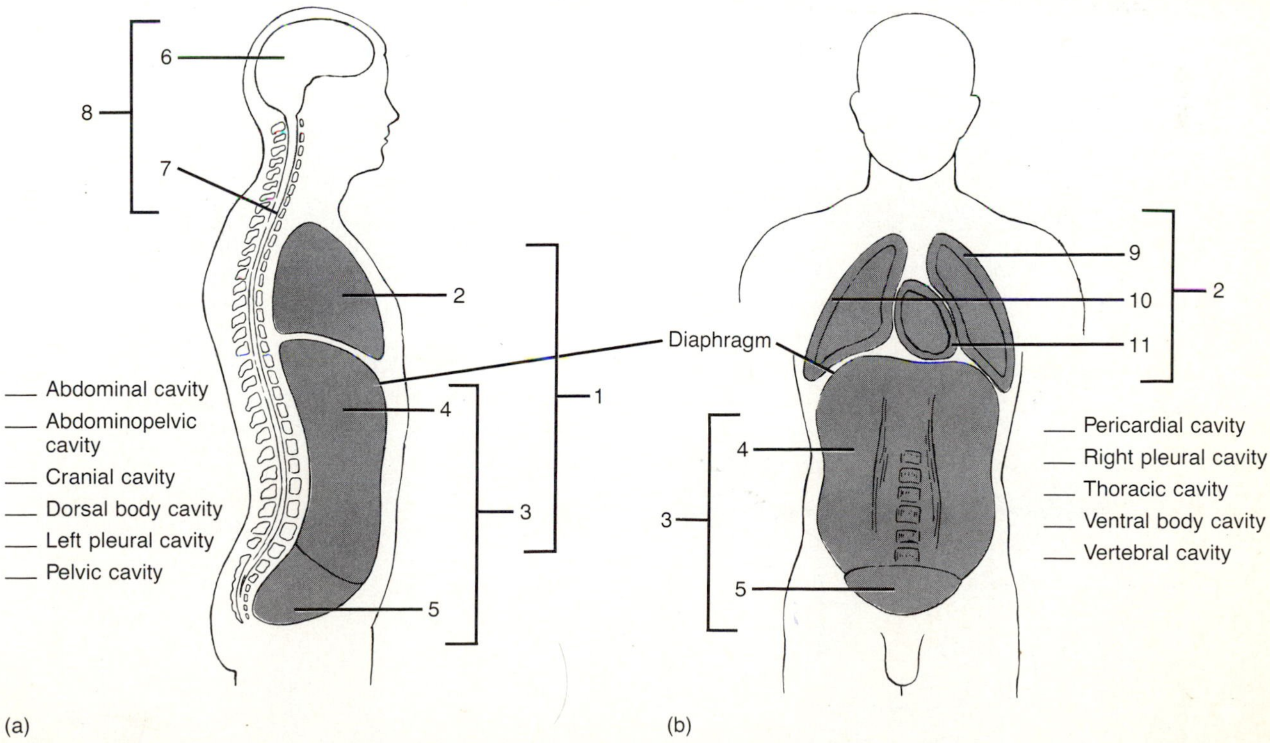

FIGURE 2.3 Body cavities. (a) Median section. (b) Anterior view.

midclavicular line drawn through the midpoint of the left clavicle slightly medial to the left nipple.

The four imaginary lines divide the abdominopelvic cavity into the following nine regions: (1) ***umbilical*** (um-BIL-i-kul) ***region***; which is centrally located; (2) ***left lumbar*** (*lumbus* = loin) ***region***, to the left of the umbilical region; (3) ***right lumbar***, to the right of the umbilical region; (4) ***epigastric*** (ep-i-GAS-trik; *epi* = above; *gaster* = stomach) ***region***, directly above the umbilical region; (5) ***left hypochondriac*** (hī′-pō-KON-drē-ak; *hypo* = under; *chondro* = cartilage) ***region***, to the left of the epigastric region; (6) ***right hypochondriac region***, to the right of the epigastric region; (7) ***hypogastric (pubic) region***, directly below the umbilical region; (8) ***left iliac*** (IL-ē-ak; *iliacus* = superior part of hipbone) or ***inguinal region***, to the left of the hypogastric (pubic) region; and (9) ***right iliac (inguinal) region***, to the right of the hypogastric (pubic) region.

Label Figure 2.4 by indicating the names of the four imaginary lines and the nine abdominopelvic regions.

Examine a torso and determine which organs or parts of organs lie within each of the nine abdominopelvic regions.

I. ABDOMINOPELVIC QUADRANTS

Another way to divide the abdominopelvic cavity is into ***quadrants*** by passing one hor-

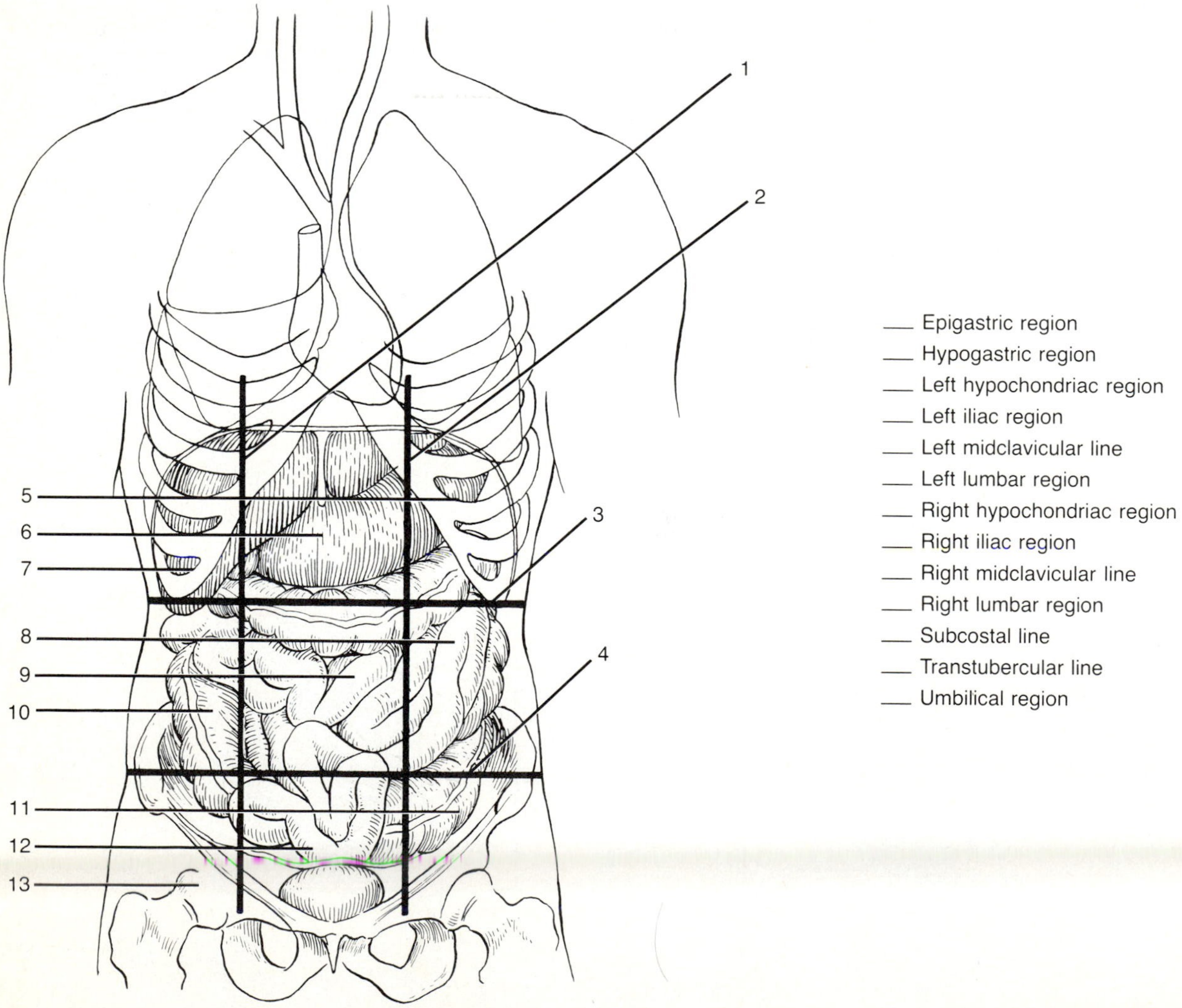

FIGURE 2.4 Abdominopelvic regions.

izontal line and one vertical line through the umbilicus. The two lines thus divide the abdominopelvic cavity into a ***right upper quadrant (RUQ), left upper quadrant (LUQ), right lower quadrant (RLQ),*** and ***left lower quadrant (LLQ).*** Quadrant names are frequently used by health care professionals for locating the site of an abdominopelvic pain, tumor, or other abnormality.

Examine a torso or wall chart, or both, and determine which organs or parts of organs lie within each of the abdominopelvic quadrants.

J. DISSECTION OF WHITE RAT

Now that you have some idea of the names of the various body systems and the principal organs that comprise each, you can actually observe some of these organs by dissecting a white rat. Dissect means "to separate." This dissection gives you an excellent opportunity to see the different sizes, shapes, locations, and relationships of organs and to compare the different textures and external features of organs. In addition, this exercise will introduce you to the general procedure for dissection before you dissect the cat in later exercises.

CAUTION! *Please reread Section D. "Precautions Related to Dissection" at the beginning of the laboratory manual on page xii before you begin your dissection.*

PROCEDURE

1. Place the rat on its backbone on a wax dissecting pan (tray). Using dissecting pins, anchor each of the four extremities to the wax (Figure 2.5a).

2. To expose the contents of the thoracic, abdominal, and pelvic cavities, you will have to first make a midline incision. This is done by lifting the abdominal skin with a forceps to separate the skin from the underlying connective tissue and muscle. While lifting the abdominal skin, cut through it with scissors and make an incision that extends from the lower jaw to the anus (Figure 2.5a).

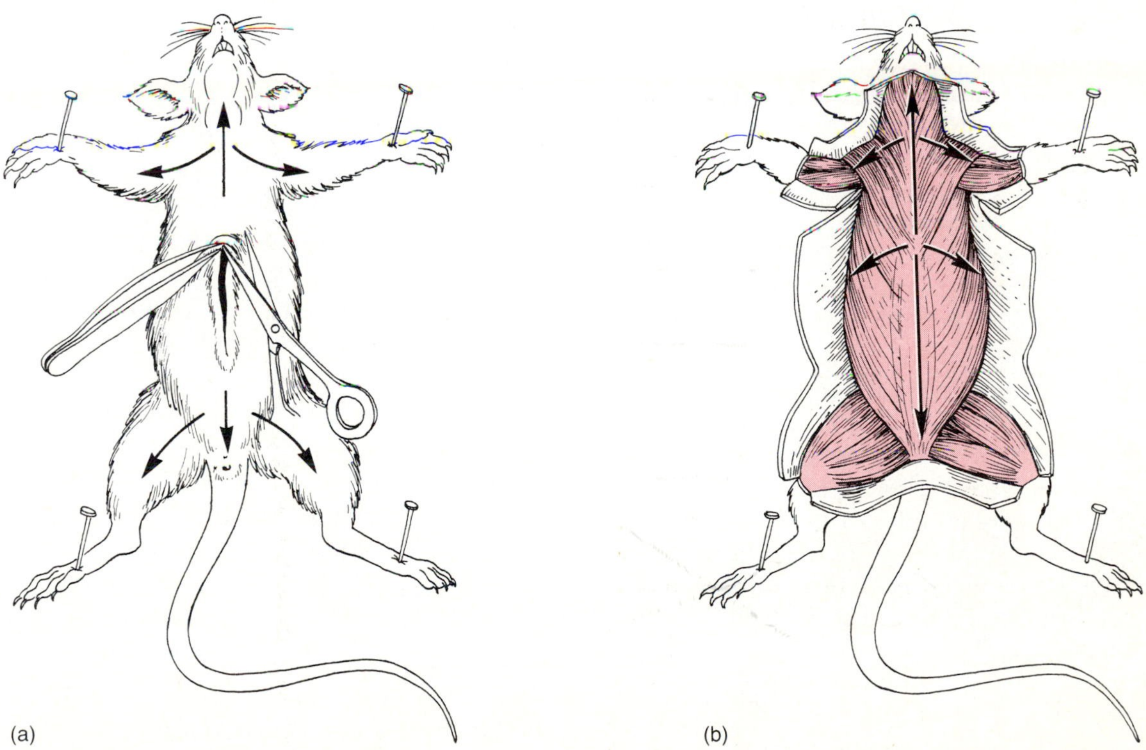

FIGURE 2.5 Dissection procedure for exposing thoracic and abdominopelvic viscera of the white rat for examination. (a) Lines of incision in skin. (b) Peeling back skin and lines of incision in muscles.

3. Now make four lateral incisions that extend from the midline incision into the four extremities (Figure 2.5a).

4. Peel the skin back and pin the flaps to the wax to expose the superficial muscles (Figure 2.5b).

5. Next, lift the abdominal muscles with a forceps and cut through the muscle layer, being careful not to damage any underlying organs. Keep the scissors parallel to the rat's backbone. Extend this incision from the anus to a point just below the bottom of the rib cage (Figure 2.5b). Make two lateral incisions just below the rib cage and fold back the muscle flaps to expose the abdominal and pelvic viscera (Figure 2.5b).

6. To expose the thoracic viscera, cut through the ribs on either side of the sternum. This incision should extend from the diaphragm to the neck (Figure 2.5b). The ***diaphragm*** is the thin muscular partition that separates the thoracic from the abdominal cavity. Again make lateral incisions in the chest wall so that you can lift the ribs to view the thoracic contents.

1. Examination of Thoracic Viscera

You will first examine the thoracic viscera (large internal organs). As you dissect and observe the various structures, palpate (feel with the hand) them so that you can compare their texture. Use Figure 2.6 as a guide.

a. ***Thymus gland***—An irregular mass of glandular tissue superior to the heart and superficial to the trachea. Push the thymus gland aside or remove it.

b. ***Heart***—A structure located in the midline, inferior to the thymus gland and between the lungs. The sac covering the heart is the ***pericardium***, which may be removed. The large vein that returns blood to the heart from the lower regions of the body is the ***inferior vena cava***, while the large vein that returns blood from the upper regions of the body is the ***superior vena cava***. The large artery that carries blood from the heart to most parts of the body is the ***aorta***.

c. ***Lungs***—Reddish, spongy structures on either side of the heart. Note that the lungs are divided into regions called ***lobes***.

d. ***Trachea***—A tubelike passageway superior to the heart and deep to the thymus gland. Note that the wall of the trachea consists of rings of cartilage. Identify the ***larynx*** (voice box) at the superior end of the trachea and the ***thyroid gland***, a bilobed structure on either side of the larynx. The lobes of the thyroid gland are connected by a band of thyroid tissue, the isthmus.

e. ***Bronchial tubes***—Trace the trachea inferiorly and note that it divides into bronchial tubes that enter the lungs and continue to divide within them.

f. ***Esophagus***—A muscular tube posterior to the trachea that transports food from the throat into the stomach. Trace the esophagus inferiorly to see where it passes through the diaphragm to join the stomach.

2. Examination of Abdominopelvic Viscera

You will now examine the principal viscera of the abdomen and pelvis. As you do so, again refer to Figure 2.6.

a. ***Stomach***—An organ located on the left side of the abdomen and in contact with the liver. The digestive organs are attached to the posterior abdominal wall by a membrane called the ***mesentery***. Note the blood vessels in the mesentery.

b. ***Small intestine***—An extensively coiled tube that extends from the stomach to the first portion of the large intestine called the cecum.

c. ***Large intestine***—A wider tube than the small intestine that begins at the cecum and ends at the rectum. The cecum is a large, saclike structure. In humans, the appendix arises from the cecum.

d. ***Rectum***—A muscular passageway, located on the midline in the pelvic cavity, that terminates in the anus.

e. ***Anus***—Terminal opening of the digestive system to the exterior.

f. ***Pancreas***—A pale gray, glandular organ posterior and inferior to the stomach.

g. ***Spleen***—A small, dark red organ lateral to the stomach.

h. ***Liver***—A large, brownish-red organ directly inferior to the diaphragm. The rat does not have a gallbladder, a structure associated with the liver. To locate the remaining viscera, either move the superficial viscera aside or remove them. Use Figure 2.7 as a guide.

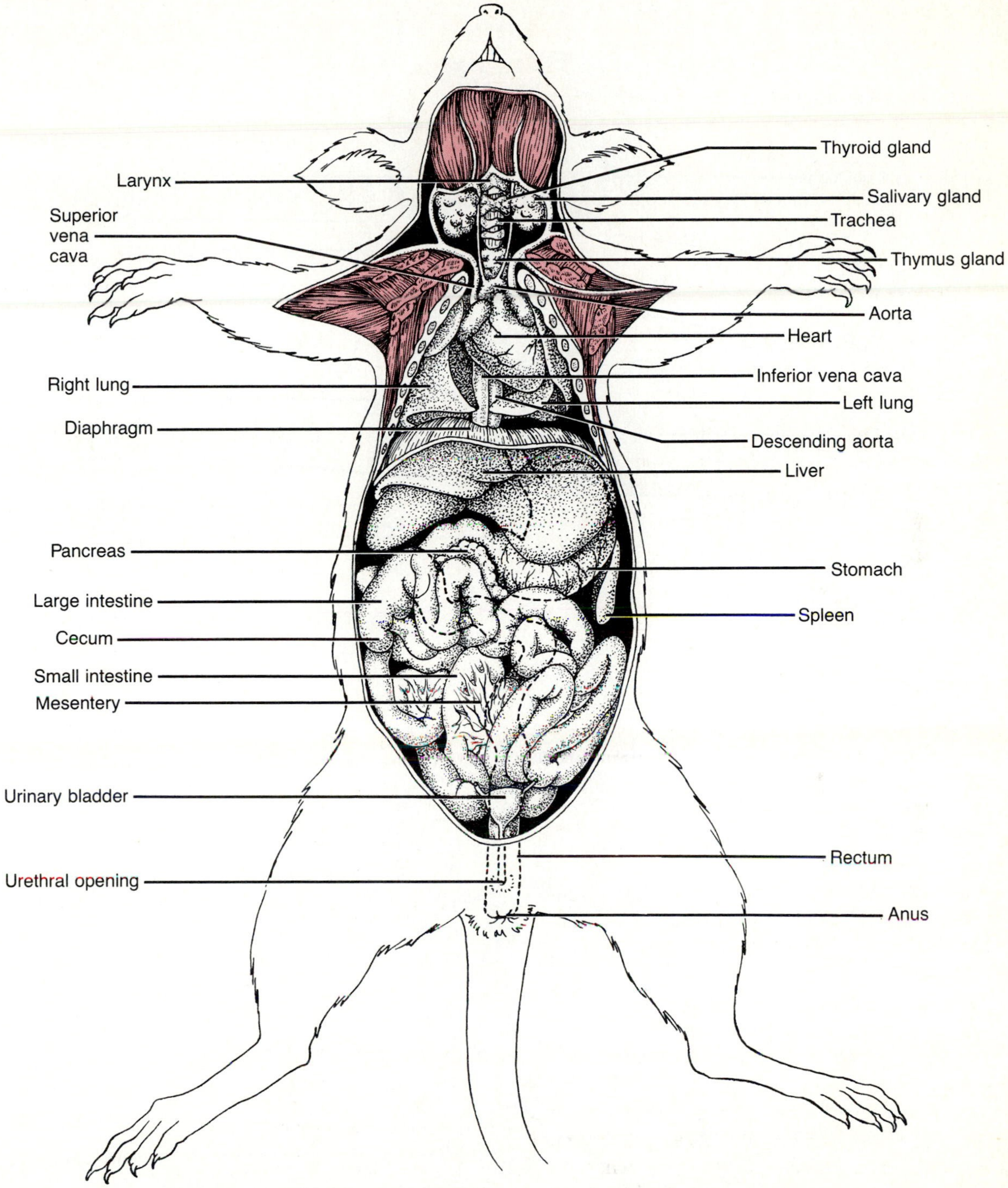

FIGURE 2.6 Superficial structures of the thoracic and abdominopelvic cavities of the white rat.

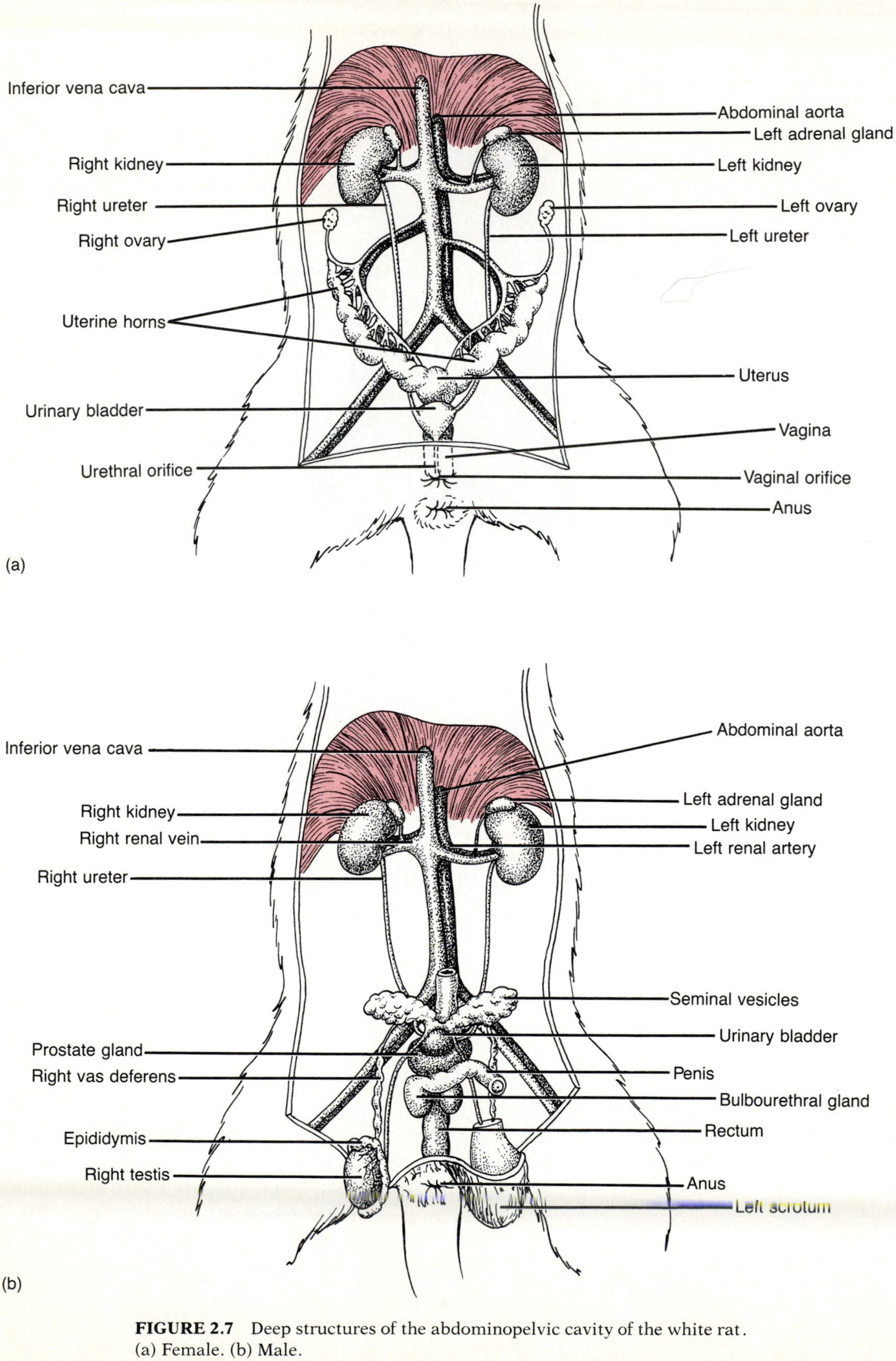

FIGURE 2.7 Deep structures of the abdominopelvic cavity of the white rat. (a) Female. (b) Male.

i. ***Kidneys***—Bean-shaped organs embedded in fat and attached to the posterior abdominal wall on either side of the backbone. As will be explained later, the kidneys and a few other structures are behind the membrane that lines the abdomen ***(peritoneum)***. Such structures are referred to as ***retroperitoneal*** and are not actually within the abdominal cavity. See if you can find the ***abdominal aorta***, the large artery located along the midline behind the inferior vena cava. Also, locate the ***renal arteries*** branching off the abdominal aorta to enter the kidneys.

j. ***Adrenal (suprarenal) glands***—Glandular structures. One is located on top of each kidney.

k. ***Ureters***—Tubes that extend from the medial surface of the kidney inferiorly to the urinary bladder.

l. ***Urinary bladder***—A saclike structure in the pelvic cavity that stores urine.

m. ***Urethra***—A tube that extends from the urinary bladder to the exterior. Its opening to the exterior is called the ***urethral orifice***. In male rats, the urethra extends through the penis; in female rats, the tube is separate from the reproductive tract.

If your specimen is female (no visible scrotum anterior to the anus), identify the following:

n. ***Ovaries***—Small, dark structures inferior to the kidneys.

o. ***Uterus***—An organ located near the urinary bladder consisting of two sides (horns) that join separately into the vagina.

p. ***Vagina***—A tube that leads from the uterus to the external vaginal opening, the ***vaginal orifice***. This orifice is in front of the anus and behind the urethral orifice.

If your specimen is male, identify the following:

q. ***Scrotum***—Large sac anterior to the anus that contains the testes.

r. ***Testes***—Egg-shaped glands in the scrotum. Make a slit into the scrotum and carefully remove one testis. See if you can find a coiled duct attached to the testis ***(epididymis)*** and a duct that leads from the epididymis into the abdominal cavity ***(vas deferens)***.

s. ***Penis***—Organ of copulation medial to the testes.

When you have finished your dissection, store or dispose of your specimen. Wash your dissecting pan and dissecting instruments with laboratory detergent, dry them, and return them to their storage areas.

ANSWER THE LABORATORY REPORT QUESTIONS AT THE END OF THE EXERCISE

LABORATORY REPORT QUESTIONS

INTRODUCTION TO THE HUMAN BODY 2

Student ______________________ **Date** ______________________

Laboratory Section ______________________ **Score/Grade** ______________________

PART I. MULTIPLE CHOICE

d 1. The directional term that best describes the eyes in relation to the nose is (a) distal (b) superficial (c) anterior (d) lateral

_____ 2. Which does *not* belong with the others? (a) right pleural cavity (b) pericardial cavity (c) vertebral cavity (d) left pleural cavity

a 3. Which plane divides the brain into an anterior and a posterior portion? (a) frontal (b) median (c) sagittal (d) horizontal

_____ 4. The urinary bladder lies in which region? (a) umbilical (b) hypogastric (c) epigastric (d) left iliac

c 5. Which is *not* a characteristic of the anatomical position? (a) the subject is erect (b) the subject faces the observer (c) the palms face backward (d) the arms are at the sides

_____ 6. The abdominopelvic region that is bordered by all four imaginary lines is the (a) hypogastric (b) epigastric (c) left hypochondriac (d) umbilical

b 7. Which directional term best describes the position of the phalanges with respect to the carpals? (a) lateral (b) distal (c) anterior (d) proximal

_____ 8. The pancreas is found in which body cavity? (a) abdominal (b) pericardial (c) pelvic (d) vertebral

c 9. The anatomical term for the leg is (a) brachial (b) tarsal (c) crural (d) sural

_____ 10. In which abdominopelvic region is the spleen located? (a) left lumbar (b) right lumbar (c) epigastric (d) left hypochondriac

_____ 11. Which of the following represents the most complex level of structural organization? (a) organ (b) cellular (c) tissue (d) chemical

_____ 12. Which body system is concerned with support, protection, leverage, blood cell production, and mineral storage? (a) cardiovascular (b) lymphatic (c) skeletal (d) digestive

_____ 13. The skin and structures derived from it, such as nails, hair, sweat glands, and oil glands, are components of which system? (a) respiratory (b) integumentary (c) muscular (d) lymphatic

_____ 14. Hormone-producing glands belong to which body system? (a) cardiovascular (b) lymphatic (c) endocrine (d) digestive

_____ 15. Which body system brings about movement, maintains posture, and produces heat? (a) skeletal (b) respiratory (c) reproductive (d) muscular

_____ 16. Which abdominopelvic quadrant contains most of the liver? (a) RUQ (b) RLQ (c) LUQ (d) LLQ

_____ 17. The physical and chemical breakdown of food for use by body cells and the elimination of solid wastes are accomplished by which body system? (a) respiratory (b) urinary (c) cardiovascular (d) digestive

PART II. COMPLETION

18. The tibia is ____________ to the fibula.

19. The ovaries are found in the ____________ body cavity.

20. The upper horizontal line that helps divide the abdominopelvic cavity into nine regions is the ____________ line.

21. The anatomical term for the hollow behind the knee is ____________ .

22. A plane that divides the stomach into a superior and an inferior portion is a(n) Horizontal plane.

23. The wrist is described as distal to the elbow.

24. The heart is located in the ____________ cavity within the thoracic cavity.

25. The abdominopelvic region that contains the rectum is the ____________ region.

26. A plane that divides the body into unequal left and right sides is the Parasagittal plane.

27. The spinal cord is located within the ____________ cavity.

28. The body system that removes carbon dioxide from body cells, delivers oxygen to body cells, helps maintain acid-base balance, helps protect against disease, helps regulate body temperature, and prevents hemorrhage by forming clots is the ____________ system.

29. The ____________ abdominopelvic quadrant contains the descending colon of the large intestine.

30. Which body system returns proteins and plasma to the cardiovascular system, transports fats from the digestive system to the cardiovascular system, filters blood, protects against disease, and produces white blood cells? ____________

PART III. MATCHING

____ 31. Right hypochondriac region
____ 32. Hypogastric region
____ 33. Left iliac region
____ 34. Right lumbar region
____ 35. Epigastric region
____ 36. Left hypochondriac region
____ 37. Right iliac region
____ 38. Umbilical region
____ 39. Left lumbar region

A. Junction of descending and sigmoid colons of large intestine
B. Descending colon of large intestine
C. Spleen
D. Most of right lobe of liver
E. Appendix
F. Ascending colon of large intestine
G. Middle of transverse colon of large intestine
H. Adrenal (suprarenal) glands
I. Sigmoid colon of large intestine

PART IV. MATCHING

R 40. Anterior
H 41. Thumb
___ 42. Transtubercular line
F 43. Armpit
___ 44. Umbilical region
A 45. Medial
___ 46. Cranial cavity
P 47. Front of knee
K 48. Breast
D 49. Chest
J 50. Buttock
Q 51. Superior
U 52. Groin
B 53. Vertebral cavity
N 54. Cheek
E 55. Neck
O 56. Distal
G 57. Pericardial cavity
I 58. Forearm
L 59. Plantar
S 60. Mouth

A. Passes through iliac crests
B. Contains spinal cord
C. Nearer the midline
D. Thoracic
E. Cervical
F. Axillary
G. Contains the heart
H. Pollex
I. Antebrachial
J. Gluteal
K. Mammary
L. Sole of foot
M. Contains navel
N. Buccal
O. Farther from the attachment of an extremity
P. Patellar
Q. Toward the head
R. Nearer to or at the front of the body
S. Oral
T. Contains brain
U. Inguinal

CELLS 3

A ***cell*** is the basic living structural and functional unit of the body. The study of cells is called ***cytology*** (sī-TOL-ō-jē; *cyto* = cell; *logos* = study of). The different kinds of cells (blood, nerve, bone, muscle, epithelial, and others) perform specific functions and differ from one another in shape, size, and structure. You will start your study of cytology by learning the important components of a theoretical, generalized cell.

A. CELL PARTS

Refer to Figure 3.1, a generalized cell based on electron micrograph studies. With the aid

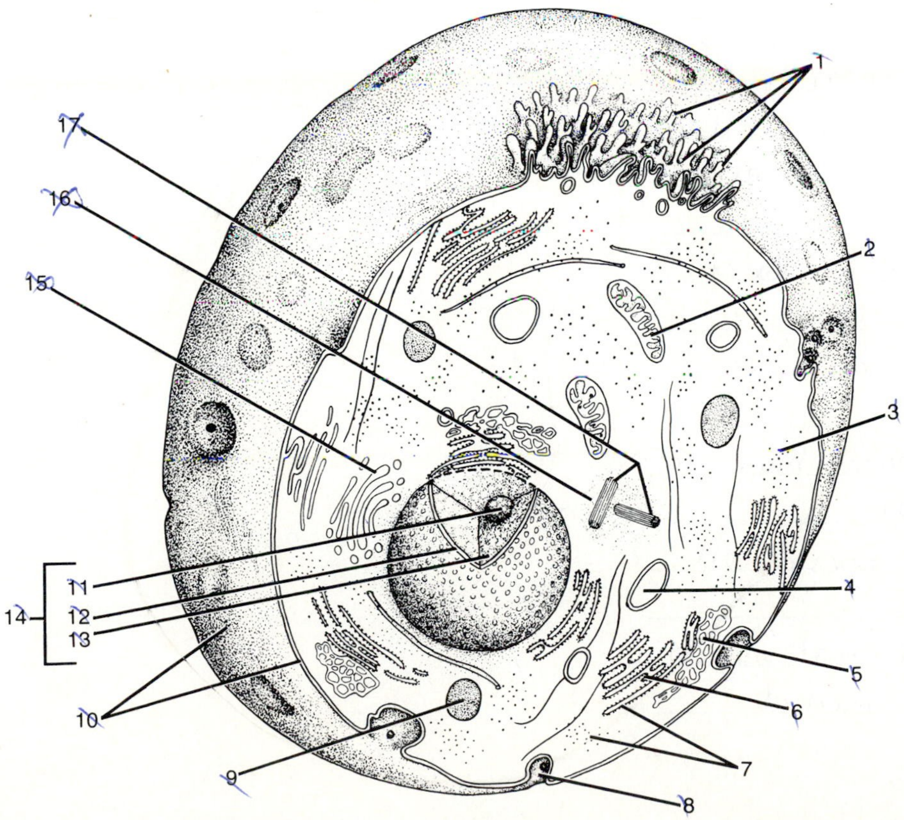

___ Centrioles
___ Centrosome
___ Cytoplasm
___ Golgi complex
___ Lysosome
___ Microvilli
___ Mitochondrion
___ Nuclear membrane
___ Nucleolus
___ Nucleoplasm
___ Nucleus
___ Pinocytic vesicle
___ Plasma (cell) membrane
___ Ribosomes
___ Rough (granular) endoplasmic reticulum
___ Smooth (agranular) endoplasmic reticulum
___ Vacuole

FIGURE 3.1 Generalized animal cell.

of your textbook and any other items made available by your instructor, label the parts of the cell indicated. In the spaces that follow, describe the function of the following cellular structures:

1. ***Plasma (cell) membrane*** outer covering
2. ***Cytoplasm*** (SĪ-tō-plazm′) fluid w/in cell & surrounds organelles
3. ***Nucleus*** (NOO-klē-us) contains genetic material
4. ***Endoplasmic reticulum*** (en′-dō-PLAS-mik re-TIK-yoo-lum) or ***ER*** serves in intracellular transportation
5. ***Ribosome*** Synthesize proteins
6. ***Golgi*** (GOL-jē) ***complex*** packages protein & helps to deliver to correct location
7. ***Mitochondria*** (mī′-tō-KON-drē-a) produces ATP
8. ***Lysosome*** (LĪ-so-sōm) contains digestive enzymes metabolism / liver cells
9. ***Peroxisome*** (pe-ROKS-i-sōm) contains enzymes related to hydrogen peroxide
10. ***Cytoskeleton*** micro tubules & filament / actin
11. ***Centriole*** (SEN-trē-ōl) play a role in cell division
12. ***Cilium*** (SIL-ē-um) short, hair-like projectiles (lungs)
13. ***Flagellum*** (fla-JEL-um) sperm cell the tail
14. ***Pinocytic vesicle*** cell that ingests a particle & bulges inward

B. DIVERSITY OF CELLS

Now obtain prepared slides of the following types of cells and examine them under the magnifications suggested:

1. Ciliated columnar epithelial cells (high power)
2. Sperm cells (oil immersion)
3. Nerve cells (high power)
4. Muscle cells (high power)

After you have made your examination, draw an example of each of these kinds of cells in the spaces provided and under each cell indicate how each is adapted to its particular function.

Ciliated columnar epithelial cell

Sperm cell

Nerve cell

Muscle cell

C. MOVEMENT OF SUBSTANCES ACROSS PLASMA MEMBRANES

In general, substances move across plasma membranes by two principal kinds of processes—passive and active. In ***passive (physical) processes***, substances move because of differences in concentration (or pressure) from regions of higher concentration (or pressure) to regions of lower concentration (or pressure). The movement usually continues until an equilibrium (even distribution) of substances is accomplished. Passive processes are the result of the kinetic energy (energy of motion) of the substances themselves and the cell does not expend energy to move the substances. Examples of passive processes are diffusion, facilitated diffusion, osmosis, filtration, and dialysis.

In ***active (physiological) processes***, substances may move from areas of lower to higher concentration. Moreover, cells must expend energy (ATP) to carry on active processes. Examples of active processes are active transport and endocytosis (phagocytosis and pinocytosis).

CAUTION! *Please reread Section A. "General Safety Precautions and Procedures" on page xi and Section C. "Precautions Related to Working with Reagents" on page xii at the beginning of the laboratory manual before you begin any of the following experiments. You should also read the experiments before you perform them to be sure that you understand all the procedures and safety cautions.*

1. Passive Processes

a. BROWNIAN MOVEMENT

At temperatures above absolute zero (−273°C or −460°F), all molecules are in constant random motion because of their inherent kinetic energy. This phenomenon is called ***Brownian movement***. Less energy is required to move smaller molecules or particles than larger ones. Since all molecules are constantly bombarded by the molecules surrounding them, the smaller the particle, the greater its random motion in terms of speed and distance.

PROCEDURE

1. With a medicine dropper place a drop of dilute detergent solution in a depression (concave) slide.
2. Using another medicine dropper, add a drop of dilute India ink to the first drop in the depression slide.
3. Stir the two solutions with a toothpick and cover the depression with a cover glass.
4. Let the slide stand for ten minutes and then place it on your microscope stage and observe it under high power.
5. Using forceps, place the slide on a hot plate and warm it for 15 seconds. Then remove the slide with forceps and observe it again under high power.
6. Record your observations in Section C.1.a of the LABORATORY REPORT RESULTS at the end of the exercise.

b. DIFFUSION

Diffusion is the net (greater) movement of molecules or ions from a region of higher concentration to a region of lesser concentration until they are evenly distributed (equilibrium). An example of diffusion in the human body is the movement of oxygen and carbon dioxide between body cells and blood.

PROCEDURE

1. To demonstrate diffusion of a solid in a liquid, *using forceps, carefully* place a large crystal of potassium permanganate ($KMnO_4$) into a test tube filled with water. CAUTION! *Avoid contact of $KMnO_4$ with your skin by using acid- or caustic-resistant gloves.*
2. Place the tube in a rack against a white background where it will not be disturbed.
3. Note the diffusion of the crystal material through the water at 15-minute intervals for 120 minutes (2 hours).
4. Record the diffusion of the crystal in millimeters (mm) per minute at 15-minute intervals in Section C.1.b of the LABORATORY REPORT RESULTS at the end of the exercise. Simply measure the distance of diffusion using a millimeter ruler.

PROCEDURE

1. To demonstrate diffusion of a solid in a solid, *using forceps, carefully* place a large crystal of methylene blue on the surface of agar in the center of a petri plate.

2. Note the diffusion of the crystal through the agar at 15-minute intervals for 120 minutes (2 hours).

3. Record the diffusion of the crystal in millimeters (mm) per minute at 15-minute intervals, using a millimeter ruler in Section C.1.b of the LABORATORY REPORT RESULTS at the end of the exercise.

c. FACILITATED DIFFUSION

Facilitated diffusion is the movement of a substance across a selectively permeable membrane from a region of higher to lower concentration with the assistance of integral proteins and/or lipids that serve as carriers in the plasma membrane. Different sugars, especially glucose, cross plasma membranes by facilitated diffusion.

In this experiment, the substance undergoing facilitated diffusion is neutral red which is red at a pH just below 7 but is yellow at a pH between 7–8.

PROCEDURE

1. In a 125 ml flask, combine one gram (1 g) bakers' yeast with 25 ml 0.75% Na_2CO_3 (sodium carbonate) solution. Swirl until the yeast is evenly suspended in the solution.

2. Divide the suspension into two large test tubes marked "U1" (unboiled) and "B1" (boiled) with a wax pencil.

3. Place tube "B1" in a boiling water bath for 2–3 minutes. Be sure that all the suspension is below water level so that all the yeast cells will be killed. CAUTION! *Make sure that the mouth of the test tube is pointed away from you and all other persons in the area.*

4. Using a test tube holder, place both test tubes into a rack and add 7.5 ml of 0.02% neutral red to each.

5. Record the color of the solution in both test tubes in Section C.1.c of the LABORATORY REPORT RESULTS at the end of the exercise.

6. After 15 minutes, place 1 ml from each test tube into test tubes marked "U2" and "B2."

7. Add enough drops of 0.75% acetic acid to tube "B2" to make its color identical to that of tube "U2."

8. Mark two microscope slides "U" and "B"; examine a sample from both tubes under high power (using a cover slip) and note which cells are stained. ____________________

____

9. Filter half of the solution remaining in tube "B1" and examine the filtrate. Repeat for tube "U1."

10. Observe the color of the filtrate and that of the yeast cells on the filter paper. ____________

____

11. To the suspension remaining in tubes "U1" and "B1" add about 9 ml 0.75 acetic acid. Filter again according to the instructions in step 9.

12. Record the color of these "U1" and "B1" filtrates and cells in Section C.1.c of the LABORATORY REPORT RESULTS at the end of the exercise.

13. How does boiling affect the facilitated diffusion of neutral red by the cells? ____________

____

Describe the membrane's permeability to neutral red. ____________________

____

Did acetic acid enter the living cells? ________

____

Did sodium carbonate enter the living cells?

____

d. OSMOSIS

Osmosis is the net movement of *water* through a selectively permeable membrane from a region of higher water (lower solute) concentration to a region of lower water (higher solute) concentration. In the body, fluids move between cells as a result of osmosis.

PROCEDURE

1. Refer to the osmosis apparatus in Figure 3.2.

2. Tie a knot very tightly at one end of a 4-in. piece of cellophane dialysis tubing that has been soaking in water for a few minutes and fill

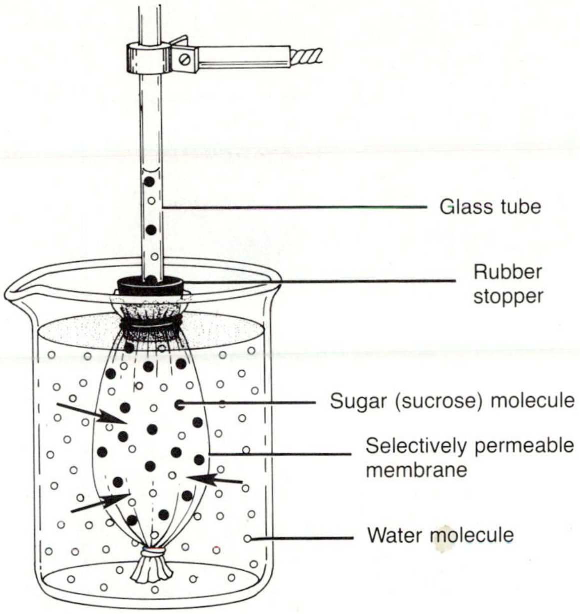

FIGURE 3.2 Osmosis apparatus.

the dialysis tubing with a 10% sugar (sucrose) solution that has been colored with Congo red (red food coloring may also be used).

3. Close the open end of the dialysis tubing with a one-hole rubber stopper into which a glass tube has already been inserted by a laboratory assistant or your instructor.

CAUTION! *If you are unfamiliar with the procedure, do not attempt to insert the glass tube into the stopper yourself since it may break and result in serious injury.*

4. Tie a piece of string tightly around the dialysis tubing to secure it to the stopper.

5. Secure and suspend the glass tube and dialysis tubing by means of a clamp attached to a ring stand.

6. Insert the dialysis tubing into a beaker or flask of water until the water comes up to the bottom of the rubber stopper.

7. As soon as the sugar solution becomes visible in the glass tube, mark it with a wax pencil and note the time.

8. Mark the height of liquid in the glass tube after 10-, 20-, and 30-minute intervals by using your millimeter (mm) ruler and record your results in Section C.1.d of the LABORATORY REPORT RESULTS at the end of the exercise.

e. HEMOLYSIS AND CRENATION

Osmosis can also be understood by noting the effects of different water concentrations on red blood cells. Red blood cells maintain their normal shape when placed in an ***isotonic solution*** (that is, one having the same salt concentration; 0.90% solution of NaCl [sodium chloride] is isotonic to red blood cells). If, however, red blood cells are placed in a ***hypotonic solution*** (concentration of NaCl lower than 0.90%), a net movement of water into the cells occurs, causing the cells to swell and possibly burst. The rupture of blood cells in this manner with the loss of hemoglobin into the surrounding liquid is called ***hemolysis*** (hē-MOL-i-sis). If, instead, red blood cells are placed in a ***hypertonic solution*** (concentration of NaCl higher than 0.90%), a net movement of water out of the cells occurs, causing them to shrink. This shrinkage is known as ***crenation*** (kre-NĀ-shun).

PROCEDURE

CAUTION! *Please reread Section B. "Precautions Related to Working with Blood, Blood Products, or Other Body Fluids" on page xi at the beginning of the laboratory manual before you begin any of the following experiments. You should also read the experiments before you perform them to be sure that you understand all the procedures and safety precautions. When you finish this part of the exercise, place the toothpicks, medicine droppers, microscope slides, and cover slips in a fresh bleach solution.*

1. With a wax marking pencil, mark three microscope slides as follows: 0.90%, DW (distilled water), and 3%.

2. Using a medicine dropper, place three drops of fresh (uncoagulated) ox blood on a microscope slide that contains 2 milliliters (ml) of a 0.90% NaCl solution (isotonic solution). Mix gently and thoroughly with a clean toothpick.

3. Using a medicine dropper, place three drops of fresh blood on a microscope slide that contains 2 ml of distilled water (hypotonic solution). Mix gently and thoroughly with a clean toothpick.

4. Now, using a medicine dropper, add three drops of fresh blood to a microscope slide that contains 2 ml of a 3% NaCl solution (hypertonic solution). Mix gently and thoroughly with a clean toothpick.

5. Using a medicine dropper, place two drops of the red blood cells in the isotonic solution on another microscope slide, cover with a cover slip, and examine the red blood cells under high power. Reduce your illumination.

What is the shape of the cells? round, little

.9% dots

Explain their shape. round

6. Using a medicine dropper, place two drops of the red blood cells in the hypotonic solution on another microscope slide, cover with a cover slip, and examine the red blood cells under high power. Reduce your illumination.

What is the shape of the cells? mass of

red.

.3 Explain their shape. non-distinguishable

7. Using a medicine dropper, place two drops of the red blood cells in the hypertonic solution on another microscope slide, cover with a cover slip, and examine the red blood cells under high power. Reduce your illumination.

What is the shape of the cells? small

H_2O circular

Explain their shape.

f. FILTRATION

Filtration is the movement, under the influence of gravity, of solvents and dissolved substances across a selectively permeable membrane from regions of higher pressure to regions of lower pressure. The pressure is called hydrostatic pressure and is the result of the solvent, usually water. In general, any substance having a molecular weight of less than 100 is filtered because the pores in the filter paper (membrane) are larger than the molecules of the substance. Filtration is one mechanism by which the kidneys regulate the chemical composition of the blood.

PROCEDURE

1. Refer to Figure 3.3, the filtration apparatus.
2. Fold a piece of filter paper in half and then in half again.

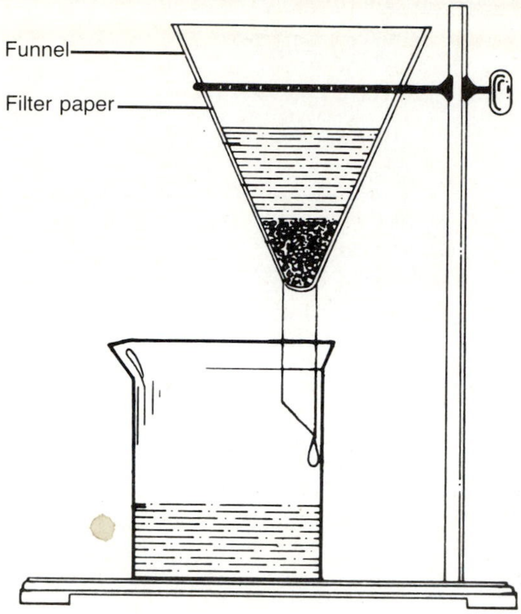

FIGURE 3.3 Filtration apparatus.

3. Open it into a cone, place it in a funnel, and place the funnel over the beaker.
4. Shake a mixture of a few particles of powdered wood charcoal (black), 1% copper sulfate (blue), boiled starch (white), and water and pour it into the funnel until the mixture almost reaches the top of the filter paper. Gravity will pull the particles through the pores of the membrane (that is, the filter paper).
5. Count the number of drops passing through the funnel for the following time intervals (10, 30, 60, 90, 120 seconds) and record your observations in Section C.1.f of the LABORATORY REPORT RESULTS at the end of the exercise.
6. Observe which substances passed through the filter paper by noting their color in the filtered fluid in the beaker.
7. Examine the filter paper to determine whether any colored particles were not filtered.
8. To determine if any starch is in the filtrate (liquid in the beaker), add several drops of 0.01 M IKI solution. A blue-black color reaction indicates the presence of starch.

g. DIALYSIS

Dialysis (dī-AL-i-sis) is the separation of smaller molecules from larger ones by using a selectively permeable membrane that permits diffusion of the smaller molecules but not the larger ones. The principle of dialysis is employed in artificial kidneys.

PROCEDURE

1. Refer to the dialysis apparatus in Figure 3.4.
2. Tie off one end of a piece of dialysis tubing that has been soaking in water. Place a prepared solution containing starch, sodium chloride, 5% glucose, and albumin into the dialysis tubing.
3. Tie off the other end of the dialysis tubing and immerse into a beaker of distilled water.
4. After 1 hour, test the solution in the beaker for the presence of each of the substances in the tubing, as follows, and record your observations in Section C.1.g of the LABORATORY REPORT RESULTS at the end of the exercise.

a. ***Albumin***—*Carefully* add several drops of concentrated nitric acid to a test tube containing 2 ml of the solution in the beaker. CAUTION! *Be extremely careful using nitric acid. It can severely damage your skin. Use acid- or caustic-resistant gloves*. Positive reaction = white coagulate.

b. ***Sugar***—Test 5 ml of the solution in the beaker in a test tube with 5 ml of Benedict's solution. Place the test tube in a boiling water bath for three minutes. CAUTION! *Make sure that the mouth of the test tube is pointed away from you and all other persons in the area.*

Using a test tube holder, remove the test tube from the water bath. Note the color. Positive reaction = green, yellow, orange, or red precipitate.

c. ***Starch***—Add several drops of IKI solution to 2 ml of the solution in the beaker in a test tube. Note the color. Positive reaction = blue-black color.

d. ***Sodium chloride***—Place 2 ml of the solution in the beaker in a test tube and add several drops of 1% silver nitrate. Note the color. Positive reaction = white precipitate.

2. Active Processes

a. ACTIVE TRANSPORT

Active transport is a process by which substances, usually ions, are transported across a plasma membrane from an area of lower to one of higher concentration. An example is the movement of sodium (Na^+) and potassium (K^+) ions across nerve cell membranes during nerve impulse conduction. Because reliable results are difficult to demonstrate simply, you will not be asked to demonstrate active transport.

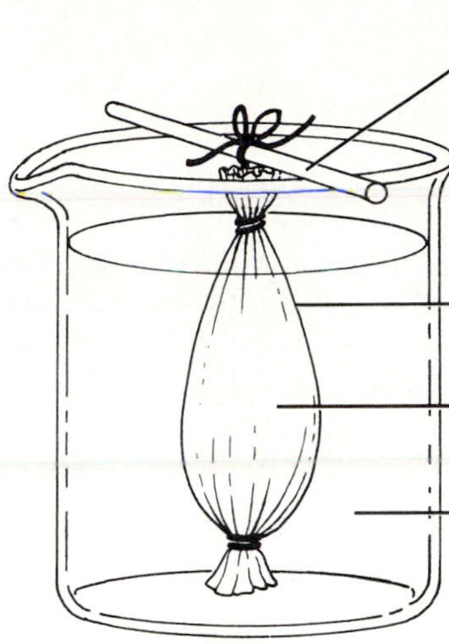

FIGURE 3.4 Dialysis apparatus.

b. ENDOCYTOSIS

Endocytosis refers to the passage of large molecules and particles across a plasma membrane, in which a segment of the membrane surrounds the substance, encloses it, and brings it into the cell. Here we will consider two types of endocytosis—phagocytosis and pinocytosis.

(1) Phagocytosis

Phagocytosis (fag′-ō-sī-TŌ-sis), or cell "eating," is the engulfment of solid particles or organisms by *pseudopodia* (temporary fingerlike projections of cytoplasm) of the cell. Once the particle is surrounded by the membrane, the membrane folds inward, pinches off from the rest of the plasma membrane, and forms a vacuole around the particle. Enzymes are secreted into the vacuole or the vacuole combines with a lysosome and the particle is digested.

Phagocytosis can be demonstrated by observing the feeding of an amoeba, a unicellular organism whose movement and ingestion are similar to those of human white blood cells (leucocytes).

PROCEDURE

1. Using a medicine dropper, place a drop of culture containing amoebae that have been starved for 48 hours into the well of a depression slide and cover the well with a cover slip. Cultures containing *Chaos chaos* or *Amoeba proteus* should be used. (Your instructor may wish to use the hanging-drop method instead. If so, she or he will give you verbal instructions.)
2. Examine the amoebae under low power, and be sure that your light is reduced considerably.
3. Observe the locomotion of an amoeba for several minutes. Pay particular attention to the pseudopodia that appear to flow out of the cell.
4. To observe phagocytosis, use a medicine dropper and add a drop containing small unicellular animals called *Tetrahymena pyriformis* to the culture containing the amoebae.
5. Examine under low power, and observe the ingestion of *Tetrahymena pyriformis* by an amoeba. Note the action of the pseudopodia and the formation of the vesicle around the ingested organism.

(2) Pinocytosis

Pinocytosis (pi-nō-sī-TŌ-sis), or cell "drinking," is the engulfment of a liquid. The liquid is attracted to the surface of the membrane, and the membrane folds inward, surrounds the liquid, and detaches from the rest of the intact membrane.

D. CELL INCLUSIONS

Cell inclusions are a large and diverse group of mostly organic substances produced by cells, that may appear or disappear at various times in the life of a cell. Using your textbook as a reference, indicate the function of the following cell inclusions:

1. ***Melanin*** ______________________
2. ***Glycogen*** ______________________
3. ***Lipids*** ______________________

E. EXTRACELLULAR MATERIALS

Substances that lie outside the plasma membranes of body cells are referred to as ***extracellular materials***. They include body fluids, such as interstitial fluid and plasma, which provide a medium for dissolving, mixing, and transporting substances. Extracellular materials also include special substances that form the matrix (substance between cells) in which some cells are embedded.

Matrix materials are produced by certain cells and deposited outside their plasma membranes where they support cells, bind them together, and provide strength and elasticity. Some matrix materials have no definite shape and are referred to as *amorphous*. These include hyaluronic (hī′-a-loo-RON-ik) acid and chondroitin (kon-DROY-tin) sulfate. Other matrix materials are *fibrous* (threadlike). Examples include collagenous, reticular, and elastic fibers.

Using your textbook as a reference, indicate the location and function for each of the following matrix materials:

Hyaluronic acid

Location ______________________

____

____

____

Function ______________________

____

____

____

Chondroitin sulfate

Location ______________________

____

____

____

Function ______________________

____

____

____

Collagenous fibers

Location ________________________________

____

____

____

Function ________________________________

____

____

____

Reticular fibers

Location ________________________________

____

____

____

Function ________________________________

____

____

____

Elastic fibers

Location ________________________________

____

____

____

Function ________________________________

____

____

____

F. CELL DIVISION

Cell division is the basic mechanism by which cells reproduce themselves. It consists of a nuclear division and a cytoplasmic division. Since nuclear division may be of two types, two kinds of cell division are recognized. In the first type, called ***somatic cell division***, a single starting cell called a ***parent cell*** duplicates itself and the result is two identical cells called ***daughter cells***. Somatic cell division consists of a nuclear division called ***mitosis*** and a cytoplasmic division called ***cytokinesis*** (sī-tō-ki-NĒ-sis) and provides the body with a means of growth and of replacement of diseased or damaged cells (Figure 3.5). The second type of cell division is called ***reproductive cell division*** and is the mechanism by which sperm and eggs are produced (Exercise 25). Reproductive cell division consists of a nuclear division called ***meiosis*** and two cytoplasmic divisions (cytokinesis) and results in the development of four nonidentical daughter cells.

In order to study somatic cell division, obtain a prepared slide of a whitefish blastula and examine it under high power.

A cell between divisions is said to be in ***interphase*** of the cell cycle. Interphase is the longest part of the cell cycle and is the period of time during which a cell carries on its physiological activities. One of the most important activities of interphase is the replication of DNA so that the two daughter cells that eventually form will each have the same kind and amount of DNA as the parent cell. Scan your slide and find a cell in interphase. Such a parent cell is characterized by a clearly defined nuclear membrane. Within the nucleus look for the nucleolus, the karyolymph or nucleoplasm (amorphous nuclear material), and DNA (which is associated with protein in the form of a granular substance called ***chromatin***). Also locate a pair of centrioles. Make a labeled diagram of an interphase cell in the space provided.

Once a cell completes its interphase activities, mitosis begins. Mitosis is the distribution of two sets of chromosomes into two separate and equal nuclei after replication of the chromosomes of the parent cell, an event that takes place in the interphase preceding mitosis. Although a continuous process, mitosis is divided into four stages for purposes of study: prophase, metaphase, anaphase, and telophase.

1. ***Prophase***—The initial stage of mitosis takes about one-half the total time required for cell

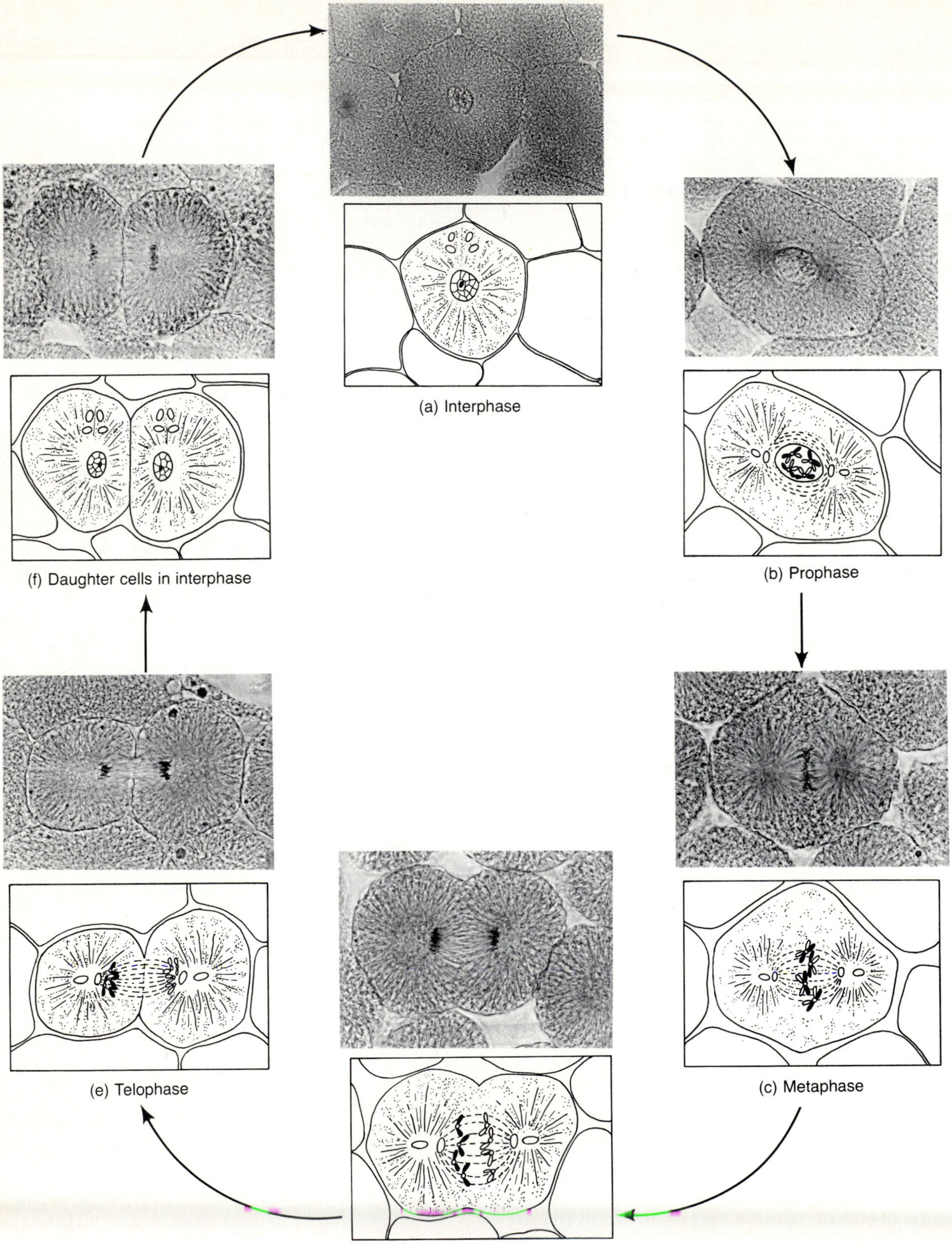

FIGURE 3.5 Cell division: mitosis and cytokinesis. Photomicrographs and diagrams of the various stages of cell division in whitefish eggs.

division and is characterized by a series of rather conspicuous events. The paired centrioles separate and move to opposite poles (ends) of the cell. Once in position they project a series of ***continuous microtubules*** that grow toward each other, connect, and thus extend from one pole of the cell to another. As this is happening, ***chromosomal microtubules*** grow out of each centriole and extend to both poles of the cell. Together, the continuous and chromosomal microtubules constitute the ***mitotic spindle*** and, with the centrioles, are referred to as the ***mitotic apparatus***. At the same time the microtubules are forming, the chromatin shortens and thickens into distinct rod-shaped bodies, the ***chromosomes***; nucleoli become less distinct; and the nuclear membrane disappears. Careful examination of prophase chromosomes reveals that each consists of two separate units called ***chromatids***, joined at a point called a ***centromere***. At the end of prophase, the chromatid pairs move toward the center of a cell in a region called the ***equatorial plane***. Isolate a cell in prophase, study it carefully, and draw a labeled diagram in the space provided.

2. ***Metaphase***—In the second stage of mitosis, the centromeres of the chromatid pairs line up along the equatorial plane of the cell. The centromere of each chromatid pair forms chromosomal microtubules that attach the centromere to a pole of the cell. Next, the lengthwise separation of the chromatids takes place. After separation, each chromatid is referred to as a chromosome. Draw and label a cell in metaphase in the space provided.

3. ***Anaphase***—The third stage of mitosis is characterized by the division of the centromeres and the movement of complete identical sets of chromosomes to opposite poles of the cell. During this movement, the centromeres attached to chromosomal microtubules lead, while the rest of the chromosome trails like a streamer. Draw and label a cell in anaphase in the space provided.

4. ***Telophase***—The final stage of mitosis consists of a series of events nearly the reverse of prophase. When the chromosome sets reach their respective poles, new nuclear membranes enclose them, chromosomes uncoil and lengthen into their chromatin form, nucleoli reappear, and the mitotic spindle disappears. The formation of two nuclei identical to those in cells of interphase terminates telophase and completes a mitotic cycle. Draw and label a cell in telophase in the space provided.

Cytokinesis begins in late anaphase and terminates in telophase. You can recognize cytokinesis in animal cells by the formation of a ***cleavage furrow*** that runs around the equatorial plane of the cell. The furrow spreads inward as a constricting ring and cuts completely through the cell, forming two portions of cytoplasm. At this point, two identical daughter cells are formed.

Following cytokinesis, each daughter cell returns to interphase. Each cell in most tissues of the body eventually grows and undergoes mitosis and cytokinesis, and a new divisional cycle begins. Examine your telophase cell again and be sure that it contains a cleavage furrow.

Using high power and starting at 12 o'clock, move around the blastula and count the number of cells in interphase and in each mitotic phase. It will be easier to do this if you imagine lines dividing the blastula into quadrants. Count the interphase cells in each quadrant, then assign each dividing cell to a specific mitotic stage. It will be hard to assign some cells to a phase, e.g., to distinguish late anaphase from early telophase. If you cannot make a decision, assign one cell to the earlier phase in question, the next cell to the later phase.

Divide the number of cells in each stage by the total number of cells counted and multiply by 100 to determine the percent of the cells in each mitotic stage at a given point in time. Record your results in Section F of the LABORATORY REPORT RESULTS at the end of the exercise.

ANSWER THE LABORATORY REPORT QUESTIONS AT THE END OF THE EXERCISE

Student ______________________ Date ______________________

Laboratory Section ______________________ Score/Grade ______________________

SECTION C. MOVEMENT OF SUBSTANCES ACROSS PLASMA MEMBRANES

1. Passive Processes

a. BROWNIAN MOVEMENT

Describe the movement of the India ink particles on the unheated slide. ______________________

____

How does this movement differ from that on the heated slide? ______________________

b. DIFFUSION

Solid in Liquid

Time	Distance in mm
15 minutes	
30 minutes	
45 minutes	
60 minutes	
75 minutes	
90 minutes	
105 minutes	
120 minutes	

Solid in Solid

Time	Distance in mm (blue)	(pink)
15 minutes	1mm	4mm
30 minutes	2mm	4mm
45 minutes	2mm	5mm
60 minutes	2mm	6mm
75 minutes		
90 minutes		
105 minutes		
120 minutes		

c. FACILITATED DIFFUSION

Neutral Red

Tube "U1" ______________________

Tube "B1" ______________________

Acetic Acid

Tube "U1" ______________________

Tube "B1" ______________________

d. OSMOSIS

Time	Height of liquid (mm)
10 minutes	______
20 minutes	______
30 minutes	blue = 4.6cm green = 13.9 yellow = 1.1

Explain what happened. ______

f. FILTRATION

10 seconds clear 14
30 seconds clear 30
60 seconds clear 50
90 seconds blue 63
120 seconds blue 64

g. DIALYSIS

Place a check in the appropriate place to indicate if the following tests are positive (+) or negative (−).

	(+)	(−)
Albumin		−
Sugar	+	
Starch		−
Sodium chloride	+	

SECTION F. CELL DIVISION

Percent of cells in interphase ______
Percent of cells in prophase ______
Percent of cells in metaphase ______
Percent of cells in anaphase ______
Percent of cells in telophase ______

CELLS

Student ______________________ Date ______________________

Laboratory Section ______________________ Score/Grade ______________________

PART I. MULTIPLE CHOICE

______ 1. The portion of the cell that forms part of the mitotic spindle during division is the (a) endoplasmic reticulum (b) Golgi complex (c) cytoplasm (d) centriole

______ 2. Movement of molecules or ions from a region of higher concentration to a region of lower concentration until they are evenly distributed is called (a) phagocytosis (b) diffusion (c) active transport (d) pinocytosis

______ 3. If red blood cells are placed in a hypertonic solution of sodium chloride, they will (a) swell (b) burst (c) shrink (d) remain the same

______ 4. The reagent used to test for the presence of sugar is (a) silver nitrate (b) nitric acid (c) IKI (d) Benedict's solution

______ 5. A cell that carries on a great deal of digestion also contains a large number of (a) lysosomes (b) centrioles (c) mitochondria (d) nuclei

______ 6. Which process does *not* belong with the others? (a) active transport (b) dialysis (c) phagocytosis (d) pinocytosis

______ 7. Movement of oxygen and carbon dioxide between blood and body cells is an example of (a) osmosis (b) active transport (c) diffusion (d) pinocytosis

______ 8. Which type of solution will cause hemolysis? (a) isotonic (b) hypotonic (c) isometric (d) hypertonic

______ 9. One process by which kidneys regulate the chemical composition of blood is (a) diffusion (b) active transport (c) filtration (d) osmosis

______ 10. Engulfment of solid particles or organisms by pseudopodia is called (a) active transport (b) dialysis (c) phagocytosis (d) filtration

______ 11. Rupture of red blood cells with subsequent loss of hemoglobin into the surrounding medium is called (a) hemolysis (b) plasmolysis (c) plasmoptysis (d) hemoglobinuria

______ 12. The area of the cell between the plasma membrane and nuclear membrane where chemical reactions occur is the (a) centrosome (b) vacuole (c) peroxisome (d) cytoplasm

______ 13. The "powerhouses" of the cell where ATP is produced are the (a) ribosomes (b) mitochondria (c) centrioles (d) lysosomes

______ 14. The sites of protein synthesis in the cell are (a) peroxisomes (b) flagella (c) ribosomes (d) centrosomes

______ 15. Which process does *not* belong with the others? (a) diffusion (b) phagocytosis (c) active transport (d) pinocytosis

______ 16. A cell inclusion that is a pigment in skin and hair is (a) glycogen (b) melanin (c) mucus (d) collagen

_____ 17. Which extracellular material is found in ligaments and tendons? (a) elastic fibers (b) chondroitin sulfate (c) collagenous fibers (d) mucus

_____ 18. The organelles that contain enzymes for the metabolism of hydrogen peroxide are (a) lysosomes (b) mitochondria (c) vacuoles (d) peroxisomes

_____ 19. The framework of cilia, flagella, centrioles, and spindle fibers is formed by (a) endoplasmic reticulum (b) collagenous fibers (c) chondroitin sulfate (d) microtubules

_____ 20. A viscous fluidlike substance that binds cells together, lubricates joints, and maintains the shape of the eyeballs is (a) elastin (b) hyaluronic acid (c) mucus (d) plasmin

PART II. COMPLETION

21. The external boundary of the cell through which substances enter and exit is called the _____________ .

22. The cytoskeleton is formed by microtubules, intermediate filaments, and _____________ .

23. The portion of the cell that contains hereditary information is the _____________ .

24. The tail of a sperm cell is a long whiplash structure called a(n) _____________ .

25. Cells placed in a _____________ solution will undergo hemolysis.

26. Division of cytoplasm is referred to as _____________ .

27. Lipid and protein secretion, carbohydrate synthesis, and assembly of glycoproteins are functions of the _____________ .

28. Storage of digestive enzymes is accomplished by the _____________ of a cell.

29. Projections of cells that move substances along their surfaces are called _____________ .

30. The _____________ provides a surface area for chemical reactions, a pathway for transporting molecules, and a storage area for synthesized molecules.

31. _____________ is a cell inclusion that represents stored glucose in the liver and skeletal muscles.

32. A jellylike substance that supports cartilage, bone, heart valves, and the umbilical cord is _____________ .

33. The framework of many soft organs is formed by _____________ fibers.

34. The net movement of water through a selectively permeable membrane from a region of higher concentration of water to a region of lower concentration of water is known as _____________ .

35. The principle of _____________ is employed in the operation of an artificial kidney.

36. In an interphase cell, DNA is in the form of a granular substance called _____________ .

37. Distribution of chromosomes into separate and equal nuclei is referred to as _____________ .

38. The constant random motion of molecules due to their inherent kinetic energy is called _____________ .

PART III. MATCHING

_____ 39. Anaphase
_____ 40. Metaphase
_____ 41. Interphase
_____ 42. Telophase
_____ 43. Prophase

A. Mitotic spindle appears
B. Movement of chromosome sets to opposite poles of cell
C. Centromeres line up on equatorial plane
D. Formation of two identical nuclei
E. Phase between divisions

TISSUES 4

A *tissue* is a group of similar cells and their intercellular substance that have a similar embryological origin and operate together to perform a specific function. The cells constituting a tissue are usually found near one another. The study of tissues is called ***histology*** (hiss-TOL-ō-jē; *histio* = tissue; *logos* = study of) or ***microscopic anatomy***. The various body tissues can be categorized into four principal kinds: (1) epithelial, (2) connective, (3) muscular, and (4) nervous. In this exercise you will examine the structure and functions of epithelial and connective tissues, except for bone or blood. Other tissues will be studied later as parts of the systems to which they belong.

A. COVERING AND LINING EPITHELIUM

Epithelial (ep′-i-THĒ-lē-al) ***tissue*** is avascular (without blood vessels) and covers and lines body surfaces, including hollow organs and body cavities. Epithelium also forms glands and is the tissue from which gametes (sperm and eggs) develop. Epithelial tissue consists entirely or largely of closely packed cells with little or no intercellular material (matrix) and the cells are arranged in continuous sheets, either single (simple) or multilayered (stratified). Also, epithelial tissue is attached to an underlying connective tissue by a ***basement membrane***. Finally, epithelium retains the ability to renew itself by cell division.

Before you start your microscopic examination of epithelial tissues, refer to Figure 4.1. Study the tissues carefully to familiarize yourself with their general structural characteristics. For each of the types of epithelium listed, obtain a prepared slide and, unless otherwise specified by your instructor, examine each under high power. In conjunction with your examination, consult a textbook of anatomy.

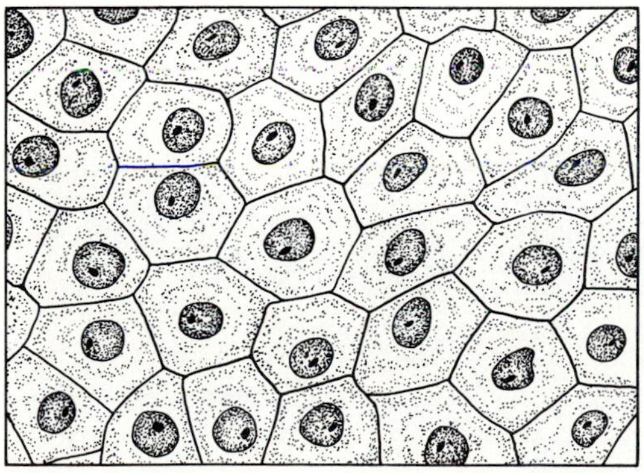

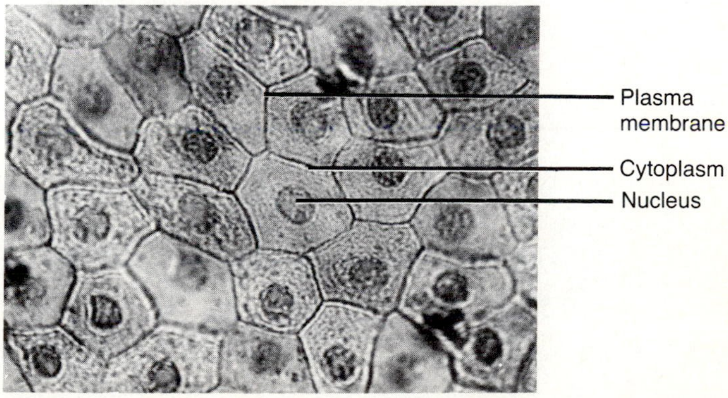

(a)

FIGURE 4.1 Epithelial tissues. Photomicrographs are labeled, while the line drawing of the same tissue is unlabeled. (a) Simple squamous epithelium, surface view.

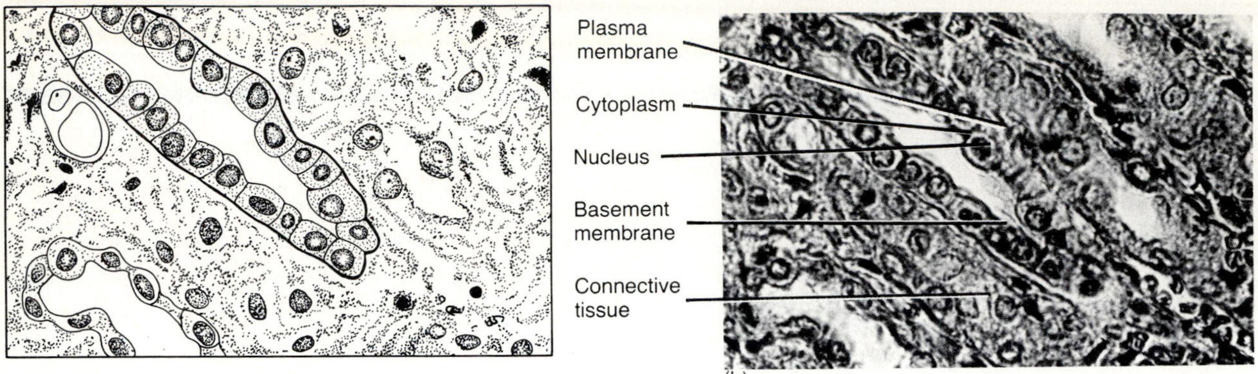

(b)

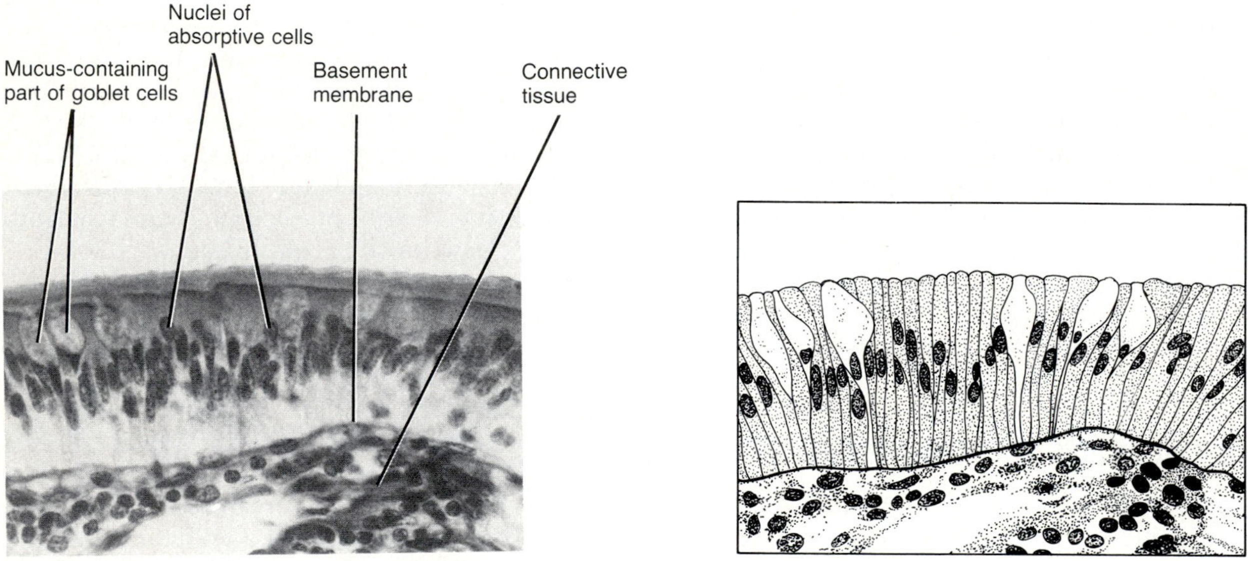

(c)

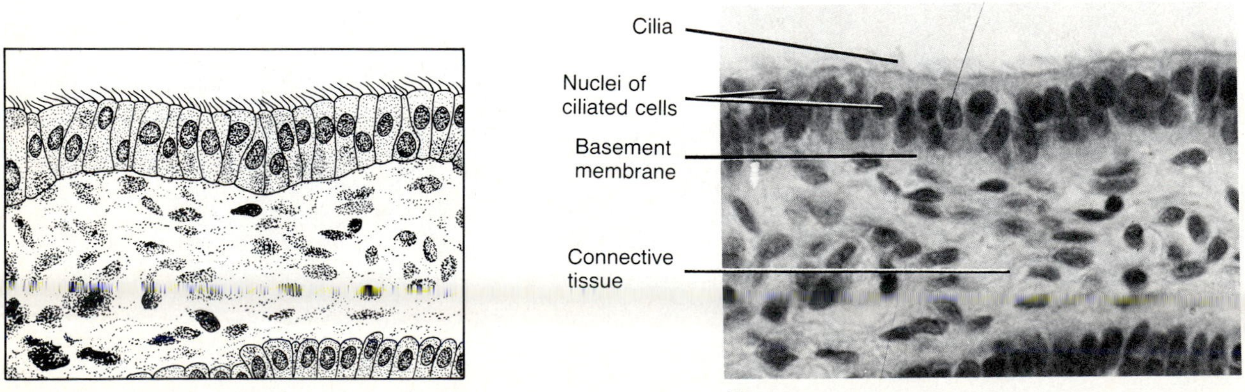

(d)

FIGURE 4.1 ***(Continued)*** Epithelial tissues. (b) Simple cuboidal epithelium. (c) Simple columnar (nonciliated) epithelium. (d) Simple columnar (ciliated) epithelium.

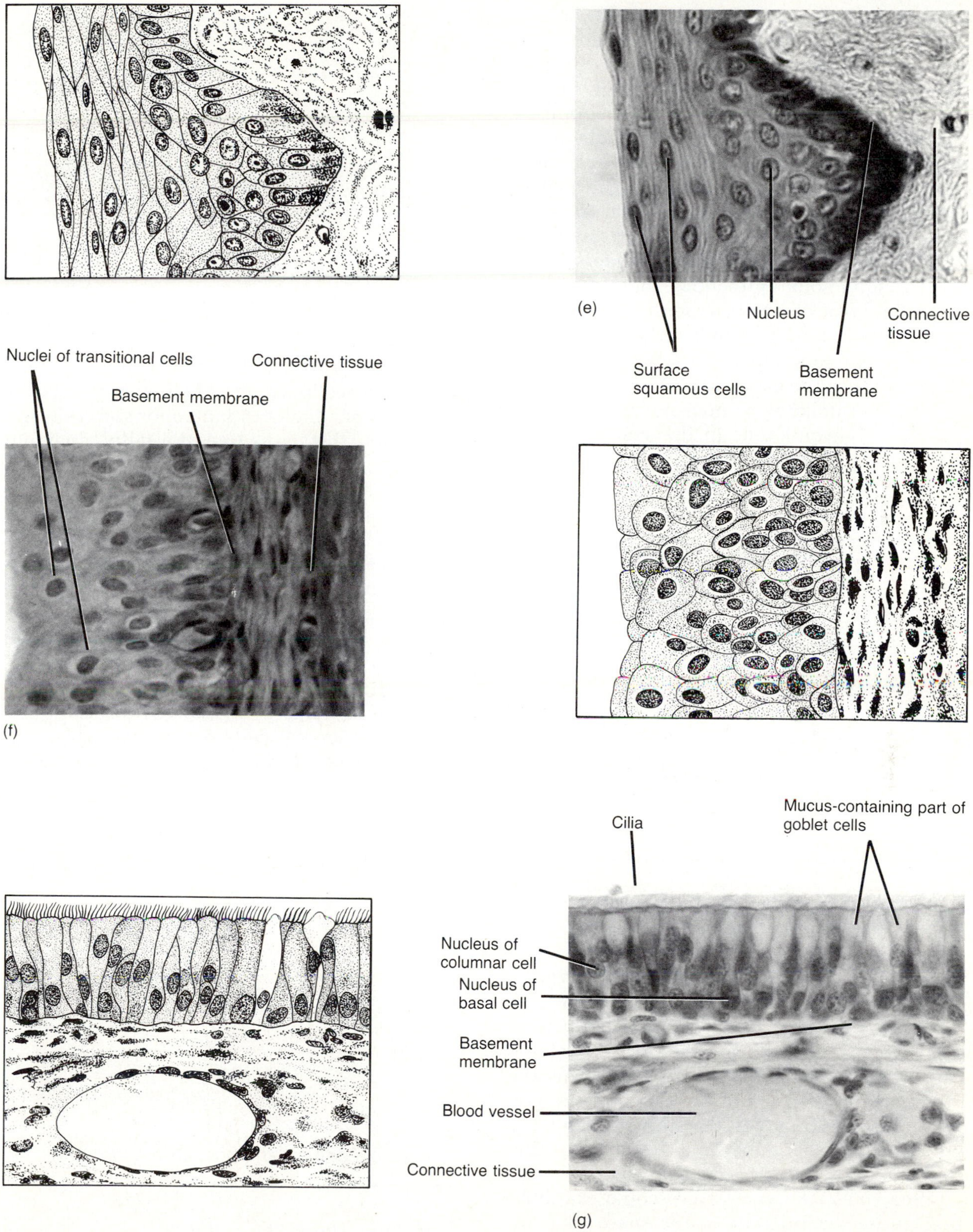

FIGURE 4.1 ***(Continued)*** Epithelial tissues. (e) Stratified squamous epithelium. (f) Stratified transitional epithelium. (g) Pseudostratified epithelium.

1. ***Simple squamous*** (SKWĀ-mus)—This tissue consists of a single layer of flat, scale-like cells and is highly adapted for diffusion, osmosis, and filtration because of its thinness. Simple squamous tissue lines the air sacs of the lungs, glomerular capsule (filtering unit) of the kidneys, inner surface of the wall of the membranous labyrinth of the inner ear, and inner surface of the tympanic membrane of the ear. Simple squamous tissue that lines the heart, blood vessels, and lymphatic vessels and forms capillary walls is called ***endothelium***. Simple squamous epithelium that forms the epithelial layer of a serous membrane is called ***mesothelium***. Serous membranes line the thoracic and abdominopelvic cavities and cover viscera within the cavities.

After you make your microscopic examination, draw several cells in the space that follows and label plasma membrane, cytoplasm, and nucleus.

Simple squamous epithelium

2. ***Simple cuboidal***—This tissue consists of a single layer of cube-shaped cells. When the tissue is sectioned at right angles, its cuboidal nature is obvious. Highly adapted for secretion and absorption, cuboidal tissue covers the surface of the ovaries; lines the smaller ducts of some glands and the anterior surface of the capsule of the lens of the eye; and forms the pigmented epithelium of the retina of the eye, part of the tubules of the kidneys, and the secreting units of other glands, such as the thyroid gland.

After you make your microscopic examination, draw several cells in the space that follows and label plasma membrane, cytoplasm, nucleus, basement membrane, and connective tissue layer.

Simple cuboidal epithelium

3. ***Simple columnar (nonciliated)***—This tissue consists of a single layer of columnar cells and, when sectioned at right angles, these cells appear as rectangles. Adapted for secretion and absorption, this tissue lines the gallbladder and the gastrointestinal tract from the cardia of the stomach to the anus and is found in the excretory ducts of many glands. In these locations, the cells protect underlying tissues. Some of the cells are modified in that the plasma membranes are folded into ***microvilli*** that increase the surface area for absorption. Other cells are modified as ***goblet cells*** that store and secrete mucus to protect the lining of the gastrointestinal tract.

After you make your microscopic examination, draw several cells in the space that follows and label plasma membrane, cytoplasm, nucleus, goblet cell, absorptive cell, basement membrane, and connective tissue layer.

Simple columnar nonciliated epithelium

4. ***Simple columnar (ciliated)***—This type of epithelium consists of a single layer of columnar absorptive, goblet, and ciliated cells. It lines some portions of the upper respiratory

tract, uterine (Fallopian) tubes, uterus, some paranasal sinuses, and the central canal of the spinal cord. Mucus produced by goblet cells forms a thin film over the surface of the tissue, and movements of the cilia propel substances over the surface of the tissue.

After you make your microscopic examination, draw several cells in the space that follows and label plasma membrane, cytoplasm, nucleus, cilia, goblet cell, basement membrane, and connective tissue layer.

Simple columnar ciliated epithelium

5. ***Stratified squamous***—This tissue consists of several layers of cells. The superficial cells are flat whereas cells of the deep layers vary in shape from cuboidal to columnar. The basal (bottom) cells continually multiply by cell division. As surface cells are sloughed off, new cells replace them from the basal layer. The surface cells of ***keratinized stratified squamous*** contain a waterproofing protein called ***keratin*** that also resists friction and bacterial invasion. The keratinized variety forms the outer layer of the skin. Surface cells of ***nonkeratinized stratified squamous*** do not contain keratin. The nonkeratinized variety lines wet surfaces such as the tongue, mouth, esophagus, part of the epiglottis, and vagina. Both varieties of stratified squamous afford considerable protection against friction.

After you make your microscopic examination, draw several cells in the space that follows and label plasma membrane, cytoplasm, nucleus, squamous surface cells, basal cells, basement membrane, and connective tissue layer.

Stratified squamous epithelium

6. ***Stratified squamous (student prepared)***—Before examining the next slide, prepare a smear of cheek cells from the epithelial lining of the mouth. As noted previously, epithelium that lines the mouth is nonkeratinized stratified squamous epithelium. However, you will be examining surface cells only, and these will appear similar to simple squamous epithelium.

PROCEDURE

a. Using the blunt end of a toothpick, *gently* scrape the lining of your cheek several times to collect some surface cells of the stratified squamous epithelium.

b. Now move the toothpick across a clean glass microscope slide until a thin layer of scrapings is left on the slide.

c. Allow the preparation to air dry.

d. Next, cover the smear with several drops of 1% methylene blue stain. After about 1 minute, gently rinse the slide in cold tap water or distilled water to remove excess stain.

e. Gently blot the slide dry using a paper towel.

f. Examine the slide under low and high power. See if you can identify the plasma membrane, cytoplasm, nuclear membrane, and nucleoli. Some bacteria are commonly found on the slide and usually appear as very small rods or spheres.

7. ***Stratified transitional***—This tissue resembles nonkeratinized stratified squamous, except that the superficial cells are larger and

more rounded. When stretched, the surface cells are drawn out into squamouslike cells. This drawing out permits the tissue to stretch without the outer cells breaking apart from one another. The tissue lines parts of the urinary system, such as the urinary bladder, parts of the ureters, and urethra that are subject to expansion from within.

After you have made your microscopic examination, draw several cells in the space that follows and label plasma membrane, cytoplasm, nucleus, surface cells, basement membrane, and connective tissue layer.

Stratified transitional epithelium

8. ***Pseudostratified***—Nuclei of columnar cells in this tissue are at varying depths, and, although all of the cells are attached to the basement membrane in a single layer, some do not reach the surface. This arrangement gives the impression of multilayered tissue, thus the name pseudostratified. The nonciliated type lines large excretory ducts of many glands, epididymis, parts of the male urethra, and parts of the auditory (Eustachian) tubes. The ciliated type contains goblet cells and lines most of the upper respiratory tract and certain ducts of the male reproductive system. Ciliated tissue is highly adapted for secretion and movement of substances by ciliary action.

After you have made your microscopic examination, draw several cells in the space that follows and label plasma membrane, cytoplasm, nucleus, cilia, goblet cell, basement membrane, and connective tissue layer.

Pseudostratified epithelium

B. GLANDULAR EPITHELIUM

A ***gland*** may consist of a single epithelial cell or a group of highly specialized epithelial cells that secrete various substances. Glands that have no ducts (ductless), secrete hormones, and release their secretions into the blood are called ***endocrine glands***. Examples include the pituitary, thyroid, and adrenals (Exercise 15). Glands that secrete their products into ducts are called ***exocrine glands***. Examples include sweat glands and salivary glands.

1. Structural Classification of Exocrine Glands

Based on the shape of the secretory portion and the degree of branching of the duct, exocrine glands can be structurally classified as follows:

a. ***Unicellular***—One-celled glands that secrete mucus. An example is the goblet cell (see Figure 4.1c). These cells line portions of the respiratory and digestive systems.

b. ***Multicellular***—Many-celled glands that occur in several different forms (see Figure 4.2).

1. ***Simple***—Single, nonbranched duct.
   Tubular—Secretory portion is straight and tubular (intestinal glands).
   Branched tubular—Secretory portion is branched and tubular (gastric and uterine glands).
   Coiled tubular—Secretory portion is coiled (sudoriferous [soo-dor-IF-er-us], or sweat, glands).

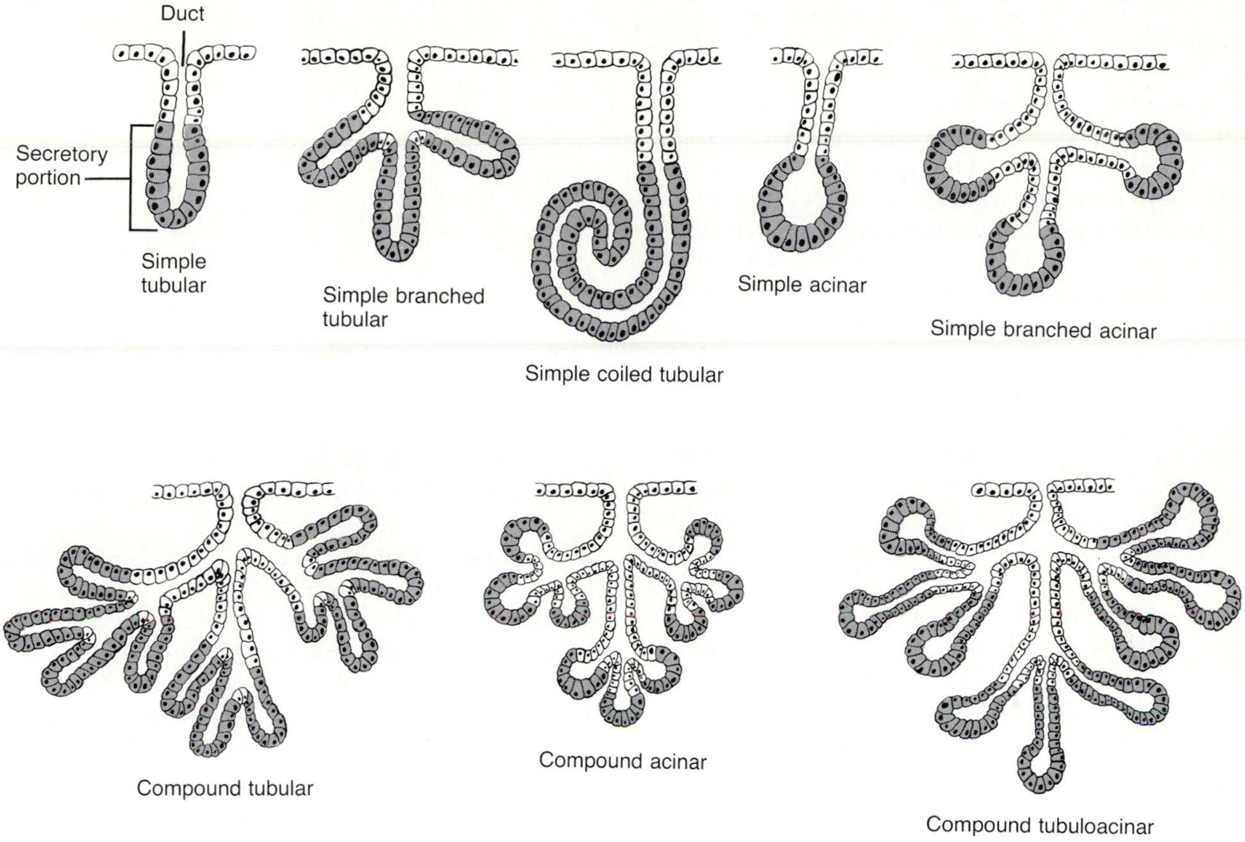

FIGURE 4.2 Structural classification of multicellular exocrine glands.

Acinar (AS-i-nar)—Secretory portion is flasklike (seminal vesicle glands).

Branched acinar—Secretory portion is branched and flasklike (sebaceous [se-BĀ-shus], or oil, glands).

2. *Compound*—Branched duct.

   Tubular—Secretory portion is tubular (bulbourethral or Cowper's glands, testes, liver).

   Acinar—Secretory portion is flasklike (sublingual and submandibular salivary glands).

   Tubuloacinar—Secretory portion is both tubular and flasklike (parotid salivary glands, pancreas).

Obtain a prepared slide of a representative of each of the types of multicellular exocrine glands just described. As you examine each slide, compare your observations to the diagrams of the glands in Figure 4.2.

2. Functional Classification of Exocrine Glands

Functional classification is based on whether a secretion is a product of a cell or consists of entire or partial glandular cells themselves. ***Holocrine glands***, such as sebaceous (oil) glands, accumulate their secretory product in their cytoplasm. The cell then dies and is discharged with its contents as the glandular secretion; the discharged cell is replaced by a new one. ***Apocrine glands***, such as the sudoriferous (sweat) glands in the axilla, accumulate secretory products at the outer margins of the secreting cells. The margins pinch off as the secretion and the remaining portions of the cells are repaired so that the process can be repeated. ***Merocrine (eccrine) glands***, such as the pancreas and salivary glands, produce secretions that are simply formed by the secretory cells and then discharged into a duct.

C. CONNECTIVE TISSUE

Connective tissue, the most abundant tissue in the body, functions by protecting, supporting, separating structures (e.g., skeletal muscles), and binding structures together. It is typically well vascularized, except for cartilage. Connective tissue contains a large amount of intercellular substance (matrix) that surrounds widely scattered cells. This tissue usually does not occur on free surfaces.

Before you start your microscopic examination of connective tissues, refer to Figure 4.3. Study the tissues carefully to familiarize yourself with their general structural characteristics. For each type of connective tissue listed, obtain a prepared slide and, unless otherwise specified by your instructor, examine each under high power.

1. ***Loose*** or ***areolar*** (a-RĒ-ō-lar)—One of the most widely distributed connective tissues in the body, this type consists of a semifluid intercellular substance containing hyaluronic acid, chondroitin and other sulfates, three kinds of fibers, and several types of cells. ***Collagenous fibers*** are synthesized from the protein *collagen* and often occur in parallel bundles to provide strength; ***elastic fibers*** are smaller than collagenous, branch freely, consist of the protein *elastin*, have nonparallel construction, and provide elasticity; and ***reticular fibers*** are very thin, branch extensively, consist of collagen plus *glycoprotein*, have nonparallel construction, and provide the stroma (framework) of many soft organs. The cells in loose connective tissue include ***fibroblasts*** that form fibers and intercellular substance; ***fibrocytes*** that are inactive fibroblasts; ***macrophages*** (MAK-rō-fā-jez) that are capable of phagocytosis; ***plasma cells*** that develop from B lymphocytes and produce antibodies; ***mast cells*** that form the anticoagulant heparin and the vasodilator histamine; ***adipocytes*** (fat cells); and ***leucocytes*** (white blood cells). Loose connective tissue is present in many mucous membranes, around blood vessels, nerves, and organs, and, together with adipose tissue, forms the ***subcutaneous*** (sub′-kyoo-TĀ-nē-us) ***layer*** or superficial fascia (FASH-ē-a). This layer is located between the skin and underlying tissues.

After you make your microscopic examination, draw a small area of the tissue in the space that follows and label the elastic fibers, collagenous fibers, fibroblasts, and mast cells.

Loose (areolar) connective tissue

2. ***Adipose***—This is fat tissue in which cells derived from fibroblasts, called ***adipocytes***, are modified for fat storage. Because the cytoplasm and nuclei of the cells are pushed to the side, the cells resemble signet rings. The tissue is found wherever loose connective tissue is located and around the kidneys and heart, in the marrow of long bones, and behind the eyeball, where it provides insulation, energy reserve, support, and protection.

After your microscopic examination, draw several cells in the space that follows and label the fat storage area, cytoplasm, nucleus, and plasma membrane.

Adipose tissue

3. ***Dense (collagenous)***—This tissue has a predominance of closely packed collagenous fibers with fibroblasts arranged between the bundles of fibers. Its principal function is strength. It forms the tendons, aponeuroses (ap′-ō-noo-RŌ-sēz), and ligaments; the membrane capsules around the kidneys, heart, testes, liver, and lymph nodes; and, finally, the deep fasciae around muscles.

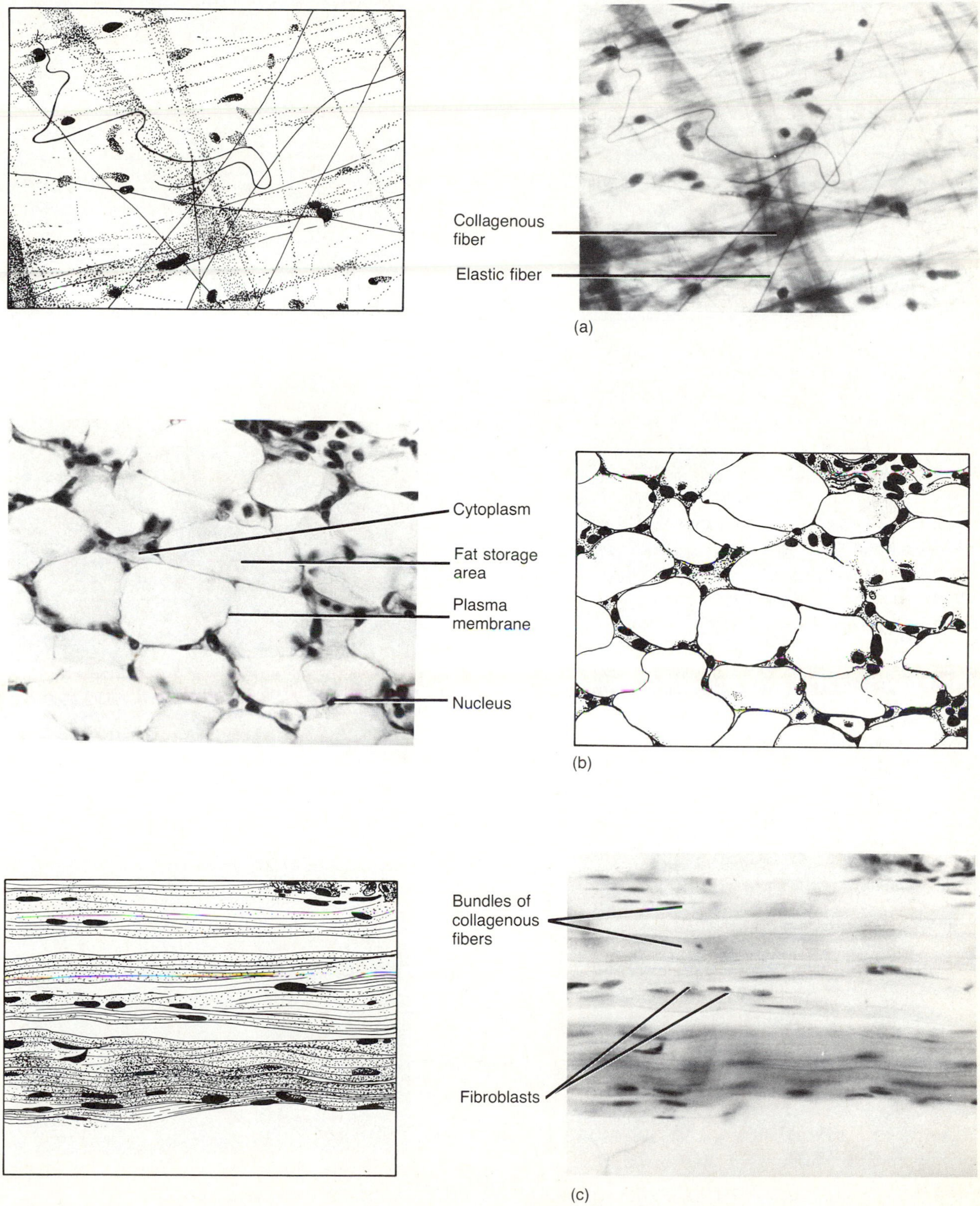

FIGURE 4.3 Connective tissues. Photomicrographs are labeled, while the line drawing of the same tissue is unlabeled. (a) Loose (areolar). (b) Adipose. (c) Dense (collagenus).

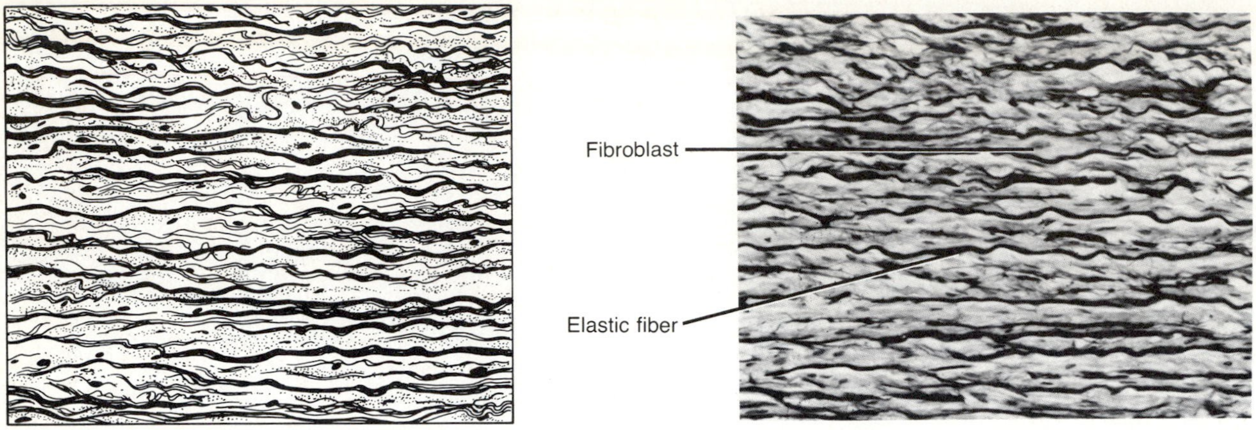

(d)

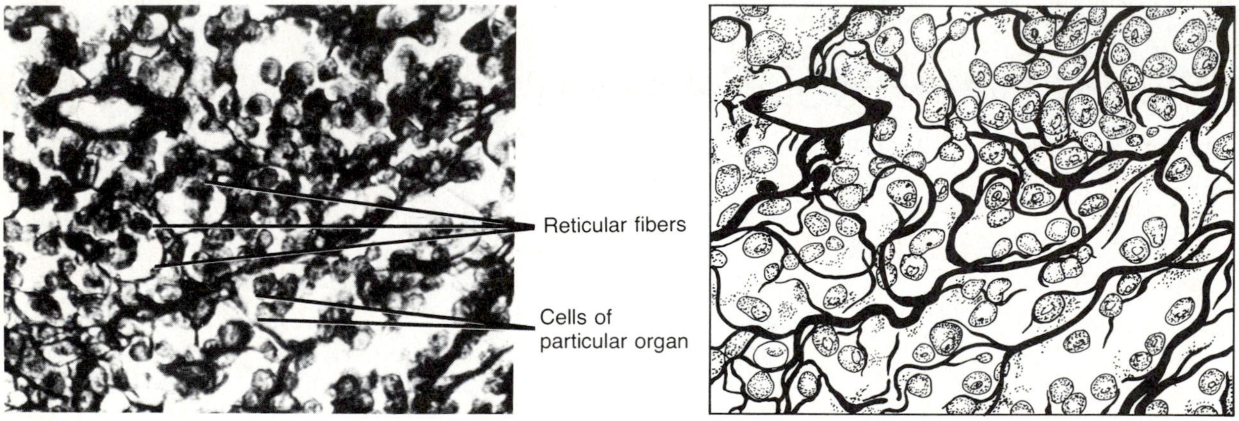

(e)

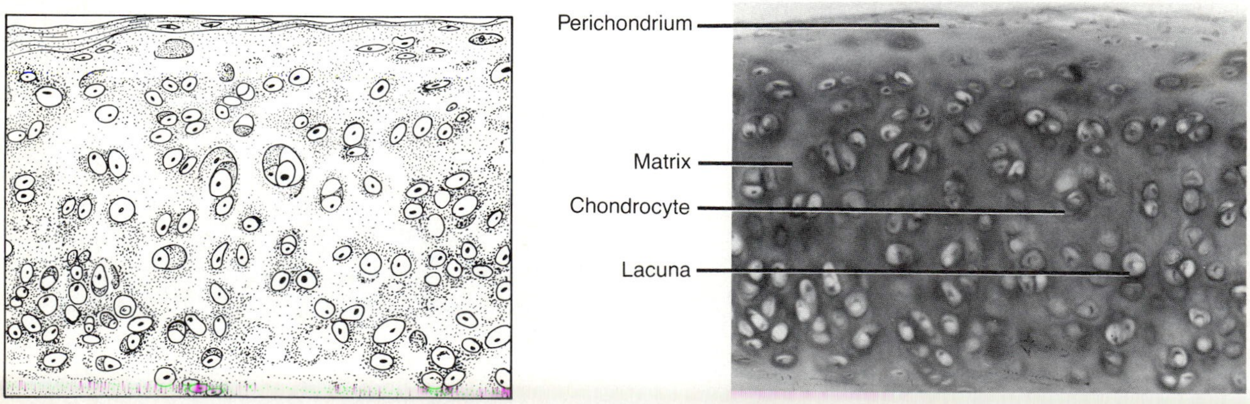

(f)

FIGURE 4.3 ***(Continued)*** Connective tissues. (d) Elastic. (e) Reticular. (f) Hyaline cartilage.

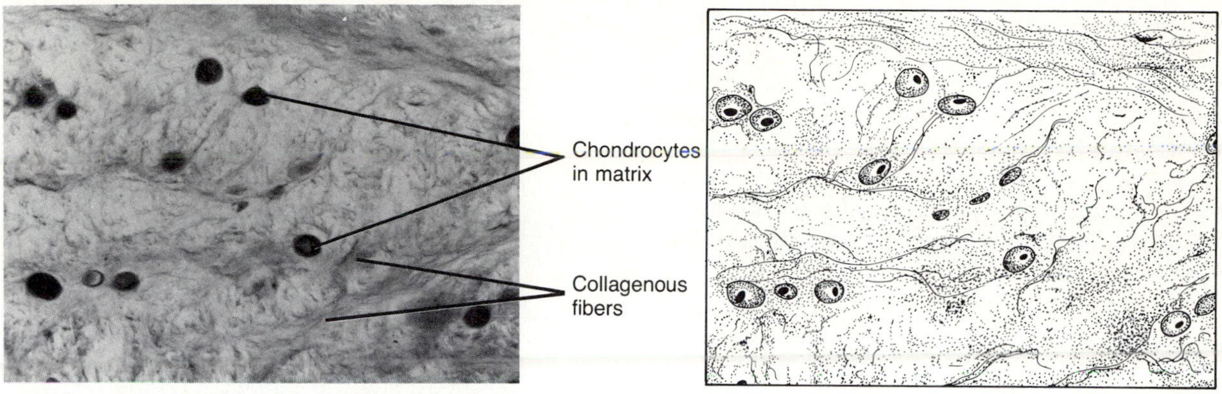

(g)

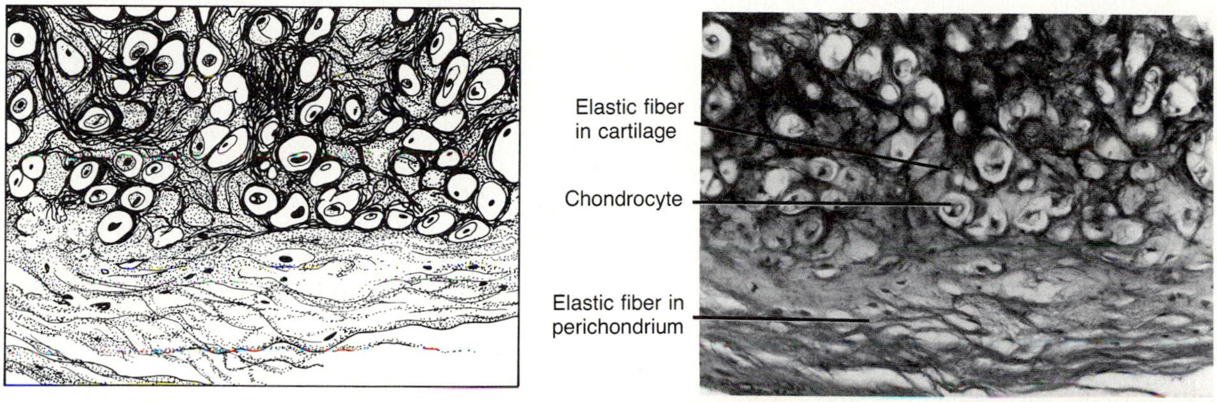

(h)

FIGURE 4.3 ***(Continued)*** Connective tissues. (g) Fibrocartilage. (h) Elastic cartilage.

After you make your microscopic examination, draw a sample of the tissue in the space that follows and label the collagenous fibers and fibroblasts.

Dense (collagenous) connective tissue

4. ***Elastic***—This tissue has a predominance of elastic fibers with fibroblasts in the spaces between fibers. It provides strength and elasticity and is found in the walls of elastic arteries, bronchial tubes, lungs, ligamenta flava of the vertebrae, suspensory ligament of the penis, and true vocal cords.

After you make your microscopic examination, draw a sample of the tissue in the space that follows and label the elastic fibers and fibroblasts.

Elastic connective tissue

5. ***Reticular***—This tissue consists of interlacing reticular fibers in which the cells are interspersed between the fibers. It provides strength and support, forms the stroma (framework) of the liver, spleen, and lymph nodes, and binds smooth muscle fibers (cells) together.

After you make your microscopic examination, draw a sample of the tissue in the space that follows and label the reticular fibers and cells of the organ.

Reticular connective tissue

6. ***Hyaline cartilage***—Cartilage consists of a dense network of collagenous and elastic fibers firmly embedded in a gel-like substance (chondroitin sulfate). The fibers are not usually visible in hyaline cartilage. The cells of cartilage are called ***chondrocytes*** (KON-drō-sīts) and occur singly or in groups in spaces called ***lacunae*** (la-KOO-nē) in the intercellular substance. The surface of cartilage is surrounded by a connective tissue membrane called the ***perichondrium*** (per′-i-KON-drē-um). Cartilage, like epithelium, is avascular, except for the perichondrium.

Hyaline cartilage, also called gristle, is the most abundant cartilage in the body. It provides strength and support and is found at joints over the ends of long bones ***(articular cartilage)*** and at the ventral ends of the ribs ***(costal cartilage)***. It helps form the nose, larynx, trachea, bronchi, bronchial tubes, and most of the embryonic skeleton.

After you make your microscopic examination, draw a sample of the tissue in the space that follows and label the perichondrium, chondrocytes, lacunae, and intercellular substance (matrix).

Hyaline cartilage

7. ***Fibrocartilage***—Unlike those in hyaline cartilage, collagenous fibers between chondrocytes are visible in fibrocartilage. This tissue combines strength and rigidity. It is found in the symphysis pubis joint in the discs between vertebrae, and menisci (cartilages) of the knee.

After you make your microscopic examination, draw a sample of the tissue in the space that follows and label the chondrocytes, lacunae, intercellular substance, and collagenous fibers.

Fibrocartilage

8. ***Elastic cartilage***—In this tissue, the chondrocytes are located in a threadlike network of elastic fibers. The tissue provides strength and maintains the shape of organs such as the epiglottis of the larynx, external part of the ear (pinna), and auditory (Eustachian) tubes.

After you make your microscopic examination, draw a sample of the tissue in the space that follows and label the perichondrium, chondrocytes, lacunae, intercellular substance, and elastic fibers.

Elastic cartilage

D. MEMBRANES

The combination of an epithelial layer and an underlying layer of connective tissue constitutes an ***epithelial membrane***. Examples are mucous, serous, and cutaneous membranes. Another kind of membrane, a synovial membrane, has no epithelium. ***Mucous membranes***, also called the ***mucosa***, line body cavities that open directly to the exterior, such as the gastrointestinal, respiratory, urinary, and reproductive tracts. The surface tissue of a mucous membrane consists of epithelium and has a variety of functions, depending on location. Accordingly, the epithelial layer secretes mucus but may also secrete enzymes, filter dust, and have a protective and absorbent action. The underlying connective tissue layer of a mucous membrane, called the *lamina propria*, binds the epithelial layer in place, protects underlying tissues, provides the epithelium with nutrients and oxygen and removes wastes, and holds blood vessels in place.

Serous membranes, also called the ***serosa***, line body cavities that do not open to the exterior and cover organs that lie within the cavities. Serous membranes consist of a surface layer of mesothelium and an underlying layer of loose connective tissue. The mesothelium secretes a lubricating fluid. Serous membranes consist of two portions. The part attached to the cavity wall is called the ***parietal layer***; the part that covers the organs in the cavity is called the ***visceral layer***. Examples of serous membranes are the pleurae, pericardium, and peritoneum.

The ***cutaneous membrane*** or skin is the principal component of the integumentary system, which will be considered in the next exercise.

Synovial membranes line joint cavities. They do not contain epithelium but rather consist of loose connective tissue, adipose tissue, and elastic fibers. Synovial membranes produce synovial fluid, which lubricates the ends of bones as they move at joints and nourishes the articular cartilage around the ends of bones.

ANSWER THE LABORATORY REPORT QUESTIONS AT THE END OF THE EXERCISE

TISSUES

Student ______________ Date ______________

Laboratory Section ______________ Score/Grade ______________

PART I. MULTIPLE CHOICE

_____ 1. In parts of the body such as the urinary bladder, where considerable distension occurs, you can expect to find which epithelial tissue? (a) pseudostratified (b) cuboidal (c) columnar (d) transitional

_____ 2. Stratified epithelium is usually found in areas of the body where the principal activity is (a) filtration (b) absorption (c) protection (d) diffusion

_____ 3. Ciliated epithelium destroyed by disease would cause malfunction in which system? (a) digestive (b) respiratory (c) skeletal (d) cardiovascular

_____ 4. The tissue that provides the skin with resistance to wear and tear and serves to waterproof it is (a) keratinized stratified squamous (b) pseudostratified (c) transitional (d) simple columnar

_____ 5. The connective tissue cell that would most likely increase its activity during an infection is the (a) melanocyte (b) macrophage (c) adipocyte (d) fibroblast

_____ 6. Torn ligaments would involve damage to which tissue? (a) collagenous (b) reticular (c) elastic (d) loose

_____ 7. Simple squamous tissue that lines the heart, blood vessels, and lymphatic vessels is called (a) transitional (b) adipose (c) endothelium (d) mesothelium

_____ 8. Microvilli and goblet cells are associated with which tissue? (a) hyaline cartilage (b) simple columnar nonciliated (c) transitional (d) stratified squamous

_____ 9. Superficial fascia contains which tissue? (a) elastic (b) reticular (c) fibrocartilage (d) loose (areolar) connective tissue

_____ 10. Which tissue forms articular cartilage and costal cartilage? (a) fibrocartilage (b) elastic cartilage (c) adipose (d) hyaline cartilage

_____ 11. Because the sublingual gland contains a branched duct and flasklike secretory portions, it is classified as (a) simple coiled tubular (b) compound acinar (c) simple acinar (d) compound tubular

_____ 12. Which glands discharge an entire dead cell and its contents as their secretory products? (a) merocrine (b) apocrine (c) endocrine (d) holocrine

_____ 13. Membranes that line cavities that open directly to the exterior are called (a) synovial (b) serous (c) mucous (d) cutaneous

INTEGUMENTARY SYSTEM 5

The skin and its derivatives (hair, nails, and glands), and several specialized receptors constitute the ***integumentary*** (in-teg-yoo-MEN-tar-ē) ***system***, which you will study in this exercise. An ***organ*** is an aggregation of tissues of definite form that performs a definite function; a ***system*** is a group of organs that operate together to perform specialized functions.

A. SKIN

The ***skin*** is one of the larger organs of the body in terms of surface area, occupying a surface area of about 19,355 cm^2 (3000 $in.^2$). Among the functions performed by the skin are maintenance of body temperature; protection of underlying tissues from physical abrasion, microorganisms, dehydration, and ultraviolet (UV) radiation; excretion of water and salts and several organic compounds, synthesis of vitamin D in the presence of sunlight, reception of stimuli for touch, pressure, pain, and temperature change sensations; and immunity.

The skin consists of an outer, thinner ***epidermis***, which is avascular, and an inner, thicker ***dermis***, which is vascular. Below the dermis is the ***subcutaneous layer (superficial fascia*** or ***hypodermis)*** that attaches the skin to underlying tissues and organs.

1. Epidermis

The epidermis consists of four principal kinds of cells. ***Keratinocytes*** are the most numerous cells and they undergo keratinization, that is, newly formed cells produced in the basal layers are pushed up to the surface and in the process take on a waterproofing chemical, keratin. ***Melanocytes*** are pigment cells that impart color to the skin. The third and fourth types of cells in the epidermis are now called ***nonpigmented granular dendrocytes***, two distinct cell types, formerly known as ***Langerhans cells*** and ***Granstein cells***. These cells differ both in their sensitivity to damage by ultraviolet (UV) radiation and their functions in immunity. Langerhans cells are a small population of dendrocytes that arise from bone marrow and migrate to the epidermis and other stratified squamous epithelial tissue in the body. These cells are sensitive to UV radiation and lie above the basal layer of keratinocytes. Langerhans cells interact with cells called helper T cells to assist in the immune response. Granstein cells are dendrocytes that are more resistant to UV radiation and interact with cells called suppressor T cells to assist in the immune response. At this point, we will concentrate only on keratinocytes.

Obtain a prepared slide of human skin and carefully examine the epidermis. Identify the following layers from the outside inward:

a. ***Stratum corneum***—25 to 30 rows of flat, dead cells that are filled with ***keratin***, a waterproofing protein.

b. ***Stratum lucidum***—Several rows of clear, flat cells that contain ***eleidin*** (el-Ē-i-din), a precursor of keratin; found only in the palms and soles.

c. ***Stratum granulosum***—3 to 5 rows of flat cells that contain ***keratohyalin*** (ker′-a-tō-HĪ-a-lin), a precursor of eleidin.

d. ***Stratum spinosum***—8 to 10 rows of polyhedral (many-sided) cells.

e. ***Stratum basale (germinativum)***—Single layer of cuboidal to columnar cells that constantly undergo division. In hairless skin, this layer contains tactile (Merkel's) discs, receptors sensitive to touch.

Label the epidermal layers in Figures 5.1 and 5.2.

2. Skin Color

The color of skin results from (1) ***blood in capillaries*** of the dermis (beneath the epidermis); (2) ***carotene***, a yellow/orange pigment in the stratum corneum of the epidermis and fatty areas of the dermis; and (3) ***melanin***, a pale yellow to black pigment found primarily

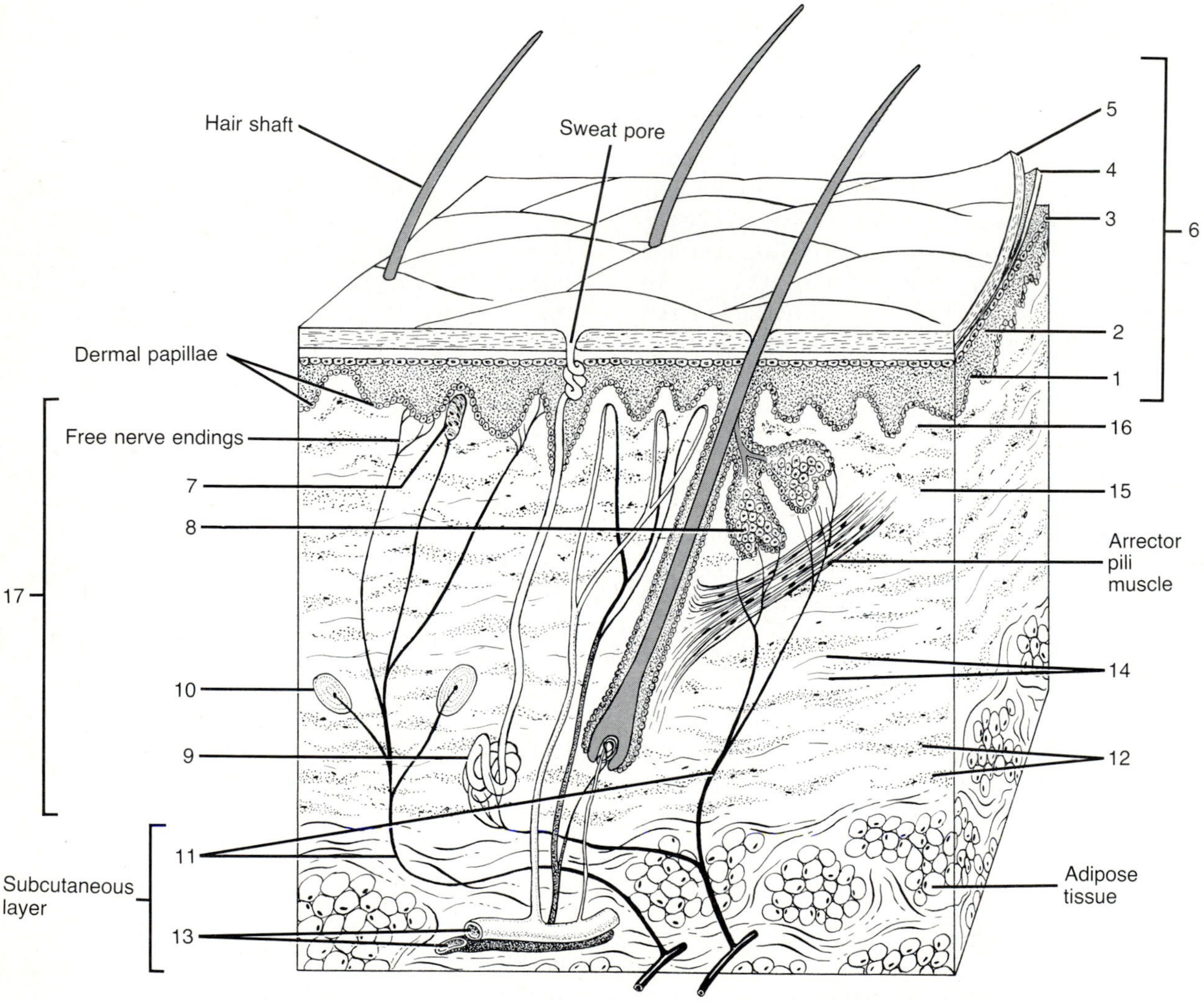

13 Blood vessels
___ Collagenous fibers
___ Corpuscle of touch (Meissner's corpuscle)
17 Dermis
___ Elastic fibers
6 Epidermis
___ Lamellated (Pacinian) corpuscle
11 Nerves
16 Papillary region
15 Reticular region
8 Sebaceous (oil) gland
___ Stratum basale
___ Stratum corneum
___ Stratum granulosum
___ Stratum lucidum
___ Stratum spinosum
___ Sudoriferous (sweat) gland

FIGURE 5.1 Structure of the skin.

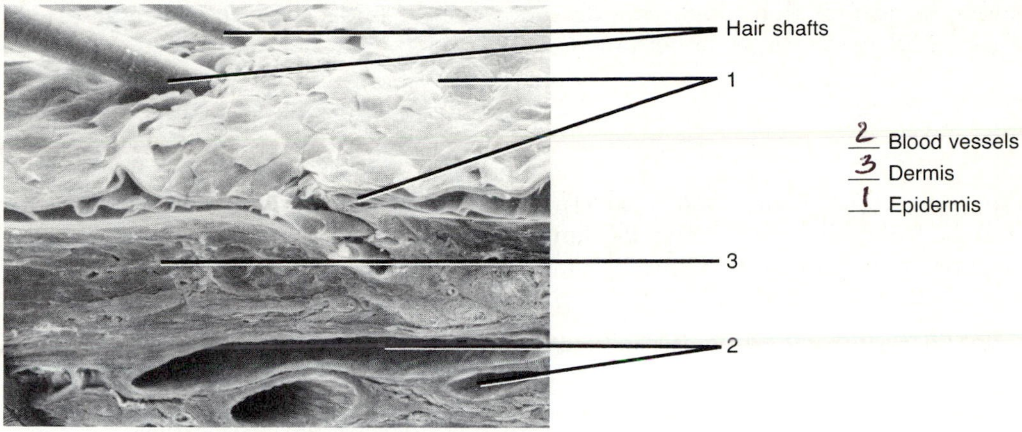

FIGURE 5.2 Scanning electron micrograph of the skin and several hairs at a magnification of 260 ×. (From *Tissues and Organs: A Text-Atlas of Scanning Electron Microscopy* by Richard G. Kessel and Randy H. Kardon. W. H. Freeman and Company. Copyright © 1979.)

in the melanocytes in the stratum basale and spinosum of the epidermis. Whereas blood in capillaries imparts a pink color to Caucasian skin, carotene imparts the characteristic yellow color to Asian skin. Because the number of ***melanocytes*** (me-LAN-ō-sīts) or melanin-producing cells is about the same in all races, most differences in skin color are due to the amount of melanin that the melanocytes synthesize and disperse. Exposure to ultraviolet (UV) radiation increases melanin synthesis, resulting in darkening (tanning) of the skin to protect the body against further UV radiation.

An inherited inability of a person of any race to produce melanin results in ***albinism*** (AL-bi-nizm). The pigment is absent from the hair and eyes as well as from the skin, and the individual is referred to as an ***albino***. In some people, melanin tends to form in patches called ***freckles***. Others inherit patches of skin that lack pigment, a condition called ***vitiligo*** (vit-i-LĪ-gō).

3. Dermis

The dermis is divided into two regions and is composed of connective tissue containing collagenous and elastic fibers and a number of other structures. The upper region of the dermis (***papillary region***) is loose areolar connective tissue containing fine elastic fibers. This layer contains fingerlike projections, the ***dermal papillae*** (pa-PIL-ē). Some papillae enclose blood capillaries; others contain ***corpuscles of touch (Meissner's corpuscles)***, nerve endings sensitive to touch. The lower region of the dermis ***(reticular region)*** consists of denser areolar connective tissue with interlacing bundles of larger collagenous and coarse elastic fibers. Spaces between the fibers may be occupied by ***hair follicles, sebaceous (oil) glands, bundles of smooth muscle (arrector pili muscle), sudoriferous (sweat) glands, blood vessels***, and ***nerves***.

The reticular region of the dermis is attached to the underlying structures (bones and muscles) by the subcutaneous layer. This layer also contains nerve endings sensitive to deep pressure called ***lamellated (Pacinian) corpuscles***.

Carefully examine the dermis and subcutaneous layer on your microscope slide. Label the following structures in Figure 5.1: papillary region, reticular region, corpuscle of touch, blood vessels, nerves, elastic fibers, collagenous fibers, sebaceous gland, sudoriferous gland, and lamellated corpuscle. Also label the dermis and blood vessels in Figure 5.2.

If a model of the skin and subcutaneous layer is available, examine it to see the three-dimensional relationship of the structures to each other.

B. HAIR

Hairs (pili) develop from the epidermis and are variously distributed over the body. Each hair consists of a ***shaft***, most of which is visible above the surface of the skin, and a ***root***, the portion below the surface that penetrates deep into the dermis and even into the subcutaneous layer.

The shaft of a coarse hair consists of the following parts:

1. ***Medulla***—Inner region composed of several rows of polyhedral cells containing eleidin; poorly developed or not present in fine hairs.

2. ***Cortex***—Several rows of dark cells surrounding the medulla; contains pigment in dark hair and pigment and air in gray hair.

3. ***Cuticle of the hair***—Outermost layer; consists of a single layer of flat, scalelike, keratinized cells arranged like shingles on a house.

The root of a hair also contains a medulla, cortex, and cuticle of the hair along with the following associated parts:

1. ***Hair follicle***—Structure surrounding the root that consists of an external root sheath, an internal root sheath, and a bulb. These three epidermally derived layers are surrounded by a dermal layer of connective tissue.

2. ***External root sheath***—Downward continuation of strata basale and spinosum of the epidermis.

3. ***Internal root sheath***—Cellular tubular sheath that separates the hair from the external root sheath; consists of (1) the ***cuticle of the internal root sheath***, an inner single layer of

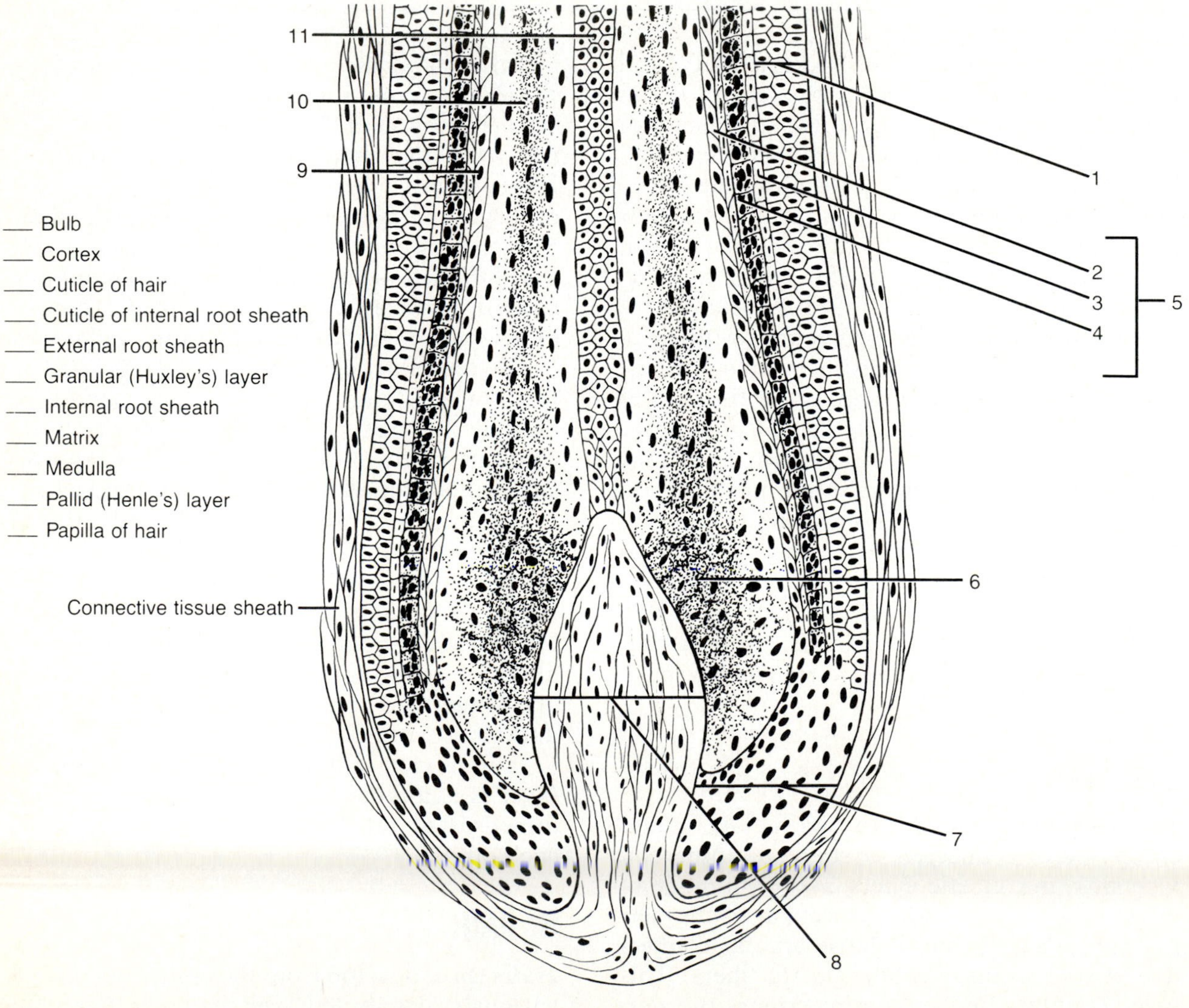

FIGURE 5.3 Longitudinal section of a hair root.

flattened cells with atrophied nuclei, (2) ***granular (Huxley's) layer***, a middle layer of one to three rows of cells with flattened nuclei, and (3) ***pallid (Henle's) layer***, an outer single layer of cuboidal cells with flattened nuclei.

4. ***Bulb***—Enlarged, onion-shaped structure at the base of the hair follicle.

5. ***Papilla of the hair***—Dermal indentation into the bulb; contains loose connective tissue and blood vessels to nourish the hair.

6. ***Matrix***—Region of cells at the base of the bulb that divides to produce new hair.

7. ***Arrector pili muscle***—Bundle of smooth muscle extending from the dermis of the skin to the side of the hair follicle; its contraction, under the influence of fright or cold, causes the hair to move into a vertical position, producing "goose bumps."

8. ***Root hair plexus***—Nerve endings around each hair follicle that are sensitive to touch and respond when the hair shaft is moved.

Obtain a prepared slide of a cross section and a longitudinal section of a hair root and identify as many parts as you can. Using your textbook as a reference, label Figure 5.3. Also label the parts of a hair shown in Figure 5.4.

C. GLANDS

Sebaceous (se-BĀ-shus) or ***oil glands***, with few exceptions, are connected to hair follicles (see Figures 5.1 and 5.4). They secrete an oily substance called ***sebum***, a mixture of fats, cholesterol, proteins, and salts. Sebaceous glands are absent in the skin of the palms and soles but

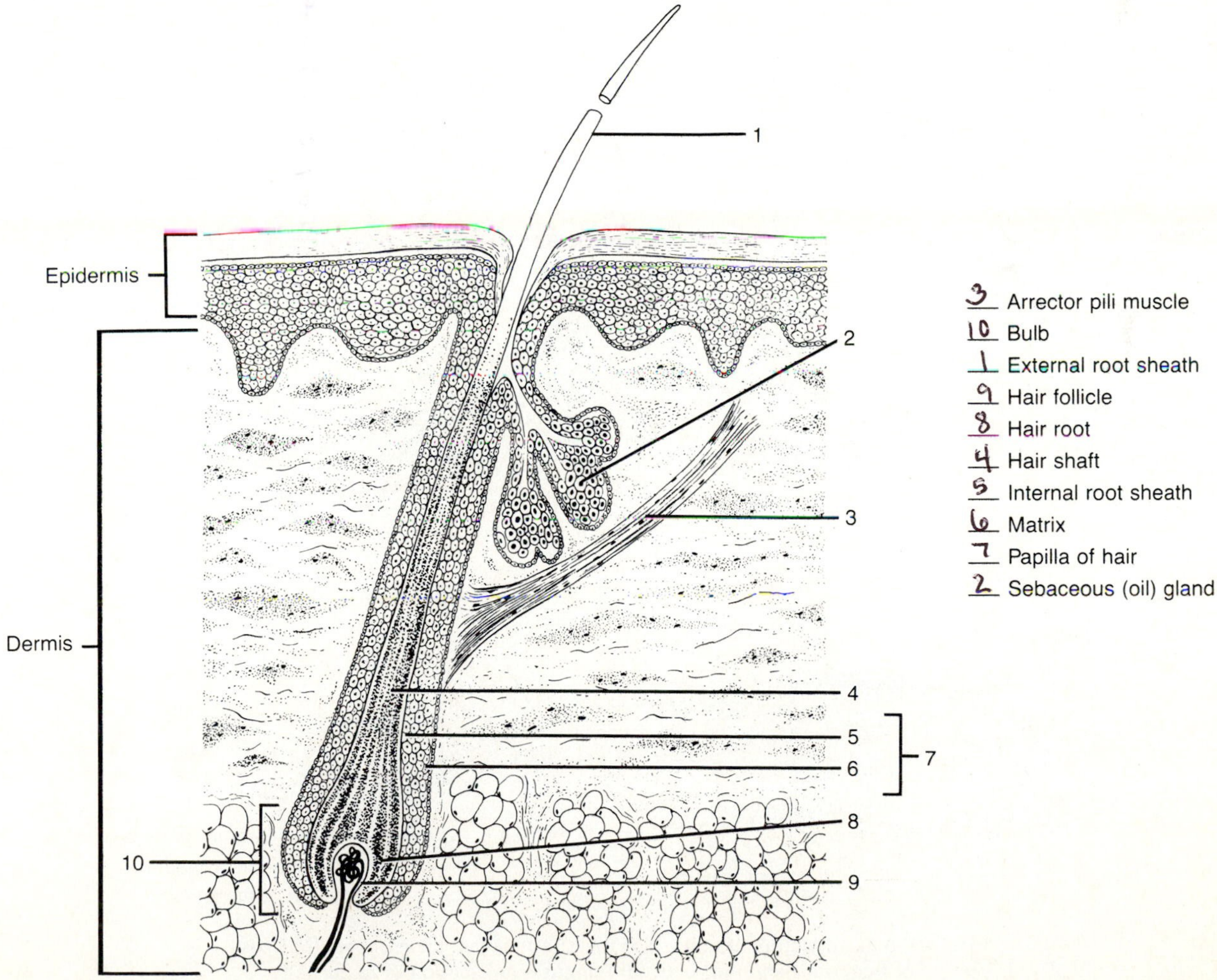

FIGURE 5.4 Parts of a hair and associated structures.

are numerous in the skin of the face, neck, upper chest, and breasts. Sebum helps prevent hair from drying and forms a protective film over the skin that prevents excessive evaporation and keeps the skin soft and pliable.

Sudoriferous (soo′-dor-IF-er-us) or ***sweat glands*** are separated into two principal types on the basis of structure, location, and secretion. ***Apocrine sweat glands*** are simple, branched tubular glands found primarily in the skin of the axilla, pubic region, and areolae (pigmented areas) of the breasts. Their secretory portion is located in the dermis; the excretory duct opens into hair follicles. Apocrine sweat glands begin to function at puberty and produce a more viscous secretion than the other type of sweat gland. ***Eccrine sweat glands*** are simple, coiled tubular glands found throughout the skin, except for the margins of the lips, nail beds of the fingers and toes, glans penis, glans clitoris, and eardrums. The secretory portion of these glands is in the subcutaneous layer; the excretory duct projects upward and terminates at a pore at the surface of the epidermis (see Figure 5.1). Eccrine sweat glands function throughout life and produce a more watery secretion than the apocrine glands. Sudoriferous glands produce ***perspiration***, a mixture of water, salt, urea, uric acid, amino acids, ammonia, sugar, lactic acid, and ascorbic acid. The evaporation of perspiration helps to maintain normal body temperature.

Ceruminous (se-ROO-mi-nus) ***glands*** are modified sudoriferous glands in the external auditory meatus. The combined secretion of ceruminous and sudoriferous glands is called ***cerumen*** (earwax). Cerumen, together with hairs in the external auditory meatus, provides

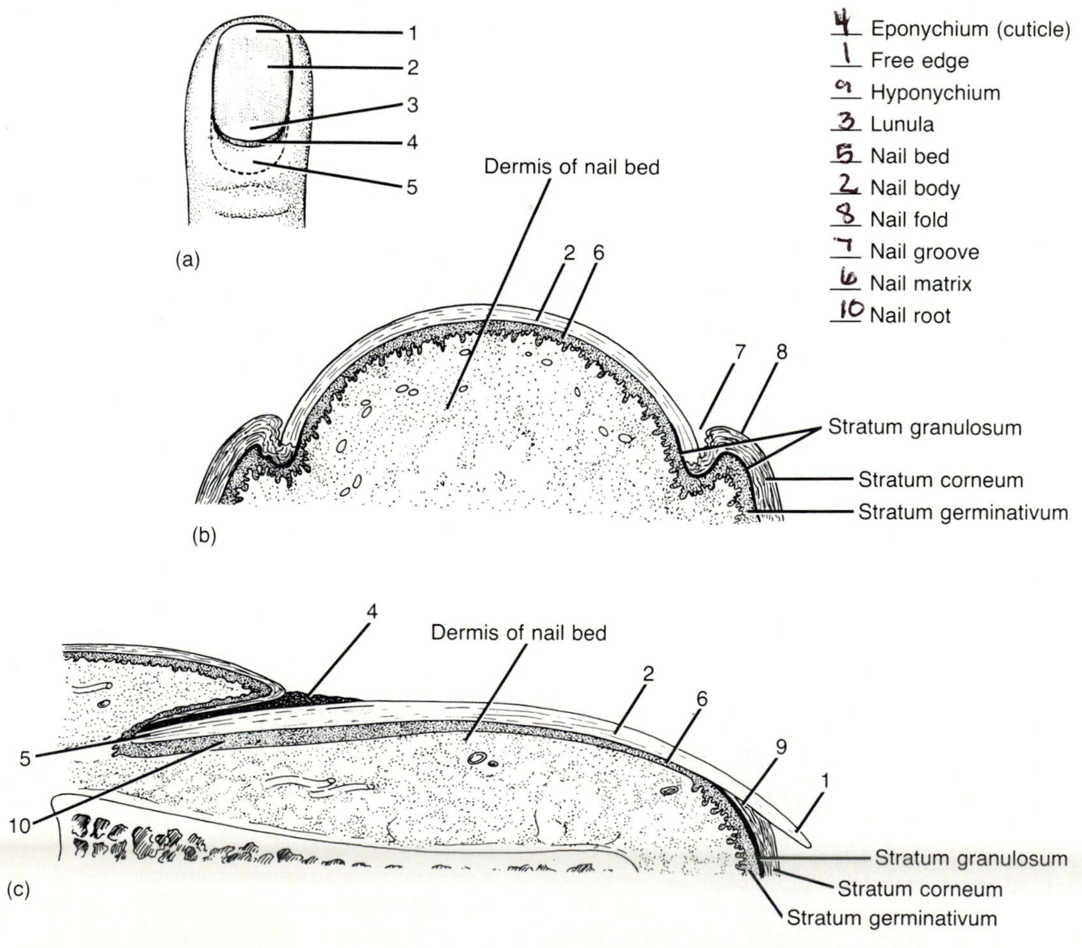

FIGURE 5.5 Structure of nails. (a) Fingernail viewed from above. (b) Cross section of fingernail and nail bed. (c) Longitudinal section of fingernail.

a sticky barrier that prevents foreign bodies from reaching the eardrum.

D. NAILS

Nails are hard, keratinized epidermal cells that form a clear covering over the dorsal surfaces of the terminal portions of the fingers and toes. Each nail consists of the following parts:

1. ***Nail body***—Portion that is visible.
2. ***Free edge***—Part that may project beyond the distal end of the digit.
3. ***Nail root***—Portion hidden in nail groove.
4. ***Lunula*** (LOO-nyoo-la)—Whitish semilunar area at proximal end of body.
5. ***Nail fold***—Fold of skin that extends around the proximal end and lateral borders of the nail.
6. ***Nail bed***—Strata basale and spinosum of the epidermis beneath the nail.
7. ***Nail groove***—Furrow between the nail fold and nail bed.
8. ***Eponychium*** (ep′-ō-NIK-ē-um)—Cuticle; a narrow band of epidermis.
9. ***Hyponychium***—Thickened area of stratum corneum below the free edge of the nail.
10. ***Nail matrix***—Epithelium of the proximal part of the nail bed; division of the cells brings about growth of nails.

Using your textbook as a reference, label the parts of a nail shown in Figure 5.5. Also, identify the parts that are visible on your own nails.

ANSWER THE LABORATORY REPORT QUESTIONS AT THE END OF THE EXERCISE

LABORATORY REPORT QUESTIONS
INTEGUMENTARY SYSTEM 5

Student ______________________ **Date** ______________________

Laboratory Section ______________________ **Score/Grade** ______________________

PART I. MULTIPLE CHOICE

c 1. The waterproofing quality of skin is due to the presence of (a) melanin (b) carotene (c) keratin (d) receptors

b 2. Sebaceous glands (a) produce a watery solution called sweat (b) produce an oily substance that prevents excessive water evaporation from the skin (c) are associated with mucous membranes (d) are part of the subcutaneous layer

c 3. Which of the following is the proper sequence of layering of the epidermis, going from the free surface toward the underlying tissues? (a) basale, spinosum, granulosum, corneum (b) spinosum, basale, granulosum, corneum (c) corneum, lucidum, granulosum, spinosum, basale (d) corneum, granulosum, lucidum, spinosum

c 4. Skin color is *not* determined by the (a) presence or absence of melanin (b) presence or absence of carotene (c) presence or absence of keratin (d) presence or absence of blood vessels in the dermis

d 5. Destruction of what part of a single hair would result in its inability to grow? (a) sebaceous gland (b) arrector pili muscles (c) matrix (d) bulb

a 6. One would expect to find relatively few, if any, sebaceous glands in the skin of the (a) palms (b) face (c) neck (d) upper chest

d 7. Which of the following sequences, from outside to inside, is correct? (a) epidermis, reticular region, papillary region, subcutaneous layer (b) epidermis, subcutaneous layer, reticular region, papillary region (c) epidermis, reticular region, subcutaneous layer, papillary region (d) epidermis, papillary region, reticular region, subcutaneous layer

c 8. The attached visible portion of a nail is called the (a) nail bed (b) nail root (c) nail fold (d) nail body

a 9. Nerve endings sensitive to touch are called (a) corpuscles of touch (Meissner's corpuscles) (b) papillae (c) lamellated (Pacinian) corpuscles (d) follicles

b 10. The cuticle of a nail is referred to as the (a) matrix (b) eponychium (c) hyponychium (d) fold

d 11. One would *not* expect to find sudoriferous glands associated with the (a) forehead (b) axilla (c) palms (d) nail beds

a 12. Fingerlike projections of the dermis that contain loops of capillaries and receptors are called (a) dermal papillae (b) nodules (c) polyps (d) pili

d 13. Which of the following statements about the function of skin is *not* true? (a) it helps control body temperature (b) it prevents excessive water loss (c) it synthesizes several compounds (d) it absorbs water and salts

______ 14. Which is *not* part of the internal root sheath? (a) granular (Huxley's) layer (b) cortex (c) pallid (Henle's) layer (d) cuticle of the internal root sheath

__b__ 15. Growth in the length of nails is the result of the activity of the (a) eponychium (b) nail matrix (c) hyponychium (d) nail fold

PART II. COMPLETION

16. A group of tissues that performs a definite function is called a(n) ~~system~~ organ?
17. The outer, thinner layer of the skin is known as the epidermis .
18. The skin is attached to underlying tissues and organs by the CT (basale) .
19. A group of organs that operate together to perform a specialized function is called a(n) system .
20. The epidermal layer that contains eleidin is the stratum lucidum .
21. The epidermal layers that produce new cells are the stratum spinosum and stratum basale .
22. The smooth muscle attached to a hair follicle is called the arretor pili muscle.
23. An inherited inability to produce melanin is called albinism .
24. Nerve endings sensitive to deep pressure are referred to as Meissner's corpuscles.
25. The inner region of a hair shaft and root is the bulb .
26. The portion of a hair containing loose connective tissue and blood vessels is the shaft .
27. Modified sweat glands that line the external auditory meatus are called ceruminous glands.
28. The whitish semilunar area at the proximal end of the nail body is referred to as the lucala .
29. The secretory product of sudoriferous glands is called sweat .
30. Melanin is synthesized in cells called melanocytes .

OSSEOUS TISSUE 6

Structurally, the ***skeletal system*** consists of two types of connective tissue: cartilage and bone. The microscopic structure of cartilage has been discussed in Exercise 4. In this exercise the gross structure of a typical bone and the histology of ***osseous (bone) tissue*** will be studied.

The skeletal system has the following basic functions:

1. ***Support***—It provides a supporting framework for the soft tissues, maintaining the body's shape and posture.
2. ***Protection***—It protects delicate structures such as the brain, spinal cord, heart, lungs, major blood vessels in the chest, and pelvic viscera.
3. ***Movement***—When muscles contract, bones serve as levers to help produce body movements.
4. ***Storage***—Bones store mineral salts, especially calcium and phosphorus.
5. ***Blood cell production*** or ***hematopoiesis*** (hē′-mat-ō-poy-Ē-sis)—***Red marrow*** consists of primitive blood cells in immature stages, fat cells, and macrophages. It is responsible for producing red blood cells, platelets, and some white blood cells.

A. GROSS STRUCTURE OF A LONG BONE

Examine the external features of a fresh long bone and locate the following structures:

1. ***Periosteum*** (per′-ē-OS-tē-um)—Dense white fibrous membrane covering the surface of the bone, except for areas covered by articular cartilage; the outer ***fibrous layer*** is composed of connective tissue and contains blood vessels, lymphatics, nerves that pass into the bone; the inner ***osteogenic layer*** contains elastic fibers, blood vessels, ***osteoprogenitor cells*** (stem cells derived from mesenchyme that differentiate into osteoblasts), and ***osteoblasts***, cells that secrete the organic components and mineral salts involved in bone formation. The periosteum functions in bone growth, nutrition, and repair, and as an attachment site for tendons and ligaments.
2. ***Articular cartilage***—Thin layer of hyaline cartilage covering the ends of the bone where joints are formed.
3. ***Epiphysis*** (e-PIF-i-sis)—End or extremity of a bone; the epiphyses are referred to as proximal and distal.
4. ***Diaphysis*** (dī-AF-i-sis)—Elongated shaft of a bone between the epiphyses.
5. ***Metaphysis*** (me-TAF-i-sis)—In mature bone, the region where the diaphysis joins the epiphysis; in growing bone, the region near the epiphyseal plate where calcified cartilage is replaced by bone as the bone lengthens.

Now examine a mature long bone that has been sectioned longitudinally and locate the following structures:

1. ***Medullary*** (MED-yoo-lar′-ē) or ***marrow cavity***—Cavity within the diaphysis that contains yellow marrow (primarily fat cells and a few scattered cells that can produce blood cells).
2. ***Endosteum*** (end-OS-tē-um)—Membrane that lines the medullary cavity and contains osteoprogenitor cells, osteoblasts, and scattered

osteoclasts (cells that remove bone); it coats spongy bone and separates it from marrow.

3. ***Compact (dense) bone***—Bone that forms the bulk of the diaphysis and covers the epiphyses.

4. ***Spongy (cancellous) bone***—Bone that forms a small portion of the diaphysis at the ends of the medullary cavity and the bulk of the epiphysis; spongy bone contains red marrow.

Label the parts of a long bone indicated in Figure 6.1.

B. CHEMISTRY OF BONE

Unlike other connective tissues, the intercellular substance of bone is very hard. This hardness results from the presence of mineral salts, mainly calcium phosphate ($Ca_3(PO_4)_2 \cdot [OH]_2$) and some calcium carbonate ($CaCO_3$). Mineral salts comprise about 67% of the weight of bone. Despite its hardness, bone is also flexible, a characteristic that enables it to resist various forces. The flexibility of bone comes from organic substances in its intercellular

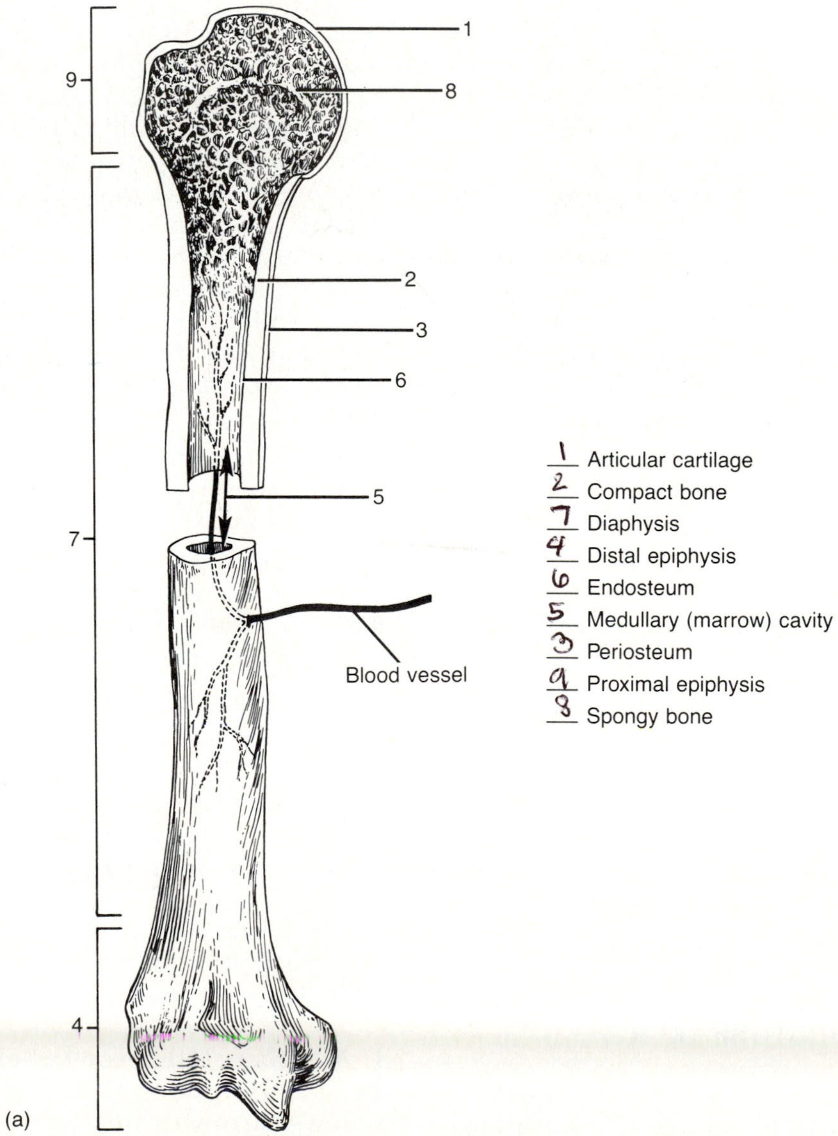

FIGURE 6.1 Parts of a long bone. (a) Diagram.

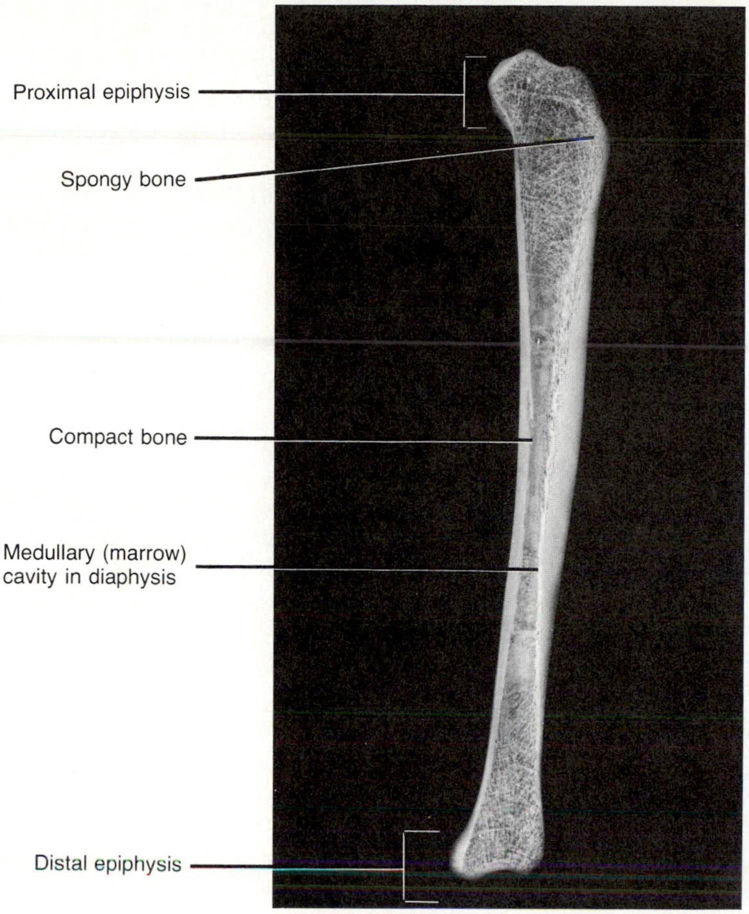

(b)

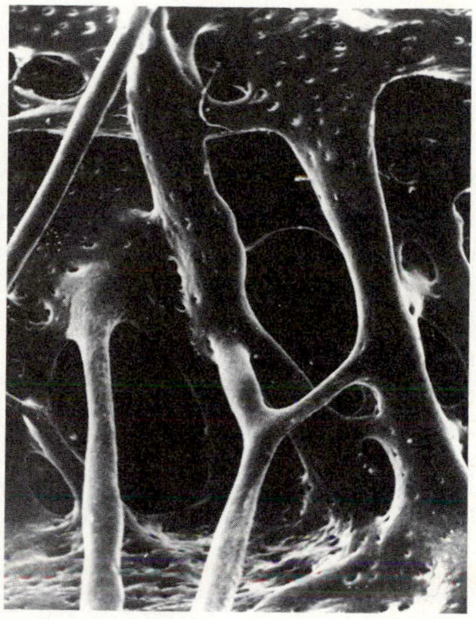

(c)

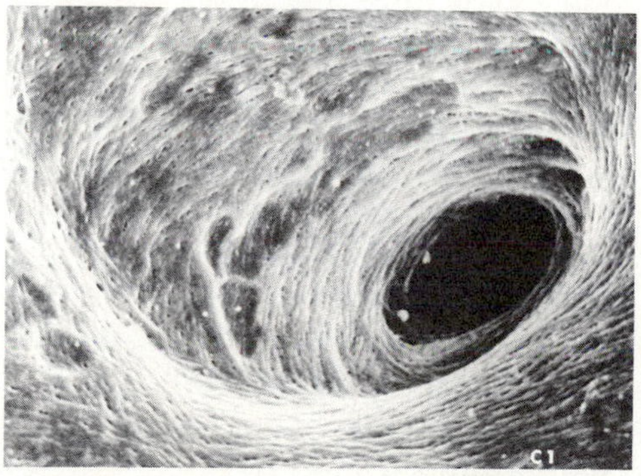

(d)

FIGURE 6.1 ***(Continued)*** Parts of a long bone. (b) Photograph of a section through the tibia. (c) Scanning electron micrograph of spongy bone trabeculae at a magnification of 25×. (d) Scanning electron micrograph of a view inside a central (Haversian) canal at a magnification of 250×.

substance, especially collagenous fibers. Organic materials comprise about 33% of the weight of bone.

PROCEDURE

1. Obtain a bone that has been baked. How does this compare to an untreated one?

____

What substances does baking remove from the bone (inorganic or organic)? _______________

____

2. Now obtain a bone that has already been soaked in nitric acid by your instructor. How does this bone compare to an untreated one?

____

What substances does nitric acid treatment remove from the bone (inorganic or organic)?

____

C. HISTOLOGY OF BONE

Bone tissue, like other connective tissues, contains a large amount of intercellular substance that surrounds widely separated cells. This intercellular substance consists of (1) collagenous fibers and (2) abundant mineral salts. The cells include osteoprogenitor cells, osteoblasts, osteocytes (mature bone cells that maintain daily activities of bone tissue), and osteoclasts.

Depending on the size and distribution of spaces, bone may be categorized as spongy (cancellous) or compact (dense). ***Spongy bone***, which contains many large spaces that store mainly red marrow, composes most of the tissue of short, flat, and irregularly shaped bones and most of the epiphyses of long bones. ***Compact bone***, which contains few spaces and is deposited in layers over spongy bone, is thicker in the diaphysis than in the epiphyses.

Obtain a prepared slide of compact bone in which several osteons (Haversian systems) are shown in cross section. Observe under high power. Look for the following structures:

1. ***Central (Haversian) canal***—Circular canal in the center of an osteon that runs longitudinally through the bone; the canal contains blood vessels, lymphatics, and nerves.

2. ***Lamellae*** (la-MEL-ē)—Concentric layers of calcified intercellular substance.

3. ***Lacunae*** (la-KOO-nē)—Spaces or cavities between lamellae that contain osteocytes.

4. ***Canaliculi*** (kan′-a-LIK-yoo-lē)—Minute canals that radiate in all directions from the lacunae and interconnect with each other; contain slender processes of osteocytes; canaliculi provide routes so that nutrients can reach osteocytes and wastes can be removed from them.

5. ***Osteocyte***—Mature bone cell located within a lacuna.

6. ***Osteon (Haversian system)***—Central (Haversian) canal plus its surrounding lamellae, lacunae, canaliculi, and osteocytes; structural unit of compact bone.

Label the parts of the osteon (Haversian system) shown in Figure 6.2.

Now obtain a prepared slide of a longitudinal section of compact bone and examine under high power. Locate the following:

1. ***Perforating (Volkmann's) canals***—Canals that extend obliquely or horizontally inward from the bone surface and contain blood vessels, lymphatics, and nerves; extend into central (Haversian) canals and medullary cavity.

2. ***Endosteum***—Layer of osteoprogenitor cells, osteoblasts, and scattered osteoclasts lining medullary (marrow) cavity and spongy bone.

3. ***Medullary (marrow) cavity***—Space within diaphysis that contains yellow marrow in adult.

4. ***Lamellae***—Concentric rings of hard, calcified, intercellular substance.

5. ***Lacunae***—Spaces that contain osteocytes.

6. ***Canaliculi***—Minute canals that radiate from lacunae and contain slender processes of osteocytes.

7. ***Osteocytes***—Mature osteoblasts in lacunae that are no longer capable of producing new bone tissue but maintain daily activities of bone tissue.

Label the parts indicated in the microscopic view of bone in Figure 6.3.

D. TYPES OF BONES

The 206 named bones of the body may be classified by shape into four principal types:

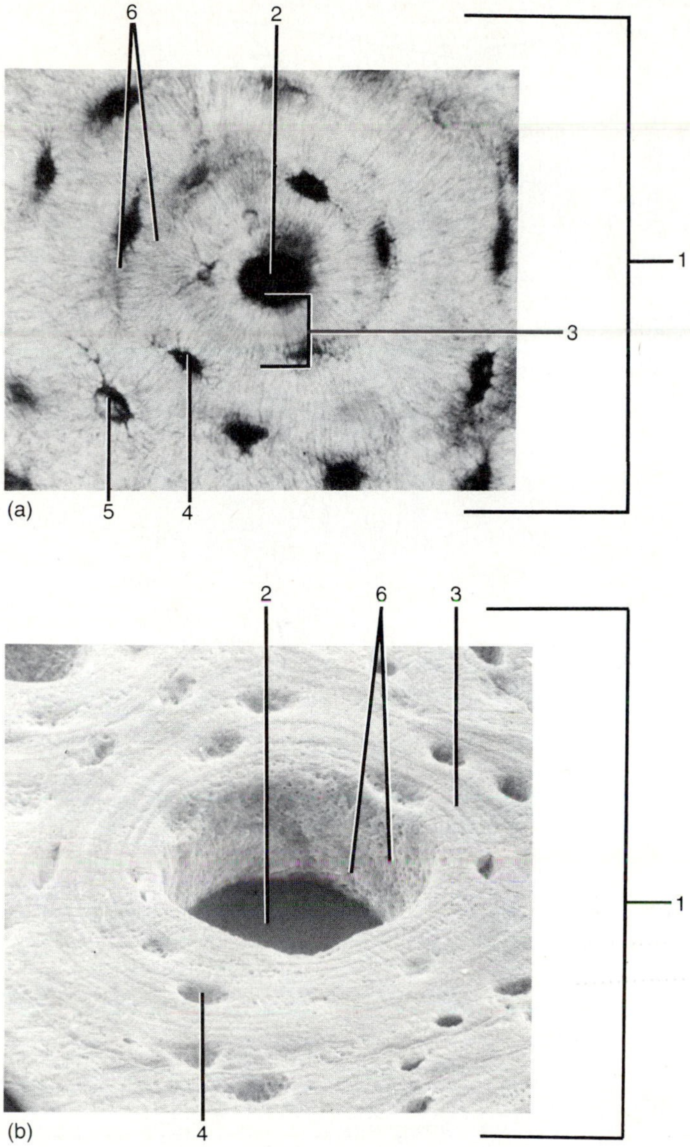

FIGURE 6.2 Osteon (Haversian system). (a) Transmission electron micrograph. (b) Scanning electron micrograph at a magnification of 1040×. ((b) from *Tissues and Organs: A Text-Atlas of Scanning Electron Microscopy* by Richard G. Kessel and Randy H. Kardon. W. H. Freeman and Company. Copyright © 1979.)

1. ***Long***—Have greater length than width, consist of a diaphysis and a variable number of epiphyses, and have a marrow cavity; contain more compact than spongy bone. Example: humerus.

2. ***Short***—Somewhat cube-shaped, and contain more spongy than compact bone. Example: wrist bones.

3. ***Flat***—Generally thin and flat and composed of two more-or-less parallel plates of compact bone enclosing a layer of spongy bone. Example: sternum.

4. ***Irregular***—Very complex shapes; cannot be grouped into any of the three categories just described. Example: vertebrae.

Other bones that are recognized, although not considered in the structural classification, include the following:

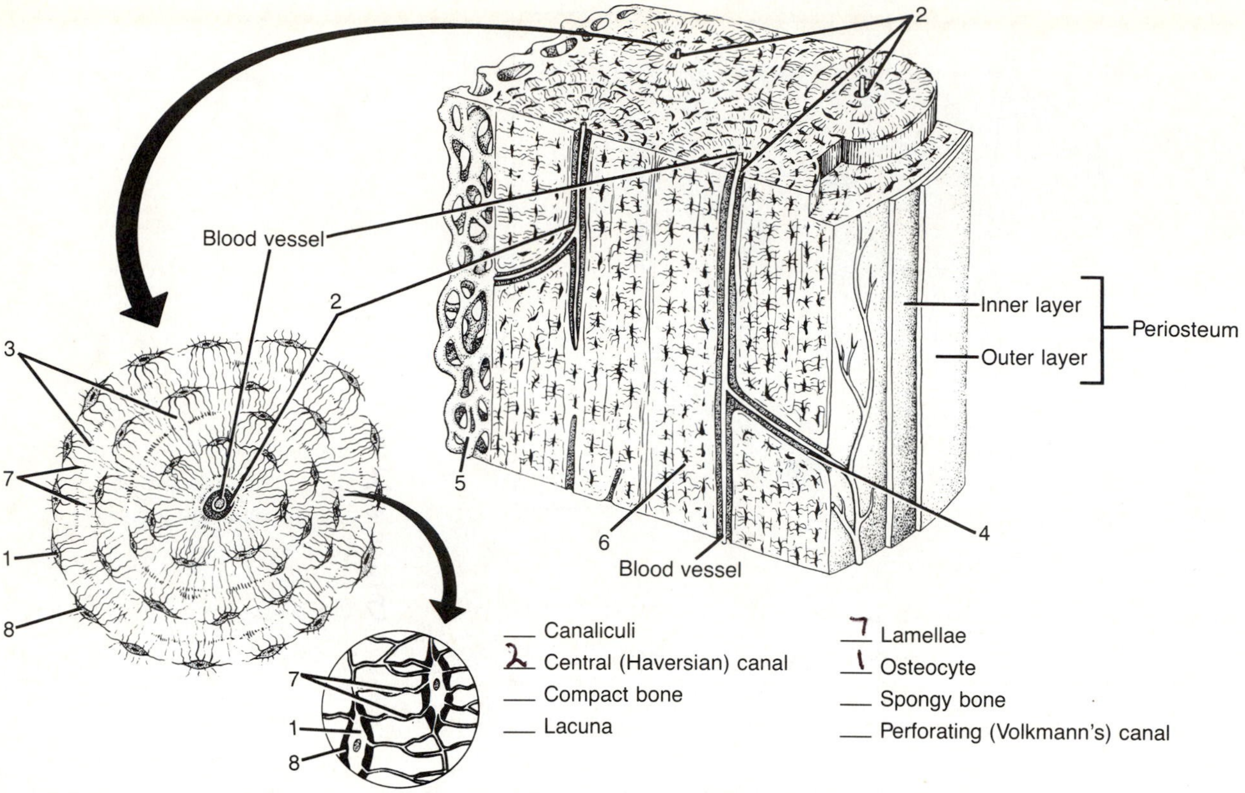

FIGURE 6.3 Cross section and longitudinal section of several osteons (Haversian systems).

1. ***Sutural (Wormian)***—Small bones between certain cranial bones; variable in number.
2. ***Sesamoid***—Small bones found in tendons; variable in number; the only constant sesamoid bones are the paired kneecaps; *sesamoid* means "resembling a grain of sesame."

Examine the disarticulated skeleton, Beauchene (disarticulated) skull, and articulated skeleton and find several examples of long, short, flat, and irregular bones. List examples of each type you find.

1. ***Long*** ______________________________

2. ***Short*** ______________________________

3. ***Flat*** ______________________________

4. ***Irregular*** ______________________________

E. BONE MARKINGS

The surfaces of bones contain various structural features that have specific functions. These features are called ***bone markings*** and are listed in Table 6.1. Knowledge of the bone markings will be very useful when you learn the bones of the body in Exercise 7.

Next to each marking listed in Table 6.1, write its definition, using your textbook as a reference.

ANSWER THE LABORATORY REPORT QUESTIONS AT THE END OF THE EXERCISE

TABLE 6.1
BONE MARKINGS

Marking	Description
DEPRESSIONS AND OPENINGS	
Fissure (FISH-ur)	
Foramen (Fō-RA-men; *foramen* = hole)	small hole that blood vessels / nerve go in
Meatus (mē-A-tus; *meatus* = canal)	canal or tube like passage way in bone
Paranasal sinus (*sin* = cavity)	
Groove or **sulcus** (*sulcus* = ditchlike groove)	
Fossa (*fossa* = basinlike depression)	depression or socket in bone
PROCESSES (prominent PROJECTIONS) THAT FORM JOINTS	
Condyle (KON-dīl; *condylus* = knucklelike process)	rounded that articulates w/ another bone 
Head	head of femur 
Facet	

tubercle — sm. rounded knob that connect w/ connective tissue

trochanter — big. rounded knob that connect w/

TABLE 6.1 ***(Continued)***

Marking	Description
PROCESSES (PROJECTIONS) TO WHICH TENDONS, LIGAMENTS, AND OTHER CONNECTIVE TISSUES ATTACH	
Tubercle (TOO-ber-kul; *tube* = knob)	
Tuberosity	
Trochanter (trō-KAN-ter)	
Crest	
Line	
Spinous process or **spine**	
Epicondyle (*epi* = above)	

JOINTS IN SKULL

sutures — seams that are immovable

4 major

Fontanels

only in infant skull * soft spots

Figure 7-4

LABORATORY REPORT QUESTIONS
OSSEOUS TISSUE 6

Student ______________________ **Date** ______________________

Laboratory Section ______________________ **Score/Grade** ______________________

PART I. COMPLETION

1. Small clusters of bones between certain cranial bones are referred to as ______________________ bones.
2. The technical name for a bone cell is a(n) ______________________ .
3. Canals that extend obliquely inward or horizontally from the bone surface and contain blood vessels and lymphatics are ______________________ canals.
4. The end, or extremity, of a bone is referred to as the ______________________ .
5. Cube-shaped bones that contain more spongy than compact bone are known as ______________________ bones.
6. The cavity within the shaft of a bone that contains marrow is the ______________________ cavity.
7. The thin layer of hyaline cartilage covering the end of a bone where joints are formed is called ______________________ cartilage.
8. Minute canals that radiate from lacuna to lacuna are called ______________________ .
9. The white fibrous covering around the surface of a bone, except for the areas covered by cartilage, is the ______________________ .
10. The shaft of a bone is referred to as the ______________________ .
11. The ______________________ are concentric layers of calcified intercellular substance.
12. The membrane that lines the medullary cavity and contains osteoblasts and a few osteoclasts is the ______________________ .
13. The technical name for bone tissue is ______________________ tissue.
14. In a mature bone, the region where the shaft joins the extremity is called the ______________________ .
15. The microscopic structural unit of compact bone is called a(n) ______________________ .
16. The hardness of bone is primarily due to the mineral salt ______________________ .

BONES 7

The 206 named bones of the adult skeleton are grouped into two divisions: axial and appendicular. The ***axial skeleton*** consists of bones that compose the axis of the body. The axis is a straight line that runs through the center of gravity of the body, through the head, and down to the space between the feet. The ***appendicular skeleton*** consists of the bones of the extremities or limbs (upper and lower) and the girdles (pectoral and pelvic), which connect the extremities to the axial skeleton.

In this exercise you will study the names and locations of bones and their markings by examining various regions of the skeleton:

Region	Number of bones
Axial skeleton	
Skull	
Cranium	8
Face	14
Hyoid (above the larynx)	1
**Auditory ossicles,* 3 in each ear	6
Vertebral column	26
Thorax	
Sternum	1
Ribs	24
	80
Appendicular skeleton	
Pectoral (shoulder) girdles	
Clavicle	2
Scapula	2
Upper extremities	
Humerus	2
Ulna	2
Radius	2
Carpals	16
Metacarpals	10
Phalanges	28
Pelvic (hip) girdles	
Coxal (hip or pelvic) bone	2
Appendicular skeleton (Continued)	
Lower extremities	
Femur	2
Fibula	2
Tibia	2
Patella	2
Tarsals	14
Metatarsals	10
Phalanges	28
	126

*Not actually considered part of the axial or appendicular skeleton, but are placed in the axial skeleton for convenience.

A. BONES OF ADULT SKULL

The ***skull*** is composed of two sets of bones—***cranial*** and ***facial***. The 8 cranial bones are 1 ***frontal***, 2 ***parietals*** (pa-RĪ-i-talz), 2 ***temporals***, 1 ***occipital*** (ok-SIP-i-tal), 1 ***sphenoid*** (SFĒ-noyd), and 1 ***ethmoid***. The 14 facial bones include 2 ***nasals***, 2 ***maxillae*** (mak-SIL-ē), 2 ***zygomatics***, 1 ***mandible***, 2 ***lacrimals*** (LAK-ri-mals), 2 ***palatines*** (PAL-a-tīns), 2 ***inferior conchae*** (KONG-kē), and 1 ***vomer***. These bones are indicated by *arrows* in Figure 7.1. Using the series of photographs in Figure 7.1 for reference, locate the cranial and facial bones on both a Beauchene (disarticulated) and an articulated skull.

Obtain a Beauchene skull and an articulated skull and observe them in anterior view. Using Figure 7.1a for reference, locate the ***frontal eminence, frontal squama (vertical plate), glabella, superciliary arch, supraorbital margin, superior orbital fissure, mental foramen*** (fō-RĀ-men), ***infraorbital foramen***, and ***supraorbital foramen***.

Turn the Beauchene skull and articulated skull so that you are looking at the right side. Using Figure 7.1b for reference, locate the ***zygomatic process, temporal process, mastoid process, mastoid portion, external auditory meatus***, and ***temporal squama***.

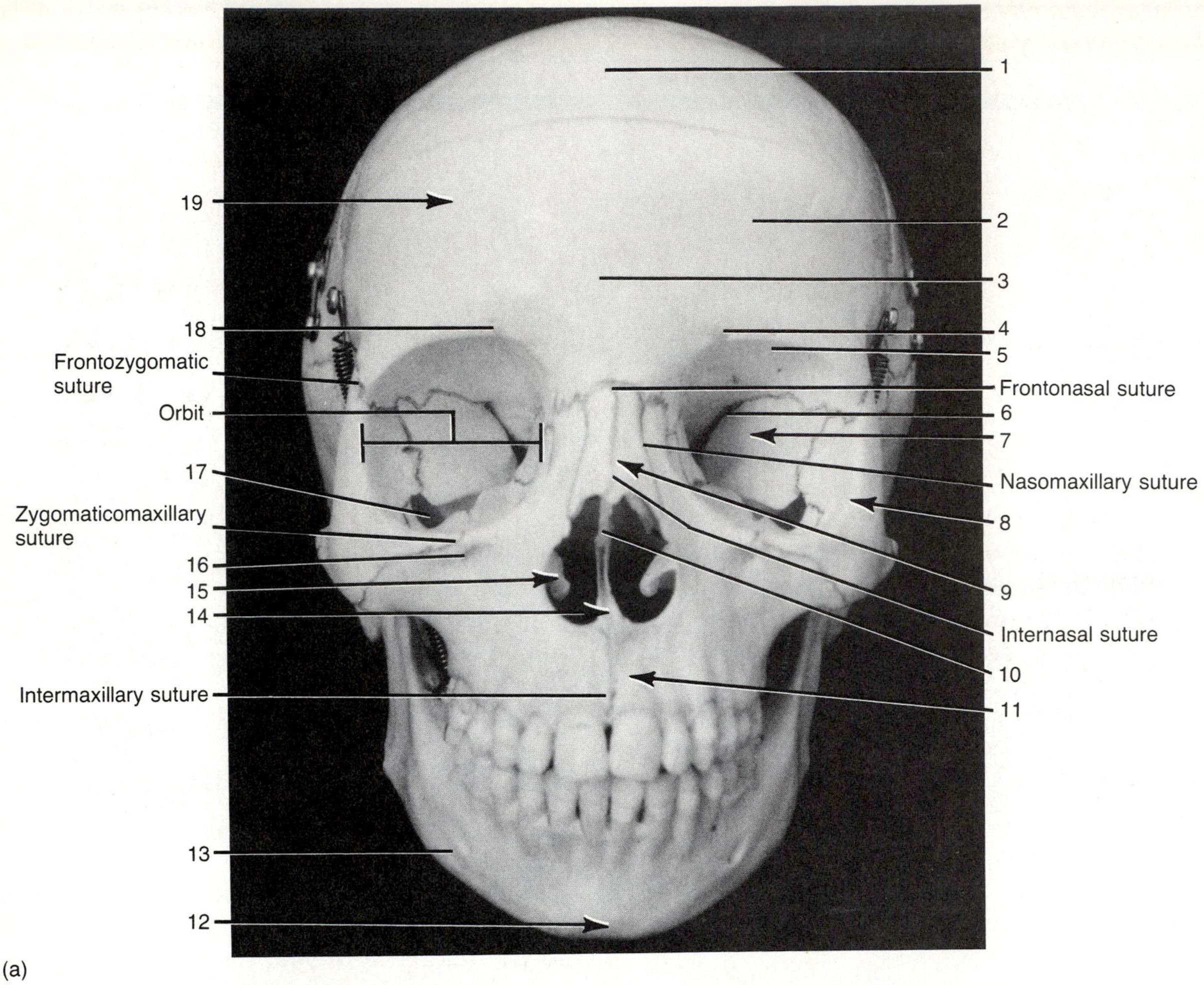

1. Frontal eminence
2. Frontal squama
3. Glabella
4. Superciliary arch
5. Supraorbital margin
6. Superior orbital fissure
7. Sphenoid bone
8. Zygomatic bone
9. Nasal bone
10. Perpendicular plate
11. Maxilla
12. Mandible
13. Mental foramen
14. Vomer
15. Inferior nasal concha
16. Infraorbital foramen
17. Inferior orbital fissure
18. Supraorbital foramen
19. Frontal bone

FIGURE 7.1 The skull. (a) Anterior view.

If a skull in median section is available, use Figure 7.1c for reference and locate the ***sella turcica*** (TUR-si-ka), ***frontal sinus, crista galli, cribriform plate, sphenoidal sinus, superior nasal concha*** (KONG-ka), ***middle nasal concha, inferior nasal concha, pterygoid*** (TER-i-goyd) ***process, styloid process, palatine process,*** and ***external occipital protuberance***. Note also the ***hyoid bone*** below the mandible. This is not a bone of the skull, but is noted here because of its proximity to the skull.

Take an articulated skull and turn it upside down so that you are looking at the inferior surface. Using Figure 7.1d for reference, locate the ***palatine process, zygomatic arch, foramen spinosum, mandibular fossa, foramen lacerum, carotid foramen, mastoid process, stylomastoid foramen, foramen magnum, external occipital protuberance, condylar fossa, occipital condyle, jugular foramen, styloid process, foramen ovale, inferior orbital fissure, middle nasal concha, lesser palatine foramina,***

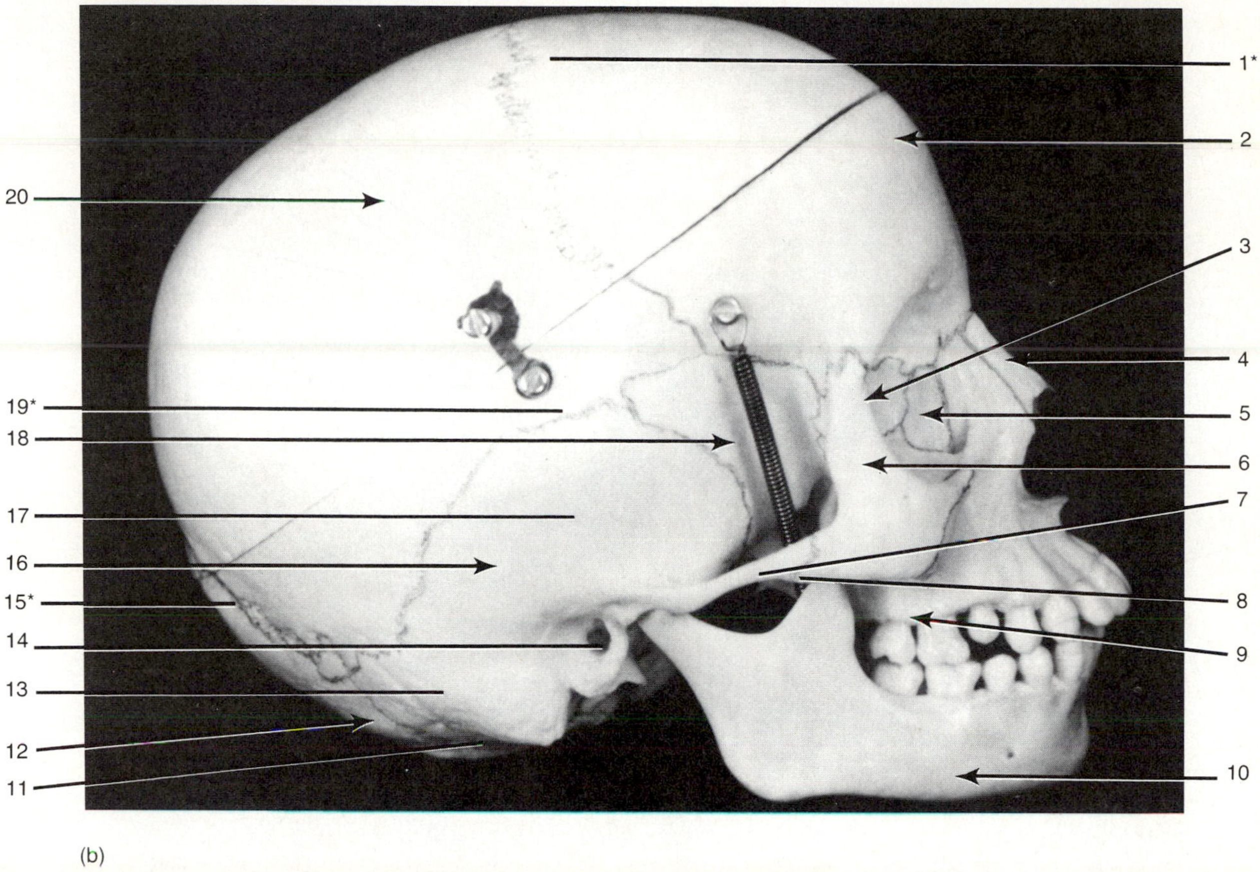

1. Coronal suture
2. Frontal bone
3. Ethmoid bone
4. Nasal bone
5. Lacrimal bone
6. Zygomatic bone
7. Zygomatic process
8. Temporal process
9. Maxilla
10. Mandible
11. Mastoid process
12. Occipital bone
13. Mastoid portion
14. External auditory meatus
15. Lambdoidal suture
16. Temporal bone
17. Temporal squama
18. Sphenoid bone
19. Squamosal suture
20. Parietal bone

FIGURE 7.1 ***(Continued)*** The skull. (b) Right lateral view

greater palatine foramen, horizontal plane, and ***incisive fossa***.

Obtain an articulated skull with a removable crown. Using Figure 7.1e for reference, locate the ***olfactory foramina, crista galli, lesser wing, optic foramen, greater wing, sella turcica, foramen ovale, foramen spinosum, foramen lacerum, petrous portion, foramen magnum, foramen rotundum, superior orbital fissure,*** and ***cribriform plate***.

Examine the right orbit of an articulated skull. Using Figure 7.2 for reference, locate the ***frontal bone, nasal bone, ethmoid bone, lacrimal bone, maxilla, zygomatic bone, sphenoid bone, optic foramen, infraorbital foramen, supraorbital foramen, inferior orbital fissure***, and ***superior orbital fissure***.

Obtain a mandible and, using Figure 7.3 for reference, identify the ***body, ramus, angle, condylar process, coronoid process, mandibular notch, mental foramen, alveolar*** (al-VĒ-ō-lar) ***process***, and ***mandibular foramen***.

Before you move on, examine Table 7.1, Summary of Foramina of the Skull. It will give

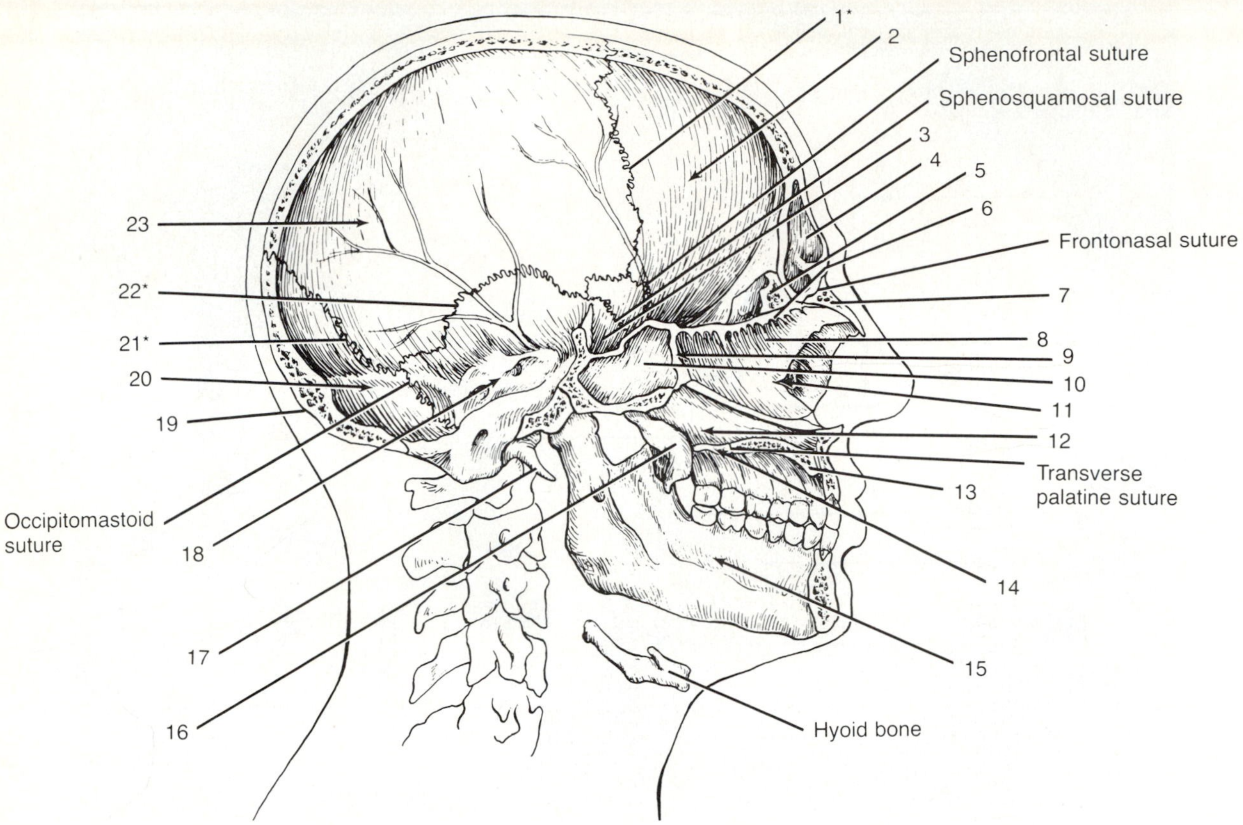

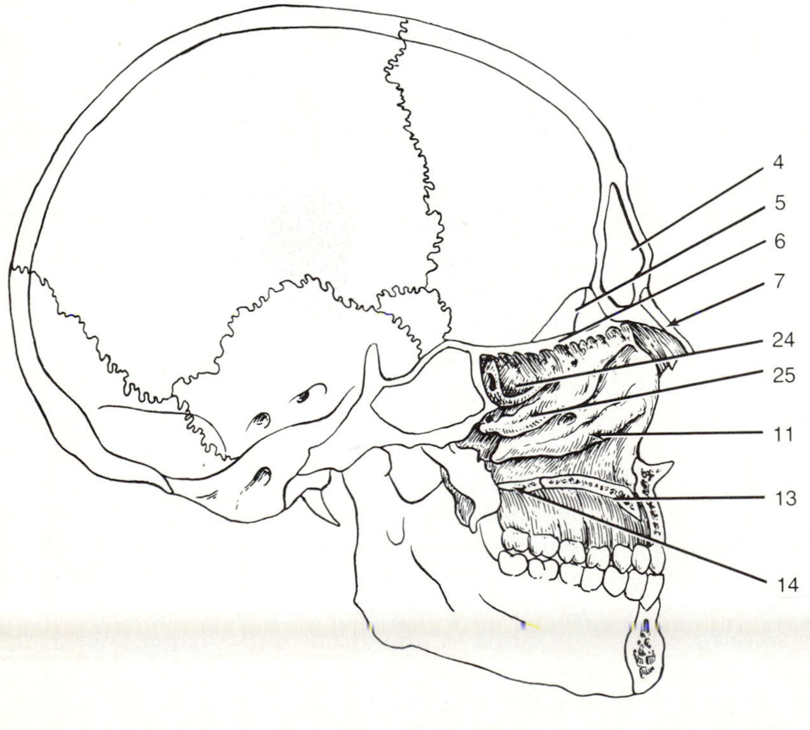

1. Coronal suture
2. Frontal bone
3. Sella turcica
4. Frontal sinus
5. Crista galli
6. Cribriform plate
7. Nasal bone
8. Perpendicular plate
9. Sphenoid bone
10. Sphenoidal sinus
11. Inferior nasal concha
12. Vomer
13. Palatine process
14. Palatine bone
15. Mandible
16. Pterygoid process
17. Styloid process
18. Temporal bone
19. External occipital protuberance
20. Occipital bone
21. Lambdoidal suture
22. Squamosal suture
23. Parietal bone
24. Superior nasal concha
25. Middle nasal concha

(c)

FIGURE 7.1 ***(Continued)*** The skull. (c) Median sections

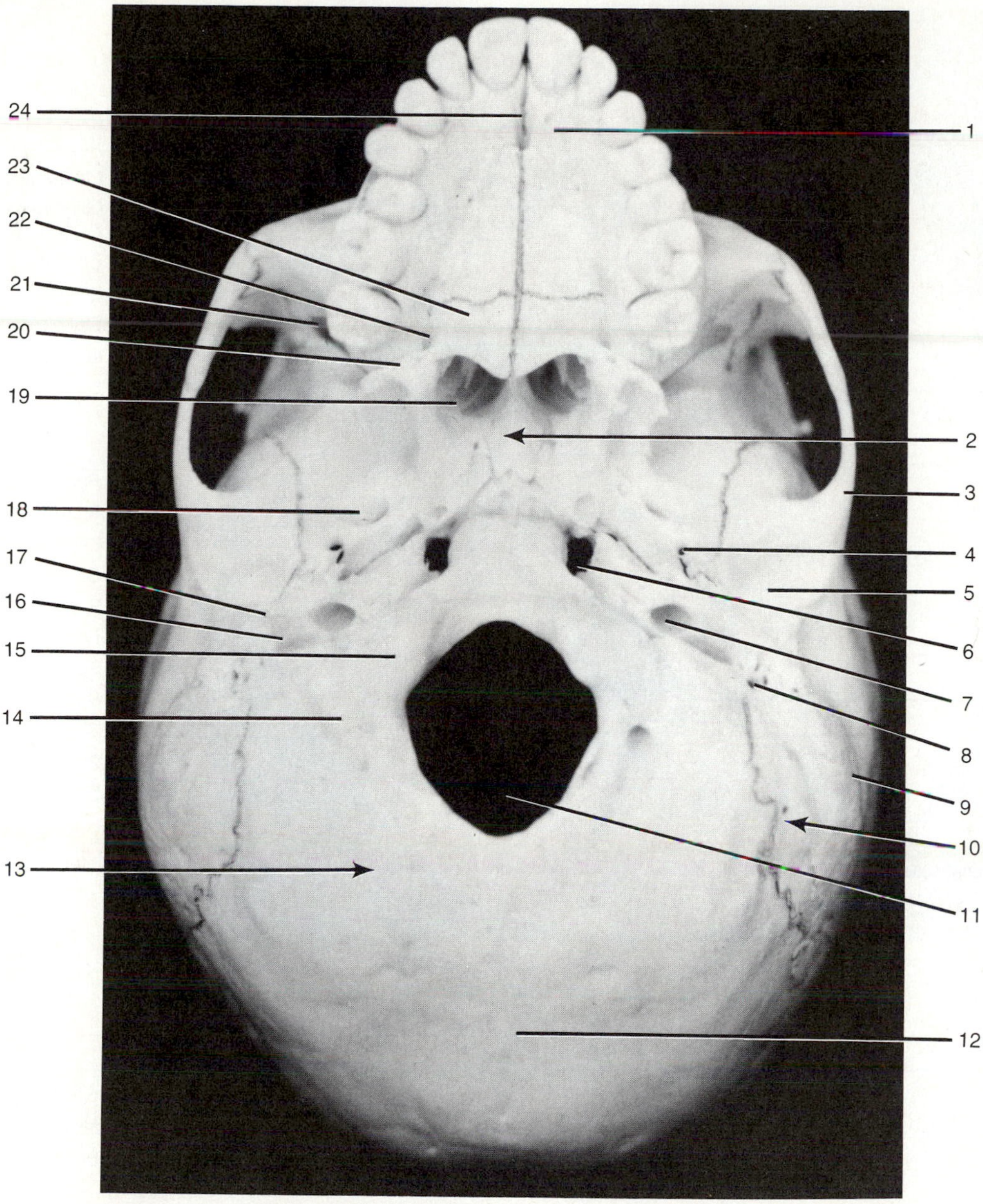

(d)

1. Palatine process
2. Vomer
3. Zygomatic arch
4. Foramen spinosum
5. Mandibular fossa
6. Foramen lacerum
7. Carotid foramen
8. Stylomastoid foramen
9. Mastoid process
10. Temporal bone
11. Foramen magnum
12. External occipital protuberance
13. Occipital bone
14. Condylar fossa
15. Occipital condyle
16. Jugular foramen
17. Styloid process
18. Foramen ovale
19. Middle nasal concha
20. Lesser palatine foramina
21. Inferior orbital fissure
22. Greater palatine foramen
23. Horizontal plate
24. Incisive fossa

FIGURE 7.1 ***(Continued)*** The skull. (d) Inferior view

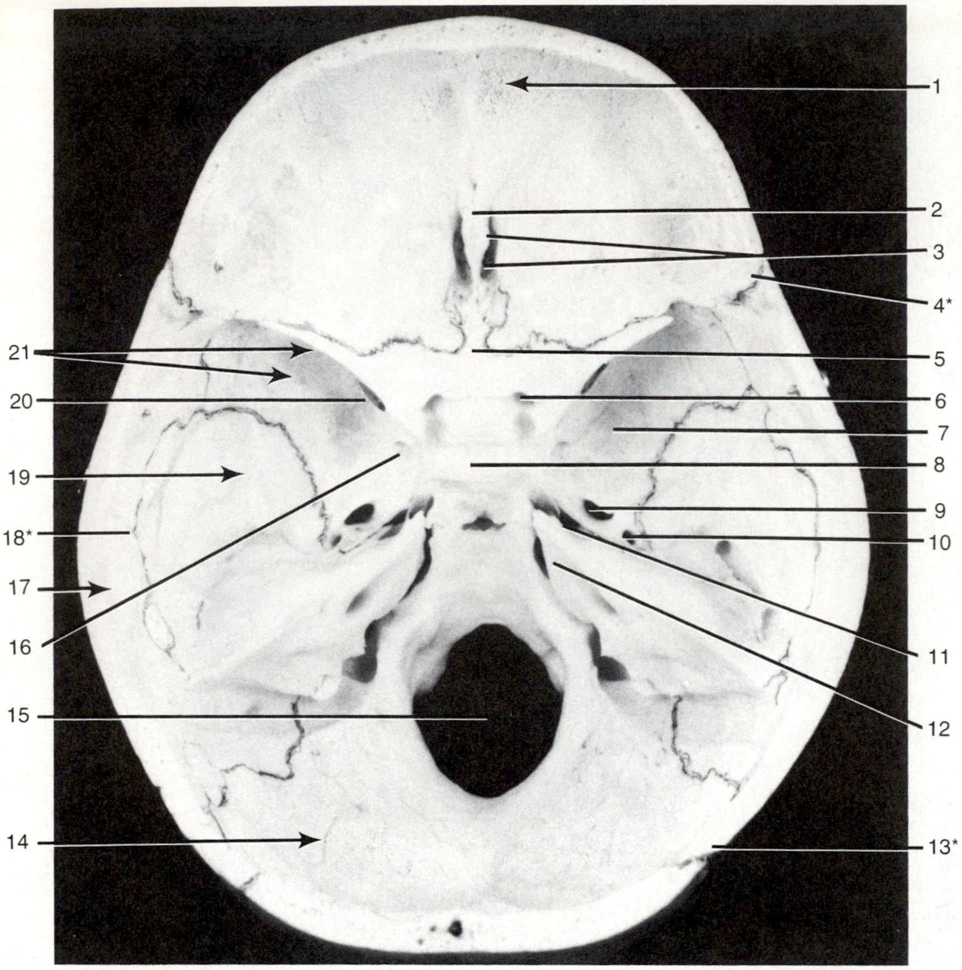

1. Frontal bone
2. Crista galli
3. Olfactory foramina in cribriform plate
4. Coronal suture
5. Lesser wing
6. Optic foramen
7. Greater wing
8. Sella turcica
9. Foramen ovale
10. Foramen spinosum
11. Foramen lacerum
12. Petrous portion
13. Lambdoidal suture
14. Occipital bone
15. Foramen magnum
16. Foramen rotundum
17. Parietal bone
18. Squamosal suture
19. Temporal bone
20. Superior orbital fissure
21. Sphenoid bone

(e)

FIGURE 7.1 ***(Continued)*** The skull. (e) Floor of cranium.

you an understanding of the structures that pass through the various foramina you just studied.

B. SUTURES OF SKULL

A ***suture*** (SOO-chur) is an immovable joint between skull bones. The four prominent sutures are the ***coronal, sagittal, lambdoidal*** (lam-DOY-dal), and ***squamosal*** (skwa-MŌ-sal). The lines identifying these sutures are marked with *asterisks* in Figure 7.1. Using Figures 7.1 and 7.4 for reference, locate these sutures on an articulated skull.

C. FONTANELS OF SKULL

At birth, the skull bones are separated by membrane-filled spaces called ***fontanels*** (fon′-ta-NELZ). They (1) enable the fetal skull to compress as it passes through the birth canal, (2) permit rapid growth of the brain during infancy, (3) facilitate determination of the degree of brain development by their state of closure, (4) serve as landmarks (anterior fontanel) for withdrawal of blood from the superior sagittal sinus, and (5) aid in determining the position of the fetal head prior to birth. The principal fontanels are ***anterior (frontal), posterior (occipital), anterolateral (sphenoidal),*** and ***posterolateral (mastoid).*** Using Figure 7.4 for reference, locate the fontanels on the skull of a newborn infant.

D. PARANASAL SINUSES OF SKULL

A ***paranasal sinus,*** or just simply ***sinus,*** is a cavity in a bone located near the nasal cavity.

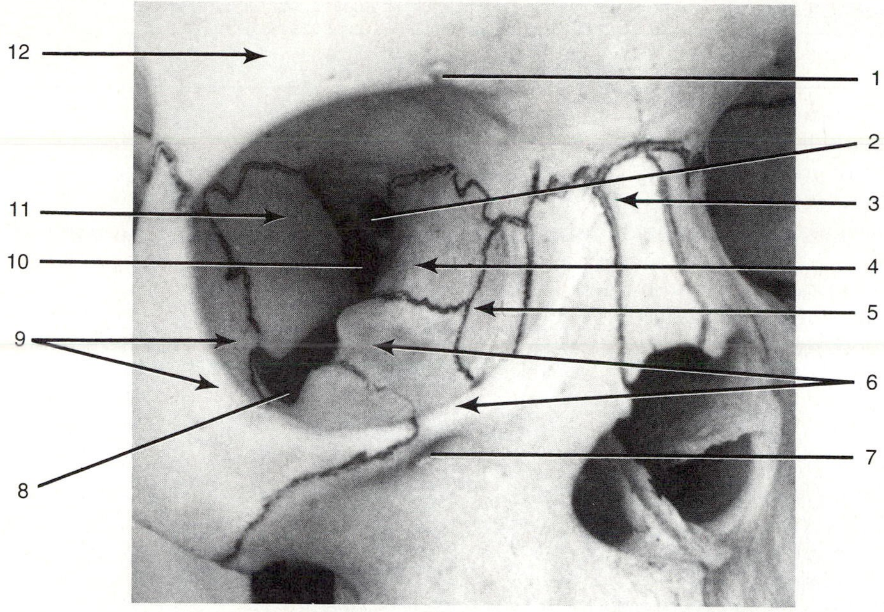

1. Supraorbital foramen
2. Optic foramen
3. Nasal bone
4. Ethmoid bone
5. Lacrimal bone
6. Maxilla
7. Infraorbital foramen
8. Inferior orbital fissure
9. Zygomatic bone
10. Superior orbital fissure
11. Sphenoid bone
12. Frontal bone

FIGURE 7.2 The right orbit.

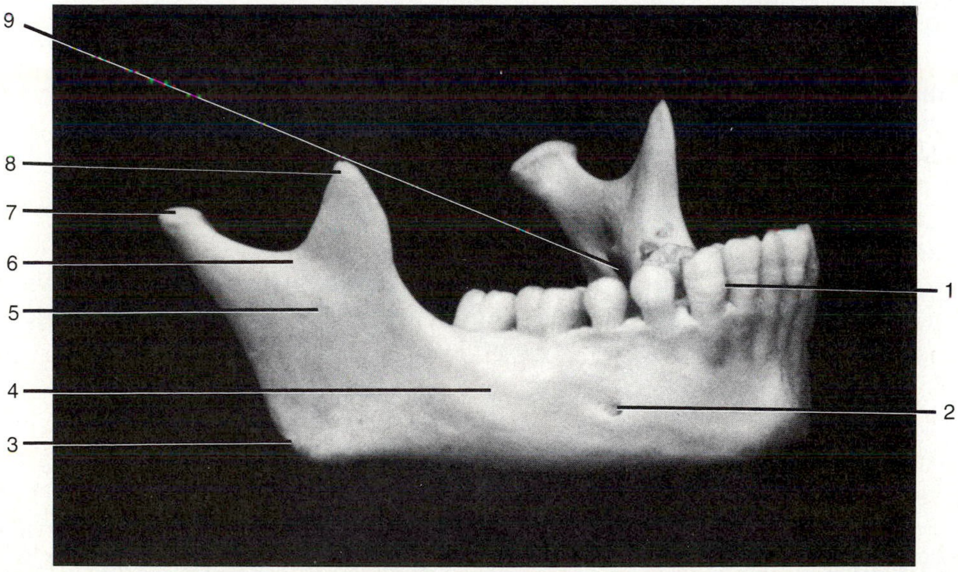

1. Alveolar process
2. Mental foramen
3. Angle
4. Body
5. Ramus
6. Mandibular notch
7. Condylar process
8. Coronoid process
9. Mandibular foramen

FIGURE 7.3 Right lateral view of the mandible.

Paired paranasal sinuses are found in the maxillae and the frontal, sphenoid, and ethmoid bones. Locate the paranasal sinuses on the Beauchene skull or other demonstration models that may be available. Label the paranasal sinuses shown in Figure 7.5.

E. VERTEBRAL COLUMN

The ***vertebral column***, together with the sternum and ribs, constitutes the skeleton of the trunk. The ***vertebrae*** of the adult column are distributed as follows: 7 ***cervical*** (SER-vi-kal)

TABLE 7.1
SUMMARY OF FORAMINA OF THE SKULL

Foramen	Structures passing through
Carotid	Internal carotid artery
Greater palatine	Greater palatine nerve and greater palatine vessels
Incisive	Branches of greater palatine vessels and nasopalatine nerve
Infraorbital	Infraorbital nerve and artery
Jugular	Internal jugular vein, glossopharyngeal (IX) nerve, vagus (X) nerve, and accessory (XI) nerve
Lacerum	Internal carotid artery and branch of ascending pharyngeal artery
Lesser palatine	Lesser palatine nerves and artery
Magnum	Medulla oblongata and its membranes, accessory (XI) nerve, and vertebral and spinal arteries and meninges
Mandibular	Inferior alveolar nerve and vessels
Mental	Mental nerve and vessels
Olfactory	Olfactory (I) nerve
Optic	Optic (II) nerve and ophthalmic artery
Ovale	Mandibular branch of trigeminal (V) nerve
Rotundum	Maxillary branch of trigeminal (V) nerve
Spinosum	Middle meningeal vessels
Stylomastoid	Facial (VII) nerve and stylomastoid artery
Supraorbital	Supraorbital nerve and artery

(neck), 12 ***thoracic*** (thō-RAS-ik) (chest), 5 ***lumbar*** (lower back), 5 ***sacral*** (fused into one bone, the ***sacrum*** between the hipbones), and 3, 4, or 5 ***coccygeal*** (kok-SIJ-ē-al) (fused into one bone, the ***coccyx*** (KOK-six), forming the tail of the column). Locate each of these regions on the articulated skeleton. Label the same regions in Figure 7.6, the anterior view.

Examine the vertebral column on the articulated skeleton and identify the cervical, thoracic, lumbar, and sacral (sacrococcygeal) curves. Label the curves in Figure 7.6, the right lateral view.

F. VERTEBRAE

A typical ***vertebra*** consists of the following portions:

1. ***Body***—Thick, disc-shaped anterior portion.
2. ***Vertebral (neural) arch***—Posterior extension from the body that surrounds the spinal cord and consists of the following parts:
    a. ***Pedicles*** (PED-i-kuls)—Two short, thick processes that project posteriorly; each has a superior and inferior notch and, when successive vertebrae are fitted together, the opposing notches form an ***intervertebral foramen*** through which a spinal nerve and blood vessels pass.
    b. ***Laminae*** (LAM-i-nē)—Flat portions that form the posterior wall of the vertebral arch.
    c. ***Vertebral foramen***—Opening through which the spinal cord passes; when all the vertebrae are fitted together, the foramina form a canal, the ***vertebral canal***.
3. ***Processes***—Seven processes arise from the vertebral arch:
    a. Two ***transverse processes***—Lateral extensions where the laminae and pedicles join.
    b. One ***spinous process***—Posterior projection of the lamina.
    c. Two ***superior articular processes***—Articulate with the vertebra above. Their top surfaces, called ***superior articular facets,*** articulate with the vertebra above.
    d. Two ***inferior articular processes***—Articulate with the vertebra below. Their bottom surfaces, called ***inferior articular facets***, articulate with the vertebra below.

Obtain a thoracic vertebra and locate each part just described. Now label the vertebra in Figure 7.7a. You should be able to distinguish the

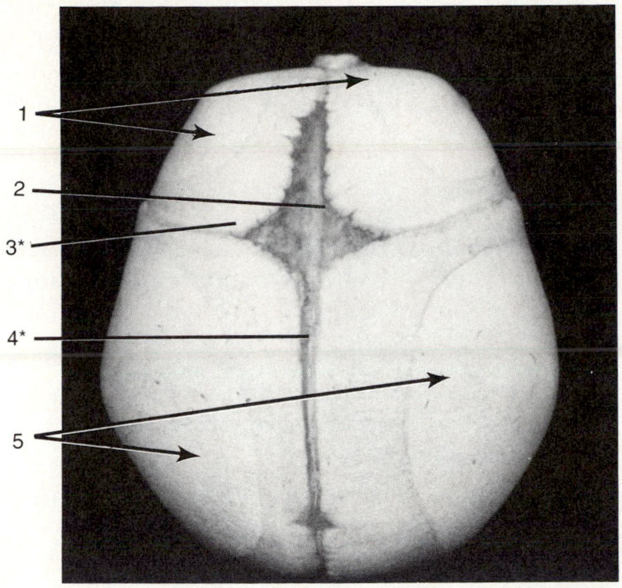

(a)

1. Frontal bone
2. Anterior fontanel
3. Coronal suture
4. Sagittal suture
5. Parietal bones

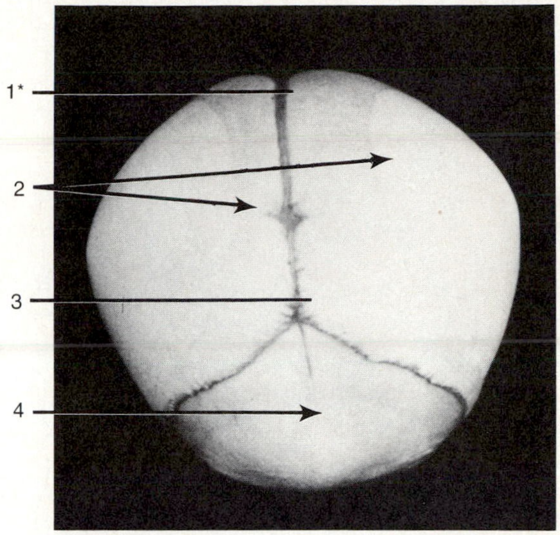

(b)

1. Sagittal suture
2. Parietal bones
3. Posterior fontanel
4. Occipital bone

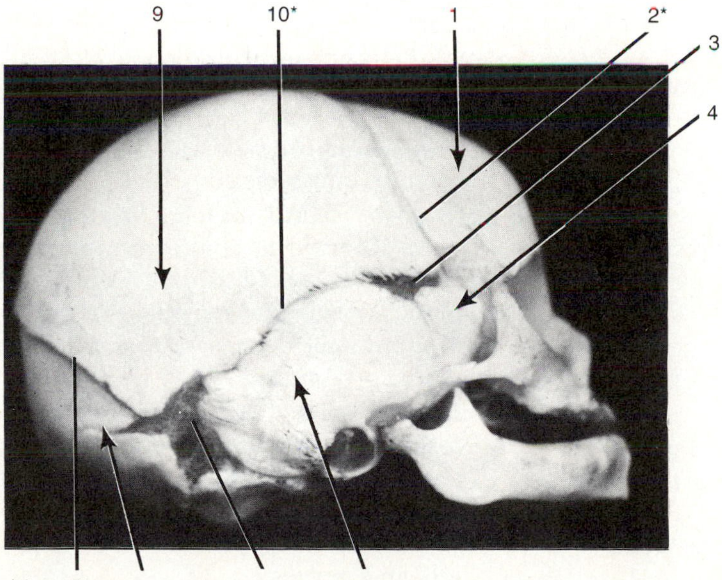

(c)

1. Frontal bone
2. Coronal suture
3. Anterolateral fontanel
4. Sphenoid bone
5. Temporal bone
6. Posterolateral fontanel
7. Occipital bone
8. Lambdoidal suture
9. Parietal bone
10. Squamosal suture

FIGURE 7.4 Fontanels. (a) Superior view. (b) Posterior view. (c) Right lateral view.

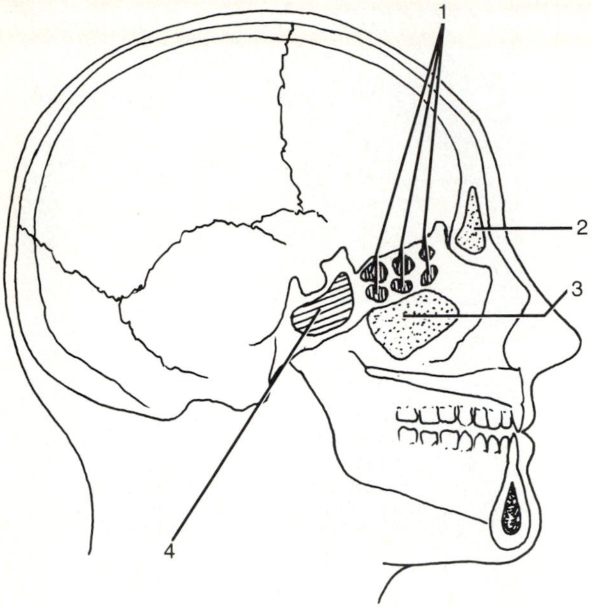

1 Ethmoidal sinus
2 Frontal sinus
3 Maxillary sinus
4 Sphenoidal sinus

FIGURE 7.5 Median section of skull showing paranasal sinuses.

general parts on all the different vertebrae that contain them.

Although vertebrae have the same basic design, those of a given region have special distinguishing features. Obtain examples of the following vertebrae and identify their distinguishing features.

1. ***Cervical vertebrae***
   a. ***Atlas***—First cervical vertebra (see Figure 7.7b)
      Transverse foramen—Opening in transverse process through which an artery, a vein, and a branch of a spinal nerve pass.
      Anterior arch—Anterior wall of vertebral foramen.
      Posterior arch—Posterior wall of vertebral foramen.
      Lateral mass—Side wall of vertebral foramen.

   Label the other indicated parts.

   b. ***Axis***—Second cervical vertebra (see Figure 7.7c)
      Dens—Superior projection of body that articulates with atlas.

   Label the other indicated parts.

   c. ***Cervicals 3 through 6*** (see Figure 7.7d)
      Bifid spinous process—Cleft in spinous processes of cervical vertebrae 2 through 6.

   Label the other indicated parts.

   d. ***Vertebra prominens***—Seventh cervical vertebra; contains a nonbifid and long spinous process.
2. ***Thoracic vertebrae*** (see Figure 7.7e)
   a. ***Facets***—For articulation with the tubercle and head of a rib; found on body and transverse processes.
   b. ***Spinous processes***—Long, pointed, downward projection.

   Label the other indicated parts.
3. ***Lumbar vertebrae*** (see Figure 7.7f)
   a. ***Spinous processes***—Broad, blunt.
   b. ***Superior articular processes***—Directed medially, not superiorly.
   c. ***Inferior articular processes***—Directed laterally, not inferiorly.

   Label the other indicated parts.
4. ***Sacrum*** (see Figure 7.7g)
   a. ***Transverse lines***—Areas where bodies of vertebrae are joined.
   b. ***Anterior sacral (pelvic) foramina***—Four pairs of foramina that communicate with dorsal sacral foramina; passages for blood vessels and nerves.
   c. ***Median sacral crest***—Spinous processes of fused vertebrae.
   d. ***Lateral sacral crest***—Transverse processes of fused vertebrae.
   e. ***Posterior sacral (dorsal) foramina***—Four pairs of foramina that communicate with pelvic foramina; passages for blood vessels and nerves.
   f. ***Sacral canal***—Continuation of vertebral canal.
   g. ***Sacral promontory***—Superior, anterior projecting border.
   h. ***Auricular surface***—Articulates with ilium of coxal bone.
   i. ***Sacral hiatus*** (hi-Ā-tus)—Inferior entrance to sacral canal where laminae of S5, and sometimes S4, fail to meet.
5. ***Coccyx*** (see Figure 7.7g)

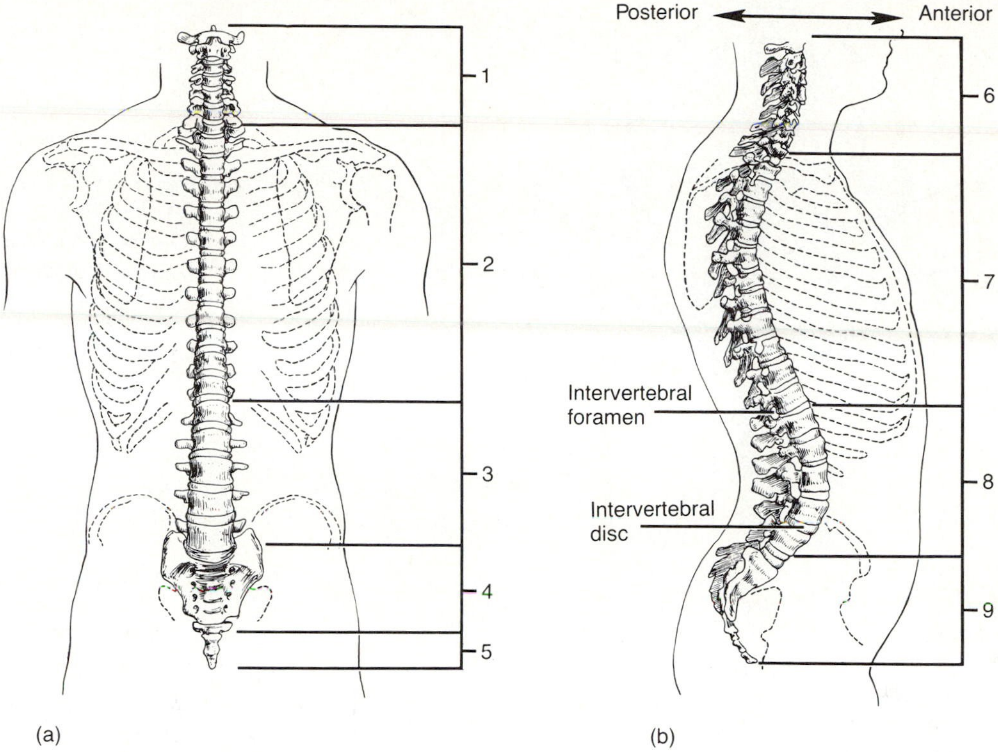

FIGURE 7.6 Vertebral column. (a) Anterior view. (b) Right lateral view.

Label the coccyx and the parts of the sacrum in Figure 7.7g.

G. STERNUM AND RIBS

The skeleton of the ***thorax*** consists of the ***sternum, costal cartilages, ribs,*** and bodies of the ***thoracic vertebrae.***

Examine the articulated skeleton and disarticulated bones and identify the following:

1. ***Sternum***—Breastbone, flat bone in midline of anterior thorax.
   a. ***Manubrium*** (ma-NOO-brē-um)—Superior portion.
   b. ***Body***—Middle, largest portion.
   c. ***Sternal angle***—Junction of the manubrium and body.
   d. ***Xiphoid*** (ZI-foyd) ***process***—Inferior, smallest portion.
   e. ***Jugular (suprasternal) notch***—Depression on the superior surface of the manubrium that may be palpated.
   f. ***Clavicular notches***—Articular surfaces for the clavicles.
   g. ***Costal notches***—Points at which ribs articulate.
2. ***Costal cartilage***—Strip of hyaline cartilage that attaches a rib to the sternum.
3. ***Ribs***—Parts of a typical rib (third through ninth) include:
   a. ***Body***—Shaft, main part of rib.
   b. ***Head***—Posterior projection.
   c. ***Neck***—Constricted portion behind head.
   d. ***Tubercle*** (TOO-ber-kul)—Knoblike elevation just below neck; consists of a ***nonarticular part*** that affords attachment for a ligament and an ***articular part*** that articulates with an inferior vertebra.

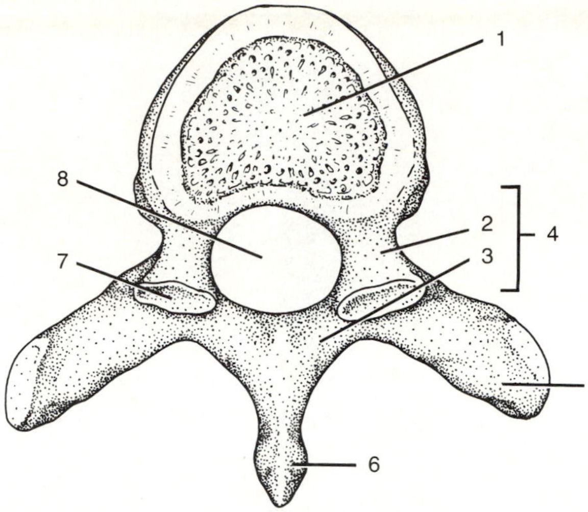

(a)

___ Body
___ Lamina
___ Pedicle
___ Spinous process
___ Superior articular facet
___ Transverse process
___ Vertebral arch
8 Vertebral foramen

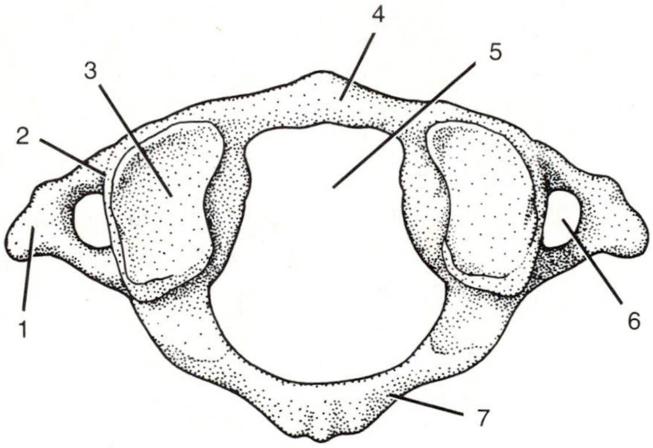

(b)

4 Anterior arch
1 Lateral mass
7 Posterior arch
2 Superior articular facet
6 Transverse foramen
3 Transverse process
5 Vertebral foramen

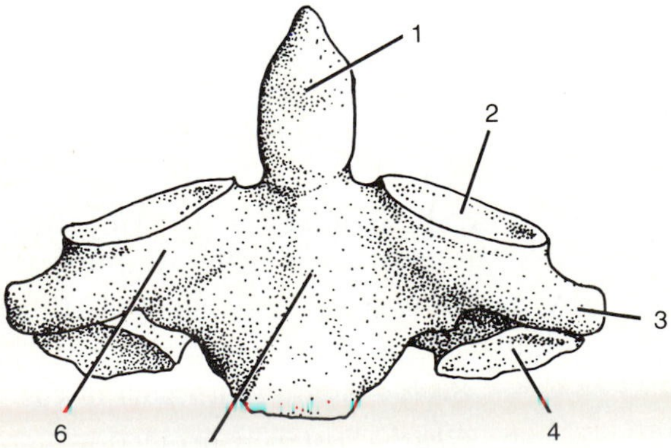

(c)

5 Body
___ Dens
3 Inferior articular facet
___ Lateral mass
6 Superior articular facet
2 Transverse process

FIGURE 7.7 Vertebrae. (a) Superior view of a typical vertebra. (b) Superior view of the atlas. (c) Anterior view of the axis.

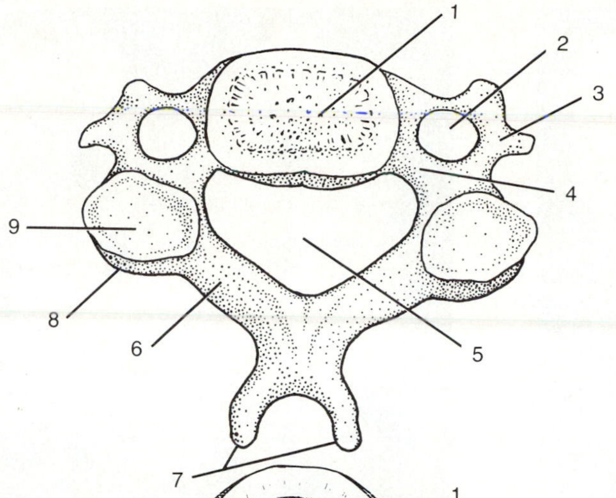

(d)

___ Bifid spinous process
___ Body
___ Inferior articular process
___ Lamina
___ Pedicle
___ Superior articular facet
___ Transverse foramen
___ Transverse process
___ Vertebral foramen

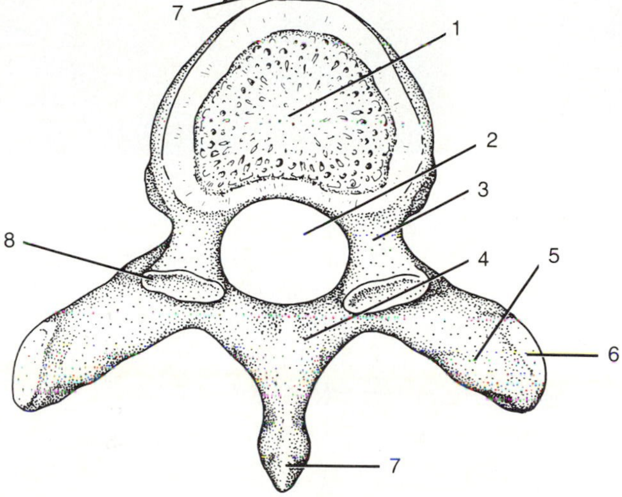

(e)

___ Body
___ Facet for tubercle of rib
___ Lamina
___ Pedicle
___ Spinous process
___ Superior articular facet
___ Transverse process
___ Vertebral foramen

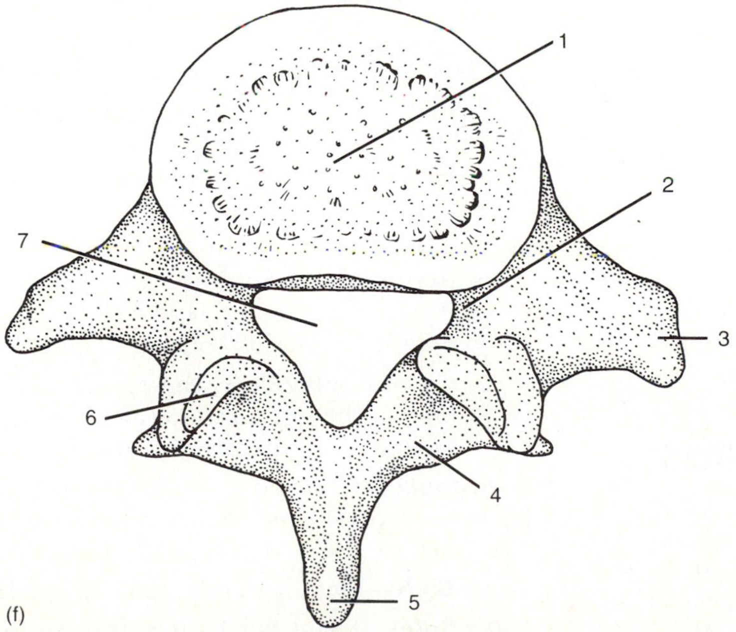

(f)

___ Body
___ Lamina
___ Pedicle
___ Spinous process
___ Superior articular facet
___ Transverse process
___ Vertebral foramen

FIGURE 7.7 ***(Continued)*** Vertebrae. (d) Superior view of a cervical vertebra. (e) Superior view of a thoracic vertebra. (f) Superior view of a lumbar vertebra.

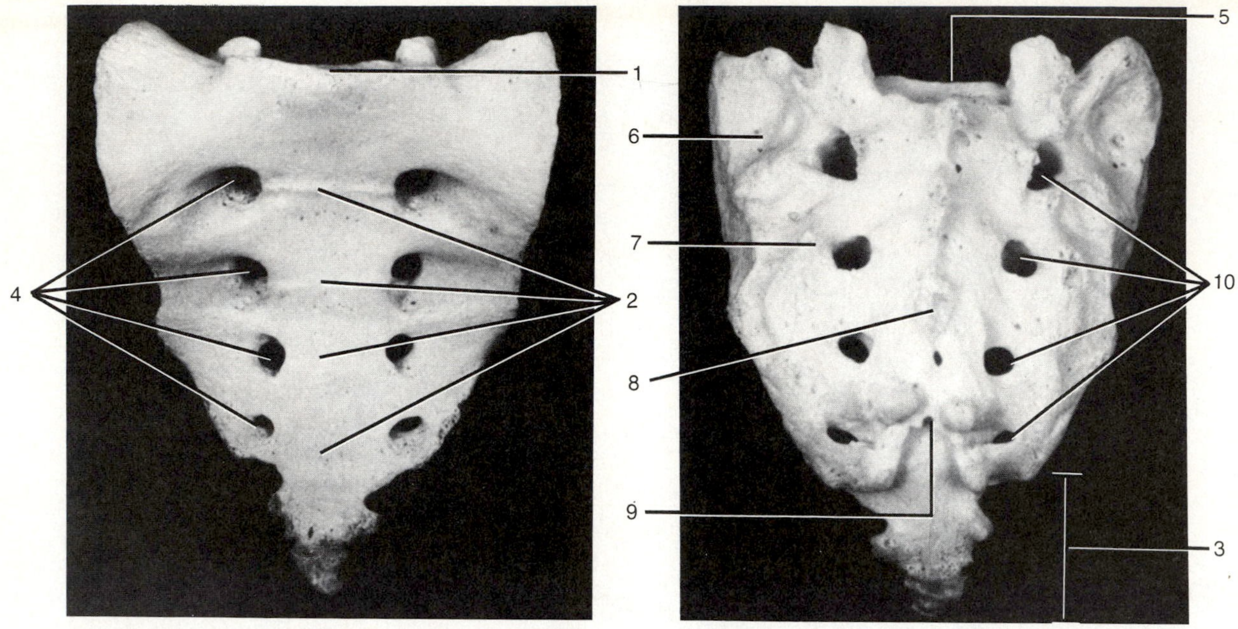

(g)

2 Auricular surface
3 Coccyx
10 Posterior sacral (dorsal) foramina
5 Lateral sacral crest
1 Median sacral crest
4 Anterior sacral (pelvic) foramina
9 Sacral canal
7 Sacral hiatus
6 Sacral promontary
8 Transverse lines

FIGURE 7.7 ***(Continued)*** Vertebrae. (g) Anterior (left) and posterior (right) views of the sacrum and the coccyx.

e. ***Costal groove***—Depression on the inner surface containing blood vessels and a nerve.

f. ***Superior facet***—Articulates with facet on superior vertebra.

g. ***Inferior facet***—Articulates with facet on inferior vertebra.

Label the parts of the sternum in Figure 7.8a and the parts of a rib in Figure 7.8b.

H. PECTORAL (SHOULDER) GIRDLES

Each ***pectoral (shoulder) girdle*** consists of two bones—***clavicle*** (collar bone) and ***scapula*** (shoulder blade). Its purpose is to attach the bones of the upper extremity to the axial skeleton.

Examine the articulated skeleton and disarticulated bones and identify the following:

1. ***Clavicle*** (KLAV-i-kul)—Slender bone with a double curvature; lies horizontally in superior and anterior part of thorax.
   a. ***Sternal extremity***—Rounded, medial end that articulates with manubrium of sternum.
   b. ***Acromial*** (a-KRŌ-mē-al) ***extremity***—Broad, flat, lateral end that articulates with the acromion of the scapula.
   c. ***Conoid tubercle*** (TOO-ber-kul)—Projection on the inferior, lateral surface for attachment of ligaments.
2. ***Scapula*** (SCAP-yoo-la)—Large, flat triangular bone in dorsal thorax between the levels of ribs 2 through 7.
   a. ***Body***—Flattened, triangular portion.
   b. ***Spine***—Ridge across posterior surface.
   c. ***Acromion*** (a-KRŌ-mē-on)—Flattened, expanded process of spine.

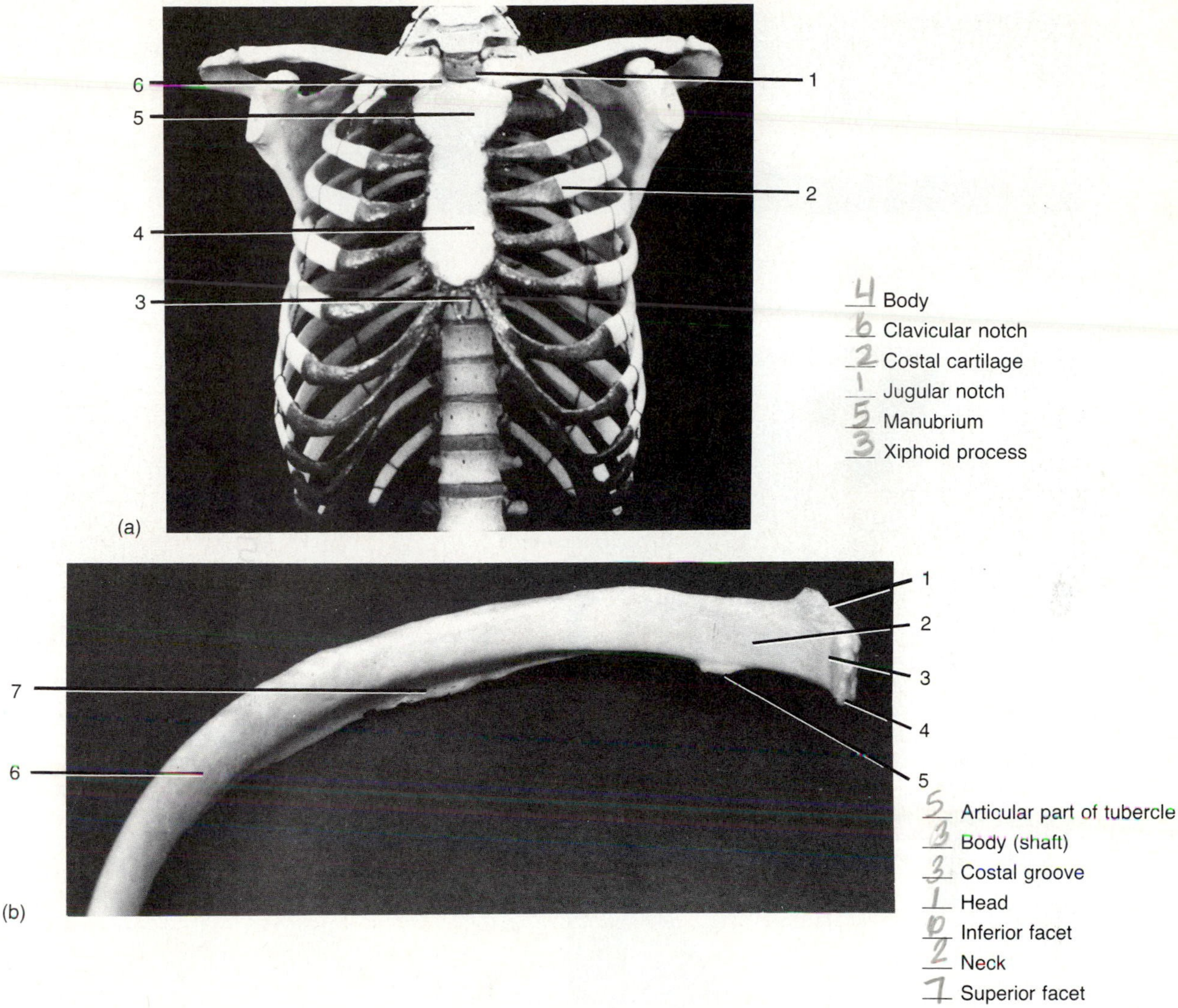

FIGURE 7.8 Bones of the thorax. (a) Anterior view of the thoracic cage. (b) Inner aspect of a portion of the fifth right rib.

d. ***Medial (vertebral) border***—Edge of body near vertebral column.

e. ***Lateral (axillary) border***—Edge of body near arm.

f. ***Inferior angle***—Bottom of body where medial and lateral borders join.

g. ***Glenoid cavity***—Depression below acromion that articulates with head of humerus to form the shoulder joint.

h. ***Coracoid*** (KOR-a-koyd) ***process***—Projection at lateral end of superior border.

i. ***Supraspinous*** (soo′-pra-SPĪ-nus) ***fossa***—Surface for muscle attachment above spine.

j. ***Infraspinous fossa***—Surface for muscle attachment below spine.

k. ***Superior border***—Superior edge of the body.

l. ***Superior angle***—Top of body where superior and medial borders join.

Label Figure 7.9.

I. UPPER EXTREMITIES

The skeleton of the ***upper extremities*** consists of a humerous in each arm, an ulna and radius in each forearm, carpals in each wrist,

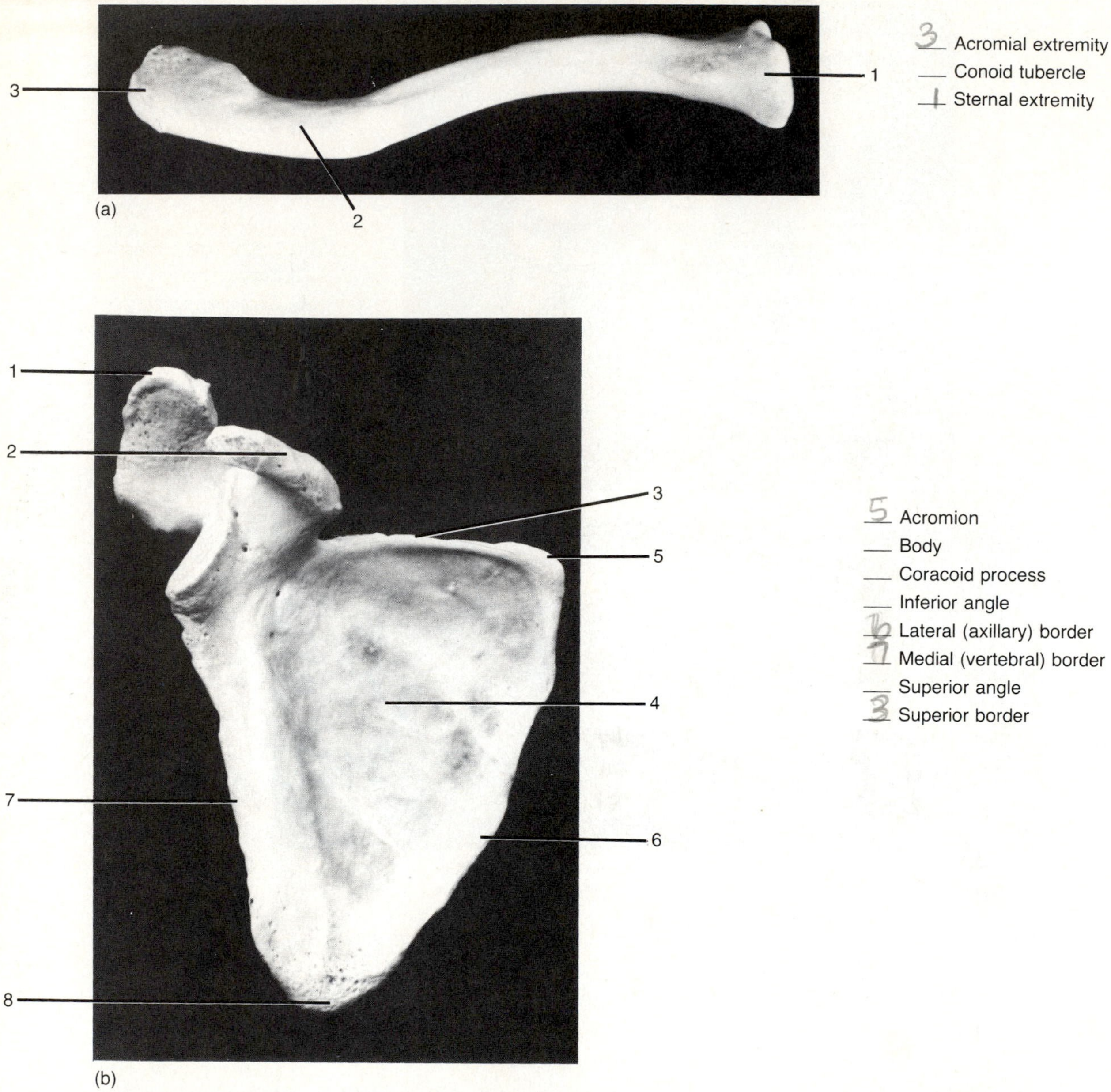

FIGURE 7.9 The pectoral (shoulder) girdle. (a) Right clavicle viewed from below. (b) Anterior view of right scapula.

metacarpals in each palm, and phalanges in the fingers.

Examine the articulated skeleton and disarticulated bones and identify the following:

1. ***Humerus*** (HYOO-mer-us)—Arm bone; longest and largest bone of the upper extremity.
   a. ***Head***—Articulates with glenoid cavity of scapula.
   b. ***Anatomical neck***—Oblique groove below head.
   c. ***Greater tubercle***—Lateral projection below anatomical neck.
   d. ***Lesser tubercle***—Anterior projection.
   e. ***Intertubercular sulcus (bicipital groove)***—Between tubercles.
   f. ***Surgical neck***—Constricted portion below turbercles.

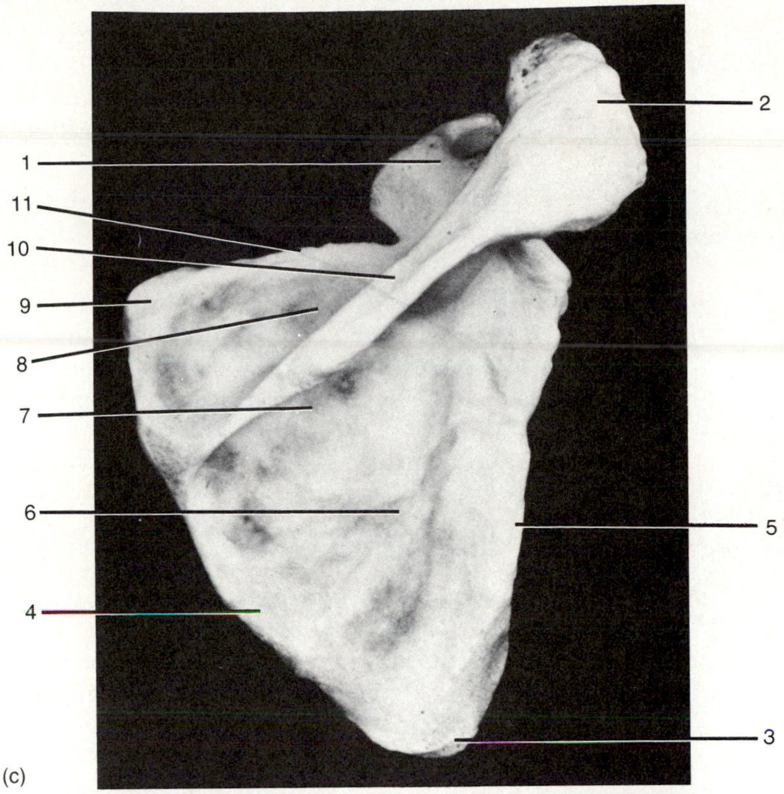

___ Acromion
___ Body
___ Coracoid process
___ Inferior angle
___ Infraspinous fossa
___ Lateral (axillary) border
___ Medial (vertrebral) border
___ Spine
___ Superior angle
___ Superior border
___ Supraspinous fossa

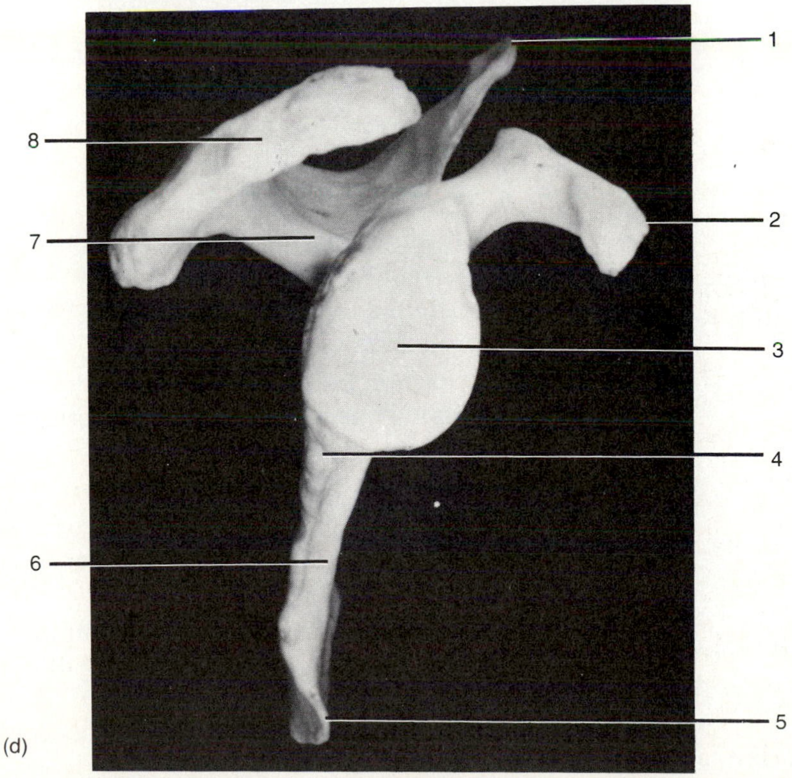

___ Acromion
___ Body
___ Coracoid process
___ Glenoid cavity
___ Inferior angle
___ Lateral (axillary) border
___ Spine
___ Superior angle

FIGURE 7.9 ***(Continued)*** The pectoral (shoulder) girdle. (c) Posterior view of right scapula. (d) Lateral border view of right scapula.

g. ***Body***—Shaft.

h. ***Deltoid tuberosity***—Roughened, V-shaped area about midway down the lateral surface of shaft.

i. ***Capitulum*** (ka-PIT-yoo-lum)—Rounded knob that articulates with head of radius.

j. ***Radial fossa***—Anterior lateral depression that receives head of radius when forearm is flexed.

k. ***Trochlea*** (TRŌK-lē-a)—Projection that articulates with the ulna.

l. ***Coronoid fossa***—Anterior medial depression that receives part of the ulna when the forearm is flexed.

m. ***Olecranon*** (ō-LEK-ra-non) ***fossa***—Posterior depression that receives the olecranon of the ulna when the forearm is extended.

n. ***Medial epicondyle***—Projection on medial side of distal end.

o. ***Lateral epicondyle***—Projection on lateral side of distal end.

2. ***Ulna***—Medial bone of forearm.

a. ***Olecranon (olecranon process)***—Prominence of elbow at proximal end.

b. ***Coronoid process***—Anterior projection that, with olecranon, receives trochlea of humerus.

c. ***Trochlear (semilunar) notch***— Curved area between olecranon and coronoid process into which trochlea of humerus fits.

d. ***Radial notch***—Depression lateral and inferior to trochlear notch that receives the head of the radius.

e. ***Head***—Rounded portion at distal end.

f. ***Styloid process***—Projection on posterior side of distal end.

3. ***Radius***—Lateral bone of forearm.

a. ***Head***—Disc-shaped process at proximal end.

b. ***Radial tuberosity***—Medial projection for insertion of the biceps brachii muscle.

c. ***Styloid process***—Projection on lateral side of distal end.

d. ***Ulnar notch***—Medial, concave depression for articulation with ulna.

4. ***Carpus***—Wrist, consists of eight small bones, called ***carpals***, united by ligaments.

a. ***Proximal row***—From medial to lateral are called ***pisiform, triquetrum, lunate,*** and ***scaphoid.***

b. ***Distal row***—From medial to lateral are called ***hamate, capitate, trapezoid,*** and ***trapezium.***

5. ***Metacarpus***—Five bones in the palm of the hand, numbered as follows beginning with thumb side: I, II, III, IV, and V metacarpals.

6. ***Phalanges*** (fa-LAN-jēz)—Bones of the fingers; two in each thumb (proximal and distal) and three in each finger (proximal, middle, and distal). The singular of phalanges is ***phalanx.***

Label Figure 7.10.

J. PELVIC (HIP) GIRDLES

The ***pelvic (hip) girdles*** consist of the two ***coxal (hip*** or ***pelvic) bones***. They provide a strong and stable support for the lower extremities on which the weight of the body is carried and attach the lower extremities to the axial skeleton.

Examine the articulated skeleton and disarticulated bones and identify the following parts of the coxal bone:

1. ***Ilium***—Superior flattened portion.

a. ***Iliac crest***—Superior border of ilium.

b. ***Anterior superior iliac spine***—Anterior projection of iliac crest.

c. ***Anterior inferior iliac spine***—Projection under anterior superior iliac spine.

d. ***Posterior superior iliac spine***—Posterior projection of iliac crest.

e. ***Posterior inferior iliac spine***—Projection below posterior superior iliac spine.

f. ***Greater sciatic*** (sī-AT-ik) ***notch***—Concavity under posterior inferior iliac spine.

g. ***Iliac fossa***—Medial concavity for attachment of iliacus muscle.

h. ***Iliac tuberosity***—Point of attachment for sacroiliac ligament posterior to iliac fossa.

i. ***Auricular surface***—Point of articulation with sacrum.

j. ***Posterior, anterior,*** and ***inferior gluteal lines***—Between these lines gluteal muscles are attached on the lateral surface.

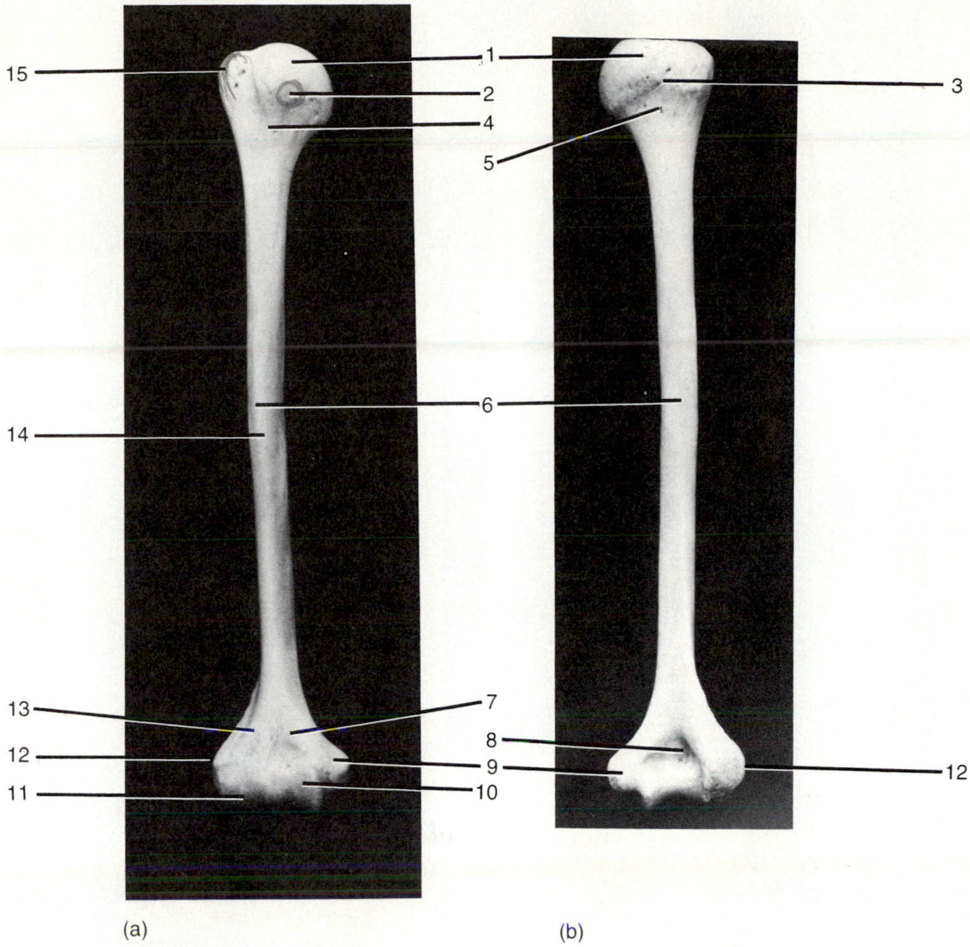

4 Anatomical neck
___ Body
12 Capitulum
7 Coronoid fossa
14 Deltoid tuberosity
15 Greater tubercle
1 Head
___ Intertubercular sulcus (bicipital groove)
___ Lateral epicondyle
2 Lesser tubercle
___ Medial epicondyle
___ Olecranon fossa
13 Radial fossa
4 Surgical neck
9 Trochlea

FIGURE 7.10 Bones of upper extremity. (a) Anterior view of right humerus. (b) Posterior view of right humerus.

2. ***Ischium***—Lower, posterior portion.
   a. ***Ischial spine***—Posterior projection of ischium.
   b. ***Lesser sciatic notch***—Concavity under ischial spine.
   c. ***Ischial tuberosity***—Roughened projection.
   d. ***Ramus***—Portion of ischium that joins the pubis and surrounds the ***obturator foramen.***
3. ***Pubis***—Anterior, inferior portion.
   a. ***Superior ramus***—Upper portion of pubis.
   b. ***Inferior ramus***—Lower portion of pubis.

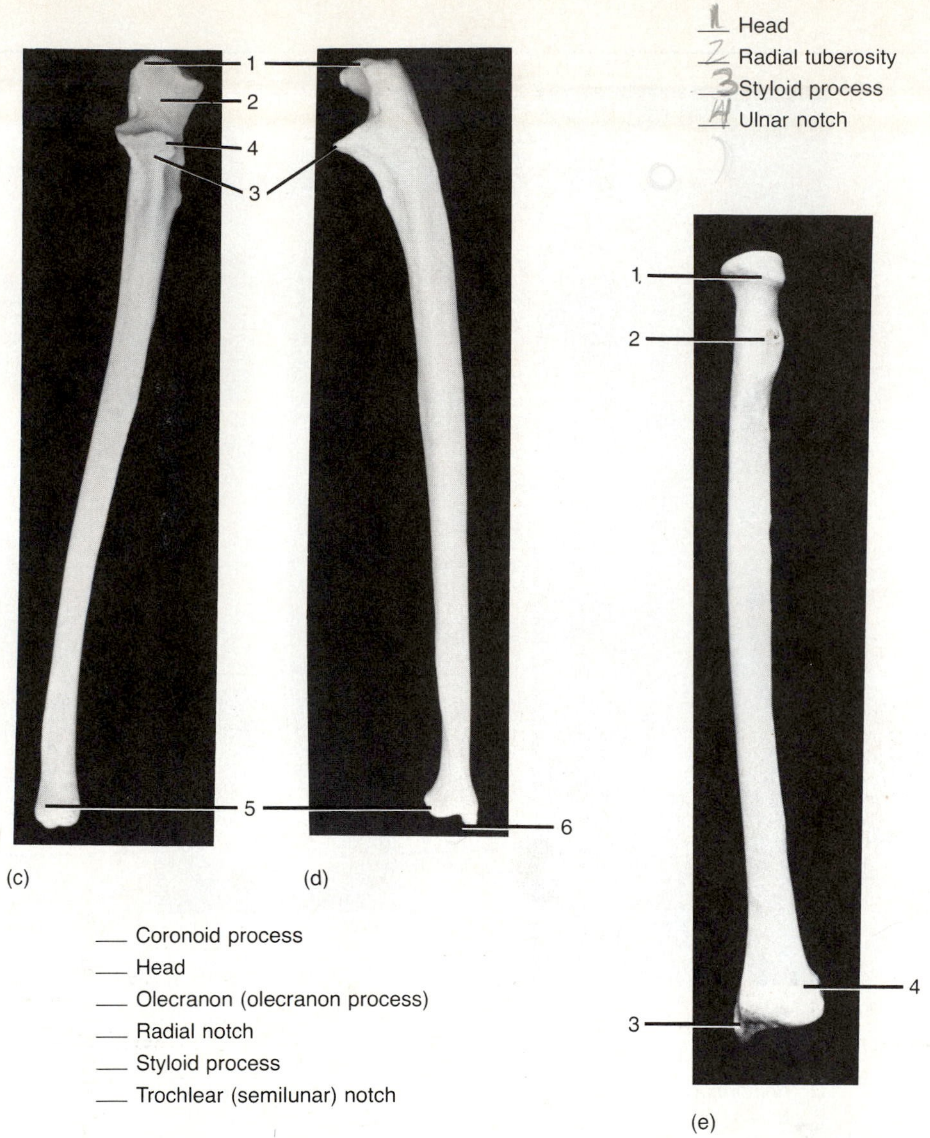

FIGURE 7.10 ***(Continued)*** Bones of upper extremity. (c) Anterior view of right ulna. (d) Lateral view of right ulna. (e) Anterior view of right radius.

c. ***Symphysis*** (SIM-fi-sis) ***pubis***—Joint between left and right hipbones.

4. ***Acetabulum*** (as′-e-TAB-yoo-lum)—Socket that receives the head of the femur to form the hip joint; two-fifths of the acetabulum is formed by the ilium, two-fifths is formed by the ischium, and one-fifth is formed by the pubis.

Label Figure 7.11.

Again, examine the articulated skeleton. This time compare the male and female pelvis. The ***pelvis*** consists of the two hipbones, sacrum, and coccyx. Identify the following:

1. ***Greater (false) pelvis***—Expanded portion situated above the brim of the pelvis; bounded laterally by the ilia and posteriorly by the upper sacrum.

2. ***Lesser (true) pelvis***—Below and behind the brim of the pelvis; constructed of parts of the ilium, pubis, sacrum, and coccyx; contains an opening above, the ***pelvic inlet,*** and an opening below, the ***pelvic outlet.***

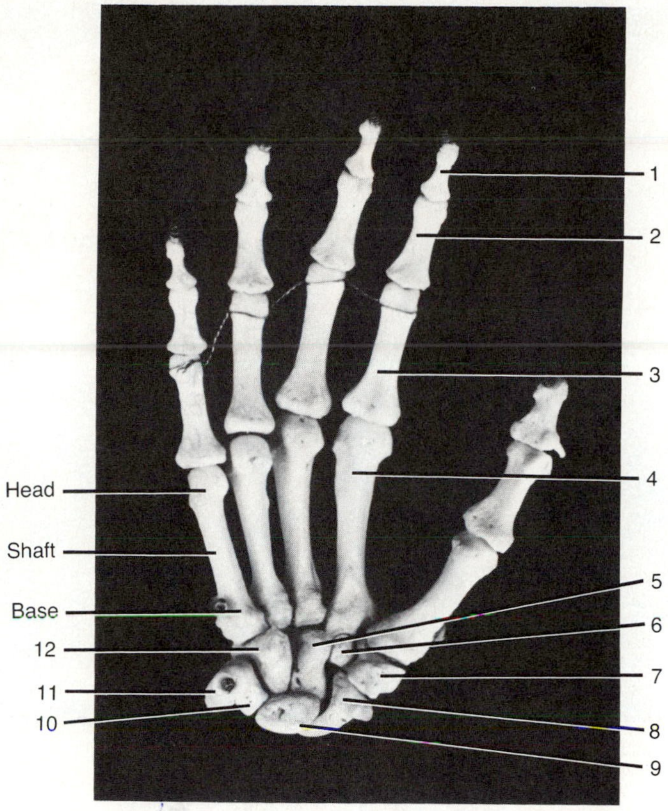

(f)

___ Capitate
___ Distal phalanx
___ Hamate
___ Lunate
___ Metacarpal
___ Middle phalanx
___ Pisiform
___ Proximal phalanx
___ Scaphoid
___ Trapezium
___ Trapezoid
___ Triquetrum

FIGURE 7.10 ***(Continued)*** Bones of upper extremity. (f) Anterior view of right wrist and hand.

K. LOWER EXTREMITIES

The bones of the ***lower extremities*** consist of a femur in each thigh, a patella in front of each knee joint, a tibia and fibula in each leg, tarsals in each ankle, metatarsals in each foot, and phalanges in the toes.

Examine the articulated skeleton and disarticulated bones and identify the following:

1. ***Femur***—Thigh bone; longest and heaviest bone in the body.
   a. ***Head***—Rounded projection at proximal end that articulates with acetabulum of hipbone.
   b. ***Neck***—Constricted portion below head.
   c. ***Greater trochanter*** (trō-KAN-ter)—Prominence on lateral side.
   d. ***Lesser trochanter***—Prominence on dorsomedial side.
   e. ***Intertrochanteric line***—Ridge on anterior surface.
   f. ***Intertrochanteric crest***—Ridge on posterior surface.
   g. ***Linea aspera***—Vertical ridge on posterior surface.
   h. ***Medial condyle***—Medial posterior projection on distal end that articulates with tibia.
   i. ***Lateral condyle***—Lateral posterior projection on distal end that articulates with tibia.
   j. ***Intercondylar*** (in′-ter-KON-di-lar) ***fossa***—Depressed area between condyles on posterior surface.
   k. ***Medial epicondyle***—Projection above medial condyle.
   l. ***Lateral epicondyle***—Projection above lateral condyle.

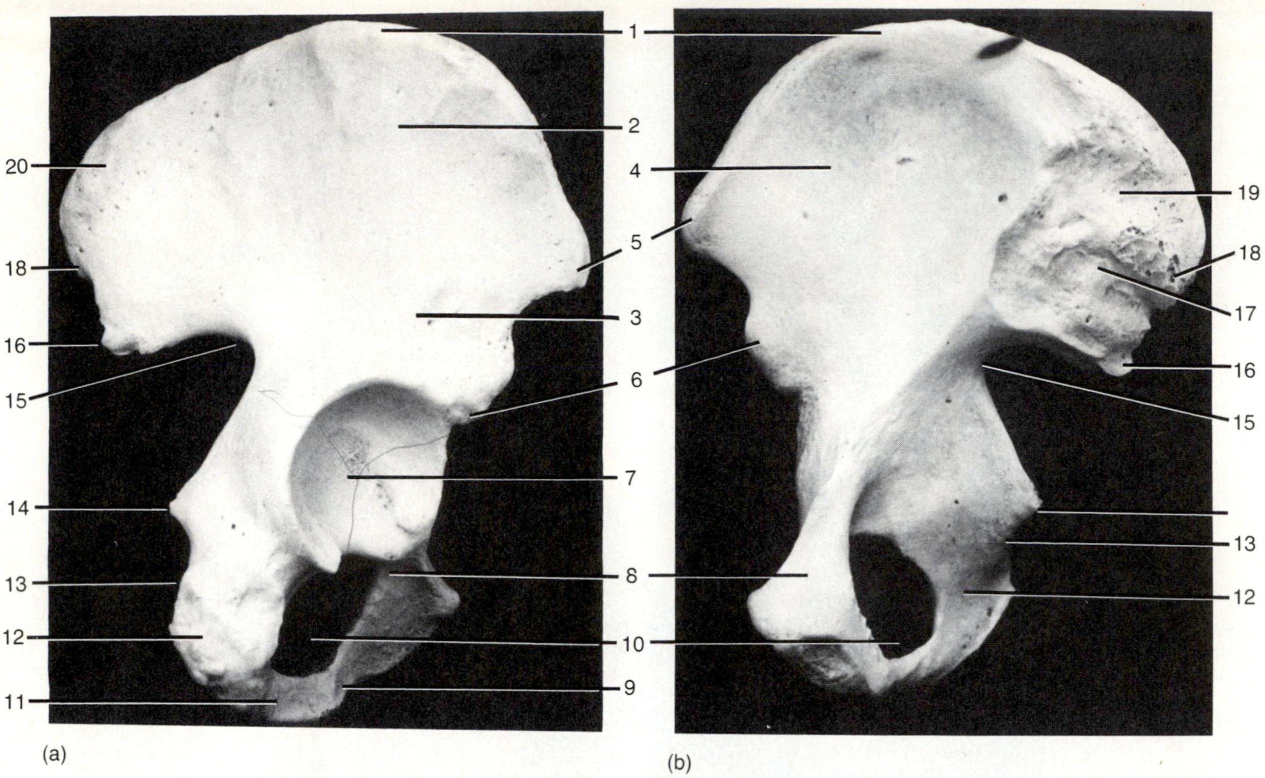

7 Acetabulum
___ Anterior gluteal line
___ Anterior inferior iliac spine
___ Anterior superior iliac spine
___ Auricular surface
15 Greater sciatic notch
1 Iliac crest
___ Iliac fossa
___ Iliac tuberosity
___ Inferior gluteal line
___ Inferior ramus
14 Ischial spine
___ Ischial tuberosity
13 Lesser sciatic notch
10 Obturator foramen
___ Posterior gluteal line
___ Posterior inferior iliac spine
___ Posterior superior iliac spine
___ Ramus
___ Superior ramus

FIGURE 7.11 Right coxal (hip) bone. (a) Lateral view. (b) Medial view.

m. ***Patellar surface***—Between condyles on the anterior surface.

2. ***Patella***—Kneecap in front of knee joint; develops in tendon of quadriceps femoris muscle.

   a. ***Base***—Broad superior portion.

   b. ***Apex***—Pointed inferior portion.

   c. ***Articular facets***—Articulating surfaces on posterior surface for medial and lateral condyles of femur.

3. ***Tibia***—Shinbone; medial bone of leg.

   a. ***Lateral condyle***—Articulates with lateral condyle of femur.

   b. ***Medial condyle***—Articulates with medial condyle of femur.

   c. ***Intercondylar eminence***—Upward projection between condyles.

   d. ***Tibial tuberosity***—Anterior projection for attachment of patellar ligament.

   e. ***Medial malleolus*** (mal-LĒ-ō-lus)—Distal projection that articules with talus bone of ankle.

f. ***Fibular notch***—Distal depression that articulates with the fibula.

4. ***Fibula***—Lateral bone of leg.

a. ***Head***—Proximal projection that articulates with tibia.

b. ***Lateral malleolus***—Projection at distal end that articulates with the talus bone of ankle.

5. ***Tarsus***—Seven bones of the ankle called ***tarsals***.

a. ***Posterior bones—talus*** and ***calcaneus*** (heel bone).

b. ***Anterior bones—cuboid, navicular (scaphoid)***, and three ***cuneiforms*** called the first (medial), second (intermediate), and third (lateral) cuneiforms.

6. ***Metatarsus***—Consists of five bones of the foot called metatarsals, numbered as follows, beginning on the medial (large toe) side: I, II, III, IV, and V metatarsals.

7. ***Phalanges***—Bones of the toes, comparable to phalanges of fingers; two in each large toe (proximal and distal) and three in each small toe (proximal, middle, and distal).

Label Figure 7.12.

L. ARTICULATED SKELETON

Now that you have studied all the bones of the body, label the entire articulated skeleton in Figure 7.13.

M. SKELETAL SYSTEM OF CAT

The ***skeletal system*** of the cat, as is that of the human, is conveniently divided into two principal divisions: axial skeleton and appendicular skeleton. The ***axial skeleton*** of the cat consists of the bones of the skull, vertebral column, ribs, sternum, and hyoid apparatus. The ***appendicular skeleton*** of the cat consists of the bones of the pectoral girdle, pelvic girdle, forelimbs, and hindlimbs.

Before you start your examination of the bones of the cat skeleton, it should be noted that the cat walks on its digits with the remainder of the palm and foot elevated. This is referred to as ***digitigrade locomotion***. Humans, by contrast, walk on the entire sole of the foot. This is called ***plantigrade locomotion***.

Refer to Figure 7.14, which shows the skeleton of the cat in lateral view. Be sure that you study the figure carefully so that you can identify the principal bones.

Obtain a mounted articulated skeleton and identify all the bones indicated in Figure 7.14. Compare the cat skeleton to a human articulated skeleton.

ANSWER THE LABORATORY REPORT QUESTIONS AT THE END OF THE EXERCISE

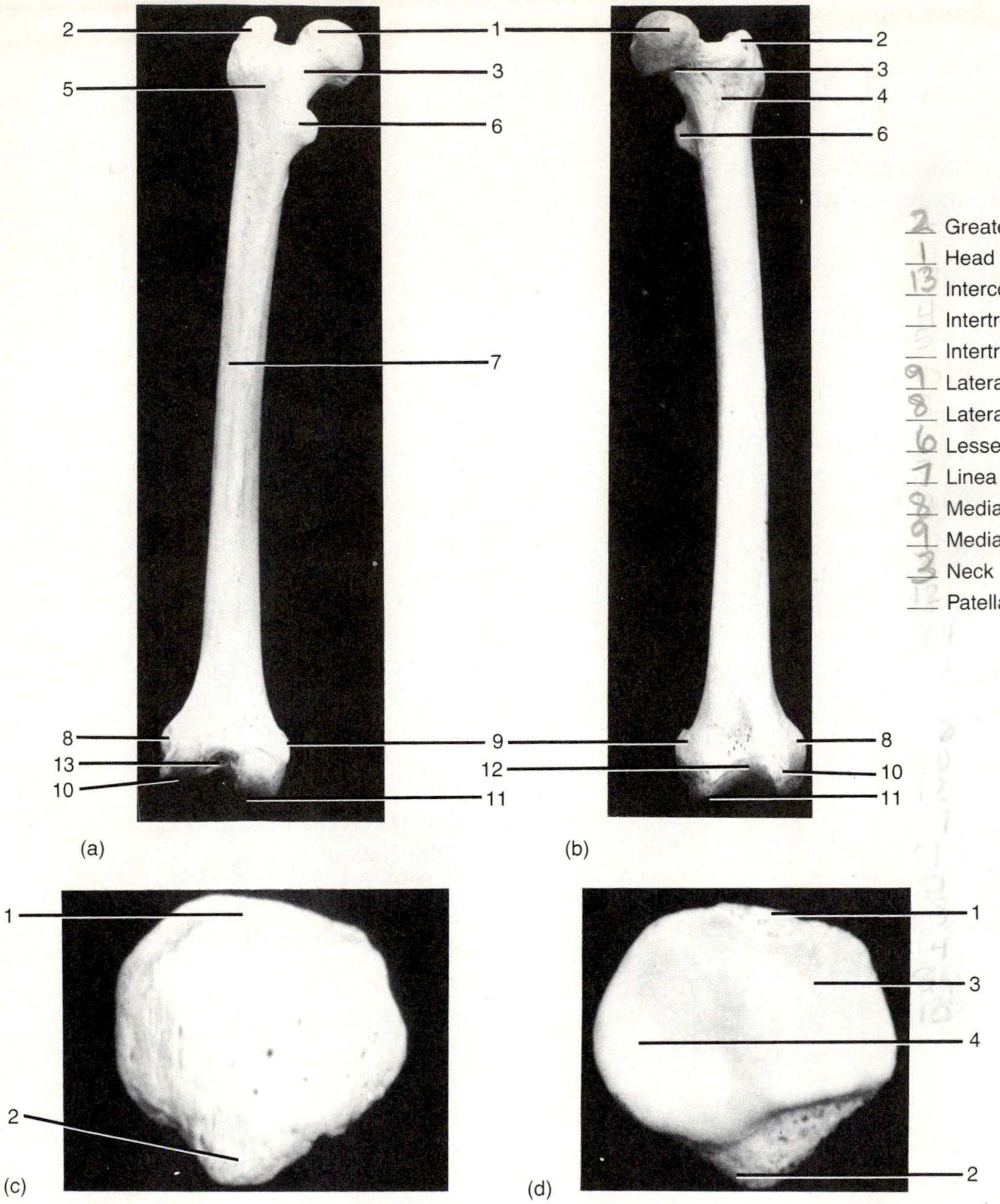

___ Greater trochanter
___ Head
___ Intercondylar fossa
___ Intertrochanteric crest
___ Intertrochanteric line
___ Lateral condyle
___ Lateral epicondyle
___ Lesser trochanter
___ Linea aspera
___ Medial condyle
___ Medial epicondyle
___ Neck
___ Patellar surface

___ Apex
___ Articular facet for lateral femoral condyle
___ Articular facet for medial femoral condyle
___ Base

FIGURE 7.12 Bones of lower extremity. (a) Posterior view of right femur. (b) Anterior view of right femur. (c) Anterior view of right patella. (d) Posterior view of right patella.

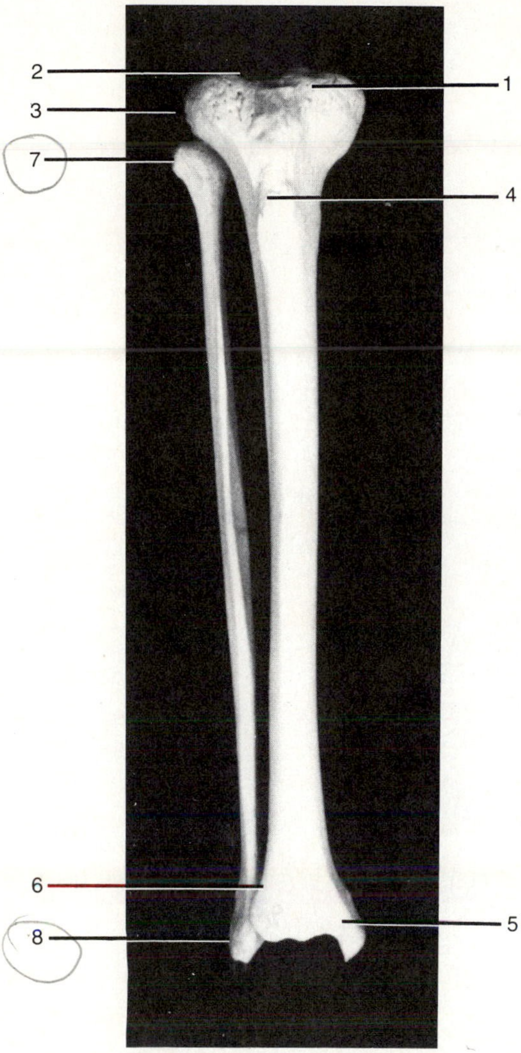

(e)

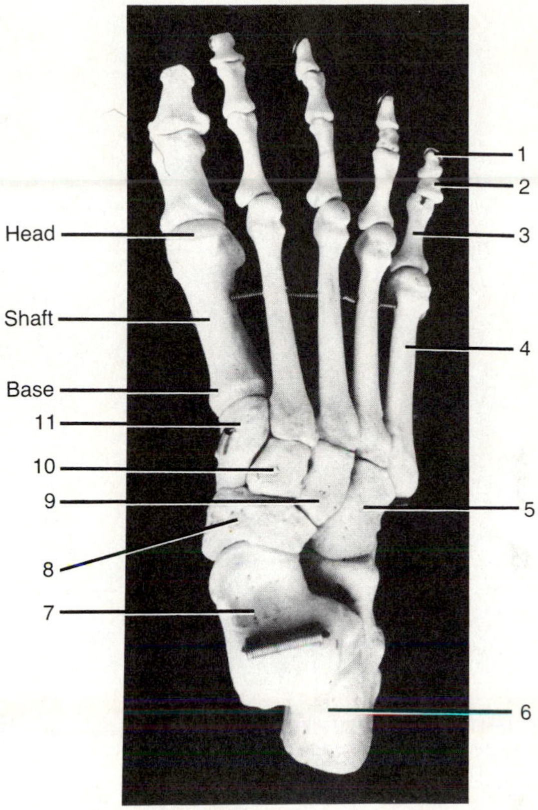

(f)

___ Fibular notch
___ Head of fibula
___ Intercondylar eminence
___ Lateral condyle
___ Lateral malleolus
___ Medial condyle
___ Medial malleolus
___ Tibial tuberosity

___ Calcaneus
___ Cuboid
___ Distal phalanx
___ First (medial) cuneiform
___ Metatarsal
___ Middle phalanx
___ Navicular
___ Proximal phalanx
___ Second (intermediate) cuneiform
___ Talus
___ Third (lateral) cuneiform

FIGURE 7.12 ***(Continued)*** Bones of lower extremity. (e) Anterior view of right fibula (left side) and right tibia (right side). (f) Superior view of right foot.

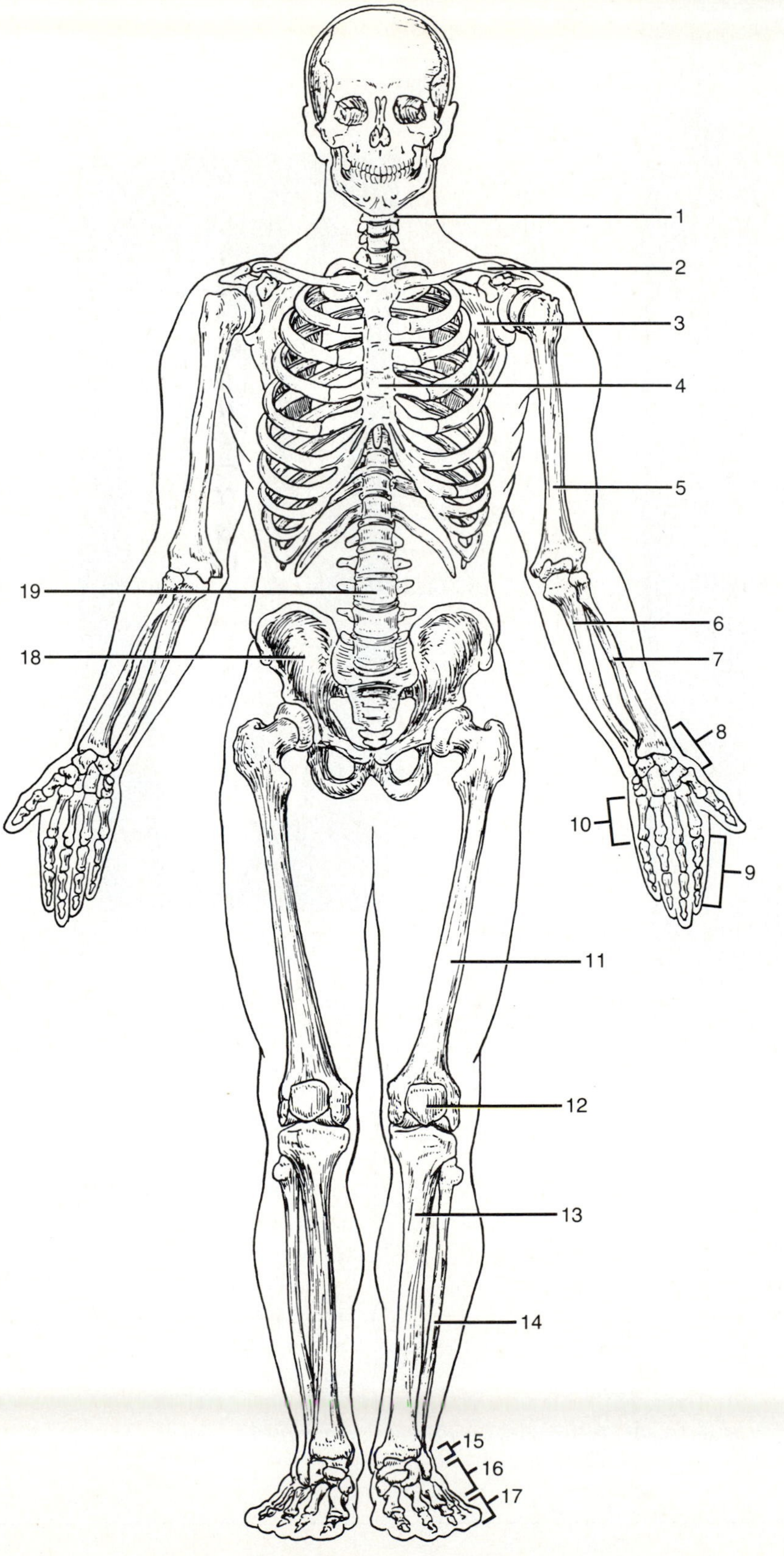

15 Carpals
2 Clavicle
18 Coxal bone
11 Femur
14 Fibula
5 Humerus
1 Hyoid bone
16 Metacarpals
10 Metatarsals
12 Patella
17 Phalanges of foot
9 Phalanges of hand
6 Radius
3 Scapula
4 Sternum
8 Tarsals
13 Tibia
7 Ulna
1 Vertebral column

FIGURE 7.13 Anterior view of entire skeleton.

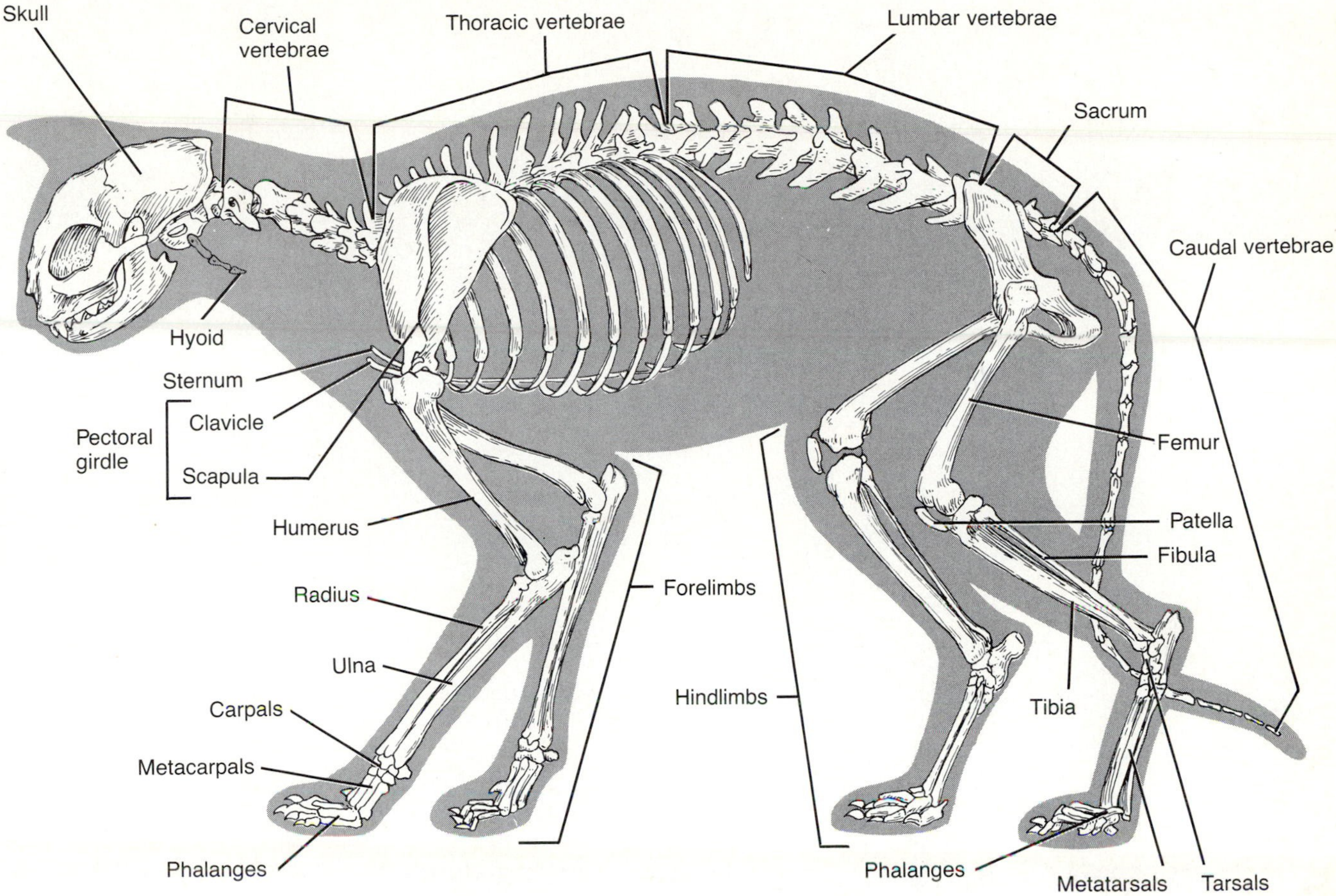

FIGURE 7.14 Skeleton of cat in left lateral view.

LABORATORY REPORT QUESTIONS

BONES 7

Student ______________________ **Date** ______________________

Laboratory Section ______________________ **Score/Grade** ______________________

PART I. MULTIPLE CHOICE

_____ 1. The suture between the parietal and temporal bone is the (a) lambdoidal (b) coronal (c) squamosal (d) sagittal

_____ 2. Which bones does *not* contain a paranasal sinus? (a) ethmoid (b) maxilla (c) sphenoid (d) sacral

_____ 3. Which is the superior, concave curve in the vertebral column? (a) thoracic (b) lumbar (c) cervical (d) sacral

_____ 4. The fontanel between the parietal and occipital bones is the (a) anterolateral (b) anterior (c) posterior (d) posterolateral

_____ 5. Which is *not* a component of the upper extremity? (a) radius (b) femur (c) carpus (d) humerus

_____ 6. All are components of the appendicular skeleton *except* the (a) humerus (b) occipital bone (c) calcaneus (d) triquetral

_____ 7. Which bone does *not* belong with the others? (a) occipital (b) frontal (c) parietal (d) mandible

_____ 8. Which region of the vertebral column is closer to the skull? (a) thoracic (b) lumbar (c) cervical (d) sacral

_____ 9. Of the following bones, the one that does *not* help form part of the orbit is the (a) sphenoid (b) frontal (c) occipital (d) lacrimal

_____ 10. Which bone does *not* form a border for a fontanel? (a) maxilla (b) temporal (c) occipital (d) parietal

PART II. IDENTIFICATION

For each surface marking listed, identify the skull bone to which it belongs:

11. Glabella ______________________
12. Mastoid process ______________________
13. Sella turcica ______________________
14. Cribriform plate ______________________
15. Foramen magnum ______________________
16. Mental foramen ______________________
17. Infraorbital foramen ______________________

18. Crista galli ____________________
19. Foramen ovale ____________________
20. Horizontal plate ____________________
21. Optic foramen ____________________
22. Superior nasal concha ____________________
23. Zygomatic process ____________________
24. Styloid process ____________________
25. Mandibular fossa ____________________

PART III. MATCHING

____ 26. Iliac crest
____ 27. Capitulum
____ 28. Medial malleolus
____ 29. Laminae
____ 30. Vertebral foramen
____ 31. Talus
____ 32. Olecranon
____ 33. Pisiform
____ 34. Acromial extremity
____ 35. Symphysis pubis
____ 36. Hamate
____ 37. Costal groove
____ 38. Xiphoid process
____ 39. Greater trochanter
____ 40. Transverse lines
____ 41. Radial tuberosity
____ 42. Greater tubercle
____ 43. Ischium
____ 44. Glenoid cavity
____ 45. Lateral malleolus

A. Inferior portion of sternum
B. Medial bone of distal carpals
C. Distal projection of tibia
D. Lateral end of clavicle
E. Points where bodies of sacral vertebrae join
F. Portion of rib that contains blood vessels
G. Prominence of elbow
H. Lateral projection of humerus
I. Medial projection for insertion of biceps brachii muscle
J. Lower posterior portion of hipbone
K. Articulates with head of radius
L. Prominence on lateral side of femur
M. Distal projection of fibula
N. Superior border of ilium
O. Articulates with head of humerus
P. Anterior joint between hipbones
Q. Opening through which spinal cord passes
R. Form posterior wall of vertebral arch
S. Medial bone of proximal carpals
T. Component of tarsus

ARTICULATIONS 8

An ***articulation*** (ar-tik′-yoo-LĀ-shun), or ***joint***, is a point of contact between bones, cartilage and bones, or teeth and bones. Some joints permit no movement, others permit a slight degree of movement, and still others permit free movement. In this exercise you will study the structure and action of joints.

A. KINDS OF JOINTS

The joints of the body may be classified into three principal kinds based on their function and structure.

The functional classification of joints is as follows:

1. ***Synarthroses*** (sin′-ar-THRŌ-sēz)—Immovable joints.
2. ***Amphiarthroses*** (am′-fē-ar-THRŌ-sēz)—Slightly movable joints.
3. ***Diarthroses*** (dī-ar-THRŌ-sēz)—Freely movable joints.

The structural classification of joints is as follows:

1. ***Fibrous***—Allow little or no movement; do not contain synovial (joint) cavity; articulating bones held together by fibrous connective tissue.
   a. ***Sutures*** (SOO-cherz)—Found between skull bones; bones separated by thin layer of fibrous tissue; synarthroses. Example: lambdoidal suture.
   b. ***Syndesmosis*** (sin′-dez-MŌ-sis)—Articulating bones united by dense fibrous tissue; usually amphiarthroses. Example: distal ends of tibia and fibula.
   c. ***Gomphosis*** (gom-FŌ-sis)—Cone-shaped peg fits into a socket and the intervening substance is the periodontal ligament; synarthroses. Example: articulation of roots of teeth with alveoli (sockets) of mandible and maxillae.
2. ***Cartilaginous***—Allow little or no movement; do not contain synovial (joint) cavity; articulating bones held together by cartilage.
   a. ***Synchondrosis*** (sin′-kon-DRŌ-sis)—Connecting cartilage is hyaline cartilage; synarthroses. Example: epiphyseal plate.
   b. ***Symphysis*** (SIM-fi-sis)—Connecting substance is a broad, flat disc of fibrocartilage; amphiarthroses. Example: intervertebral discs between vertebrae and symphysis pubis between anterior surfaces of hipbones.
3. ***Synovial*** (si-NŌ-vē-al)—Diarthroses that contain a ***synovial (joint) cavity***, a space between articulating bones (see Figure 8.1); articulating parts of the bones are covered by a thin layer of hyaline cartilage called ***articular cartilage***. The synovial cavity and articulating surfaces of bones are surrounded by a sleevelike ***articular capsule***; the outer layer of the articular capsule, the ***fibrous capsule***, consists of collagenous connective tissue and forms ligaments that are attached to the periosteum of the articulating bones; the inner layer of the articular capsule is the ***synovial membrane*** that secretes synovial fluid to lubricate the joint and nourish the articular cartilage. Many synovial joints also contain ***accessory ligaments***. Some accessory ligaments, such as the collateral ligament of the knee (see Figure 8.3), lie outside the articular capsule ***(extracapsular ligaments)***; other accessory ligaments lie within

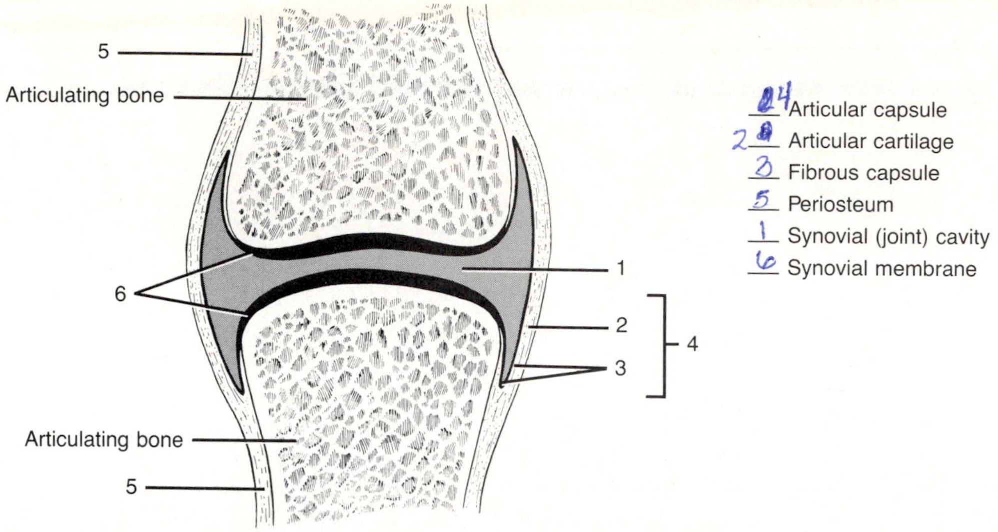

FIGURE 8.1 Parts of a synovial joint.

the articular capsule but outside the synovial cavity ***(intracapsular ligaments)***, such as the cruciate ligaments of the knee joint (see Figure 8.3). Some synovial joints contain ***articular discs (menisci)***, pads of fibrocartilage between articular surfaces of the bones that help to maintain the stability of the joint and direct the flow of synovial fluid to areas of greatest friction (see Figure 8.3). Some synovial joints also contain fluid-filled connective tissue sacs called ***bursae***, which cushion the movement of one part of the body over another (bursae are located between skin and bone, tendons and bone, muscles and bone, ligaments and bone, and between adjacent muscles [see Figure 8.3]).

Label Figure 8.1, the principal parts of a synovial joint.

B. TYPES OF SYNOVIAL JOINTS

The principal types of synovial joints are:

1. ***Gliding***—Articulating surfaces usually flat; permits movement in two planes (biaxial movement), side to side and back and forth. Example: between carpals, tarsals, sacrum and ilium, sternum and clavicle, scapula and clavicle, and articular processes of vertebrae.

2. ***Hinge***—Spool-like surface of one bone fits into the concave surface of another; movement in single plane (monaxial movement), usually flexion and extension. Example: elbow, knee, ankle, interphalangeal joints.

3. ***Pivot***—Rounded, pointed, or conical surface of one bone articulates within a ring formed partly by bone and partly by a ligament; primary movement is rotation; joint is monaxial. Example: between atlas and axis and between proximal ends of radius and ulna.

4. ***Ellipsoidal***—Oval-shaped condyle of one bone fits into an elliptical cavity of another bone; movement is biaxial, side to side and back and forth. Example: between radius and carpals.

5. ***Saddle***—Articular surface of one bone is saddle-shaped and the articular surface of the other bone is shaped like a rider sitting in the saddle; movement similar to that of an ellipsoidal joint. Example: between trapezium and metacarpal of thumb.

6. ***Ball-and-socket***—Ball-like surface of one bone fits into cuplike depression of another bone; movement is in three planes (triaxial movement), flexion-extension, abduction-adduction, and rotation. Example: shoulder and hip joint.

Label Figure 8.2, types of synovial joints.

Examine the articulated skeleton and find as many examples as you can of the joints just described. As part of your examination, be sure to note the shapes of the articular surfaces and the movements possible at each joint.

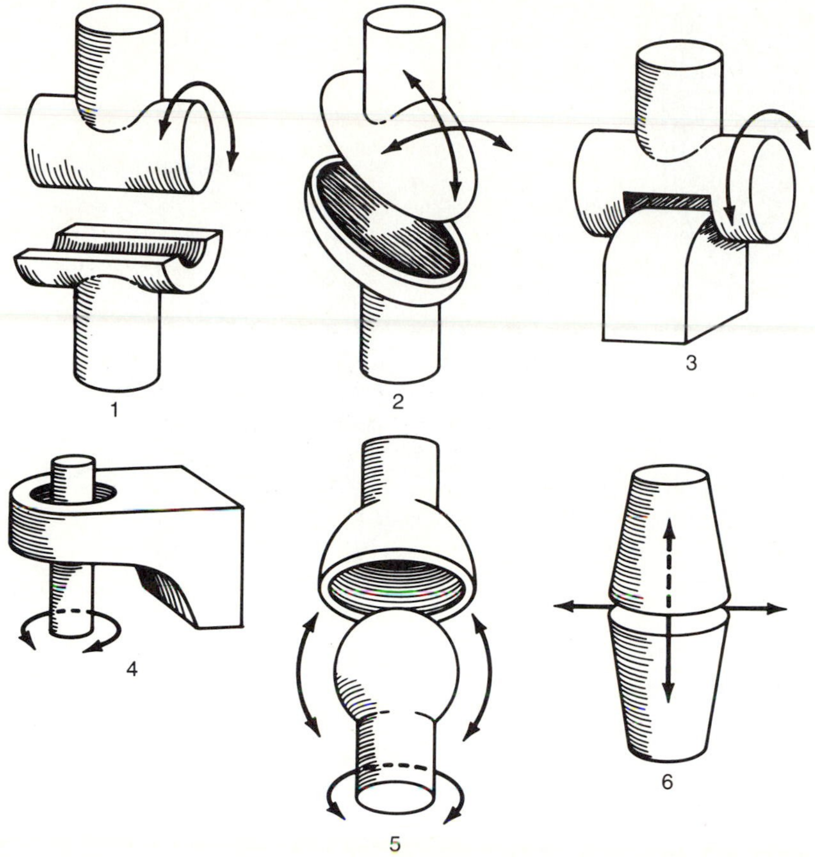

6 Gliding joint
1 Hinge joint
4 Pivot joint
2 Ellipsoidal joint
3 Saddle joint
5 Ball-and-socket joint

FIGURE 8.2 Types of synovial joints.

C. KNEE JOINT

The knee joint is one of the largest joints in the body and illustrates the basic structure of a synovial joint and the limitations on its movement. Some of the structures associated with the knee joint are as follows:

1. ***Tendon of quadriceps femoris muscle***—Strengthens joint anteriorly and externally.

2. ***Gastrocnemius muscle***—Strengthens joint posteriorly and externally.

3. ***Patellar ligament***—Strengthens anterior portion of joint and prevents leg from being flexed too far backward.

4. ***Fibular collateral ligament***—Between femur and fibula; strengthens the lateral side of the joint and prohibits side-to-side movement at the joint.

5. ***Tibial collateral ligament***—Between femur and tibia; strengthens the medial side of the joint and prohibits side-to-side movement at the joint.

6. ***Oblique popliteal ligament***—Starts in a tendon that lies over the tibia and runs upward and laterally to the lateral side of the femur; supports the back of the knee and prevents hyperextension.

7. ***Anterior cruciate ligament***—Passes posteriorly and laterally from the tibia and attaches to the femur; strengthens the joint internally and may help stabilize the knee during its movements.

8. ***Posterior cruciate ligament***—Passes anteriorly and medially from the tibia and attaches to the femur; strengthens the joint internally and may help stabilize the knee during its movement.

9. ***Menisci***—Concentric wedge-shaped pieces of fibrocartilage between the femur and tibia; called ***lateral meniscus*** and the ***medial menis***

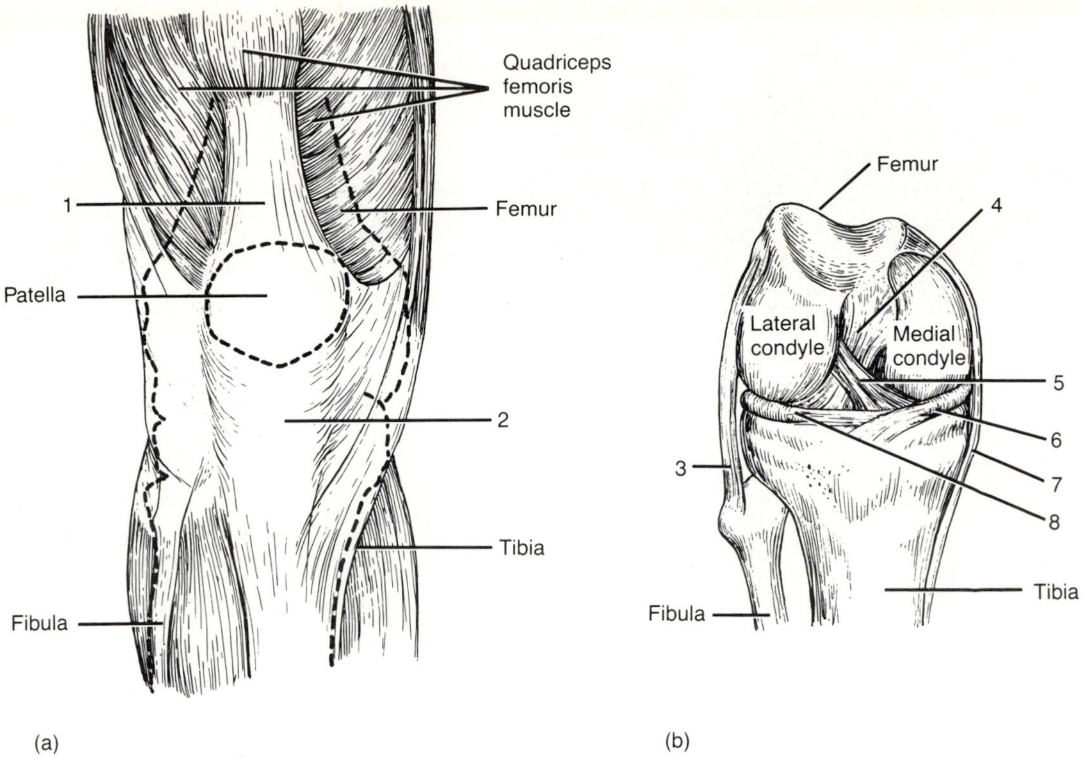

___ Anterior cruciate ligament
___ Fibular collateral ligament
___ Lateral meniscus
___ Medial meniscus

___ Patellar ligament
___ Posterior cruciate ligament
___ Tendon of quadriceps femoris muscle
___ Tibial collateral ligament

FIGURE 8.3 Ligaments, tendons, bursae, and menisci of right knee joint. (a) Anterior, superficial view. (b) Anterior view (flexed) with many superficial structures removed.

cus; provide support for the continuous weight placed on the knee joint.

Label the structures associated with a knee joint in Figure 8.3.

If a longitudinally sectioned knee joint of a cow or lamb is available, examine it and see how many structures you can identify.

D. MOVEMENTS AT SYNOVIAL JOINTS

Movements at synovial joints may be classified as follows:

1. ***Gliding***—One surface moves back and forth and from side to side without any angular or rotary movement.

2. ***Angular***—Increase or decrease the angle between bones.
   a. ***Flexion***—Involves decrease in angle between surfaces of articulating bones.
   b. ***Extension***—Involves increase in angle between surfaces of articulating bones.
   c. ***Hyperextension***—Continuation of extension beyond the anatomical position.
   d. ***Abduction***—Movement of a bone away from midline.
   e. ***Adduction***—Movement of a bone toward midline.

3. ***Rotation***—Movement of a bone around its own longitudinal axis.

4. ***Circumduction***—Movement in which the distal end of a bone moves in a circle while the

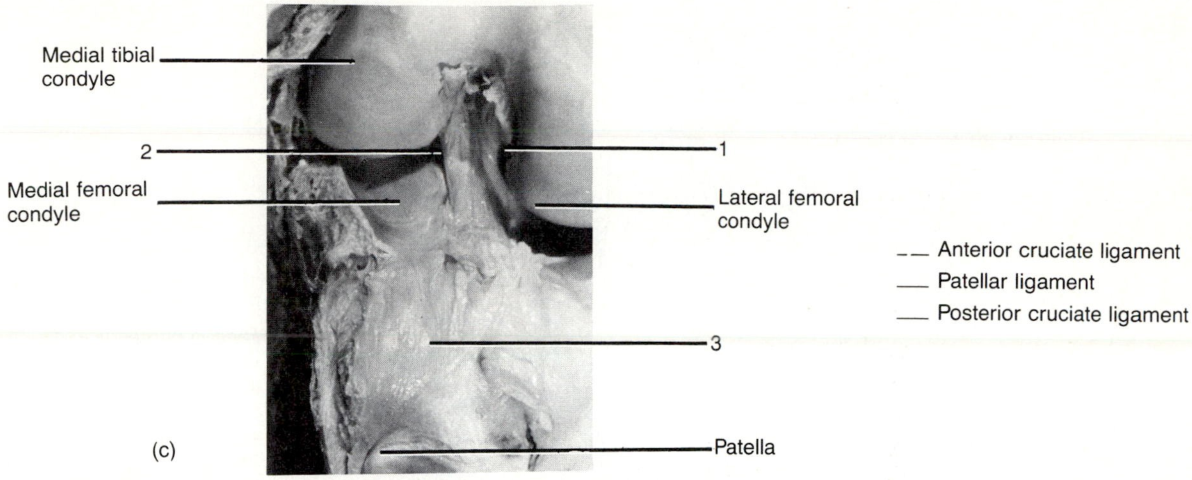

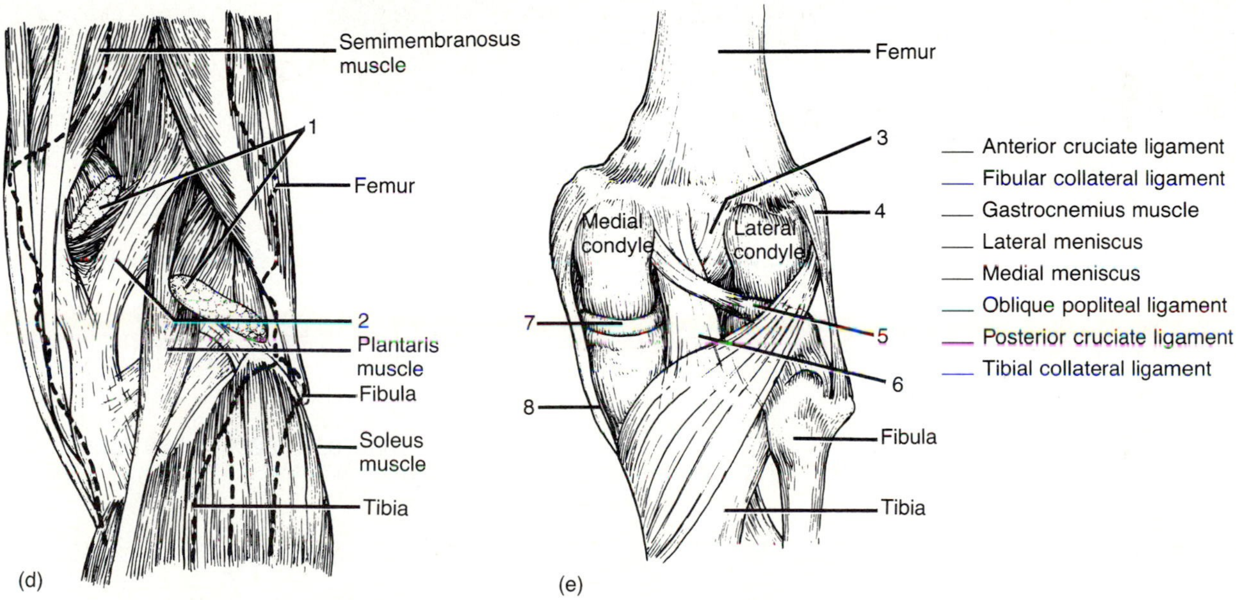

FIGURE 8.3 ***(Continued)*** Ligaments, tendons, bursae, and menisci of right knee joint. (c) Photograph of internal structure. (d) Posterior, superficial view. (e) Posterior view with many superficial structures removed.

proximal end remains relatively stable; bone describes a cone in the air.

5. ***Special***—Found only at the joints indicated:
   a. ***Inversion***—Movement of sole of foot inward.
   b. ***Eversion***—Movement of sole of foot outward.
   c. ***Dorsiflexion***—Bending the foot in the direction of the dorsum (upper surface).
   d. ***Plantar flexion***—Bending the foot in the direction of the plantar surface (sole).
   e. ***Protraction***—Movement of mandible or shoulder girdle forward on plane parallel to ground.
   f. ***Retraction***—Movement of protracted part of the body backward on plane parallel to ground.
   g. ***Supination***—Movement of forearm in

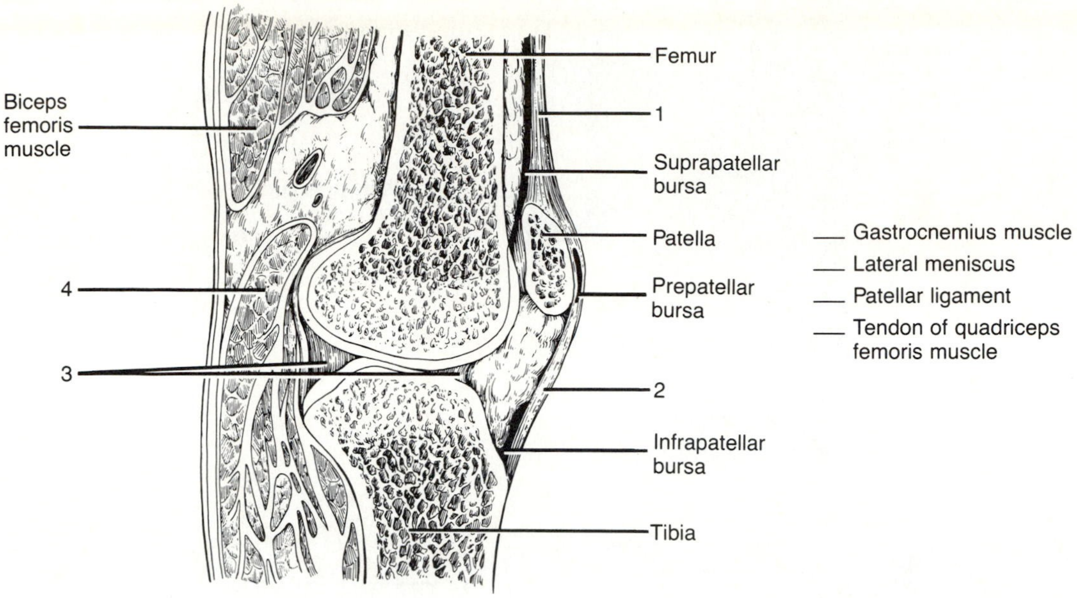

FIGURE 8.3 ***(Continued)*** Ligaments, tendons, bursae, and menisci of right knee joint. (f) Sagittal section.

which palm of hand is turned anteriorly or superiorly.

h. ***Pronation***—Movement of forearm in which palm of hand is turned posteriorly or inferiorly.

i. ***Depression***—Movement in which part of body, such as the mandible or scapula, moves inferiorly.

j. ***Elevation***—Movement in which part of body moves superiorly.

Label the various movements illustrated in Figure 8.4.

ANSWER THE LABORATORY REPORT QUESTIONS AT THE END OF THE EXERCISE

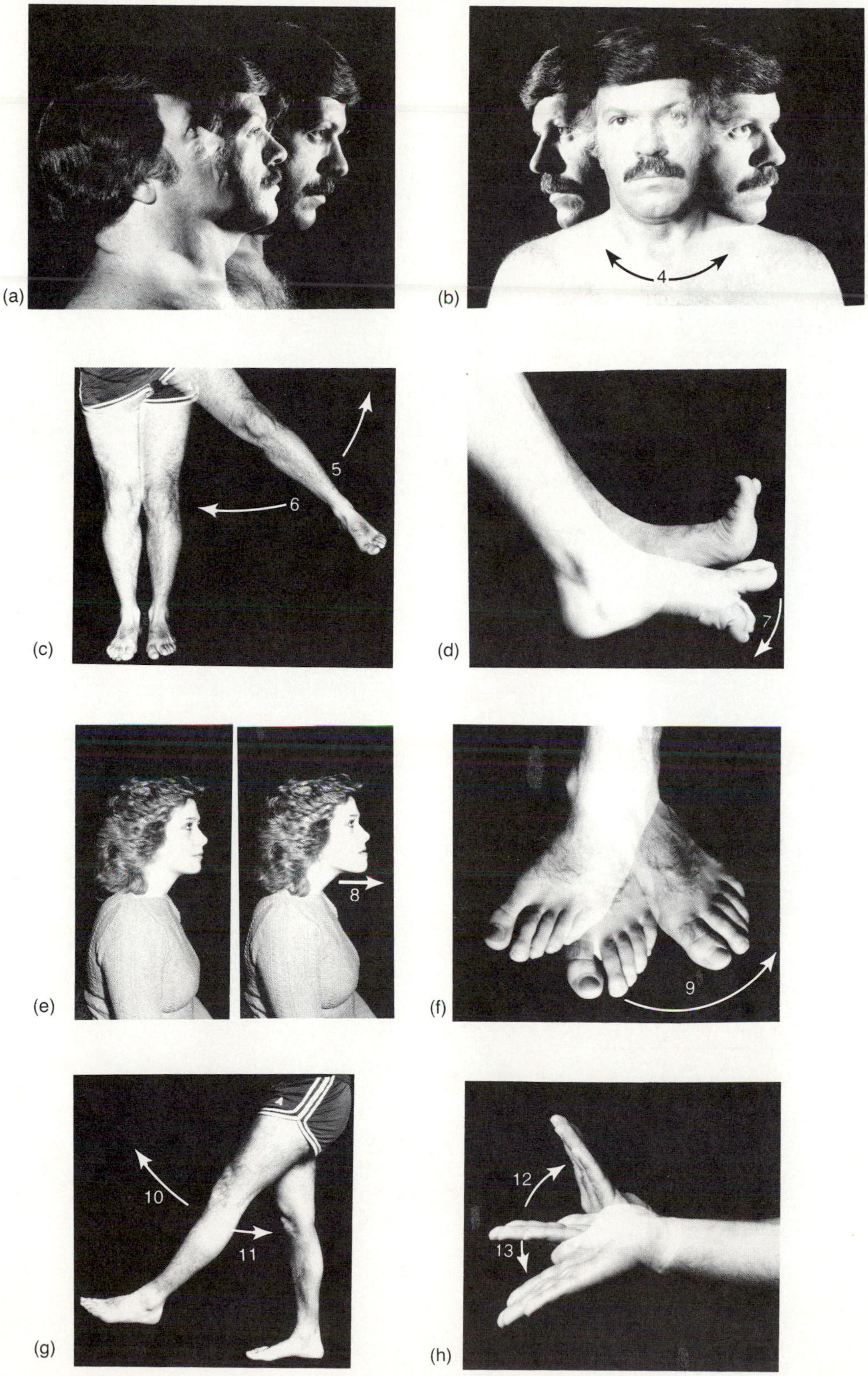

FIGURE 8.4 Movements at synovial joints.

___ Abduction
___ Adduction
___ Circumduction
___ Depression
___ Extension
___ Extension
___ Eversion
___ Flexion
___ Flexion
___ Flexion
___ Hyperextension
___ Hyperextension
___ Plantar flexion
___ Protraction
___ Rotation

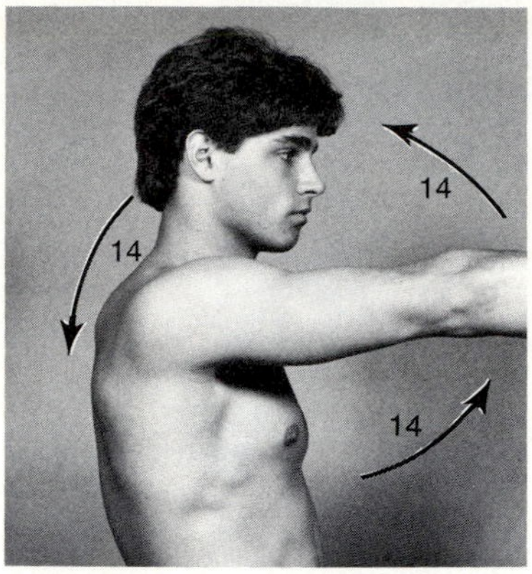

(i)

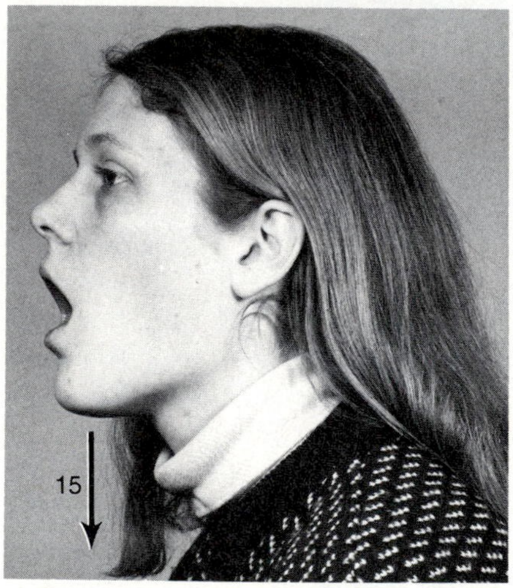

(j)

FIGURE 8.4 ***(Continued)*** Movements at synovial joints.

LABORATORY REPORT QUESTIONS

ARTICULATIONS 8

Student ______________________ Date ______________________

Laboratory Section ______________________ Score/Grade ______________________

PART I. MULTIPLE CHOICE

_____ 1. A joint united by dense fibrous tissue that permits a slight degree of movement is a (a) suture (b) syndesmosis (c) symphysis (d) synchondrosis

_____ 2. A joint that contains a broad flat disc of fibrocartilage is classified as a (a) ball-and-socket joint (b) suture (c) symphysis (d) gliding joint

_____ 3. The following characteristics define what type of joint? Presence of a synovial cavity, articular cartilage, synovial membrane, and ligaments. (a) suture (b) synchondrosis (c) syndesmosis (d) hinge

_____ 4. Which joints are slightly movable? (a) diarthroses (b) amphiarthroses (c) synovial (d) synarthroses

_____ 5. Which type of joint is immovable? (a) synarthrosis (b) syndesmosis (c) symphysis (d) diarthrosis

_____ 6. What type of joint provides triaxial movement? (a) hinge (b) ball-and-socket (c) saddle (d) ellipsoidal

_____ 7. Which ligament provides strength on the medial side of the knee joint? (a) oblique popliteal (b) posterior cruciate (c) fibular collateral (d) tibial collateral

_____ 8. On the basis of structure, which joint is fibrous? (a) symphysis (b) synchondrosis (c) pivot (d) syndesmosis

_____ 9. The elbow, knee, and interphalangeal joints are examples of which type of joint? (a) pivot (b) hinge (c) gliding (d) saddle

_____ 10. Functionally, which joint provides the greatest degree of movement? (a) diarthrosis (b) synarthrosis (c) amphiarthrosis (d) syndesmosis

PART II. COMPLETION

11. The thin layer of hyaline cartilage on articulating surfaces of bones is called ______________________ cartilage.

12. The synovial membrane and fibrous capsule together form the ______________________ capsule.

13. Pads of fibrocartilage between the articular surfaces of bones that maintain stability of the joint are called ______________________ .

14. Fluid-filled connective tissue sacs that cushion movements of one body part over another are referred to as ______________________ .

15. The ______________________ ligament supports the back of the knee and helps to prevent hyperextension.

PART III. MATCHING

_____ 16. Circumduction
_____ 17. Adduction
_____ 18. Flexion
_____ 19. Pronation
_____ 20. Elevation
_____ 21. Protraction
_____ 22. Rotation
_____ 23. Plantar flexion
_____ 24. Dorsiflexion
_____ 25. Inversion

A. Decrease in the angle, usually between the anterior surfaces of articulating bones
B. Moving a part upward
C. Bending the foot in the direction of the upper surface
D. Forward movement parallel to the ground
E. Movement of the sole of the foot inward at the ankle joint
F. Movement toward the midline
G. Bending the foot in the direction of the sole
H. Turning the palm posterior
I. Movement of a bone around its own axis
J. Distal end of a bone moves in a circle while the proximal end remains relatively stable

MUSCLE TISSUE 9

Muscle tissue constitutes 40% to 50% of total body weight and is composed of fibers (cells) that are highly specialized with respect to four characteristics: (1) ***excitability***, or ability to receive and respond to stimuli; (2) ***contractility***, or ability to contract (shorten and thicken); (3) ***extensibility (extension)***, or ability to stretch when pulled; and (4) ***elasticity***, or ability to return to original shape after contraction or extension. Through contraction, muscle performs three basic functions: motion, maintenance of posture, and heat production. In this exercise you will examine the histological structure of muscle tissue and conduct exercises on the physiology of frog muscle.

A. KINDS OF MUSCLE TISSUE

Histologically, three kinds of muscle tissue are recognized:

1. ***Skeletal muscle tissue***—Usually attached to bones; contains conspicuous striations when viewed microscopically; voluntary because it contracts under conscious control.
2. ***Cardiac muscle tissue***—Found only in the wall of the heart; striated when viewed microscopically; involuntary because it contracts without conscious control.
3. ***Smooth muscle tissue***—Located in walls of viscera and blood vessels; referred to as nonstriated because it lacks striations when viewed microscopically; involuntary because it contracts without conscious control.

B. SKELETAL MUSCLE TISSUE

Examine a prepared slide of skeletal muscle in longitudinal and cross section under high power. Look for the following:

1. ***Sarcolemma***—Plasma membrane of the muscle fiber (cell).
2. ***Sarcoplasm***—Cytoplasm of the muscle fiber.
3. ***Nuclei***—Several in each muscle fiber lying close to sarcolemma.
4. ***Striations***—Cross stripes in each muscle fiber (described shortly).
5. ***Epimysium*** (ep′-i-MĪZ-ē-um)—Fibrous connective tissue that surrounds the entire skeletal muscle.
6. ***Perimysium*** (per′-i-MĪZ-ē-um)—Fibrous connective tissue surrounding a bundle (fascicle) of muscle fibers.
7. ***Endomysium*** (en′-dō-MĪZ-ē-um)—Fibrous connective tissue surrounding individual muscle fibers.

Refer to Figure 9.1 and label the structures indicated.

With the use of an electron microscope, additional details of skeletal muscle tissue may be noted. Among these are the following:

1. ***Mitochondria***—Organelles that have smooth outer membrane and folded inner membrane in which ATP is generated.
2. ***Sarcoplasmic reticulum*** (sar′-kō-PLAZ-mik re-TIK-yoo-lum) or ***SR***—Network of membrane-enclosed tubules.

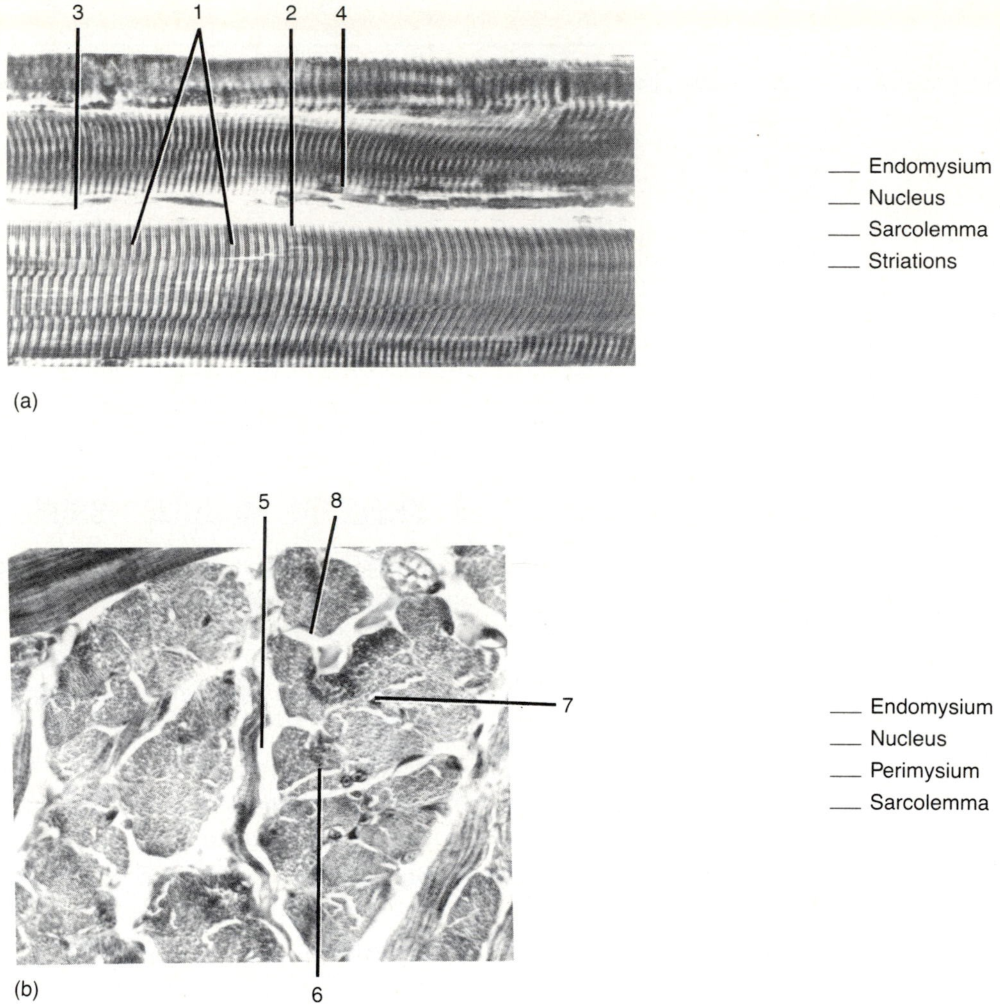

FIGURE 9.1 Histology of skeletal muscle tissue. (a) Photomicrograph of several muscle fibers in longitudinal section. (b) Photomicrograph of several muscle fibers in cross section.

3. ***T (transverse) tubules***—Run perpendicular to and connect with sarcoplasmic reticulum; open to outside of muscle fiber.

4. ***Triad***—T tubule and the segments of sarcoplasmic reticulum on both sides.

5. ***Myofibrils***—Threadlike structures that run longitudinally through a fiber and consist of ***thin myofilaments*** composed of the protein actin and ***thick myofilaments*** composed of the protein myosin.

6. ***Sarcomere***—Contractile unit of a muscle fiber; compartment within a muscle fiber separated from other sarcomeres by dense material called ***Z lines***.

7. ***A (anisotropic) band***—Dark region in a sarcomere represented by the length of thin myofilaments.

8. ***I (isotropic) band***—Light region in a sarcomere composed of thin myofilaments only. The combination of alternating dark A bands and light I bands gives the muscle fiber the striated (striped) appearance.

9. ***H zone***—Region in the A band of a sarcomere consisting of thick myofilaments only.

10. ***M line***—Series of fine threads in the center of the H zone that appear to connect the middle parts of adjacent thick myofilaments.

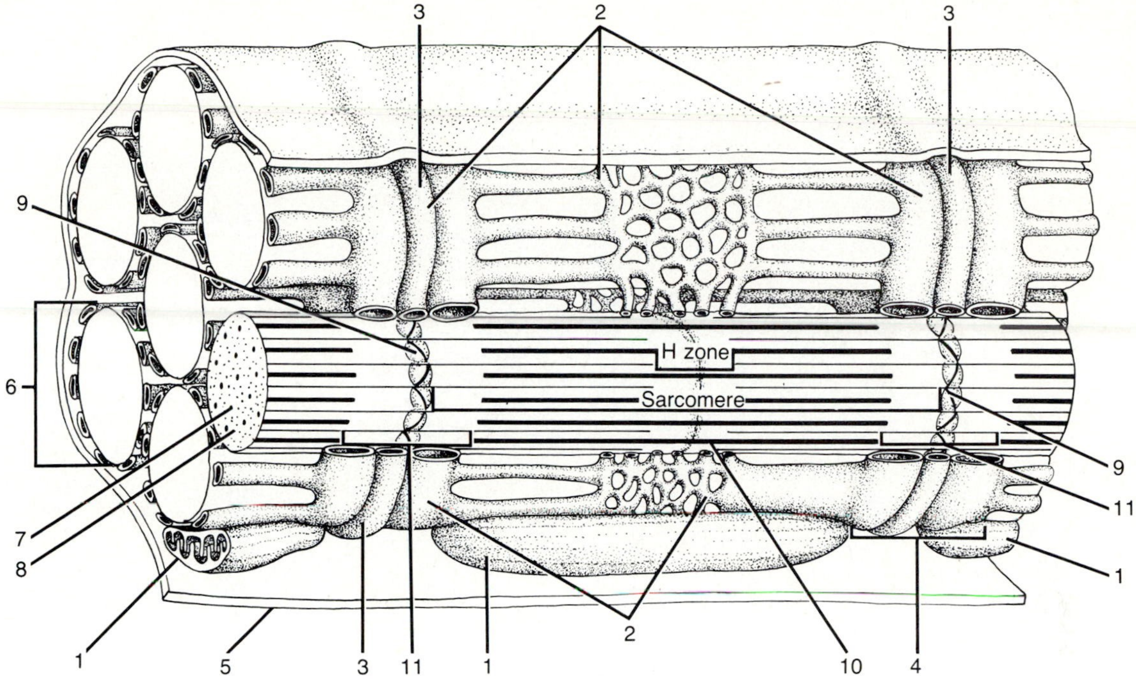

___ A (anisotropic) band
___ I (isotropic) band
___ Mitochondrion
___ Myofibril
___ Sarcolemma
___ Sarcoplasmic reticulum
___ Thick myofilament
___ Thin myofilament
___ Triad
___ T (transverse) tubule
___ Z line

FIGURE 9.2 Enlarged aspect of several myofibrils of skeletal muscle tissue based on an electron micrograph.

Figure 9.2 is a diagram of skeletal muscle tissue based on electron micrographic studies. Label the structures shown.

C. CARDIAC MUSCLE TISSUE

Examine a prepared slide of cardiac muscle tissue in longitudinal and cross section under high power. Locate the following structures: ***sarcolemma, endomysium, nuclei, striations,*** and ***intercalated discs*** (transverse thickenings of the sarcolemma that separate individual fibers). Label Figure 9.3.

D. SMOOTH MUSCLE TISSUE

Examine a prepared slide of smooth muscle tissue in longitudinal and cross section under high power. Locate and label the following structures in Figure 9.4: ***sarcolemma, sarcoplasm, nucleus,*** and ***muscle fiber***.

E. PHYSIOLOGY OF SKELETAL MUSCLE CONTRACTION

The process of skeletal muscle contraction involves a series of both electrical and chemical events. This series of events is termed ***excitation-contraction coupling***. Muscle contraction occurs only if the muscle is stimulated with a stimulus of threshold or suprathreshold intensity. A ***threshold stimulus*** is defined as the minimum strength stimulus necessary to initiate a contraction. In the body, this is accomplished by a nerve impulse being transmitted via a nerve cell called a ***motor neuron***. The portion of the motor neuron that extends from the

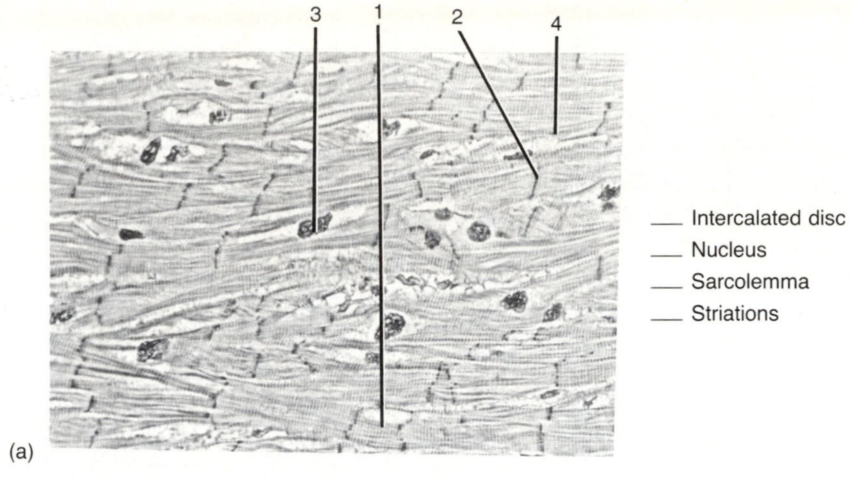

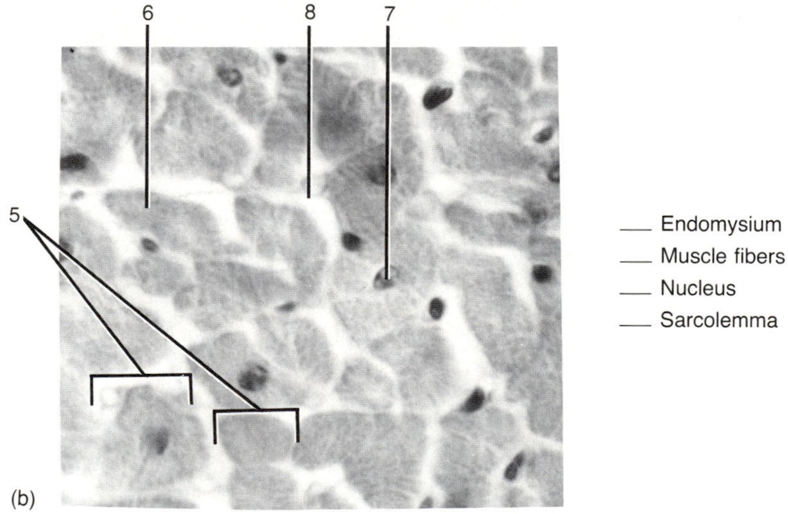

FIGURE 9.3 Histology of cardiac muscle tissue. (a) Photomicrograph of several muscle fibers in longitudinal section. (b) Photomicrograph of several muscle fibers in cross section.

nerve cell body to a muscle fiber (cell) is called an ***axon***. Upon entering the epimysium of a skeletal muscle, an axon branches into ***axon terminals (telodendria)*** that come into close approximation with the sarcolemma of a muscle fiber. Such an area of close approximation consisting of the axon terminal of a motor neuron and the portion of the sarcolemma near it is called a ***neuromuscular (myoneural) junction*** or ***motor end plate*** (Figure 9.5).

Close examination of a neuromuscular junction reveals that the distal ends of the axon terminals are expanded into bulblike structures called ***synaptic end bulbs***. The bulbs contain membrane-enclosed sacs, called ***synaptic vesicles***, that store chemicals called neurotransmitters. These chemicals allow a nerve impulse to be transmitted to a muscle. The invaginated area of the sarcolemma under the axon terminal is referred to as a ***synaptic gutter (trough)***; the space between the axon terminal and sarcolemma is known as a ***synaptic cleft***. Along the synaptic gutter are numerous folds of the sarcolemma, called ***subneural***

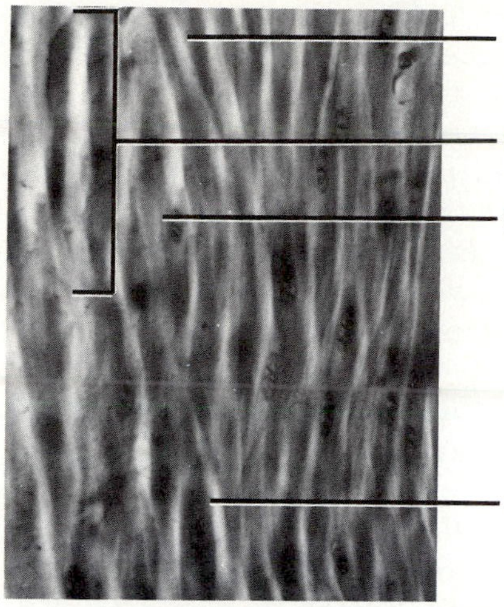

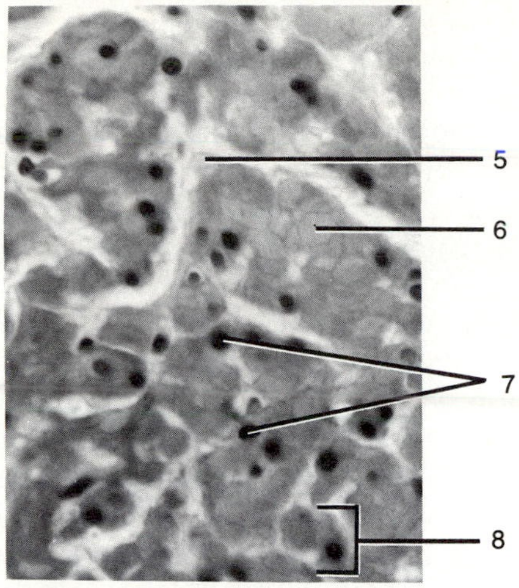

___ Muscle fiber
___ Nucleus
___ Sarcolemma
___ Sarcoplasm

___ Muscle fiber
___ Nuclei
___ Sarcolemma
___ Sarcoplasm

FIGURE 9.4 Histology of smooth muscle tissue. (a) Photomicrograph of several muscle fibers in longitudinal section. (b) Photomicrograph of several muscle fibers in cross section.

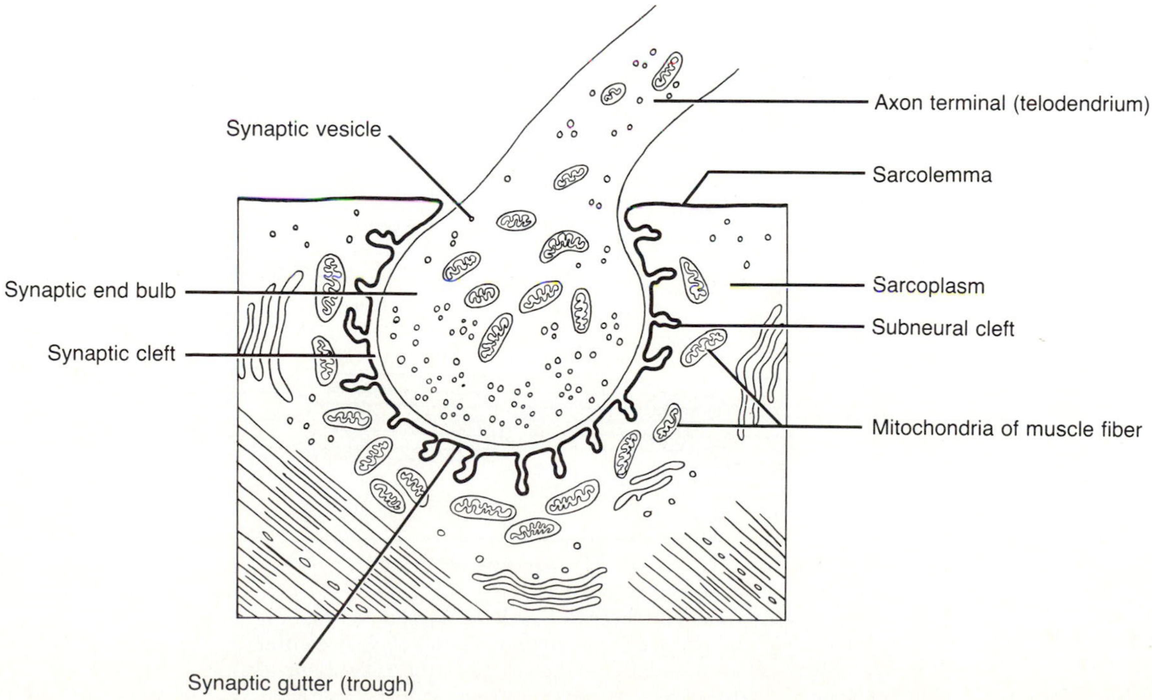

FIGURE 9.5 Details of a neuromuscular junction (motor end plate).

clefts, which greatly increase the surface area of the synaptic gutter.

In a skeletal neuromuscular junction, the nerve impulse reaches the end of an axon, causing it to depolarize. This depolarization causes the synaptic vesicles to release a neurotransmitter called ***acetylcholine*** (as′-ē-til-KŌ-lēn) or ***ACh***, which diffuses across the synaptic cleft where it binds to receptors on the sarcolemma of the muscle fiber. The binding of ACh to the sarcolemma initiates a localized depolarization called the ***end plate potential***, which achieves threshold intensity and initiates an action potential along the sarcolemma of the muscle fiber. The action potential travels along the sarcolemma and into the interior of the muscle fiber via the T tubules, causing the sarcoplasmic reticulum (SR) to release calcium ions among thin (actin) myofilaments and thick (myosin) myofilaments. As myosin cross-bridges of the thick myofilaments attach to receptors on the thin myofilaments, ATP is split. Energy thus released causes myosin cross-bridges to move, and thin myofilaments slide inward toward the H zone, causing the muscle fiber to shorten (Figure 9.6). As this is occurring, newly synthesized ATP displaces ADP from the myosin molecule. If calcium has been taken back into the SR by the calcium pump, the myosin cross-bridges release from the actin receptors and relaxation occurs. However, if sufficient calcium is still present within the sarcoplasm the entire process repeats and further contraction occurs. This theory of skeletal muscle contraction is termed the ***sliding filament theory of skeletal muscle contraction***.

A contracting skeletal muscle fiber follows the ***all-or-none principle***. Simply stated, this principle says that a skeletal muscle fiber will respond maximally or not at all to a stimulus. If the stimulus is of threshold or greater intensity the fiber will respond maximally; if the stimulus is subthreshold in intensity the fiber will not respond. This principle, however, does not imply that the entire skeletal muscle must be either fully relaxed or fully contracted because individual fibers within a mus-

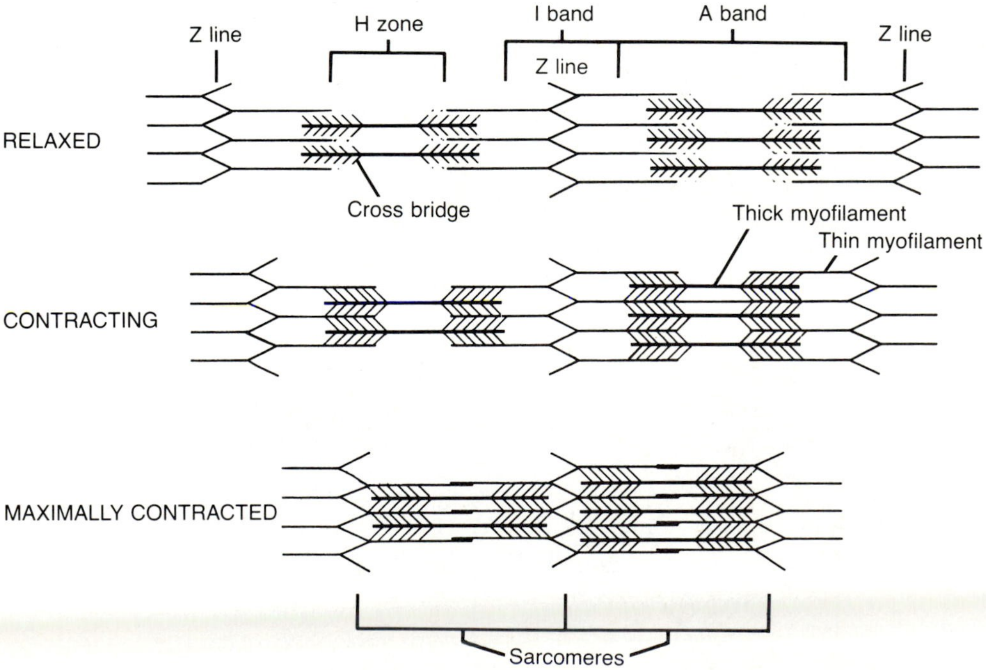

FIGURE 9.6 Sliding filament theory of skeletal muscle contraction. The positions of the various parts of two sarcomeres in relaxed, contracting, and maximally contracted states are shown. Note the movement of thin myofilaments and relative size of H zone.

cle possess varying thresholds for stimulation. Therefore, the muscle as a whole can have graded contractions.

With these facts in mind, you will now perform a few simple lab tests that will illustrate the physiology of skeletal muscle contraction.

F. LABORATORY TESTS ON SKELETAL MUSCLE CONTRACTION[1]

Skeletal muscles produce different kinds of contractions depending on the intensity and frequency of the stimulus applied. Among these contractions are isotonic, isometric, tonic, twitch, tetanic, and treppe. Twitch contractions do not occur in the body but are worth demonstrating because they show the different phases of muscle contraction quite clearly. Any record of a muscle contraction is called a ***myogram***.

You may test muscle contraction through a team exercise or by observing a demonstration prepared by your instructor. Read the sections that follow on pithing a frog, preparing muscle or nerve-muscle preparations, and using a polygraph *before* you do the exercise or observe the demonstration. Depending on the size of the class and the equipment available, you should work in teams of three to five students: one or two students should be assigned to assist the instructor to set up the physiologic apparatus, one or two other students should be assigned as "surgeons" to isolate, remove, and suspend the frog muscle, and another student should act as the recorder and coordinate the work.

1. Polygraph

The polygraph (Figure 9.7) records physiological events and works by receiving a signal from a sensing device, amplifying it, and changing it into a perceptible form. For example, a sensing device called a myograph transducer measures the force exerted by a contracting muscle and converts this mechanical force into an electrical signal. This signal is magnified by the amplifier and then changed into perceptible form by a writing pen that makes a permanent, visible record of the contraction.

A polygraph usually has a number of recording channels. Each is used to record a particular physiological event as it occurs. Different physiological events may be recorded simultaneously by using different channels. Each channel consists of a sensing device to detect the event, a coupler to join the sensing device to the amplifier, an amplifier, and a writing pen to make a record of what happened. A fifth pen, marked TIME AND EVENTS, is used to keep track of time and to mark the points at which stimuli are delivered. This time pen is the lowest pen on the instrument. The other pens are numbered to correspond to the recording channel with which they work.

Either electrodes or a transducer can serve as the sensing device for the polygraph. ***Electrodes*** are made of metal plates or small rods that can detect electrophysiological phenomena such as changes in electrical potential of the skin or nerves. Electrodes transmit the original electrical signal without change to the amplifier. In many cases a transducer must be used as the sensing device. The ***transducer*** converts nonelectrical signals into electrical signals: for example, the myograph transducer already mentioned converts mechanical force from a muscle contraction into an electrical signal; a heat transducer converts heat into an electrical signal; and a pressure transducer converts pressure changes into an electrical signal.

PROCEDURE

a. ORIENTATION TO THE POLYGRAPH AND RELATED ACCESSORIES

Plug the polygraph into an electrical outlet. Turn on the master switch and the power switch on one of the amplifiers. (If you have a Model 7000 Amplifier, the amplifiers are on when the master switch is on.) The power switch is located at the bottom left corner of the amplifier.

(1) Ink Supply

1. Check to see that the ink wells are no more than half full. Position the ink wells so that the

[1]These tests involve skeletal muscle tissue only. Tests on cardiac muscle tissue are included in Exercise 19 (Cardiovascular Physiology) and tests on smooth muscle in Exercise 22 (Digestive System). At the discretion of the instructor, muscle tissue tests may be done in individual exercises or combined into a single exercise.

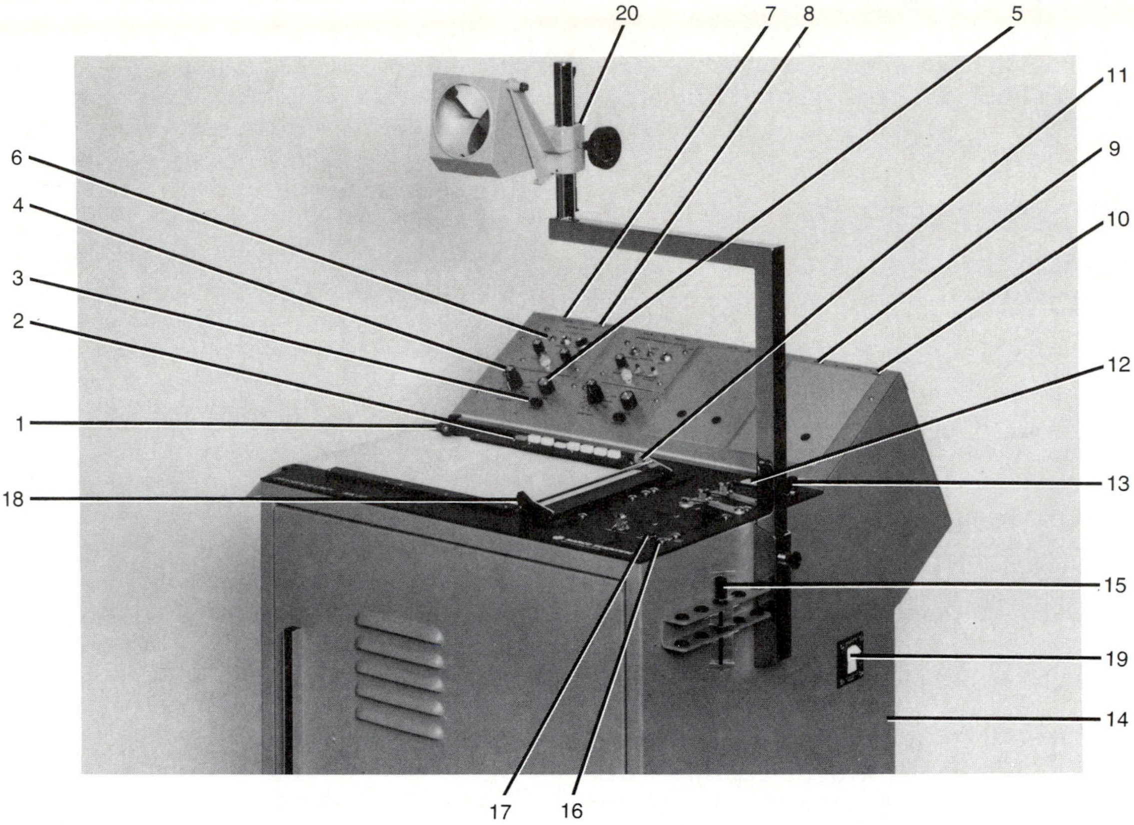

1. Chart drive tension wheel
2. Paper speed switches
3. Record
4. Sensitivity
5. Position
6. Coupler compartment
7. Auxiliary input
8. Monitor output
9. Event input
10. Event output
11. Timer switch
12. Power switch
13. Ground terminal
14. Power connector
15. Inkwell
16. External marker
17. Event marker
18. Pen lifter assembly
19. Projector power switch
20. Projector lens assembly

FIGURE 9.7 A polygraph.

ink level is even with the level of the flat paper table of the polygraph.

2. Send ink to the pens by placing a finger over the hole in the top of the ink bottle and squeezing very gently until a drop of ink appears at the pen point. You must remove your finger before you stop squeezing the bottle or the ink will not remain in the pen assembly.

3. At the conclusion of the exercise, empty the pens by reversing the above sequence, that is, squeeze the bottle, place a finger over the hole, stop squeezing, and remove the finger from the hole when the ink has been entirely withdrawn from the tubing assembly feeding the pens.

4. A watch glass containing water should be placed under the pen point when you empty the pens so that water will be drawn into the pen assembly as the ink is being withdrawn. Ink is fed by gravity pull. Therefore, if the ink is not flowing adequately (is too light), raise the ink well 3 mm.

(2) Paper and Paper Controls

The paper drive control along the back edge of the flat paper table sets the paper in motion and lowers the pens.

1. Move the paper drive control to the left. Note that the paper will move only if the leading edge of the paper is under the turning wheel.

2. The speed of the paper can be controlled by the knob marked CM/SEC—PAPER SPEED. Change the settings and observe the speed. Reuse this paper in the next steps.

3. The paper can be pulled back into the body of the unit if the cabinet door below the flat paper table is opened.

4. While you have the cabinet door open, observe the manner in which the paper is guided through the slot onto the flat paper table of the polygraph.

5. Pull the paper all the way out and practice reloading. You should observe that the polygraph paper is ruled in metric units, usually in either 5-mm squares or 1-mm squares.

(3) Time and Event Marker

A knob to the right of the paper speed knob controls the time and event marker.

1. Turn it on at the lowest setting, 1 SEC. Observe the fifth pen and determine how often the pen deflects and in what direction the pen moves.

2. Depress the red button (EVENT MARKER) at the lower right corner of the polygraph.

What is the result of this action? ____________

____

In what direction does the pen move? ________

This button can be depressed to mark the time at which either an event occurred or an external stimulus was delivered to the subject.

3. A stimulator (to be discussed later) properly connected to the polygraph will automatically activate the event marker when each single stimulus is delivered. (Some models of stimulator also mark when continuous stimuli are delivered.)

4. The time marker can be set at intervals of 0 (time marker is off), 1, 5, 30, or 60 seconds. Every 60 seconds the time marker skips a mark, which indicates that a minute has passed during the recording of the experiment.

5. Practice using the timer and the external event markers.

(4) Amplifiers

The amplifier units are located in the upper portion of the polygraph. Each amplifier controls one specific pen, which is numbered identically to the amplifier controlling it. Examples of solid state amplifiers are shown in Figure 9.8. Each amplifier has several switches.

1. The on-off switch (in lower left corner) must be ON for the unit to be operational. This will be indicated by the red signal light.

2. The record switch must be placed in RECORD phase to record data. When out of RECORD phase, the record switch disconnects the amplifier from the sensing device. *The record switch must be disengaged before any connections are made to the polygraph. Otherwise, fuses will blow and light bulbs will burn out*. The record switch must also be disengaged unless actual recording is taking place. In solid state models, engage the record switch by depressing the clear button; it will light.

3. The position control situates the pen on the paper. Rotate this control and observe the movement of the pen.

4. The sensitivity (amplitude) control determines the amount of amplification of the incoming signal. If the amplification is too low, the magnitude of the pen deflection will be small. If the amplification is increased, the magnitude of the pen deflection will also increase. This control is the one used to calibrate the size of pen deflection when a known signal is presented. As you increase the sensitivity, you also increase the susceptibility of the system to outside disturbances.

5. The filter control (Model 7070 only) is generally only used in research work and should be left in the 10K position.

6. The polarity switch (Model 7070 only) should be set to positive (+) unless measuring electrophysiological events (for example, electrocardiography, electroencephalography). When measuring electrophysiological events, the switch should be set to the negative (−) position.

(5) Preamplifiers and Couplers

A coupler is a module in a solid state system that links a sensing device to its amplifier. If preamplification is necessary, the coupler also acts in this capacity. In solid state systems the sensitivity control on the amplifier remains operational because the signal often must be amplified a second time.

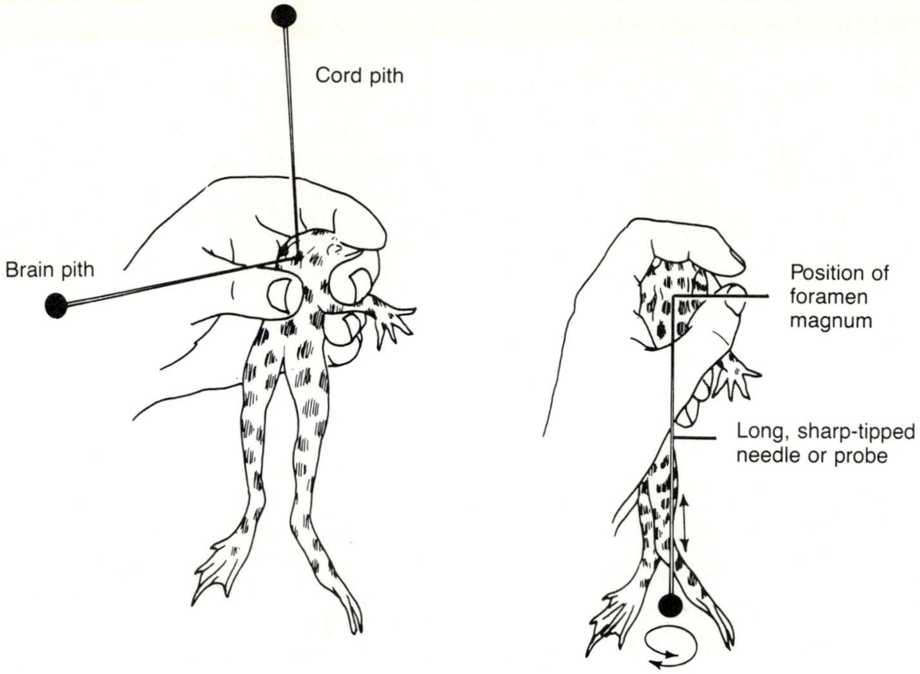

FIGURE 9.8 Procedure for pithing a frog.

(6) Transducers

Transducers are elements of the system that convert nonelectrical signals into electrical signals. Transducers are usually connected to the amplifier by a cable; however, they are occasionally connected directly to the amplifier.

1. Look at the 9-pin cable and note that when the pins of the cable are aligned with the input socket, the yellow dot on the pin is aligned with the yellow dot on the socket.
2. If the yellow dots are not properly aligned, a broken connector may result.

b. CALIBRATION OF THE SYSTEM

1. Suspend a 100-g weight from the muscle transducer and adjust the sensitivity of the amplifier so that a 5-cm pen deflection is obtained.
2. If at any time this calibration proves unsatisfactory, change the sensitivity to best fit the situation. *Record the calibration factor* for future reference during this experiment.

2. Pithing of Frogs

By definition, *pithing* is destruction of the central nervous system by piercing the brain or spinal cord. This procedure is used in animal experimentation to render the animal unconscious so that it feels no pain. A single-pithed frog is one in which only the brain is destroyed, whereas a double-pithed frog has its spinal cord destroyed too. Usually the double-pithing procedure is used.

NOTE. *It is strongly suggested that either your instructor or a laboratory assistant perform the following procedure for pithing a frog. The procedure requires some practice and may prove difficult for inexperienced individuals.*

PROCEDURE

1. There are two accepted methods for pithing a frog.
2. One method of single pithing a frog entails holding the animal in a paper tower with its dorsal side up and with the index finger pressing the nose down so that the head makes a right angle with the trunk (Figure 9.8).

3. Locate the slight depression formed by the first vertebra and skull about 3 mm behind the line joining the posterior borders of the tympanic membrane. This groove represents the area of the foramen magnum.

4. Carefully insert a long sharp-tipped needle or probe into the foramen magnum and direct it forward and a little downward.

5. Exert a steady pressure and rotate it, moving it from side to side in the cranial cavity to destroy the brain.

6. To double pith the frog, insert the needle into the vertebral canal, directing it downward until it has reached the end of the canal. Move the needle from side to side as you go.

7. A second acceptable method to single pith a frog involves holding the animal in a paper towel with its dorsal side up and inserting one blade of a pair of scissors into the animal's mouth and cutting off the top of the head just posterior to the eyes.

8. To double pith the animal insert a long sharp-tipped needle or probe into the exposed vertebral column and direct it downward until it has reached the end of the canal, moving the needle from side to side as you go.

3. Experimental Preparation

PROCEDURE

a. ISOLATED MUSCLE PREPARATION

The gastrocnemius muscle of the frog is commonly used in the laboratory to demonstrate muscle contraction.

1. After double-pithing the frog, cut the leg off two-thirds of the way up the thigh. Remove the skin by pulling it down over the ankle.

2. Cut the calcaneal (Achilles) tendon as far down over the ankle as possible and then gently pull the gastrocnemius free from the lower leg.

3. Cut through the lower leg so as to leave the gastrocnemius muscle attached to the femur, and carefully cut away the severed muscle of the upper leg.

4. The muscle preparation consists of the gastrocnemius muscle, with its calcaneal tendon, attached to the bare lower half of the femur. The other gastrocnemius muscle should not be removed until needed. The gastrocnemius muscle is shown in Figure 9.9.

5. Place the muscle preparation in the apparatus with the bone held firmly in a femur clamp and the tendon connected to a muscle transducer by means of an S-pin stuck through the tendon (Figure 9.10).

6. Remember than an isolated skeletal muscle has no blood supply. Therefore moisten the muscle with Ringer's solution frequently and follow procedures carefully.

7. Adjust the height of the femur clamp so that the muscle has a slight amount of tension on it.

b. NERVE-MUSCLE PREPARATION

Another preparation that may be substituted for the isolated muscle preparation procedure is the nerve-muscle preparation. In this preparation the nerve innervating the gastrocnemius muscle is stimulated rather than directly stimulating the muscle.

NOTE. *It is strongly suggested that either your instructor or a laboratory assistant perform the following procedure for pithing a frog. The procedure requires some practice and may prove difficult for inexperienced individuals.*

1. Double-pith the frog and pin it down onto a frog board in the prone position. Remove the skin from the entire lower appendage by first making a cut around the upper leg and pulling the skin down over the ankle.

2. Cut the calcaneal tendon as far down over the ankle as possible and then gently pull the gastrocnemius free from the lower leg.

3. Cut through the lower leg so as to leave the gastrocnemius muscle attached to the femur. Gently spread the gluteus muscle from the semimembranosus muscle and expose the sciatic nerve (see Figure 9.9).

4. *Do not stretch the nerve, and avoid any unnecessary contact between the nerve, metal instruments, and other tissues.*

5. Gently place a ligature under the nerve. *Do not tie.* Utilize this ligature to place the stimulating electrodes under the nerve. Gently lower the nerve to lie over the electrode tips.

6. Attach the muscle to a transducer via an S-pin stuck through the calcaneal tendon.

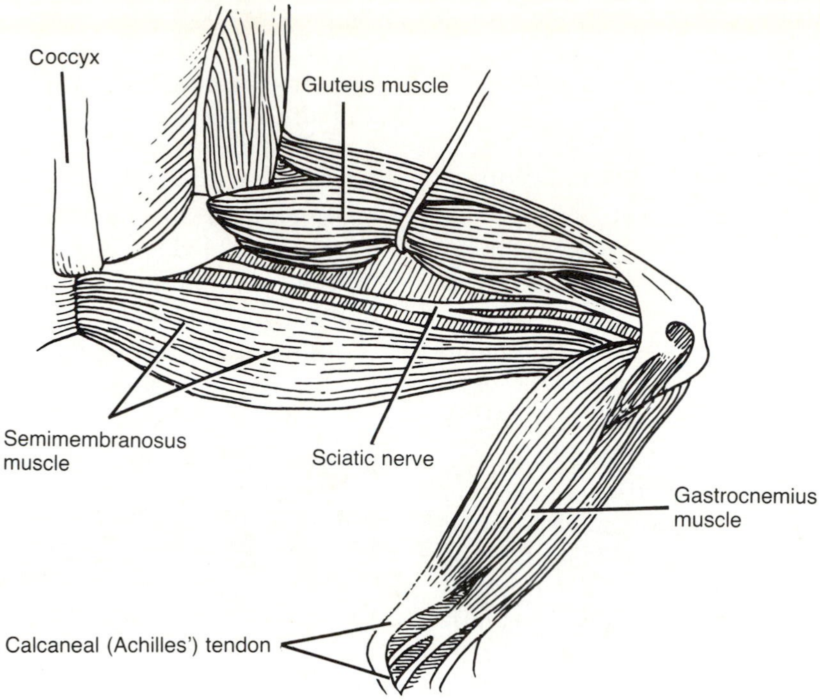

FIGURE 9.9 Structures in the lower extremity of the frog seen in posterior view.

Periodically moisten the gastrocnemius muscle and sciatic nerve with Ringer's solution (Figure 9.10).

7. Adjust the height of the muscle transducer so that a slight amount of tension is placed upon the muscle.

4. Demonstration of Threshold Stimulus

PROCEDURE

1. With the polygraph set at its lowest paper speed and the stimulus intensity knob placed at its minimum setting, apply a single stimulus to the muscle by utilizing the single stimulus key.

2. Record the stimulus strength on the polygraph paper by the pen mark made by the event marker.

3. Soak the muscle preparation with Ringer's solution between each stimulus.

4. In a stepwise manner increase the stimulus intensity and determine when a threshold strength stimulus has been reached.

5. Once threshold has been determined, continue to increase the stimulus strength in a stepwise manner until a maximum strength contraction has been obtained.

6. If skeletal muscle obeys the all-or-none principle, are the results obtained in this procedure a contradiction to this principle?

7. Include your results and response in Section F.1 of the LABORATORY REPORT RESULTS at the end of the exercise.

5. Single Muscle Twitch

PROCEDURE

1. With the polygraph set at its fastest paper speed, record a single muscle contraction resulting from a suprathreshold stimulus.

2. Calculate the latent period, contraction period, and relaxation period.

3. After rinsing the muscle with Ringer's solution, repeat this procedure several times.

4. Rinse the muscle after each procedure. Calculate the average for each value.

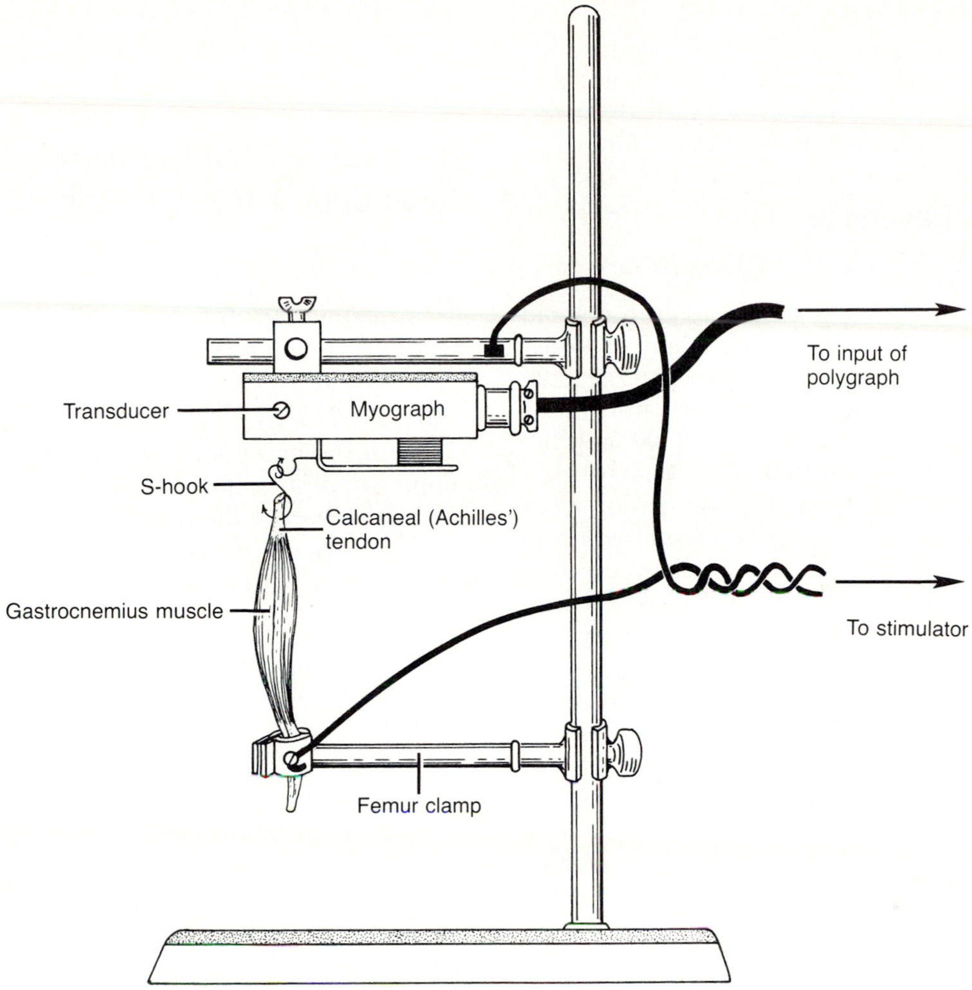

FIGURE 9.10 Frog gastrocnemius muscle preparation.

5. Place one tracing in Section F.3 of the LABORATORY REPORT RESULTS at the end of the exercise and record the average values.

6. Effect of Fatigue on a Single Muscle Twitch

PROCEDURE

1. *Familiarize yourself with the entire procedure before starting.* Set the polygraph at its lowest paper speed. Using a frequency of 60 per second and the multiple stimulus setting, stimulate the muscle until it fatigues. (Fatigue is demonstrated when the contraction strength is reduced by 50% or more as compared to the initial contraction strength.)

2. Upon onset of fatigue, rapidly turn the stimulator off, set the paper speed at its highest level, and stimulate the muscle with a single suprathreshold stimulus.

3. *Do not* rinse the muscle with Ringer's solution.

4. Obtain several additional single muscle twitches, and calculate the average latent period, contraction period, and relaxation period.

5. Place one tracing in Section F.4 of the LABORATORY REPORT RESULTS at the end of the exercise and record the average values.

6. Compare these values to those obtained in Section 5, Single Muscle Twitch. Explain any differences seen between the two tracings in the space provided in Section F.5 of the LABORA-

TORY REPORT RESULTS at the end of the exercise.

7. Before proceeding to the next section discard the muscle utilized up to this point and replace it with a fresh muscle. Quickly determine threshold for this new preparation.

7. Muscle Response to Variation of Stimulus Frequency

When a single threshold or suprathreshold stimulus is applied to either a single skeletal muscle fiber or a complete muscle, the contraction response follows the all-or-none principle. However, if two stimuli are applied in rapid succession so that the second stimulus is applied before complete relaxation from the first, one notices that the second contraction is greater than the first. Such a response is seen in either single skeletal muscle fibers or a whole muscle. The extent of the increased response to the second stimulus is dependent upon the interval between the successive stimuli, as long as the interval between the two stimuli is greater than the refractory period for that fiber or muscle. This response to increased stimulus frequency, termed ***wave summation***, does not follow the all-or-none principle, and is due to increased amounts of calcium ions accumulating within the muscle fiber after successive stimuli. If you were to continue to increase the stimulus frequency you would note that the amount of relaxation between successive stimuli would progressively decrease until further stimuli would not increase the strength of contraction and no relaxation would be seen. This response would be termed ***tetanus***. During tetanus the contraction strength may be three or four times that seen with a single contraction.

PROCEDURE

1. Set the paper speed at a medium value.

2. Turn the stimulus frequency knob to 1 per second and set the stimulator mode selector switch to repetitive stimuli.

3. Progressively increase the frequency in a stepwise manner until wave summation and tetany are observed. Record the frequencies at which summation and tetany occur in Section F.6 of the LABORATORY REPORT RESULTS at the end of the exercise.

4. Turn the stimulator off and bathe the muscle with Ringer's solution frequently.

5. Allow 2 to 3 minutes of recovery time before proceeding to the next section.

8. Effect of Resting Length Upon Contraction Strength

PROCEDURE

1. Adjust either the femur clamp or muscle transducer height (depending upon the preparation used) so that a minimal amount of resting tension is applied to the muscle.

2. With the paper drive set at its lowest speed, apply a single suprathreshold stimulus to the muscle. Observe the contraction strength.

3. Rinse the muscle with Ringer's solution after each stimulation.

4. Adjust the level of the femur clamp or transducer so that the resting tension is increased slightly. Stimulate again with a single stimulus and rinse. Observe the contraction strength at this new resting tension.

5. Repeat until the contraction strength has reached a maximum and started to decline.

6. Graph your results in Section F.7 of the LABORATORY REPORT RESULTS at the end of the exercise and explain your findings in the space provided.

G. BIOCHEMISTRY OF SKELETAL MUSCLE CONTRACTION

You will examine the effect of the following solutions on the contraction of glycerinated muscle fibers.[2] (1) ATP solution, (2) mineral ion solution, and (3) ATP plus mineral ion solution.

PROCEDURE

1. Using 7× or 10× magnification and glass needles or clean stainless steel forceps, gently tease the muscle into very thin groups of myofibers. Single cells or thin groups must be used because strands thicker than a silk thread curl when they contract.

[2]Glycerinated muscle preparation and solutions are supplied by the Carolina Biological Supply Company, Burlington, North Carolina 27215.

2. Using a Pasteur pipette or medicine dropper, transfer one strand into a drop of glycerol on a clean glass slide and cover the preparation with a cover slip. Examine the strand under low and high power and note the striations. Also note that each cell has several nuclei.

3. Transfer one of the thinnest strands to a drop of glycerol on a second microscope slide. Do not add a cover slip. If the amount of glycerol on the slide is more than a small drop, soak the excess into a piece of lens paper held at the edge of the glycerol farthest from the fibers. Using a dissecting microscope and a millimeter ruler held beneath the slide, measure the length of one of the fibers.

4. Now flood the fibers with the solution containing only ATP and observe their reaction. After 30 seconds or more, remeasure the same fiber.

How much did the fiber contract? ____ mm.

5. Using clean slides and medicine droppers, and being especially sure to use clean teasing needles or forceps, transfer other fibers to a drop of glycerol on a slide. Again measure the length of one fiber. Next flood the fibers with a solution containing mineral ions, observe their reaction, and remeasure the fiber.

How much did the fiber contract? ____ mm.

6. Repeat the exercise, this time using a solution containing a combination of ATP and ions.

7. Observe a contracted fiber under low and high power, and look for differences in appearance between muscle in a contracted and that in a relaxed state (see Figure 9.6).

H. ELECTROMYOGRAPHY

When a muscle contracts, action potentials spread along the length and interior of the muscle, and a small portion of the electrical activity spreads away from the muscle, possibly to the skin, depending upon the location of the muscle. If several muscles contract simultaneously, the bioelectrical potential measured at the skin may be of considerable magnitude. By placing two electrodes on the skin or directly in the muscle an electrical recording termed an ***electromyograph (EMG)*** may be recorded when the muscle is stimulated (Figure 9.11). Abnormalities in muscle physiology may be determined through the clinical application of the electromyograph.

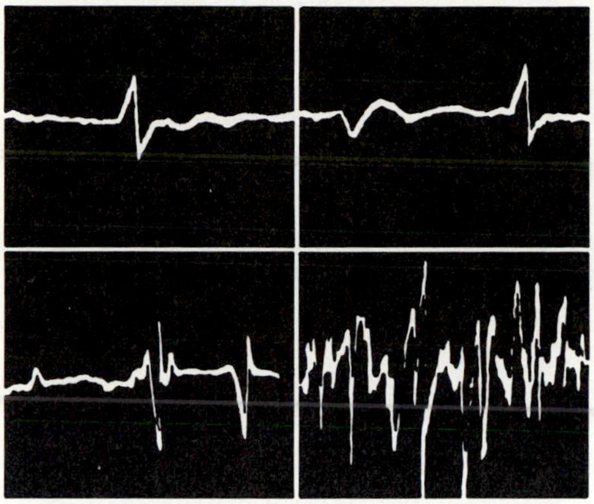

FIGURE 9.11 Diagrammatic representation of normal electromyograms. The single potential in the upper left corner has a measured amplitude of 0.8 mV and a duration of 7 msec.

1. Subject Preparation

PROCEDURE

1. Prepare to place skin EMG electrodes (or ECG electrodes) on the anterior surface of the forearm superficial to the ***flexor digitorum superficialis*** muscle (see Figure 10.15a) and on the posterior surface of the arm superficial to the ***triceps brachii*** muscle (see Figure 10.14b).

2. Prior to placing the electrodes on the skin, *gently* scrub the area with a dishwashing pad in order to remove dead epithelial cells and to facilitate recording. *Make sure that the capillaries are not damaged and no bleeding occurs.*

3. Apply a small amount of electrode gel to each electrode and apply them to the proper location.

4. Connect the two recording electrodes (placed on the anterior surface of the forearm) and the ground (placed on the posterior surface of the arm) to the preamplifier (high gain coupler) and set the gain to ×100, sensitivity at an appropriate setting between 20 and 100, time constant to 0.03, and paper speed at high.

2. Recording of Spontaneous Muscle Activity

PROCEDURE

1. With the subject completely relaxed and his/her arm placed horizontally on a labora-

tory table, record any spontaneously occurring electrical activity within the flexor digitorum superficialis over a time period of 1–3 minutes.

2. Record your observations in Section H.1 of the LABORATORY REPORT RESULTS at the end of the exercise.

3. Recruitment of Motor Units

PROCEDURE

1. Have the subject gently flex his/her digits while recording, and note the electrical activity.

2. Have the subject relax their digits completely, followed by a more forceful contraction. Note any change in the EMG.

3. Place a tennis ball in your subject's hand and ask him/her to squeeze the ball once or twice as strongly as possible. Note the EMG recording, and then let your subject rest thoroughly.

4. Record your observations in Section H.2 of the LABORATORY REPORT RESULTS at the end of the exercise.

4. Effect of Fatigue

PROCEDURE

1. *Without recording* an EMG on the polygraph, have your subject repeatedly squeeze the tennis ball as strongly as possible until no longer able to squeeze the ball.

2. When fatigue has been achieved *quickly* start recording and ask the subject to attempt to squeeze the ball as strongly as possibly four or five times in quick succession.

3. Note the EMG recording and record your observations in Section H.3 of the LABORATORY REPORT RESULTS at the end of the exercise.

ANSWER THE LABORATORY REPORT QUESTIONS AT THE END OF THE EXERCISE

MUSCLE TISSUE 9

Student ______________________ **Date** ______________________

Laboratory Section ______________________ **Score/Grade** ______________________

SECTION F. LABORATORY TESTS ON SKELETAL MUSCLE CONTRACTION

1. Complete the following table with the values obtained.

Stimulus Strength	Contraction Strength

2. The all-or-none principle states that, once threshold is reached, a muscle cell will contract maximally, and that a subthreshold stimulus will not elicit a response. Based upon the data you collected in Section F.4, did the muscle obey the all-or-none principle? ______

3. Attach and label the myogram obtained with a single stimulus. Label latent period, contraction period, and relaxation period. Include the average values obtained.

4. Attach and label the myogram obtained with a single stimulus in a fatigued muscle. Label the latent period, contraction period, and relaxation period, and include the average values obtained.

5. Explain the physiological reason for the differences observed between the two myograms.

____

____

____

____

____

6. Draw and label the myogram obtained as the stimulus frequency was increased. Label tetanus in your diagram.

7. Complete the following graph based upon your results obtained in Section F.8.

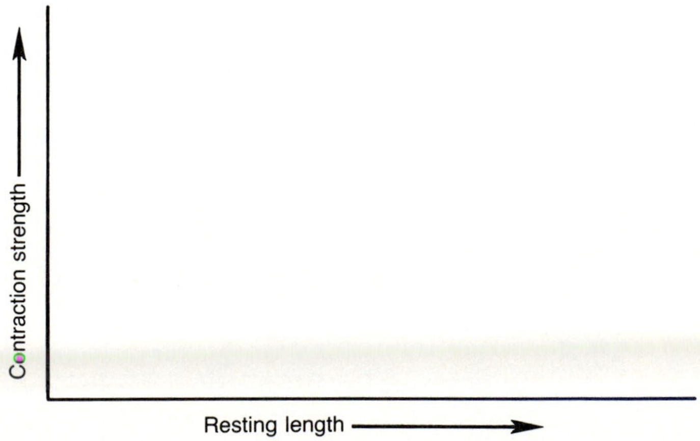

SECTION H. ELECTROMYOGRAPHY

1. Include your record of spontaneous electrical activity recorded while the subject was resting. Give a physiological explanation. ______________________________

____

2. Include your record of the EMG recorded during the recruitment exercise. Give a physiological explanation and possible function of recruitment. ______________________________

____

3. Include your record of the EMG recorded following fatigue, and give a physiological explanation of your observations. ______________________________

____

LABORATORY REPORT QUESTIONS

MUSCLE TISSUE 9

Student ______________________ **Date** ______________________

Laboratory Section ______________________ **Score/Grade** ______________________

PART I. MULTIPLE CHOICE

_____ 1. The ability of muscle tissue to return to its original shape after contraction or extension is called (a) excitability (b) elasticity (c) extension (d) tetanus

_____ 2. Which of the following is striated and voluntary? (a) skeletal muscle tissue (b) cardiac muscle tissue (c) visceral muscle tissue (d) smooth muscle tissue

_____ 3. The portion of a sarcomere composed of thin myofilaments only is the (a) H zone (b) A band (c) I band (D) Z line

_____ 4. The area of contact between a motor axon terminal and a muscle fiber sarcolemma is called the (a) synapse (b) myofilament (c) T tubule (d) neuromuscular junction

_____ 5. The connective tissue layer surrounding bundles of muscle fibers is called the (a) perimysium (b) endomysium (c) ectomysium (d) myomysium

_____ 6. Which of the following is striated and involuntary? (a) smooth muscle tissue (b) skeletal muscle tissue (c) cardiac muscle tissue (d) visceral muscle tissue

_____ 7. The portion of a sarcomere where thin and thick myofilaments overlap is called the (a) H zone (b) triad (c) A band (d) I band

_____ 8. The portion of a polygraph that converts nonelectrical signals into electrical signals is the (a) transducer (b) electrode (c) amplifier (d) coupler

_____ 9. In a skeletal muscle twitch, which period requires the *least* amount of time? (a) relaxation (b) latent (c) contraction (d) extension

_____ 10. A state of sustained muscle contraction is known as (a) tetanus (b) treppe (c) myopathy (d) hypertrophy

PART II. COMPLETION

11. The ability of muscle tissue to receive and respond to stimuli is called ______________________ .
12. Fibrous connective tissue located between muscle fibers is known as ______________________ .
13. The sections of a muscle fiber separated by Z lines are called ______________________ .
14. The plasma membrane surrounding a muscle fiber is called the ______________________ .
15. The region in a sarcomere consisting of thick myofilaments only is known as the ______________________ .
16. Muscle tissue that is nonstriated and involuntary is ______________________ .
17. The ability of muscle tissue to stretch when pulled is called ______________________ .

18. The phenomenon by which a muscle fiber contracts to its fullest or not at all is known as the

____________________ .

19. A record of muscle contraction is referred to as a(n) ____________________ .

20. A stimulus too weak to initiate a nerve impulse is called a(n) ____________________ stimulus.

SKELETAL MUSCLES 10

In this exercise you will learn the names, locations, and actions of the principal skeletal muscles of the body.

A. HOW SKELETAL MUSCLES PRODUCE MOVEMENT

1. Origin and Insertion

Skeletal muscles produce movements by exerting force on tendons, which in turn pull on bones. Most muscles cross at least one joint and are attached to the articulating bones that form the joint. When such a muscle contracts, it draws one articulating bone toward the other. The two articulating bones usually do not move equally in response to the contraction. One is held nearly in its original position because other muscles contract to pull it in the opposite direction or because its structure makes it less movable. Ordinarily, the attachment of a muscle tendon to the stationary bone is called the ***origin***. The attachment of the other muscle tendon to the movable bone is the ***insertion***. A good analogy is a spring on a door. The part of the spring attached to the door represents the insertion; the part attached to the frame is the origin. The fleshy portion of the muscle between the tendons of the origin and insertion is called the ***belly (gaster)***. The origin is usually proximal and the insertion distal, especially in the extremities. In addition, muscles that move a body part generally do not cover the moving part. For example, although contraction of the biceps brachii muscle moves the forearm, the belly of the muscle lies over the humerus.

2. Group Actions

Most movements are coordinated by several skeletal muscles acting in groups rather than individually and most skeletal muscles are arranged in opposing pairs at joints, that is, flexors–extensors, abductors–adductors, and so on. Consider flexing the forearm at the elbow, for example. A muscle that causes a desired action is referred to as the ***prime mover (agonist)***. In this instance, the biceps brachii is the prime mover (see Figure 10.14a). Simultaneously with the contraction of the biceps brachii, another muscle, called the ***antagonist***, is relaxing. In this movement, the triceps brachii serves as the antagonist (see Figure 10.14b). The antagonist has an effect opposite to that of the prime mover, that is, the antagonist relaxes and yields to the movement of the prime mover. You should not assume, however, that the biceps brachii is always the prime mover and the triceps brachii is always the antagonist. For example, when extending the forearm at the elbow, the triceps brachii serves as the prime mover and the biceps brachii functions as the antagonist; their roles are reversed. Note that if the prime mover and antagonist contracted simultaneously with equal force, there would be no movement.

In addition to prime movers and antagonists, most movements also involve muscles called ***synergists***, which serve to steady a movement, thus preventing unwanted movements and helping the prime mover function more efficiently. For example, flex your hand at the wrist and then make a fist. Note how difficult this is to do. Now extend your hand at the wrist and then make a fist. Note how much easier it is to clench your fist. In this case, the extensor muscles of the wrist act as synergists in cooperation with the flexor muscles of the fingers acting as prime movers. The extensor muscles of the fingers serve as antagonists (see Figure 10.15c,d).

Some muscles in a group also act as ***fixators***, which stabilize the origin of the prime mover so that the prime mover can act more efficiently. For example, the scapula is a freely movable bone in the pectoral (shoulder) girdle that serves as an origin for several muscles that move the arm. However, for the scapula to serve as a firm origin for muscles that move the arm, it must be held steady. This is accomplished by fixator muscles that hold the scapula firmly against the back of the chest. In abduction of the arm, the deltoid muscle serves as the prime mover, whereas fixators (pectoralis minor, rhomboideus major, rhomboideus minor, trapezius, subclavius, and serratus anterior muscles) hold the scapula firmly (see Figure 10.12). These fixators stabilize the scapula that serves as the attachment site for the origin of the deltoid muscle while the insertion of the muscle pulls on the humerus to abduct the arm. Under different conditions and depending on the movement and which point is fixed, many muscles act, at various times, as prime movers, antagonists, synergists, or fixators.

B. NAMING SKELETAL MUSCLES

Most of the almost 700 skeletal muscles of the body are named on the basis of one or more distinctive criteria. If you understand these criteria, you will find it much easier to learn and remember the names of individual muscles. Some muscles are named on the basis of the ***direction of the muscle fibers***, for example, *rectus* (meaning straight), *transverse*, and *oblique*. Rectus fibers run parallel to some imaginary line, usually the midline of the body, transverse fibers run perpendicular to the line, and oblique fibers run diagonal to the line. Muscles named according to the direction their fibers run include the rectus abdominis, transversus abdominis, and external oblique. Another criterion employed is ***location***. For example, the temporalis is so named because of its proximity to the temporal bone, and the tibialis anterior is located near the tibia. ***Size*** is also commonly employed. For instance, *maximus* means largest, *minimus* means smallest, *longus* refers to long, and *brevis* refers to short. Examples include the gluteus maximus, gluteus minimus, adductor longus, and peroneus brevis.

Some muscles, such as biceps, triceps, and quadriceps, are named on the basis of the ***number of origins*** they have. For instance, the biceps has two origins, the triceps three, and the quadriceps four. Other muscles are named on the basis of ***shape***. Common examples include the deltoid (meaning triangular) and trapezius (meaning trapezoid). Muscles may also be named after their ***insertion*** and their ***origin***. Two such examples are the sternocleidomastoid (originates on sternum and clavicle and inserts at mastoid process of temporal bone), and the stylohyoid (originates on styloid process of temporal bone and inserts at the hyoid bone).

Still another basis for naming muscles is ***action***. Listed here are the principal actions of muscles, their definitions, and examples of muscles that perform the actions.

Flexor—Usually decreases the anterior angle at a joint. Example: flexor carpi radialis.

Extensor—Usually increases the anterior angle at a joint. Example: extensor carpi ulnaris.

Abductor—Moves a bone away from the midline. Example: abductor pollicis brevis.

Adductor—Moves a bone closer to the midline. Example: adductor longus.

Levator—Produces an upward or superiorly directed movement. Example: levator scapulae.

Depressor—Produces a downward or inferiorly directed movement. Example: depressor labii inferioris.

Supinator—Turns the palm upward or to the anterior. Example: supinator.

Pronator—Turns the palm downward or to the posterior. Example: pronator teres.

Sphincter—Decreases the size of an opening. Example: external anal sphincter.

Tensor—Makes a body part more rigid. Example: tensor fasciae latae.

Rotator—Moves a bone around its longitudinal axis. Example: obturator.

C. CONNECTIVE TISSUE COMPONENTS

Skeletal muscles are protected, strengthened, and attached to other structures by several connective tissue components. For example, the entire muscle is usually wrapped with a fibrous connective tissue called the ***epimysium*** (ep′-i-MĪZ-ē-um). When the muscle is cut in cross section, invaginations of the epimysium divide the muscle into bundles called ***fasciculi (fascicles)***. These invaginations of the epimysium are called the ***perimysium*** (per′-i-MĪZ-ē-um). In turn, invaginations of the perimysium, called ***endomysium*** (en′-dō-MĪZ-ē-um), penetrate into the anterior of each fascicle and separate individual muscle fibers from each other. The epimysium, perimysium, and endomysium are all extensions of deep fascia and are all continuous with the connective tissue that attaches the muscle to another structure, such as bone or other muscle. All three elements may be extended beyond the muscle cells as a ***tendon***—a cord of connective tissue that attaches a muscle to the periosteum of bone. The connective tissue may also extend as a broad, flat band of tendons called an ***aponeurosis***. Aponeuroses also attach to the coverings of a bone or another muscle. When a muscle contracts, the tendon and its corresponding bone or muscle are pulled toward the contracting muscle. In this way skeletal muscles produce movement.

In Figure 10.1, label the epimysium, perimysium, endomysium, fasciculus, and muscle fibers.

D. PRINCIPAL MUSCLES

In the pages that follow, a series of tables has been provided for you to learn the principal skeletal muscles by region.[1] Use each table as follows:

1. Take each muscle, in sequence, and study the ***learning key*** that appears in parentheses after the name of the muscle. The learning key is a list of prefixes, suffixes, and definitions that explain the derivations of the muscles' names.

[1]A few of the muscles listed are not illustrated in the diagrams. Please consult your textbook to locate these muscles.

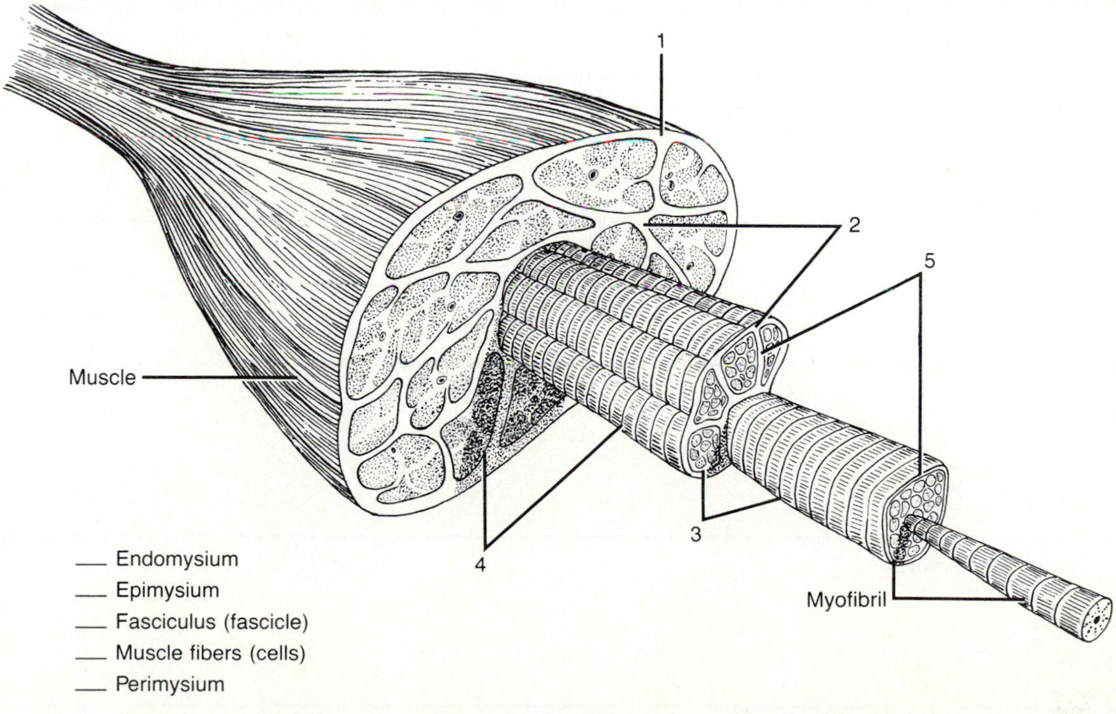

FIGURE 10.1 Connective tissue components of a skeletal muscle.

It will help you to understand the reason for giving a muscle its name.

2. As you learn the name of each muscle, determine its origin, insertion, and action and write these in the spaces provided in the table. Consult your textbook if necessary.

3. Again, using your textbook as a guide, label the diagram referred to in the table.

4. Try to visualize what happens when the muscle contracts so that you will understand its action.

5. Do steps 1 through 4 for each muscle in the table. Before moving to the next table, examine a torso or chart of the skeletal system so that you can compare and approximate the positions of the muscles.

6. When possible, try to feel each muscle on your own body.

TABLE 10.1
MUSCLES OF FACIAL EXPRESSION (After completing the table, label Figure 10.2.)

Overview: The muscles in this group provide humans with the ability to express a wide variety of signals that communicate emotions, including grief, surprise, fear, and happiness. The muscles themselves lie within the layers of superficial fascia. As a rule, they arise from the fascia or bones of the skull and insert into the skin. Because of their insertions, the muscles of facial expression move the skin rather than a joint when they contract.

Muscle	Origin	Insertion	Action
Epicranius (*epi* = over; *crani* = skull)	This muscle is divisible into two portions: the frontalis, over the frontal bone, and the occipitalis, over the occipital bone. The two muscles are united by a strong aponeurosis, the galea aponeurotica, which covers the superior and lateral surfaces of the skull.		
Frontalis (*front* = forehead)			
Occipitalis (*occipito* = base of skull)			
Orbicularis oris (*orb* = circular; *or* = mouth)			
Zygomaticus major (*zygomatic* = cheek bone; *major* = greater)			
Levator labii superioris (*levator* = raises or elevates *labii* = lip; *superioris* = upper)			
Depressor labii inferioris (*depressor* = depresses or lowers; *inferioris* = lower)			
Buccinator (*bucc* = cheek)			

TABLE 10.1 ***(Continued)***

Muscle	Origin	Insertion	Action
Mentalis (*mentum* = chin)			
Platysma (*platy* = flat, broad)			
Risorius (*risor* = laughter)			
Orbicularis oculi (*ocul* = eye)			
Corrugator supercilii (*corrugo* = to wrinkle; *supercilium* = eyebrow)			
Levator palpebrae superioris (*palpebrae* = eyelids) (See Figure 10.4)			

TABLE 10.2
MUSCLES THAT MOVE THE LOWER JAW (After completing the table, label Figure 10.3.)

Overview: Muscles that move the lower jaw are also known as muscles of mastication because they are involved in biting and chewing. These muscles also assist in speech.

Muscle	Origin	Insertion	Action
Masseter (*masseter* = chewer)			
Temporalis (*tempora* = temples)			
Medial pterygoid (*medial* = closer to midline; *pterygoid* = like a wing)			
Lateral pterygoid (*lateral* = farther from midline)			

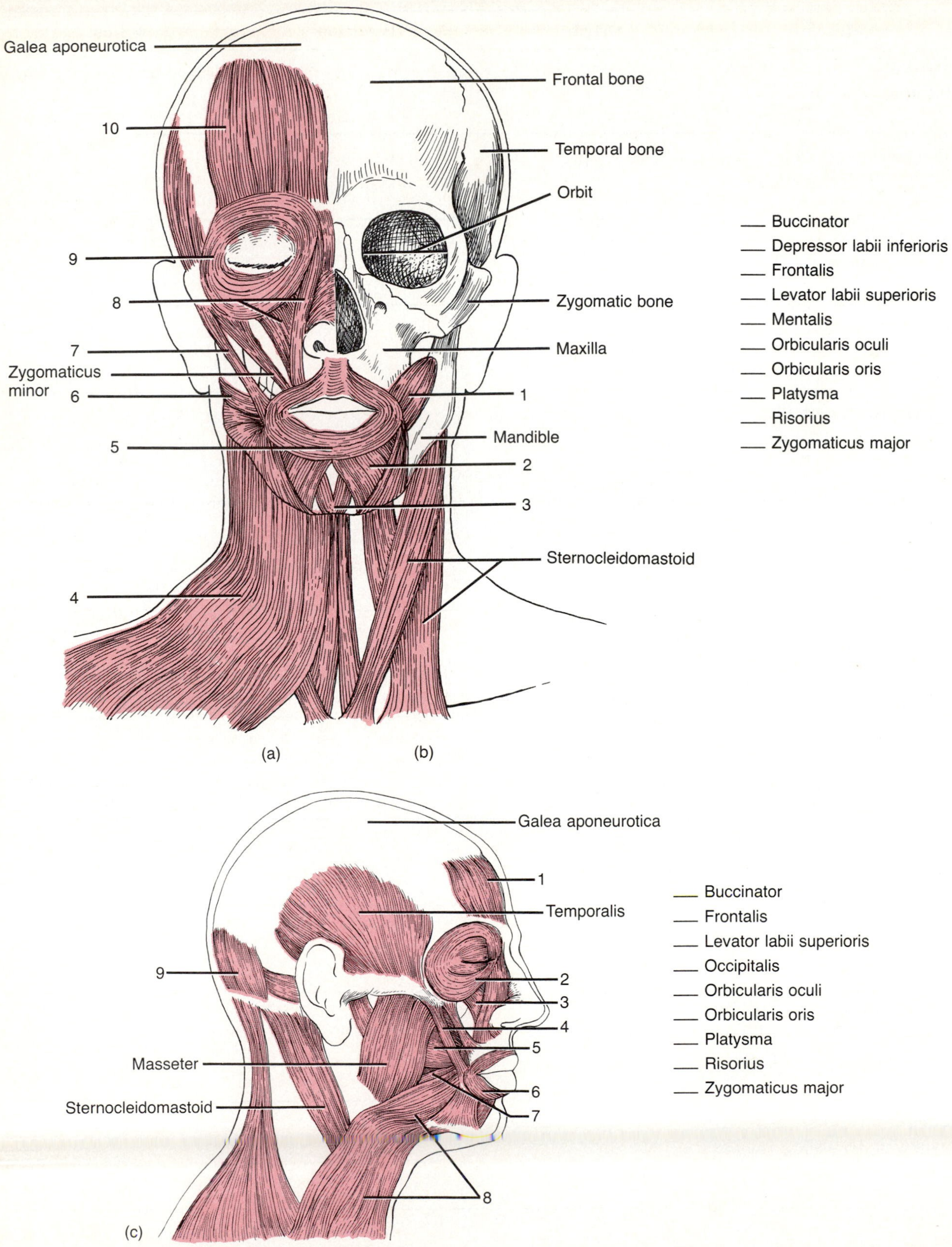

FIGURE 10.2 Muscles of facial expression. (a) Anterior superficial view. (b) Anterior deep view. Muscles of facial expression. (c) Right lateral superficial view.

TABLE 10.3
MUSCLES THAT MOVE THE EYEBALL—THE EXTRINSIC MUSCLES* **(After completing the table, label Figure 10.4.)**

Overview: Muscles associated with the eyeball are of two principal types: extrinsic and intrinsic. ***Extrinsic muscles*** originate outside the eyeball and are inserted on its outer surface (sclera). ***Intrinsic muscles*** originate and insert entirely within the eyeball.

Movements of the eyeballs are controlled by three pairs of extrinsic muscles. The two pairs of rectus muscles move the eyeball in the direction indicated by their respective names—superior, inferior, lateral, and medial. The pair of oblique muscles—superior and inferior—rotate the eyeball on its axis. The extrinsic muscles of the eyeballs are among the fastest contracting and precisely controlled skeletal muscles of the body.

Muscle	Origin	Insertion	Action
Superior rectus (*superior* = above; *rectus* = in this case, muscle fibers running parallel to long axis of eyeball)			
Inferior rectus (*inferior* = below)			
Lateral rectus			
Medial rectus			
Superior oblique (*oblique* = in this case, muscle fibers running diagonally to long axis of eyeball)			
Inferior oblique			

*Muscles situated on the outside of the eyeball.

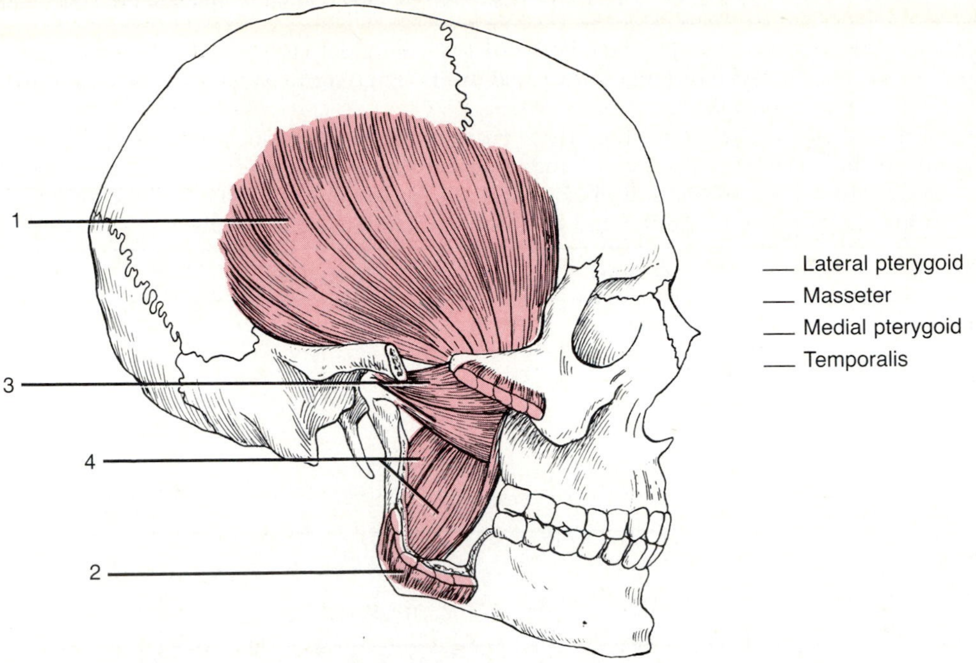

FIGURE 10.3 Muscles that move lower jaw seen in right lateral view.

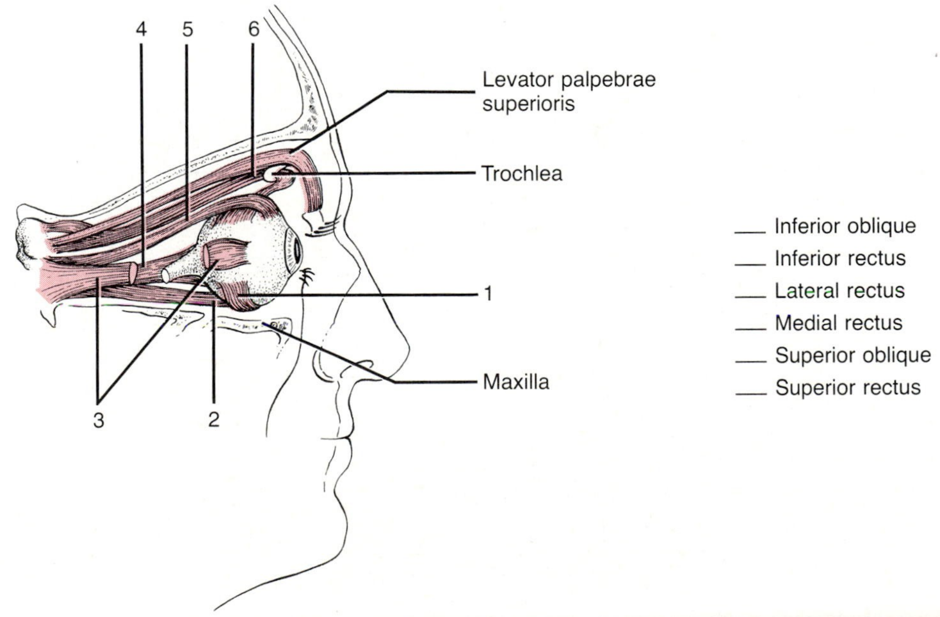

FIGURE 10.4 Extrinsic muscles of eyeball seen in right lateral view.

TABLE 10.4
MUSCLES THAT MOVE THE TONGUE—EXTRINISIC MUSCLES (After completing the table, label Figure 10.5.)

Overview: The tongue is divided into lateral halves by a median fibrous septum. The septum extends throughout the length of the tongue and is attached inferiorly to the hyoid bone. Like the muscles of the eyeball, muscles of the tongue are of two principal types—extrinsic and intrinsic. ***Extrinsic muscles*** originate outside the tongue and insert into it. ***Intrinsic muscles*** originate and insert within the tongue. The extrinsic and intrinsic muscles of the tongue are arranged in both lateral halves of the tongue.

Muscle	Origin	Insertion	Action
Genioglossus (*geneion* = chin; *glossus* = tongue)			
Styloglossus (*stylo* = stake or pole)			
Palatoglossus (*palato* = palate)			
Hyoglossus			

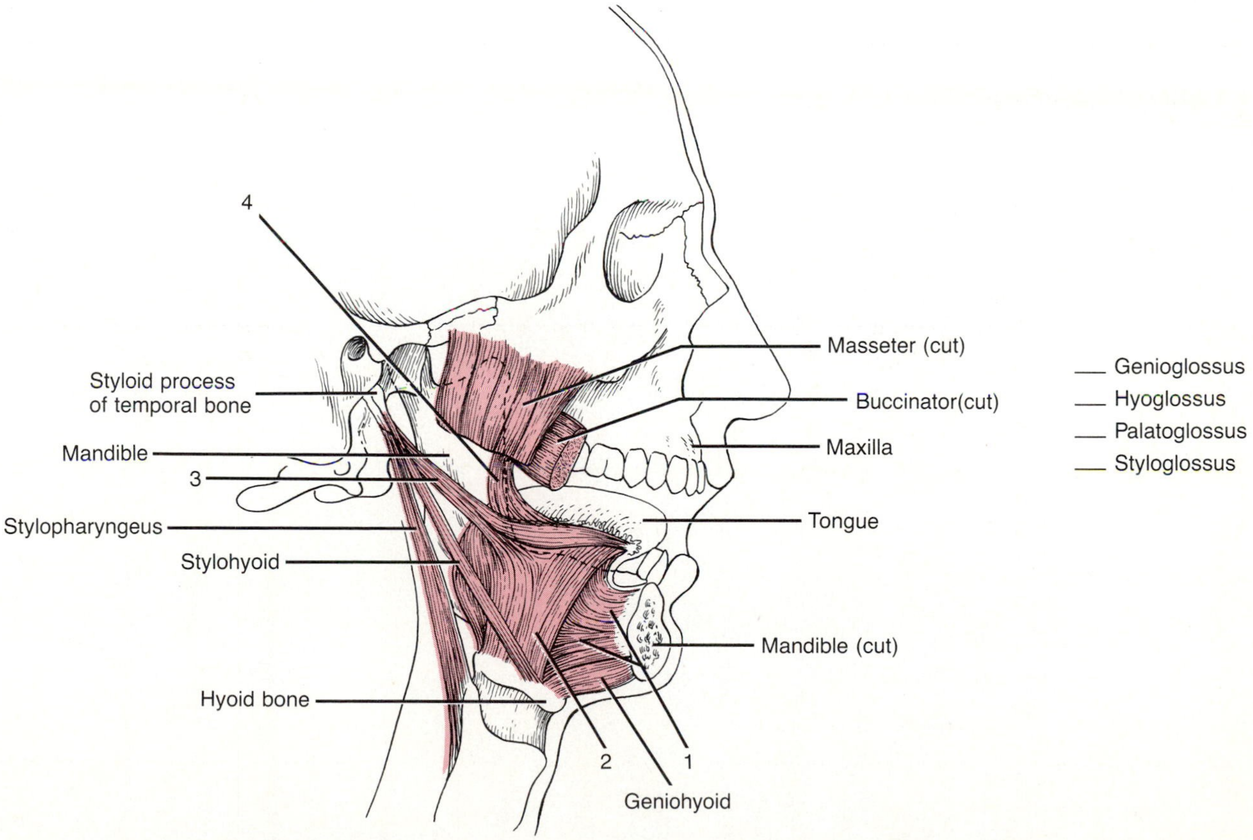

FIGURE 10.5 Muscles that move the tongue viewed from right side.

TABLE 10.5
MUSCLES OF THE PHARYNX (After completing the table, label Figure 10.6.)

Overview: The pharynx (throat) is a funnel-shaped, muscular tube located posterior to the nasal cavities, mouth, and larynx (voice box). The muscles are arranged in two layers—an outer circular layer and an inner longitudinal layer. The ***circular layer*** is composed of the three constrictors, each overlapping the one above. The remaining muscles constitute the ***longitudinal layer***.

Muscle	Origin	Insertion	Action
CIRCULAR LAYER			
Inferior constrictor (*inferior* = below; *constrictor* = decreases diameter of lumen)			
Middle constrictor			
Superior constrictor (*superior* = above)			
LONGITUDINAL LAYER			
Stylopharyngeus (*stylo* = stake or pole; *pharyngo* = pharynx)			
Salpingopharyngeus (*salping* = pertaining to auditory tube or uterine tube)			
Palatopharyngeus (*palato* = palate)			

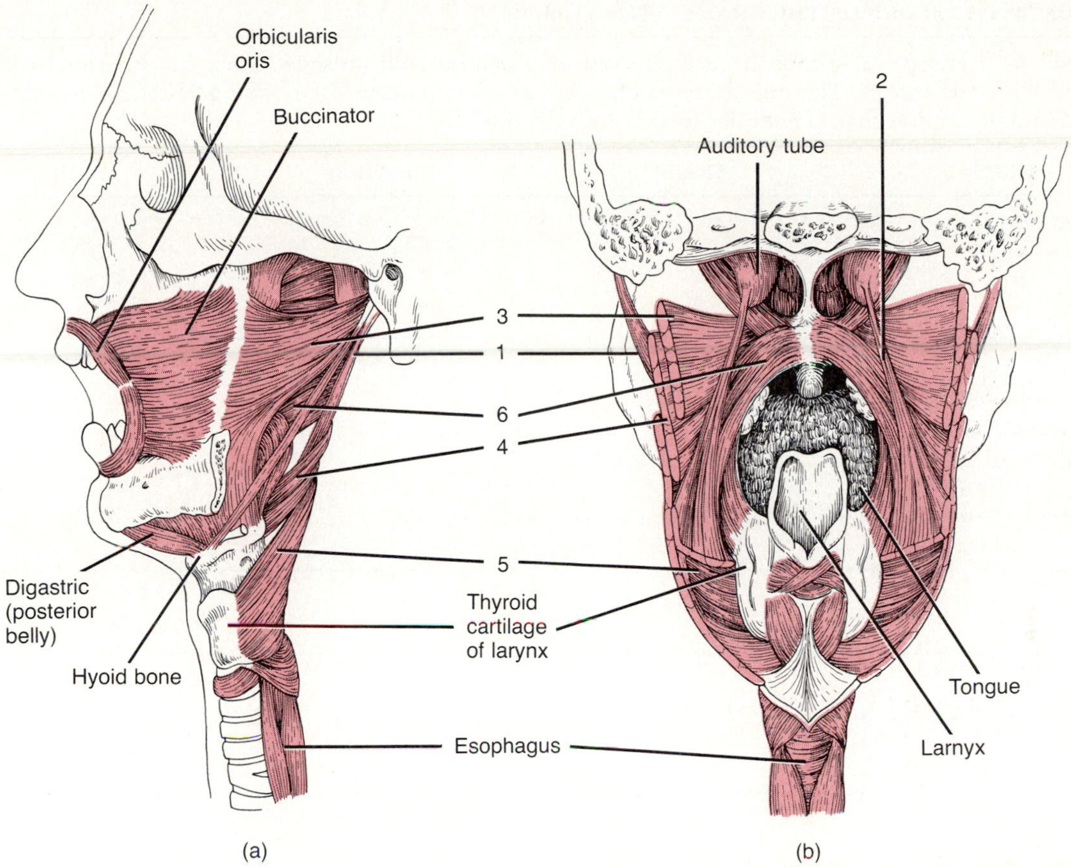

___ Inferior constrictor
___ Middle constrictor
___ Palatopharyngeus
___ Salpingopharyngeus
___ Stylopharyngeus
___ Superior constrictor

FIGURE 10.6 Muscles of pharynx. (a) Left lateral view. (b) Posterior view.

TABLE 10.6
MUSCLES OF THE FLOOR OF THE ORAL CAVITY (Figure 10.7)

Overview: As a group, these muscles are referred to as ***suprahyoid muscles***. They lie superior to the hyoid bone and all insert into it. The digastric muscle consists of an anterior belly and a posterior belly united by an intermediate tendon that is held in position by a fibrous loop.

Muscle	Origin	Insertion	Action
Digastric (*di* = two; *gaster* = belly)			
Stylohyoid (*stylo* = stake or pole, styloid process of temporal bone; *hyoedes* = U-shaped, pertaining to hyoid bone.)			
Mylohyoid			
Geniohyoid (*geneion* = chin) (See Figure 10.5)			

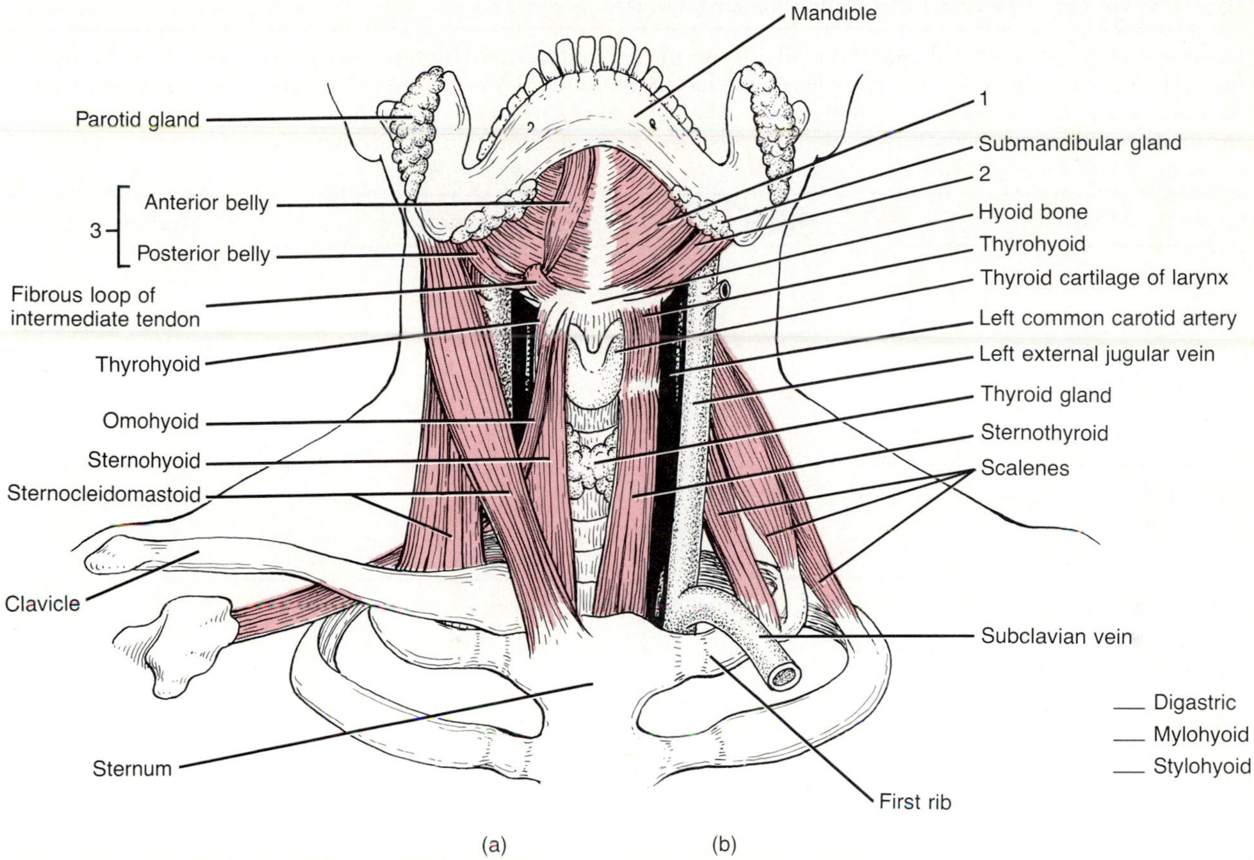

FIGURE 10.7 Muscles of the floor of the oral cavity. (a) Superficial view. (b) Deep view.

TABLE 10.7
MUSCLES OF THE LARYNX (After completing the table, label Figure 10.8.)

Overview: The muscles of the larynx, like those of the eyeballs and tongue, are grouped into ***extrinsic*** and ***intrinsic muscles***. The extrinsic muscles of the larynx marked (*) are together referred to as ***infrahyoid (strap) muscles***. They lie inferior to the hyoid bone. The omohyoid muscle, like the digastric muscle, is composed of two bellies and an intermediate tendon. In this case, however, the two bellies are referred to as superior and inferior, rather than anterior and posterior.

Muscle	Origin	Insertion	Action
EXTRINSIC			
Omohyoid* (*omo* = relationship to shoulder; *hyoedes* = U-shaped)			
Sternohyoid* (*sterno* = sternum)			
Sternothyroid* (*thyro* = thyroid gland)			
Thyrohyoid*			
Stylopharyngeus **Palatopharyngeus** **Inferior constrictor** **Middle constrictor**	**See Table 10.5 and Figure 10.6**		
INTRINSIC			
Cricothyroid (*crico* = cricoid cartilage of larynx)			
Posterior cricoarytenoid (*arytaina* = shaped like a jug)			
Lateral cricoarytenoid			
Arytenoid			
Thyroarytenoid			

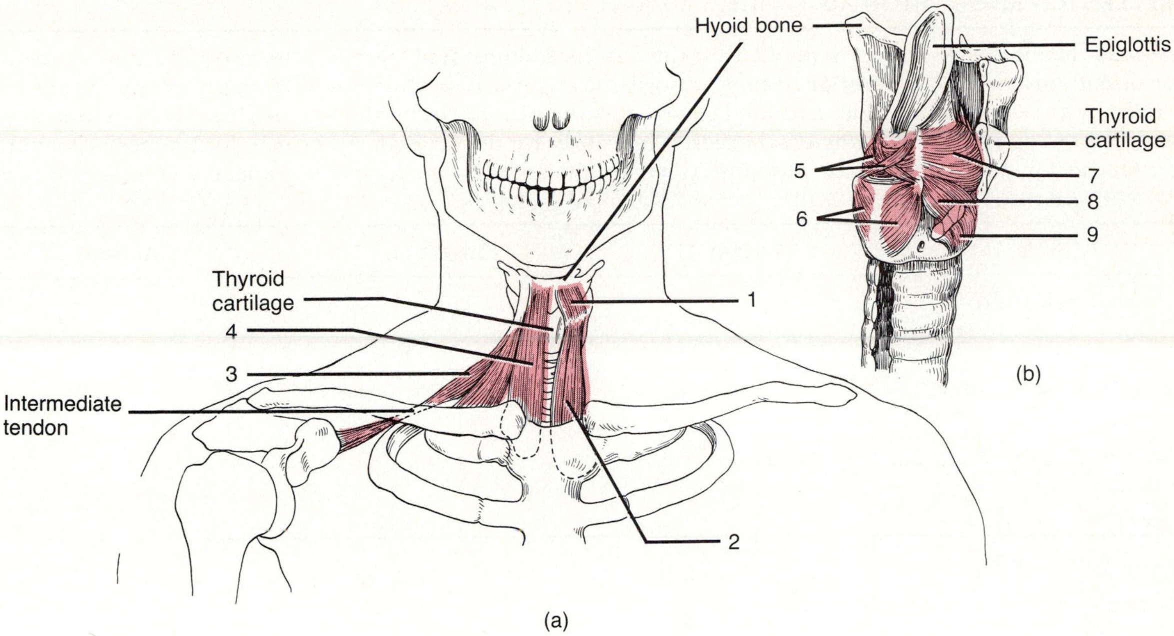

___ Arytenoid
___ Cricothyroid
___ Lateral cricoarytenoid
___ Omohyoid
___ Posterior cricoarytenoid
___ Sternohyoid
___ Sternothyroid
___ Thyroarytenoid
___ Thyrohyoid

FIGURE 10.8 Muscles of larynx. (a) Anterior view. (b) Right posterolateral view.

TABLE 10.8
MUSCLES THAT MOVE THE HEAD

Overview: The cervical region is divided by the sternocleidomastoid muscle into two principal triangles—anterior and posterior. The ***anterior triangle*** is bordered superiorly by the mandible, inferiorly by the sternum, medially by the cervical midline, and laterally by the anterior border of the sternocleidomastoid muscle (see Figure 11.2). The ***posterior triangle*** is bordered inferiorly by the clavicle, anteriorly by the posterior border of the sternocleidomastoid muscle, and posteriorly by the anterior border of the trapezius muscle (see Figure 11.2). Subsidiary triangles exist within the two principal triangles.

Muscle	Origin	Insertion	Action
Sternocleidomastoid (*sternum* = breastbone; *cleido* = clavicle; *mastoid* = mastoid process of temporal bone) (label this muscle in Figure 10.12a)			
Semispinalis capitis (*semi* = half; *spine* = spinous process; *caput* = head) (label this muscle in Figure 10.17)			
Splenius capitis (*splenion* = bandage) (label this muscle in Figure 10.17)			
Longissimus capitis (*longissimus* = longest) (label this muscle in Figure 10.17)			

TABLE 10.9
MUSCLES THAT ACT ON THE ANTERIOR ABDOMINAL WALL (After completing the table, label Figure 10.9.)

Overview: The anterolateral abdominal wall is composed of skin, fascia, and four pairs of flat, sheetlike muscles: rectus abdominis, external oblique, internal oblique, and transversus abdominis. The anterior surfaces of the rectus abdominis muscles are interrupted by transverse fibrous bands of tissue called ***tendinous intersections***, believed to be remnants of septa that separated myotomes during embryonic development. The aponeuroses of the external oblique, internal oblique, and transversus abdominis muscles meet at the midline to form the ***linea alba*** (white line), a tough fibrous band that extends from the xiphoid process of the sternum to the symphysis pubis. The inferior free border of the external oblique aponeurosis, plus some collagenous fibers, forms the ***inguinal ligament***, which runs from the anterior superior iliac spine to the pubic tubercle (see Figure 10.18). The ligament demarcates the thigh and body wall.

Just superior to the medial end of the inguinal ligament is a triangular slit in the aponeurosis referred to as the ***superficial inguinal ring***, the outer opening of the ***inguinal canal***. The canal contains the spermatic cord and inguinal nerve in males and round ligament of the uterus and ilioinguinal nerve in females.

The posterior abdominal wall is formed by the lumbar vertebrae, parts of the ilia of the hipbones, psoas major muscle (described in Table 10.19), quadratus lumborum muscle, and iliacus muscle (also described in Table 10.19). Whereas the anterolateral abdominal wall is contractile and distensible, the posterior abdominal wall is bulky and stable by comparison.

Muscle	Origin	Insertion	Action
Rectus abdominis (*rectus* = fibers parallel to midline; *abdomino* = abdomen)			
External oblique *external* = closer to the surface; *oblique* = fibers diagonal to midline)			
Internal oblique (*internal* = farther from the surface)			
Transversus abdominis (*transverse* = fibers perpendicular to midline)			
Quadratus lumborum (*quad* = four; *lumbo* = lumbar region) (See Figure 10.17)			

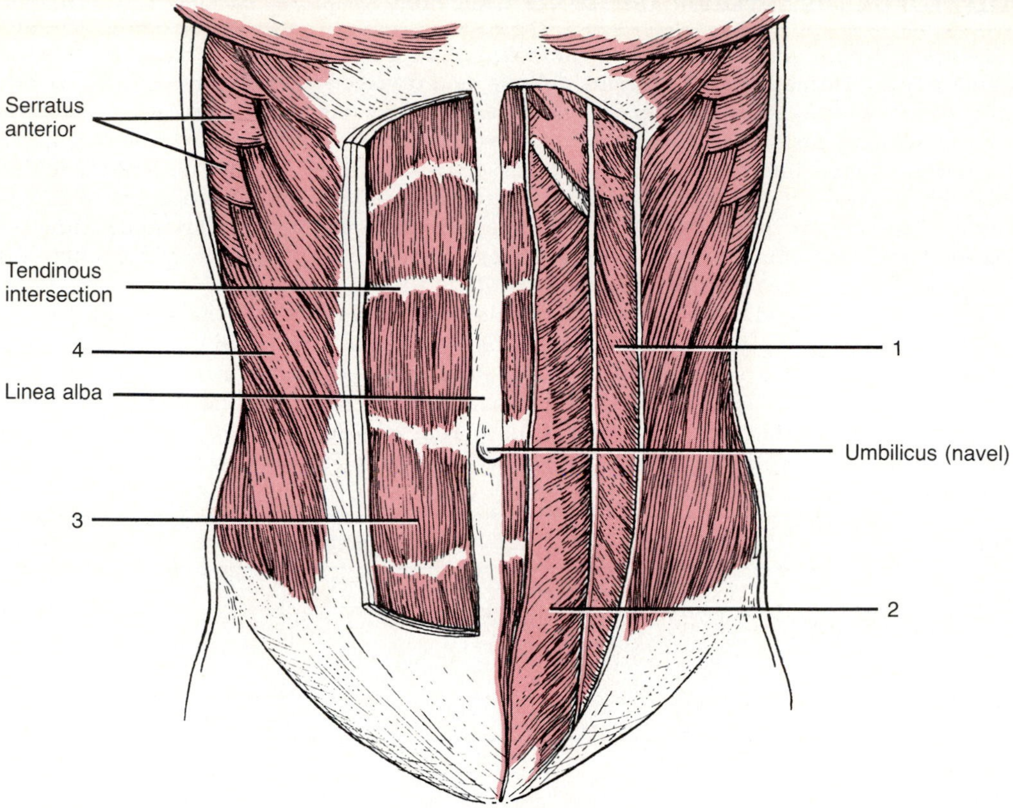

___ External oblique
___ Internal oblique
___ Rectus abdominis
___ Transversus abdominis

FIGURE 10.9 Muscles of anterior abdominal wall.

TABLE 10.10
MUSCLES USED IN BREATHING (After completing the table, label Figure 10.10.)

Overview: The muscles described here are attached to the ribs and by their contraction and relaxation alter the size of the thoracic cavity during normal breathing. In forced breathing, other muscles are involved as well. Essentially, inspiration occurs when the thoracic cavity increases in size. Expiration occurs when the thoracic cavity decreases in size.

Muscle	Origin	Insertion	Action
Diaphragm (*dia* = across, between; *phragma* = wall)			
External intercostals (*inter* = between; *costa* = rib)			
Internal intercostals (*internal* = farther from surface)			

TABLE 10.11
MUSCLES OF THE PELVIC FLOOR (After completing the table, label Figure 10.11.)

Overview: The muscles of the pelvic floor, together with the fascia covering their external and internal surfaces, are referred to as the ***pelvic diaphragm***. This diaphragm is funnel-shaped and forms the floor of the abdominopelvic cavity where it supports the pelvic viscera. It is pierced by the anal canal and urethra in both sexes and also by the vagina in the female.

Muscle	Origin	Insertion	Action
Levator ani (*levator* = raises; *ani* = anus)	This muscle is divisible into two parts, the pubococcygeus muscle and the iliococcygeus muscle.		
Pubococcygeus (*pubo* = pubis; *coccygeus* = coccyx)			
Iliococcygeus (*ilio* = ilium)			
Coccygeus*			

*Not illustrated

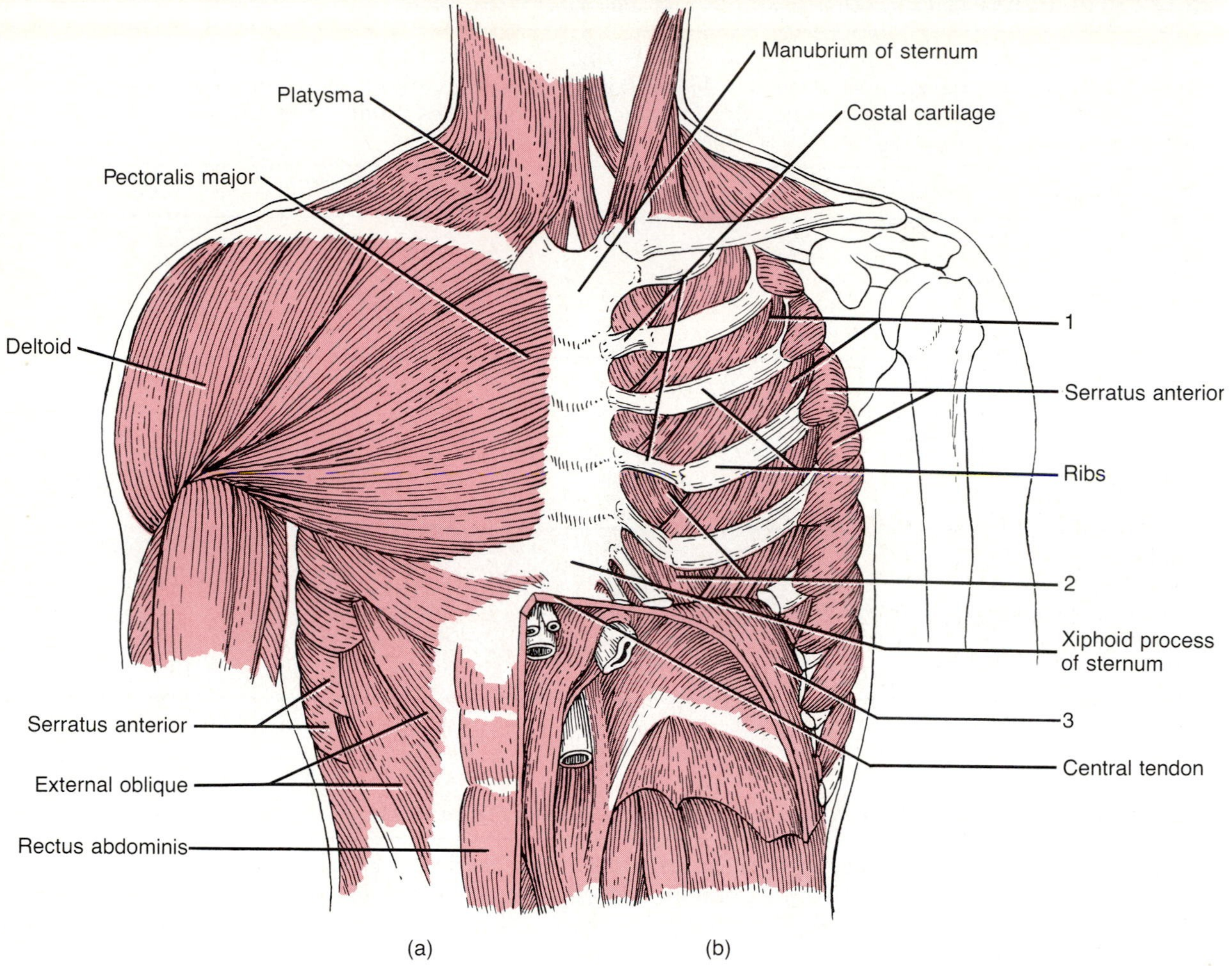

___ Diaphragm
___ External intercostals
___ Internal intercostals

FIGURE 10.10 Muscles used in breathing. (a) Superficial view. (b) Deep view.

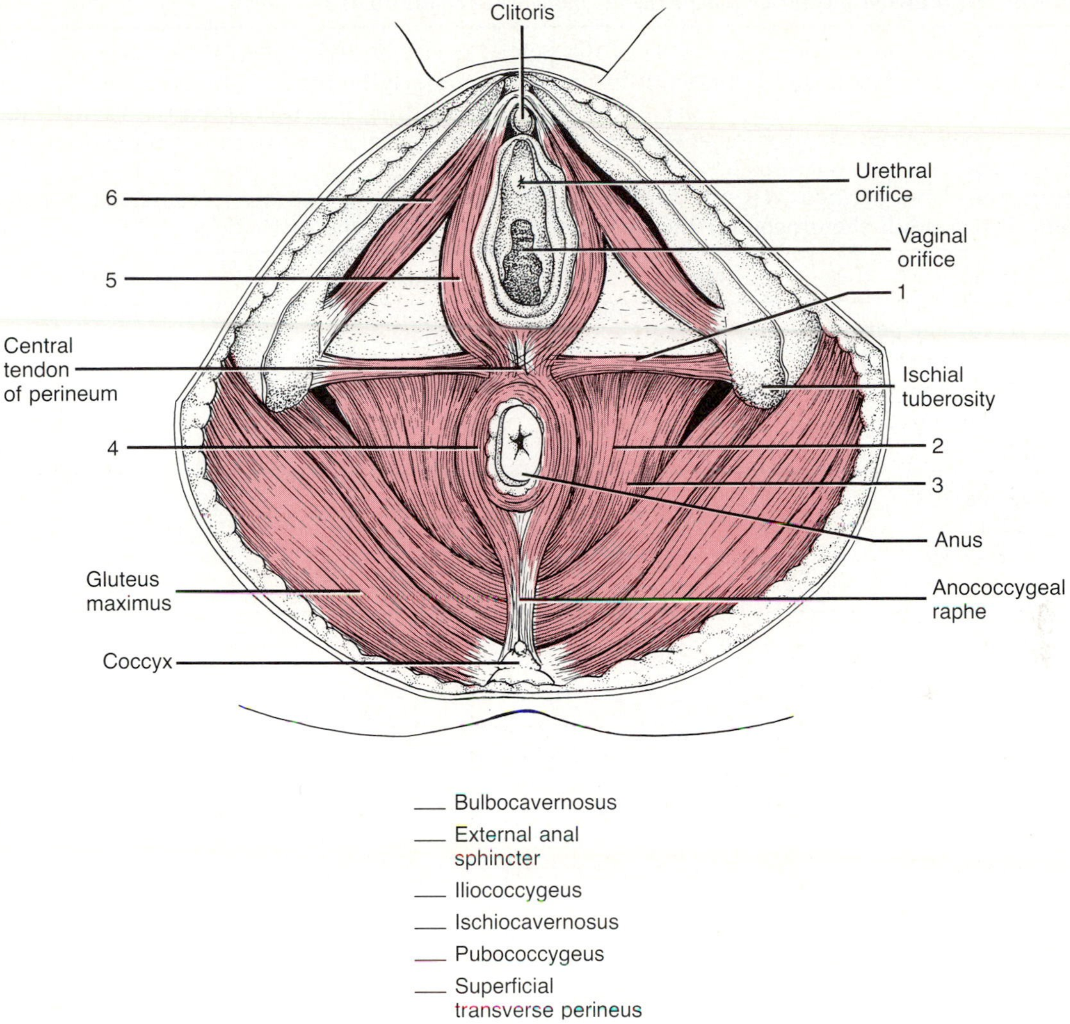

FIGURE 10.11 Muscles of female pelvic floor and perineum.

TABLE 10.12
MUSCLES OF THE PERINEUM (After completing the table, label Figure 10.11.)

Overview: The ***perineum*** is the entire outlet of the pelvis. It is a diamond-shaped area at the lower end of the trunk between the thighs and buttocks. It is bordered anteriorly by the symphysis pubis, laterally by the ischial tuberosities, and posteriorly by the coccyx. A transverse line drawn between the ischial tuberosities divides the perineum into an anterior ***urogenital triangle*** that contains the external genitals and a posterior ***anal triangle*** that contains the anus.

The deep transverse perineus, the urethral sphincter, and a fibrous membrane constitute the ***urogenital diaphragm.*** It surrounds the urogenital ducts and helps to strengthen the pelvic floor.

Muscle	Origin	Insertion	Action
Superficial transverse perineus (*superficial* = near surface; *transverse* = across; *perineus* = perineum)			
Bulbocavernosus (*bulbus* = bulb; *caverna* = hollow space)			
Ischiocavernosus (*ischion* = hip)			
Deep transverse perineus* (*deep* = farther from surface)			
Urethral sphincter* (*urethral* = pertaining to urethra; *sphincter* = circular muscle that decreases the size of an opening)			
External anal sphincter			

*Not illustrated

TABLE 10.13
MUSCLES THAT MOVE THE PECTORAL (SHOULDER) GIRDLE (After completing the table, label Figure 10.12.)

Overview: Muscles that move the pectoral (shoulder) girdle originate on the axial skeleton and insert on the clavicle or scapula. The muscles can be distinguished into ***anterior*** and ***posterior*** groups. The principal action of the muscles is to stabilize the scapula so that it can function as a stable point of origin for most of the muscles that move the arm (humerus).

Muscle	Origin	Insertion	Action
ANTERIOR MUSCLES			
Subclavius (*sub* = under; *clavius* = clavicle)			
Pectoralis minor (*pectus* = breast, chest, thorax; *minor* = lesser)			
Serratus anterior (*serratus* = serrated; *anterior* = front)			
POSTERIOR MUSCLES			
Trapezius (*trapezoides* = trapezoid shaped)			
Levator scapulae (*levator* = raises; *scapulae* = scapula)			
Rhomboideus major (*rhomboides* = rhomboid shaped or diamond-shaped)			
Rhomboideus minor			

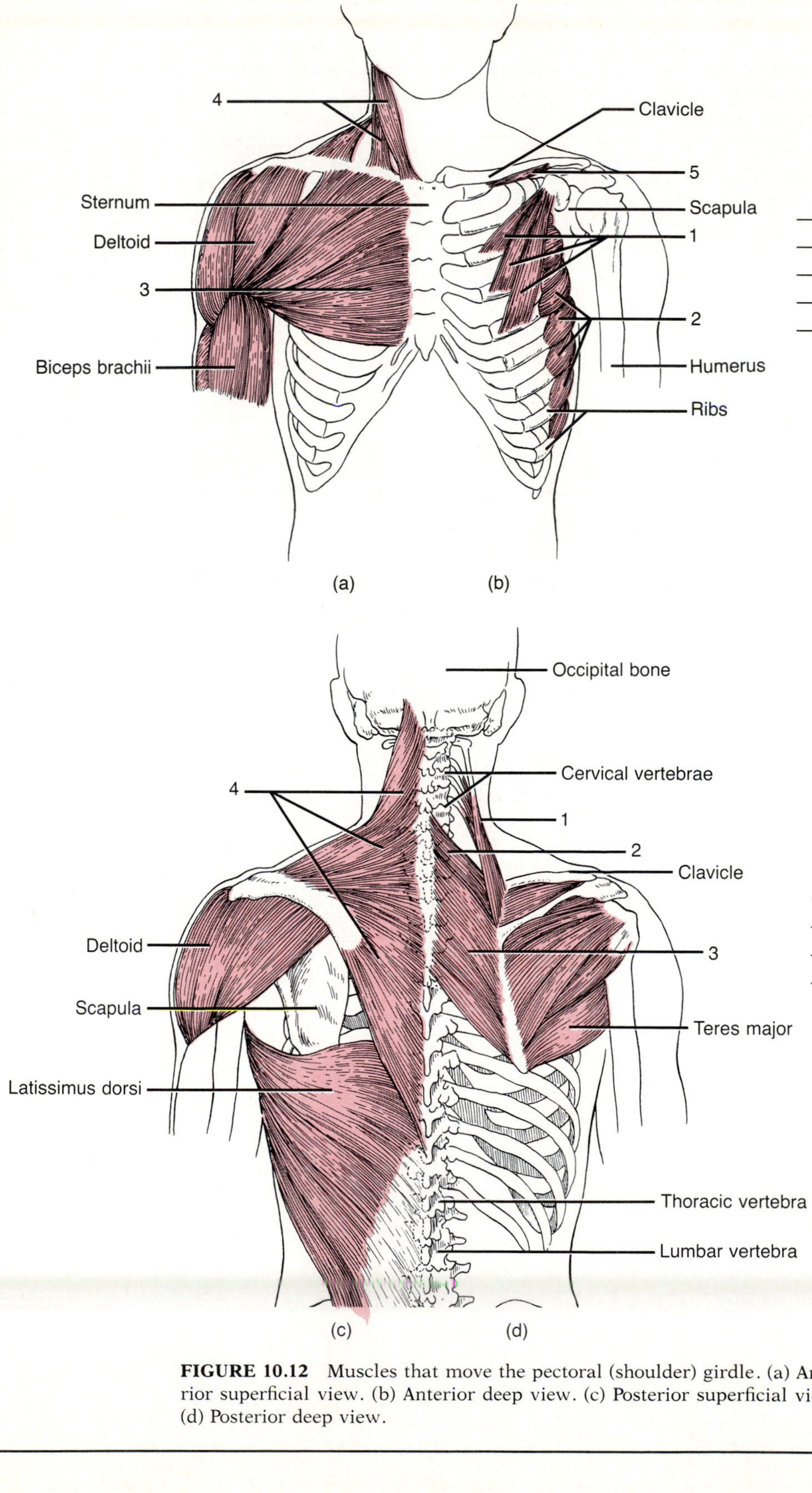

___ Pectoralis major
___ Pectoralis minor
___ Serratus anterior
___ Sternocleidomastoid
___ Subclavius

___ Levator scapulae
___ Rhomboideus major
___ Rhomboideus minor
___ Trapezius

FIGURE 10.12 Muscles that move the pectoral (shoulder) girdle. (a) Anterior superficial view. (b) Anterior deep view. (c) Posterior superficial view. (d) Posterior deep view.

TABLE 10.14
MUSCLES THAT MOVE THE ARM (After completing the table, label Figure 10.13.)

Overview: The muscles that move the arm (humerus) cross the shoulder joint. Of the nine muscles that cross the shoulder joint, only two of them (pectoralis major and latissimus dorsi) do not originate on the scapula. These two muscles are thus designated as ***axial muscles,*** since they originate on the axial skeleton. The remaining seven muscles, the ***scapular muscles,*** arise from the scapula.

The strength and stability of the shoulder joint are not provided by the shape of the articulating bones or its ligaments. Instead, four deep muscles of the shoulder and their tendons—subscapularis, supraspinatus, infraspinatus, and teres minor—strengthen and stabilize the shoulder joint. The muscles and their tendons are so arranged as to form a nearly complete circle around the joint. This arrangement is referred to as the ***rotator (musculotendinous) cuff*** and is a common site of injury to baseball pitchers, especially tearing of the supraspinatus muscle. The muscles of the rotator cuff are the principal rotators of the arm.

After you have studied the muscles in this table, arrange them according to the following actions: flexion, extension, abduction, adduction, medial rotation, and lateral rotation. (The same muscle can be used more than once.)

Muscle	Origin	Insertion	Action
AXIAL MUSCLES			
Pectoralis major (label this muscle in Figure 10.12)			
Latissimus dorsi (*dorsum* = back)			
SCAPULAR MUSCLES			
Deltoid (*delta* = triangular-shaped)			
Subscapularis (*sub* = below; *scapularis* = scapula)			
Supraspinatus (*supra* = above; *spinatus* = spine of scapula)			
Infraspinatus (*infra* = below)			
Teres major (*teres* = long and round)			
Teres minor			
Coracobrachialis (*coraco* = coracoid process)			

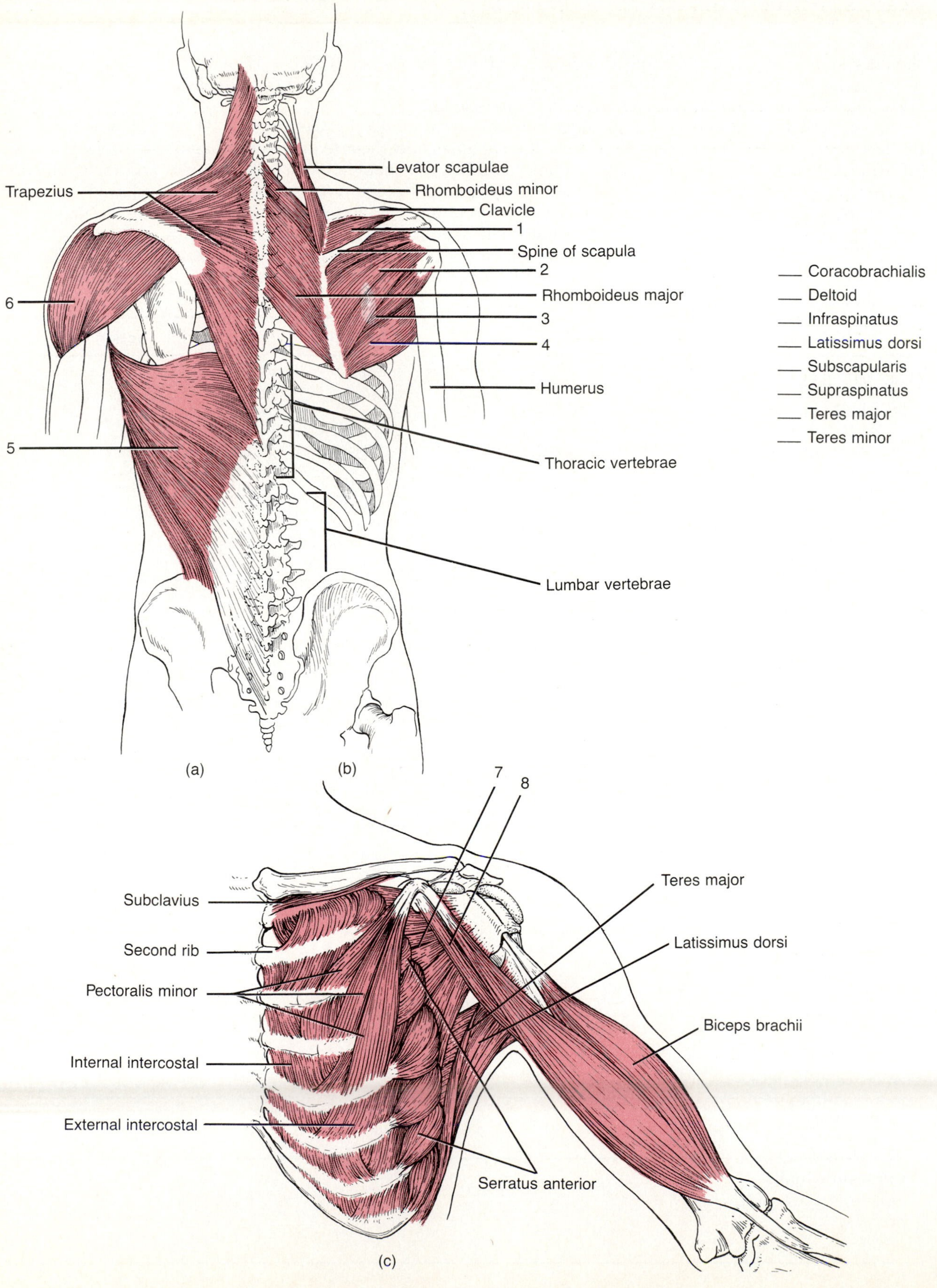

FIGURE 10.13 Muscles that move the arm. (a) Posterior superficial view. (b) Posterior deep view. (c) Anterior deep view.

TABLE 10.15
MUSCLES THAT MOVE THE FOREARM (After completing the table, label Figure 10.14.)

Overview: The muscles that move the forearm (radius and ulna) are divided into ***flexors*** and ***extensors***. Recall that the elbow joint is a hinge joint, capable only of flexion and extension under normal conditions. Whereas the biceps brachii, brachialis, and brachioradialis are flexors of the elbow joint, the triceps brachii and anconeus are extensors.

Muscle	Origin	Insertion	Action
FLEXORS			
Biceps brachii (*biceps* = two heads of origin; *brachion* = arm)			
Brachialis			
Brachioradialis (*radialis* = radius) (See Figure 10.15a also)			
EXTENSORS			
Triceps brachii (*triceps* = three heads of origin)			
Anconeus (*anconeal* = pertaining to the elbow) (label this muscle in Figure 10.15)			

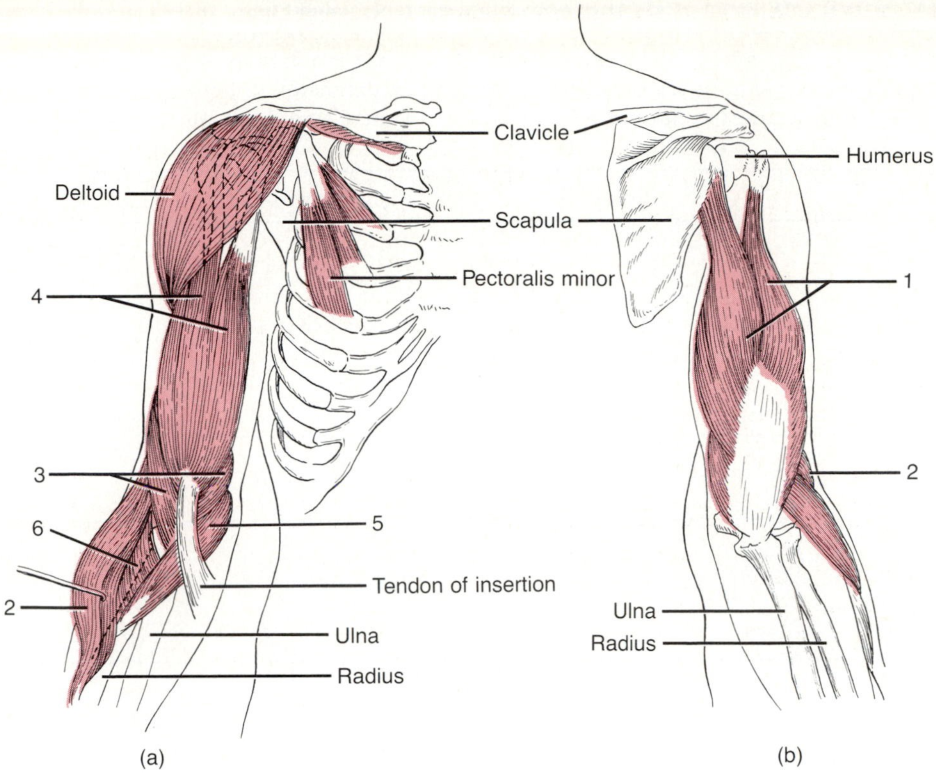

___ Biceps brachii
___ Brachialis
___ Brachioradialis
___ Pronator teres
___ Supinator
___ Triceps brachii

FIGURE 10.14 Muscles that move the forearm. (a) Anterior view. (b) Posterior view.

TABLE 10.16
MUSCLES THAT MOVE THE WRIST, HAND, AND FINGERS (Figure 10.15)

Overview: Muscles that move the wrist, hand, and fingers are many and varied. However, as you will see, their names for the most part give some indication of their origin, insertion, or action. On the basis of location and function, the muscles are divided into two groups—anterior and posterior. The ***anterior muscles*** function as flexors and pronators. They originate on the humerus and typically insert on the carpals, metacarpals, and phalanges. The bellies of these muscles form the bulk of the proximal forearm. The ***posterior muscles*** function as extensors and supinators. These muscles arise on the humerus and insert on the metacarpals and phalanges. Each of the two principal groups is also divided into superficial and deep muscles.

The tendons of the muscles of the forearm that attach to the wrist or continue into the hand, along with blood vessels and nerves, are held close to bones by strong fascial structures. The tendons are also surrounded by tendon sheaths. At the wrist, the deep fascia is thickened into fibrous bands called retinacula. The ***flexor retinaculum (transverse carpal ligament)*** is located over the palmar surface of the carpal bones. Through it pass the long flexor tendons of the digits and wrist and median nerve. The ***extensor retinaculum (dorsal carpal ligament)*** is located over the dorsal surface of the carpal bones. Through it pass the extensor tendons of the wrist and digits.

After you have studied the muscles in the table, arrange them according to the following actions: flexion, extension, abduction, adduction, supination, and pronation. (The same muscles can be used more than once.)

Muscle	Origin	Insertion	Action
ANTERIOR GROUP (flexors and pronators)			
Superficial			
Pronator teres (*pronation* = turning palm downward or posteriorly)			
Flexor carpi radialis (*flexor* = decreases angle at a joint; *carpus* = wrist; *radialis* = radius)			
Palmaris longus (*palma* = palm *longus* = long)			
Flexor carpi ulnaris (*ulnaris* = ulna)			
Flexor digitorum superficialis (*digit* = finger or toe; *superficialis* = closer to surface)			

TABLE 10.16 ***(Continued)***

Muscle	Origin	Insertion	Action
Deep			
Flexor digitorum profundus (*profundus* = deep)			
Flexor pollicis longus (*pollex* = thumb)			
Pronator quadratus (*quadratus* = squared, four-sided)			
POSTERIOR GROUP (extensors and supinators)			
Superficial			
Brachioradialis (See Table 10.15)			
Extensor carpi radialis longus (*extensor* = increases angle at a joint)			
Extensor carpi radialis brevis (*brevis* = short)			
Extensor digitorum			
Extensor digiti minimi (*minimi* = little finger)			
Extensor carpi ulnaris			
Anconeus (See Table 10.15)			

TABLE 10.16 *(Continued)*

Muscle	Origin	Insertion	Action
Deep			
Supinator (*supination* = palm upward or anteriorly)			
Abductor pollicis longus (*abductor* = moves a part away from midline)			
Extensor pollicis brevis			
Extensor pollicis longus			
Extensor indicis (*indicis* = index)			

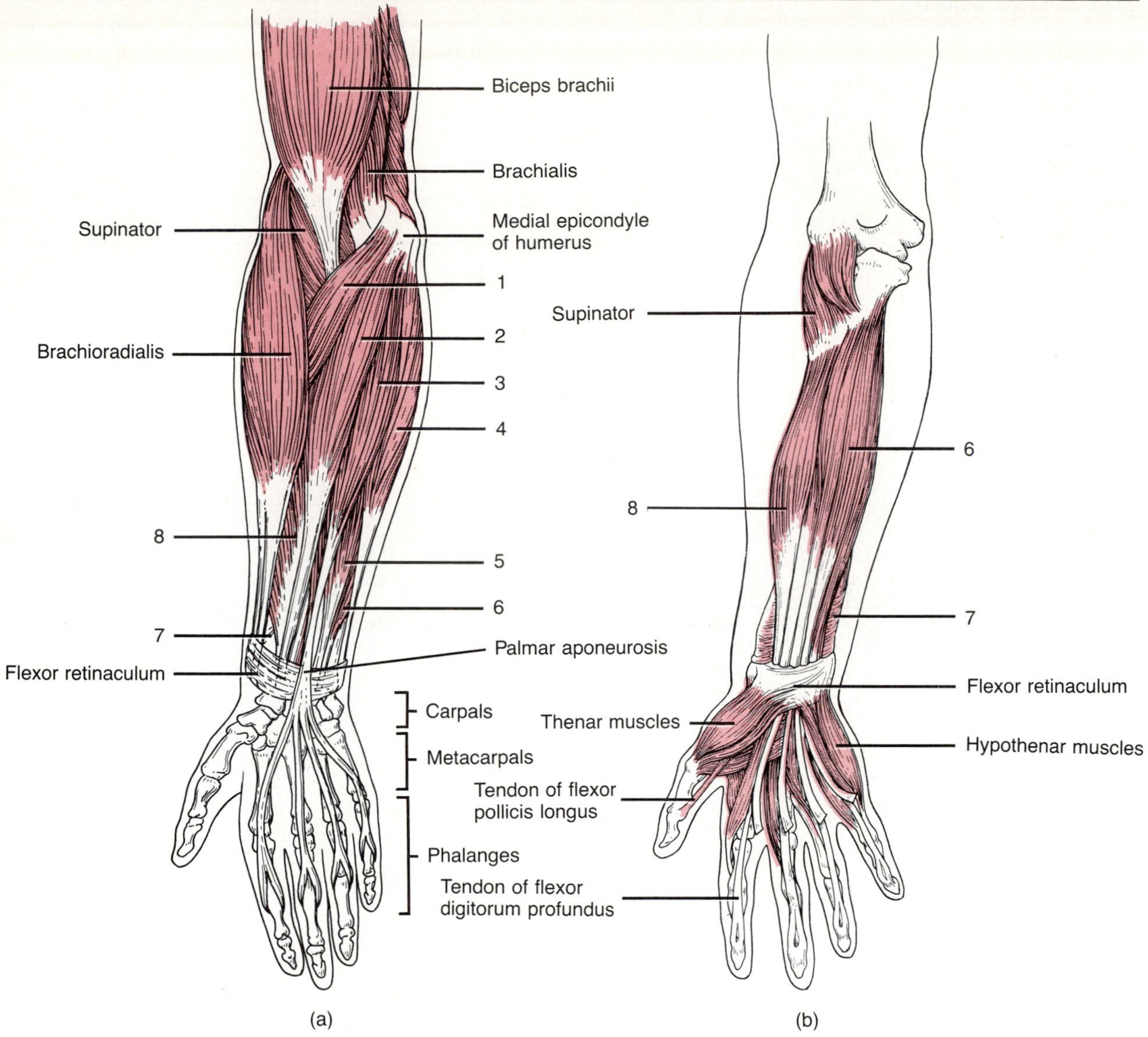

___ Flexor carpi radialis	___ Flexor pollicis longus
___ Flexor carpi ulnaris	___ Palmaris longus
___ Flexor digitorum profundus	___ Pronator quadratus
___ Flexor digitorum superficialis	___ Pronator teres

FIGURE 10.15 Muscles that move the wrist, hand, and fingers. (a) Superficial anterior view. (b) Deep anterior view.

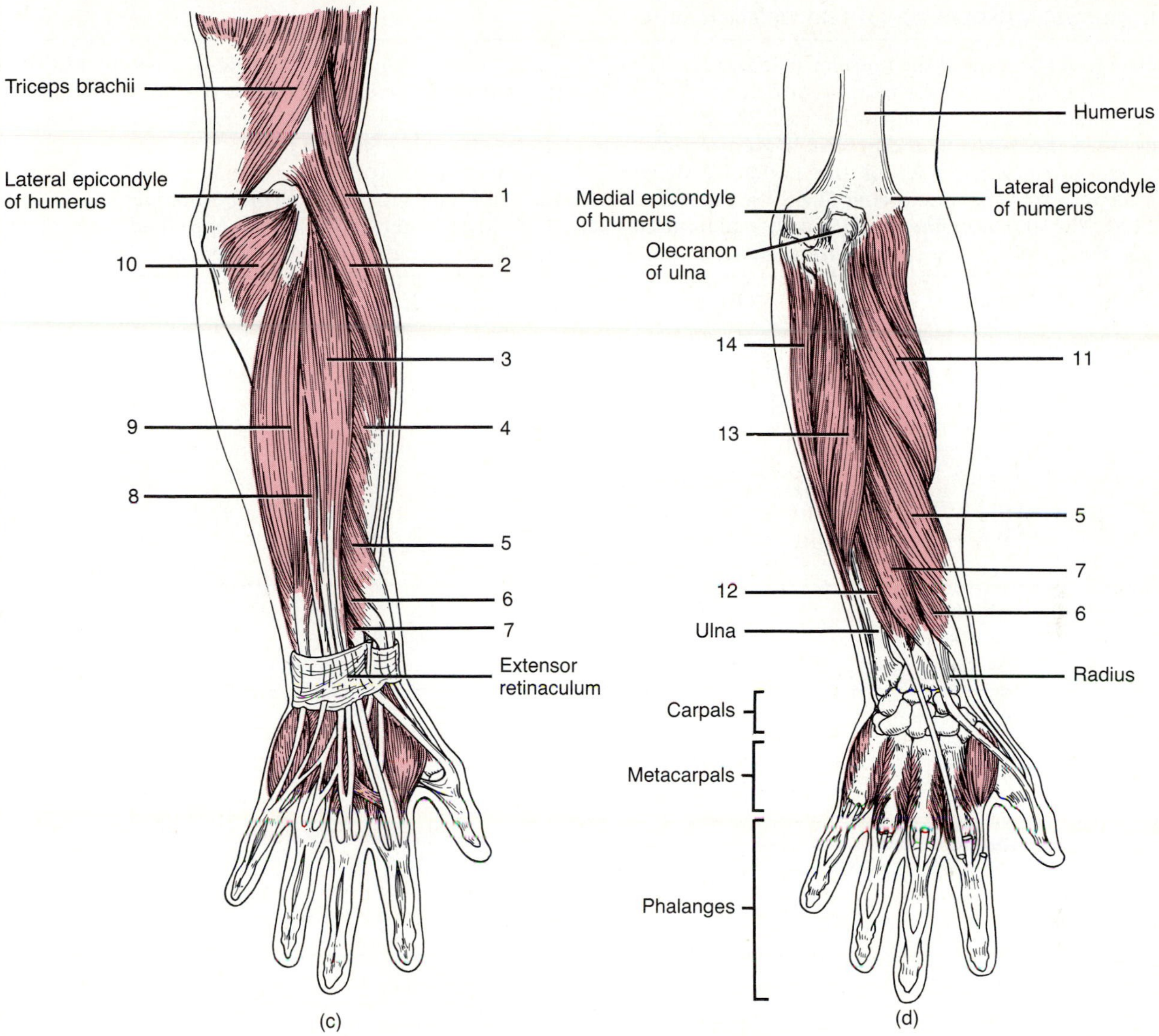

___ Abductor pollicis longus
___ Anconeus
___ Brachioradialis
___ Extensor carpi radialis brevis
___ Extensor carpi radialis longus
___ Extensor carpi ulnaris
___ Extensor digiti minimi
___ Extensor digitorum
___ Extensor indicis
___ Extensor pollicis brevis
___ Extensor pollicis longus
___ Flexor carpi ulnaris
___ Flexor digitorum profundus
___ Supinator

FIGURE 10.15 ***(Continued)*** Muscles that move the wrist, hand, and fingers. (c) Superficial posterior view. (d) Deep posterior view.

TABLE 10.17
INTRINSIC MUSCLES OF THE HAND (Figure 10.16)

Overview: Several of the muscles discussed in Table 10.16 help to move the digits in various ways. In addition, there are muscles in the palmar surface of the hand called ***intrinsic muscles*** that also help to move the digits. Such muscles are so named because their origins and insertions are both within the hands. These muscles assist in the intricate and precise movements that are charcteristic of the human hand.

The intrinsic muscles of the hand are divided into three principal groups—thenar, hypothenar, and intermediate. The four ***thenar*** (THĒ-nar) ***muscles*** act on the thumb and form the ***thenar eminence*** (see Figure 11.8). The four ***hypothenar*** (HĪ-pō-thē-nar) ***muscles*** act on the little finger and form the ***hypothenar eminence*** (see Figure 11.8). The eleven ***intermediate*** (midpalmar) ***muscles*** act on all the digits, except the thumb.

Muscle	Origin	Insertion	Action
THENAR MUSCLES			
Abductor pollicis brevis (*abductor* = moves part away from midline; *pollex* = thumb; *brevis* = short)			
Opponens pollicis (*opponens* = opposes)			
Flexor pollicis brevis (*flexor* = decreases angle at joint)			
Adductor pollicis (*adductor* = moves part toward midline)			
HYPOTHENAR MUSCLES			
Palmaris brevis (*palma* = palm)			
Abductor digiti minimi (*digit* = finger or toe; *minimi* = little finger)			
Flexor digiti minimi brevis			
Opponens digiti minimi			

TABLE 10.17 ***(Continued)***

Muscle	Origin	Insertion	Action
INTERMEDIATE (MIDPALMAR) MUSCLES			
Lumbricals (four muscles)			
Dorsal interossei (four muscles; *dorsal* = back surface; *inter* = between; *ossei* = bones)			
Palmar interossei (three muscles)			

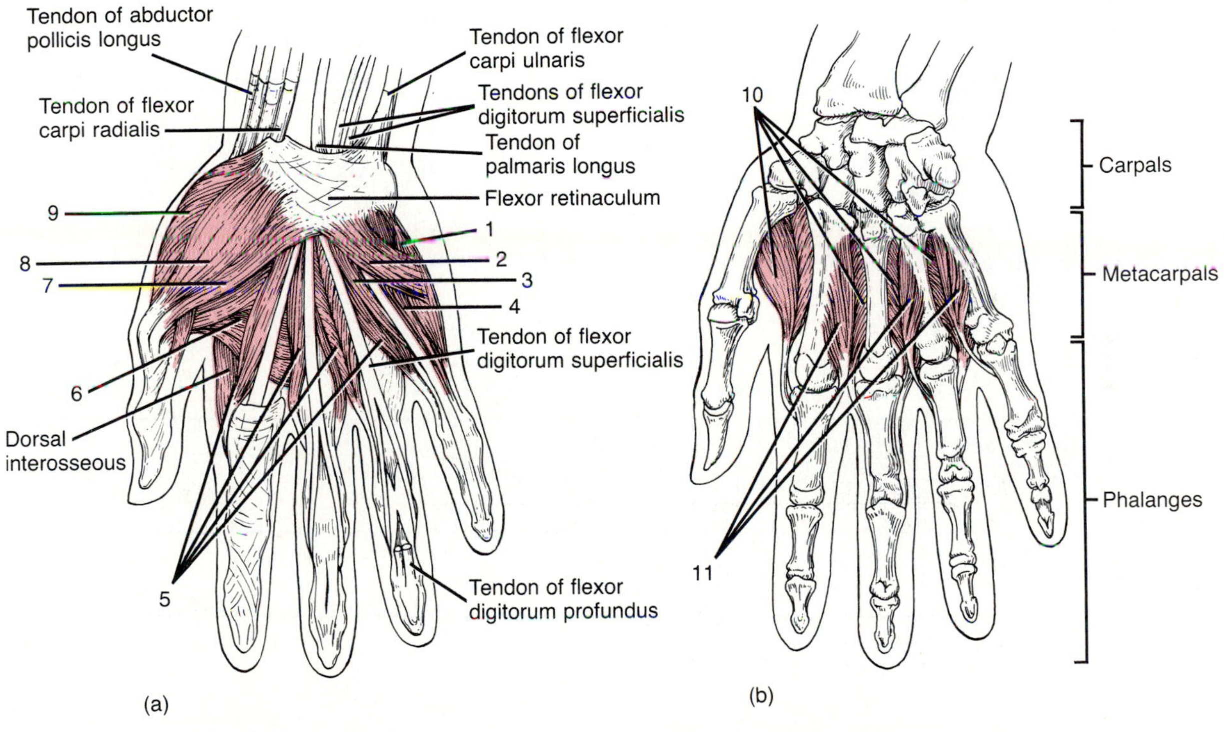

___ Abductor digiti minimi
___ Abductor pollicis brevis
___ Adductor pollicis
___ Dorsal interossei
___ Flexor digiti minimi brevis
___ Flexor pollicis brevis

___ Lumbricals
___ Opponens digiti minimi
___ Opponens pollicis
___ Palmar interossei
___ Palmaris brevis

FIGURE 10.16 Intrinsic muscles of the hand. (a) Superficial. (b) Deep.

TABLE 10.18
MUSCLES THAT MOVE THE VERTEBRAL COLUMN (Figure 10.17)

Overview: The muscles that move the vertebral column are quite complex becuase they have multiple origins and insertions and there is considerable overlapping among them. One way to group the muscles is on the basis of the direction of the muscle bundles and their approximate lengths. For example, the ***splenius muscles*** arises from the midline and run laterally and superiorly to their insertions. The ***erector spinae (sacrospinalis) muscle*** arises from either the midline or more laterally, but usually runs almost longitudinally, with neither a marked outward nor inward direction as it is traced superiorly. The ***transversospinalis muscles*** arise laterally, but run toward the midline as they are traced superiorly. Deep to these three muscle groups are small ***segmental muscles*** that run between spinous processes or transverse processes of vertebrae. Since the scalene muscles also assist in moving the vertebral column, they are included in this table.

Note in Table 10.9 that the rectus abdominis and quadratus lumborum muscles also assume a role in moving the vertebral column.

Muscle	Origin	Insertion	Action
SPLENIUS MUSCLES			
Splenius capitis (*splenium* = bandage; *caput* = head)			
Splenius cervicis (*cervix* = neck)			
ERECTOR SPINAE (SACROSPINALIS) This is the largest muscular mass of the back and consists of three groupings—iliocostalis, longissimus, and spinalis. These groups, in turn, consist of a series of overlapping muscles. The iliocostalis group is laterally placed, the longissimus group is intermediate in placement, and the spinalis group is medially placed.			
Iliocostalis (lateral) group			
Iliocostalis lumborum (*ilium* = flank; *costa* = rib)			
Iliocostalis thoracis (*thorax* = chest)			
Iliocostalis cervicis			
Longissimus (intermediate) group			
Longissimus thoracis (*longissimus* = longest)			
Longissimus cervicis			
Longissimus capitis			

TABLE 10.18 ***(Continued)***

Muscle	Origin	Insertion	Action
Spinalis (medial) group			
Spinalis thoracis (*spinalis* = vertebral column)			
Spinalis cervicis			
Spinalis capitis			
TRANSVERSOSPINALIS MUSCLES			
Semispinalis thoracis (*semi* = partially or one half)			
Semispinalis cervicis			
Semispinalis capitis			
Multifidus (*multi* = many; *findere* = to split)			
Rotatores (*rotate* = turn on an axis)			
SEGMENTAL MUSCLES			
Interspinales (*inter* = between)			
Intertransversarii			
SCALENE MUSCLES			
Anterior scalene (*anterior* = front; *skalenos* = uneven)			
Middle scalene			
Posterior scalene (*posterior* = back)			

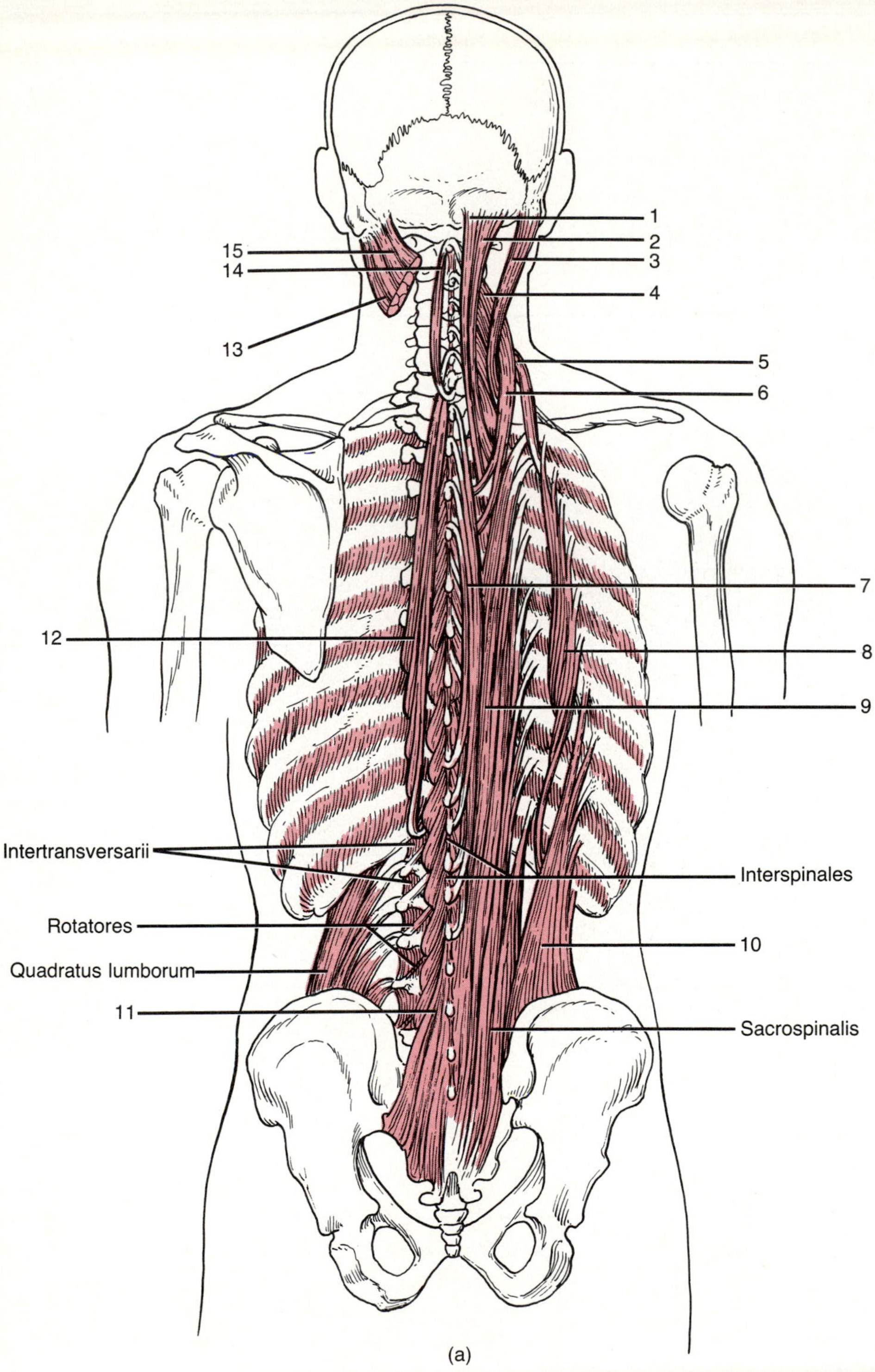

___ Iliocostalis cervicis
___ Iliocostalis lumborum
___ Iliocostalis thoracis
___ Longissimus capitis
___ Longissimus cervicis
___ Longissimus thoracis
___ Multifidus
___ Semispinalis capitis
___ Semispinalis cervicis
___ Semispinalis thoracis
___ Spinalis capitis
___ Spinalis cervicis
___ Spinalis thoracis
___ Splenius capitis
___ Splenius cervicis

FIGURE 10.17 Muscles that move the vertebral column. (a) Posterior view of back muscles. Many superficial muscles have been removed or cut to show deep muscles.

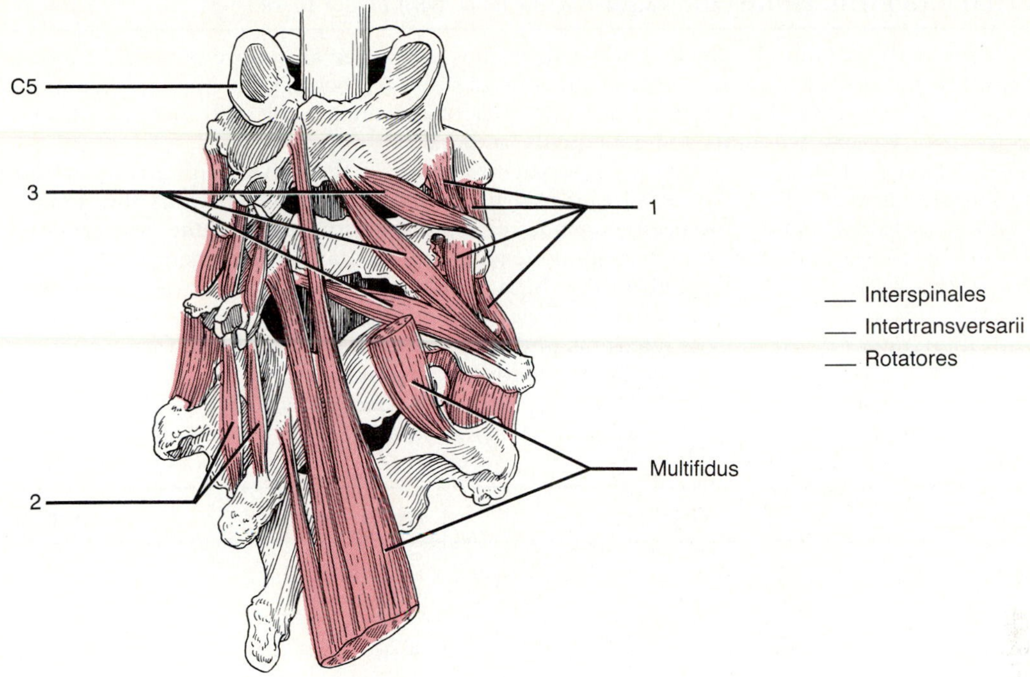

(b)

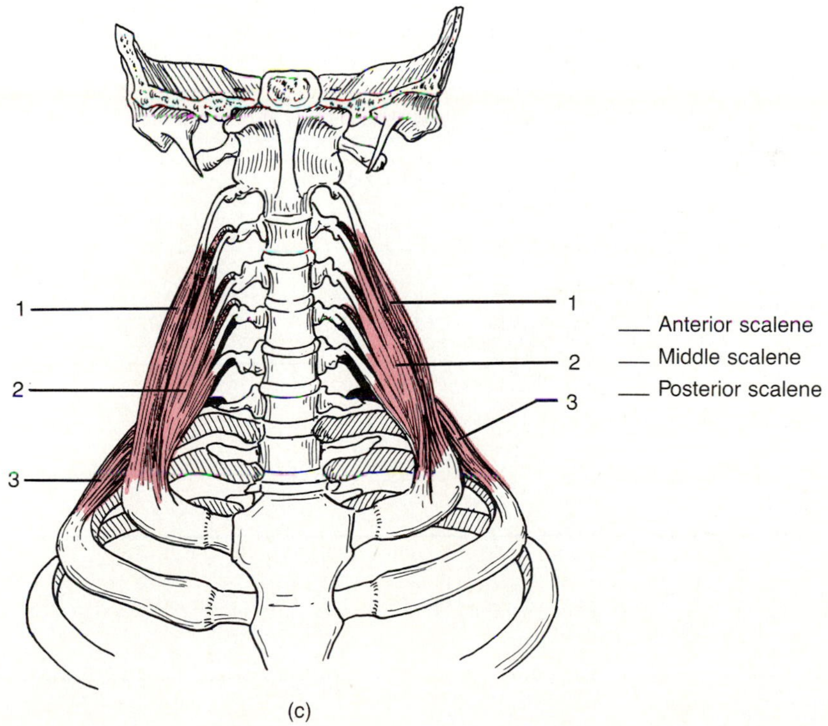

(c)

FIGURE 10.17 ***(Continued)*** Muscles that move the vertebral column. (b) Posterolateral view of several vertebrae with their associated muscles. (c) Anterior view of scalene muscles.

TABLE 10.19
MUSCLES THAT MOVE THE THIGH (After completing the table, label Figure 10.18.)

Overview: As you will see, muscles of the lower extremities are larger and more powerful than those of the upper extremities since lower extremity muscles function in stability, locomotion, and maintenance of posture. Upper extremity muscles are characterized by versatility of movement. In addition, muscles of the lower extremities frequently cross two joints and act equally on both.

The majority of muscles that act on the thigh (femur) originate on the pelvic (hip) girdle and insert on the femur. The anterior muscles are the psoas major and iliacus, together referred to as the iliopsoas muscle. The remaining muscles (except for the pectineus, adductors, and tensor fasciae latae) are posterior muscles. Technically, the pectineus and adductors are components of the medial compartment of the thigh, but they are included in this table because they act on the thigh. The tensor fasciae latae muscle is laterally placed. The ***fascia lata*** is a deep fascia of the thigh that encircles the entire thigh. It is well-developed laterally where, together with the tendons of the gluteus maximus and tensor fasciae latae muscles, forms a structure called the ***iliotibial tract***. The tract inserts into the lateral condyle of the tibia.

After you have studied the muscles in this table, arrange them according to the following actions: flexion, extension, abduction, adduction, medial rotation, and lateral rotation. (The same muscles can be used more than once.)

Muscle	Origin	Insertion	Action
Psoas major (*psoa* = muscle of loin)			
Iliacus (*iliac* = ilium)			
Gluteus maximus (*glutos* = buttock; *maximus* = largest)			
Gluteus medius (*media* = middle)			
Gluteus minimus (*minimus* = smallest)			
Tensor fasciae latae (*tensor* = makes tense; *fascia* = band; *latus* = wide)			
Piriformis (*pirum* = pear; *forma* = shape)			
Obturator internus* (*obturator* = obturator foramen; *internus* = inside)			
Obturator externus (*externus* = outside)			

TABLE 10.19 *(Continued)*

Muscle	Origin	Insertion	Action
Superior gemellus (*superior* = above; *gemellus* = twins)			
Inferior gemellus (*inferior* = below)			
Quadratus femoris (*quad* = four; *femoris* = femur)			
Adductor longus (*adductor* = moves part closer to midline; *longus* = long)			
Adductor brevis (*brevis* = short)			
Adductor magnus (*magnus* = large)			
Pectineus (*pecten* = comb-shaped)			

*Not illustrated

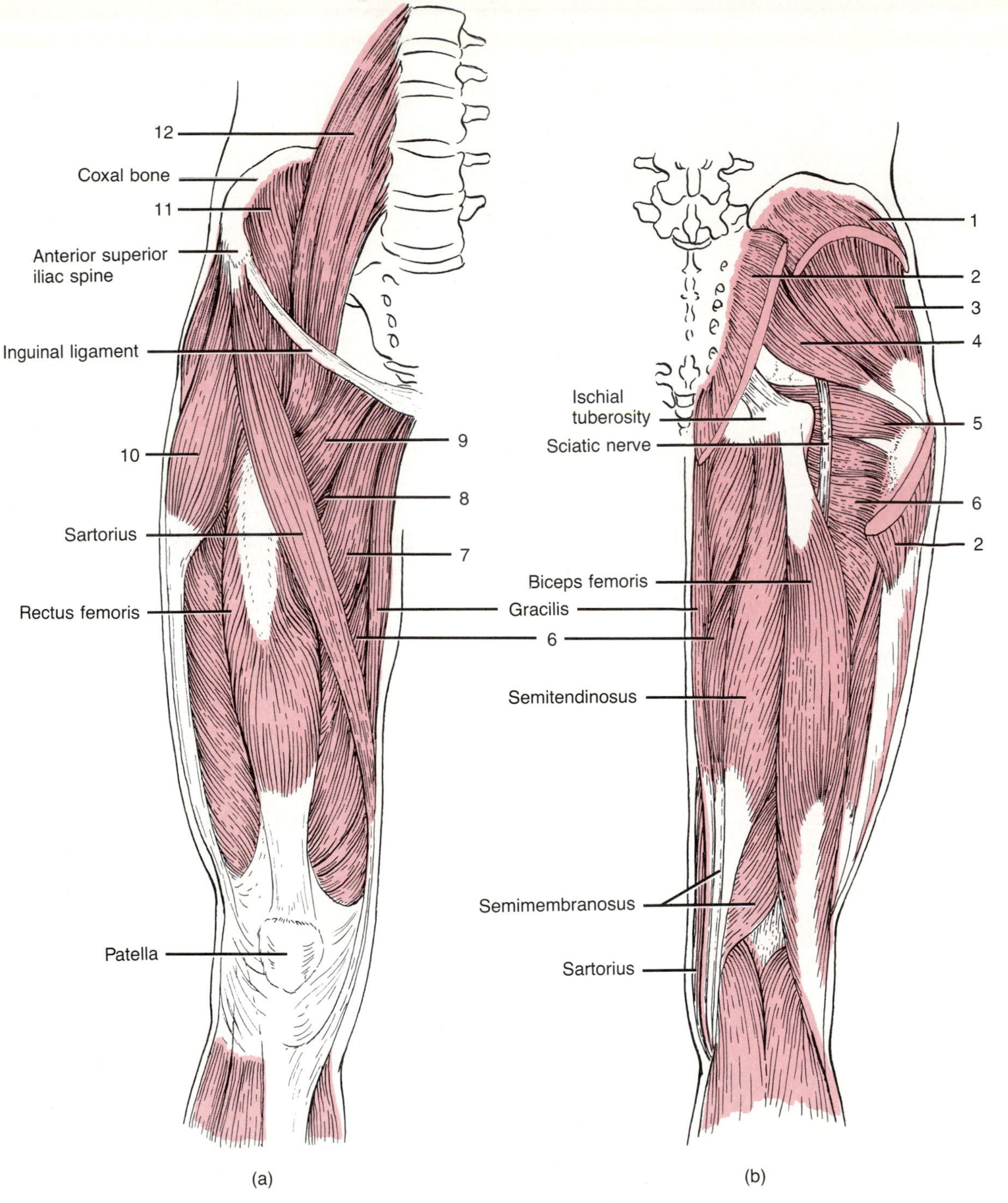

___ Adductor brevis	___ Gluteus medius	___ Obturator externus	___ Psoas major
___ Adductor longus	___ Gluteus minimus	___ Obturator internus	___ Quadratus femoris
___ Adductor magnus	___ Iliacus	___ Pectineus	___ Superior gemellus
___ Gluteus maximus	___ Inferior gemellus	___ Piriformis	___ Tensor fasciae latae

FIGURE 10.18 Muscles that move the thigh. (a) Superficial anterior view. (b) Superficial posterior view.

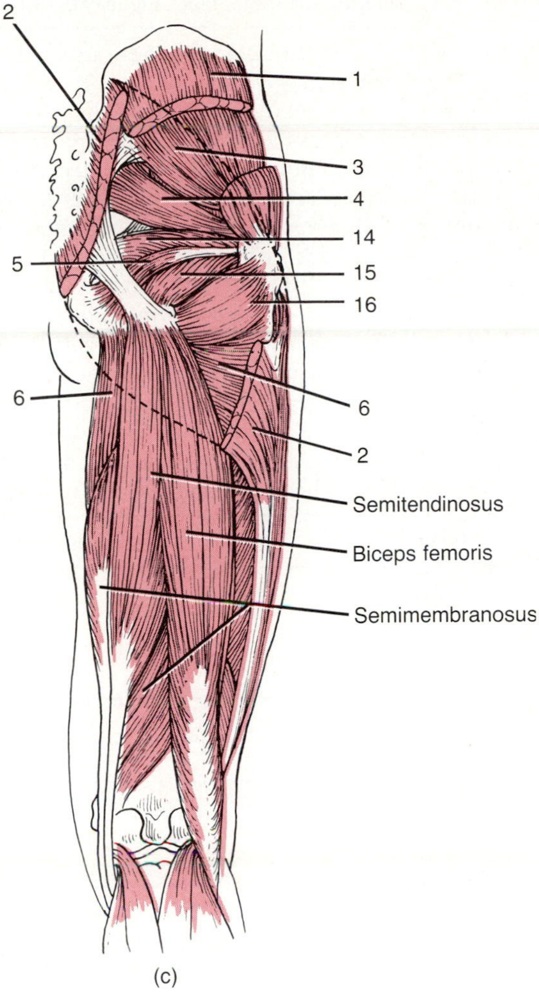

___ Adductor magnus
___ Gluteus maximus
___ Gluteus medius
___ Gluteus minimus
___ Inferior gemellus
___ Obturator internus
___ Piriformis
___ Quadratus femoris
___ Superior gemellus

FIGURE 10.18 ***(Continued)*** (c) Deep posterior view.

TABLE 10.20
MUSCLES THAT ACT ON THE LEG (After completing the table, label Figure 10.19.)

Overview: The muscles that act on the leg (tibia and fibula) originate in the hip and thigh and are separated into compartments by deep fascia. The ***medial (adductor) compartment*** is so named because its muscles adduct the thigh. It is innervated by the obturator nerve. As noted earlier, the adductor magnus, adductor longus, adductor brevis, and pectineus muscles, components of the medial compartment, are included in Table 10.19 because they act on the femur. The gracilis, the other muscle in the medial compartment, not only adducts the thigh but also flexes the leg. For this reason, it is included in this table.

The ***anterior (extensor) compartment*** is so designated because its muscles act to extend the leg. It is composed of the quadriceps femoris and sartorius muscles and is innervated by the femoral nerve. The quadriceps femoris muscle is a composite muscle that includes four distinct parts, usually described as four separate muscles (rectus femoris, vastus lateralis, vastus intermedius, and vastus medialis). The common tendon for the four muscles is known as ***patellar ligament*** and attaches to the tibial tuberosity.

The ***posterior (flexor) compartment*** is so named because its muscles flex the leg. It is innervated by branches of the sciatic nerve. Included are the hamstrings (biceps femoris, semitendinosus, and semimembranosus). The hamstrings are so named because their tendons are long and stringlike in the popliteal area. The ***popliteal fossa*** is a diamond-shaped space on the posterior aspect of the knee bordered laterally by the biceps femoris and medially by the semitendinosus and semimembranosus muscles (see also Figure 11.10b).

Muscle	Origin	Insertion	Action
MEDIAL (ADDUCTOR) COMPARTMENT			
Adductor magnus			
Adductor longus			
Adductor brevis			
Pectineus			
Gracilis (*gracilis* = slender)			
ANTERIOR (EXTENSOR) COMPARTMENT			
Quadriceps femoris (*quadriceps* = four heads of origin; *femoris* = femur)			

TABLE 10.20 ***(Continued)***

Muscle	Origin	Insertion	Action
Rectus femoris (*rectus* = fibers parallel to midline)			
Vastus lateralis (*vastus* = large; *lateralis* = lateral)			
Vastus medialis (*medialis* = medial)			
Vastus intermedius (*intermedius* = middle)			
Sartorius (*sartor* = tailor; refers to cross-legged position of tailors)			
POSTERIOR (FLEXOR) COMPARTMENT			
Hamstrings			
Biceps femoris (*biceps* = two heads of origin)			
Semitendinosus (*semi* = half; *tendo* = tendon)			
Semimembranosus (*membran* = membrane)			

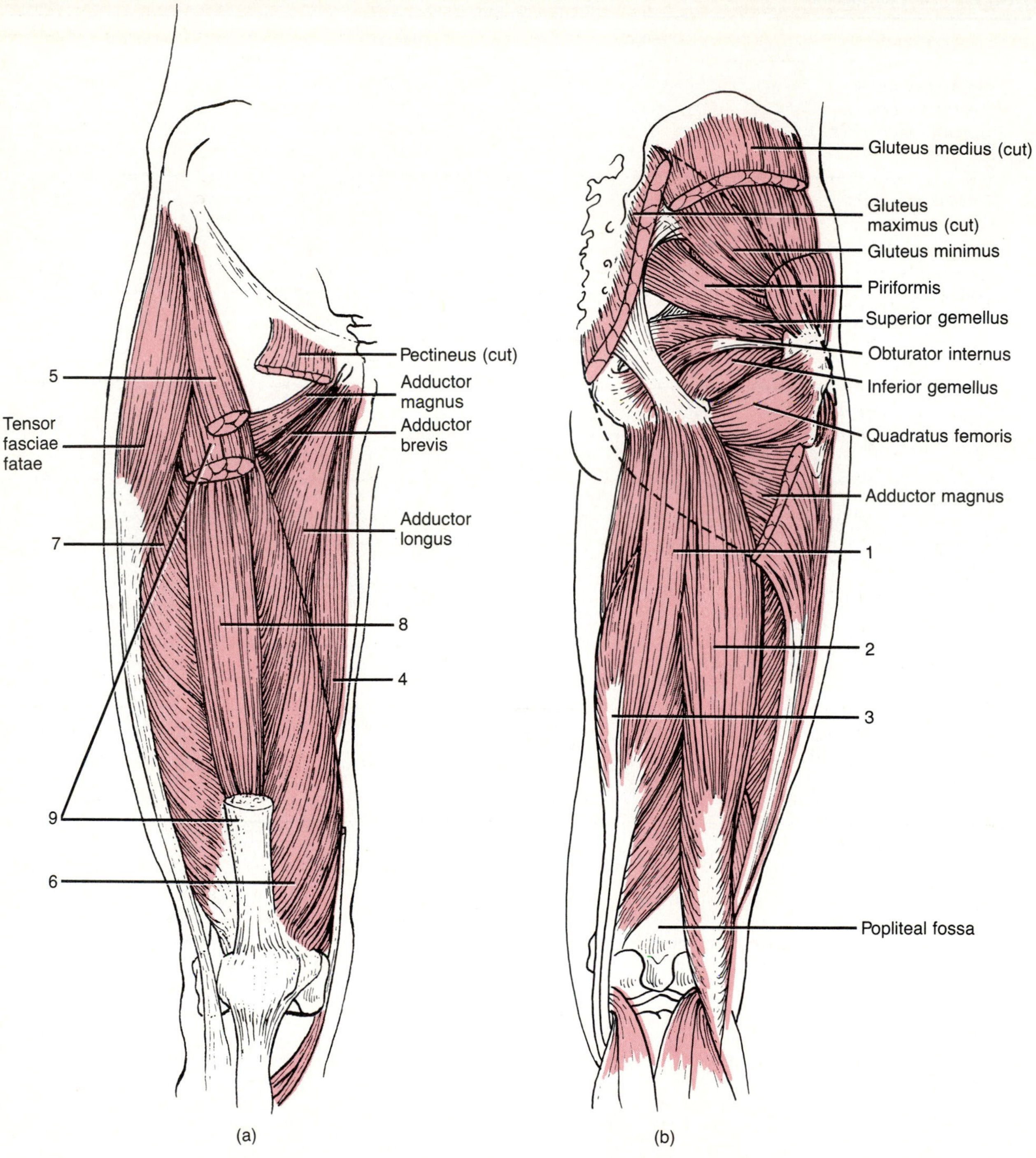

___ Biceps femoris
___ Gracilis
___ Rectus femoris
___ Sartorius
___ Semimembranosus
___ Semitendinosus
___ Vastus intermedius
___ Vastus lateralis
___ Vastus medialis

FIGURE 10.19 Muscles that act on the leg. (a) Anterior deep view. (b) Posterior deep view.

TABLE 10.21
MUSCLES THAT MOVE THE FOOT AND TOES (Figure 10.20)

Overview: The musculature of the leg, like that of the thigh, can be distinguished into three compartments by deep fascia. In addition, all the muscles in the respective compartments are innervated by the same nerve. The ***anterior compartment*** consists of muscles that dorsiflex the foot and are innervated by the deep peroneal nerve. In a situation analogous to the wrist, the tendons of the muscles of the anterior compartment are held firmly to the ankle by thickenings of deep fascia called the ***superior extensor retinaculum*** (***transverse ligament of the ankle***) and ***inferior extensor retinaculum*** (***cruciate ligament of the ankle***).

The ***lateral*** (***peroneal***) ***compartment*** contains two muscles that plantar flex and evert the foot. They are supplied by the superficial peroneal nerve.

The ***posterior compartment*** consists of muscles that are divisible into superficial and deep groups. All are innervated by the tibial nerve. All three superficial muscles share a common tendon of insertion, the calcaneal (Achilles) tendon that inserts into the calcaneal bone of the ankle. The superficial muscles are plantar flexors of the foot. Of the four deep muscles, three plantar flex the foot.

Muscle	Origin	Insertion	Action
ANTERIOR COMPARTMENT			
Tibialis anterior (*tibialis* = tibia; *anterior* = front)			
Extensor hallucis longus (*extensor* = increases angle at joint; *hallucis* = hallux or great toe; *longus* = long)			
Extensor digitorum longus			
Peroneus tertius (*perone* = fibula; *tertius* = third)			
LATERAL (PERONEAL) COMPARTMENT			
Peroneus longus			
Peroneus brevis (*brevis* = short)			

TABLE 10.21 *(Continued)*

Muscle	Origin	Insertion	Action
POSTERIOR COMPARTMENT			
Superficial			
Gastrocnemius (*gaster* = belly; *kneme* = leg)			
Soleus (*soleus* = sole of foot)			
Plantaris (*plantar* = sole of foot)			
Deep			
Popliteus (*poples* = posterior surface of knee)			
Flexor hallucis longus (*flexor* = decreases angle at joint)			
Flexor digitorum longus (*digitorum* = finger or toe)			
Tibialis posterior (*posterior* = back)			

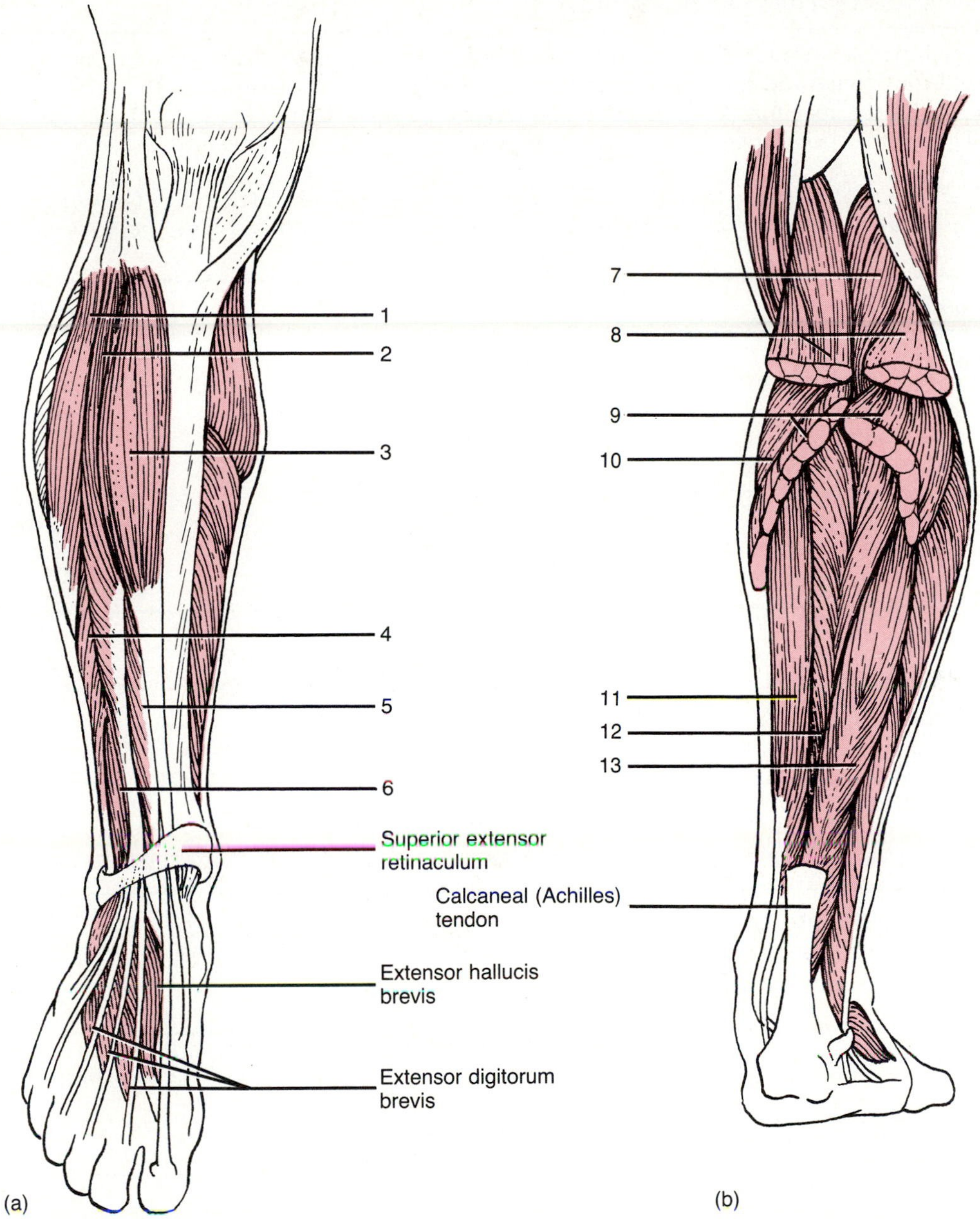

___ Extensor digitorum longus
___ Extensor hallucis longus
___ Flexor digitorum longus
___ Flexor hallucis longus
___ Gastrocnemius
___ Peroneus brevis

___ Peroneus longus
___ Peroneus tertius
___ Plantaris
___ Popliteus
___ Soleus
___ Tibialis anterior
___ Tibialis posterior

FIGURE 10.20 Muscles that move the foot and toes. (a) Anterior superficial view. (b) Posterior deep view.

TABLE 10.22
INTRINSIC MUSCLES OF THE FOOT (Figure 10.21)

Overview: The intrinsic muscles of the foot are comparable to those in the hand. Where the muscles of the hand are specialized for precise and intricate movements, those of the foot are limited to support and locomotion. The deep fascia of the foot forms the ***plantar aponeurosis (fascia),*** that extends from the calcaneus to the phalanges. The aponeurosis supports the longitudinal arch of the foot and transmits the flexor tendons of the foot.

The intrinsic musculature of the foot is divided into two groups—dorsal and plantar. There is only one ***dorsal muscle***. The ***plantar muscles*** are arranged in four layers, the most superficial layer being referred to as the first layer.

Muscle	Origin	Insertion	Action
DORSAL MUSCLE			
Extensor digitorum brevis (*extensor* = increases angle at joint; *digit* = finger or toe; *brevis* = short) (Illustrated in Figure 10.20a)			
PLANTAR MUSCLES			
First (superficial) layer			
Abductor hallucis (*abductor* = moves part away from midline; *hallucis* = hallux or great toe)			
Flexor digitorum brevis (*flexor* = decreases angle at joint)			
Abductor digiti minimi (*minimi* = small toe)			
Second layer			
Quadratus plantae (*quad* = four; *planta* = sole of foot)			
Lumbricals (four muscles)			

TABLE 10.22 *(Continued)*

Muscle	Origin	Insertion	Action
Third layer			
Flexor hallucis brevis			
Adductor hallucis (*adduct* = moves part closer to midline)			
Flexor digiti minimi brevis			
Fourth (deep) layer			
Dorsal interossei (four muscles; *dorsal* = back surface; *inter* = between; *ossei* = bone)			
Plantar interossei			

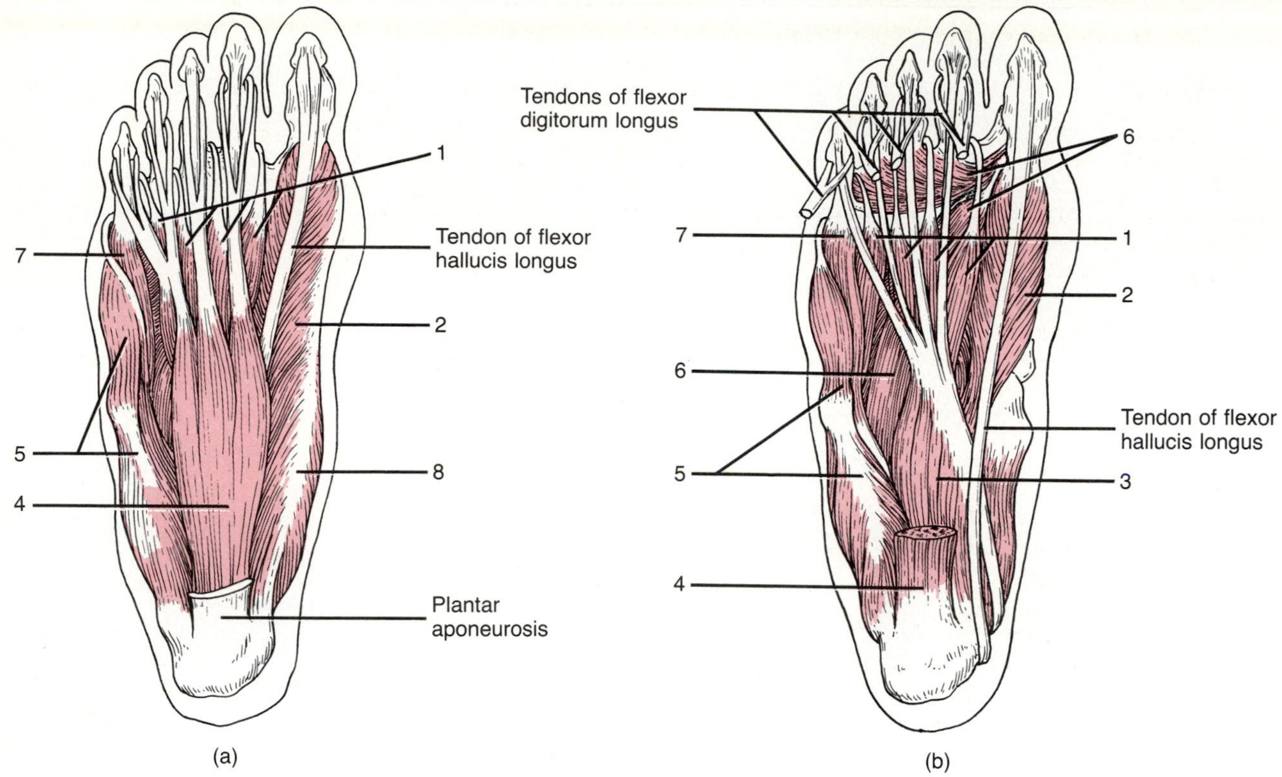

___ Abductor digiti minimi
___ Abductor hallucis
___ Adductor hallucis
___ Flexor digitorum brevis
___ Flexor digiti minimi brevis
___ Flexor hallucis brevis
___ Lumbricals
___ Quadratus plantae

FIGURE 10.21 Intrinsic muscles of the foot. (a) Plantar view showing some superficial and deeper muscles. (b) Plantar view showing deeper muscles.

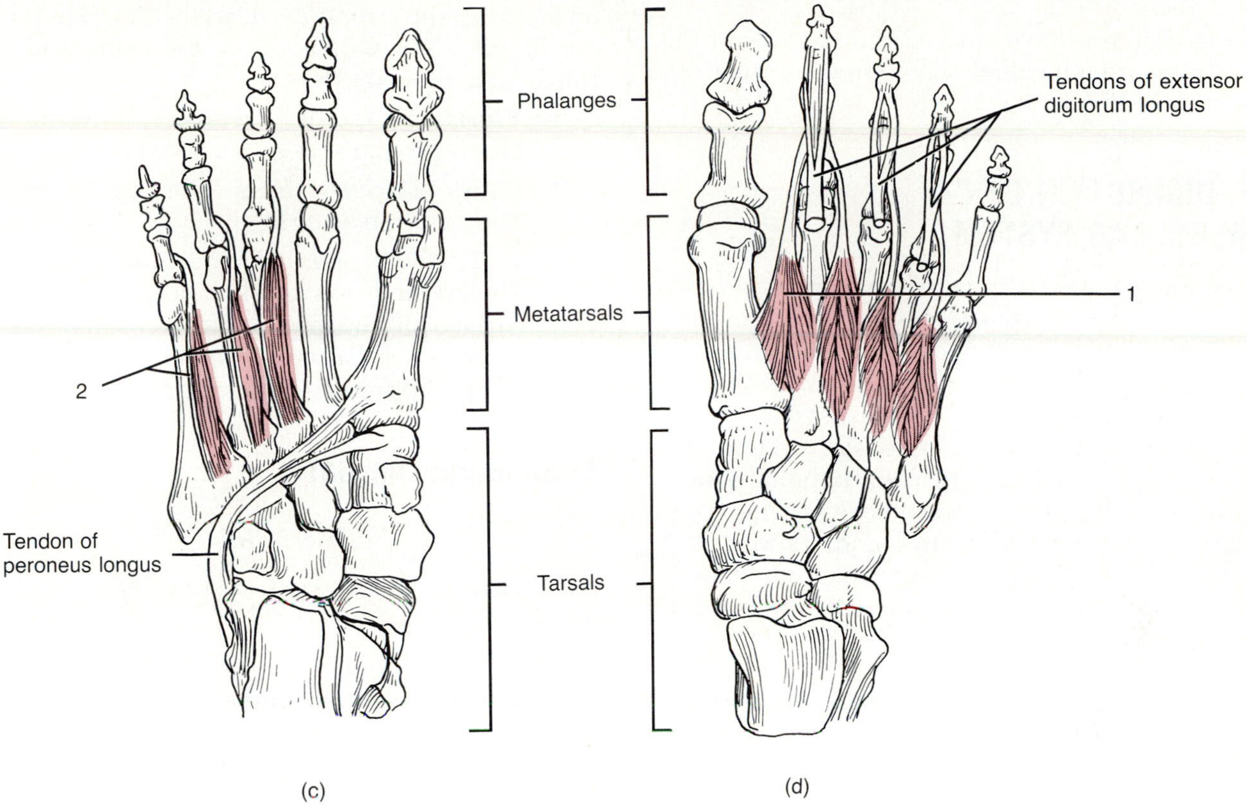

___ Dorsal interossei
___ Plantar interossei

FIGURE 10.21 ***(Continued)*** Intrinsic muscles of the foot. (c) Plantar deep view. (d) Dorsal deep view.

E. COMPOSITE MUSCULAR SYSTEM

Now that you have studied the muscles of the body by region, label the composite diagram shown in Figure 10.22.

F. DISSECTION OF CAT MUSCULAR SYSTEM

Assuming that you have completed your study of the human muscular system, you can now begin your examination of the cat musculature. The principal objective of dissecting the cat muscular system is to increase your understanding of the human muscular system. Do not be concerned that the cat and human muscular systems differ somewhat. Instead, concentrate on the similarities and, as you perform the dissection, try to relate your knowledge of the human muscular system to that of the cat.

CAUTION! *Please reread Section D. "Precautions Related to Dissection" at the beginning of the laboratory manual on p. xii before you begin your dissection.*

PROCEDURE

1. After obtaining a cat from your instructor, remove it from the plastic bag and place it on your dissecting tray ventral side up. Keep the liquid preservative in the plastic bag.
2. Determine the sex of your cat. A male has a ***scrotum*** ventral to the base of the tail. The ***prepuce*** appears as a slight elevation anterior to the scrotum. In contrast to the human male, the cat penis is usually retracted.
3. A female cat has a ***urogenital aperture***, the external opening of the vagina and the urethra, just ventral to the anus. Both male and female cats have four or five ***nipples (teats)*** on either side of the midline, so their presence does not necessarily mean that your cat is female. Consult your instructor if you have difficulty determining the sex of your cat. Compare your cat with one of the opposite sex and learn to identify both sexes externally.
4. Before skinning your cat, review Section F in Exercise 2, in which you were introduced to the planes of the body. Identify the planes of the cat's body and compare their orientation to those in the human (Figures 2.2 and 10.23). Also review Section D in Exercise 2 and locate, on your cat, the structures described therein for the human. You will note the following additional structures on the cat:
   a. ***Vibrissae***—Tactile whiskers located around the mouth and above the eyes.
   b. ***Claws***—Derived from the epidermis. They are retractable.
   c. ***Tori (foot pads*** or ***friction pads)***—Thickened areas of skin on the ventral surface of the paws. They cushion the body when the cat walks. Count the tori on the front and the back paws.

1. Skinning the Cat

PROCEDURE

Note: *Please read this entire section before starting.*

1. Grasp the skin at the ventral surface of the neck, pinch it up, and make a small longitudinal incision at the midline with the scissors. The incision, as well as all others you will make in skinning, should not be too deep because the skin is only about 1/8 in. thick and underlying muscles are easily damaged.
2. Now continue the incision upward along the midline to the lower lip and downward along the midline to the genital region. Do not remove the skin from the genital region.
3. If your specimen is female, the mammary glands may be removed. These appear as large masses along the ventral surface of the abdomen. Do not damage the underlying muscles as you remove the glands.
4. Next make incisions from the midline incision on the ventral side to the wrists and from the midline incision to the ankles. Cut all the way around the wrists and ankles.
5. Now you can begin to separate the skin from the underlying muscles to expose the muscles. This can be done by pulling the skin away from the body, using your fingers as much as possible. If you must use a scalpel, make sure that the sharp edge is directed away from the muscle and toward the skin.
6. Continue peeling the skin until only two points of attachment remain, one at the base of the tail and the other at the sides and top of the head.

7. Cut across the skin at the base of the tail to free the skin from the tail. The skin around the tail may be left in position.

8. Make a final incision from the left side of the lower lip, up to the top of the left eye, across to the top of the right eye, and down to the right side of the lower lip.

9. Remove the skin from the head, cutting around the bases of the ears as you come to them.

10. As part of the skinning procedure, be careful of the following:

- □ While skinning the neck, do not damage the external jugular veins in the neck (see Figure 10-24a).
- □ While skinning the forelimbs, do not damage the cephalic veins (see Figure 18.21).
- □ While skinning the hindlimbs, do not damage the greater saphenous veins (see Figure 18.19).
- □ While skinning the dorsal surface of the abdomen, do not cut through the deep fascia (lumbodorsal fascia) along the dorsal midline (see Figure 10.26a).
- □ The edge of the latissimus dorsi muscle (see Figure 10.26a) has a tendency to tear.

11. Before continuing, find the ***cutaneous muscles***, which are attached to the undersurface of the skin and are generally removed with the skin. One of these, the ***cutaneous maximus***, is closely bound to the underside of the skin in the trunk area. Another is the ***platysma***, which is closely bound to the skin on either side of the neck. The cutaneous muscles move the skin.[2]

12. Clean off as much fat as possible. Large fat deposits are usually found in the neck, axillary regions, and inguinal regions. As you pick away the fat with forceps, do not damage blood vessels or nerves. Also remove as much of the superficial fascia covering muscles as possible. This can be done with your forceps. If your specimen has been properly cleaned, the individual fibers in a muscle will be visible. Usually, the fibers of one muscle run in one direction, while those of another muscle run in another direction.

13. Save the skin and wrap it around the cat at the end of each laboratory period to prevent your specimen from drying out. Moist paper towels or an old cotton towel may also be used.

2. Dissecting Skeletal Muscles

Keep in mind that *dissect* means to separate. Dissection of a skeletal muscle includes separating the body of the muscle from surrounding tissues, but points of attachment are left intact. These are the ***origin*** (attachment to a fixed bone) and the ***insertion*** (attachment to a movable bone). When a muscle contracts, the insertion moves toward the origin. You can pull the insertion toward the origin to observe the action of the muscle.

Most muscles are attached to bones by bands of white fibrous connective tissue called ***tendons***. Some muscles are attached by white, flattened, sheetlike tendons called ***aponeuroses***.

Skeletal muscles are distinguished from each other by normal lines of cleavage. These lines consist of deep fascia and may be exposed by slightly pulling adjacent muscles apart. The muscles can be separated from each other by picking the fascia out with your forceps. If, after removing the fascia, the muscles separate as distinct units, your procedure is correct. If, instead, the separated muscles have a ragged appearance, you have damaged a muscle by tearing it into two portions.

Sometimes superficial muscles must be severed in order to examine deep muscles. When you cut a muscle, ***transect*** it (i.e., cut it at right angles to the fibers at about the center of the belly of the muscle) using a scissors. Then ***reflect*** the cut ends (i.e., pull them back, one toward the origin and one toward the insertion). As you reflect the ends, carefully break the underlying fascia with your scalpel to separate the superficial muscle from the deep muscle. Transection keeps the origin and insertion of the superficial muscle intact and preserves the original shape and position of the superficial muscle.

3. Dissection, Transection, and Reflection of a Single Skeletal Muscle

Before you start to identify skeletal muscles, you should dissect a single muscle to familiarize yourself with the procedure and the

[2]If preskinned cats are used the instructor may wish to demonstrate the cutaneous maximus and the platysma.

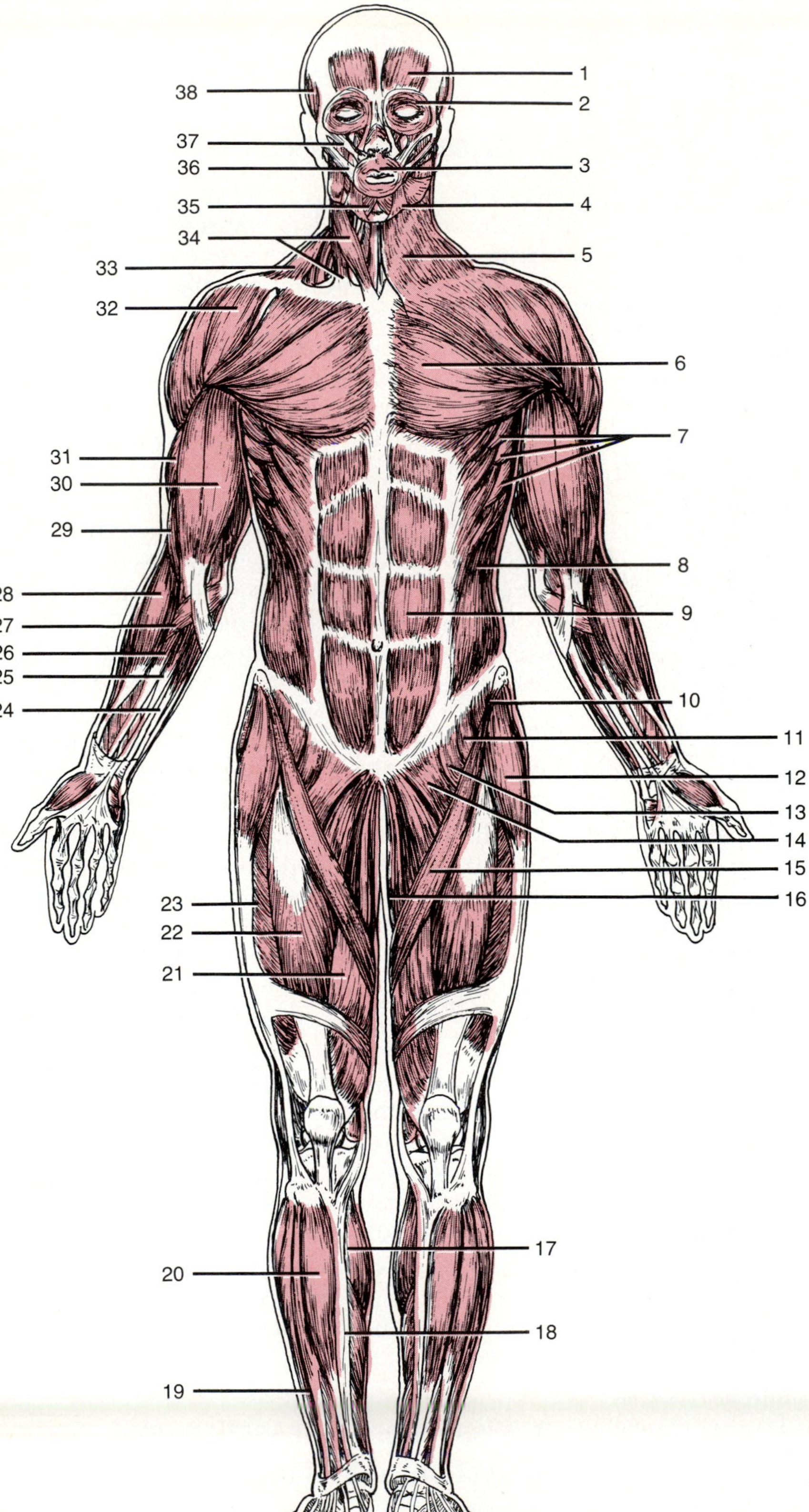

(a)

___ Adductor longus
___ Biceps brachii
___ Brachialis
___ Brachioradialis
___ Deltoid
___ Depressor labii inferioris
___ External oblique
___ Flexor carpi radialis
___ Flexor carpi ulnaris
___ Frontalis
___ Gastrocnemius
___ Gracilis
___ Iliacus
___ Mentalis
___ Orbicularis oculi
___ Orbicularis oris
___ Palmaris longus
___ Pectineus
___ Pectoralis major
___ Peroneus longus
___ Platysma
___ Pronator teres
___ Psoas major
___ Rectus abdominis
___ Rectus femoris
___ Sartorius
___ Serratus anterior
___ Soleus
___ Sternocleidomastoid
___ Temporalis
___ Tensor fasciae latae
___ Tibialis anterior
___ Trapezius
___ Triceps brachii
___ Vastus lateralis
___ Vastus medialis
___ Zygomaticus major
___ Zygomaticus minor

FIGURE 10.22 Principal superficial muscles. (a) Anterior view.

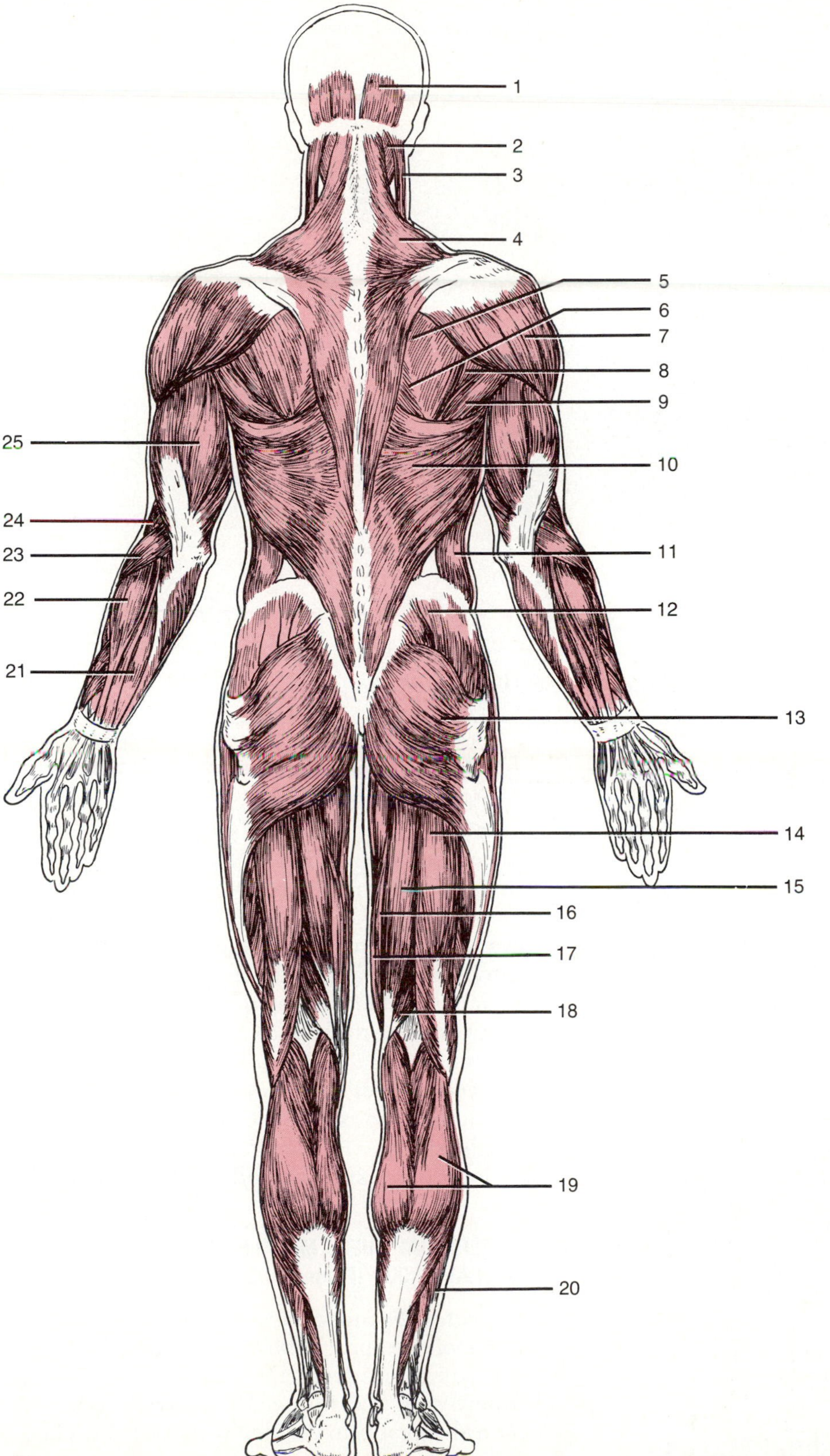

(b)

___ Biceps femoris
___ Brachioradialis
___ Deltoid
___ Extensor carpi radialis longus
___ Extensor carpi ulnaris
___ Extensor digitorum
___ External oblique
___ Gastrocnemius
___ Gluteus maximus
___ Gluteus medius
___ Gracilis
___ Infraspinatus
___ Latissimus dorsi
___ Occipitalis
___ Rhomboideus major
___ Sartorius
___ Semimembranosus
___ Semispinalis capitis
___ Semitendinosus
___ Soleus
___ Sternocleidomastoid
___ Teres major
___ Teres minor
___ Trapezius
___ Triceps brachii

FIGURE 10.22 ***(Continued)*** Principal superficial muscles. (b) Posterior view.

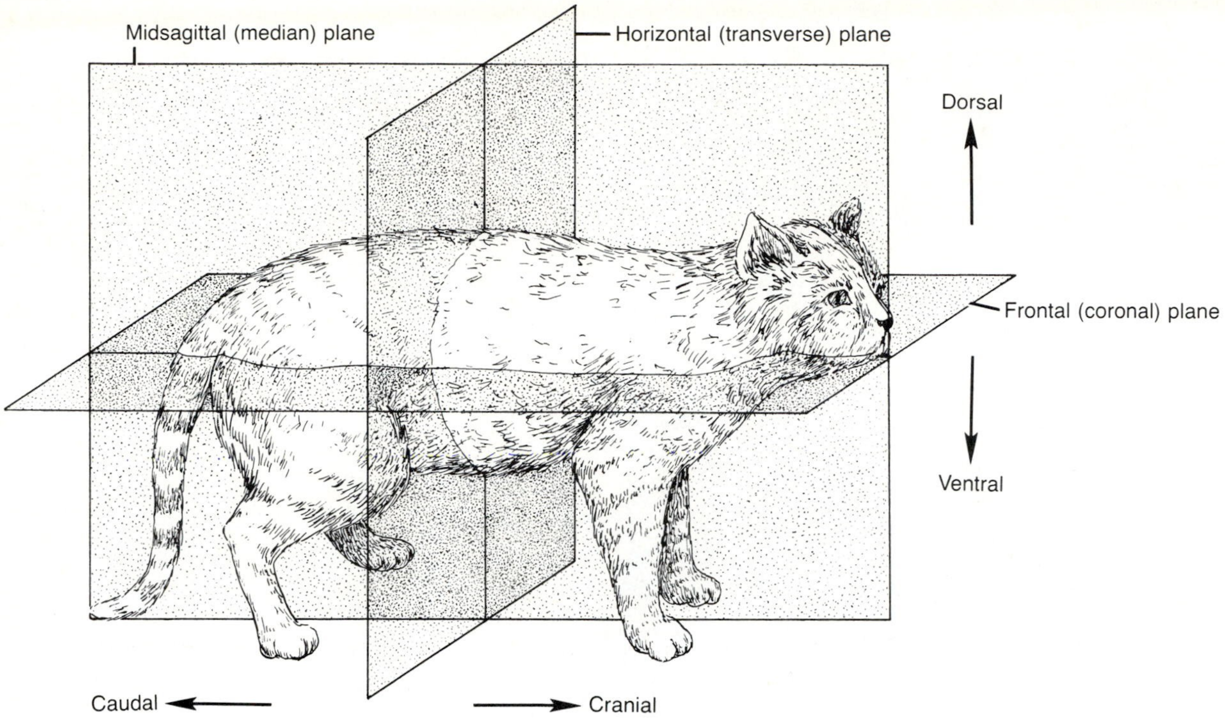

FIGURE 10.23 Anatomical planes and directional terms of the cat.

terminology. The sample muscle that has been selected is the pectoantebrachialis (see Figure 10.24a).

PROCEDURE

1. Place the cat on its dorsal surface and find the pectoantebrachialis and pectoralis major muscles. Separate the muscles from each other.

2. Dissect the body of the pectoantebrachialis muscle toward its origin and insertion. Its origin is the manubrium of the sternum and it inserts on the fascia on the proximal end of the forearm. Leave the origin and insertion intact.

3. By drawing the insertion toward the origin you can note the action of the muscle—adduction of the arm.

4. Transect the pectoantebrachialis muscle and reflect the cut ends to expose the underlying pectoralis major. As you reflect, break the fascia under the pectoantebrachialis muscle.

4. Identification of Skeletal Muscles

You are now ready to begin your identification of selected skeletal muscles of the cat. The muscles you will study are grouped according to region. Each muscle discussed is shown in the corresponding figure. Locate each muscle discussed and try to determine its origin and insertion. Then dissect it and pull its insertion toward its origin to observe its action. If you are told that a particular muscle is a deep muscle, you will also have to transect and reflect the superficial muscle to locate and study the deep muscle.

As you dissect, transect, and reflect muscles of various regions, work only on either the left or right side of the cat. You will use the other side later on to study blood vessels and nerves.

a. SUPERFICIAL CHEST MUSCLES: PECTORAL GROUP (Figure 10.24a)

The ***pectoralis*** muscle group covers the ventral surface of the chest. The large, triangular muscle group arises from the sternum and inserts primarily on the humerus. On each side, the pectoralis group is divided into four muscles.

1. ***Pectoantebrachialis***—A narrow ribbonlike muscle superficial to the pectoralis major

muscle. It consists of a narrow band (½ in.) of fibers that run from the sternum to the forearm. Dissect, transect, and reflect.

2. ***Pectoralis major***—A muscle 2 to 3 in. wide that lies immediately deep to the pectoantebrachialis and the clavobrachialis. Its fibers run roughly parallel to those of the pectoantebrachialis. Separate the pectoralis major from the pectoralis minor and clavobrachialis, transect, and reflect.

3. ***Pectoralis minor***—A muscle lying dorsal and caudal to the pectoralis major. Dissect, transect, and reflect.

4. ***Xiphihumeralis***—The most caudad of the pectoral group. Its fibers run parallel to pectoralis minor and eventually pass deep to the pectoralis minor. Dissect, transect, and reflect.

In humans, the pectoral group has only two muscles—the pectoralis major and minor.

b. DEEP CHEST MUSCLES (Figure 10.24b)

1. ***Serratus ventralis***—Fan-shaped muscle extending obliquely upward from the ribs to the shoulder blade. Dissect.

2. ***Levator scapulae***—Continuous with the serratus ventralis at its cranial border. Dissect.

3. ***Transversus costarum***—Small, thin muscle that covers cranial end of rectus abdominis muscle. Dissect.

4. ***Scalenus***—Complex muscle that extends longitudinally along the ventrolateral thorax and neck. Its three main parts are called the scalenus dorsalis, scalenus medius, and scalenus ventralis. The scalenus medius is easiest to recognize. Dissect.

5. ***External intercostals***—Series of muscles[3] in the intercostal spaces. Their fibers run in same direction (caudoventrally) as those of the external oblique muscle (next section). Dissect, transect, and reflect.

6. ***Internal intercostals***—Series of muscles deep to the external intercostals. Their fibers run in same direction (caudodorsally) as those of the internal oblique muscle (next section).

c. MUSCLES OF THE ABDOMINAL WALL (Figure 10.24a and b)

Before examining the muscles of the abdominal wall, locate the ***linea alba***, a midventral fascial line formed by fusion of the aponeuroses of the ventrolateral abdominal muscles. The muscles of the abdominal wall include the following:

1. ***External oblique***—Most superficial, lateral abdominal muscle. Its dorsal half is fleshy; its ventral half is an aponeurosis. Its fibers run dorsally and caudally. Make a very shallow longitudinal incision in the external oblique from the ribs to the pelvis. Transect and reflect. Find the aponeurosis of the external oblique.

2. ***Internal oblique***—Deep to the external oblique. Its fibers run ventrally and cranially, opposite those of the external oblique. Make a very shallow longitudinal incision in the internal oblique muscle until you can see transverse fibers. They belong to the transversus abdominis muscle. Reflect the internal oblique and find its aponeurosis.

3. ***Transversus abdominis***—This is the third and deepest layer of the abdominal wall and is deep to the internal oblique. Its fibers run transversely. Locate the aponeurosis of the transversus abdominis.

4. ***Rectus abdominis***—Longitudinal band of muscle just lateral to the linea alba. The muscle is visible when the external oblique aponeurosis is reflected.

The abdominal muscles are similar to those in humans.

d. MUSCLES OF THE NECK (Figure 10.25a and b)

1. ***Sternomastoid***—Band of muscle extending from the manubrium to the mastoid portion of the skull. The muscle is deep to the external jugular vein. Dissect.

2. ***Cleidomastoid***—Narrow band of muscle that is between the sternomastoid and clavotrapezius. Most of the cleidomastoid is deep to the clavotrapezius. Dissect.

3. ***Sternohyoid***—Narrow, straight muscle on either side of the midline covering the trachea. Caudal end partially covered by the sternomastoid. Dissect.

4. ***Sternothyroid***—Narrow band of muscle lateral and dorsal to the sternohyoid. The ster-

[3]Not illustrated.

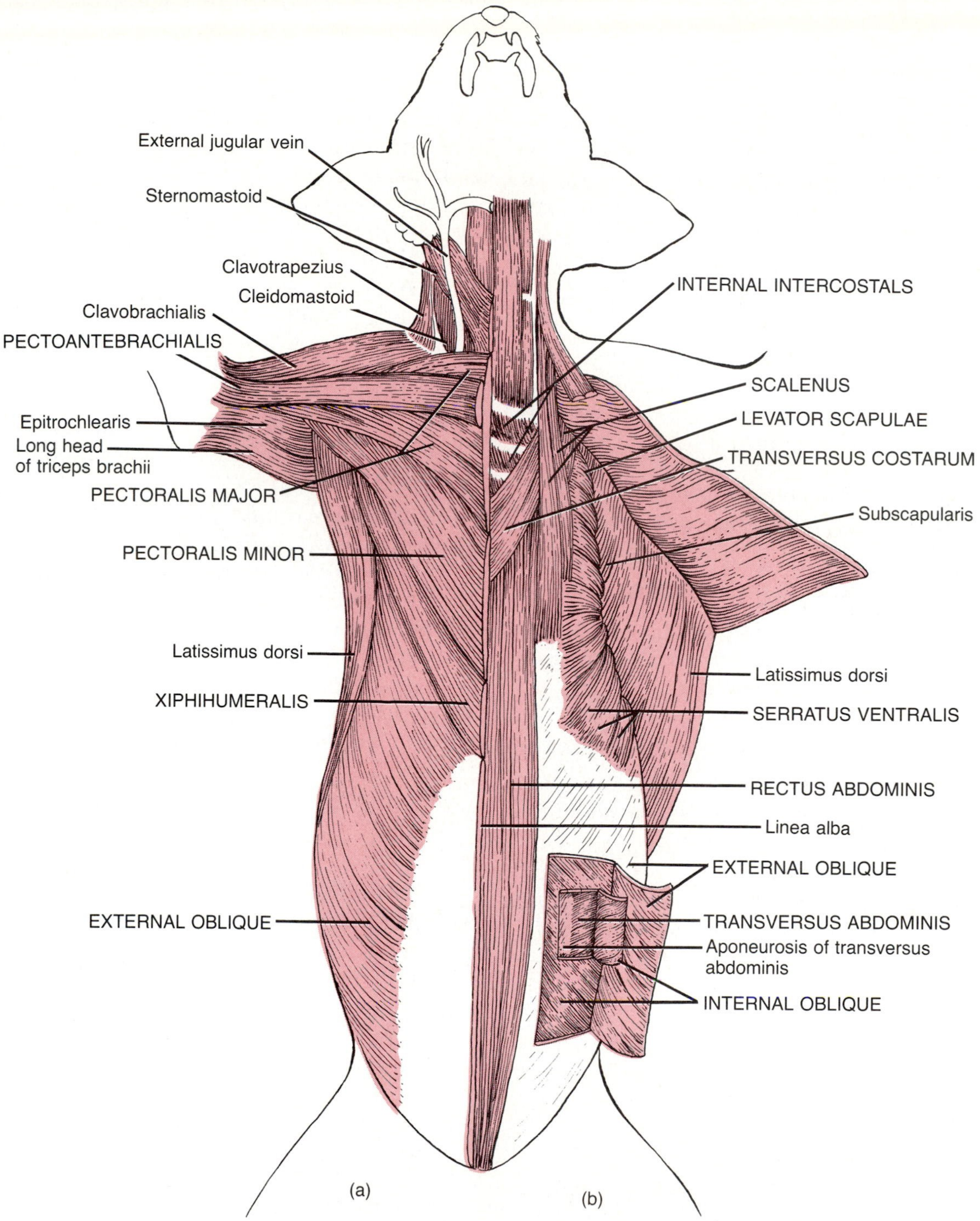

FIGURE 10.24 Ventral view of chest and abdominal muscles. (a) Superficial view. (b) Deep view.

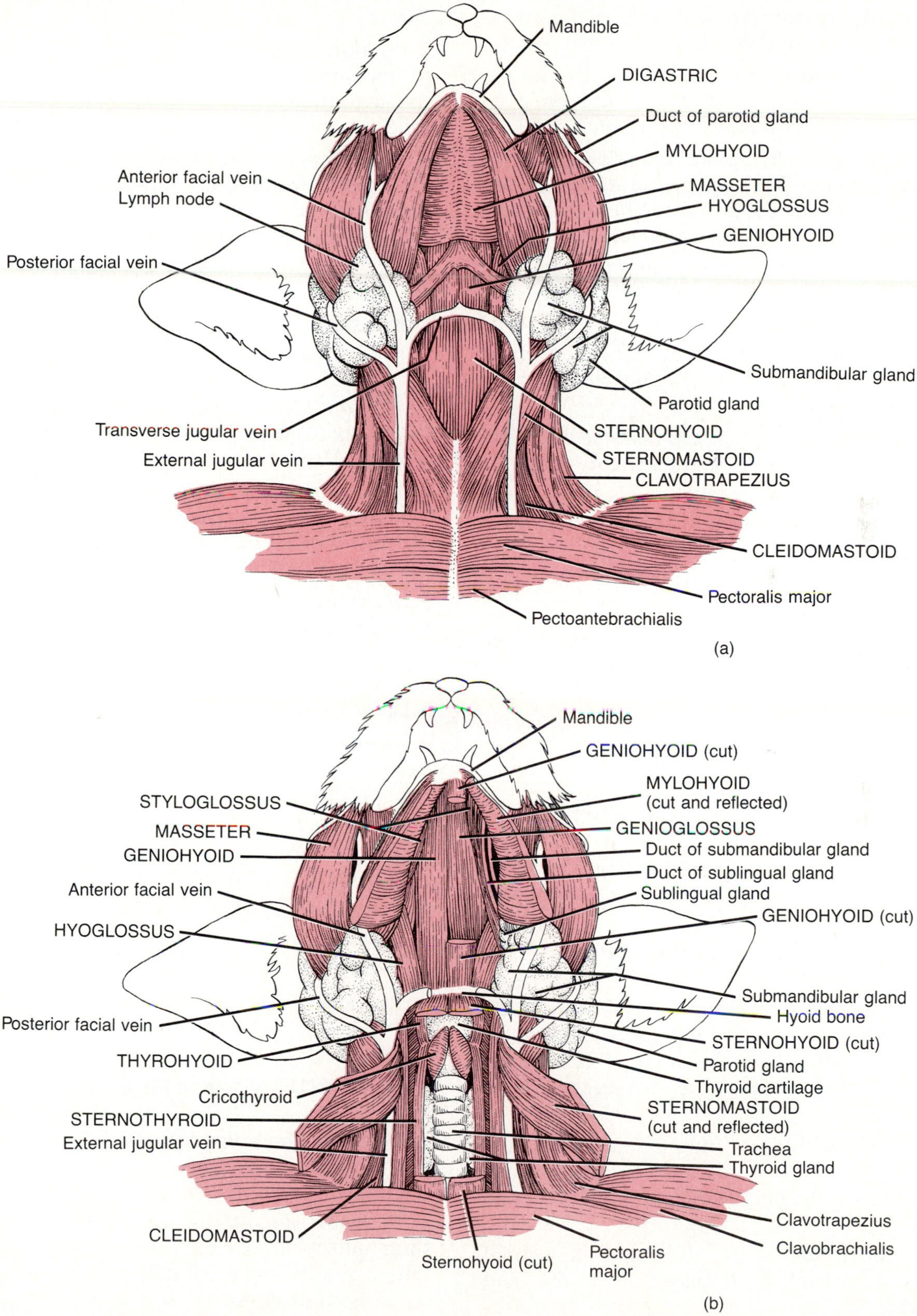

FIGURE 10.25 Ventral view of neck muscles. (a) Superficial view. (b) Deep view.

nohyoid must be lifted and pushed aside to expose the sternothyroid. The thyroid gland is deep to the sternothyroid.

5. ***Thyrohyoid***—Small muscle on the lateral side of the thyroid cartilage of the larynx. Its location is just cranial to the sternothyroid.

6. ***Geniohyoid***—Straplike muscle cranial to the sternohyoid. The mylohyoid must be transected and reflected to locate the geniohyoid. Transect and reflect the geniohyoid.

7. ***Mylohyoid***—Thin, sheetlike muscle seen after transection and reflection of the digastric muscle. Transect and reflect.

8. ***Digastric***—Superficial muscle attached to the ventral border of the mandible. Transect and reflect.

9. ***Masseter***—Large muscle mass at the angle of the jaw behind and below the ear.

10. ***Temporalis***—A muscle just above and caudal to the eye and medial to the ear (Figure 10.26a).

11. ***Hyoglossus***—Short, obliquely directed muscle lateral to the geniohyoid. Dissect.

12. ***Styloglossus***—Long muscle that runs parallel to the inside of the body of the mandible. Dissect.

13. ***Genioglossus***—A muscle lying deep to the geniohyoid.

e. SUPERFICIAL SHOULDER MUSCLES (Figure 10.26a)

1. ***Clavotrapezius***—Large, broad muscle on the back and side of the neck. Dissect, transect, and reflect.

2. ***Levator scapulae ventralis***—A muscle lying caudal to the clavotrapezius and passing underneath the clavotrapezius. Dissect.

3. ***Clavobrachialis (clavodeltoid)***—Triangular muscle extending onto the shoulder as a continuation of the clavotrapezius. Dissect, transect, and reflect.

4. ***Acromiodeltoid***—A muscle caudal to the clavobrachialis and lateral to the levator scapulae ventralis. Dissect, transect, and reflect.

5. ***Spinodeltoid***—Short, thick muscle caudal to the acromiodeltoid. Dissect, transect, and reflect.

6. ***Acromiotrapezius***—Thin, flat muscle extending from the middle of the back to the shoulder. The muscle is caudal to the clavotrapezius. Dissect, transect, and reflect.

7. ***Spinotrapezius***—A muscle lying caudal to the acromiotrapezius, partly covered by the acromiotrapezius. Dissect, transect, and reflect.

8. ***Latissimus dorsi***—Large, triangular muscle caudal to the spinotrapezius. This muscle extends from the middle of the back to the humerus. Note its attachment to the lumbodorsal fascia. Dissect, transect, and reflect.

f. DEEP SHOULDER MUSCLES (Figure 10.26b)

1. ***Rhomboideus capitis***—Narrow muscle extending from the back of the skull to the scapula. Dissect.

2. ***Rhomboideus***—A muscle lying deep to medial portions of the acromiotrapezius and the spinotrapezius, between the scapula and the body wall. Dissect.

3. ***Splenius***—Broad muscle deep to the clavotrapezius and the rhomboideus capitis. Dissect.

4. ***Supraspinatus***—A muscle deep to the acromiotrapezius, located in the supraspinous fossa of the scapula. Dissect.

5. ***Infraspinatus***—A muscle caudal to the supraspinatus, occupying the infraspinous fossa of the scapula. Dissect.

6. ***Teres major***—Band of thick muscle caudal to the infraspintuas muscle and covering the axillary border of the scapula. Dissect.

7. ***Teres minor***—Tiny, round muscle caudal and adjacent to the insertion of the infraspinatus.[4] Dissect.

8. ***Subscapularis***—Large, flat muscle in the subscapular fossa. Dissect (see Figure 10.29b).

g. BACK MUSCLES (Figure 10.26a and b)

1. ***Serratus dorsalis cranialis***—Four or five short muscle slips above and caudal to the scapula.

2. ***Serratus dorsalis caudalis***—Four or five short muscle slips caudal to the serratus dorsalis cranialis and medial to the upper part of the latissimus dorsi.

[4]Not illustrated.

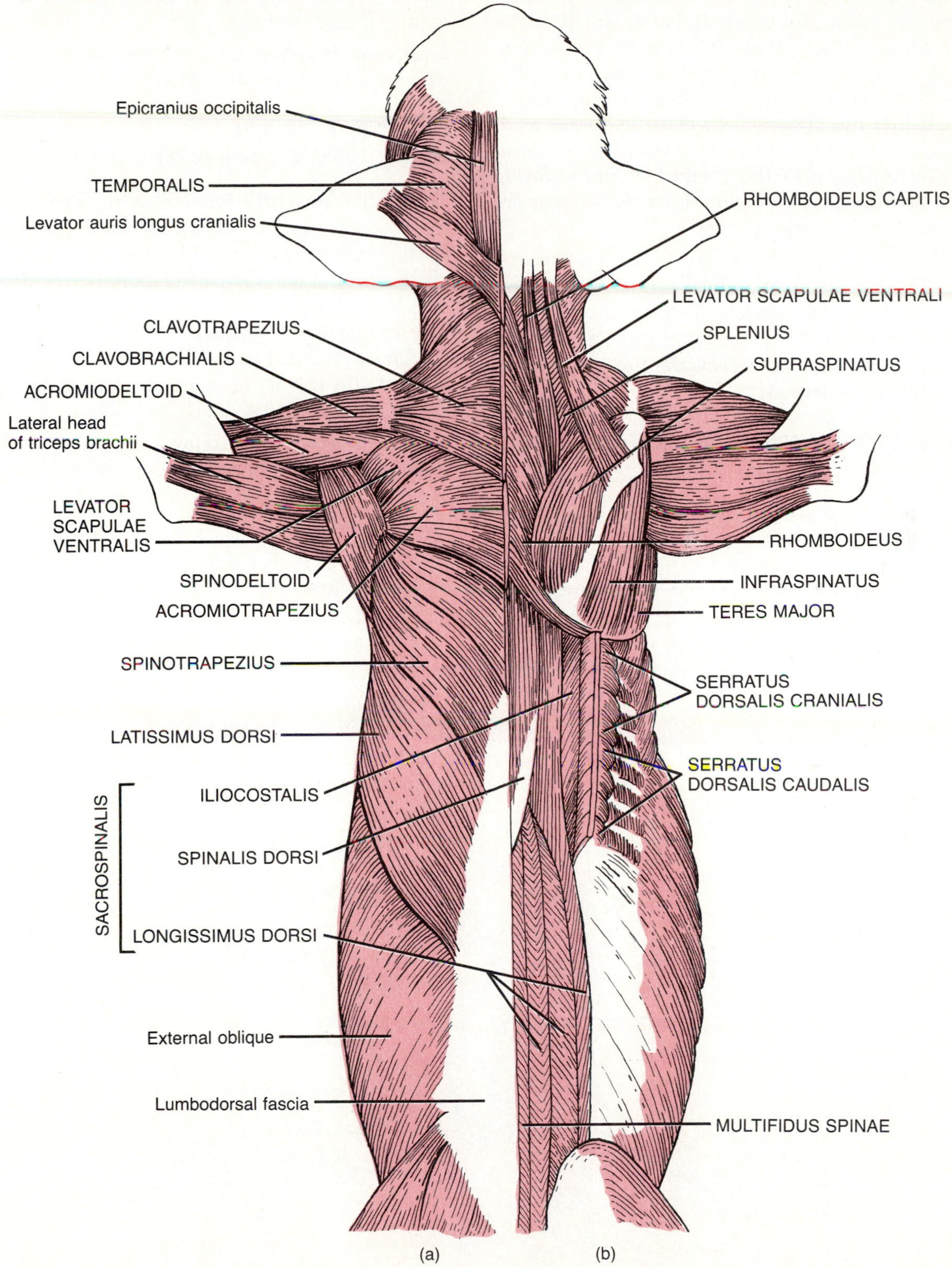

FIGURE 10.26 Dorsal view of shoulder and back muscles. (a) Superficial view. (b) Deep view.

3. ***Sacrospinalis***—Complex muscle extending along the spine. Its three divisions are the lateral ***iliocostalis***, which is deep to the serratus dorsalis cranialis and the serratus dorsalis caudalis; the intermediate ***longissimus dorsi***, which fills the spaces between the spines and transverse processes of the thoracic and lumbar vertebrae; and the medial ***spinalis dorsi***, next to the spines of the thoracic vertebrae. Dissect these components.

4. ***Multifidus spinae***—Narrow band of muscle adjacent to the vertebral spines extending the length of the vertebral column and best seen in the sacral region. Its cranial portion is called the ***semispinalis cervicis***[5] muscle, which is under the splenius.

h. MUSCLES OF THE ARM: MEDIAL SURFACE (Figure 10.27)

1. ***Epitrochlearis***—Flat band of muscle along the medial surface of the arm between the biceps brachii and the triceps brachii. Dissect, transect, and reflect.

2. ***Biceps brachii***—Located on the ventromedial surface of the arm. Dissect.

i. MUSCLES OF THE ARM: LATERAL SURFACE (Figure 10.28)

1. ***Triceps brachii***—Muscle with three main heads covering the caudal surface and much of the sides of the arm. The ***long head***, located on the caudal surface of the arm, is the largest. The ***lateral head*** covers much of the lateral surface of the arm. Dissect, transect, and reflect the lateral head. The ***medial head***, found on the medial surface of the arm, is located between the long and lateral heads.

2. ***Anconeus***—A tiny triangular muscle lying deep to the distal end of the lateral head of the

[5]Not illustrated.

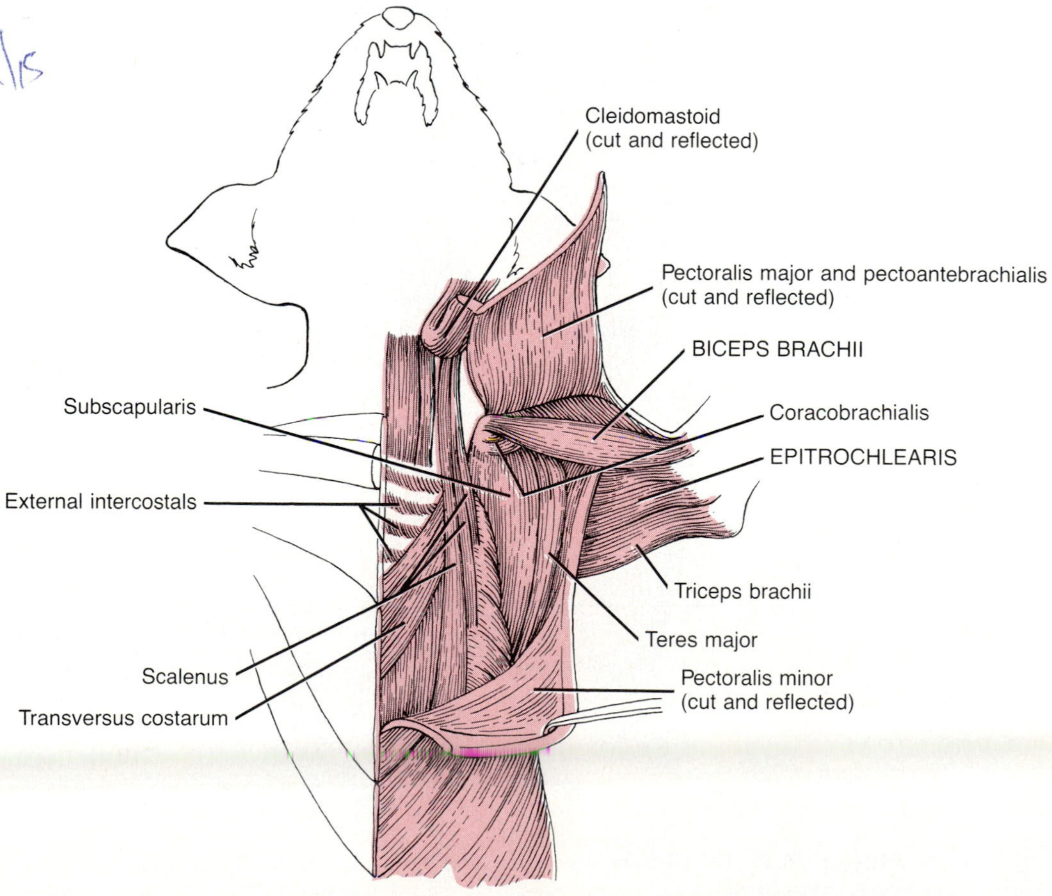

FIGURE 10.27 Muscles of medial surface of arm.

triceps brachii. This is a fourth head of the triceps brachii. Dissect.

3. ***Brachialis***—A muscle located on the ventrolateral surface of the arm, cranial to the lateral head of the triceps brachii. Dissect.

j. MUSCLES OF THE FOREARM: VENTRAL SURFACE (Figure 10.29a and b)

1. ***Pronator teres***—Small, triangular muscle that originates from the medial epicondyle of the humerus. Dissect.

2. ***Flexor carpi radialis***—Long, spindle-shaped muscle above the pronator teres. Dissect.

3. ***Palmaris longus***—Large, flat, broad muscle near the center of the medial surface. Dissect, transect, and reflect.

4. ***Flexor digitorum sublimis (superficialis)***—A muscle deep to the palmaris longus. Dissect.

5. ***Flexor carpi ulnaris***—Band of muscle above the palmaris longus consisting of two heads. Dissect.

6. ***Flexor digitorum profundus***—A muscle with five heads of origin, deep to the palmaris longus. Dissect.

7. ***Pronator quadratus***—Flat, quadrilateral muscle extending across the ventral surface of about the distal half of the ulna and radius. Dissect.

k. MUSCLES OF THE FOREARM: DORSAL SURFACE (Figure 10.30a and b)

1. ***Brachioradialis***—Narrow, ribbonlike muscle on the lateral surface of the forearm. Dissect.

2. ***Extensor carpi radialis longus***—Larger muscle adjacent to the brachioradialis. Dissect, transect, and reflect.

3. ***Extensor carpi radialis brevis***—Muscle usually covered by the extensor carpi radialis longus at the proximal end. Dissect.

4. ***Extensor digitorum communis***—Large muscle running entire length of the lateral surface of forearm. Dissect, transect, and reflect.

5. ***Extensor digitorum lateralis***—Muscle with about the same structure as the extensor digitorum communis muscle. It lies lateral to the ulna. Dissect, transect, and reflect.

6. ***Extensor carpi ulnaris***—Muscle adjacent to

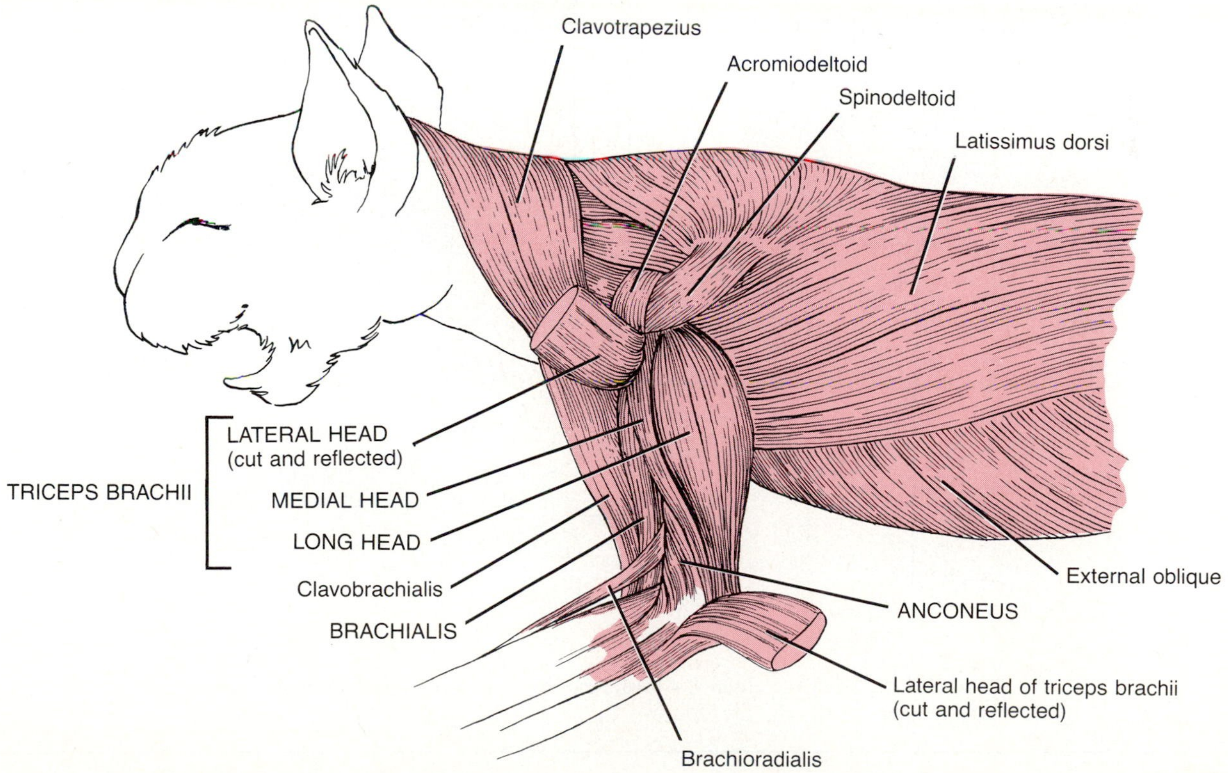

FIGURE 10.28 Muscles of lateral surface of arm.

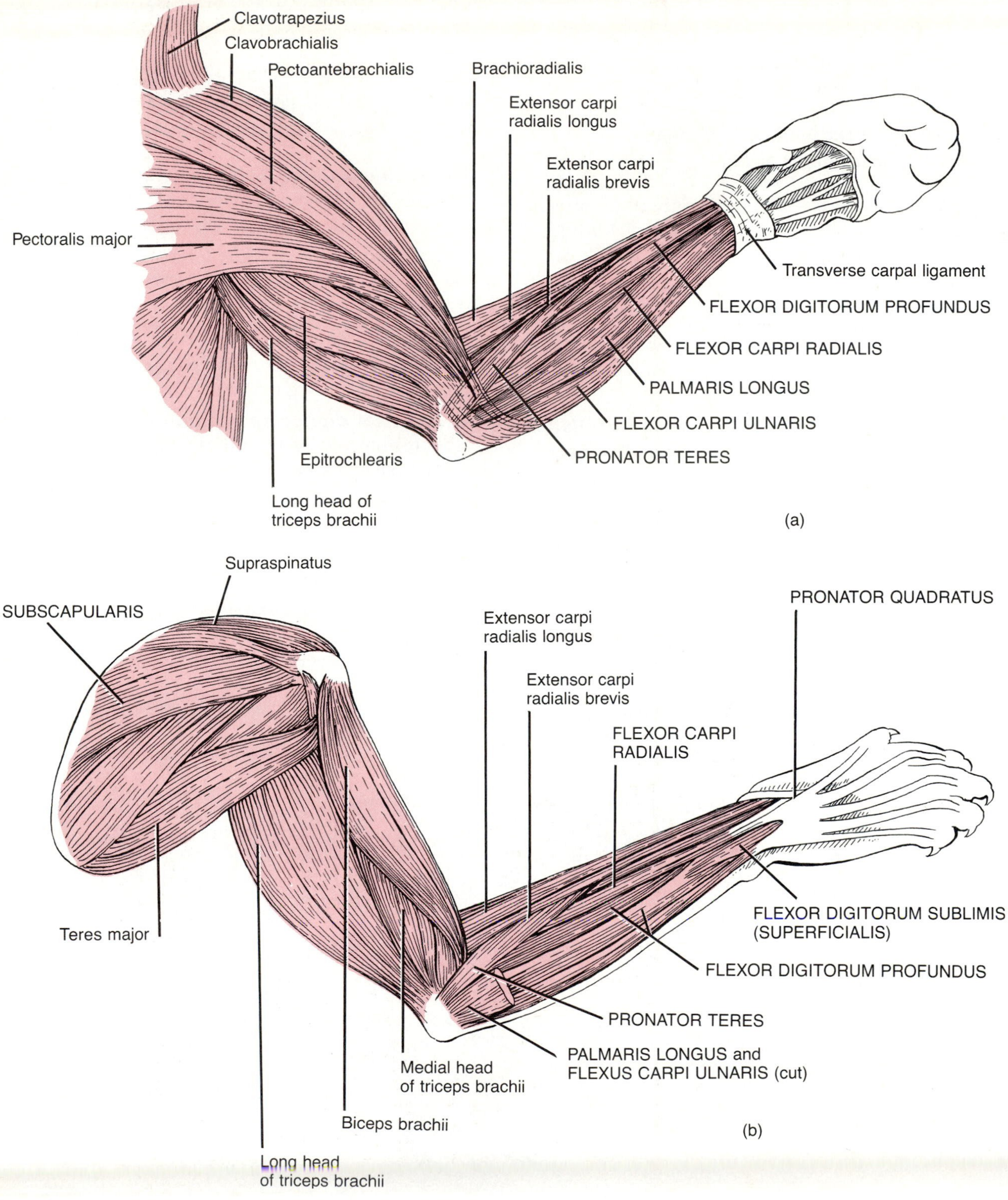

FIGURE 10.29 Ventral view of muscles of forearm. (a) Superficial view. (b) Deep view.

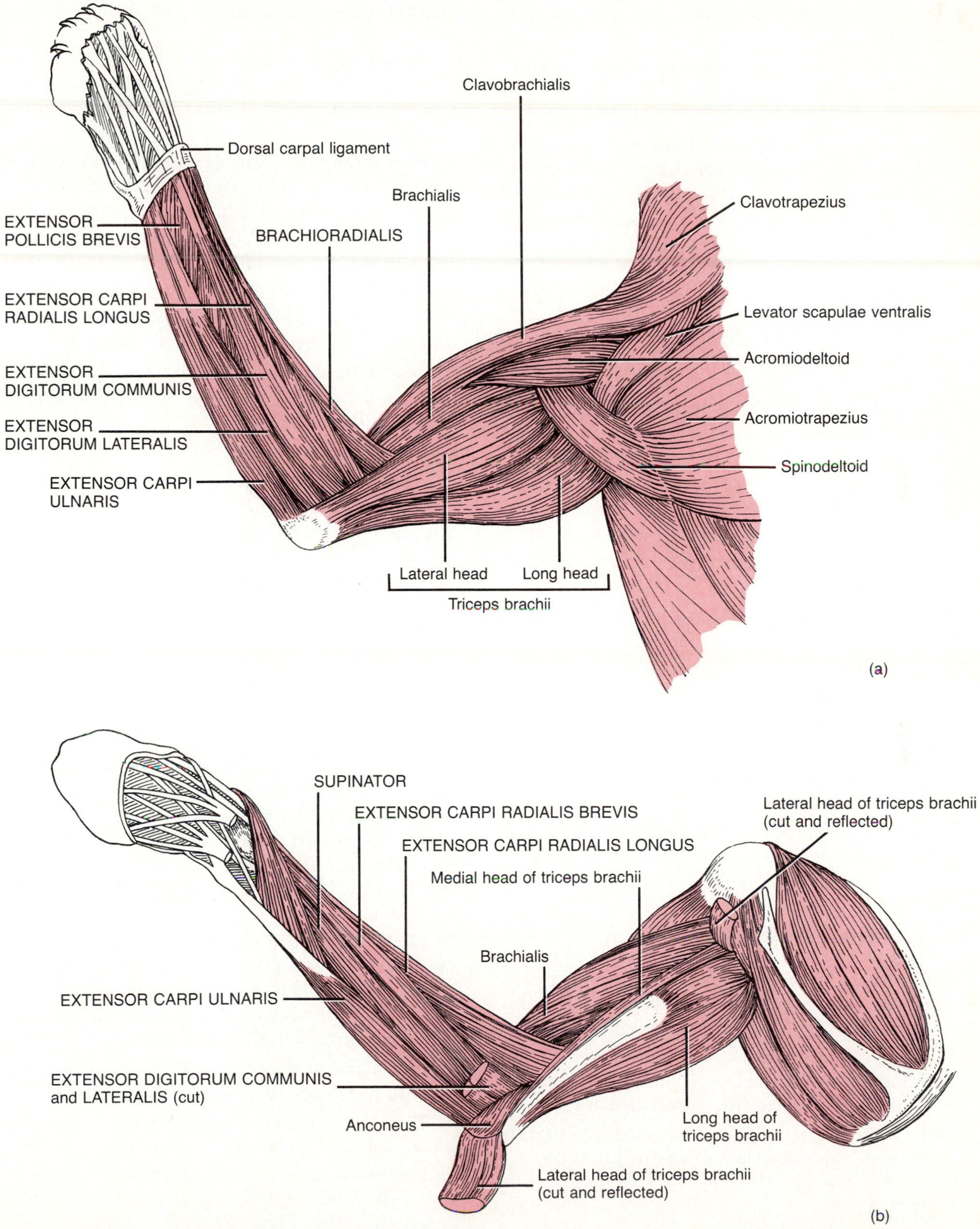

FIGURE 10.30 Dorsal view of muscles of forearm. (a) Superficial view. (b) Deep view.

the extensor digitorum lateralis, extending down the forearm laterally. Dissect, transect, and reflect.

7. ***Extensor indicis proprius***—Narrow muscle deep to the extensor carpi ulnaris.[6] Dissect.

8. ***Extensor pollicis brevis***—Muscle extending obliquely across the forearm deep to the extensor carpi ulnaris, extensor digitorum lateralis, and extensor digitorum communis. Dissect.

9. ***Supinator***—Muscle that passes obliquely across radius deep to the extensor digitorum communis.

l. MUSCLES OF THE THIGH: MEDIAL SURFACE (Figure 10.31a and b)

1. ***Sartorius***—Straplike muscle on the cranial half of the medial side of the thigh. Dissect, transect, and reflect.

2. ***Gracilis***—Wide, flat muscle covering most of the caudal portion of the medial side of the thigh. Dissect, transect, and reflect.

3. ***Semimembranosus***—Thick muscle deep to the gracilis. Dissect.

4. ***Adductor femoris***—Triangular muscle deep to the gracilis and lateral to the semimembranosus. Dissect.

5. ***Adductor longus***—Small, triangular muscle cranial to the adductor femoris. Dissect.

6. ***Pectineus***—Very small triangular muscle cranial to the adductor longus. Dissect.

7. ***Iliopsoas***—Long, cylindrical muscle that originates on the last two thoracic and lumbar vertebrae and passes along the medial surface of the thigh adjacent to the pectineus. Dissect.

8. ***Rectus femoris***—Narrow band of muscle deep to the sartorius, one of four components of the ***quadriceps femoris*** muscle. The others are the vastus medialis, vastus intermedius, and vastus lateralis. Dissect the rectus femoris.

9. ***Vastus medialis***—Large muscle on the medial surface of the thigh caudal to the rectus femoris and deep to the sartorius. Dissect.

10. ***Vastus intermedius***—Best seen by separating the rectus femoris and vastus lateralis.[7]

[6,7]Not illustrated.

m. MUSCLES OF THE THIGH: LATERAL SURFACE (Figure 10.32a and b)

1. ***Tensor fasciae latae***—Triangular mass of muscle cranial to the biceps femoris. Dissect, transect, and reflect. Reflect the tough, white fasciae latae.

2. ***Biceps femoris***—Muscle caudal to the tensor fasciae latae, covers most of the lateral surface of the thigh. Dissect, transect, and reflect.

3. ***Semitendinosus***—Large band of muscle on the ventral border of the thigh between the biceps femoris and semimembranosus. Dissect.

4. ***Caudofemoralis***—Band of muscle cranial and dorsal to the biceps femoris. Dissect.

5. ***Tenuissimus***—Very slender band of muscle deep to the biceps femoris. The sciatic nerve runs parallel to it. Dissect.

6. ***Gluteus maximus***—Triangular mass of muscle just cranial to the caudofemoralis. Dissect.

7. ***Gluteus medius***—Relatively large, somewhat triangular muscle cranial to the gluteus maximus. Dissect.

8. ***Vastus lateralis***—Large muscle deep to the tensor fasciae latae that occupies the craniolateral surface of the thigh. Dissect.

n. MUSCLES OF THE SHANK: MEDIAL SURFACE (Figure 10.31a and b)

1. ***Gastrocnemius (medial head)***—Muscle originating on medial epicondyle of femur. The medial and lateral heads unite to form a large muscle mass that inserts into the calcaneus by the ***calcaneal (Achilles) tendon***. Separate the medial head from the lateral head and reflect the medial head.

2. ***Flexor digitorum longus***—Muscle consisting of two heads. The first is partly covered by the soleus muscle. The second is between the first head and the medial head of the gastrocnemius. Dissect, transect, and reflect.

3. ***Flexor hallucis longus***—Long, heavy muscle deep to the flexor digitorum longus. Dissect.

4. ***Tibialis anterior***—Tapered band of muscle on the anterior aspect of the tibia. Dissect.

5. ***Plantaris***—Strong, round muscle between the two heads of the gastrocnemius. Also inserts into the calcaneal tendon. Dissect.

o. MUSCLES OF THE SHANK: LATERAL SURFACE (Figure 10.32a and b)

1. ***Gastrocnemius (lateral head)***—Muscle originating on the lateral epicondyle of the femur. Dissect, transect, and reflect.

2. ***Soleus***—Small band of muscle next to the lateral head of the gastrocnemius and inserting into the calcaneal (Achilles) tendon. Dissect.

3. ***Extensor digitorum longus***—Muscle posterior to the tibialis anterior. Dissect.

4. ***Peroneus longus***—Slender, superficial muscle caudal to the extensor digitorum longus. Dissect, transect, and reflect.

5. ***Peroneus brevis***—Short, thick muscle with the lateral surface covered by the peroneus longus. Dissect.

6. ***Peroneus tertius***—Slender muscle adjacent to the extensor digitorum longus. Dissect.

ANSWER THE LABORATORY REPORT QUESTIONS AT THE END OF THE EXERCISE

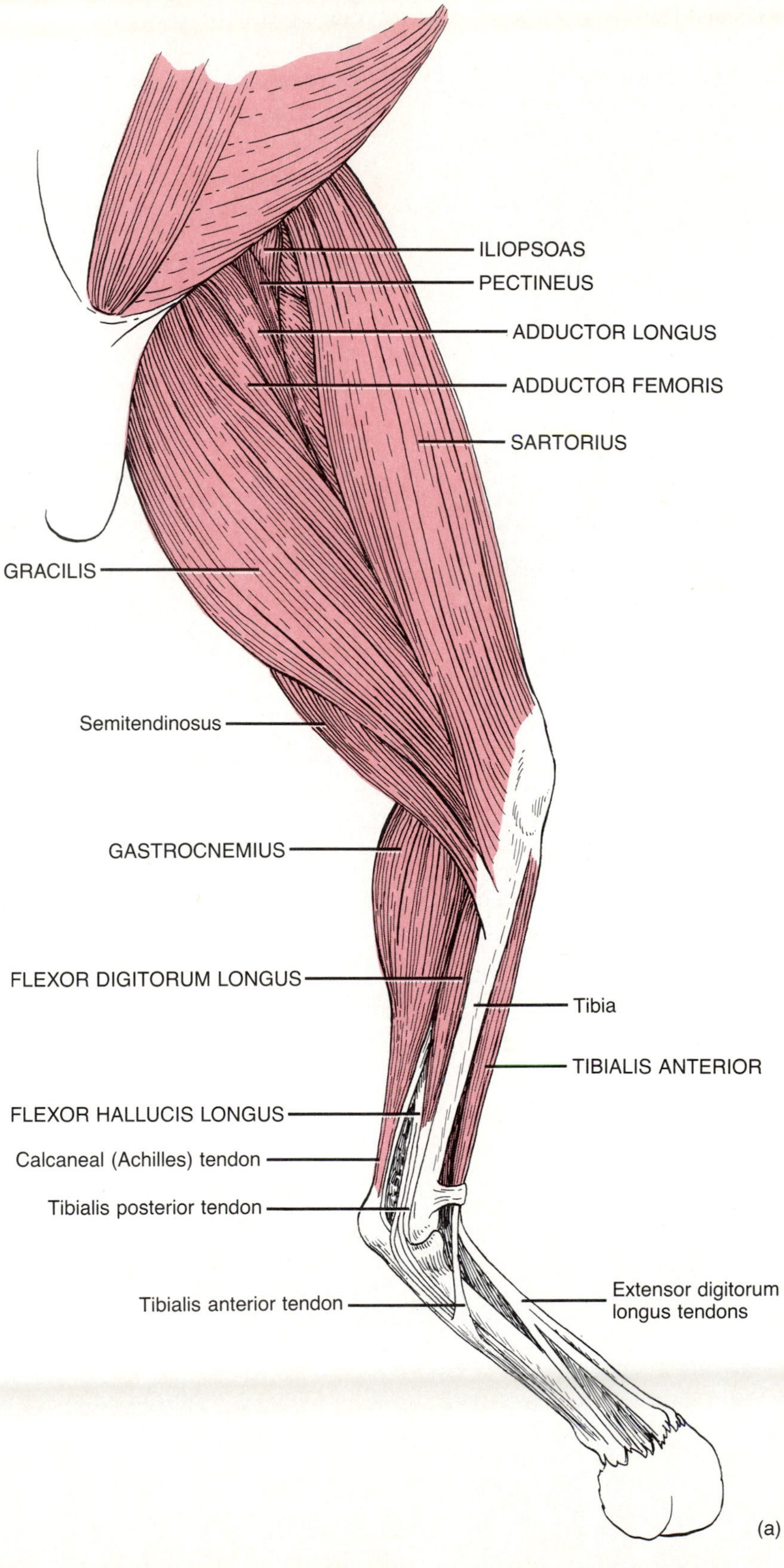

FIGURE 10.31 Muscles of medial surface of thigh and shank. (a) Superficial view.

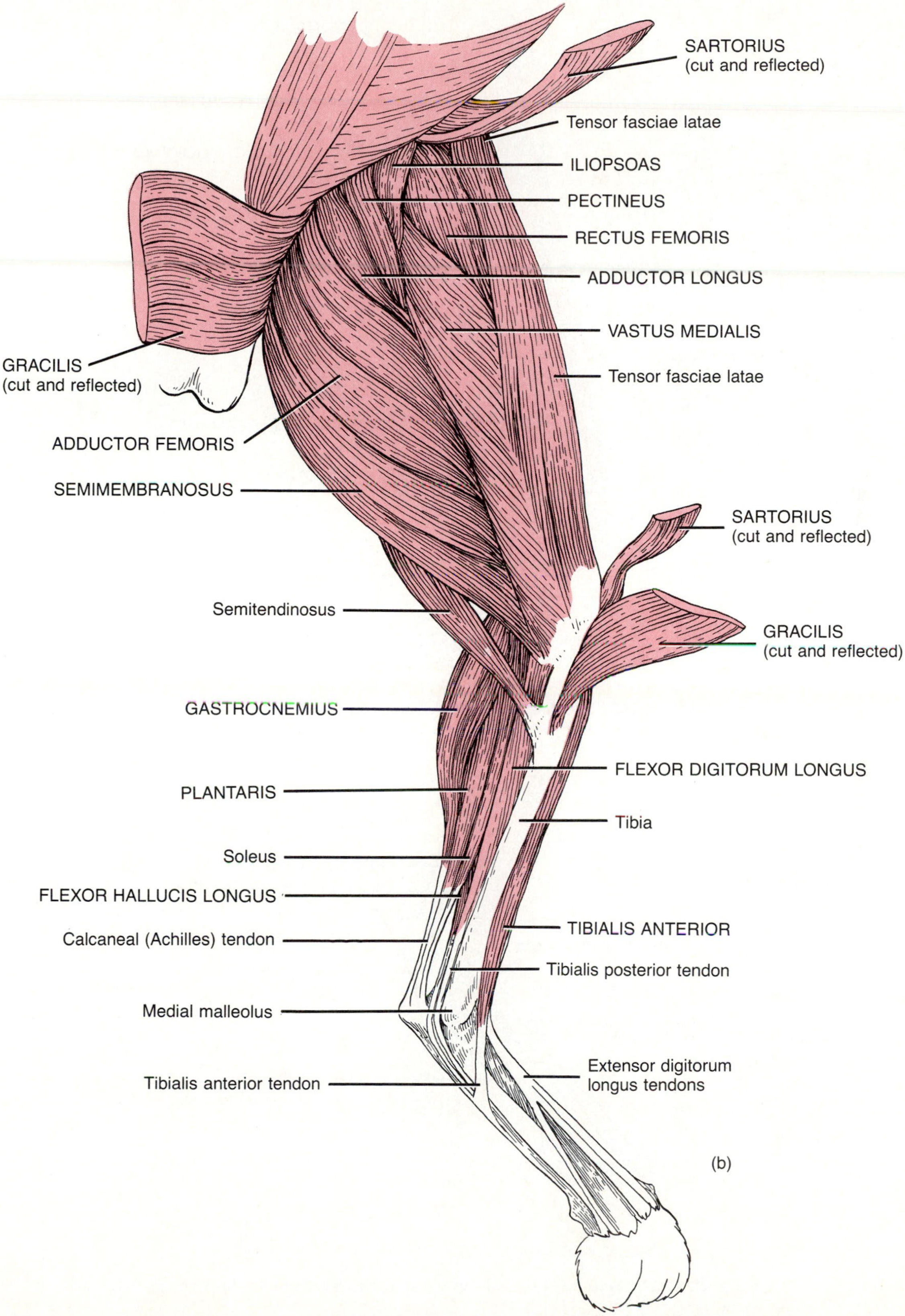

FIGURE 10.31 ***(Continued)*** Muscles of medial surface of thigh and shank. (b) Deep view.

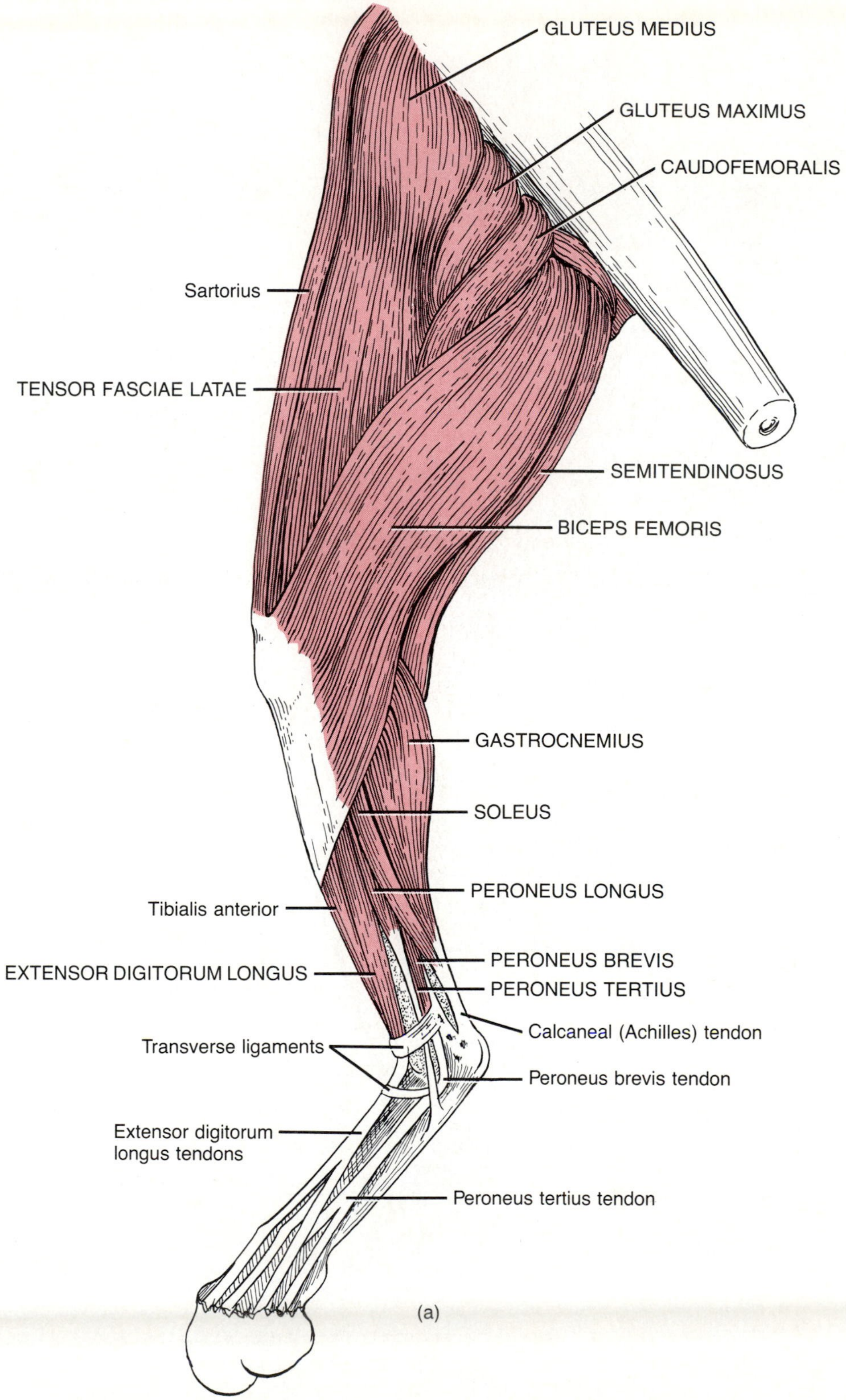

FIGURE 10.32 Muscles of lateral surface of thigh and shank. (a) Superficial view.

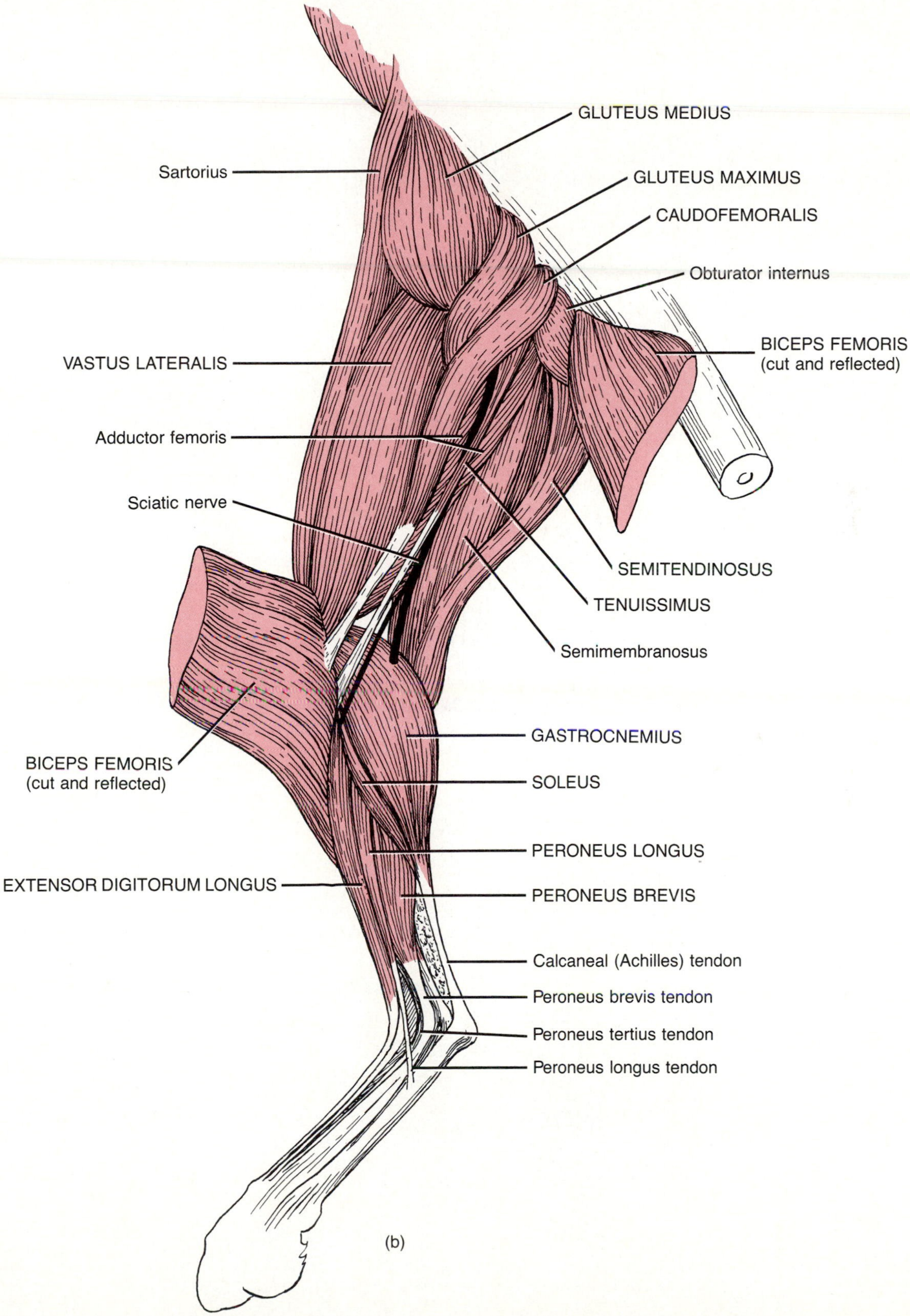

FIGURE 10.32 ***(Continued)*** Muscles of lateral surface of thigh and shank. (b) Deep view.

LABORATORY REPORT QUESTIONS

SKELETAL MUSCLES 10

Student ______________________ Date ______________________

Laboratory Section ______________________ Score/Grade ______________________

PART I. MULTIPLE CHOICE

_____ 1. The connective tissue covering that encloses the entire skeletal muscle is the (a) perimysium (b) endomysium (c) epimysium (d) mesomysium

_____ 2. A cord of connective tissue that attaches a skeletal muscle to the periosteum of bone is called a(n) (a) ligament (b) aponeurosis (c) perichondrium (d) tendon

_____ 3. A skeletal muscle that decreases the angle at a joint is referred to as a(n) (a) flexor (b) abductor (c) pronator (d) evertor

_____ 4. The name *abductor* means that a muscle (a) produces a downward movement (b) moves a part away from the midline (c) elevates a body part (d) increases the angle at a joint

_____ 5. Which connective tissue layer directly encircles the fasciculi of skeletal muscles? (a) epimysium (b) endomysium (c) perimysium (d) mesomysium

_____ 6. Which muscle is *not* associated with a movement of the eyeball? (a) superior rectus (b) superior oblique (c) medial rectus (d) genioglossus

_____ 7. Of the following, which muscle is involved in compression of the abdomen? (a) external oblique (b) superior oblique (c) medial rectus (d) genioglossus

_____ 8. A muscle directly concerned with breathing is the (a) sternocleidomastoid (b) mentalis (c) brachialis (d) external intercostal

_____ 9. Which muscle is *not* related to movement of the wrist? (a) extensor carpi ulnaris (b) flexor carpi radialis (c) supinator (d) flexor carpi ulnaris

_____ 10. A muscle that helps move the thigh is the (a) piriformis (b) triceps brachii (c) hypoglossus (d) peroneus tertius

_____ 11. Which muscle is *not* related to mastication? (a) temporalis (b) masseter (c) lateral rectus (d) medial pterygoid

_____ 12. All of these muscles form the circular layer of the musculature of the pharynx *except* the (a) middle constrictor (b) inferior gemellus (c) inferior constrictor (d) superior constrictor

_____ 13. Which muscle does *not* belong with the others because of its relationship to the hyoid bone? (a) digastric (b) mylohyoid (c) geniohyoid (d) thyrohyoid

_____ 14. Which muscle is *not* a component of the anterolateral abdominal wall? (a) external oblique (b) psoas major (c) rectus abdominis (d) internal oblique

_____ 15. All are components of the pelvic diaphragm *except* the (a) anconeus (b) coccygeus (c) iliococcygeus (d) pubococcygeus

_____ 16. Which is *not* a flexor of the forearm? (a) biceps brachii (b) brachialis (c) brachioradialis (d) triceps brachii

_____ 17. The abductor pollicis brevis, opponeus pollicis, flexor pollicis brevis, and adductor pollicis are all components of the (a) thenar eminence (b) midpalmar muscles (c) hypothenar eminence (d) erector spinae

_____ 18. Which muscle is *not* involved in moving the vertebral column? (a) splenius (b) longissimus (c) spinalis (d) sartorius

_____ 19. Which muscle flexes the wrists? (a) palmaris longus (b) extensor carpi radialis longus (c) supinator (d) extensor indicis

_____ 20. Which muscle is *not* involved in flexion of the thigh? (a) rectus femoris (b) sartorius (c) biceps femoris (d) vastus intermedius

_____ 21. Of the muscles that move the foot and toes, the anterior compartment muscles are involved in (a) plantar flexion (b) dorsiflexion (c) abduction (d) adduction

_____ 22. Which muscle is deepest? (a) plantar interosseous (b) adductor hallucis (c) quadratus plantae (d) abductor hallucis

PART II. MATCHING

Identify the criterion (or criteria) used to name the following muscles:

_____ 23. Supinator
_____ 24. Deltoid
_____ 25. Stylohyoid
_____ 26. Flexor carpi radialis
_____ 27. Gluteus maximus
_____ 28. External oblique
_____ 29. Triceps brachii
_____ 30. Adductor longus
_____ 31. Temporalis
_____ 32. Trapezius

A. Location
B. Shape
C. Size
D. Direction of fibers
E. Action
F. Number of origins
G. Insertion and origin

PART III. COMPLETION

33. The principal cheek muscle used in compression of the cheek is the _____________ .

34. The muscle that protracts the tongue is the _____________ .

35. The eye muscle that rolls the eyeball downward is the _____________ .

36. The _____________ muscle flexes the neck on the chest.

37. The abdominal muscle that flexes the vertebral column is the _____________ .

38. The muscle of the pectoral (shoulder) girdle that depresses the clavicle is the _____________ .

39. Flexion, adduction, and medial rotation of the arm are accomplished by the _____________ muscle.

40. The _____________ muscle is the most important extensor of the forearm.

41. The muscle that flexes and abducts the wrist is the _____________ .

42. The four muscles that extend the legs are the vastus lateralis, vastus medialis, vastus intermedius, and _____________ .

43. Muscles that lie inferior to the hyoid bone are referred to as _____________ muscles.

44. The neck region is divided into two principal triangles by the _____________ muscle.

45. Muscles that move the pectoral (shoulder) girdle originate on the axial skeleton and insert on the clavicle or ______________________ .
46. The muscles that move the arm (humerus) and do not originate on the scapula are called ______________________ muscles.
47. Together, the subscapularis, supraspinatus, infraspinatus, and teres major muscles form the ______________________ .
48. The posterior muscles involved in moving the wrist, hand, and fingers function in extension and ______________________ .
49. The posterior muscles of the thigh are involved in ______________________ of the leg.
50. Together, the biceps femoris, semitendinosus, and semimembranosus are referred to as the ______________________ muscles.

SURFACE ANATOMY 11

Now that you have studied the skeletal and muscular systems, you will be introduced to the study of ***surface anatomy***, in which you will study the form and markings of the surface of the body.[1] A knowledge of surface anatomy will help you identify certain superficial structures by visual inspection or palpation (to feel with the hand).

A convenient way to study surface anatomy is first to divide the body into its principal regions: head, neck, trunk, and upper and lower extremities. These may be reviewed in Figure 2.1.

A. HEAD

The ***head*** (cephalic region or caput) is divisible into the cranium and face. Several surface features of the head are:

1. ***Cranium*** (brain case or skull)
   a. ***Frontal region***—Front of skull (sinciput) that includes forehead.
   b. ***Parietal region***—Crown of skull (vertex).
   c. ***Temporal region***—Side of skull (tempora).
   d. ***Occipital region***—Base of skull (occiput).
2. ***Face*** (Facies)
   a. ***Orbital*** or ***ocular region***—Includes eyeballs (bulbi oculorum), eyebrows (supercilia), and eyelids (palpebrae).
   b. ***Nasal region***—Nose (nasus).
   c. ***Infraorbital region***—Inferior to orbit.
   d. ***Oral region***—Mouth.
   e. ***Mental region***—Anterior part of mandible.
   f. ***Buccal region***—Cheek.
   g. ***Parotid-masseteric region***—External to parotid gland and masseter muscle.
   h. ***Zygomatic region***—Inferolateral to orbit.
   i. ***Auricular region***—Ear.

Using your textbook as an aid, label Figure 11.1.

Using a mirror, examine the various features of the head just described. Working with a partner, be sure that you can identify the regions by both common *and* anatomical names.

The surface anatomy features of the eyeball and accessory structures are presented in Figure 14.10, of the ear in Figure 14.19, and of the nose in Figure 21.3.

B. NECK

The ***neck*** (collum) can be divided into an anterior cervical region, two lateral cervical regions, and a posterior (nuchal) region. Among the surface features of the neck are:

[1]At the discretion of your instructor, surface anatomy may be studied either before or in conjunction with your study of various body systems. This exercise can be used as an excellent review of many topics already studied.

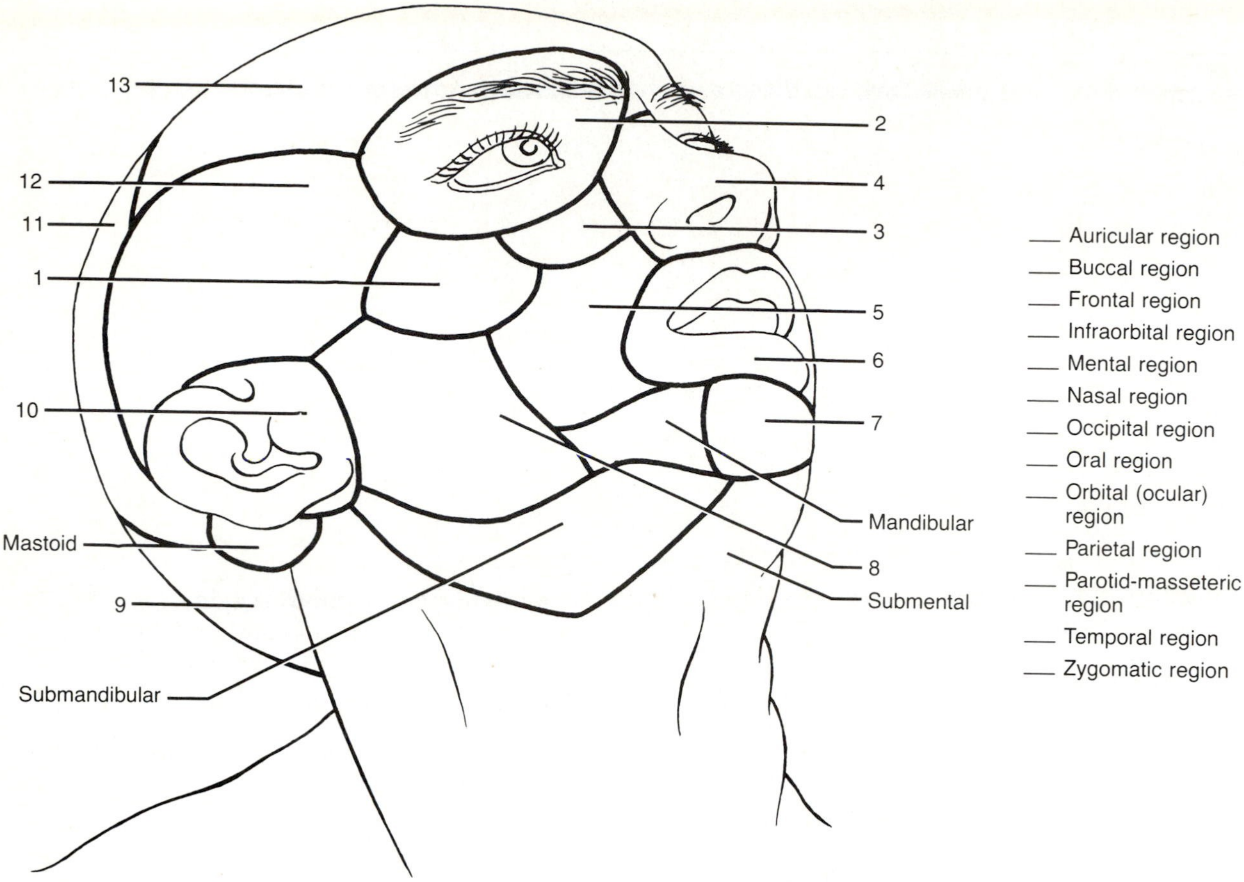

FIGURE 11.1 Regions of the cranium and face.

1. ***Thyroid cartilage (Adam's apple)***—Triangular laryngeal cartilage in the midline of the anterior cervical region.

2. ***Hyoid bone***—First resistant structure palpated in the midline below the chin, lying just above thyroid cartilage.

3. ***Cricoid cartilage***—Inferior laryngeal cartilage that attaches larynx to trachea. This structure can be palpated by running your fingertip down from your chin over the thyroid cartilage. (After you pass the cricoid cartilage, your fingertip sinks in.)

4. ***Sternocleidomastoid muscles***—Form major portion of lateral cervical regions, extending from mastoid process of temporal bone (felt as bump behind pinna of ear) to sternum and clavicle. Each muscle divides the neck into an anterior and posterior (lateral) triangle.

5. ***Trapezius muscles***—Form portion of lateral cervical region, extending downward and outward from base of skull. "Stiff neck" is frequently associated with inflammation of these muscles.

6. ***Anterior triangle of neck***—Bordered superiorly by mandible, inferiorly by sternum, medially by cervical midline, and laterally by anterior border of sternocleidomastoid muscle.

7. ***Posterior (lateral) triangle of neck***—Bordered inferiorly by clavicle, anteriorly by posterior border of sternocleidomastoid muscle, and posteriorly by anterior border of trapezius muscle.

8. ***External jugular veins***—Prominent veins along lateral cervical regions, readily seen when a person is angry or a collar fits too tightly.

Using a mirror and working with a partner, use your textbook as an aid in identifying the surface features of the neck just described. Then label Figure 11.2.

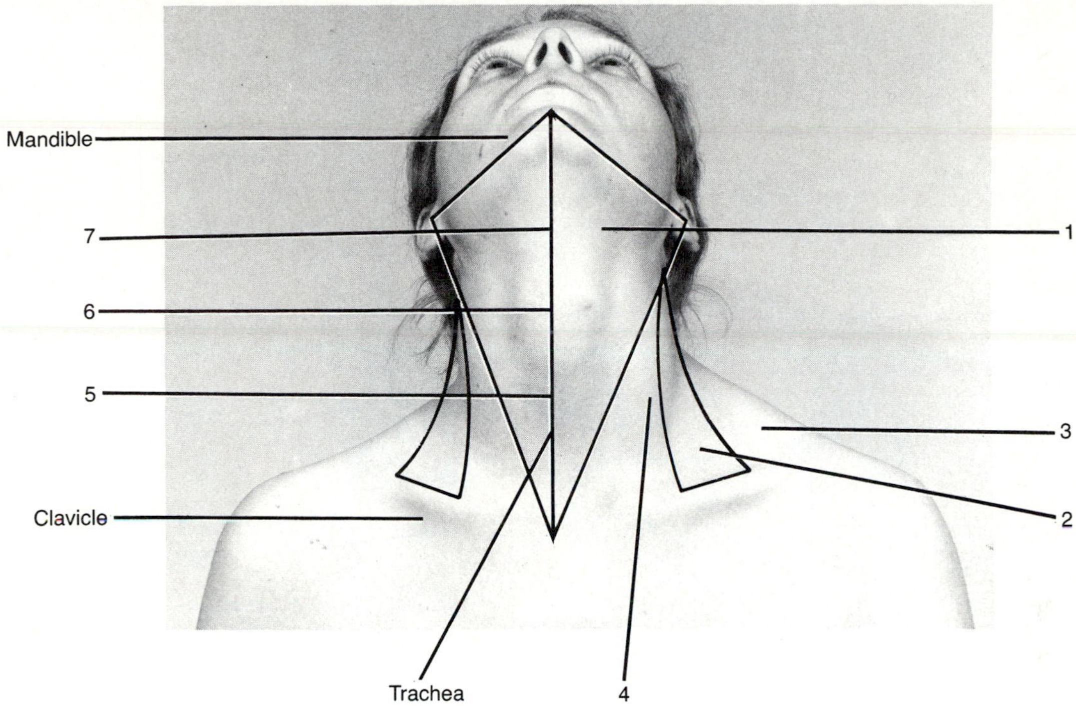

___ Anterior triangle of neck
___ Cricoid cartilage
___ Hyoid bone
___ Posterior triangle of neck
___ Sternocleidomastoid muscle
___ Thyroid cartilage
___ Trapezius muscle

FIGURE 11.2 Surface anatomy of neck seen in anterior view.

C. TRUNK

The ***trunk*** is divided into the back, chest, abdomen, and pelvis. Its surface features include:

1. ***Back (dorsum)***
   a. ***Vertebral spines***—Dorsally pointed projections of vertebrae. Very prominent is the ***vertebrae prominens***, the spine of C7, easily palpated when the head is flexed.
   b. ***Scapula***—Shoulder blade. Several parts of the scapula, such as the medial (axillary) border, lateral (vertebral) border, inferior angle, spine, and acromion, may be observed or palpated.
   c. ***Ribs***—These may be seen in thin individuals.
   d. ***Muscles***—Among the visible superficial back muscles are the ***latissimus dorsi*** covers lower half of back), ***erector spinae*** (on either side of vertebral column), ***infraspinatus*** (inferior to spine of scapula), ***trapezius***, and ***teres major*** (inferior to infraspinatus).

Using your textbook as an aid, label Figure 11.3.

2. ***Chest (thorax)***
   a. ***Clavicles***—Collarbones. These lie in superior region of thorax and can be palpated along their entire length.
   b. ***Sternum***—Breastbone, lies in midline of chest. The following parts of the sternum are important surface features:
      1. ***Jugular (suprasternal) notch***—Depression on superior surface of sternum between medial ends of clavicles. The trachea can be palpated in the notch.

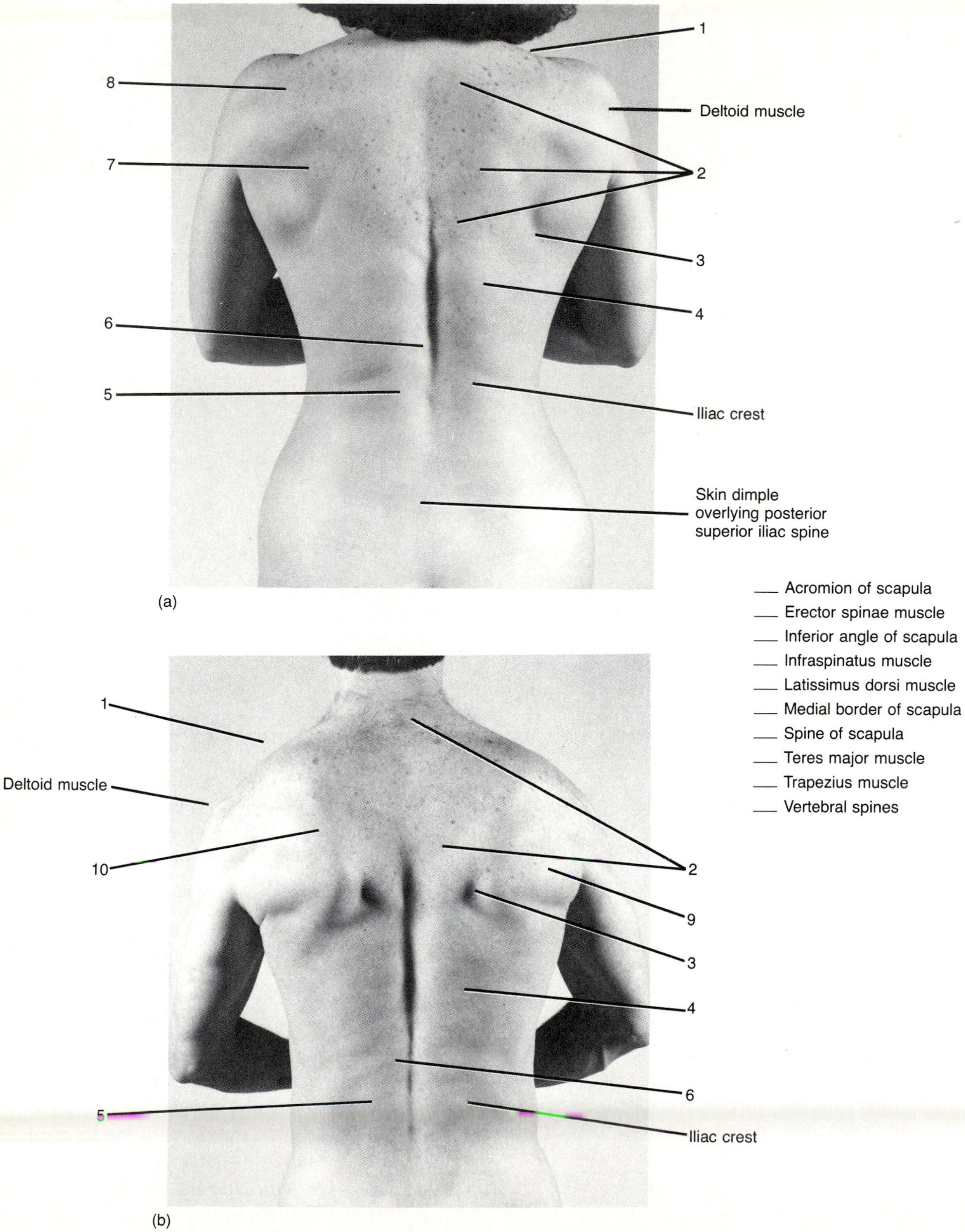

FIGURE 11.3 Surface anatomy of back. (a) Female. (b) Male.

2. ***Sternal angle***—Formed by junction of manubrium and body of sternum. This is palpable under the skin and locates the sternal end of second rib.
3. ***Xiphoid process***—Inferior portion of sternum.

c. ***Ribs***

d. ***Costal margins***—Inferior edges of costal cartilages of ribs 7 through 10.

e. ***Muscles***—Among the superficial chest muscles that can be seen are the ***pectoralis major*** (principal upper chest muscle) and ***serratus anterior*** (inferior and lateral to pectoralis major).

Using your textbook as an aid, label Figure 11.4.

3. ***Abdomen*** and ***Pelvis***

a. ***Umbilicus***—Also called navel, site of attachment of umbilical cord in fetus. Umbilicus is most obvious surface marking on the abdomen of most individuals.

b. ***Linea alba***—Slight groove in midline extending from xiphoid process to symphysis pubis. This is particularly obvious in thin, muscular individuals.

c. ***Tendinous intersections***—Fibrous bands that run transversely across the rectus

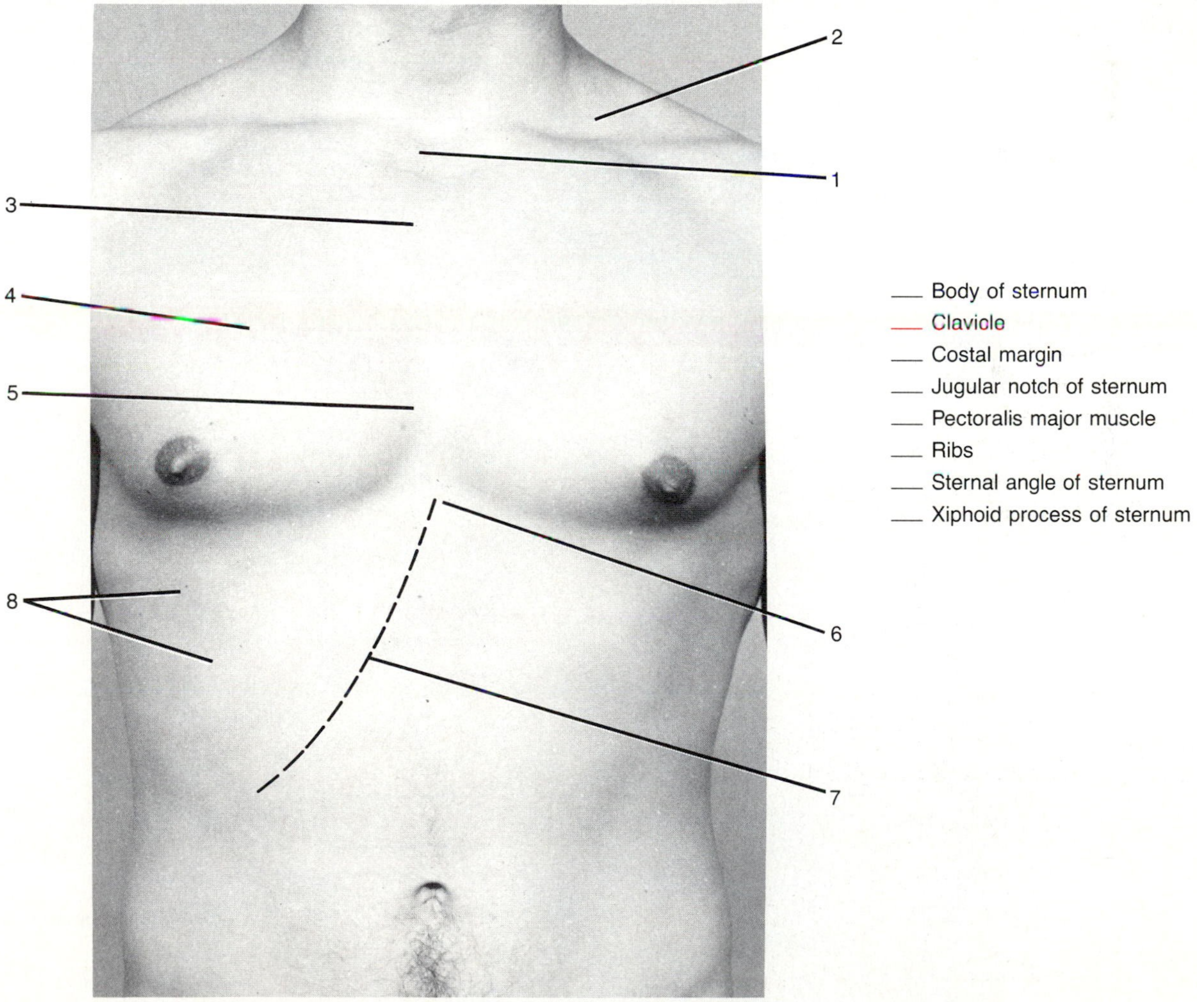

FIGURE 11.4 Surface anatomy of chest. (The serratus anterior muscle is shown in Figure 11.5.)

abdominis muscle. Three or more are visible in muscular individuals.

d. ***Muscles***—Among the superficial abdominal muscles are the ***external oblique*** (inferior to serratus anterior) and ***rectus abdominis*** (just lateral to midline of abdomen).

e. ***Symphysis pubis***—Anterior joint of hipbones. This structure is palpated as a firm resistance in the midline at the inferior portion of the anterior abdominal wall.

Using your textbook as an aid, label Figure 11.5.

D. UPPER EXTREMITY

The ***upper extremity*** consists of the armpit, arm, shoulder, elbow, forearm, wrist, and hand (palm and fingers).

1. Major surface features of the ***shoulder (acromial)*** region are:

   a. ***Acromioclavicular joint***—Slight elevation at lateral end of clavicle.

   b. ***Acromion***—Expanded end of spine of scapula. This is clearly visible in some individuals and can be palpated about 2.5 cm (1 in.) distal to acromioclavicular joint (see Figure 11.3).

   c. ***Deltoid muscle***—Triangular muscle that forms rounded prominence of shoulder. This is a frequent site for intramuscular injections (see Figure 11.3).

2. Major surface features of the ***arm (brachium)*** and ***elbow (cubitus)*** are:

   a. ***Biceps brachii muscle***—Forms bulk of anterior surface of arm.

   b. ***Triceps brachii muscle***—Forms bulk of posterior surface of arm.

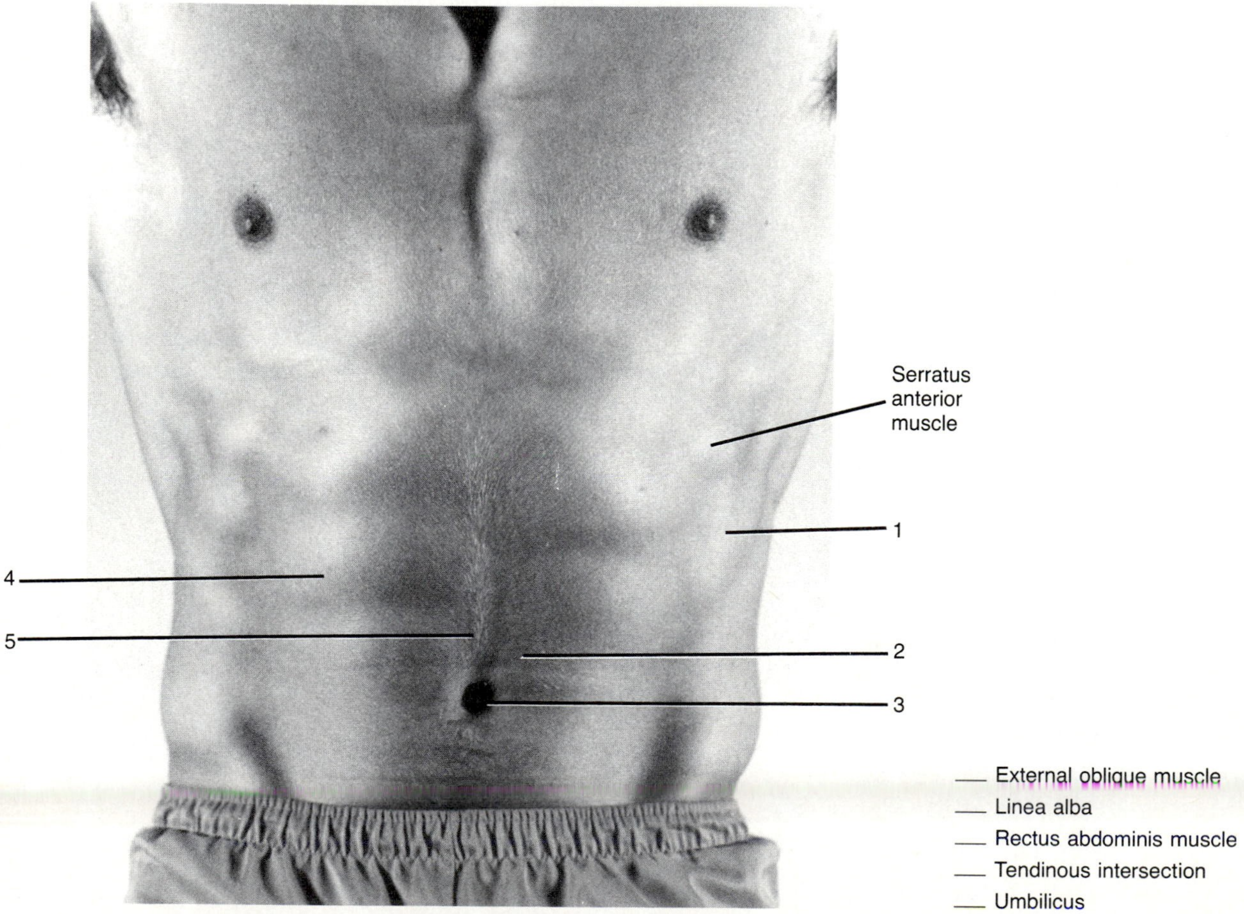

FIGURE 11.5 Surface anatomy of the anterior abdomen.

c. ***Medial epicondyle***—Medial projection at distal end of humerus.

d. ***Ulnar nerve***—Can be palpated as a rounded cord in a groove behind the medial epicondyle.

e. ***Lateral epicondyle***—Lateral projection at distal end of humerus.

f. ***Olecranon***—Projection of proximal end of ulna that lies between and slightly superior to epicondyles when forearm is extended; forms elbow.

g. ***Cubital fossa***—Triangular space in anterior region of elbow. This area contains biceps brachii tendon, brachial artery and its terminal branches (radial and ulnar arteries), and parts of median and radial nerves.

h. ***Median cubital vein***—Crosses cubital fossa obliquely. This vein is frequently selected for removal of blood.

Using your textbook as an aid, label Figure 11.6.

3. Major surface features of the ***forearm (antebrachium)*** and ***wrist (carpus)*** are:

a. ***Styloid process of ulna***—Projection of distal end of ulna at medial side of wrist.

b. ***Styloid process of radius***—Projection of distal end of radius at lateral side of wrist.

c. ***Brachioradialis muscle***—Located at superior and lateral aspect of forearm.

d. ***Flexor carpi radialis muscle***—Located along midportion of forearm.

e. ***Flexor carpi ulnaris muscle***—Located at medial aspect of forearm.

f. ***Tendon of palmaris longus muscle***—Located on anterior surface of wrist near ulna. When you make a fist, you can see this tendon. The palmaris longus muscle is frequently absent on one or both sides.

g. ***Tendon of flexor carpi radialis muscle***—Tendon on anterior surface of wrist lateral to tendon of palmaris longus.

h. ***Radial artery***—Can be palpated just medial to styloid process of radius; this artery is frequently used to take pulse.

i. ***Pisiform bone***—Medial bone of proximal carpals. The bone is easily palpated as a projection distal to styloid process of ulna.

j. ***Tendon of extensor pollicis brevis muscle***—Tendon close to styloid process of radius along posterior surface of wrist, best seen when thumb is bent backward.

k. ***Tendon of extensor pollicis longus muscle***—Tendon closer to styloid process of ulna along posterior surface of wrist, best seen when thumb is bent backward.

l. ***"Anatomical snuffbox"***—Depression between tendons of extensor pollicis brevis and extensor pollicis longus muscles. Radial artery can be palpated in the depression.

m. ***Wrist creases***—Three more or less constant lines on anterior aspect of wrist where skin is firmly attached to underlying deep fascia.

Using your textbook as an aid, label Figure 11.7.

4. Major surface features of the ***hand (manus)*** are:

a. ***"Knuckles"***—Commonly refers to dorsal aspects of distal ends of metacarpals II, III, IV, and V. Term also includes dorsal aspects of metacarpophalangeal and interphalangeal joints.

b. ***Thenar eminence***—Lateral rounded contour on anterior surface of hand formed by muscles of thumb.

c. ***Hypothenar eminence***—Medial rounded contour on anterior surface of hand formed by muscles of little finger.

d. ***Skin creases***—Several more or less constant lines on the anterior aspect of the palm and digits where skin is firmly attached to underlying deep fascia. These are generally divided into ***palmar flexion creases*** and ***digital flexion creases***.

e. ***Extensor tendons***—Besides the tendons of the extensor pollicis brevis and extensor pollicis longus muscles associated with the thumb, the following extensor tendons are also visible on the posterior aspect of the hand: ***extensor digiti minimi tendon*** in line with phalanx V (little finger) and ***extensor digitorum*** in line with phalanges II, III, and IV.

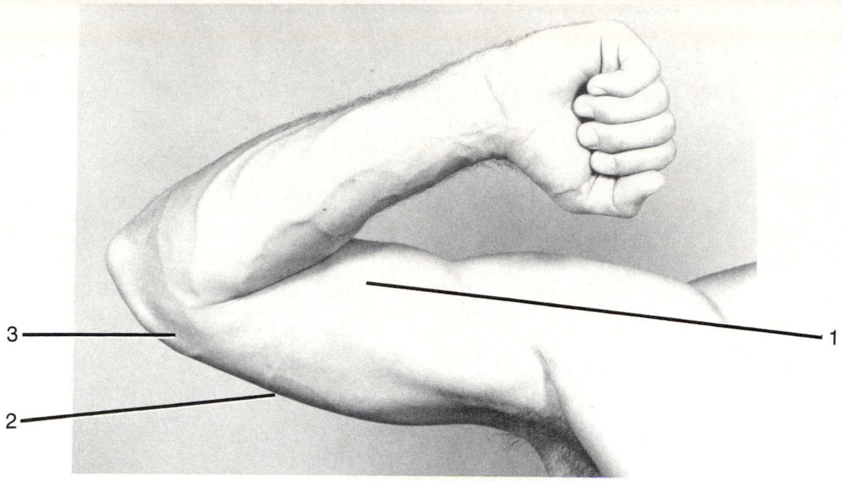

(a)

___ Biceps brachii muscle
___ Cubital fossa
___ Lateral epicondyle of humerus
___ Medial epicondyle of humerus
___ Median cubital vein
___ Olecranon of ulna
___ Triceps brachii muscle

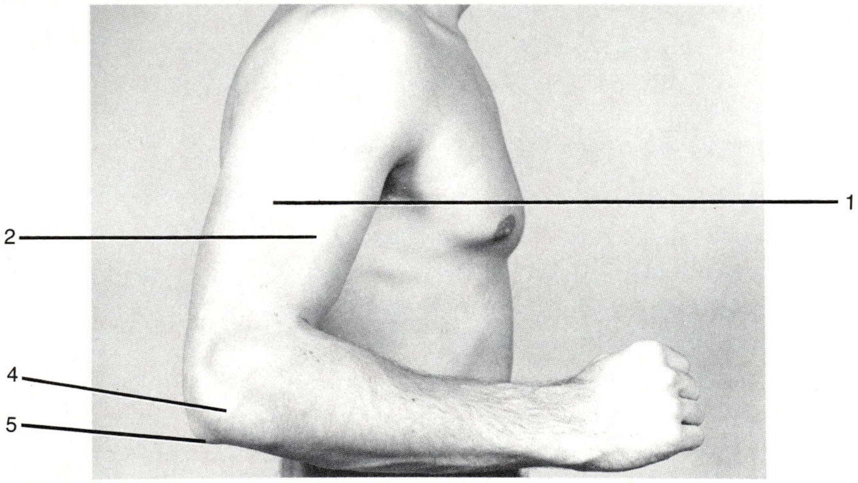

(b)

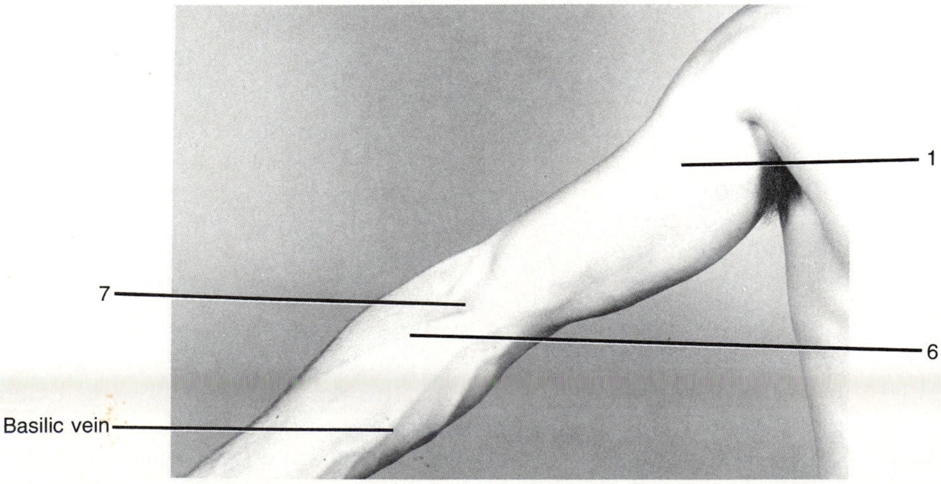

(c)

FIGURE 11.6 Surface anatomy of arm and elbow. (a) Medial view. (b) Lateral view. (c) Anterior view.

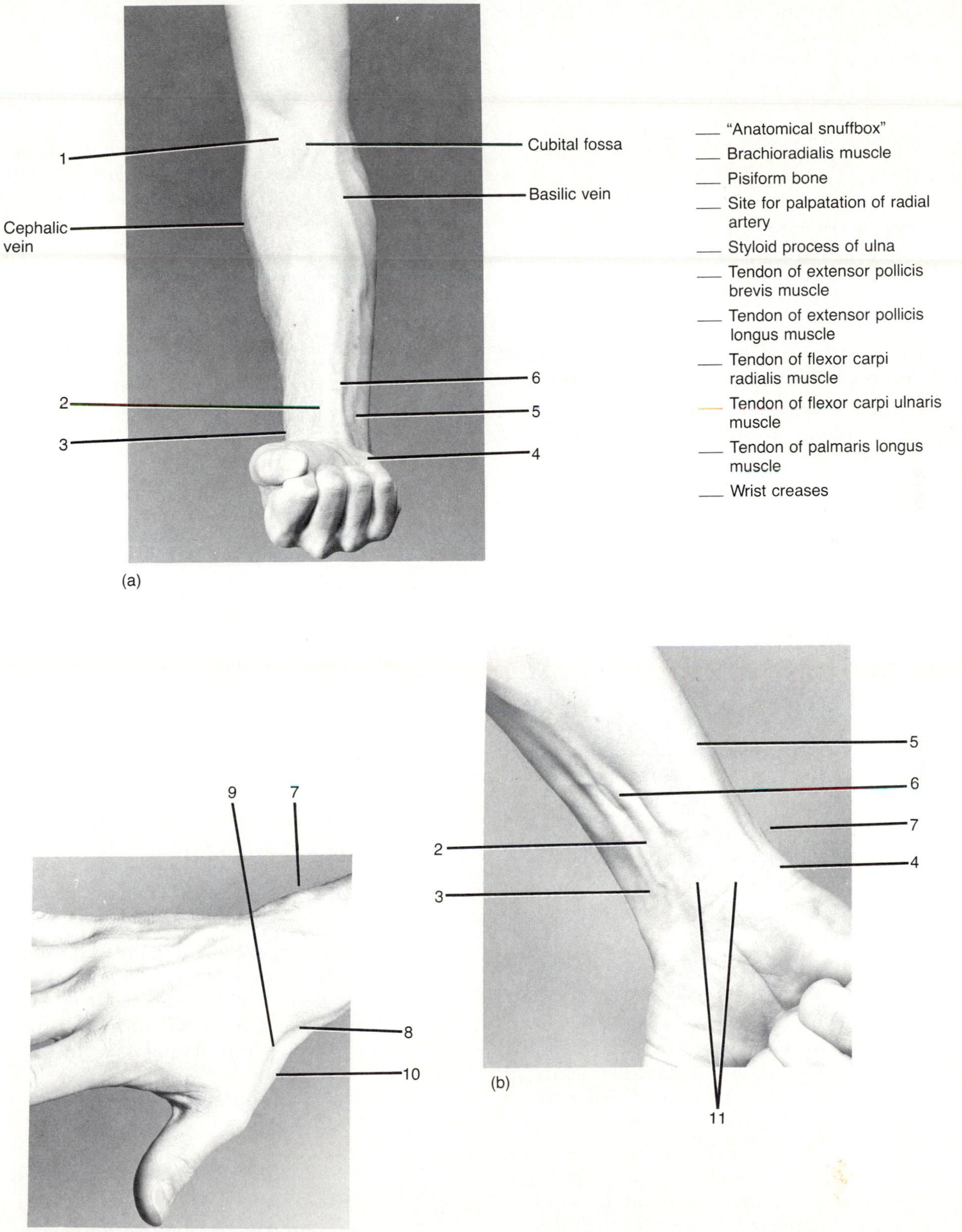

FIGURE 11.7 Surface anatomy of forearm and wrist. (a) and (b) Anterior views. (c) Posterolateral view.

f. ***Dorsal venous arch***—Superficial veins on posterior surface of hand; displayed by compressing the blood vessels at the wrist for a few minutes as the hand is opened and closed.

With the aid of your textbook, label Figure 11.8.

E. LOWER EXTREMITY

The ***lower extremity*** consists of the buttocks, thigh, knee, leg, ankle, and foot.

1. Major surface features of the ***buttocks (gluteal region)*** and ***thigh (femoral region)*** are:

a. ***Iliac crest***—Superior margin of ilium of hipbone, forming outline of superior border of buttock. When you rest your hands on your hips, they rest on the iliac crests.

b. ***Posterior superior iliac spine***—Posterior termination of iliac crest; lies deep to a dimple (skin depression) about 4 cm (1.5 in.) lateral to midline. Dimple forms because the skin and underlying fascia are attached to bone.

c. ***Gluteus maximus muscle***—Forms major portion of prominence of buttock.

d. ***Gluteus medius muscle***—Superolateral to gluteus maximus. This is a frequent site for intramuscular injections.

e. ***Gluteal cleft***—Depression along midline that separates the buttocks.

f. ***Gluteal fold***—Inferior limit of buttock formed by inferior margin of gluteus maximus muscle.

g. ***Ischial tuberosity***—Bony prominence of ischium of hipbone, bears weight of body when seated.

h. ***Greater trochanter***—Projection of proximal end of femur on lateral surface of thigh. This can be palpated about 20 cm (8 in.) inferior to iliac crest.

i. ***Anterior thigh muscles***—Include ***sartorius*** (runs obliquely across thigh) and ***quadriceps femoris***, which consists of ***rectus femoris*** (midportion of thigh), ***vastus lateralis*** (anterolateral surface of thigh), ***vastus medialis*** (medial inferior portion of thigh), and ***vastus intermedius*** (deep to rectus femoris). Vastus lateralis is frequent injection site for diabetics.

j. ***Posterior thigh muscles***—Include the ***hamstrings (semitendinosus, semimembranosus,*** and ***biceps femoris).***

k. ***Medial thigh muscles***—Include ***adductor magnus, adductor brevis, adductor longus, gracilis, obturator externus,*** and ***pectineus***.

With the aid of your textbook, label Figure 11.9.

2. Major surface anatomy features of the ***knee (genu)*** are as follows:

a. ***Patella***—Kneecap, located within quadriceps femoris tendon on anterior surface of knee along midline. Margins of condyles (described shortly) can be felt on either side of it.

b. ***Patellar ligament***—Continuation of quadriceps femoris tendon inferior to patella.

c. ***Popliteal fossa***—Diamond-shaped space on posterior aspect of knee visible when knee is flexed. Fossa is bordered superolaterally by the biceps femoris muscle, superomedially by the semimembranosus and semitendinosus muscles, and inferolaterally and inferomedially by the lateral and medial heads of the gastrocnemius muscle.

d. ***Medial condyles of femur and tibia***—Medial projections just below patella. Upper part of projection belongs to distal end of femur, lower part of projection belongs to proximal end of tibia.

e. ***Lateral condyles of femur and tibia***—Lateral projections just below patella. Upper part of projections belongs to distal end of femur, lower part of projections belongs to proximal end of tibia.

With the aid of your textbook, label Figure 11.10.

3. Major surface anatomy features of the ***leg (crus)*** and ***ankle (tarsus)*** are:

a. ***Tibial tuberosity***—Bony prominence of tibia below patella into which patellar ligament inserts (see Fig. 11.10a).

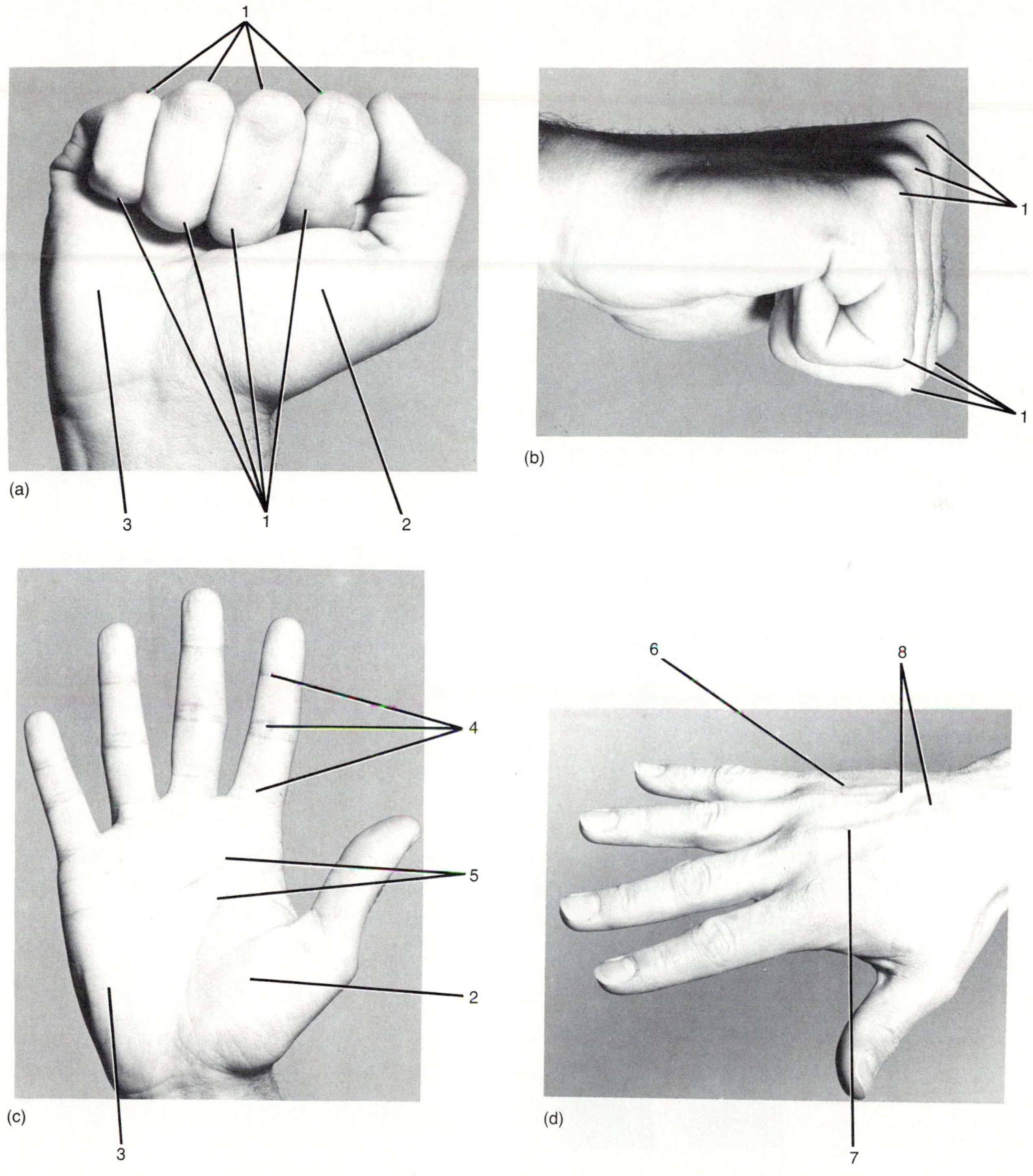

___ Digital flexion creases
___ Dorsal venous arch
___ Hypothenar eminence
___ "Knuckles"
___ Palmar flexion creases
___ Tendon of extensor digiti minimi muscle
___ Tendon of extensor digitorum muscle
___ Thenar eminence

FIGURE 11.8 Surface anatomy of hand. (a) Anterior view. (b) Medial view. (c) Anterior view. (d) Posterior view.

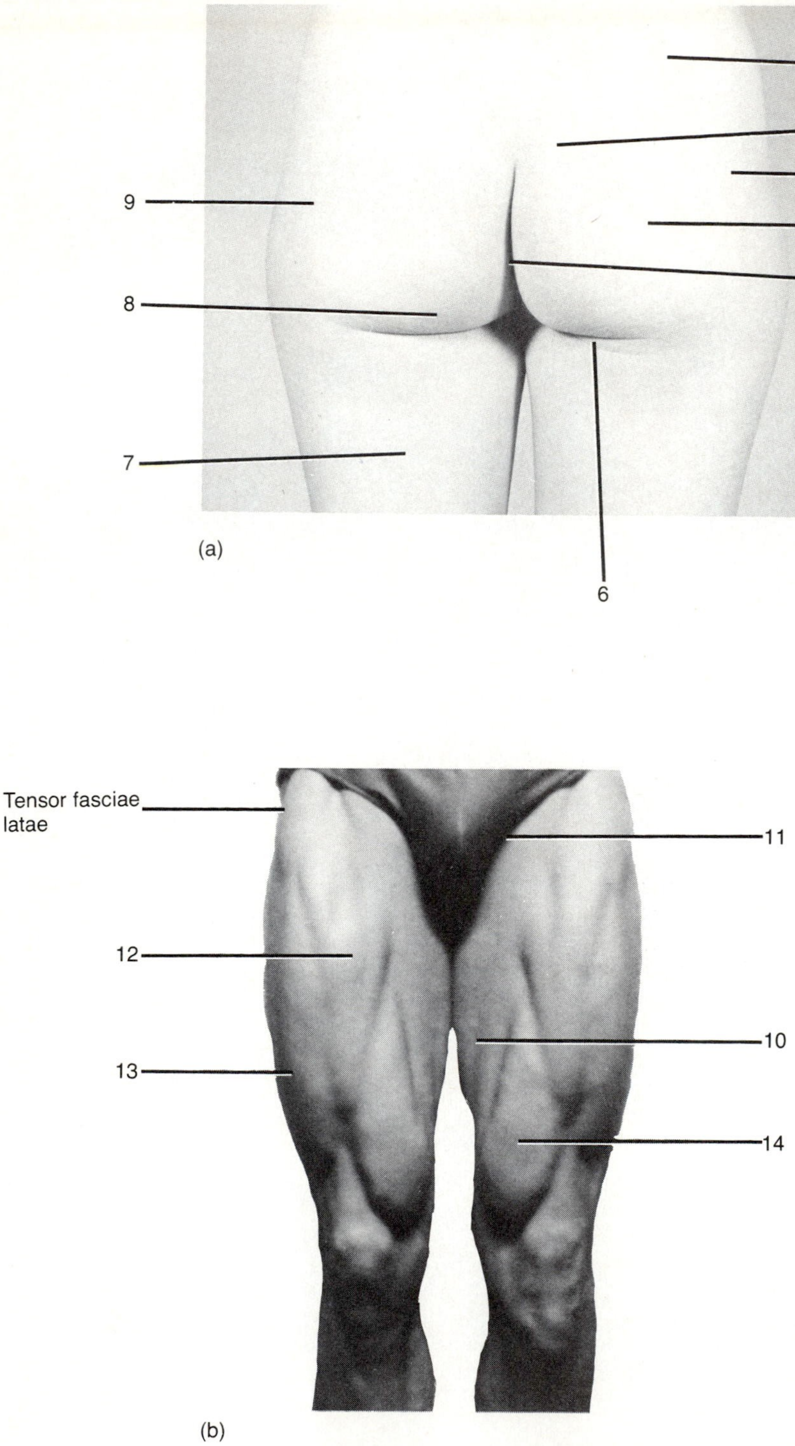

___ Adductor magnus muscle
___ Gluteal cleft
___ Gluteal fold
___ Gluteus maximus muscle
___ Gluteus medius muscle
___ Greater trochanter of femur
___ Hamstrings
___ Iliac crest
___ Posterior superior iliac spine
___ Rectus femoris muscle
___ Sartorius muscle
___ Site of ischial tuberosity
___ Vastus lateralis muscle
___ Vastus medialis muscle

FIGURE 11.9 Surface anatomy of buttocks and thigh. (a) Posterior view. (b) Anterior view.

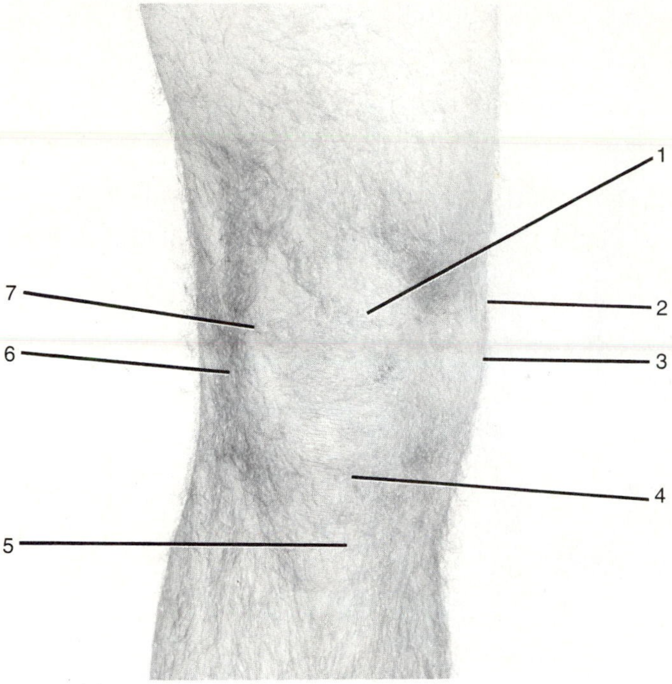

(a)

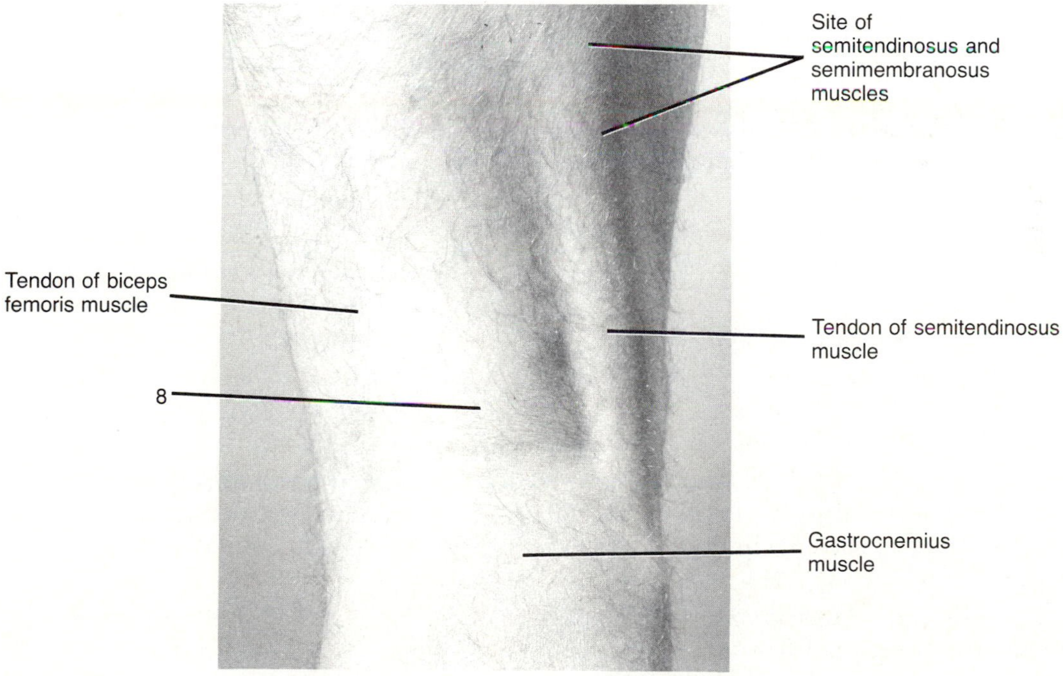

(b)

___ Lateral condyle of femur
___ Lateral condyle of tibia
___ Medial condyle of femur
___ Medial condyle of tibia
___ Patella
___ Patellar ligament
___ Popliteal fossa
___ Tibial tuberosity

FIGURE 11.10 Surface anatomy of knee. (a) Anterior view. (b) Posterior view.

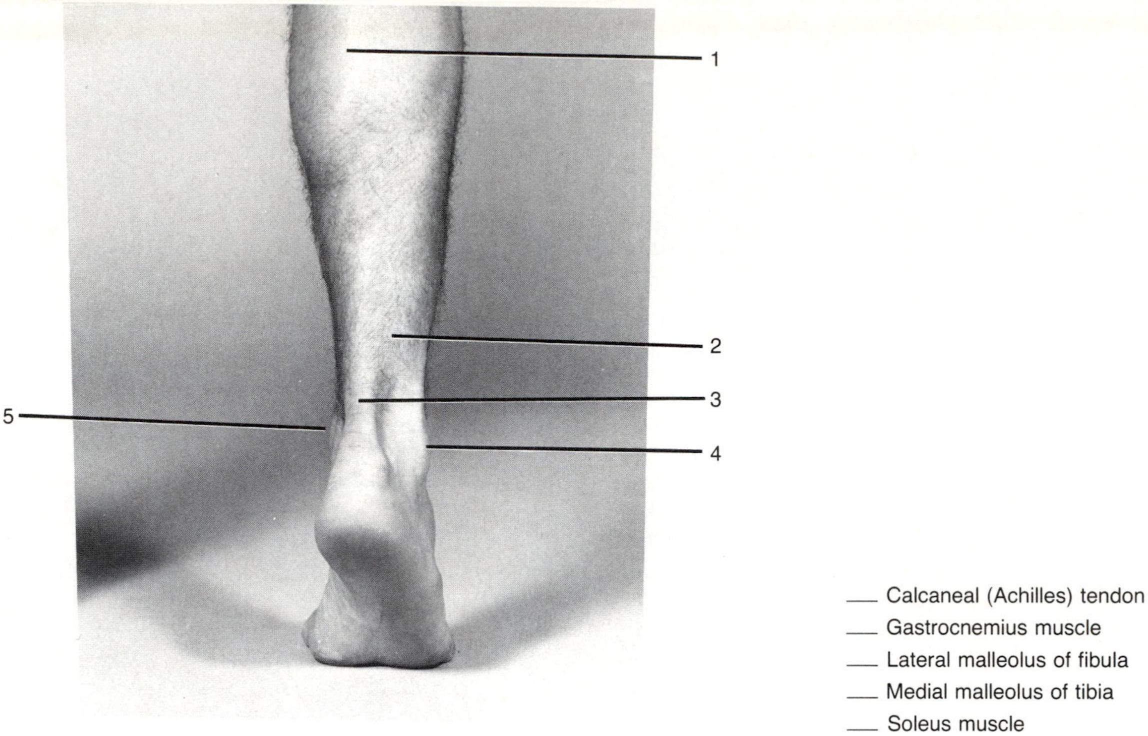

FIGURE 11.11 Surface anatomy of leg seen in posterior view.

b. ***Medial malleolus***—Projection of distal end of tibia that forms medial prominence of ankle.

c. ***Lateral malleolus***—Projection of distal end of fibula that forms lateral prominence of ankle.

d. ***Calcaneal (Achilles) tendon***—Tendon of insertion into calcaneus for gastrocnemius and soleus muscles.

e. ***Tibialis anterior muscle***—Located on anterior surface of leg.

f. ***Gastrocnemius muscle***—Forms bulk of middle and superior portions of posterior surface of leg.

g. ***Soleus muscle***—Mostly deep to gastrocnemius, visible on either side of gastrocnemius below middle of leg.

With the aid of your textbook, label Figure 11.11.

4. Major surface anatomy features of the ***foot (pes)*** are:

a. ***Calcaneus***—Heel bone to which calcaneal (Achilles) tendon inserts.

b. ***Tendon of extensor hallucis longus muscle***— Visible in line with phalanx I (great toe).

c. ***Tendons of extensor digitorum longus muscles*** —Visible in line with phalanges II through V.

d. ***Dorsal venous arch***—Superficial veins on dorsum of foot that unite to form small and great saphenous veins.

With the aid of your textbook, label Figure 11.12.

ANSWER THE LABORATORY REPORT QUESTIONS AT THE END OF THE EXERCISE

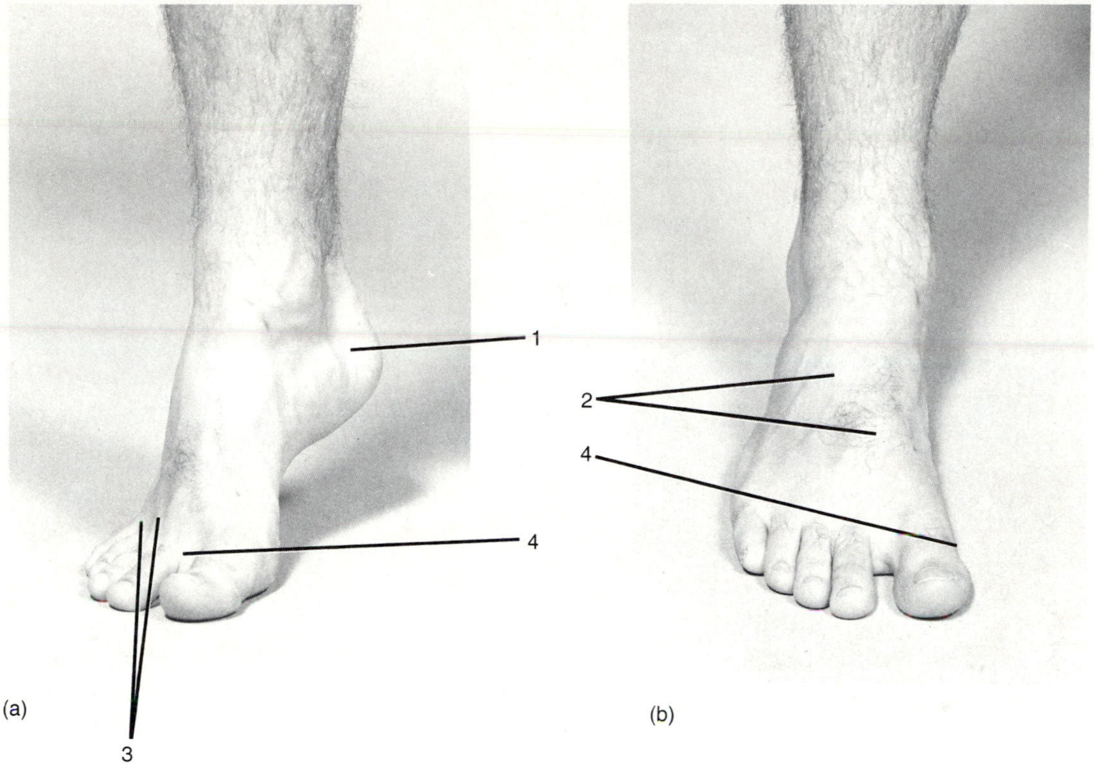

___ Calcaneus
___ Dorsal venous arch
___ Tendon of extensor digitorum longus muscle
___ Tendon of extensor hallucis longus muscle

FIGURE 11.12 Surface anatomy of foot. (a) Anteromedial view. (b) Anterior view.

LABORATORY REPORT QUESTIONS

SURFACE ANATOMY 11

Student ______________________ Date ______________________

Laboratory Section ______________________ Score/Grade ______________________

PART I. MULTIPLE CHOICE

_____ 1. The term used to refer to the crown of the skull (vertex) is (a) occipital (b) mental (c) parietal (d) zygomatic

_____ 2. The laryngeal cartilage in the midline of the anterior cervical region known as the Adam's apple is the (a) cricoid (b) epiglottis (c) arytenoid (d) thyroid

_____ 3. Inflammation of which muscle is associated with "stiff neck"? (a) teres major (b) trapezius (c) deltoid (d) cervicalis

_____ 4. The skeletal muscle located directly on either side of the vertebral column is the (a) serratus anterior (b) infraspinatus (c) teres major (d) erector spinae

_____ 5. The jugular notch and xiphoid process are associated with the (a) sternum (b) scapula (c) clavicle (d) ribs

_____ 6. The expanded end of the spine of the scapula is the (a) acromion (b) linea alba (c) olecranon (d) superior angle

_____ 7. Which nerve can be palpated as a rounded cord in a groove behind the medial epicondyle? (a) median (b) radial (c) ulnar (d) brachial

_____ 8. Which carpal bone can be palpated as a projection distal to the styloid process of the ulna? (a) trapezoid (b) trapezium (c) hamate (d) pisiform

_____ 9. The lateral rounded contour on the anterior surface of the hand (at the base of the thumb) formed by the muscles of the thumb is the (a) "anatomical snuffbox" (b) thenar eminence (c) hypothenar eminence (d) dorsal venous arch

_____ 10. The superior margin of the hipbone is the (a) symphysis pubis (b) iliac spine (c) acetabulum (d) iliac crest

_____ 11. Which bony structure bears the weight of the body when a person is seated? (a) greater trochanter (b) iliac crest (c) ischial tuberosity (d) gluteal fold

_____ 12. Which muscle is *not* a component of the quadriceps femoris group? (a) biceps femoris (b) vastus lateralis (c) vastus medialis (d) rectus femoris

_____ 13. The diamond-shaped space on the posterior aspect of the knee is the (a) cubital fossa (b) posterior triangle (c) popliteal fossa (d) nuchal groove

_____ 14. The projection of the distal end of the tibia that forms the prominence on one side of the ankle is the (a) medial condyle (b) medial malleolus (c) lateral condyle (d) lateral malleolus

_____ 15. The tendon that can be seen in line with the great toe belongs to which muscle? (a) extensor digiti minimi (b) extensor digitorum longus (c) extensor hallucis longus (d) extensor carpi radialis

PART II. COMPLETION

16. The laryngeal cartilage that connects the larynx to the trachea is the ____________ cartilage.

17. The ____________ triangle is bordered by the mandible, sternum, cervical midline, and sternocleidomastoid muscle.

18. The depression on the superior surface of the sternum between the medial ends of the clavicle is the ____________ .

19. The principal superficial chest muscle is the ____________ .

20. Tendinous intersections are associated with the ____________ muscle.

21. The muscle that forms the rounded prominence of the shoulder is the ____________ muscle.

22. The triangular space in the anterior aspect of the elbow is the ____________ .

23. The "anatomical snuffbox" is bordered by the tendons of the extensor pollicis brevis muscle and the ____________ muscle.

24. The dorsal aspects of the distal ends of metacarpals II through V are commonly referred to as ____________ .

25. The tendon of the ____________ muscle is in line with phalanx V.

26. The dimple that forms about 4 cm lateral to the midline just above the buttocks lies superficial to the ____________ .

27. The femoral projection that can be palpated about 20 cm (8 in.) inferior to the iliac crest is the ____________ .

28. The continuation of the quadriceps femoris tendon inferior to the patella is the ____________ .

29. The tendon of insertion for the gastrocnemius and soleus muscles is the ____________ tendon.

30. Superficial veins in the dorsum of the foot that unite to form the small and great saphenous veins belong to the ____________ .

31. The prominent veins along the lateral cervical regions are the ____________ veins.

32. The pronounced vertebral spine of C7 is the ____________ .

33. The most reliable surface anatomy feature of the chest is the ____________ .

34. A slight groove extending from the xiphoid process to the symphysis pubis is the ____________ .

35. The vein that crosses the cubital fossa and is frequently used to remove blood is the ____________ vein.

PART III. MATCHING

_____ 36. Mental region
_____ 37. Xiphoid process
_____ 38. Arm
_____ 39. Nucha
_____ 40. Shoulder
_____ 41. Gluteus maximus muscle
_____ 42. Costal margin
_____ 43. Olecranon
_____ 44. Wrist
_____ 45. Auricular region
_____ 46. Semitendinosus muscle
_____ 47. Leg
_____ 48. Cranium
_____ 49. Ankle
_____ 50. Hand

A. Inferior edges of costal cartilages of ribs 7 through 10
B. Forms elbow
C. Inferior portion of sternum
D. Manus
E. Crus
F. Brachium
G. Tarsus
H. Anterior part of mandible
I. Brain case
J. Component of hamstrings
K. Forms main part of prominence of buttock
L. Carpus
M. Posterior neck region
N. Acromial (omos)
O. Ear

NERVOUS TISSUE AND PHYSIOLOGY 12

In this exercise you will identify the parts of a neuron and the components of a reflex arc. You will also perform several experiments on reflexes in the frog.

A. HISTOLOGY OF NERVOUS TISSUE

Despite its complexity, the nervous system consists of only two principal kinds of cells: neurons and neuroglia. ***Neurons*** (nerve cells) constitute the nervous tissue and are highly specialized for nerve impulse conduction. ***Neuroglia*** (noo-ROG-lē-a) bind together nervous tissue, form myelin sheaths around axons within the central nervous system, and carry on phagocytosis. They do not transmit nerve impulses.

A neuron consists of the following parts:

1. ***Cell body (perikaryon)*** (per′-i-KAR-ē-on)—Contains a nucleus, cytoplasm, mitochondria, Golgi apparatus, chromatophilic substance (Nissl bodies), and neurofibrils.
2. ***Dendrites***—Usually short, highly branched extensions of the cell body that conduct nerve impulses toward the cell body.
3. ***Axon***—Single, usually relatively long process that conducts nerve impulses away from the cell body. An axon, in turn, consists of the following:
   a. ***Axon hillock***—The origin of an axon from the cell body represented as a small conical elevation.
   b. ***Axoplasm***—Cytoplasm of an axon.
   c. ***Axolemma***—Plasma membrane around the axoplasm.
   d. ***Axon collateral***—Side branch of an axon.
   e. ***Axon terminals*** or ***telodendria*** (tel′-ō-DEN-drē-a)—Fine, branching filaments of an axon or axon collateral.
   f. ***Synaptic end bulbs***—Bulblike structures at distal end of telodendria that contain storage sacs (***synaptic vesicles***) for neurotransmitters.
   g. ***Myelin sheath***—Phospholipid, segmented covering of many axons, especially large peripheral ones; the myelin sheath is produced by peripheral nervous system neuroglia called ***neurolemmocytes (Schwann cells)***.
   h. ***Neurolemma (sheath of Schwann)***—Peripheral, nucleated cytoplasmic layer of the neurolemmocyte (Schwann cell) that encloses the myelin sheath.
   i. ***Neurofibral node (node of Ranvier)***—Unmyelinated gap between segments of the myelin sheath.

Using your textbook and models of neurons as a guide, label Figure 12.1.

Now obtain a prepared slide of an ox spinal cord (cross section), human spinal cord (cross and longitudinal sections), nerve endings in skeletal muscle, and a nerve trunk (cross and longitudinal sections). Examine each under

___ Axolemma
___ Axon
___ Axon collateral
___ Axon hillock
___ Axon terminals
___ Cell body
___ Dendrites
___ Myelin sheath
___ Neurofibral node (node of Ranvier)
___ Neurolemma (sheath of Schwann)
___ Synaptic end bulbs

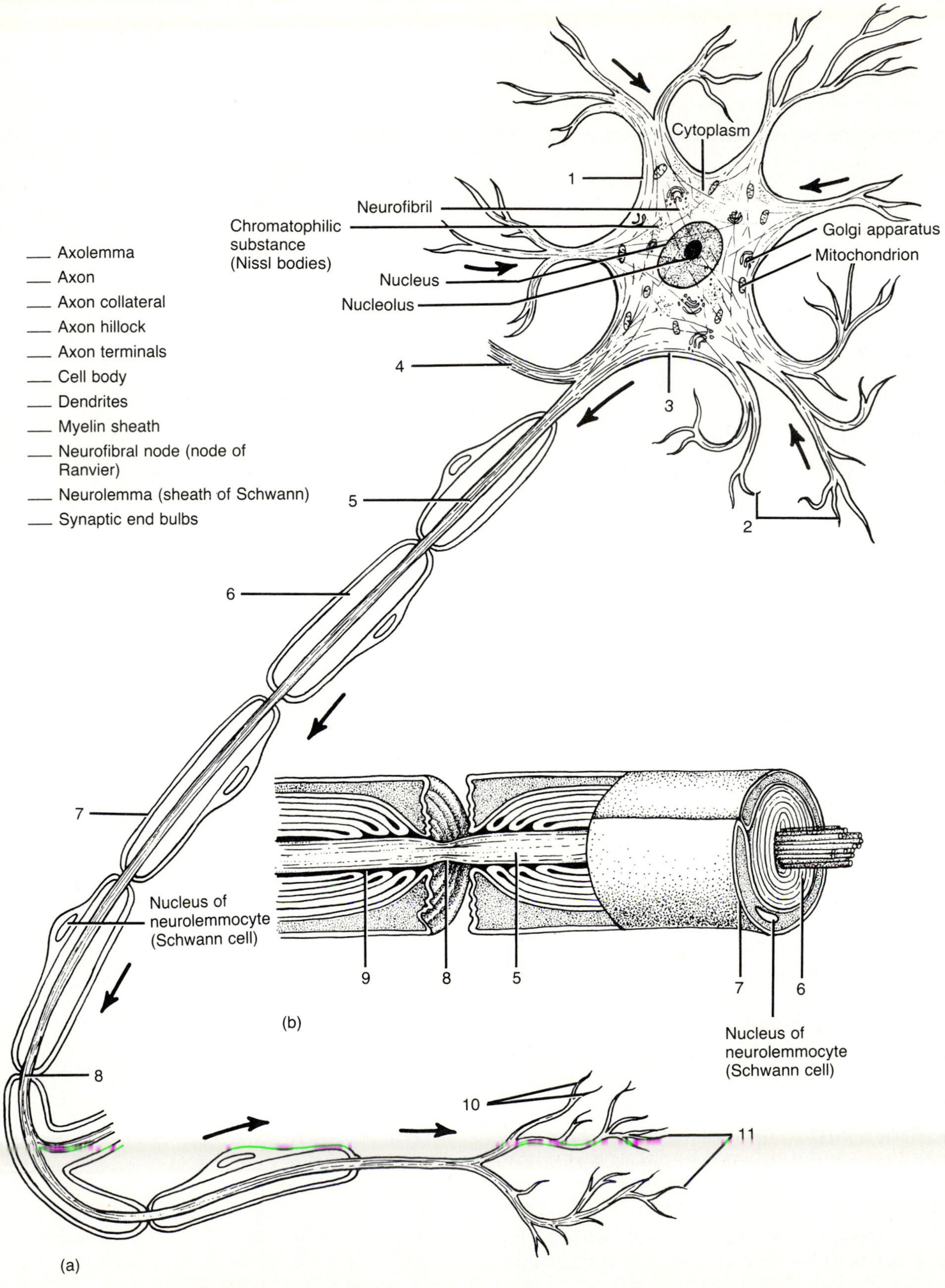

FIGURE 12.1 Structure of a neuron. (a) Shown is an entire multipolar neuron. Arrows indicate direction in which the nerve impulse travels. (b) Cross section and longitudinal section through a myelinated fiber.

high power and identify as many parts of the neuron as you can.

Neurons may be classified on the basis of structure and function. Structurally, neurons are ***multipolar*** (several dendrites and one axon), ***bipolar*** (one dendrite and one axon), and ***unipolar*** (a single process that branches into an axon and a dendrite). Functionally, neurons are classified as (1) ***sensory (afferent)***, which carry nerve impulses toward the central nervous system; (2) ***motor (efferent)***, which carry nerve impulses away from the central nervous system; and (3) ***association (connecting*** or ***interneurons)***, neurons that are located in the central nervous system and carry nerve impulses between sensory and motor neurons.

The neuron you have already labeled in Figure 12.1 is a motor neuron.

B. HISTOLOGY OF NEUROGLIA

Among the types of neuroglial cells are the following:

1. ***Astrocytes*** (AS-trō-sīts)—Star-shaped cells having numerous processes; they twine around neurons to form a supporting network in the central nervous system, attach neurons to blood vessels, and together with tightly packed endothelial cells of brain capillaries, form part of blood-brain barrier (BBB), a selective barrier that protects brain cells from harmful substances.
   a. ***Protoplasmic astrocytes*** are found in the gray matter of the central nervous system.
   b. ***Fibrous astrocytes*** are found in the white matter of the central nervous system.
2. ***Oligodendrocytes*** (ol′-i-gō-DEN-drō-sīts)—Resemble astrocytes but with fewer and shorter processes; they provide support in the central nervous system and produce a myelin sheath on axons of neurons of the central nervous system.
3. ***Microglia*** (mī-KROG-lē-a)—Small cells with few processes, although normally stationary, they can migrate to damaged nervous tissue and there carry on phagocytosis; they are also called ***brain macrophages***.
4. ***Ependyma*** (e-PEN-di-ma)—Epithelial cells arranged in a single layer that range from squamous to columnar in shape; many are ciliated; they form a continuous epithelial lining for the central canal of the spinal cord and for the ventricles of the brain, spaces that form and circulate cerebrospinal fluid; ependyma probably assist in circulating cerebrospinal fluid in these areas.

Obtain prepared slides of astrocytes (protoplasmic and fibrous), oligodendrocytes, microglia, and ependyma. Using your textbook, Figure 12.2, and models of neuroglia as a guide, identify the various kinds of cells. In the spaces provided, draw each of the cells.

Protoplasmic astrocyte

Fibrous astrocyte

Oligodendrocyte

Microglial cell

Ependymal cell

C. REFLEX ARC

Reflexes are rapid responses to changes in the internal or external environment that attempt

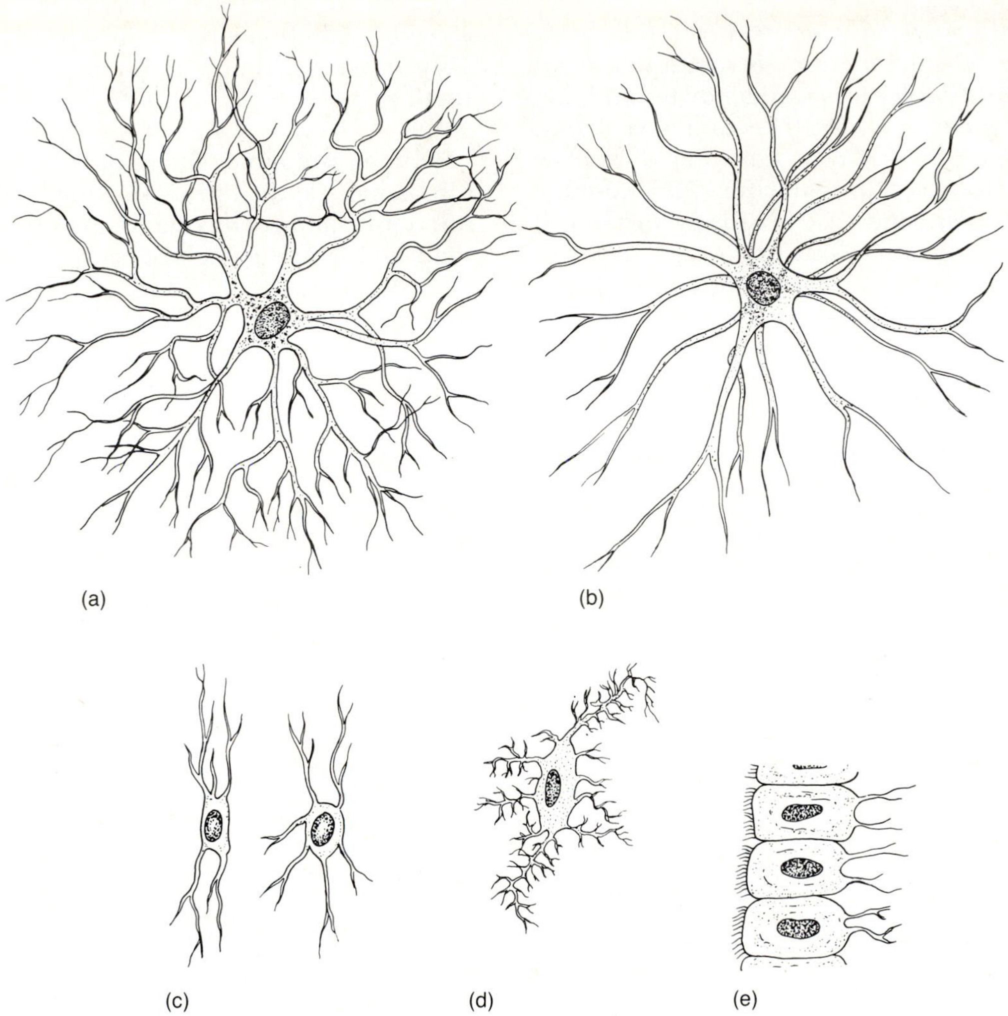

Figure 12.2 Neuroglia. (a) Protoplasmic astrocyte. (b) Fibrous astrocyte. (c) Oligodendrocytes. (d) Microglial cell. (e) Ependyma.

to restore body functions to homeostasis. For a reflex to occur, a nerve impulse must travel over a neural pathway called a ***reflex arc***, which consists of the following components:

1. ***Receptor***—The distal end of a dendrite or sensory structure associated with the distal end of a neuron; it responds to a stimulus by initiating a localized depolarization that, if it is of threshold value, initiates a nerve impulse in a sensory neuron.

2. ***Sensory neuron***—Nerve cell that carries the nerve impulse from the receptor to the central nervous system.

3. ***Center***—Region within the central nervous system where an incoming sensory impulse is processed. The center may generate an outgoing motor impulse; the center may also contain an association neuron between the sensory neuron and the motor neuron. The number of synapses in the center is directly proportional to the complexity of the reflex.

4. ***Motor neuron***—Nerve cell that transmits the nerve impulse generated by the sensory neuron or the association neuron in the center to the part of the body that will respond.

5. ***Effector***—The part of the body, either a muscle or a gland, that responds to the motor neuron impulse and thus the stimulus.

Label the components of a reflex arc in Figure 12.3.

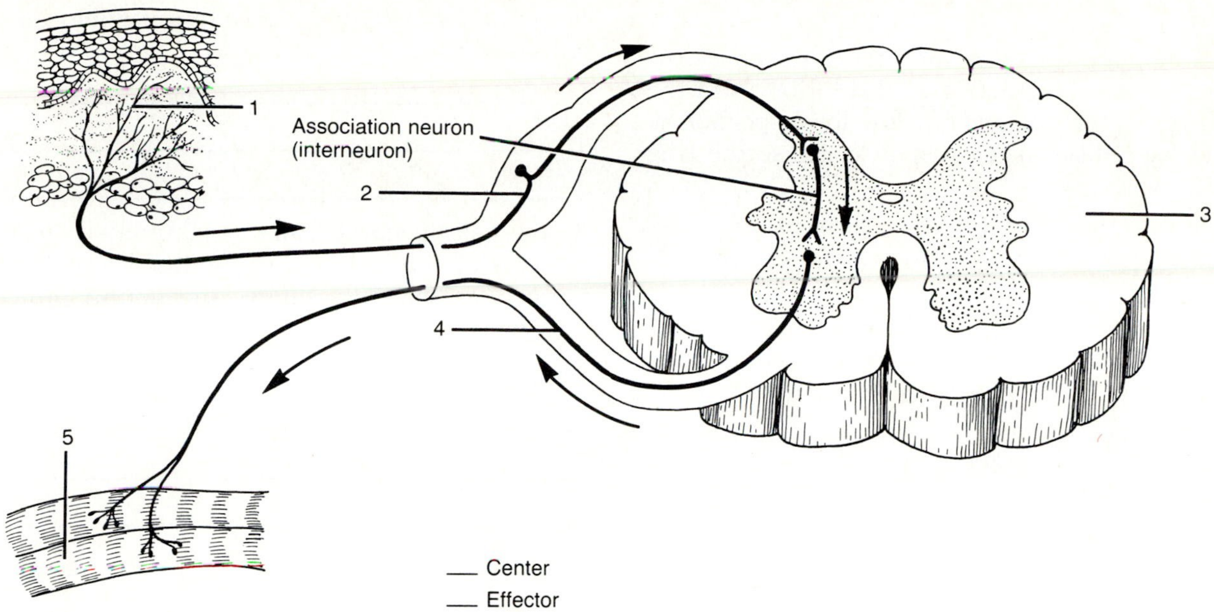

FIGURE 12.3 Components of a reflex arc.

D. SPINAL REFLEXES OF FROG

The following exercise will demonstrate the varying levels of complexity encountered in several common reflexes. Through the results obtained in this experiment you should be able to determine which reflexes involve synapses within the brain and which involve synapses within the spinal cord only. You should also be able to determine the relative complexities of the reflexes. Refer to Exercise 9 for pithing instructions and for the nerve-muscle preparation.

All test results should be recorded in the table provided in Section D of the LABORATORY REPORT RESULTS at the end of the exercise.

PROCEDURE

1. Place the frog on the lab bench and observe the position of the head and legs.
2. Rate of respiration can be determined by counting the number of times that the frog raises and lowers the floor of its mouth or opens and closes its nostrils per minute.
3. Place the frog on it back and see if it rights itself (righting reflex).
4. Place the frog on a turtle board and then slowly tilt the board and observe the animal's response (horizon reflex).
5. Place the frog in water and observe whether or not it stays afloat or swims.
6. Pinch the toes of the hind legs and observe if the frog withdraws its foot (withdrawal reflex).
7. Touch the frog's cornea and observe the presence or absence of a corneal reflex.
8. Slap the table sharply, close to the frog, and observe its response to this loud noise and table vibration.

NOTE. *It is strongly suggested that either your instructor or a laboratory assistant perform the following procedure for pithing a frog. The procedure requires some practice and may prove difficult for inexperienced individuals.*

Single-pith the frog and repeat the same 8 procedures. Upon completion of these procedures, the following tests should also be conducted on a single-pithed frog:

PROCEDURE

1. Insert a small wire through the anterior area of both jaws. Attach the wire so that the frog

is suspended over a pan or sink with its legs hanging freely.

2. Using a piece of cotton that has been moistened with a 30% acetic acid solution, hold the cotton by forceps and rub the lower portion of one leg. Observe the response. Rinse the frog with water.

3. Place the acid on the chest of the frog, observe the response, and wash the frog as before.

Repeat all 11 procedures with a double-pithed frog and record the observations.

PROCEDURE

1. Prepare the sciatic nerve and muscle preparation of one leg of the frog as shown in Figure 9.9. Separate the gluteus and semimembranosus muscles to expose the sciatic nerve.

2. Ascertain reactivity of the sciatic nerve by determining the threshold of the nerve. What response should you be observing in order to determine that threshold, and therefore a nerve action potential (nerve impulse) has been obtained in the nerve?

___

___

___

If the sciatic nerve can be stimulated, what aspects of the reflex are functional and what aspects are not in a single- and double-pithed frog?

___

___

___

___

___

___

ANSWER THE LABORATORY REPORT QUESTIONS AT THE END OF THE EXERCISE

NERVOUS TISSUE AND PHYSIOLOGY 12

Student ______________________ Date ______________________

Laboratory Section ______________________ Score/Grade ______________________

SECTION D. SPINAL REFLEXES OF FROG

Record the results of your observations of frog reflexes in the following table.

Reflex activity	Normal frog	Single-pithed frog	Double-pithed frog
1. Head and leg position			
2. Rate of respiration			
3. Righting reflex			
4. Horizon reflex			
5. Swimming			
6. Withdrawal reflex			
7. Corneal reflex			
8. Response to noise			
9. Acetic acid on leg			
10. Acetic acid on chest			

What conclusions may be drawn as to the levels of complexity of the above reflexes?

____

____

____

What conclusions may be drawn as to the reactivity of peripheral nerves in a single- or double-pithed frog? ____

____

____

LABORATORY REPORT QUESTIONS

NERVOUS TISSUE AND PHYSIOLOGY 12

Student ______________________ Date ______________________

Laboratory Section ______________________ Score/Grade ______________________

PART I. MULTIPLE CHOICE

_____ 1. The portion of a neuron that conducts nerve impulses away from the cell body is the (a) dendrite (b) axon (c) receptor (d) effector

_____ 2. The fine branching filaments of an axon are called (a) myelin sheaths (b) axolemmas (c) telodendria (d) axon hillocks

_____ 3. The component of a reflex arc that responds to a motor impulse is the (a) center (b) receptor (c) sensory neuron (d) effector

_____ 4. Which type of neuron conducts nerve impulses toward the central nervous system? (a) afferent (b) association (c) internuncial (d) efferent

_____ 5. In a reflex arc, the nerve impulse is transmitted directly to the effector by the (a) sensory neuron (b) motor neuron (c) center (d) receptor

_____ 6. Bulblike structures at the distal ends of telodendria that contain storage sacs for neurotransmitters are called (a) dendrites (b) synaptic end bulbs (c) axon collaterals (d) neurofibrils

_____ 7. Which neuroglial cell is phagocytic? (a) oligodendrocyte (b) protoplasmic astrocyte (d) microglial cell (d) fibrous astrocyte

PART II. COMPLETION

8. A neuron that contains several dendrites and one axon is classified as ______________________ .

9. The portion of a neuron that contains the nucleus and cytoplasm is the ______________________ .

10. The phospholipid covering around many peripheral axons is called the ______________________ .

11. The two types of cells that compose the nervous system are neurons and ______________________ .

12. The peripheral, nucleated layer of the neurolemmocyte (Schwann cell) that encloses the myelin sheath is the ______________________ .

13. The side branch of an axon is referred to as the ______________________ .

14. The part of a neuron that conducts nerve impulses toward the cell body is the ______________________ .

15. Neurons that carry nerve impulses between sensory neurons and motor neurons are called ______________________ neurons.

16. Neurons with one dendrite and one axon are classified as ______________ .

17. Unmyelinated gaps between segments of the myelin sheath are known as ______________ .

18. The neuroglial cell that produces a myelin sheath around axons of neurons of the central nervous system is called a(n) ______________ .

19. In a reflex arc, the muscle or gland that responds to a motor impulse is called the ______________ .

NERVOUS SYSTEM 13

The nervous system has two principal divisions: central and peripheral. The ***central nervous system (CNS)*** consists of the brain and spinal cord and is the control center of the entire nervous system. The ***peripheral nervous system (PNS)*** consists of the various processes that connect the CNS with receptors, muscles, and glands. The PNS is divided into afferent and efferent systems. The ***afferent system*** consists of neurons that convey information from receptors to the CNS. The ***efferent system*** consists of nerve cells that convey information from the CNS to muscles and glands. Finally, the efferent system is subdivided into a somatic nervous system and an autonomic nervous system. The ***somatic nervous system (SNS)*** consists of efferent neurons that convey information from the CNS to skeletal muscles and is under conscious control. The ***autonomic nervous system (ANS)*** contains efferent neurons that convey information from the CNS to smooth muscle, cardiac muscle, and glands and is usually involuntary.

In this exercise, you will examine the principal structural features of the spinal cord and spinal nerves, perform several experiments on reflexes, identify the principal structural features of the brain, trace the course of cerebrospinal fluid, identify the cranial nerves, perform several experiments designed to test for cranial nerve function, and examine the structure and function of the autonomic nervous system. You will also dissect and study the cat nervous system and sheep brain.

A. SPINAL CORD AND SPINAL NERVES

1. General Features

Obtain a model or preserved specimen of the spinal cord and identify the following general features:

a. ***Cervical enlargement***—Between vertebrae C4 and T1; origin of nerves to upper extremities.

b. ***Lumbar enlargement***—Between vertebrae T9 and T12; origin of nerves to lower extremities.

c. ***Conus medullaris*** (KŌ-nus med-yoo-LAR-is)—Tapered conical portion of spinal cord near vertebra L1 or L2.

d. ***Filum terminale*** (FĪ-lum ter-mi-NAL-ē)—Nonnervous fibrous tissue arising from the conus medullaris and extending inferiorly to the coccyx; consists mostly of pia mater.

e. ***Cauda equina*** (KAW-da ē-KWĪ-na)—Spinal nerves that angle inferiorly in the vertebral canal giving the appearance of wisps of coarse hair.

f. ***Anterior median fissure***—Deep, wide groove on the anterior surface of the spinal cord.

g. ***Posterior median sulcus***—Shallow, narrow groove on the posterior surface of the spinal cord.

After you have located the parts on a model or preserved specimen of the spinal cord, label the spinal cord in Figure 13.1.

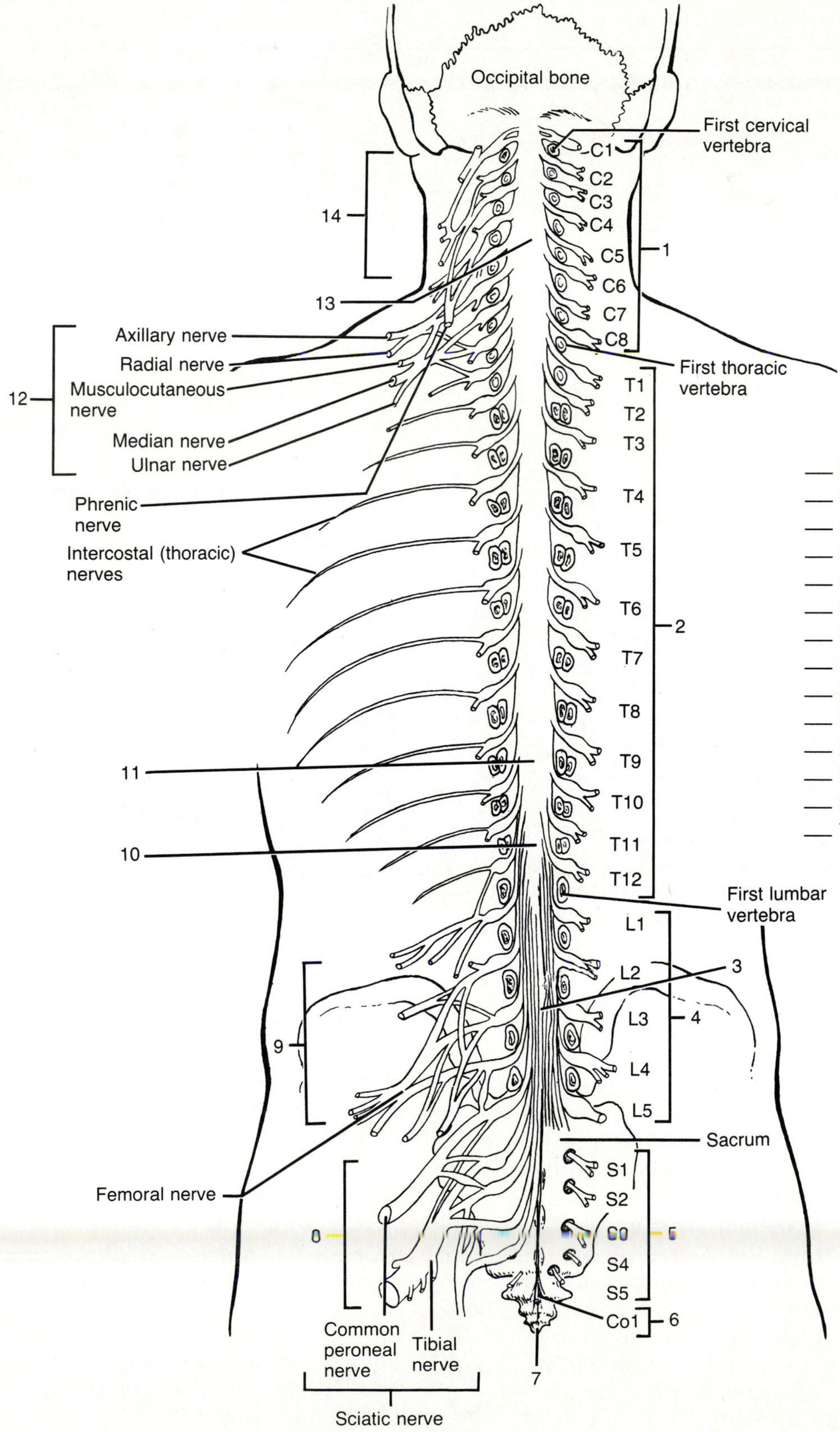

FIGURE 13.1 Spinal cord in posterior view.

___ Brachial plexus
___ Cauda equina
___ Cervical enlargement
___ Cervical nerves
___ Cervical plexus
___ Coccygeal nerves
___ Conus medullaris
___ Filum terminale
___ Lumbar enlargement
___ Lumbar nerves
___ Lumbar plexus
___ Sacral nerves
___ Sacral plexus
___ Thoracic nerves

2. Meninges

The *meninges* (me-NIN-jēz) are coverings that run continuously around the spinal cord and brain (*meninx* is singular). They protect the central nervous system. The spinal meninges are:

a. ***Dura mater*** (DYOO-ra MĀ-ter)—The outer meninx composed of dense fibrous connective tissue. Between the wall of the vertebral canal and the dura mater is the ***epidural space***, which is filled with fat, connective tissue, and blood vessels.

b. ***Arachnoid*** (a-RAK-noyd)—The middle meninx composed of very delicate connective tissue. Between the arachnoid and the dura mater is a spaced called the ***subdural space***, which contains serous fluid.

c. ***Pia mater*** (PĒ-a MĀ-ter)—The inner meninx is a transparent fibrous membrane that contains blood vessels. Between the pia mater and the arachnoid is a space called the ***subarachnoid space*** where cerebrospinal fluid circulates. Extensions of the pia mater called ***denticulate*** (den-TIK-yoo-lāt) ***ligaments*** are attached laterally to the dura mater along the length of the spinal cord and suspend the spinal cord and afford protection against shock and sudden displacement.

Label the meninges, subarachnoid space, and denticulate ligament in Figure 13.2.

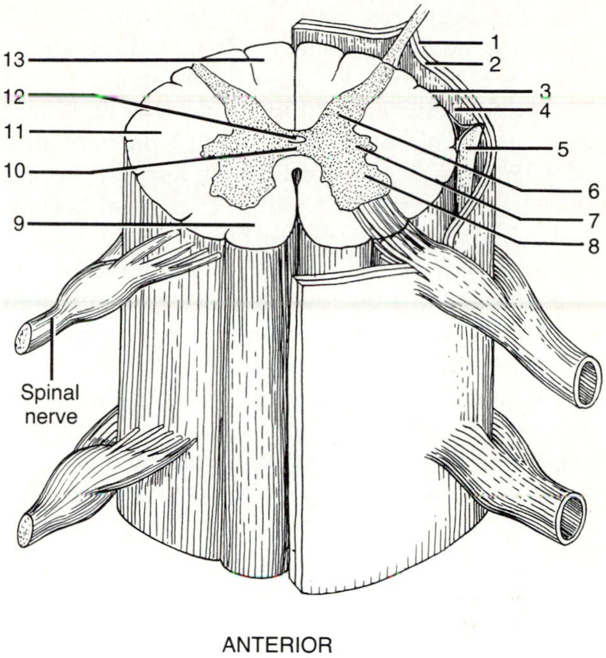

___ Anterior gray horn
___ Anterior white column
___ Arachnoid
___ Central canal
___ Denticulate ligament
___ Dura mater
___ Gray commissure
___ Lateral gray horn
___ Lateral white column
___ Pia mater
___ Posterior gray horn
___ Posterior white column
___ Subarachnoid space

FIGURE 13.2 Cross section of spinal cord showing meninges on right side of figure.

3. Cross Section of Spinal Cord

Obtain a model or specimen of the spinal cord in cross section and note the gray matter, shaped like a letter H or a butterfly. Identify the following parts:

a. ***Gray commissure*** (KOM-mi-shur)—Cross bar of the letter H.

b. ***Central canal***—Small space in the center of the gray commissure that contains cerebrospinal fluid.

c. ***Anterior gray horn***—Anterior region of the upright portion of the H.

d. ***Posterior gray horn***—Posterior region of the upright portion of the H.

e. ***Lateral gray horn***—Intermediate region between the anterior and posterior gray horns present in the thoracic, upper lumbar, and sacral segments of the spinal cord.

f. ***Anterior white column***—Anterior region of white matter.

g. ***Posterior white column***—Posterior region of white matter.

h. ***Lateral white column***—Intermediate region of white matter between the anterior and posterior white columns.

Label these parts of the spinal cord in cross section in Figure 13.2.

Examine a prepared slide of a spinal cord in cross section and see how many structures you can identify.

4. Spinal Nerve Attachments

Spinal nerves are the paths of communication between the spinal cord tracts and the periphery. The 31 pairs of spinal nerves are named and numbered according to the region of the spinal cord from which they emerge. The first cervical pair emerges between the atlas and occipital bone; all other spinal nerves leave the vertebral column from intervertebral foramina between adjoining vertebrae. There are 8 pairs of cervical nerves, 12 pairs of thoracic nerves, 5 pairs of lumbar nerves, 5 pairs of sacral nerves, and 1 pair of coccygeal nerves. Label the spinal nerves in Figure 13.1.

Each pair of spinal nerves is connected to the spinal cord by two points of attachment called roots. The ***posterior (sensory) root*** contains sensory nerve fibers only and conducts nerve impulses from the periphery to the spinal cord. Each posterior root has a swelling, the ***posterior (sensory) root ganglion***, which contains the cell bodies of the sensory neurons from the periphery. Fibers extend from the ganglion into the posterior gray horn. The other point of attachment, the ***anterior (motor) root***, contains motor nerve fibers only and conducts nerve impulses from the spinal cord to the periphery. The cell bodies of the motor neurons are located in lateral or anterior gray horns.

Label the posterior root, posterior root ganglion, anterior root, spinal nerve, cell body of sensory neuron, axon of sensory neuron, cell body of motor neuron, and axon of motor neuron in Figure 13.3.

5. Structure of Spinal Nerves

The posterior and anterior roots unite to form a spinal nerve at the intervertebral foramen. Because the posterior root contains sensory nerve fibers and the anterior root contains motor nerve fibers, all spinal nerves are mixed nerves.

Individual nerve fibers within a nerve, whether myelinated or unmyelinated, are wrapped in a connective tissue covering called the ***endoneurium*** (en′-dō-NOO-rē-um). Groups of fibers with their endoneurium are arranged in bundles called ***fascicles***, and each bundle is wrapped in a connective tissue covering called the ***perineurium*** (per′-i-NOO-rē-um). All the fascicles, in turn, are wrapped in a connective tissue covering called the ***epineurium*** (ep′-i-NOO-rē-um). This is the outermost covering around the entire nerve.

Obtain a prepared slide of a nerve in cross section and identify the fibers, endoneurium, perineurium, epineurium, and fascicles. Now label Figure 13.4.

6. Branches of Spinal Nerves

Shortly after leaving its intervertebral foramen, a spinal nerve divides into several branches called rami:

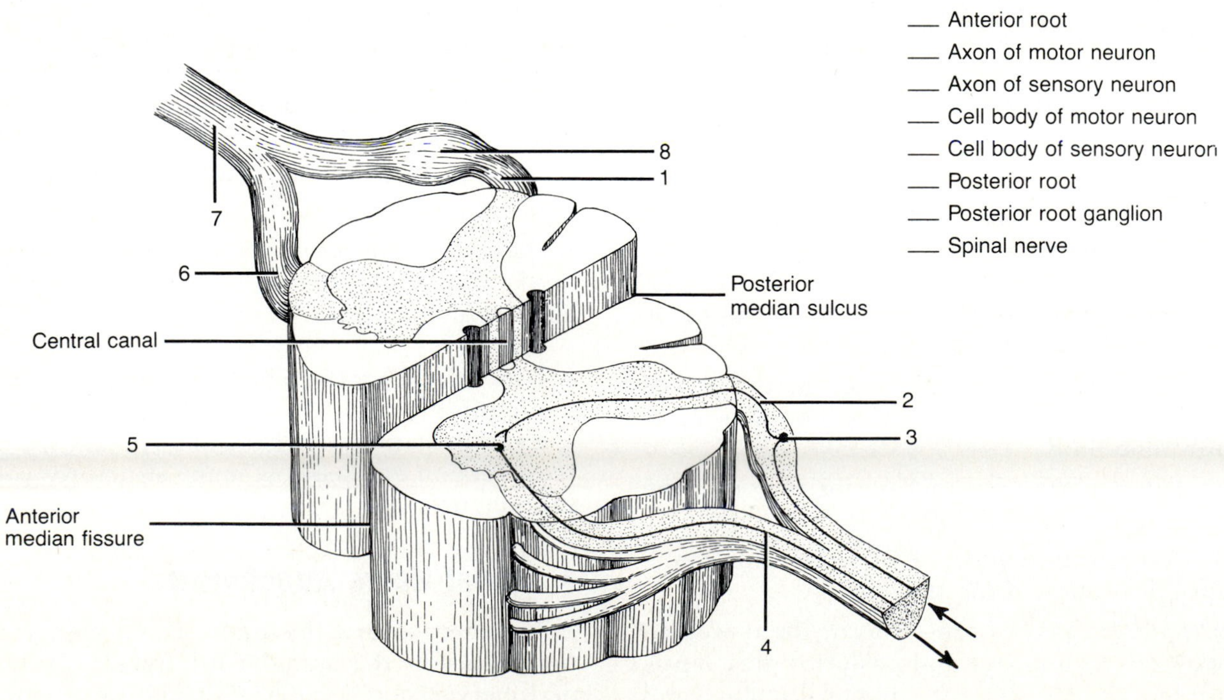

FIGURE 13.3 Spinal nerve attachments.

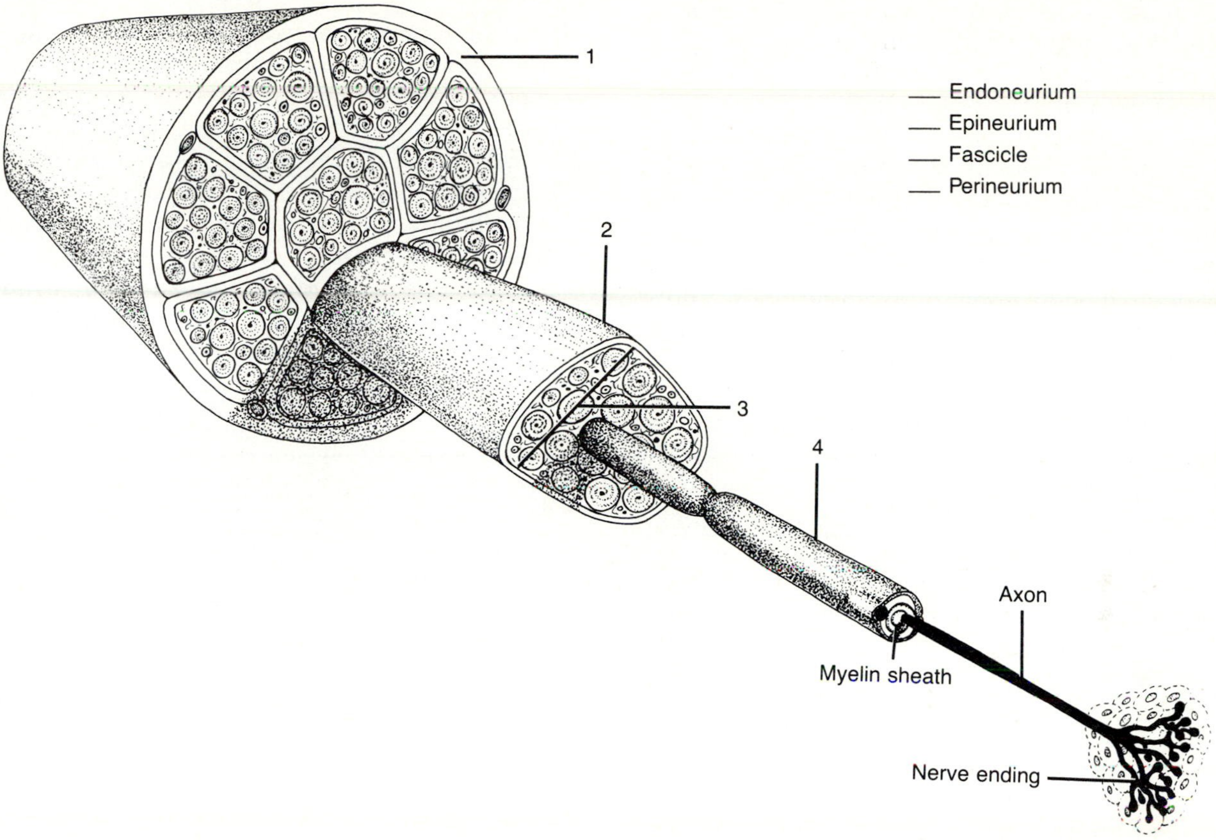

FIGURE 13.4 Coverings of a spinal nerve.

a. ***Dorsal ramus*** (RĀ-mus)—Innervates deep muscles and skin of the dorsal surface of the back.

b. ***Ventral ramus***—Innervates superficial back muscles and all structures of the extremities and lateral and ventral trunk; except for thoracic nerves T2–T11, the ventral rami of the other spinal nerves form plexuses before innervating their structures.

c. ***Meningeal branch***—Innervates vertebrae, vertebral ligaments, blood vessels of the spinal cord, and meninges.

d. ***Rami communicantes*** (RĀ-mē ko-myoo-nē-KAN-tēz)—Gray and white rami communicantes are components of the autonomic nervous system; they connect the ventral rami with sympathetic trunk ganglia.

7. Plexuses

The ventral rami of spinal nerves, except for T2–T11, join with adjacent nerves on either side of the body to form networks called ***plexuses*** (PLEK-sus-ēz).

a. ***Cervical plexus*** (PLEK-sus)—Formed by the ventral rami of the first four cervical nerves (C1–C4) with contributions from C5; one is located on each side of the neck alongside the first four cervical vertebrae; the plexus supplies the skin and muscles of the head, neck, and upper part of shoulders.

b. ***Brachial plexus***—Formed by the ventral rami of spinal nerves C5–C8 and T1 with contributions from C4 and T2; each is located on either side of the last four cervical and first thoracic vertebrae and extends downward and laterally, over the first rib behind the clavicle, and into the axilla; the plexus constitutes the entire nerve supply for the upper extremities and shoulder region.

c. ***Lumbar plexus***—Formed by the ventral rami of spinal nerves L1–L4; each is located on either side of the first four lumbar vertebrae posterior to the psoas major muscle and anterior to the quadratus lumborum muscle; the plexus supplies the anterolateral abdominal wall, external genitals, and part of the lower extremity.

d. ***Sacral plexus***—Formed by the ventral rami of spinal nerves L4–L5 and S1–S4; each is located largely anterior to the sacrum; the plexus supplies the buttocks, perineum, and lower extremities.

Label the cervical, brachial, lumbar, and sacral plexuses in Figure 13.1. Also note the names of some of the major peripheral nerves that arise from the plexuses.

For each nerve listed in the table below, indicate the plexus to which it belongs and the structure(s) it innervates.

8. Spinal Cord Tracts

The vital function of conveying sensory and motor information to and from the brain is carried out by ascending (sensory) and descending (motor) tracts in the spinal cord. The names of the tracts indicate the white column in which the tract travels, where the cell bodies of the tract originate, and where the axons of the tract terminate. For example, the anterior spinothalamic tract is located in the ***anterior*** white column, it originates in the ***spinal*** cord, and it terminates in the ***thalamus*** of the brain. Since

Nerve	Plexus	Innervation
Musculocutaneous		
Femoral		
Phrenic		
Pudendal		
Axillary		
Transverse cervical		
Radial		
Obturator		
Tibial		
Thoracodorsal		
Perforating cutaneous		
Ulnar		
Long thoracic		
Median		
Iliohypogastric		
Deep peroneal		

it conveys nerve impulses from the spinal cord upward to the brain, it is an ascending (sensory) tract.

Label the ascending and descending tracts shown in Figure 13.5.

For each of the ascending tracts in the table below, indicate the origin, termination, and function.

For each of the descending tracts in the table below, indicate the origin, termination, and action.

9. Reflex Experiments

In this section, you will determine the response obtained in various common human reflexes. Reflexes that result in contraction of skeletal

Ascending tracts	Origin	Termination	Function
Anterior (ventral) spinothalamic			
Lateral spinothalamic			
Fasciculus gracilis and fasciculus cuneatus			
Posterior (dorsal) spinocerebellar			
Anterior (ventral) spinocerebellar			

Descending tracts	Origin	Termination	Function
Lateral corticospinal			
Anterior (ventral) corticospinal			
Rubrospinal			
Tectospinal			
Vestibulospinal			
Lateral reticulospinal			
Anterior (ventral) or medial reticulospinal			

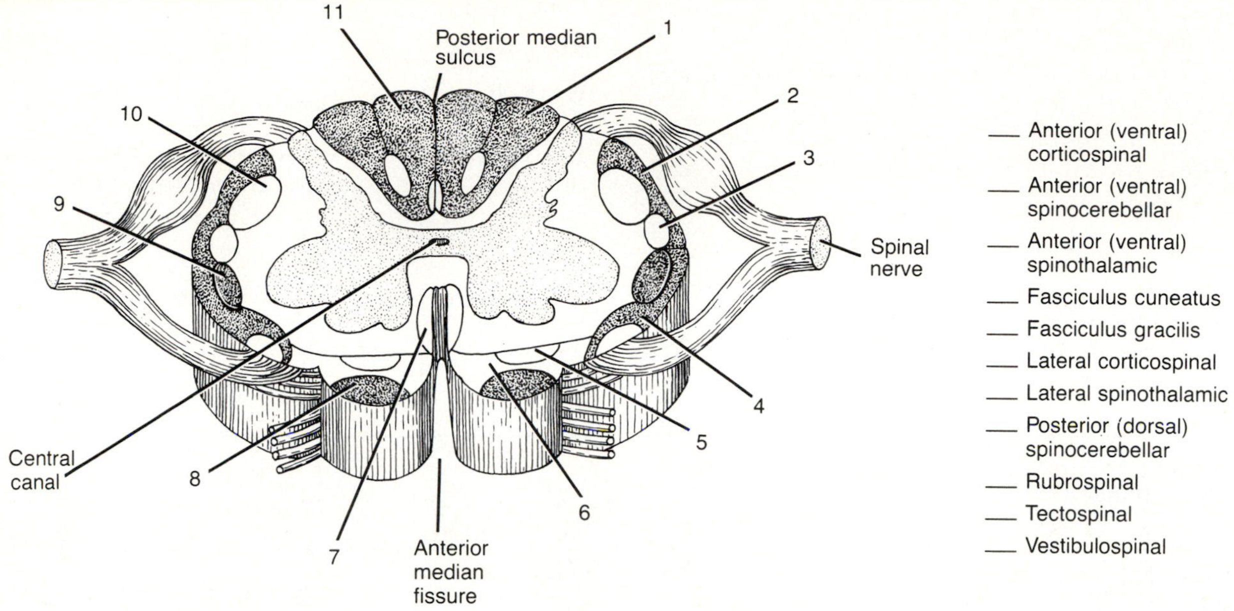

FIGURE 13.5 Selected ascending and descending tracts of the spinal cord.

muscles are called ***somatic reflexes***. Those that cause contraction of cardiac or smooth muscle or secretion by glands are known as ***visceral (autonomic) reflexes***.

a. ***Corneal reflex***—*Very slowly and carefully*, approach, but *do not touch*, the cornea of your partner's eye with lens paper.

What happens? ______________________

____

What is the purpose of this reflex? ______________________

____

b. ***Light (photopupil) reflex***—Have your partner close his or her eyes for about 2 minutes while facing a moderate (*not bright*) light. Now have your partner open his or her eyes and note the size of the pupils.

Describe the response. ______________________

____

Which nerves and muscles act in this reflex?

____

What is the purpose of this reflex? ______________________

____

c. ***Accommodation pupil reflex***—Have your partner focus on an object at a distance of 6.10 m (20 ft) or more.

Describe the size of the subject's pupils.

____

Now have the subject focus on an object 25.4 cm (10 in.) away. Make sure that there has been no change in illumination. Note the size of the pupils again.

Has there been any change? ______________________

____

____

What is the purpose of this reflex? ______________________

____

____

What nerve and muscles act in this reflex?

____

____

d. ***Convergence reflex***—Observe the position of your partner's eyeballs while he or she looks at a distant object. Now have your partner look at a nearby object.

What changes do you observe in the eyeballs?

What is the purpose of this reflex?

e. ***Swallowing reflex***—Swallow the saliva in your mouth and immediately swallow again and again for 20 seconds. Swallow again for 20 seconds but this time drink a small amount of water each time you swallow.

How do the results of the first and second experiment compare?

What is the stimulus for the receptors to initiate swallowing?

What muscles are involved in swallowing?

f. ***Patellar reflex***—Have your partner sit on a table so that the leg hangs freely. Strike the patellar ligament just below the kneecap with the reflex hammer.

What is the response?

Test the patellar reflex while the subject is adding a column of numbers. Test the reflex again while the subject interlocks his or her fingers and pulls one hand against the other.

Compare the results of these two experiments.

What is the purpose of the patellar reflex?

g. ***Achilles reflex***—Have the subject kneel on a chair and let the feet hang freely over the edge of the chair. Bend one foot to increase the tension on the gastrocnemius muscle. Tap the calcaneal (Achilles) tendon with the reflex hammer.

What is the result?

h. ***Plantar reflex***—Scratch the sole of your partner's foot by moving a blunt object along the sole toward the toes.

What is the response?

What is the Babinski sign?

Why is it normal in children under the age of 18 months?

What does the Babinski sign indicate in an adult?

B. BRAIN

1. Parts

The brain may be divided into four principal parts: (1) ***brain stem***, which consists of the medulla oblongata, pons, and midbrain; (2) ***diencephalon*** (dī-en-SEF-a-lon), which consists primarily of the thalamus and hypothalamus; (3) ***cerebellum***, which is posterior to the brain stem; and (4) ***cerebrum***, which is superior to the brain stem and comprises about seven-eighths of the total weight of the brain.

Examine a model and preserved specimen of the brain and identify the parts just described. Then refer to Figure 13.6 and label the parts of the brain.

2. Meninges

As in the spinal cord, the brain is protected by ***meninges***. The cranial meninges are continuous with the spinal meninges. The cranial meninges are the outer ***dura mater***, the middle ***arachnoid***, and the inner ***pia mater***. The cranial dura mater consists of two layers called the periosteal layer (adheres to the cranial bones

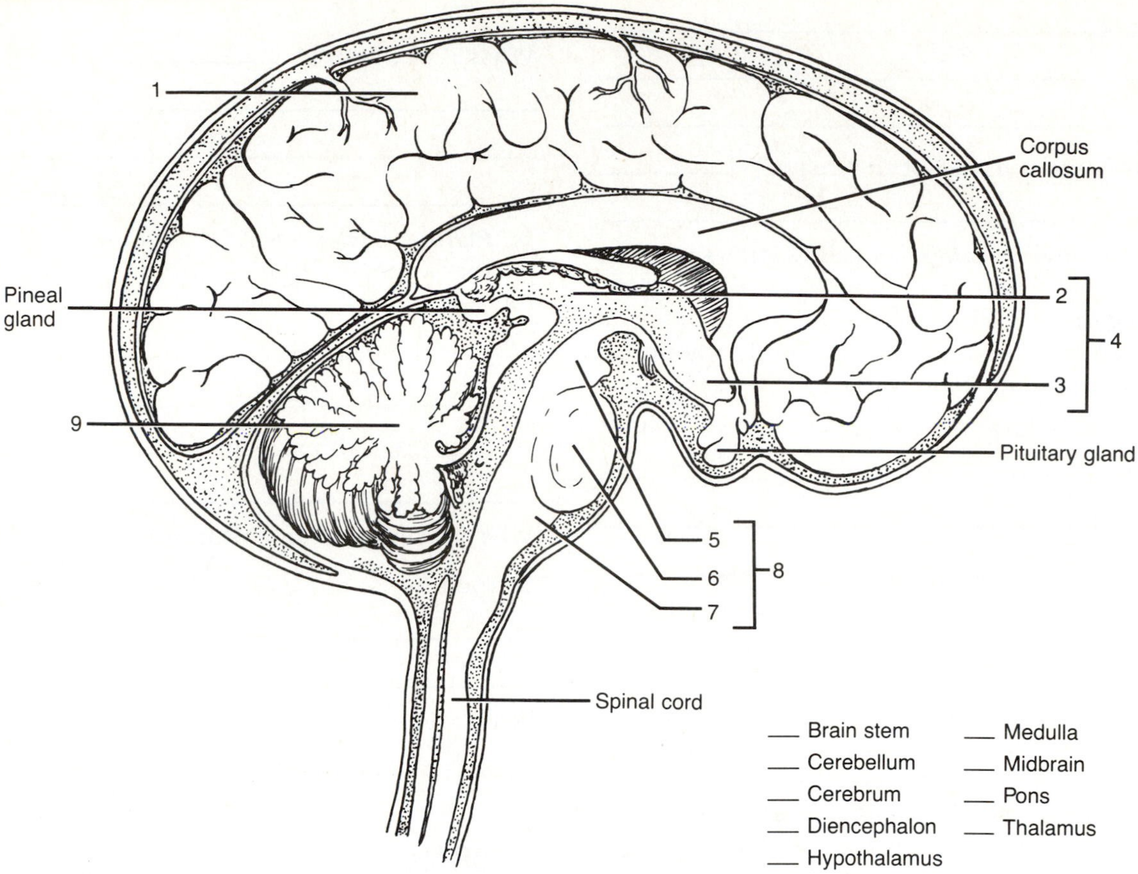

FIGURE 13.6 Principal parts of the brain in midsagittal section.

and serves as a periosteum) and the meningeal layer (thinner, inner layer that corresponds to the spinal dura mater).

Refer to Figure 13.7 and label all of the meninges.

3. Cerebrospinal Fluid (CSF)

The central nervous system is protected by ***cerebrospinal fluid (CSF)***. The fluid circulates through the subarachnoid space around the brain and spinal cord and through the ventricles of the brain. The ventricles are cavities in the brain that communicate with each other, with the central canal of the spinal cord, and with the subarachnoid space.

Cerebrospinal fluid is formed primarily by filtration and secretion from networks of capillaries and ependymal cells in the ventricles, called ***choroid*** (KŌ-royd) ***plexuses*** (see Figure 13.7). Each of the two ***lateral ventricles*** (VEN-tri-kuls) is located within a hemisphere (side) of the cerebrum under the corpus callosum. The fluid formed in the choroid plexuses of the lateral ventricles circulates through an opening called the ***interventricular foramen*** into the third ventricle. The ***third ventricle*** is a slit between and inferior to the right and left halves of the thalamus and between the lateral ventricles. More fluid is added by the choroid plexus of the third ventricle. Then the fluid circulates through a canal-like structure called the ***cerebral aqueduct*** (AK-we-dukt) into the fourth ventricle. The ***fourth ventricle*** lies between the inferior brain stem and the cerebellum. More fluid is added by the choroid plexus of the fourth ventricle. The roof of the fourth ventricle has three openings: one ***median aperture*** (AP-e-chur) and two ***lateral apertures***. The fluid circulates through the apertures into the subarachnoid space around the back of the brain and downward through the central canal of the spinal cord to the subarachnoid space around the posterior surface

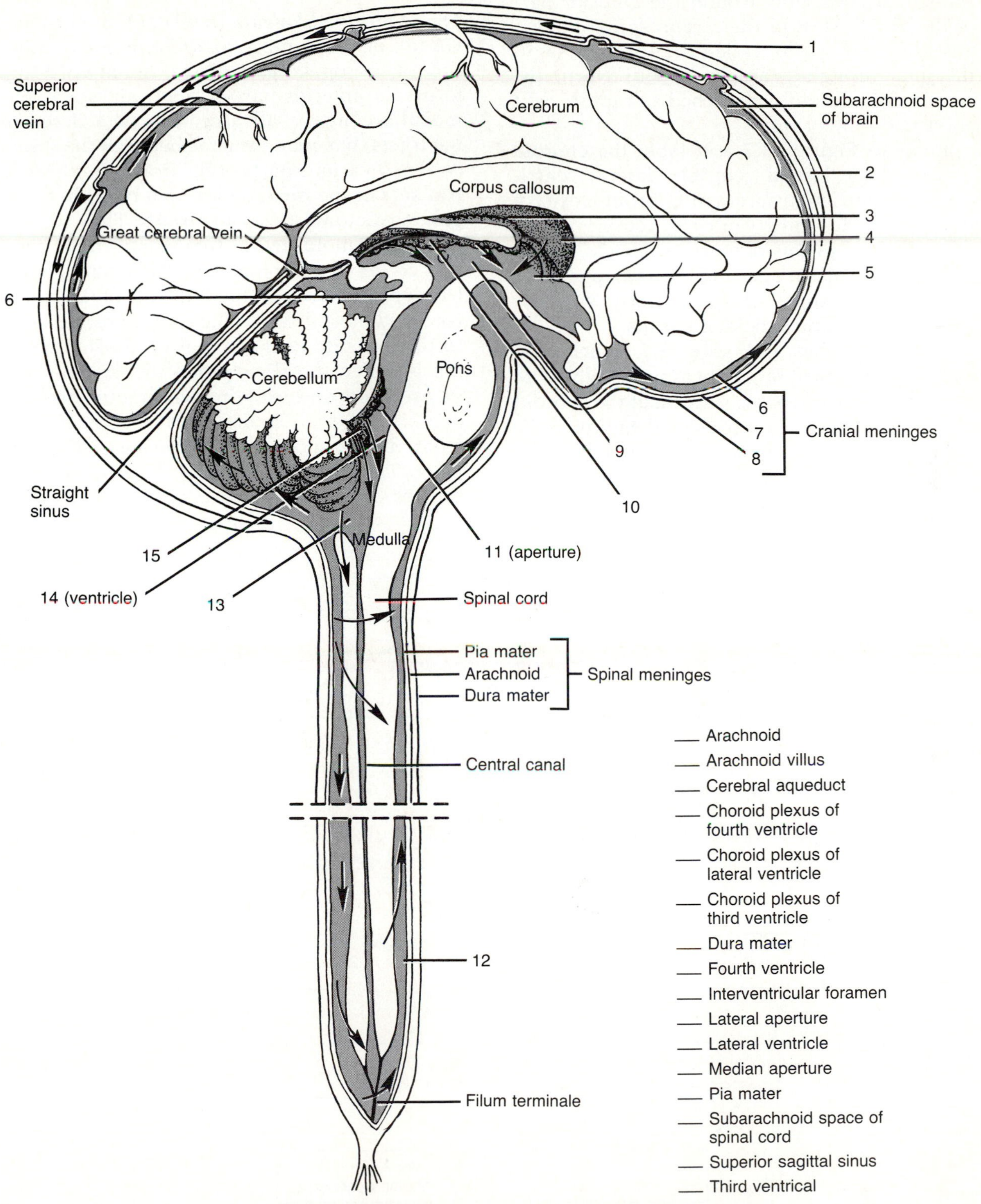

___ Arachnoid
___ Arachnoid villus
___ Cerebral aqueduct
___ Choroid plexus of fourth ventricle
___ Choroid plexus of lateral ventricle
___ Choroid plexus of third ventricle
___ Dura mater
___ Fourth ventricle
___ Interventricular foramen
___ Lateral aperture
___ Lateral ventricle
___ Median aperture
___ Pia mater
___ Subarachnoid space of spinal cord
___ Superior sagittal sinus
___ Third ventrical

FIGURE 13.7 Brain and spinal cord in midsagittal section.

of the spinal cord, up the anterior surface of the spinal cord, and around the anterior part of the brain. Most of the cerebrospinal fluid is absorbed into a vein called the superior sagittal sinus through its arachnoid villi. Normally, cerebrospinal fluid is absorbed as rapidly as it is formed.

Refer to Figure 13.7 and label the choroid plexus of the lateral ventricle, lateral ventricle, interventricular foramen, choroid plexus of third ventricle, third ventricle, cerebral aqueduct, choroid plexus of fourth ventricle, fourth ventricle, median aperture, lateral aperture, subarachnoid space of spinal cord, superior sagittal sinus, and arachnoid villus.

Note the arrows in Figure 13.7, which indicate the path taken by cerebrospinal fluid. With the aid of your textbook, starting at the choroid plexus of the lateral ventricle and ending at the superior sagittal sinus, see if you can follow the remaining path of the fluid.

Now complete Figure 13.8.

4. Medulla Oblongata

The ***medulla oblongata*** (me-DULL-la ob′-long-GA-ta), or just simply ***medulla***, is a continuation of the upper part of the spinal cord and forms the inferior part of the brain stem. The medulla contains all ascending and descending tracts that communicate between the spinal cord and various parts of the brain. On the ventral side of the medulla are two roughly triangular structures called ***pyramids***. They contain the largest motor tracks that run from the cerebral cortex to the spinal cord. Most fibers in the left pyramid cross to the right side of the spinal cord and most fibers in the right pyramid cross to the left side of the spinal cord. This crossing is called the ***decussation*** (dē′-ku-SĀ-shun) ***of pyramids*** and explains why motor areas of one side of the cerebral cortex control muscular movements on the opposite side of the body. The medulla contains three vital reflex centers called the ***cardiac center***, ***respiratory***

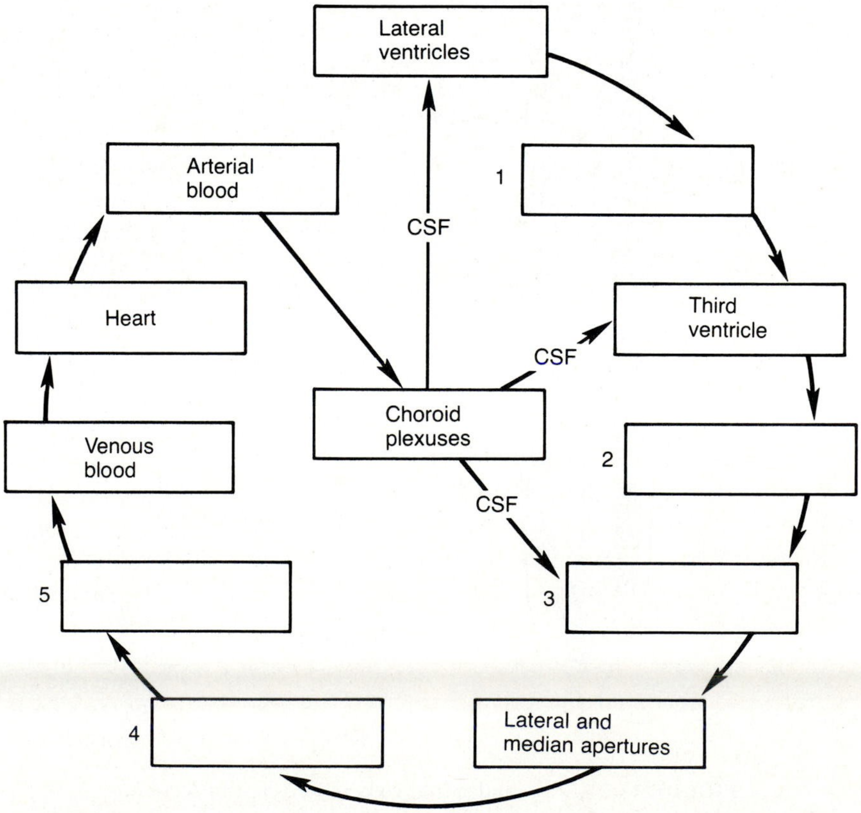

FIGURE 13.8 Formation, circulation, and absorption of cerebrospinal fluid (CSF).

center, and ***vasomotor center***. The medulla also contains the nuclei of origin of several cranial nerves. These are the cochlear and vestibular branches of the vestibulocochlear (VIII) nerves, glossopharyngeal (IX) nerves, vagus (X) nerves, spinal portions of the accessory (XI) nerves, and hypoglossal (XII) nerves. Examine a model or specimen of the brain and identify the parts of the medulla. Then refer to Figure 13.6 and locate the medulla.

5. Pons

The ***pons*** lies directly above the medulla and anterior to the cerebellum. The pons contains fibers that connect parts of the cerebellum and medulla with the cerebrum. The pons contains the nuclei or origin of the following pairs of cranial nerves: trigeminal (V) nerves, abducens (VI) nerves, facial (VII) nerves, and vestibular branch of the vestibulocochlear (VIII) nerves. Other important nuclei in the pons are the ***pneumotaxic*** (noo-mō-TAK-sik) ***area*** and ***apneustic*** (ap-NOO-stik) ***area*** that help control respiration.

Identify the pons on a model or specimen of the brain. Locate the pons in Figure 13.6.

6. Midbrain

The ***midbrain*** extends from the pons to the lower portion of the cerebrum. The ventral portion of the midbrain contains the paired ***cerebral peduncles*** (pe-DUNG-kulz), which connect the upper parts of the brain to lower parts of the brain and spinal cord. The dorsal part of the midbrain contains four rounded eminences called ***corpora quadrigemina*** (KOR-por-ra kwad-ri-JEM-in-a). Two of the eminences, the ***superior colliculi*** (ko-LIK-yoo-lī), serve as reflex centers for movements of the eyeballs and the head in response to visual and other stimuli. The other two eminences, the ***inferior colliculi***, serve as reflex centers for movements of the head and trunk in response to auditory stimuli. The midbrain contains the nuclei of origin for two pairs of cranial nerves: oculomotor (III) and trochlear (IV).

Identify the parts of the midbrain on a model or specimen of the brain. Locate the midbrain in Figure 13.6.

7. Thalamus

The ***thalamus*** (THAL-a-mus) is a large oval structure, located above the midbrain, that consists of two masses of gray matter covered by a layer of white matter. The two masses are joined by a bridge of gray matter called the ***intermediate mass***. The thalamus contains numerous nuclei that serve as relay stations for all sensory impulses, except smell. The most prominent are the ***medial geniculate*** (je-NIK-yoo-lāt) ***nuclei*** (hearing), ***lateral geniculate nuclei*** (vision), ***ventral posterior nuclei*** (general sensations and taste), ***ventral lateral nuclei*** (voluntary motor actions), and the ***ventral anterior nuclei*** (voluntary motor actions and arousal).

Identify the thalamic nuclei on a model or specimen of the brain. Then refer to Figure 13.9 and label the nuclei.

8. Hypothalamus

The ***hypothalamus*** is located inferior to the thalamus and forms the floor and part of the wall of the third ventricle. Among the functions served by the hypothalamus are control and integration of the autonomic nervous system and parts of the endocrine system, reception and integration of sensory impulses from the viscera, secretion of regulating hormones (or factors), and control of body temperature. The hypothalamus also assumes a role in feelings of rage and aggression, food intake, thirst, the waking state and sleep patterns, and circadian rhythms.

Identify the hypothalamus on a model or specimen of the brain. Locate the hypothalamus in Figure 13.6.

9. Cerebrum

The ***cerebrum*** is the largest portion of the brain and is supported on the brain stem. Its outer surface consists of gray matter and is called the ***cerebral cortex***. Beneath the cerebral cortex is the cerebral white matter. The upfolds of the cerebral cortex are termed ***gyri*** (JĪ-rī) or ***convolutions***, the deep downfolds are termed ***fissures***, and the shallow downfolds are termed ***sulci*** (SUL-sī).

The ***longitudinal fissure*** separates the cerebrum into right and left halves called ***hemispheres***. Each hemisphere is further divided into lobes by sulci or fissures. The ***central sulcus*** separates the ***frontal lobe*** from the ***parietal lobe***. The ***lateral cerebral sulcus*** separates the frontal lobe from the ***temporal lobe***. The ***parietooccipital sulcus*** separates the

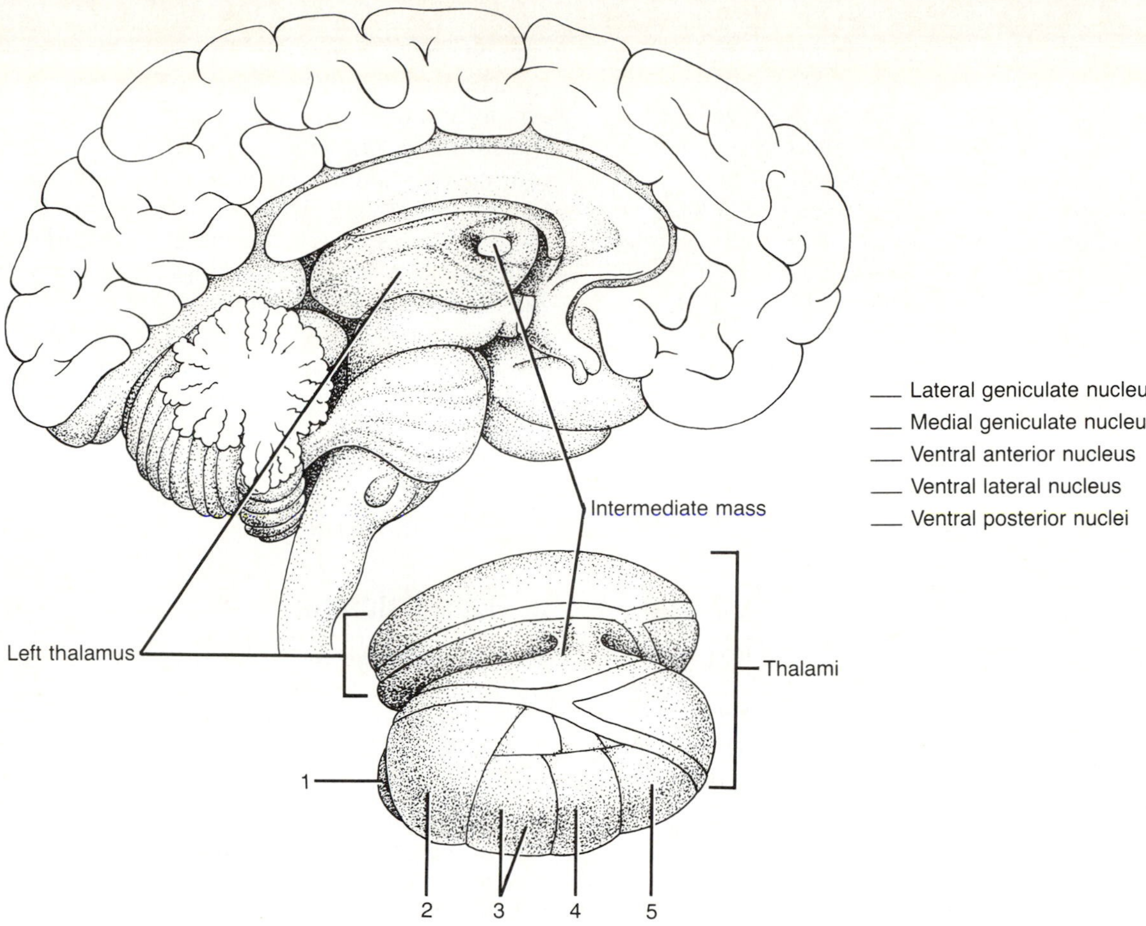

FIGURE 13.9 Thalamic nuclei.

parietal lobe from the ***occipital lobe***. Another prominent fissure, the ***transverse fissure***, separates the cerebrum from the cerebellum. Another lobe of the cerebrum, the ***insula***, lies deep within the lateral cerebral fissure under the parietal, frontal, and temporal lobes. It cannot be seen in external view. Two important gyri on either side of the ***central sulcus*** are the ***precentral gyrus*** and the ***postcentral gyrus***. The olfactory (I) and optic (II) cranial nerves are associated with the cerebrum.

Examine a model of the brain and identify the parts of the cerebrum just described. Refer to Figure 13.10 and label the lobes.

10. Basal Ganglia

Basal ganglia (GANG-lē-a) or ***cerebral nuclei*** are paired masses of gray matter, with one member of each pair in each cerebral hemisphere. The largest of the basal ganglia is the ***corpus striatum*** (strī-Ā-tum), which consists of the ***caudate nucleus*** and the ***lentiform nucleus***. The lentiform nucleus, in turn, is subdivided into a lateral ***putamen*** (pu-TĀ-men) and a medial ***globus pallidus***. The basal ganglia control large, subconscious movement of skeletal muscles. An example is swinging the arms while walking.

Examine a model or specimen of the brain and identify the basal ganglia described. Now refer to Figure 13.11 and label the basal ganglia shown.

11. Functional Areas of Cerebral Cortex

The functions of the cerebrum are numerous and complex. In a general way, the cerebral cortex can be divided into sensory, motor, and

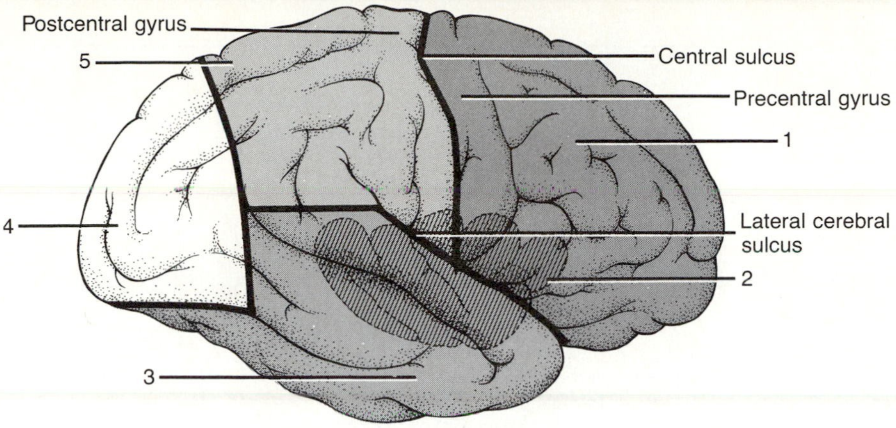

___ Frontal lobe
___ Insula
___ Occipital lobe
___ Parietal lobe
___ Temporal lobe

FIGURE 13.10 Lobes of cerebrum.

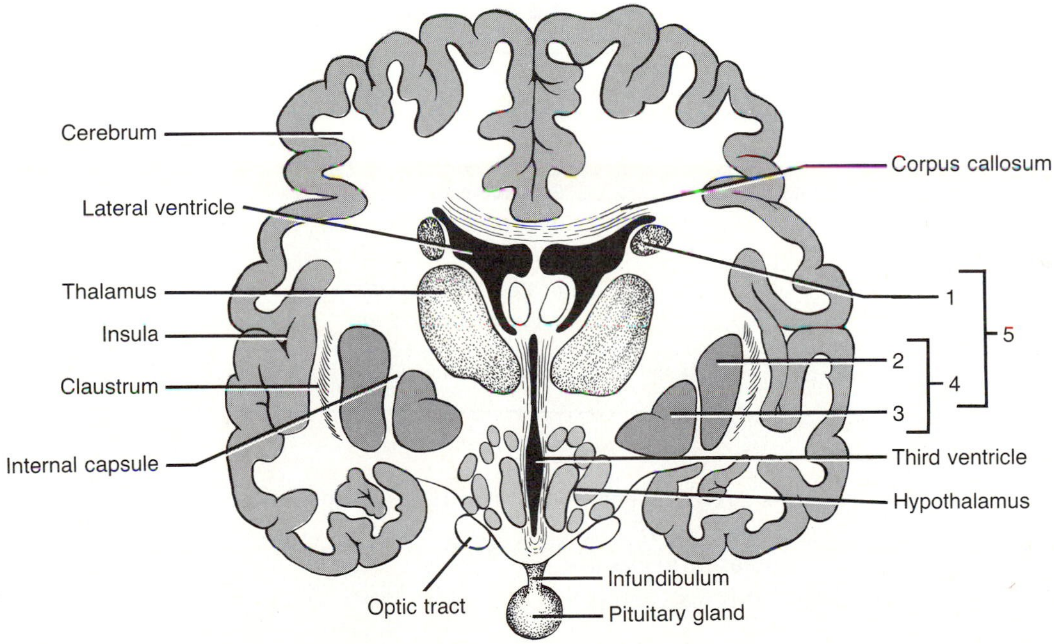

___ Caudate nucleus
___ Corpus striatum
___ Globus pallidus
___ Lentiform nucleus
___ Putamen

FIGURE 13.11 Basal ganglia in frontal section of cerebrum.

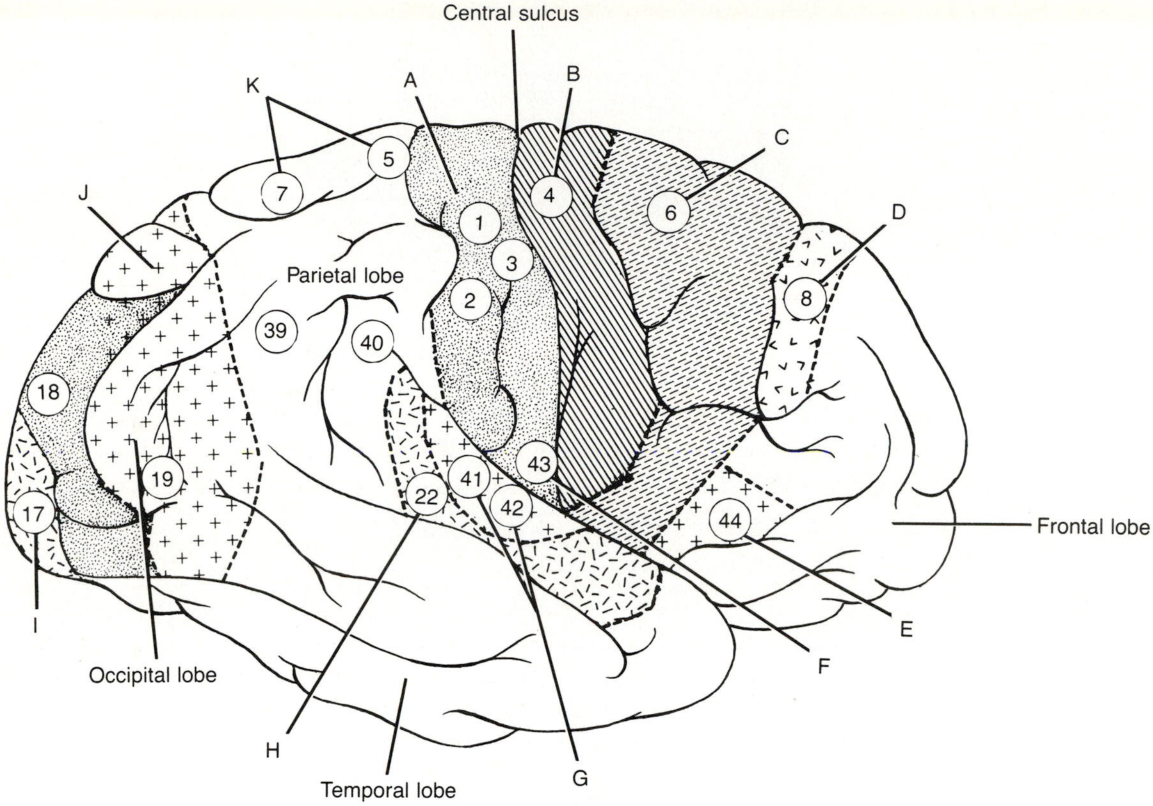

___ Auditory association area
___ Frontal eye field area
___ Motor speech (Broca's) area
___ Premotor area
___ Primary auditory area
___ Primary gustatory area
___ Primary motor area
___ Primary somesthetic area
___ Primary visual area
___ Somesthetic association area
___ Visual association area

FIGURE 13.12 Functional areas of the cerebrum seen in right lateral view.

association areas. The ***sensory areas*** interpret sensory impulses, the ***motor areas*** control muscular movement, and the ***associated areas*** are concerned with emotional and intellectual processes.

The principal sensory and motor areas of the cerebral cortex are indicated by numbers based on K. Brodmann's map of the cerebral cortex. His map, first published in 1909, attempts to correlate structure and function. Refer to Figure 13.12 and match the name of the sensory or motor area next to the appropriate letter (A, B, C, etc.).

12. Cerebellum

The ***cerebellum*** is inferior to the posterior portion of the cerebrum and separated from it by the transverse fissure. The central constricted area of the cerebellum is called the ***vermis*** and the lateral portions are referred to as ***hemispheres***. The surface of the cerebellum, called the ***cerebellar cortex***, consists of gray matter thrown into a series of slender parallel ridges called ***folia***. Beneath the gyri are tracts of white matter called ***arbor vitae***.

The cerebellum is attached to the brain stem by three paired bundles of fibers called ***cerebellar peduncles***. The ***inferior cerebellar peduncles*** connect the cerebellum with the medulla at the base of the brain stem and with the spinal cord. The ***middle cerebellar peduncles*** connect the cerebellum with the pons. The ***superior cerebellar peduncles*** connect the cerebellum with the midbrain.

The cerebellum is a motor area of the brain

that coordinates certain subconscious movements in skeletal muscles required for coordination, posture, and balance.

Examine a model or specimen of the brain and locate the parts of the cerebellum. Identify the cerebellum in Figure 13.6.

13. Determination of Cerebellar Function

Working with your lab partner, perform the following tests to determine cerebellar function:

1. While looking straight ahead, walk heel to toe for 20 ft without losing your balance.
2. Stand with your feet together and your eyes closed without losing your balance.
3. While looking straight ahead, move the heel of one foot down the shin of your other leg.
4. Move your hands and fingers as quickly as you can.
5. With your eyes open, touch your partner's fingers.
6. With your eyes closed, touch your nose with the index finger of each hand.

C. ELECTROENCEPHALOGRAM (EEG)

Electrical activity of the brain may be demonstrated via direct brain surface recordings, or through indirect recordings from the outer surface of the head. The pattern and level of activity observed are determined, to a great extent, by the overall excitation level of the brain. Such brain excitation levels are, in turn, largely determined by activity of the ***reticular activating system***. The electrical potential variations recorded are called ***brain waves*** (Figure 13.13), and an entire record is termed an ***electroencephalogram (EEG)***.

The intensity of brain waves varies from 0 to 300 microvolts (μV), while frequencies range from 0.25/sec to 50/sec or more. Brain waves vary depending upon cerebral cortex activity levels, and will vary greatly among coma, sleep, and wakefulness. Although most EEG's are irregular and without any pattern, specific neurological abnormalities such as epilepsy will present characteristic and repeatable patterns. Normal repeatable patterns also appear during certain conditions in individuals without any neurological abnormalities.

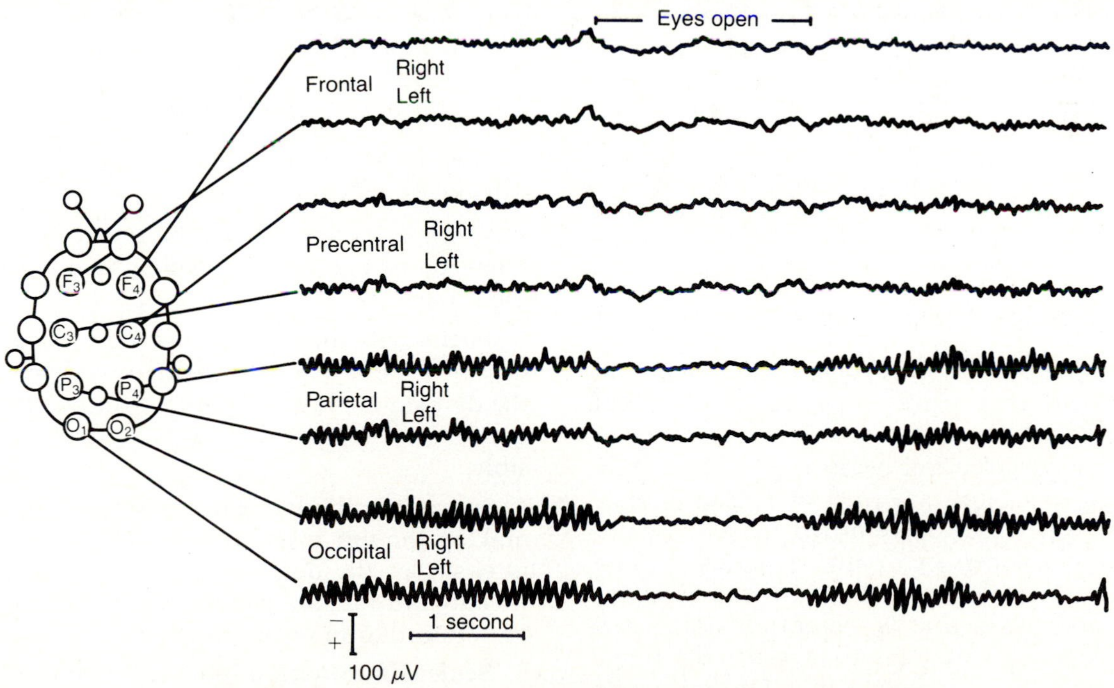

FIGURE 13.13 Diagrammatic representation of a normal EEG of a resting, awake individual. The record represents a simultaneous eight-channel unit recording from the external surface of the skull with a ground electrode placed behind the ear on the temporal lobe. (From Schmidt, R. F., editor: *Fundamentals of Neurophysiology*, ed. 2, New York, 1978, Springer-Verlag)

Alpha waves are rhythmic waves with a frequency of 8–13/sec and are found in the EEG of almost all normal adults during quiet and restful relaxation. These waves are normally 50 microvolts in intensity, and are usually recorded from the occipital region, although they may also be obtained from electrodes placed upon the frontal or parietal regions of the skull. Such waves are replaced by brain waves of lower voltage and higher frequency during directed mental activity, and disappear entirely during sleep.

Beta waves have a frequency between 14–25/sec and are most frequently seen during times of tension (Beta II waves) or activation of the central nervous system (Beta I waves). Beta waves are usually obtained by recording from either the parietal or frontal regions of the scalp.

Theta waves are normally seen in children, and may also be seen in normal adults during emotional stress (i.e., frustration and/or disappointment) or in adults with certain brain disorders. Theta waves usually exhibit a frequency of 4–7/sec. Such waves are usually recorded from the parietal or temporal regions of the scalp.

Delta waves include all waves of the EEG with a frequency less than 3.5/sec. Such waves occur during sleep, in infancy, and in serious organic brain disorders. Such brain waves may also be produced experimentally when the cerebral cortex is separated from lower brain centers, indicating that such waves may originate in the cerebral cortex independent of lower brain center influence.

1. Use of the Oscilloscope

A ***cathode ray oscilloscope*** is an electrical instrument that traces a graph of electrical potential difference on a screen in a left to right direction. An electron beam is produced by a cathode tube and is aimed at a phosphorescent screen. Electrons emitted from the cathode pass between two pairs of plates enroute to the screen (Figure 13.14a). Because ***bioelectrical potentials***, that is, potentials associated with the activity of living tissues, usually range between a few microvolts to a few millivolts (mV), some electronic device is necessary to amplify the potentials. The amplified potentials are directed to the ***Y plates*** of the cathode tube, making one plate positive and the other negative, thereby deflecting the electron beam prior to its reaching the screen. The amount of vertical deflection of the spot produced on the screen by the electron beam is determined by the magnitude of the potential difference across the Y plates and the sensitivity of the instrument.

In order to record an event in time, a ***sweep circuit*** is present within the instrument, which is able to sweep the electron beam from left to right by applying current to the ***X plates***. *Therefore the image produced on the screen of an oscilloscope is actually a plot of voltage (y axis) versus time (x axis)*.

While looking at the front panel of the oscilloscope familiarize yourself with all of the various functions of the instrument by following the procedures outlined below.

Some oscilloscopes may produce two simultaneous displays, and are termed ***dual beam oscilloscopes*** (Figure 13.14b) while others produce only one beam and are termed ***single beam oscilloscopes***. If your instrument is a dual beam oscilloscope, notice that all of the input jacks and dials except for those immediately adjacent to the screen are enclosed within continuous rectangular lines termed ***Upper Beam Vertical*** and ***Lower Beam Vertical, Time Base*** and ***Horizontal Display***. A description of the various functions follows.

PROCEDURE

1. Turn on the instrument and allow time for warm-up.

2. When a dot appears on the screen turn your attention to the dials immediately adjacent to the screen.

3. ***Focus*** will increase or decrease the sharpness of the dot as demonstrated while rotating the dial slowly to the right and then to the left. Set the dial so that the dot is as sharp as possible.

4. ***Intensity*** will alter the intensity of the beam, making the dot brighter or softer. Change the intensity of the dot, and then set it in such a way so that the dot is easily seen in contrast to the illumination of the room.

5. ***Scale illumination*** will alter the illumination of the background screen of the oscilloscope. Set the scale illumination such that it will not hinder observing the electron dot on the screen.

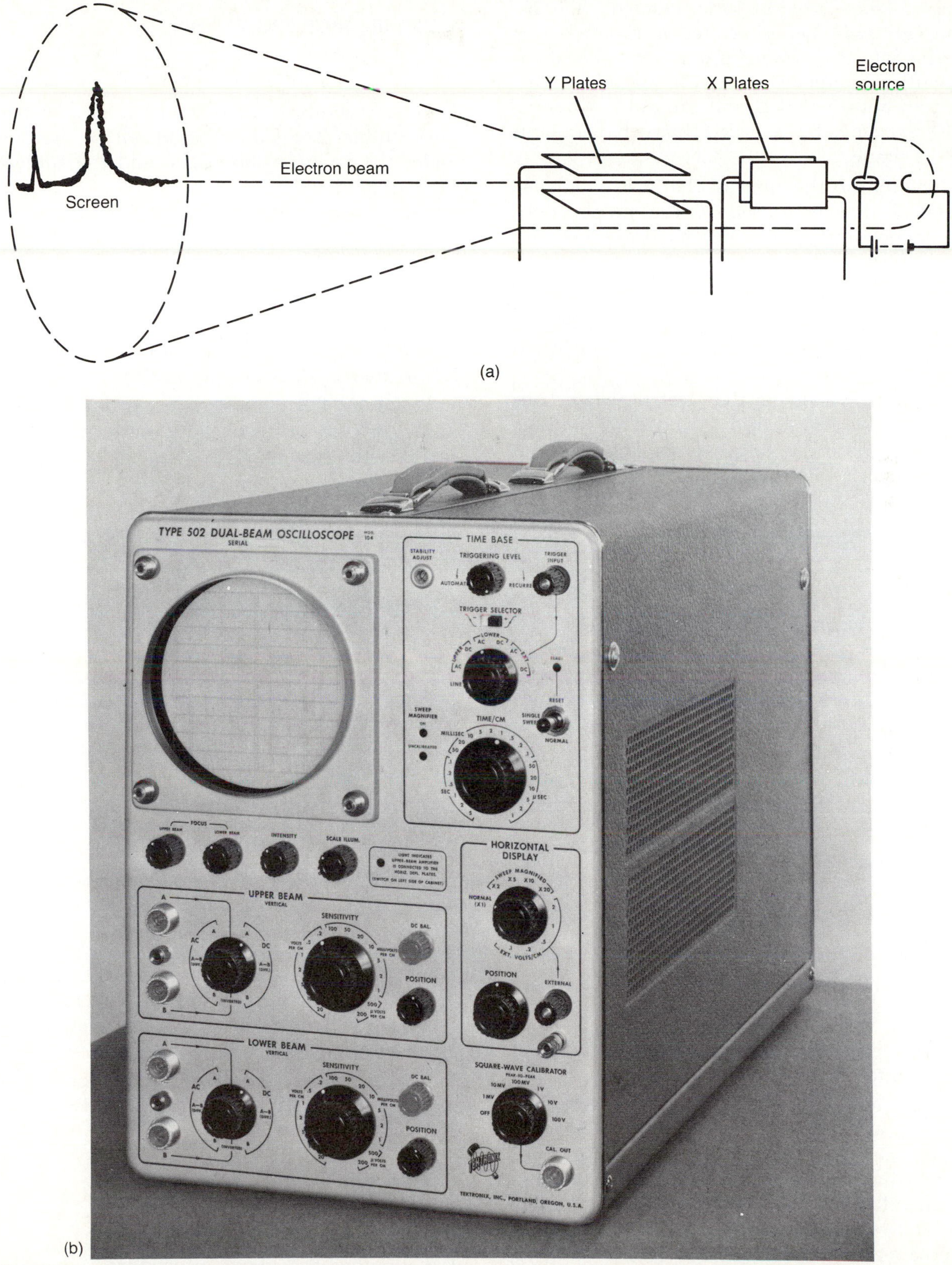

FIGURE 13.14 Oscilloscope. (a) Schematic diagram. (b) Photograph.

6. Within the enclosure marked Upper Beam you will note several jacks labeled ***"A,"*** ***"B,"*** and ***"ground"***. Leads are fed into either A or B and into the ground if one is to record the potential difference between tissue and ground, or into both A and B and ground if potential difference is to be recorded between two points of the tissue.

7. The dial to the right of the jacks is labeled ***Selector*** and is turned to the proper position depending upon which jacks are receiving input from the tissue.

8. Also within this box is a dial labeled ***Sensitivity***. This dial indicates the potential required to deflect the electron beam vertically one centimeter.

9. ***Position*** will bring the electron beam back onto the screen should it move off for any reason. Ignore the ***DC Balance*** knob for this experiment.

2. Preparation of Subject

PROCEDURE

1. For the purpose of recording an EEG in this experiment, place one electrode on the frontal region of the subject's skull immediately below the hair line and place another at the hair line on the temporal region of the skull posterior to the ear.

2. Brain waves will be detected from the frontal electrode, while the temporal electrode will serve as the ground.

3. With a dishwashing pad *gently* rub the areas where the electrodes will be placed. This will remove dead epithelial cells that will hamper recording.

CAUTION! *Be sure that the capillaries are not damaged and that bleeding does not occur.*

4. Place electrode jelly on each electrode and fasten into place with a stretchable Band-Aid.

5. Connect the electrode selector box to the "A" and "ground" input jacks of the oscilloscope. Since the EEG activity of each subject will vary, adjust the sweep and sensitivity to a midpoint value and observe the results obtained in the following procedures. If necessary slowly increase the sensitivity until definitive brain waves are obtained with the following procedures.

3. EEG Recording

a. CONTROL BRAIN WAVE PATTERN

PROCEDURE

1. Have the subject lie down on a cot or laboratory table. The leads should not be binding under the head or shoulders, and the subject must lie perfectly still.

2. Alpha waves will be difficult to demonstrate in a laboratory setting, as common disturbances may disrupt the subject and prevent the recording of alpha waves.

3. With the classroom dimly lit and all observers being absolutely still, ask your subject to close his/her eyes and relax completely.

4. Following five or more minutes of such conditions a control brain wave pattern will be established that may be composed of alpha waves.

5. Record your observations in Section C.1 of the LABORATORY REPORT RESULTS at the end of the exercise.

b. EFFECT OF HYPERVENTILATION

PROCEDURE

1. After the subject has relaxed completely for an adequate amount of time, have the person *slowly* inhale deeply and *forcibly* exhale deeply (hyperventilate).

2. Have the subject continue until there is a noticeable change in brain wave pattern.

CAUTION! *The subject should stop hyperventilating at the first sign of dizziness or lightheadedness.*

3. Once a brain wave pattern change is detected, allow the subject to breath normally and relax completely until the brain wave pattern again returns to normal.

4. Record your observations in Section C.2 of the LABORATORY REPORT RESULTS at the end of the exercise.

c. ALPHA BLOCK

PROCEDURE

1. Following the above procedure, have the subject open his/her eyes and allow them to adjust to the lighting in the room.

2. Quickly increase the lighting in the room and note any changes in the EEG.

3. Following 2–3 minutes of observing the EEG have the subject *mentally* complete a long division problem and observe any changes in the EEG during this process.

4. Record your observations in Section C.3 of the LABORATORY REPORT RESULTS at the end of the exercise.

D. CRANIAL NERVES

Of the 12 pairs of ***cranial nerves***, 10 originate from the brain stem, but all 12 pairs leave the skull through foramina in the base of the skull. The cranial nerves are designated by Roman numerals and names. The Roman numerals indicate the order in which the nerves arise from the brain, from front to back. The names indicate the distribution or function of the nerves.

Obtain a model of the brain and, using your text and any other aids available, identify the 12 pairs of cranial nerves. Now refer to Figure 13.15 and label the cranial nerves.

E. TESTS OF CRANIAL NERVE FUNCTION

The following simple tests may be performed to determine cranial nerve function. Although they provide only superficial information, they

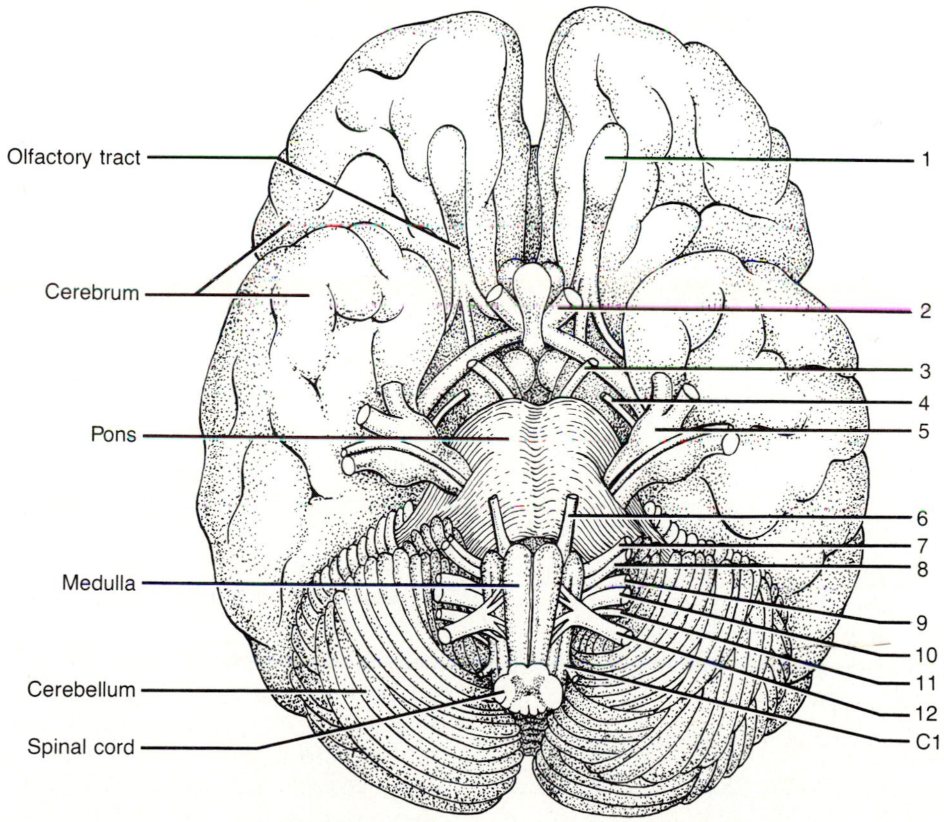

___ Abducens (VI) nerve
___ Accessory (XI) nerve
___ Facial (VII) nerve
___ Glossopharyngeal (IX) nerve
___ Hypoglossal (XII) nerve
___ Oculomotor (III) nerve
___ Olfactory bulb
___ Optic (II) nerve
___ Trigeminal (V) nerve
___ Trochlear (IV) nerve
___ Vagus (X) nerve
___ Vestibulocochlear (VIII) nerve

FIGURE 13.15 Cranial nerves of human brain seen in ventral view.

will help you to understand how the various cranial nerves function. Perform each of the tests with your partner.

1. ***Olfactory (I) nerve***—Have your partner smell several familiar substances such as spices, first using one nostril and then the other. Your partner should be able to distinguish the odors equally with each nostril.

What structures could be malfunctioning if the ability to smell is lost? ______

______

______

______

______

2. ***Optic (II) nerve***—Have your partner read a portion of a printed page using each eye. Do the same while using a Snellen chart at a distance of 6.10 m (20 ft).

Describe the visual pathway from the optic nerve to the cerebral cortex. ______

______

______

3. ***Oculomotor (III), Trochlear*** (TROK-lē-ar) ***(IV)***, and ***Abducens*** (ab-DOO-sens) ***(VI) nerves***—To test the motor responses of these nerves, have your partner follow your finger with his or her eyes without moving the head. Move your finger up, down, medially, and laterally.

Which nerves control which movements of the eyeball? ______

______

Look for signs of ptosis (drooping of one or both eyelids).

Which cranial nerve innervates the upper eyelid? ______

To test for pupillary light reflex, shine a small flashlight into each eye from the side.

CAUTION! *Do not make contact with the eyes.*

Observe the pupil.

What cranial nerve controls this reflex? ______

______

4. ***Trigeminal*** (trī-JEM-i-nal) ***(V) nerve***—To test the motor responses of this nerve, have your partner close his or her jaws tightly.

What muscles are used? ______

______

Now while holding your hand under your partner's lower jaw to provide resistance, ask your partner to open his or her mouth. To test the sensory responses of this nerve, have your partner close his or her eyes and lightly whisk a piece of dry cotton over the mandibular, maxillary, and ophthalmic areas on each side of the face. Do the same with cotton that has been moistened with cold water.

What cranial nerves bring about this response?

______

5. ***Facial (VII) nerve***—To test the motor responses of this nerve, ask your partner to bring the corners of the mouth straight back, smile while showing his or her teeth, whistle, puff his or her cheeks, frown, raise his or her eyebrows, and wrinkle his or her forehead.

Define Bell's palsy. ______

______

To test the sensory responses of this nerve, touch the tip of your partner's tongue with an applicator that has been dipped into a salt solution. Rinse the mouth out with water and now place an applicator dipped into a sugar solution along the anterior surface of the tongue.

What are the four basic tastes that can be distinguished? ______

______

6. ***Vestibulocochlear*** (ves-tib′-yoo-lō-KŌK-lē-ar) ***(VIII) nerve***—To test for the functioning of the vestibular portion of this nerve, have your partner sit on a swivel stool and turn him or her *slowly and very carefully* about 10 times (about one turn every two seconds). Stop the stool suddenly and look for nystagmus (rapid movement or quivering of the eyeballs). This is a normal response when a person is dizzy.

CAUTION! *Do not allow the subject to stand or walk until dizziness is completely gone.*

What cranial nerves would be affected if nystagmus occurs when the person is not dizzy?

______

______

To test for the functioning of the cochlear portion of the nerve, have your partner close his or her eyes and determine his or her ability to hear a ticking watch.

Measure the distance at which sound can no longer be heard for each ear. ______________

____

7. ***Glossopharyngeal*** (glos-ō-fa-RIN-jē-al) ***(IX)*** and ***Vagus (X) nerves***—The palatal (gag) reflex can be used to test the functioning of both of these nerves. Using a cotton-tipped applicator, *very slowly and gently* touch your partner's uvula.

What is the purpose of this reflex? __________

____

____

To test the sensory responses of these nerves, have your partner swallow. Does swallowing occur easily? Now *gently* hold your partner's tongue down with a tongue depressor and ask him or her to say "ah." Does the uvula move? Are the movements on both sides of the soft palate the same? The sensory function of the glossopharyngeal nerve may be tested by *lightly* applying a cotton-tipped applicator dipped in quinine to the top, sides, and back of the tongue.

In which area of the tongue was the quinine tasted? ______________________

What taste sensation is located there? ______

____

8. ***Accessory (XI) nerve***—The strength and muscle tone of the sternocleidomastoid and trapezius muscles indicate the proper functioning of the accessory nerve. To ascertain the strength of the sternocleidomastoid muscle, have your partner turn his or her head from side to side against *slight* resistance that you supply by placing your hands on either side of your partner's head. To ascertain the strength of the trapezius muscle, place your hands on your partner's shoulders and while *gently* pressing down firmly ask him or her to shrug his or her shoulders.

Do both muscles appear to be reasonably strong?

____

9. ***Hypoglossal (XII) nerve***—Have your partner protrude his or her tongue. It should protrude without deviation. Now have your partner protrude his or her tongue and move it from side to side while you attempt to *gently* resist the movements with a tongue depressor.

F. DISSECTION OF NERVOUS SYSTEM

CAUTION! *Please reread Section D. "Precautions Related to Dissection" at the beginning of the laboratory manual on page xii before you begin your dissection.*

The following instructions for dissection of the mammalian brain can be used in cats, fetal pigs, and sheep, because their brains are structurally similar. One main difference is the surface (gyri and sulci) of the cerebrum. Another difference is the proportions of certain parts. For example, the human brain is basically the same as those of sheep, pig, and cat, but the cerebrum, the center of intelligence, is disproportionately larger.

Figures 13.16 through 13.19 are based on the sheep brain, but they can be used as references if you are to dissect the cat brain instead. Because the only major dissections needed of the brain are a cross-sectional cut through the cerebrum and a midsagittal cut, brains can be returned to their preservative when you finish with them and used again. The instructor should provide brains that show cranial nerves, spinal cord, and meninges. Even if the cat brain is not to be dissected, the instructor might want to demonstrate a newly dissected one to the class.

PROCEDURE

1. If you are to dissect the brain from your cat, start by removing all of the muscles from the dorsal and lateral surfaces of the head.
2. *Very carefully*, using either bone shears or a bone saw, make an opening in the parietal bone. Do not damage the brain below.
3. Slowly chip away all the bone from the dorsal and lateral surfaces of the skull until the brain is exposed.
4. Between the cerebrum and the cerebellum is a transverse bony partition, the tentorium cerebelli, which should be removed carefully.
5. Cut the spinal cord transversely at the foramen magnum.
6. Gently lift the brain out of the floor of the cranium. As you do so, sever each cranial nerve as far from the brain as possible.

Study the following sections and dissect either the cat brain (Section E.2) or sheep brain (Section E.6) as your instructor directs.

1. Meninges

On the brain that you are to use, locate the three layers of the meninges that cover the brain and the spinal cord: outer ***dura mater***, middle ***arachnoid***, and inner ***pia mater***. Refer to pages 255 and 261 as you examine them. Note that the dura mater has two folds, the ***falx cerebri*** located between the cerebral hemispheres, and the ***tentorium cerebelli*** between the cerebral hemispheres and the cerebellum (the latter is ossified and fused to the parietal bones in the cat).

2. Dissection of Cat Brain

PROCEDURE

1. First identify on the dorsal surface of the brain the two large, cranial ***cerebral hemispheres*** (Figure 13.16). Together they constitute the ***cerebrum***. The cerebral hemispheres are

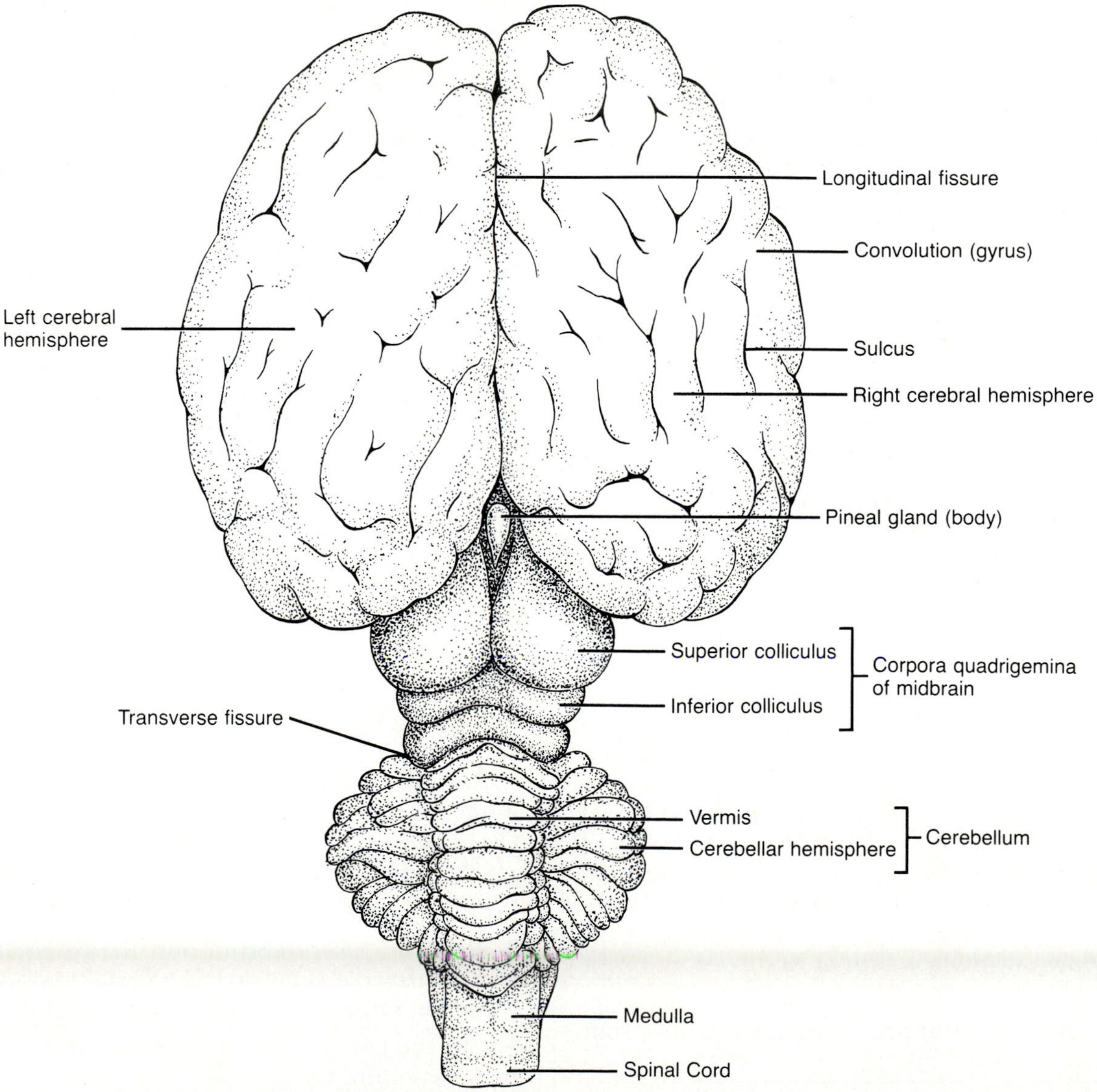

FIGURE 13.16 Dorsal view of sheep brain in which cerebellum has been spread apart from the cerebrum.

separated from each other by the deep ***longitudinal fissure***. The surface of each hemisphere consists of upward folds called ***convolutions (gyri)*** and grooves called ***sulci***.

2. Spread the hemispheres and look into the longitudinal fissure. Note the ***corpus callosum***, a band of transverse fibers that connects the two hemispheres internally.

3. Make a frontal section through a cerebral hemisphere. Note the ***gray matter*** near the surface of the cerebral cortex and the ***white matter*** beneath the gray matter.

4. Now locate the ***cerebellum*** caudal to the cerebral hemisphere and dorsal to the medulla oblongata. It is separated from the cerebral hemispheres by a deep ***transverse fissure***. The cerebellum consists of a median ***vermis*** and two lateral ***cerebellar hemispheres***. As does the cerebrum, the cerebellum has folia (convolutions), fissures, and sulci. The treelike arrangement of white matter tracts in the gray matter in the cerebellum is the ***arbor vitae*** ("tree of life").

5. Section a portion of the cerebellum longitudinally to examine the arbor vitae.

6. The cerebellum is connected to the brain stem by three prominent fiber tracts called ***peduncles***. The ***superior cerebellar peduncle*** connects the cerebellum with the midbrain, the ***inferior cerebellar peduncle*** connects the cerebellum with the medulla, and the ***middle cerebellar peduncle*** connects the cerebellum with the pons.

7. The dorsal portion (roof) of the ***midbrain*** can be examined if you spread the cerebral hemispheres apart from the cerebellum. The root of the midbrain is formed by four rounded bulges, the ***corpora quadrigemina***. The two larger cranial bulges are called the ***superior colliculi*** and the two smaller caudal bulges are called the ***inferior colliculi***. If you look between the superior colliculi you should see the ***pineal gland (body)***.

8. The ***medulla oblongata***, or just simply medulla, is the most caudal portion of the brain and is continuous with the spinal cord.

9. To examine other parts of the brain, you will have to make your observations from its ventral surface (Figure 13.17). At the cranial end of the brain are the ***olfactory bulbs***. The olfactory (I) nerves terminate on the bulbs after passing through the cribriform plate of the ethmoid bone. Bands of nerve fibers extending caudally and ventrally from the olfactory bulbs are the ***olfactory tracts***. The tracts terminate in the cranial part of the cerebrum.

10. Caudal to the olfactory tracts is the ***hypothalamus***. At the cranial border of the hypothalamus, the optic (II) nerves come together and partially cross. This point of crossing is called the ***optic chiasma***. The rest of the hypothalamus is caudal to the optic chiasma, covered mostly by the ***pituitary gland (hypophysis)***. The pituitary gland is attached to a small round elevation on the hypothalamus called the ***tuber cinereum*** by a narrow stalk called the ***infundibulum***.

11. Now examine the ventral portion (floor) of the midbrain called the ***cerebral peduncles***. These bundles of white fibers connect the pons to the cerebrum. They run obliquely and cranially on either side of the hypophysis.

12. ***Caudal*** to the cerebral peduncles is the ***pons***, which consists of transverse fibers that connect the cerebellar hemispheres.

13. Caudal to the pons is the ***medulla oblongata***. Identify the ***ventral median fissure***, a midline groove. The longitudinal bundles of fibers on either side of the ventral medial fissure are the ***pyramids***.

14. Lateral to the cranial end of the pyramids are the ***trapezoid bodies***. These are narrow bands of transverse fibers that cross the cranial end of the medulla.

15. The remaining parts of the brain may be located by cutting the brain longitudinally into halves. Cut through the midsagittal plane. A long sharp knife may be used here.

16. Using Figure 13.18 as a guide, identify the ***corpus callosum***, a transverse band of white fibers at the bottom of the cerebral hemispheres. The cranial portion of the corpus callosum is called the ***genu***, the middle part is known as the ***body*** or ***trunk***, and the caudal portion is called the ***splenium***. The ventral deflection of the genu is known as the ***rostrum***. Just beneath the splenium identify the ***pineal gland (body)***.

17. Ventral to the corpus callosum is the ***fornix***, a band of white fibers connecting the cerebral hemispheres. It is cranial to the third ventricle.

18. Between the fornix and corpus callosum is a thin vertical membrane, the ***septum pellucidum***, which separates the two lateral ventricles of the brain. If you break the membrane,

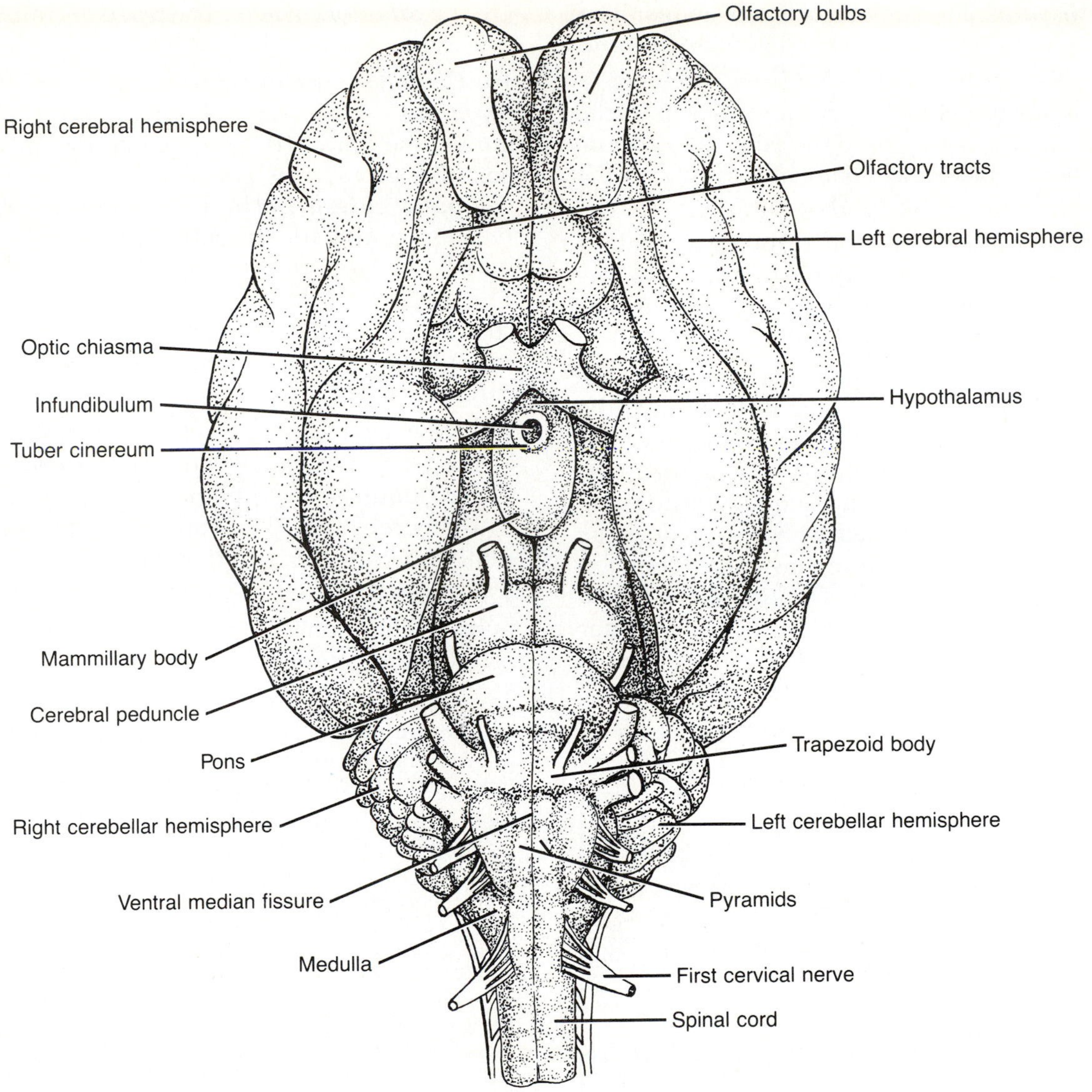

FIGURE 13.17 Ventral view of sheep brain.

you can expose the ***lateral ventricle***, a cavity in each cerebral hemisphere that contains cerebrospinal fluid in the living animal.

19. Identify the ***anterior commissure***, a small group of fibers ventral to the fornix. Extending from the anterior commissure to the optic chiasma is the ***lamina terminalis***. It forms the cranial wall of the third ventricle.

20. The ***thalamus*** forms the lateral walls of the ***third ventricle***, another cavity containing cerebrospinal fluid. The thalamus is caudal to the fornix and extends ventrally to the infundibulum and caudally to the corpora quadrigemina. The third ventricle lies mostly in the thalamus and hypothalamus, and, in the living animal, is filled with cerebrospinal fluid. The lateral ventricles communicate with the third ventricle by an opening called the ***interventricular foramen***.

21. The ***intermediate mass*** is a circular area that extends across the third ventricle, connecting the two sides of the thalamus. The interventricular foramen lies in the depression cranial to the intermediate mass. The intermediate mass is absent in about 30% of human brains and, when present, is proportionately much smaller.

22. Identify the ***posterior commissure***, a small

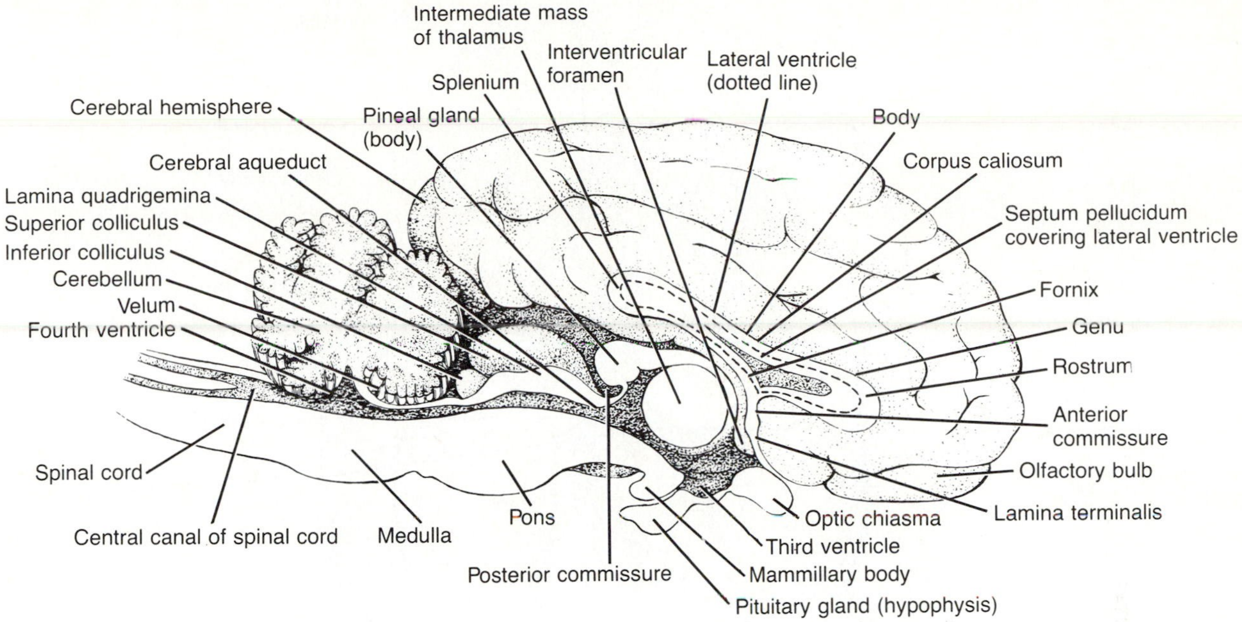

FIGURE 13.18 Midsagittal section of sheep brain.

round mass of fibers that crosses the midline. It is located just ventral to the pineal gland (body).

23. Now find the ***cerebral aqueduct***, a narrow longitudinal channel that passes through the midbrain and connects the third and fourth ventricles. The roof of the cerebral aqueduct is called the ***lamina quadrigemina***.

24. The ***fourth ventricle*** is a cavity superior to the pons and medulla and inferior to the cerebellum. It is continuous with the cerebral aqueduct and contains cerebrospinal fluid in a living animal. A membrane that forms the roof of the fourth ventricle is called the ***velum***. The fourth ventricle leads into the ***central canal*** of the spinal cord.

3. Cranial Nerves

Briefly review Section D, Cranial Nerves (page 273). Using Figure 13.19 as a guide, identify the cranial nerves.

a. ***Olfactory (I) nerve***—Consists of fibers that pass from the olfactory bulb, through the foramina of the cribriform plate, to the mucous membrane of the nasal cavity. It functions in smell.

b. ***Optic (II) nerve***—Arises from the optic chiasma and passes through the optic foramen to the retina of the eye. It functions in vision.

c. ***Oculomotor (III) nerve***—Arises from the ventral surface of the cerebral peduncle, passes through the orbital fissure, and supplies the superior rectus, medial rectus, inferior rectus, inferior oblique, and levator palpebrae eye muscles.

d. ***Trochlear (IV) nerve***—Arises just caudal to the corpora quadrigemina, curves around the lateral aspect of the cerebral peduncle, passes through the orbital fissure, and supplies the superior oblique muscle of the eyeball.

e. ***Trigeminal (V) nerve***—Originates from the lateral surface of the pons by a large sensory root and a small motor root. It exits through the orbital fissure and foramen ovale. The three branches of the trigeminal nerve are the ***ophthalmic, maxillary,*** and ***mandibular***. All three branches supply the skin of the head and face, the epithelium of the oral cavity, the anterior two-thirds of the tongue, and the teeth. The mandibular branch also supplies the muscles of mastication.

f. ***Abducens (VI) nerve***—Arises from between the trapezoid body and pyramid, passes through the orbital fissure, and supplies the lateral rectus muscle of the eyeball.

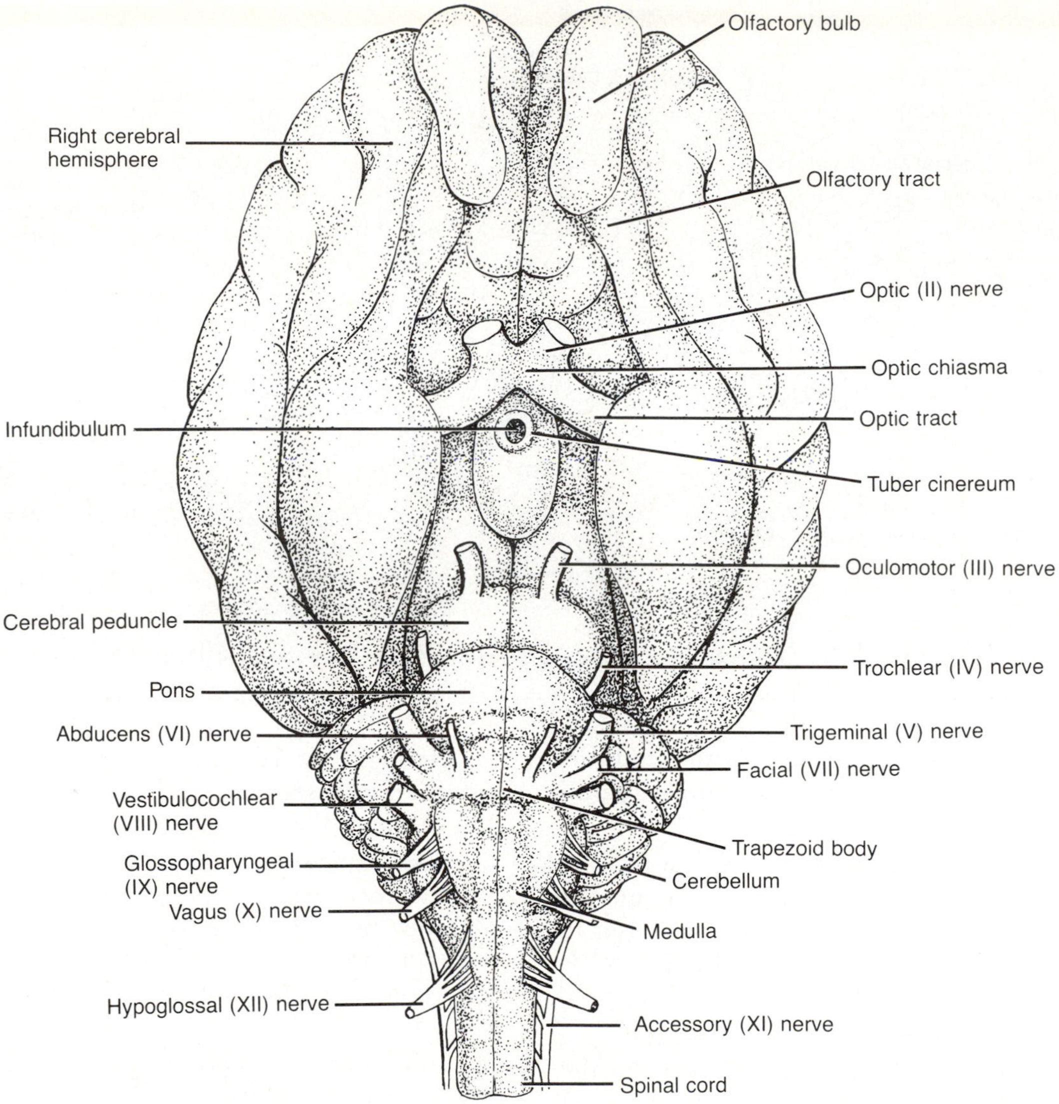

FIGURE 13.19 Cranial nerves of sheep brain seen in ventral view.

g. ***Facial (VII) nerve***—Arises caudal to the trigeminal nerve between the pons and trapezoid body, passes through the middle ear, and emerges at the stylomastoid foramen. It supplies many head and facial muscles and the mucosa of the tongue.

h. ***Vestibulocochlear (VIII) nerve***—Arises just caudal to the facial nerve, passes into the internal auditory meatus, and divides into two branches. One branch transmits auditory sensations from the cochlea. The other branch transmits sensations for equilibrium from the semicircular canals.

i. ***Glossopharyngeal (IX) nerve***—Arises just caudal to the vestibulocochlear nerve, emerges from the jugular foramen, and supplies the muscles of the pharynx and the taste buds.

j. ***Vagus (X) nerve***—Arises just caudal to the glossopharyngeal nerve and exits through the jugular foramen. It supplies most of the thoracic and abdominal viscera.

k. ***Accessory (XI) nerve***—Arises from the medulla and spinal cord, passes through the jugular foramen, and supplies the neck and shoulder muscles.

l. ***Hypoglossal (XII) nerve***—Arises from the ventral medulla, passes through the hypoglossal foramen, and supplies tongue muscles.

4. Dissection of Cat Spinal Cord

PROCEDURE

1. Remove the muscles dorsal to the vertebral column.

2. Using small bone clippers, cut through the laminae of the vertebrae and remove the neural arches of all the vertebrae starting at the foramen magnum.

3. Once the entire spinal cord is exposed, identify the following (Figure 13.20a):

   a. ***Cervical enlargement***—Between the fourth cervical and first thoracic vertebrae.

   b. ***Lumbar enlargement***—Between the third and seventh lumbar vertebrae.

   c. ***Conus medullaris***—Tapered conical tip of the spinal cord caudal to the lumbar enlargement.

   d. ***Filum terminale***—A slender strand continuous with the conus medullaris, which extends into the caudal region. The filum and caudal spinal nerves are referred to as the ***cauda equina***.

   e. ***Dorsal median sulcus***—Shallow groove on the dorsal surface of the spinal cord along the midline.

   f. ***Ventral median fissure***—Deeper groove on the ventral surface of the spinal cord along the midline (Figure 13.20b).

   g. ***Dorsolateral sulcus***—Groove parallel and lateral to the dorsal median sulcus. The dorsal roots of spinal nerves attach to this sulcus (see Figure 13.20b).

   h. ***Meninges***—These are similar in structure and location to those of the brain (see Figure 13.20b).

4. If you examine the cut end of the spinal cord at the foramen magnum with a hand lens, you can see the gray matter arranged like a letter H in the interior and the white matter external to the gray matter. Also identify the dorsal median sulcus and the ventral median fissure.

5. Spinal Nerves of Cat

The cat has 38 or 39 pairs of ***spinal nerves***. These include 8 pairs of ***cervical***, 13 pairs of ***thoracic***, 7 pairs of ***lumbar***, 3 pairs of ***sacral***, and 7 or 8 pairs of ***caudal (coccygeal)*** (see Figure 13.20a).

Each spinal nerve arises by a ***dorsal (sensory) root***, which originates from the dorsolateral sulcus, and a ***ventral root*** (Figure 13.20b). The dorsal root has an oval swelling called the ***dorsal root ganglion***. The first cervical nerve exits from the vertebral canal through the atlantal foramen. The second cervical nerve exits between the arches of the atlas and axis. All other spinal nerves exit through intervertebral foramina.

Just outside the intervertebral foramina, a spinal nerve divides into a ventral ramus and a dorsal ramus. The ***dorsal rami*** are small and are distributed to the skin and muscles of the back. The ***ventral rami*** are large and supply the skin and muscles of the ventral part of the trunk.

Expose the roots of several spinal nerves and identify the parts just described.

6. Dissection of Sheep Brain

CAUTION! *Please reread Section D. "Precautions Related to Dissection" at the beginning of the laboratory manual on page xii before you begin your dissection.*

The brains of the fetal pig, sheep, and human show many similarities. They possess the protective membranes called the ***meninges***, which can easily be seen as you proceed with the dissections. The outermost layer is the ***dura mater***. It is the toughest, protective one, and may be missing in the preserved sheep brain. The middle membrane is the ***arachnoid***, and the inner one containing blood vessels and adhering closely to the surface of the brain itself is the ***pia mater***.

PROCEDURE

1. Use Figures 13.16 through 13.18 as references for this dissection.

2. The most prominent external parts of the brain are the pair of large ***cerebral hemispheres*** and the posterior ***cerebellum*** on the dorsal surface. These large hemispheres are separated from each other by the ***longitudinal fissure***. The ***transverse fissure*** separates the hemispheres from the cerebellum. The surfaces of these hemispheres form many ***gyri***, or raised ridges, that are separated by grooves, or ***sulci***.

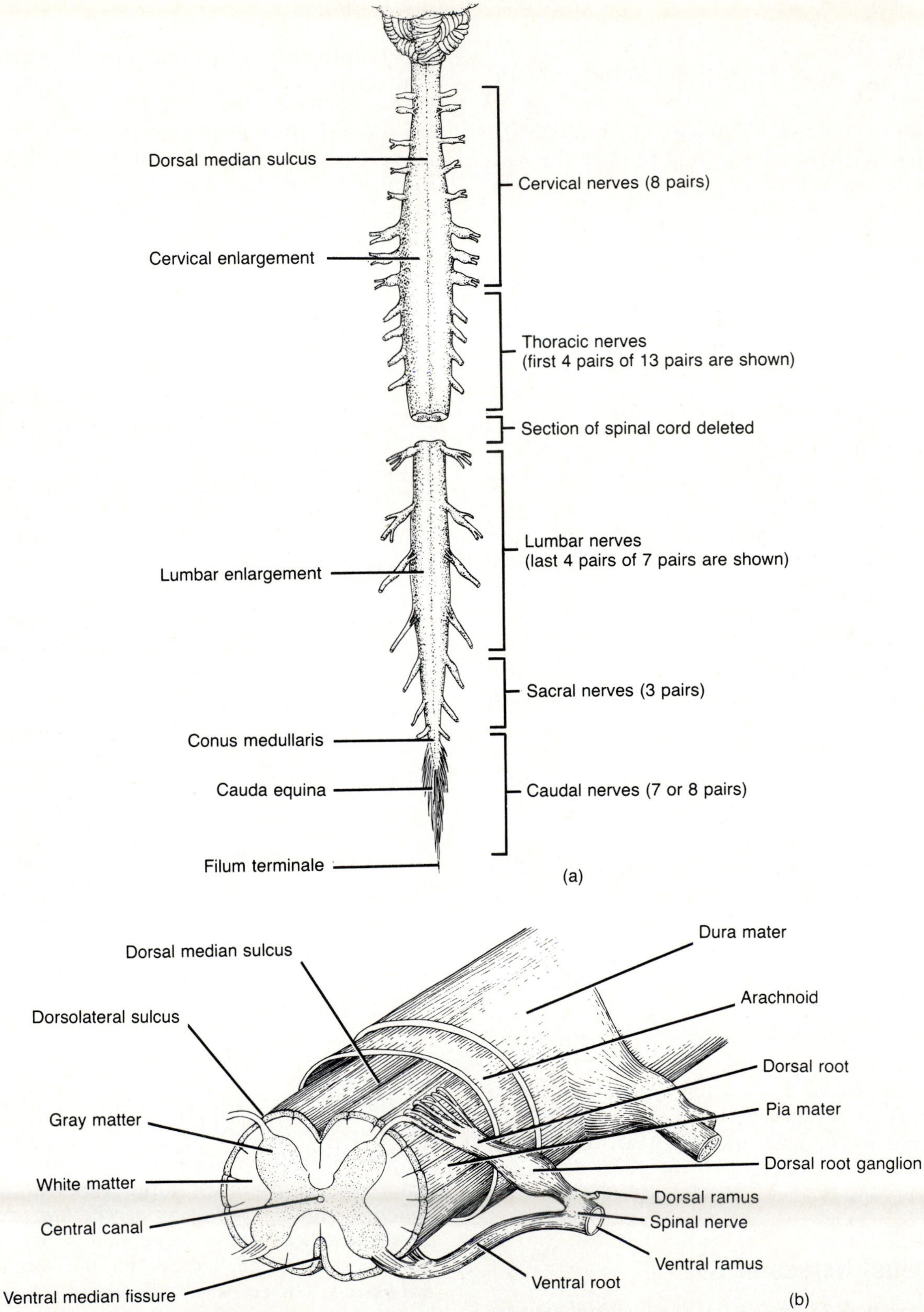

FIGURE 13.20 The spinal cord. (a) Dorsal view. (b) Cross section.

3. If you spread the hemispheres gently apart you can see, deep in the longitudinal fissure, thick bundles of white transverse fibers. These bundles form the ***corpus callosum***, which connects the hemispheres.

4. Most of the following structures can be identified by examining a midsagittal section of the sheep brain, or by cutting an intact brain along the longitudinal fissure completely through the corpus callosum.

5. If you break through the thin ventral wall, the ***septum pellucidum*** of the corpus callosum, you can see part of a large chamber, the ***lateral ventricle***, inside the hemisphere.

6. Each hemisphere has one of these ventricles. Ventral to the septum pellucidum, locate a smaller band of white fibers called the ***fornix***. Close by where the fornix disappears is a small, round bundle of fibers called the ***anterior commissure***.

7. The ***third ventricle*** and the ***thalamus*** are located ventral to the fornix. The third ventricle is outlined by its shiny epithelial lining, and the thalamus forms the lateral walls of this ventricle. This ventricle is crossed by a large circular mass of tissue, the ***intermediate mass***, which connects the two sides of the thalamus. Each lateral ventricle communicates with the third ventricle through an opening, the ***interventricular foramen***, which lies in a depression anterior to the intermediate mass and can be located with a dull probe.

8. Spreading the cerebral hemispheres and the cerebellum apart reveals the roof of the midbrain (mesencephalon), which is seen as two pairs of round swellings collectively called the ***corpora quadrigemina***. The larger, more anterior pair are the ***superior colliculi***. The smaller posterior pair are the ***inferior colliculi***. The ***pineal gland (body)*** is seen directly between the superior colliculi. Just posterior to the inferior colliculi, appearing as a thin white strand, is the ***trochlear (IV) nerve***.

9. The cerebellum is connected to the brain stem by three prominent fiber tracts called ***peduncles***. The ***superior cerebellar peduncle*** connects the cerebellum with the midbrain, the ***inferior cerebellar peduncle*** connects the cerebellum with the medulla, and the ***cerebellar peduncle*** connects the cerebellum with the pons.

10. Most of the following parts can be located on the ventral surface of the intact brain.

11. Just beneath the cerebral hemispheres are two ***olfactory bulbs***, which continue posteriorly as two ***olfactory tracts***. Posterior to these tracts, the ***optic (II) nerves*** undergo a crossing (decussation) known as the ***optic chiasma***.

12. Locate the ***pituitary gland (hypophysis)*** just posterior to the chiasma. This gland is connected to the ***hypothalamus*** portion of the diencephalon by a stalk called the ***infundibulum***. The ***mammillary body*** appears immediately posterior to the infundibulum.

13. Just posterior to this body are the paired ***cerebral peduncles***, from which arise the large ***oculomotor (III) nerves***. They may be partially covered by the pituitary gland.

14. The ***pons*** is a posterior extension of the hypothalamus and the ***medulla oblongata*** is a posterior extension of the pons.

15. The ***cerebral aqueduct*** dorsal to the peduncles runs posteriorly and connects the third ventricle with the ***fourth ventricle***, which is located dorsal to the medulla and ventral to the cerebellum.

16. The medulla merges with the ***spinal cord***, and is separated by the ***ventral median fissure***. The ***pyramids*** are the longitudinal bands of tissue on either side of this fissure.

17. Identify the remaining ***cranial nerves*** on the ventral surface of the brain. They are the trigeminal (V), abducens (VI), facial (VII), vestibulocochlear (VIII), glossopharyngeal (IX), vagus (X), accessory (XI), and hypoglossal (XII). The previously identified cranial nerves are the olfactory (I), optic (II), oculomotor (III), and trochlear (IV), for a total of twelve.

18. A cross section through a cerebral hemisphere reveals ***gray matter*** near the surface of the ***cerebral cortex*** and ***white matter*** beneath this layer.

19. A cross section through the spinal cord reveals the ***central canal***, which is connected to the fourth ventricle and contains ***cerebrospinal fluid***.

20. A midsagittal section through the cerebellum reveals a treelike arrangement of gray and white matter called the ***arbor vitae*** (tree of life).

G. AUTONOMIC NERVOUS SYSTEM

The ***autonomic nervous system (ANS)*** regulates the activities of smooth muscle, cardiac muscle, and glands, usually involuntarily. In the somatic nervous system (SNS), which is voluntary, the cell bodies of the efferent (motor) neurons are in the CNS, and their axons extend all the way to skeletal muscles in spinal nerves. The ANS always has two efferent neurons in the pathway. The first efferent neuron, the ***preganglionic neuron***, has its cell body in the CNS. Its axon leaves the CNS and synapses in an autonomic ganglion with the second neuron called the ***postganglionic neuron***. The cell body of the postganglionic neuron is inside an autonomic ganglion, and its axon terminates in a ***visceral effector*** (muscle or gland).

Label the components of the autonomic pathway shown in Figure 13.21.

The ANS consists of two divisions: sympathetic and parasympathetic. Most viscera are innervated by both divisions. In general, nerve impulses from one division stimulate a structure, whereas nerve impulses from the other division decrease its activity (Figure 13.22).

In the ***sympathetic division***, the cell bodies of the preganglionic neurons are located in the lateral gray horns of the spinal cord in the thoracic and first two lumbar segments. The axons of preganglionic neurons are myelinated and leave the spinal cord through the ventral root of a spinal nerve. Each axon travels briefly in a ventral ramus and then through a small branch called a white ramus communicans to enter a sympathetic trunk ganglion. These ganglia lie in a vertical row, on either side of the vertebral column, from the base of the skull to the coccyx. In the ganglion, the axon may synapse with a postganglionic neuron, travel upward or downward through the sympathetic trunk ganglia to synapse with postganglionic neurons at different levels, or pass through the ganglion without synapsing to form part of the splanchnic nerves. If the preganglionic axon synapses in a sympathetic trunk ganglion, it reenters the ventral or dorsal ramus of a spinal nerve via a small branch called a gray ramus communicans. If the preganglionic axon forms part of the splanchnic nerves, it passes through the sympathetic trunk ganglion but synapses with a postganglionic neuron in a prevertebral (collateral) ganglion. These ganglia are anterior to the vertebral column close to large abdominal arteries from which their names are derived (celiac, superior mesenteric, and inferior mesenteric).

In the ***parasympathetic division***, the cell bodies of the preganglionic neurons are located in nuclei in the brain stem and lateral gray

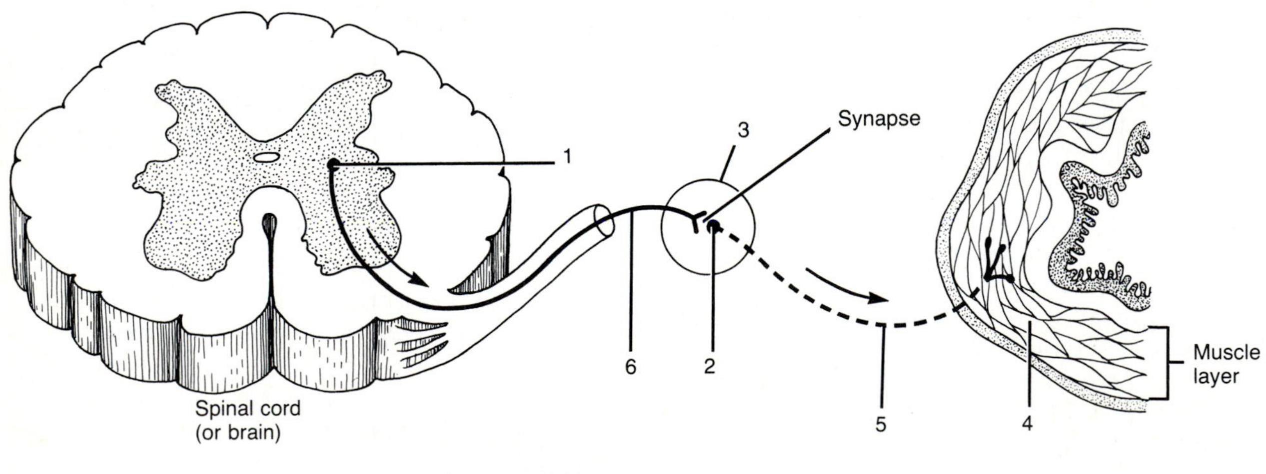

___ Autonomic ganglion
___ Cell body of postganglionic neuron
___ Cell body of preganglionic neuron
___ Axon of postganglionic neuron
___ Axon of preganglionic neuron
___ Visceral effector

FIGURE 13.21 Components of an autonomic pathway.

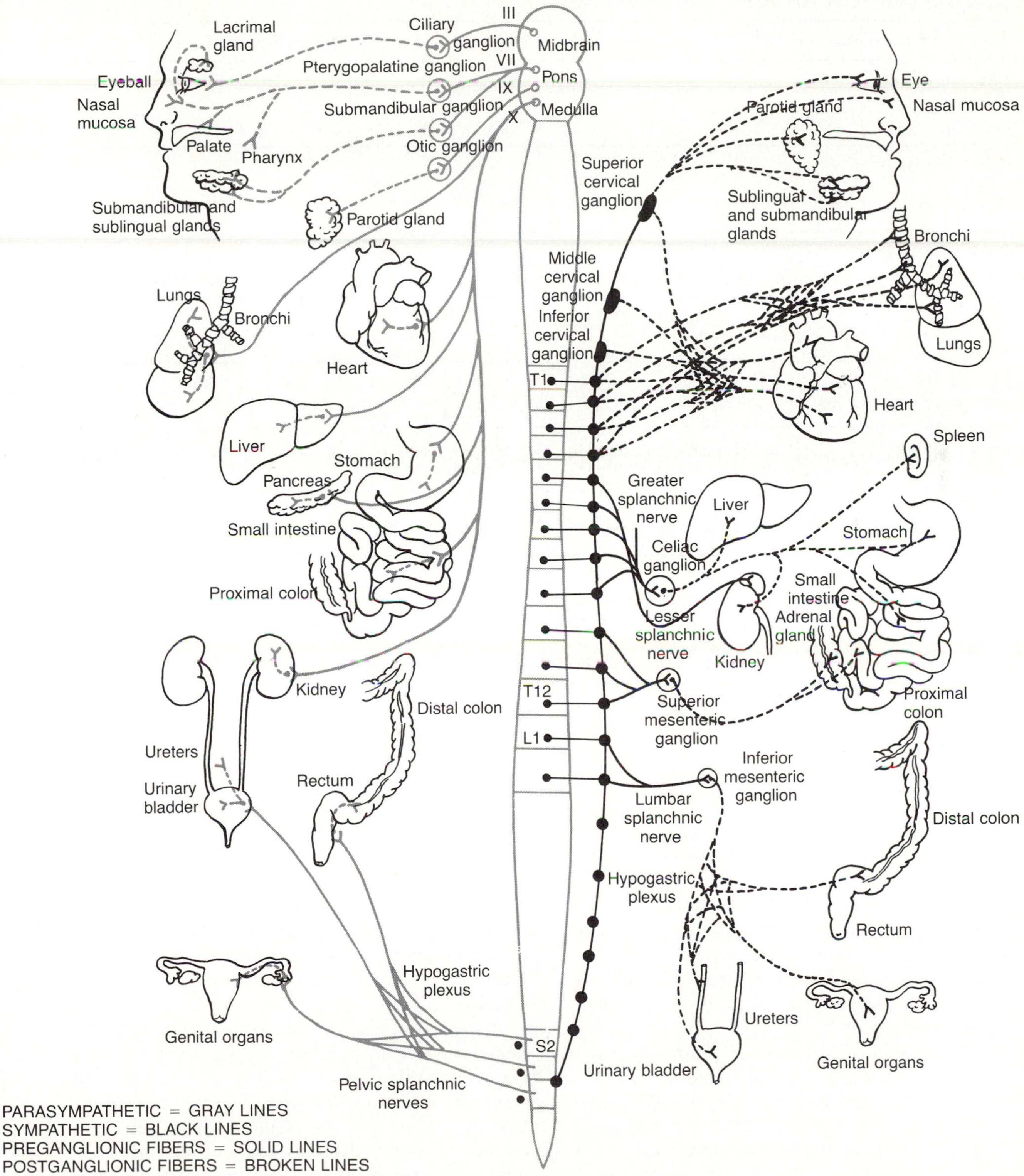

FIGURE 13.22 Structure of the autonomic nervous system.

TABLE 13.1
ACTIVITIES OF THE AUTONOMIC NERVOUS SYSTEM

Visceral effector	Sympathetic response	Parasympathetic response
Radial muscle of iris of eye		
Sphincter muscle of iris of eye		
Ciliary muscle of eye		
Sweat glands		
Lacrimal glands		
Salivary glands		
Gastric and intestinal glands		
Adrenal medulla		
Lungs		
Arterioles of skin and viscera		
Arterioles of skeletal muscles		

horn of the second through fourth sacral segments of the spinal cord. The axons emerge as part of cranial or spinal nerves. The preganglionic axons synapse with postganglionic neurons in terminal ganglia, near or within visceral effectors.

The sympathetic division is primarily concerned with processes that expend energy. During stress, the sympathetic division sets into operation a series of reactions collectively called the ***fight-or-flight response***, designed to help the body counteract the stress and return to homeostasis. During the fight-or-flight response, the heart and breathing rates increase and the blood sugar level rises, among other things.

The parasympathetic division is primarily concerned with activities that restore and conserve energy. It is thus called the ***rest-repose system***. Under normal conditions, the parasympathetic division dominates the sympathetic division in order to maintain homeostasis.

Autonomic fibers, like other axons of the nervous system, release neurotransmitters at synapses as well as at points of contact with visceral effectors ***(neuroeffector junctions)***. On the basis of the neurotransmitter produced, autonomic fibers may be classified as either cholinergic or adrenergic. ***Cholinergic*** (kō-lin-ER-jik) ***fibers*** release ***acetylcholine (ACh)*** and include the following: (1) all sympathetic and parasympathetic preganglionic axons, (2) all parasympathetic postganglionic axons, and (3) some sympathetic postganglionic axons. ***Adrenergic*** (ad′-ren-ER-jik) ***fibers***

TABLE 13.1 (***Continued***)

Visceral effector	Sympathetic response	Parasympathetic response
Heart		
Liver		
Gallbladder and ducts		
Stomach and intestines		
Kidney		
Ureter		
Urinary bladder		
Pancreas		
Spleen		
Arrector pili of hair follicle		
Uterus		
Sex organs		

produce ***norepinephrine (NE)***. Most sympathetic postganglionic axons are adrenergic.

The actual effects produced by ACh are determined by the type of receptor with which it interacts. The two types of receptors are known as nicotinic receptors and muscarinic receptors. ***Nicotinic receptors*** are found on both sympathetic and parasympathetic postganglionic neurons. These receptors are so named because the action of ACh on them are similar to those produced by nicotine. ***Muscarinic receptors*** are found on effectors innervated by parasympathetic postganglionic axons. These receptors are so named because the actions of ACh on them are similar to those produced by muscarine, a toxin produced by a mushroom.

The effects of NE and epinephrine, like those of ACh, are also determined by the type of receptor with which they interact. Such receptors are found on visceral effectors innervated by most sympathetic postganglionic axons and are referred to as ***alpha receptors*** and ***beta receptors***. Although cells of most effectors contain either alpha or beta receptors, some effector cells contain both. NE, in general, stimulates alpha receptors to a greater extent than beta receptors, and epinephrine, in general, stimulates both alpha and beta receptors.

Using your textbook as a reference, write in the sympathetic and parasympathetic responses for the visceral effectors listed in Table 13.1.

ANSWER THE LABORATORY REPORT QUESTIONS AT THE END OF THE EXERCISE

NERVOUS SYSTEM 13

Student ______________________ **Date** ______________________

Laboratory Section ______________________ **Score/Grade** ______________________

SECTION C. ELECTROENCEPHALOGRAM (EEG)

1. Did the subject demonstrate an alpha wave pattern during the control conditions? If not give a physiological explanation for your observations. ______________________

____

2. Explain the physiological reasons for the observed effect of hyperventilation upon the EEG.

____

____

3. Describe the effects of increased lighting and performance of a long division problem upon the EEG. ______________________

____

LABORATORY REPORT QUESTIONS

NERVOUS SYSTEM 13

Student ______________________ Date ______________________

Laboratory Section ______________________ Score/Grade ______________________

PART I. MULTIPLE CHOICE

_____ 1. The tapered, conical portion of the spinal cord is the (a) filum terminale (b) conus medullaris (c) cauda equina (d) lumbar enlargement

_____ 2. The outermost meninx composed of dense fibrous connective tissue is the (a) pia mater (b) arachnoid (c) dura mater (d) denticulate

_____ 3. The portion of a spinal nerve that contains motor nerve fibers only is the (a) posterior root (b) posterior root ganglion (c) lateral root (d) anterior root

_____ 4. The connective tissue covering around individual nerve fibers is the (a) endoneurium (b) epineurium (c) perineurium (d) ectoneurium

_____ 5. On the basis of organization, which does *not* belong with the others? (a) pons (b) medulla (c) thalamus (d) midbrain

_____ 6. The lateral ventricles are connected to the third ventricle by the (a) interventricular foramen (b) cerebral aqueduct (c) median aperture (d) lateral aperture

_____ 7. The vital centers for heartbeat, respiration, and blood vessel diameter regulation are found in the (a) pons (b) cerebrum (c) cerebellum (d) medulla

_____ 8. The reflex centers for movements of the head and trunk in response to auditory stimuli are located in the (a) inferior colliculi (b) medial geniculate nucleus (c) superior colliculi (d) ventral posterior nucleus

_____ 9. Which thalamic nucleus controls general sensations and taste? (a) medial geniculate (b) ventral posterior (c) ventral lateral (d) ventral anterior

_____ 10. Integration of the autonomic nervous system, secretion of regulating factors, control of body temperature, and the regulation of food intake and thirst are functions of the (a) pons (b) thalamus (c) cerebrum (d) hypothalamus

_____ 11. The left and right cerebral hemispheres are separated from each other by the (a) central sulcus (b) transverse fissure (c) longitudinal fissure (d) insula

_____ 12. Which structure does *not* belong with the others? (a) putamen (b) caudate nucleus (c) insula (d) globus pallidus

_____ 13. Which peduncles connect the cerebellum with the midbrain? (a) superior (b) inferior (c) middle (d) lateral

_____ 14. Which cranial nerve has the most anterior origin? (a) XI (b) IX (c) VII (d) IV

_____ 15. Extensions of the pia mater that suspend the spinal cord and protect against shock are the (a) choroid plexuses (b) pyramids (c) denticulate ligaments (d) superior colliculi

_____ 16. Which branch of a spinal nerve enters into formation of plexuses? (a) meningeal (b) dorsal (c) rami communicantes (d) ventral

_____ 17. Which plexus innervates the upper extremities and shoulders? (a) sacral (b) brachial (c) lumbar (d) cervical

_____ 18. How many pairs of thoracic spinal nerves are there? (a) 1 (b) 5 (c) 7 (d) 12

PART II. COMPLETION

19. The narrow, shallow groove on the posterior surface of the spinal cord is the _____________.

20. The space between the dura mater and wall of the vertebral canal is called the _____________.

21. In a spinal nerve, the cell bodies of sensory neurons are found in the _____________.

22. The outermost connective tissue covering around a spinal nerve is the _____________.

23. The middle meninx is referred to as the _____________.

24. The nuclei of origin for cranial nerves IX, X, XI, and XII are found in the _____________.

25. The portion of the brain containing the cerebral peduncles is the _____________.

26. Cranial nerves V, VI, VII, and VIII have their nuclei of origin in the _____________.

27. A shallow downfold of the cerebral cortex is called a(n) _____________.

28. The _____________ separates the frontal lobe of the cerebrum from the parietal lobe.

29. White matter tracts of the cerebellum are called _____________.

30. The space between the dura mater and the arachnoid is referred to as the _____________.

31. Together, the thalamus, hypothalamus, and pineal gland are called the _____________.

32. Cerebrospinal fluid passes from the third ventricle into the fourth ventricle through the _____________.

33. The cerebrum is separated from the cerebellum by the _____________ fissure.

34. The branches of a spinal nerve that are components of the autonomic nervous system are known as _____________.

35. The plexus that innervates the buttocks, perineum, and lower extremities is the _____________ plexus.

36. There are _____________ pairs of spinal nerves.

37. Together, the brain and spinal cord are referred to as the _____________ nervous system.

38. The cell bodies of _____________ neurons of the ANS are found inside autonomic ganglia.

39. The portion of the ANS concerned with the fight-or-flight response is the _____________ division.

40. The autonomic ganglia that are anterior to the vertebral column and close to large abdominal arteries are called _____________ ganglia.

41. The ______________________ nerve innervates the diaphragm.

42. The ______________________ tract conveys nerve impulses for touch and pressure.

43. The flexor muscles of the thigh and extensor muscles of the leg are innervated by the ______________________ nerve.

44. The ______________________ tract conveys nerve impulses related to muscle tone and posture.

45. The ______________________ nerve supplies the extensor muscles of the arm and forearm.

46. The ______________________ area of the cerebral cortex receives sensations from cutaneous, muscular, and visceral receptors in various parts of the body.

47. The portion of the cerebral cortex that translates thoughts into speech is the ______________________ area.

SENSATIONS AND SENSORY PATHWAYS 14

Now that you have studied the nervous system, you will study sensations, which are closely related to and integrated with this system. If you could not "sense" your environment and make the necessary homeostatic adjustments, you could not survive on your own. In its broadest sense, the term ***sensation*** refers to the state of awareness of external or internal conditions of the body. Sensations are picked up by ***sensory receptors*** that change stimuli into ***receptor potentials***. If the receptor potential is of threshold value or higher, it will change the environmental physical and chemical stimuli into coded nerve impulses that the brain can decode.

A. CHARACTERISTICS OF SENSATIONS

For a sensation to occur, five prerequisites must be met: (1) a ***stimulus***, or changing input of energy in the environment, capable of stimulating a receptor, must be present; (2) a ***sensory unit*** must receive the stimulus and convert it to a ***receptor potential*** (a sensory receptor or sense organ may be viewed as specialized nervous tissue that exhibits a high degree of sensitivity to specific internal or external conditions); (3) this ***receptor potential*** must be of threshold or greater value in order to initiate a nerve action potential (nerve impulse) in the sensory neuron; (4) the nerve impulse must be ***conducted*** along a nervous pathway from the sensory receptor or sense organ to the brain; (5) a ***region*** of the brain must ***interpret*** the nerve impulse into a sensation.

A sensory unit consists of a single peripheral neuron located in a spinal or cranial nerve ganglion and its central and peripheral processes. The peripheral processes may terminate either as free nerve endings or in association with a sensory receptor. Sensory receptors may be very simple or quite complex structures, such as the eye, which contains highly specialized neurons, epithelium, and connective tissues. All sensory receptors contain the terminal processes of a neuron, exhibit a high degree of excitability, and possess a specific threshold for a particular type of stimulus. Furthermore, the majority of sensory impulses are conducted to the sensory areas of the cerebral cortex, for it is in this region of the body that a stimulus produces conscious feeling.

We see with our eyes, hear with our ears, and feel pain in an injured part of our body only because the cortex interprets the sensation as coming from the stimulated sensory receptor. One characteristic of sensations, that of ***projection***, describes the process by which the brain refers sensations to their point of learned origin of the stimulation. A second characteristic of many sensations is ***adaptation***, that is, disappearance of a sensation even though a stimulus is still being applied. For example, when you first get into a tub of hot water, you might feel an intense burning sensation. But after a brief period of time the sensation decreases to one of comfortable warmth, even though the stimulus (hot water) is still present. Another characteristic is ***afterimages***, that is, the persistence of a sensation after the stimulus has been removed. One common example of an afterimage occurs when you look at a bright light and then look away. You will still see the light for several seconds afterward. The fourth characteristic of sensations is ***modality***, that is, the possession of distinct properties by which one sensation may be distinguished from another. For example, pain, pressure, touch, body position, equilibrium, hearing, vision, smell, and taste are all distinctive because the body perceives each differently.

B. CLASSIFICATION OF SENSATIONS

One convenient method of classifying sensations is by location of the receptor:

1. ***Exteroceptors*** (EKS′-ter-ō-sep′-tors), located near the body surface, are stimulated by changes in the external environment. They transmit sensations such as hearing, sight, smell, taste, touch, pressure, temperature, and pain.

2. ***Visceroceptors*** (VIS-er-ō-sep′-tors) or ***enteroceptors***, located in blood vessels and viscera, are stimulated by changes in the internal environment. These stimuli are perceived as pain, taste, fatigue, hunger, thirst, and nausea. Two specific types of visceroceptors are ***baroreceptors (pressoreceptors)***, stimulated by changes in the hydrostatic pressure of blood, and ***osmoreceptors***, which are stimulated by changes in the solute concentration of extracellular fluid.

3. ***Proprioceptors*** (PRŌ-prē-ō-sep′-tors) or stretch receptors, located in muscles, tendons, or joints, are stimulated by stretching or movement. They provide sensations of position, movement, and muscle tension.

Another classification of sensations is based on the type of stimulus received.

1. ***Mechanoreceptors*** are stimulated by mechanical deformation of the receptor itself. They provide sensations of touch, pressure, vibration, proprioception (awareness of the location and movement of body parts), hearing, and equilibrium. Mechanoreceptors are also triggered by changes in blood pressure.

2. ***Thermoreceptors*** are stimulated by changes in temperature.

3. ***Nociceptors*** (NŌ-sē-sep′-tors) are triggered by painful stimuli.

4. ***Photoreceptors (electromagnetic receptors)*** are stimulated by light energy.

5. ***Chemoreceptors*** are stimulated by changes in the chemical environment of cells such as the presence of certain molecules on the tongue or nasal mucosa (taste and smell), and the presence of oxygen, carbon dioxide, hydrogen ions, certain other electrolytes, hormones, and glucose.

Sensations may also be classified according to the simplicity or complexity of the receptor and the neural pathway involved:

1. ***General senses*** involve simple receptors and neural pathways. Taken together, a, b, and c below are referred to as ***cutaneous sensations***.
   a. ***Tactile*** (TAK-tīl) ***sensations*** (touch, pressure, vibration).
   b. ***Thermoreceptive sensations*** (hot and cold).
   c. ***Pain sensations***.
   d. ***Proprioceptive*** (prō′-prē-ō-SEP-tiv) ***sensations*** (awareness of the activities of muscles, tendons, and joints, and equilibrium).

2. ***Special senses*** involve complex receptors and neural pathways.
   a. ***Olfactory*** (ol-FAK-tō-rē) ***sensations*** (smell).
   b. ***Gustatory*** (GUS-ta-tō-rē) ***sensations*** (taste).
   c. ***Visual sensations*** (sight).
   d. ***Auditory sensations*** (hearing).
   e. ***Equilibrium*** (ē′-kwi-LIB-rē-um) ***sensations*** (orientation of the body).

C. RECEPTORS FOR GENERAL SENSES

1. Tactile Receptors

Although touch, pressure, and vibration are classified as separate sensations, all are detected by the same type of receptor, that is, one that responds to tissue deformation.

Touch sensations generally result from stimulation of tactile receptors in the skin or in tissues immediately beneath the skin. *Light touch* refers to the ability to perceive that something has touched the skin, although the size, or texture cannot be determined. *Discriminative touch* refers to the ability to recognize exactly what point of the body is touched. Receptors for touch include corpuscles of touch, tactile discs, root hair plexuses, and free (naked) nerve endings. ***Corpuscles of touch (Meissner's corpuscles)*** are receptors for discriminative touch that are found in dermal papillae. They have already been discussed in Exercise 5. ***Tactile discs***, also called ***Merkel's*** (MER-kelz) ***discs***, consist of disc-like formations of dendrites attached to deeper layers of the epidermis. They also function in discriminative touch. ***Root hair plexuses*** are dendrites arranged in networks around the roots of hairs that detect movement when hairs are disturbed. ***Free (naked) nerve endings*** are found everywhere in the skin and in many other tissues (see Figure 5.1). They are important pain receptors and also respond to objects, such as clothing, in contact with the skin.

Receptors for touch, like other cutaneous receptors, are not randomly distributed; some areas contain many receptors, others contain few. Such a clustering of receptors is called ***punctate distribution***. Touch receptors are most numerous in fingertips, palms of hands, and soles of feet. They are also abundant in the eyelids, tip of the tongue, lips, nipples, clitoris, and tip of the penis.

Examine prepared slides of corpuscles of touch (Meissner's corpuscles), tactile (Merkel's) discs, root hair plexuses, and free nerve endings. With the aid of your textbook, draw each of the receptors in the spaces that follow.

Corpuscles of touch (Meissner's corpuscles)

Tactile (Merkel's) discs

Root hair plexuses

Free (naked) nerve endings

Pressure sensations generally result from deformation of deeper tissues and are longer lasting and have less variation in intensity than touch sensations. Pressure is really sustained touch. Moreover, whereas touch is felt in a small "pinprick" area, pressure is felt over a much larger area. Receptors for pressure sensations include free nerve endings (discussed under touch sensations), lamellated corpuscles, and type II cutaneous mechanoreceptors. ***Lamellated (Pacinian) corpuscles*** are found in the subcutaneous layer and have already been discussed in Exercise 5. ***Type II cutaneous mechanoreceptors (end organs of Ruffini)*** are deeply embedded in the dermis.

Pressure receptors are found in the subcutaneous tissue under the skin, in the deep subcutaneous tissues that lie under mucous membranes, around joints and tendons, in the perimysium of muscles, in the mammary glands, in the external genitals of both sexes, and in some viscera.

Examine prepared slides of lamellated (Pacinian) corpuscles and type II cutaneous mechanoreceptors (end organs of Ruffini). With the aid of your textbook, draw the receptors in the spaces that follow.

Lamellated (Pacinian) corpuscles

Type II cutaneous mechanoreceptors (end organs of Ruffini)

Vibration sensations result from rapidly repetitive sensory signals from tactile receptors. Receptors for vibration include corpuscles of touch (Meissner's corpuscles) that detect low-frequency vibration and lamellated (Pacinian) corpuscles that detect high-frequency vibration.

2. Thermoreceptors

The cutaneous receptors for the sensation of cold are widely distributed in the dermis and subcutaneous connective tissue, and are also located in the cornea of the eye, tip of the tongue, and external genitals. The cutaneous receptors for heat are deeply embedded in the dermis and are less abundant than cold receptors. Cutaneous receptors for heat and cold have not yet been identified histologically but might be free (naked) nerve endings.

3. Pain Receptors

Receptors for ***pain***, called ***nociceptors*** (nō'-sē-SEP-tors), are free (naked) nerve endings (see Figure 5.1). Pain receptors are found in practically every tissue of the body and adapt only slightly or not at all. They may be excited by any type of stimulus. Excessive stimulation of any sense organ causes pain. For example, when stimuli for other sensations such as touch, pressure, heat, and cold reach a certain threshold, they stimulate pain receptors as well. Pain receptors, because of their sensitivity to all stimuli, have a general protective function of informing us of changes that could be potentially dangerous to health or life. Adaptation to pain does not readily occur. This low level of adaptation is important, because pain indicates disorder or disease. If we became used to it and ignored it, irreparable damage could result.

4. Proprioceptive Receptors

An awareness of the activities of muscles, tendons, and joints is provided by the ***proprioceptive (kinesthetic) sense***. It informs us of the degree to which tendons are tensed and muscles are contracted. The proprioceptive sense enables us to recognize the location and rate of movement of one part of the body in relation to other parts. It also allows us to estimate weight and to determine the muscular work necessary to perform a task. With the proprioceptive sense, we can judge the position and movements of our limbs without using our eyes when we walk, type, play a musical instrument, or dress in the dark.

Proprioceptive receptors are located in skeletal muscles and tendons, in and around joints, and in the internal ear. Proprioceptors adapt only slightly. This slight adaptation is beneficial because the brain must be apprised of the status of different parts of the body at all times so that adjustments can be made to ensure coordination.

Receptors for proprioception are as follows. The ***joint kinesthetic*** (kin'-es-THET-ik) ***receptors*** are located in the articular capsules of joints and ligaments about joints. These receptors provide feedback information on the degree and rate of angulation (change of position) of a joint. ***Muscle spindles*** consist of the endings of sensory neurons that are wrapped around specialized muscle fibers (cells). They are located in nearly all skeletal muscles and are more numerous in the muscles of the extremities. Muscle spindles provide feedback information on the degree of muscle stretch. This information is relayed to the central nervous system to assist in the coordination and efficiency of muscle contraction. ***Tendon organs (Golgi tendon organs)*** are located at the junction of skeletal muscle and tendon. They function by sensing the tension applied to a tendon. The degree of tension is related to the degree of muscle contraction and is translated by the central nervous system.

Proprioceptors in the internal ear are the maculae and cristae that function in equilibrium. These are discussed at the end of the exercise.

Examine prepared slides of joint kinesthetic receptors, muscle spindles, and tendon organs (Golgi tendon organs). With the aid of your textbook, draw the receptors in the spaces that follow.

Joint kinesthetic receptors

Muscle spindles

Tendon organs (Golgi tendon organs)

D. TESTS FOR GENERAL SENSES

Areas of the body that have few cutaneous receptors are relatively insensitive, whereas those regions that contain large numbers of cutaneous receptors are quite sensitive. This difference can be demonstrated by the ***two-point discrimination test*** for touch. In the following tests, students can work in pairs, with one acting as subject and the other as experimenter. The subject will keep his or her eyes closed during the experiments.

1. Two-Point Discrimination Test

In this test, the two points of a measuring compass are applied to the skin and the distance in millimeters (mm) between the two points is varied. The subject indicates when he or she feels two points and when he or she feels only one.

PROCEDURE

CAUTION! *Wash the compass in a fresh bleach solution before using it on another subject to prevent transfer of saliva.*

1. *Very gently*, place the compass on the tip of the tongue, an area where receptors are very densely packed.

2. Narrow the distance between the two points to 1.4 mm. At this distance, the points are able to stimulate two different receptors, and the subject feels that he or she is being touched by two objects.

3. Decrease the distance to less than 1.4 mm. The subject feels only one point, even though both points are touching the tongue, because the points are so close together that they reach only one receptor.

4. Now *gently* place the compass on the back of the neck, where receptors are relatively few and far between. Here the subject feels two distinctly different points only if the distance between them is 36.2 mm or more.

5. The two-point discrimination test shows that the more sensitive the area, the closer the compass points can be placed and still be felt separately.

6. The following order, from greatest to least sensitivity, has been established from the test: tip of tongue, tip of finger, side of nose, back of hand, and back of neck.

7. Test the tip of finger, side of nose, and back of hand and record your results in Section D.1 of the LABORATORY REPORT RESULTS at the end of the exercise.

2. Identifying Touch Receptors

PROCEDURE

1. Using a water-soluble colored felt marking pen, draw a 1-in. square on the back of the forearm and divide the square into 16 smaller squares.

2. With the subject's eyes closed, press a Von Frey hair or bristle against the skin, just enough to cause the hair to bend, once in each of the 16 squares. The pressure should be applied in the same manner each time.

3. The subject should indicate when he or she experiences the sensation of touch, and the experimenter should make dots in Square 1 at the places corresponding to the points at which the subject feels the sensations.

4. The subject and the experimenter should switch roles and repeat the test.

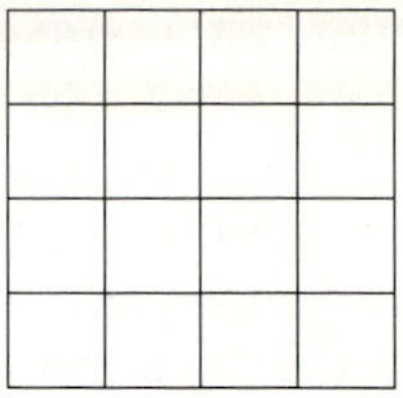

Square 1. Touch, forearm.

The pair of students working as a team should examine their 1-in. squares after the test is done. They should compare the number of positive and negative responses in each of the 16 small squares, and see how uniformly the touch receptors are distributed throughout the entire 1-in. square. Other general areas used for locating touch receptors are the arm and the back of the hand.

3. Identifying Pressure Receptors

PROCEDURE

1. The experimenter touches the skin of the subject (whose eyes are closed) with the point of a piece of colored chalk.
2. With eyes still closed, the subject then tries to touch the same spot with a piece of differently colored chalk. The distance between the two points is then measured.
3. Proceed using various parts of the body, such as palm of hand, arm, forearm, and back of neck.
4. Record your results in Section D.3 of the LABORATORY REPORT RESULTS at the end of the exercise.

4. Identifying Thermoreceptors

PROCEDURE

1. Draw a 1-in. square on the back of the wrist.
2. Place a forceps or other metal probe in ice-cold water for a minute, dry it quickly, and, with the *dull* point, explore the area in the square for the presence of cold spots.
3. Keep the probe cold and, using ink, mark the position of each spot that you find.
4. Mark each corresponding place in Square 2 with the letter "c."
5. Immerse the forceps in hot water so that it will give a sensation of warmth when removed and applied to the skin, but *avoid having it so hot that it causes pain*.
6. Proceeding as before, locate the position of the warm spots in the same area of the skin.
7. Mark these spots with ink of a different color, and then mark each corresponding place in Square 2 with the letter "h."
8. Repeat the entire procedure, using both cold forceps and warm forceps on the back of the hand and the palm of the hand, respectively, and mark Squares 3 and 4 as you did Square 2.

Square 2. Cold and heat, back of wrist.

Square 3. Cold and heat, back of hand.

Square 4. Cold and heat, palm of hand.

9. In order to demonstrate adaptation of thermoreceptors, fill three 1 liter beakers with 700 ml of: (1) ice water, (2) water at room temperature, and (3) water at 45°C. Immerse your left hand in the ice water and your right hand in the water at 45°C for 1 minute.

Now move the left hand to the beaker with water at room temperature and record the sensation. ______________________________

Move the right hand to the beaker with the water at room temperature and record the sensation. ______________________________

How do you explain the experienced difference in the temperature of the water in the beaker with the water at room temperature? ________

____

5. Identifying Pain Receptors

PROCEDURE

1. Using the same 1-in. square of the forearm that was previously used for the touch test in Section D.2, perform the following experiment.
2. Apply a piece of absorbent cotton soaked with water to the area of the forearm for 5 minutes to soften the skin.

3. Add water to the cotton as needed.

4. Place the blunt end of a probe to the surface of the skin and press enough to produce a sensation of pain. Explore the marked area systematically.

5. Using dots, mark the places in Square 5 that correspond to the points that give pain sensation when stimulated.

6. Distinguish between sensations of pain and touch. Are the areas for touch and pain identical? ________________

7. At the end of the test, compare your squares as you did in Section D.2.

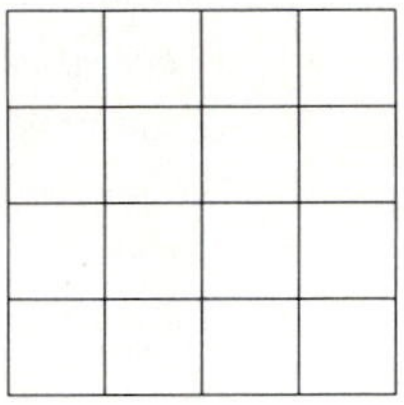

Square 5. Pain, forearm.

Perform the following test to demonstrate the phenomenon of ***referred pain***. Place your elbow in a large shallow pan of ice water, and note the progression of sensation that you experience. At first, you will feel some discomfort in the region of the elbow. Later, pain sensations will be felt elsewhere.

Where do you feel the referred pain? ________

____

Label the areas of referred pain indicated in Figure 14.1.

6. Identifying Proprioceptors

PROCEDURE

1. Face a blackboard close enough so that you can easily reach to mark it. Mark a small X on the board in front of you and keep the chalk on the X for a moment. Now close your eyes, raise your right hand above your head and then, with your eyes still closed, mark a dot as near as possible to the X. Repeat the procedure by placing your chalk on the X, closing your eyes, raising your arm above your head, and then marking another dot as close as possible to the X. Repeat the procedure a third time. Record your results by estimating or measuring how far you missed the X for each trial in Section D.6 of the LABORATORY REPORT RESULTS at the end of the exercise.

2. Write the word "physiology" on the left line that follows. Now, with your *eyes closed*, write the same word immediately to the right. How do the samples of writing compare?

________________ ________________

Explain your results. ________________

____

____

____

3. The following experiments demonstrate that kinesthetic sensations facilitate repetition of certain acts involving muscular coordination.

Students work in pairs for these experiments.

a. The experimenter asks the subject to carry out certain movements with his or her eyes closed, for example, point to the middle finger of the subject's left hand with the index finger of the subject's right hand.

b. With his or her eyes closed, the subject extends the right arm as far as possible behind the body, and then brings the index finger quickly to the tip of his or her nose.

How accurate is the subject in doing this?

____

c. Ask the subject, with eyes shut, to touch the named fingers of one hand with the index finger of the other hand.

How well does the subject carry out the directions? ________________

____

____

E. SENSORY PATHWAYS

Sensory information transmitted from the spinal cord to the brain is conducted along two general pathways: the posterior column pathway and the spinothalamic pathway.

1. Posterior Column Pathway

The ***posterior column pathway*** conducts nerve impulses related to proprioception, discrim-

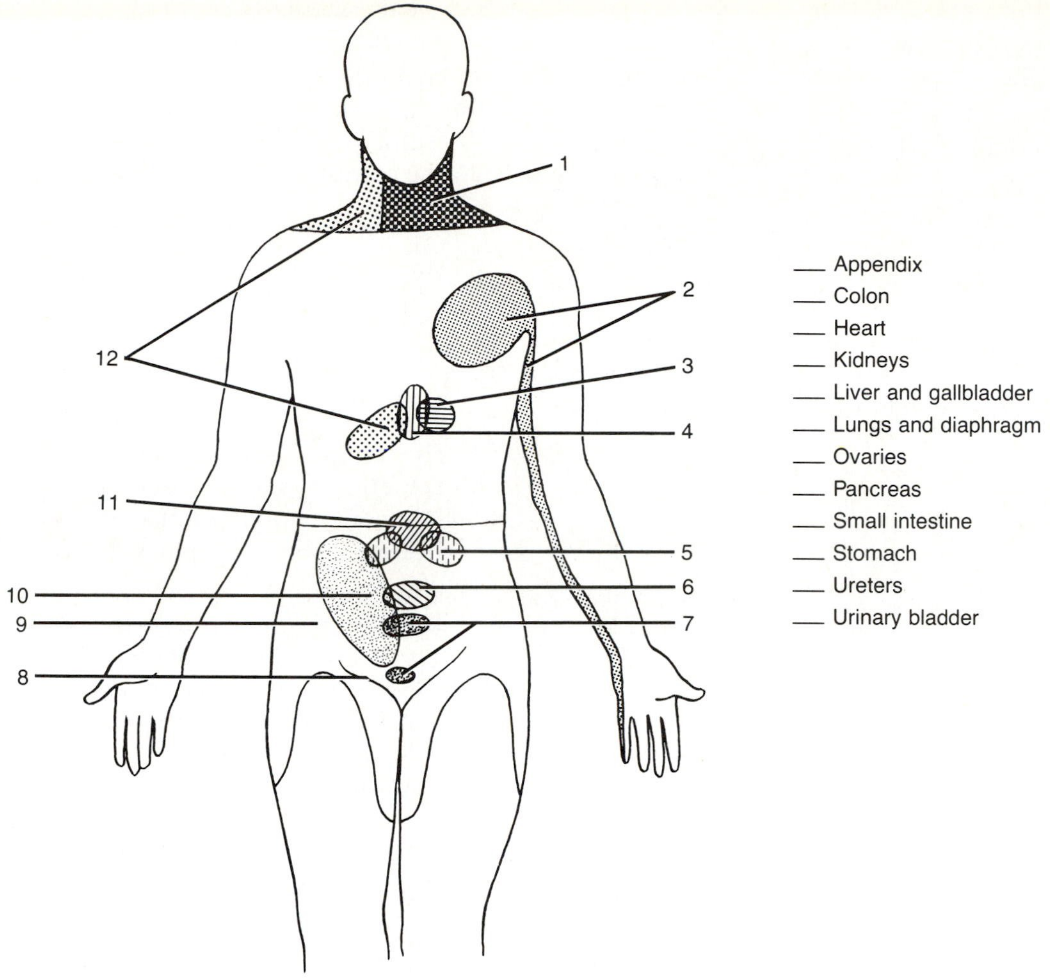

FIGURE 14.1 Referred pain.

inative touch, two-point discrimination, and vibrations.

Label the components of the posterior column pathway in Figure 14.2.

2. Spinothalamic Pathway

The spinothalamic pathway conveys nerve impulses for pain and temperature as well as light touch and pressure. The sensory pathway for pain and temperature is called the ***lateral spinothalamic pathway***.

Label the components of the lateral spinothalamic pathway in Figure 14.3.

The neural pathway that conducts nerve impulses for light touch and pressure is the ***anterior (ventral) spinothalamic pathway***.

Label the components of the anterior spinothalamic pathway in Figure 14.4.

F. OLFACTORY SENSATIONS

1. Olfactory Receptors

The receptors for the ***olfactory*** (ol-FAK-tō-rē) ***sense*** are found in the nasal epithelium in the superior portion of the nasal cavity on either side of the nasal septum. The nasal epithelium consists of three principal kinds of cells: supporting, olfactory, and basal. The ***supporting (sustentacular) cells*** are columnar epithelial cells of the mucous membrane that lines the nose. The ***olfactory cells*** are bipolar neurons. Their cell bodies lie between the supporting cells. The distal (free) end of each olfactory cell contains a dendrite that terminates in a swelling, the ***olfactory vesicle***, from which six to eight cilia, called ***olfactory hairs***, terminate. ***Basal cells*** lie between the bases of the supporting cells and are believed to produce new

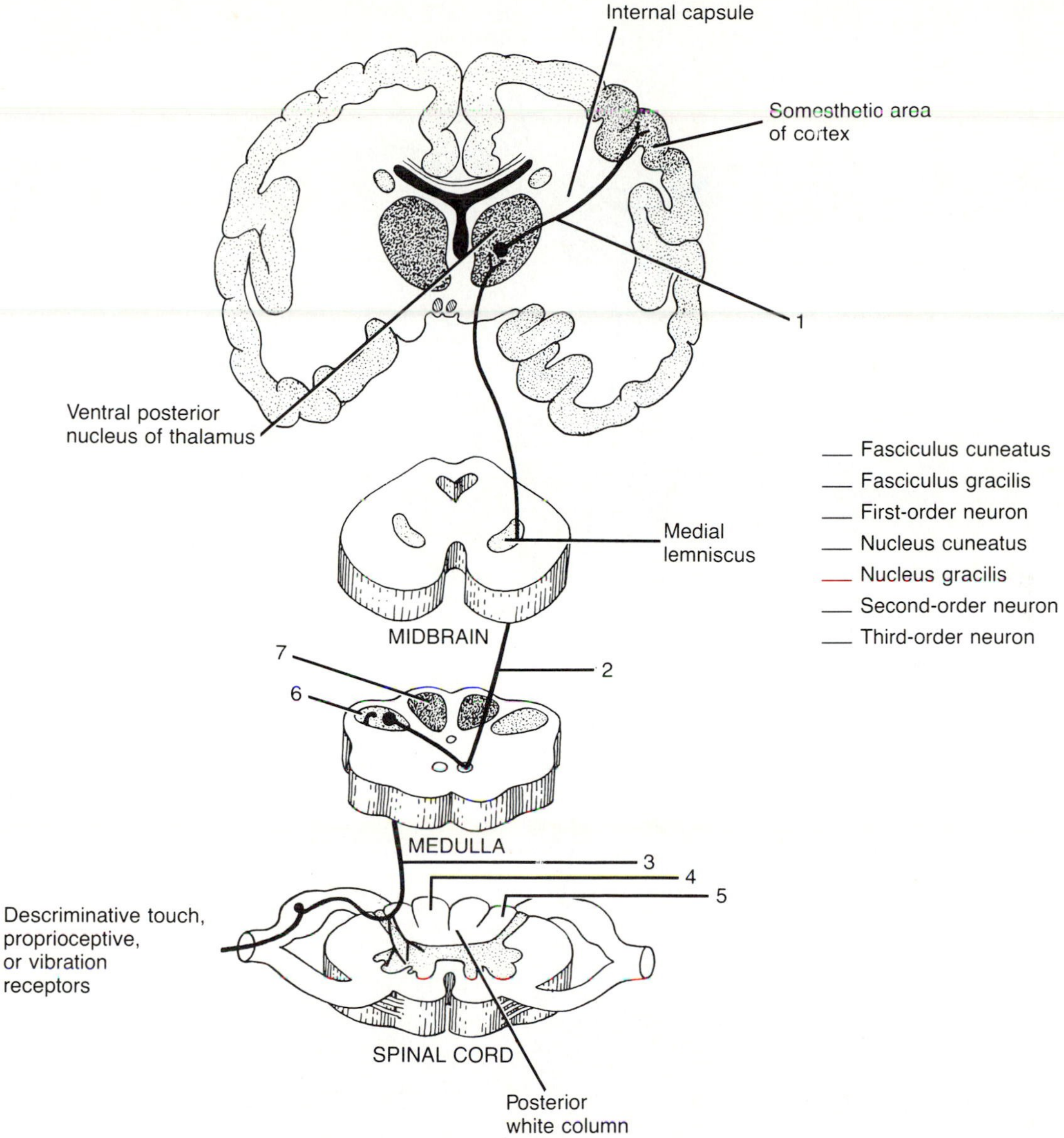

FIGURE 14.2 The posterior column pathway.

supporting cells. Within the connective tissue beneath the olfactory epithelium are ***olfactory (Bowman's) glands*** that secrete mucus.

The unmyelinated axons of the olfactory cells unite to form the ***olfactory (I) nerves***, which pass through foramina in the cribriform plate of the ethmoid bone. The olfactory nerves terminate in paired masses of gray matter, the ***olfactory bulbs***. They lie beneath the frontal lobes of the cerebrum on either side of the crista galli of the ethmoid bone. The first synapse of the olfactory neural pathway occurs in the olfactory bulbs between the axons of the olfactory (I) nerves and the dendrites of neurons inside the olfactory bulbs. Axons of these neurons run posteriorly to form the ***olfactory tract***. From here, nerve impulses are conveyed to the olfactory area of the cerebral cortex. In the cortex, the nerve impulses are interpreted as odor and give rise to the sensation of smell.

Adaptation happens quickly, especially adaptation to odors. For this reason, we become accustomed to some odors and are also able to endure unpleasant ones. Rapid adaptation also accounts for the failure of a person to detect gas that accumulates slowly in a room.

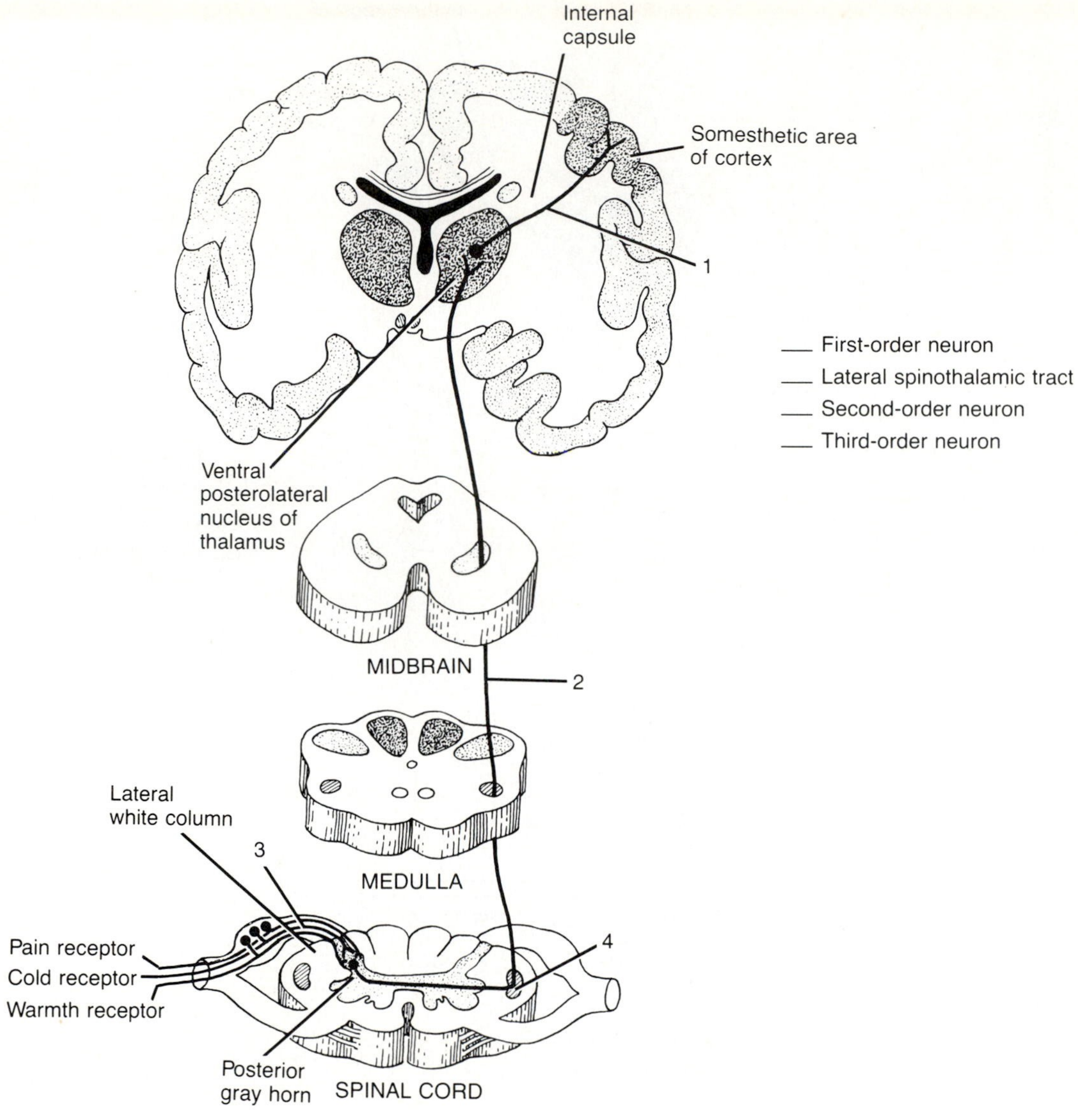

FIGURE 14.3 The lateral spinothalamic pathway.

Label the structures associated with olfaction in Figure 14.5.

Now examine a slide of the olfactory epithelium under high power. Identify the olfactory cells and supporting cells and label Figure 14.6.

2. Olfactory Adaptation

PROCEDURE

1. The subject should close his or her eyes after plugging one nostril with cotton.

2. Hold a bottle of oil of cloves, or other substance having a distinct odor, under the open nostril.

3. The subject breathes in through the open nostril, and exhales through the mouth. Note the time required for the odor to disappear, and repeat with the other nostril.

4. As soon as olfactory adaptation has occurred, test an entirely different substance.

5. Compare results for the various materials tested.

Olfactory stimuli, such as pepper, onions, ammonia, ether, and chloroform, are irritating and may cause tearing because they stimulate the receptors of the trigeminal (V) nerve as well as the olfactory neurons.

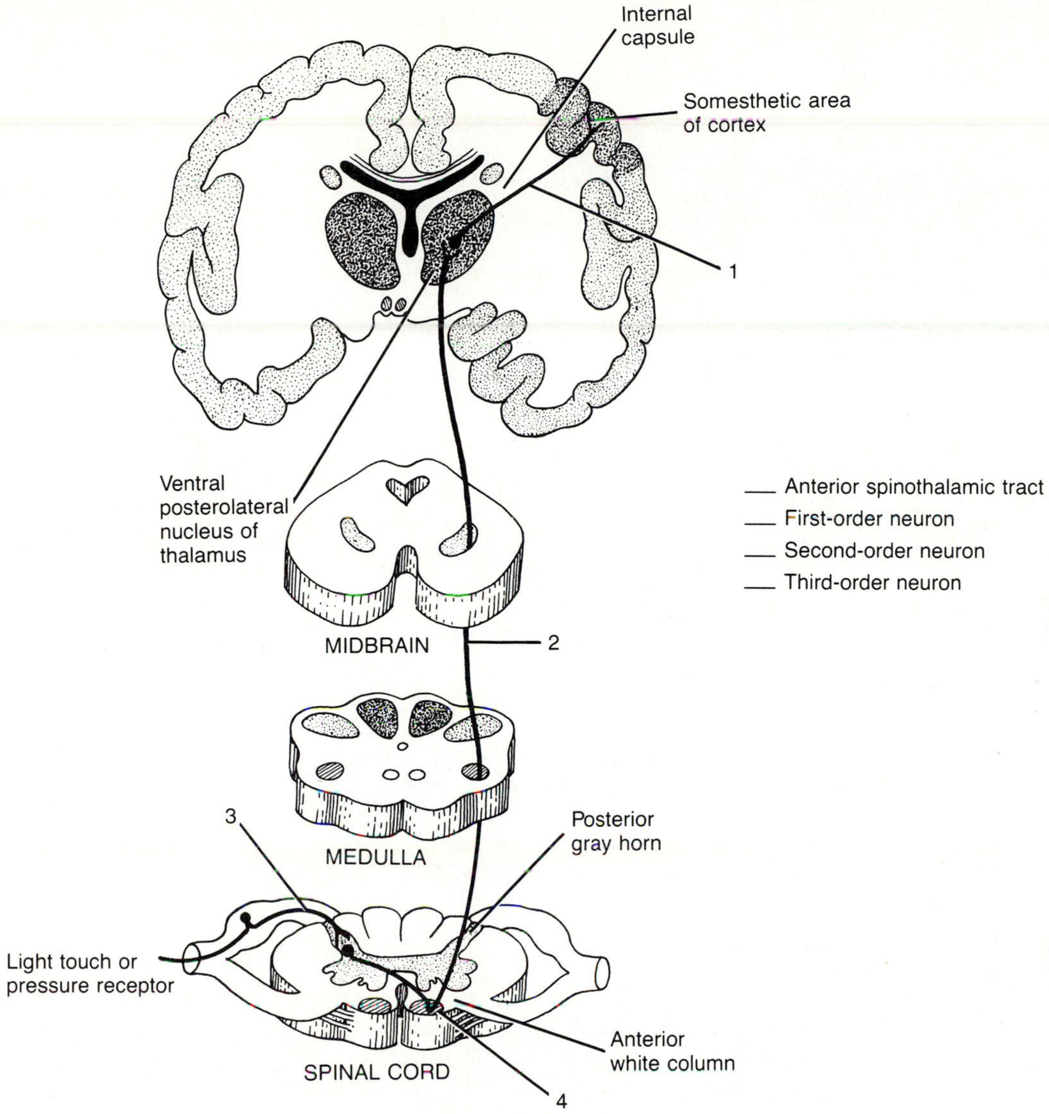

FIGURE 14.4 The anterior spinothalamic pathway.

G. GUSTATORY SENSATIONS

1. Gustatory Receptors

The receptors for ***gustatory*** (GUS-ta-tō-rē) ***sensations***, or sensations of taste, are located in the taste buds. Although taste buds are most numerous on the tongue, they are also found on the soft palate and in the pharynx. ***Taste buds*** are oval bodies consisting of three kinds of cells: supporting, gustatory, and basal. The ***supporting (sustentacular) cells*** are a specialized epithelium that forms a capsule. Inside each capsule are 4 to 20 ***gustatory cells***. Each gustatory cell contains a hairlike process ***(gustatory hair)*** that projects to the surface through an opening in the taste bud called the ***taste pore***. Gustatory cells make contact with taste stimuli through the taste pore. ***Basal cells*** are found at the periphery of the taste bud and produce new supporting and gustatory cells.

Examine a slide of taste buds and label the structures associated with gustation in Figure 14.7.

Taste buds are located in some connective tissue elevations on the tongue called ***papillae*** (pa-PILL-ē). They give the upper surface of the tongue its rough texture and appearance. ***Circumvallate*** (ser-kum-VAL-āt) ***papillae*** are circular and form an inverted V-shaped row at the posterior portion of the tongue. ***Fungiform***

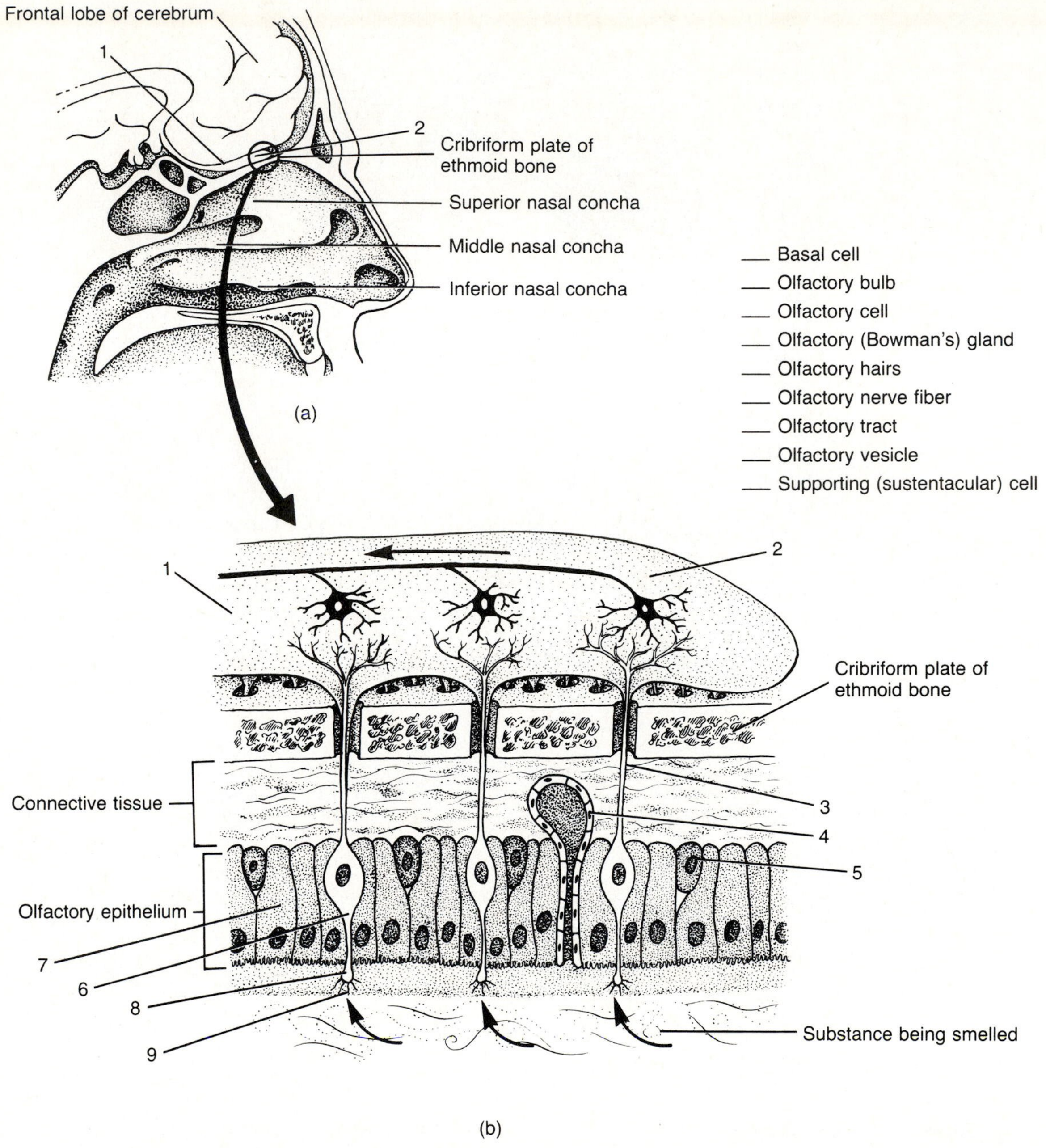

FIGURE 14.5 Olfactory receptors. (a) In relation to nasal cavity. (b) Enlarged aspect.

(FUN-ji-form) ***papillae*** are knoblike elevations found primarily on the tip and sides of the tongue. All circumvallate and most fungiform papillae contain taste buds. ***Filiform*** (FIL-i-form) ***papillae*** are threadlike structures that cover the anterior two-thirds of the tongue.

Have your partner protrude his or her tongue and examine its surface with a hand lens to identify the shape and position of the papillae.

2. Identifying Taste Zones

For gustatory cells to be stimulated, substances tasted must be in solution in the saliva in order to enter the taste pores in the taste buds.

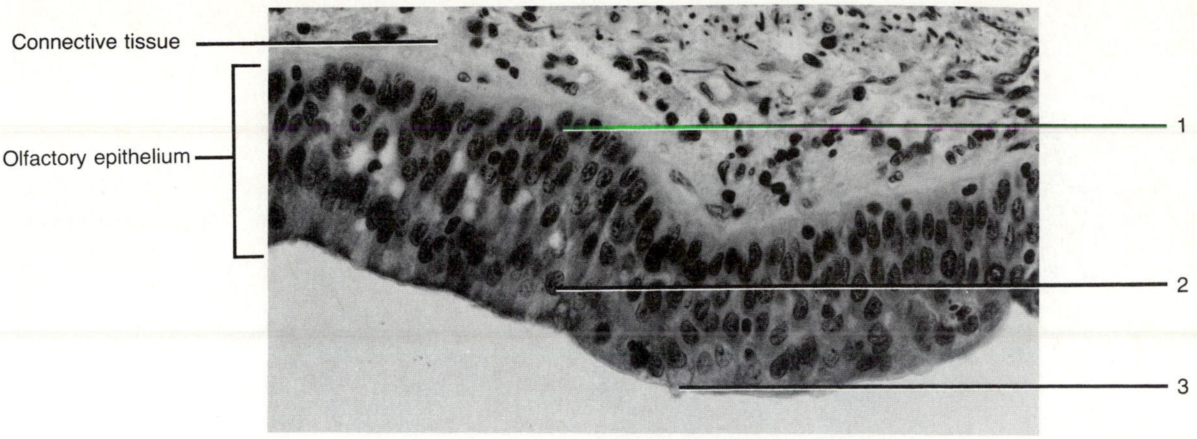

___ Basal cell
___ Olfactory cell
___ Supporting cell

FIGURE 14.6 Photomicrograph of olfactory epithelium.

Despite the many substances tasted, there are basically only four taste sensations: sour, salty, bitter, and sweet. Each taste is due to a different response to different chemicals. Some regions of the tongue react more strongly than others to particular taste sensations.

To identify the taste zones for the four taste sensations, perform the following steps and record the results in Section G.2 of the LABORATORY REPORT RESULTS at the end of the exercise by inserting a "+" (taste detected) or a "−" (taste not detected) where appropriate.

PROCEDURE

1. The subject thoroughly dries his or her tongue (use a clean paper towel). The experimenter places some granulated sugar on the tip of the tongue and notes the time. The subject indicates when he or she tastes sugar by raising his or her hand. The experimenter notes the time again and records how long it takes for the subject to taste the sugar.

2. Repeat the experiment, but this time use a drop of sugar solution. Again record how long it takes for the subject to taste the sugar. How do you explain the difference in time periods?

3. The subject rinses his or her mouth again. The experiment is then repeated using the quinine solution (bitter taste), and then the salt solution.

4. After rising yet again, the experiment is repeated using the acetic acid solution or vinegar (sour) placed on the tip and *sides* of the tongue.

3. Taste and Inheritance

Taste for certain substances is inherited, and geneticists for many years have been using the chemical ***phenylthiocarbamide (PTC)*** to test taste. To some individuals this substance tastes bitter, to others it is sweet, and some cannot taste it at all.

PROCEDURE

1. Place a few crystals of PTC on the subject's tongue. Does he or she taste it? If so, describe the taste.

2. Special paper that is flavored with this chemical may be chewed and mixed with saliva and tested in the same manner.

3. Record on the blackboard your response to the PTC test. Usually about 70% of the people tested can taste this compound; 30% cannot. Compare this percentage with the class results.

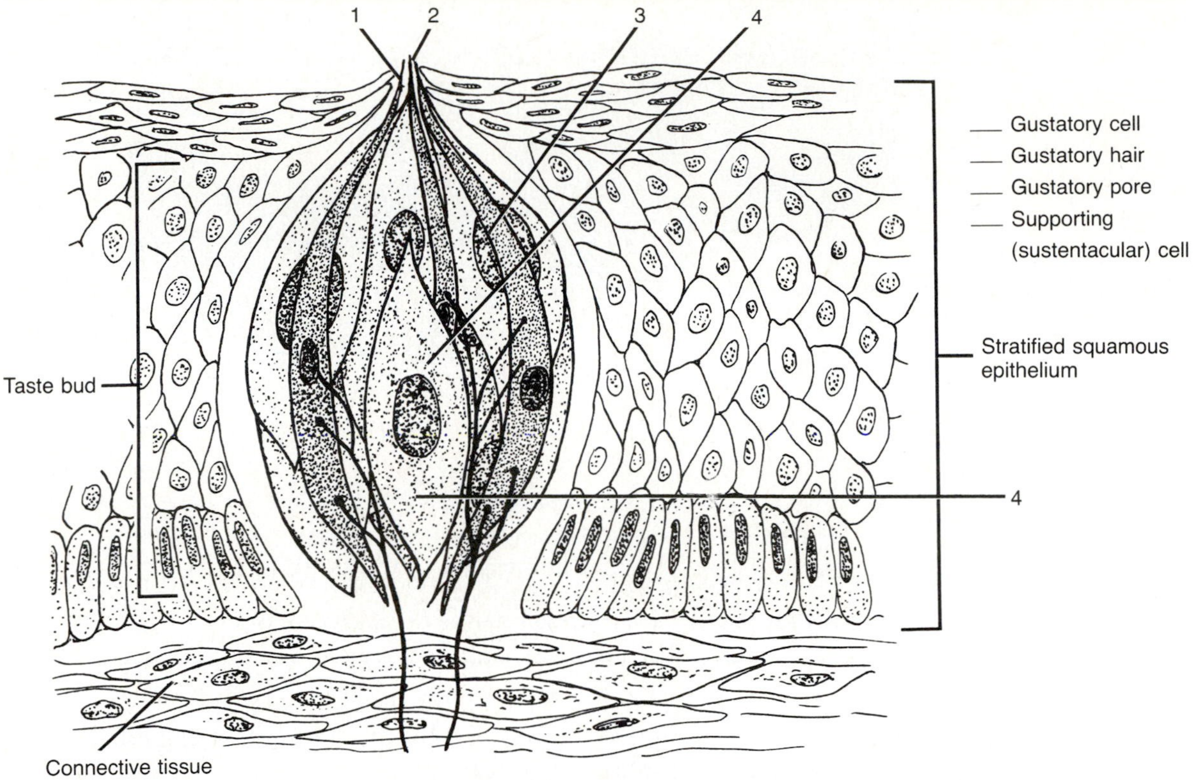

(a)

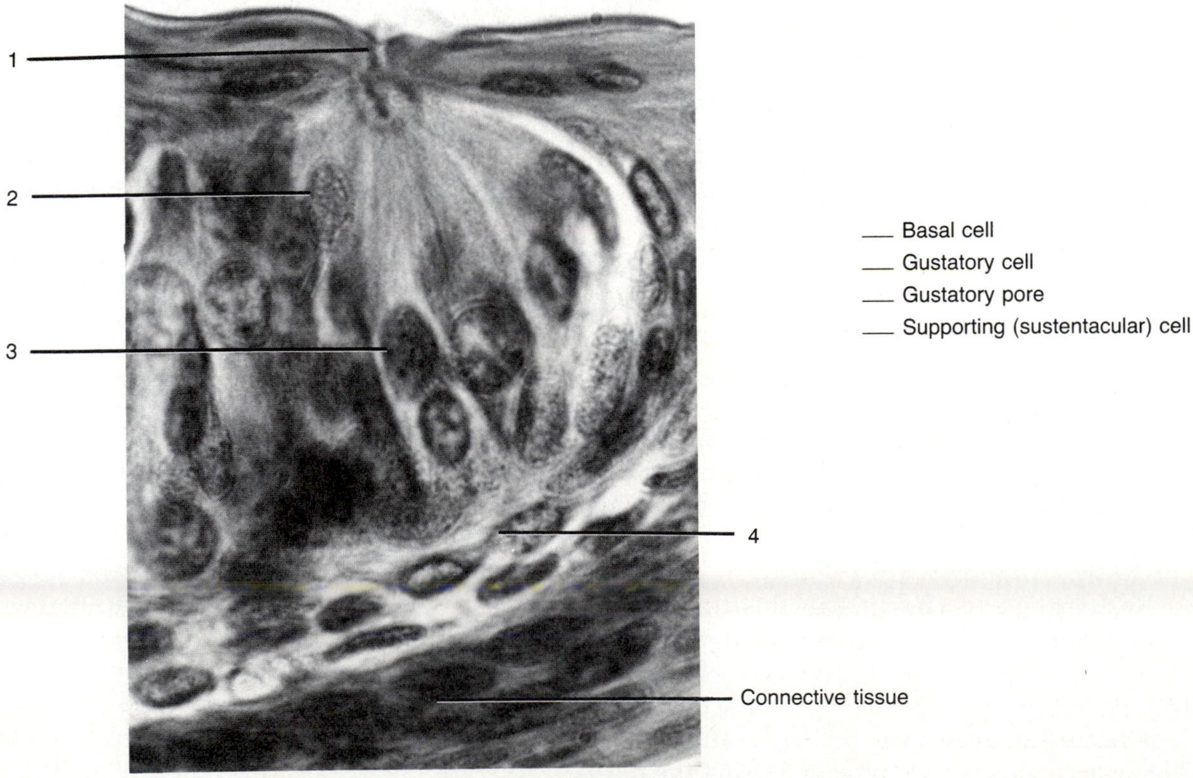

(b)

FIGURE 14.7 Structure of a taste bud. (a) Diagram. (b) Photograph.

4. Taste and Smell

This test combines the effect of smell on the sense of taste.

PROCEDURE

1. Obtain small cubes of carrot, onion, potato, and apple.

2. The subject dries the tongue with a clean paper towel, closes the eyes, and pinches the nostrils shut. The experimenter places the cubes, one by one, on the subject's tongue.

3. The subject attempts to identify each cube in the following sequences: (1) immediately, (2) after chewing (nostrils closed), and (3) after opening the nostrils.

4. Record your results in Section G.4 of the LABORATORY REPORT RESULTS at the end of the exercise.

H. VISUAL SENSATIONS

Structures related to ***vision*** are the eyeball (which is the receptor organ for visual sensations), optic (II) nerve, brain, and accessory structures. The extrinsic muscles of the eyeball may be reviewed in Figure 10.4.

1. Accessory Structures

Among the ***accessory structures*** are the eyebrows, eyelids, eyelashes, and lacrimal, or tearing, apparatus. ***Eyebrows*** protect the eyeball from falling objects, prevent perspiration from getting into the eye, and shade the eye from the direct rays of the sun. ***Eyelids (palpebrae)*** (PAL-pe-brē) consist primarily of muscle covered externally by skin. The underside of the muscle is lined by a mucous membrane called the ***conjunctiva*** (kon-junk-TĪ-va), which also covers the surface of the eyeball. Also within eyelids are ***tarsal (Meibomian) glands***, modified sebaceous glands whose oily secretion keeps the eyelids from adhering to each other. Infection of these glands produces a ***chalazion*** (cyst) in the eyelid. Eyelids shade the eye during sleep, protect the eye from light rays and foreign objects, and spread lubricating secretions over the surface of the eyeball. Projecting from the border of each eyelid is a row of short, thick hairs, the ***eyelashes***. Sebaceous glands at the base of the hair follicles of the eyelashes, called ***sebaceous ciliary glands (glands of Zeis)***, pour a lubricating fluid into the follicles. An infection of these glands is called a ***sty***.

The ***lacrimal*** (LAK-ri-mal) ***apparatus*** consists of a group of structures that manufactures and drains tears. Each ***lacrimal gland*** is located at the superior lateral portion of both orbits. Leading from the lacrimal glands are 6 to 12 ***excretory lacrimal ducts*** that empty tears onto the surface of the conjunctiva of the upper lid. From here, the tears pass medially and enter two small openings called ***puncta lacrimalia*** that appear as two small pores, one in each papilla of the eyelid, at the medial commissure of the eye. The tears then pass into two ducts, the ***lacrimal canals***, and are next conveyed into the lacrimal sac. The ***lacrimal sac*** is the superior expanded portion of the ***nasolacrimal duct***, a canal that carries the tears into the inferior meatus of the nose. Tears clean, lubricate, and moisten the external surface of the eyeball.

Label the parts of the lacrimal apparatus in Figure 14.8.

2. Structure of the Eyeball

The eyeball can be divided into three principal layers: (1) fibrous tunic, (2) vascular tunic, and (3) retina (nervous tunis). See Figure 14.9.

The ***fibrous tunic*** is the outer coat. It is divided into the posterior sclera and the anterior cornea. The ***sclera*** (SKLE-ra), called the "white of the eye," is a white coat of fibrous tissue that covers all the eyeball except the anterior colored portion (iris). The sclera gives shape to the eyeball and protects its inner parts. The anterior portion of the fibrous tunic is known as the ***cornea*** (KOR-nē-a). This nonvascular, transparent fibrous coat covers the iris. The cornea is also composed of fibrous tissue. The outer surface of the cornea contains epithelium that is continuous with the epithelium of the bulbar conjunctiva. At the junction of the sclera and cornea is the ***scleral venous sinus (canal of Schlemm)***.

The ***vascular tunic*** is the middle layer of the eyeball and consists of three portions: posterior choroid, anterior ciliary body, and iris. The ***choroid*** (KŌ-royd) is a thin, dark brown membrane that lines most of the internal surface of the sclera and contains blood vessels and pigment. The choroid absorbs light rays so they are not reflected back out of the eyeball and maintains the nutrition of the retina. The ante-

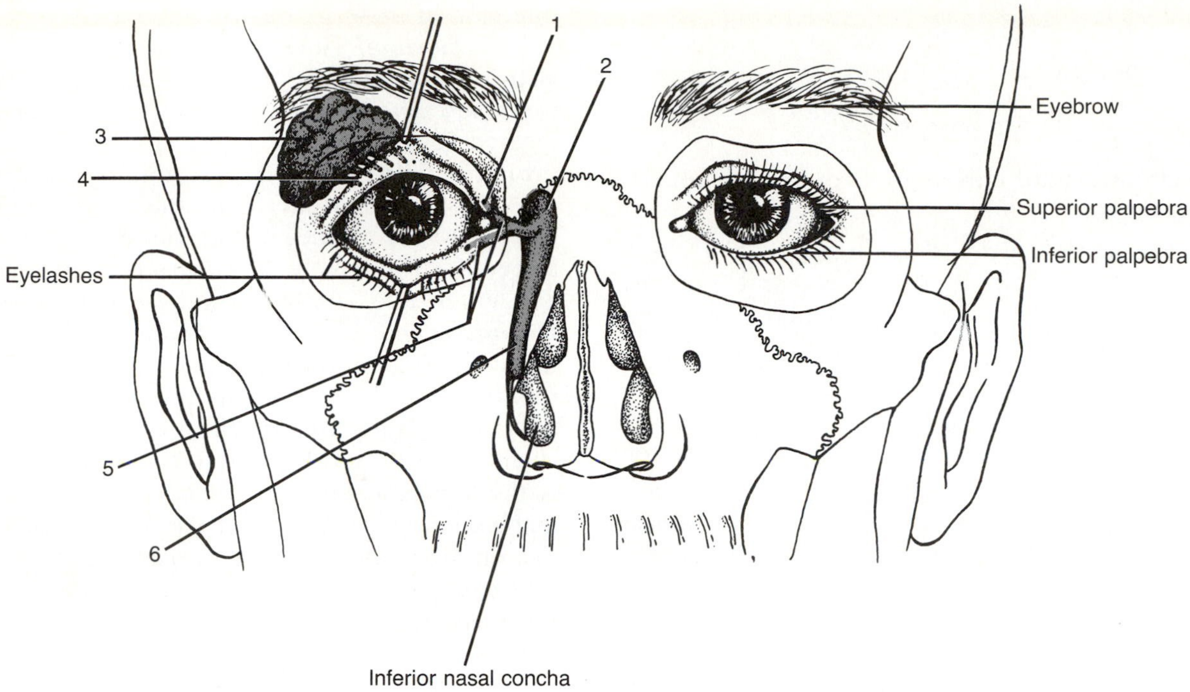

FIGURE 14.8 Lacrimal apparatus.

rior portion of the choroid is the ***ciliary*** (SIL-ē-ar′-ē) ***body***, the thickest portion of the vascular tunic. It extends from the ***ora serrata*** (Ō-ra ser-RĀ-ta) of the retina (inner tunic) to a point just behind the sclerocorneal junction. The ora serrata is the jagged margin of the retina. The ciliary body contains the ***ciliary muscle***, a smooth muscle that alters the shape of the lens for near or far vision. The ***iris***, the third portion of the vascular tunic, is the colored portion of the eyeball and consists of circular and radial smooth muscle fibers arranged to form a doughnut-shaped structure. The black hole in the center of the iris is the ***pupil***, through which light enters the eyeball. One function of the iris is to regulate the amount of light entering the eyeball.

The third and inner coat of the eye, the ***retina (nervous tunic)*** is found only in the posterior portion of the eye. Its primary function is image formation. It consists of a nervous tissue layer and pigmented layer. The outer ***pigmented layer*** (nonvisual portion) consists of epithelial cells in contact with the choroid. The inner ***nervous layer*** (visual portion) is composed of three zones of neurons. Named in the order in which they conduct nerve impulses, these are the ***photoreceptor neurons, bipolar neurons,*** and ***ganglion neurons***. Structurally, the photoreceptor layer is just internal to the pigmented layer, which lies adjacent to the choroid. The zone of ganglion neurons is the innermost of the neuronal layers.

The dendrites of the photoreceptor neurons are called rods and cones because of their respective shapes. ***Rods*** are specialized for vision in dim light. In addition, they allow discrimination between different shades of dark and light and permit discernment of shapes and movement. ***Cones*** are specialized for color vision and for sharpness of vision, that is, ***visual acuity***. Cones are stimulated only by bright light and are most densely concentrated in the ***central fovea***, a small depression in the center of the macula lutea. The ***macula lutea*** (MAK-yoo-la LOO-tē-a), or yellow spot, is situated in the exact center of the posterior portion of the retina and corresponds to the visual axis of the eye. The fovea is the area of sharpest vision because of the high concentration of cones. Rods are absent from the fovea and macula but increase in density toward the periphery of the retina.

When light stimulates photoreceptor neurons, impulses are conducted across synapses to the bipolar neurons in the intermediate zone of the nervous layer of the retina. From there, the impulses pass to the ganglion neurons. Axons of the ganglion neurons extend posteriorly to a small area of the retina called the ***optic disc (blind spot)***. This region contains openings through which fibers of the ganglion neurons exit as the ***optic (II) nerve***. Because this area contains neither rods nor cones, and only nerve fibers, no image is formed on it. For this reason it is called the blind spot.

The eyeball itself also contains the lens, just behind the pupil and iris. The ***lens*** is constructed of numerous layers of protein fibers arranged like the layers of an onion. Normally, the lens is perfectly transparent and is enclosed by a clear capsule and held in position by the ***suspensory ligament***. A loss of transparency of the lens is called a ***cataract***.

The interior of the eyeball contains a large cavity divided into two smaller cavities. These are called the anterior cavity and the posterior cavity and are separated from each other by the lens. The ***anterior cavity***, in turn, has two subdivisions known as the anterior chamber and the posterior chamber. The ***anterior chamber*** lies posterior to the cornea and anterior to the iris. The ***posterior chamber*** lies posterior to the iris and anterior to the suspensory ligaments and lens. The anterior cavity is filled with a clear, watery fluid known as the ***aqueous humor***. From the posterior chamber, the fluid permeates the posterior cavity and then passes forward between the iris and the lens, through the pupil into the anterior chamber. From the anterior chamber, the aqueous humor is drained off into the scleral venous sinus and passes into the blood. Pressure in the eye, called ***intraocular pressure (IOP)***, is produced mainly by the aqueous humor. Intraocular pressure keeps the retina smoothly applied to the choroid so that the retina may form clear images. Abnormal elevation of intraocular pressure, called ***glaucoma*** (glaw-KŌ-ma), results in degeneration of the retina and blindness.

The second, larger cavity of the eyeball is the ***posterior cavity***. It is located between the lens and retina and contains a soft, jellylike substance called the ***vitreous humor***. This substance contributes to intraocular pressure, helps to prevent the eyeball from collapsing, and holds the retina flush against the internal portions of the eyeball.

Label the parts of the eyeball in Figure 14.9.

3. Surface Anatomy

Refer to Figure 14.10 for a summary of several surface anatomy features of the eyeball and accessory structures of the eye.

4. Dissection of Vertebrate Eye (Beef or Sheep)

CAUTION! *Please reread Section D. "Precautions Related to Dissection" at the beginning of the laboratory manual on page xii before you begin your dissection.*

a. EXTERNAL EXAMINATION

PROCEDURE

1. Note any ***fat*** on the surface of the eyeball that protects the eyeball from shock in the orbit. Remove the fat.
2. Locate the ***sclera***, the tough external white coat, and the ***conjunctiva***, a delicate membrane that covers the anterior surface of the eyeball and is attached near the edge of the cornea. The ***cornea*** is the anterior, transparent portion of the sclera. It is probably opaque in your specimen due to the preservative.
3. Locate the ***optic (II) nerve***, a solid, white cord of nerve fibers on the posterior surface of the eyeball.
4. If possible, identify six ***extrinsic eye muscles*** that appear as flat bands near the posterior part of the eyeball.

b. INTERNAL EXAMINATION

PROCEDURE

1. With a sharp scalpel, make an incision about .6 cm (1/4 in.) lateral to the cornea (Figure 14.11).
2. Insert scissors into the incision and carefully and slowly cut all the way around the corneal region. The eyeball contains fluid, so take care

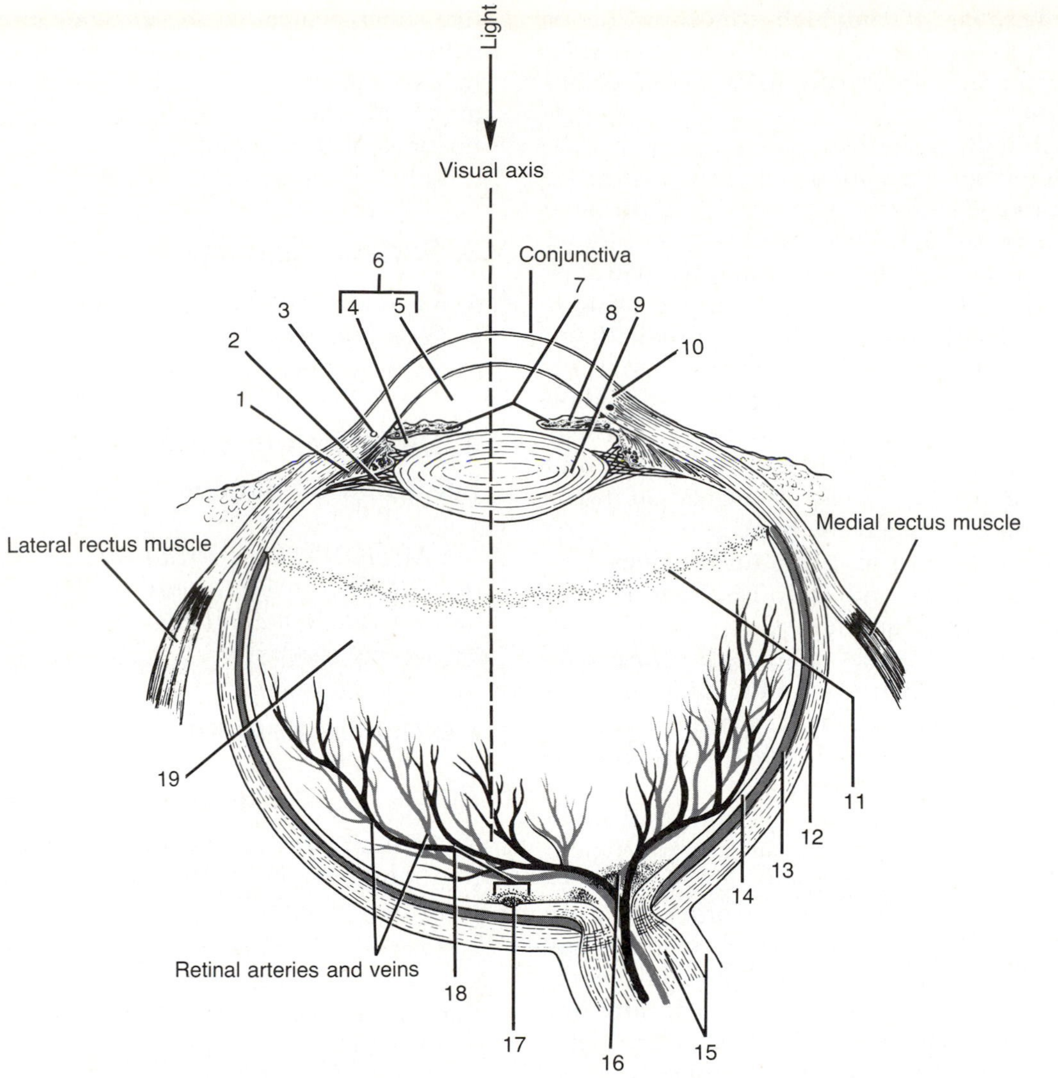

___ Anterior cavity
___ Anterior chamber
___ Central fovea
___ Choroid
___ Ciliary body and muscle
___ Cornea
___ Iris
___ Lens
___ Macula lutea
___ Optic disc (blind spot)
___ Optic (II) nerve
___ Ora serrata
___ Posterior chamber
___ Posterior cavity
___ Pupil
___ Retina (nervous tunic)
___ Sclera
___ Scleral venous sinus (canal of Schlemm)
___ Suspensory ligament

FIGURE 14.9 Parts of eyeball.

that it does not squirt out when you make your first incision. Examine the inside of the anterior part of the eyeball.

3. The ***lens*** can be seen held in position by ***suspensory ligaments***, which are delicate fibers. Around the outer margin of the lens, with a pleated appearance, is the black ***ciliary body***, which also functions to hold the lens in place. Free the lens and notice how hard it is.

4. The ***iris*** can be seen just anterior to the lens and is also heavily pigmented or black.

5. The ***pupil*** is the circular opening in the center of the iris.

6. Examine the inside of the posterior part of the eyeball, identifying the thick ***vitreous humor*** that fills the space between the lens and retina.

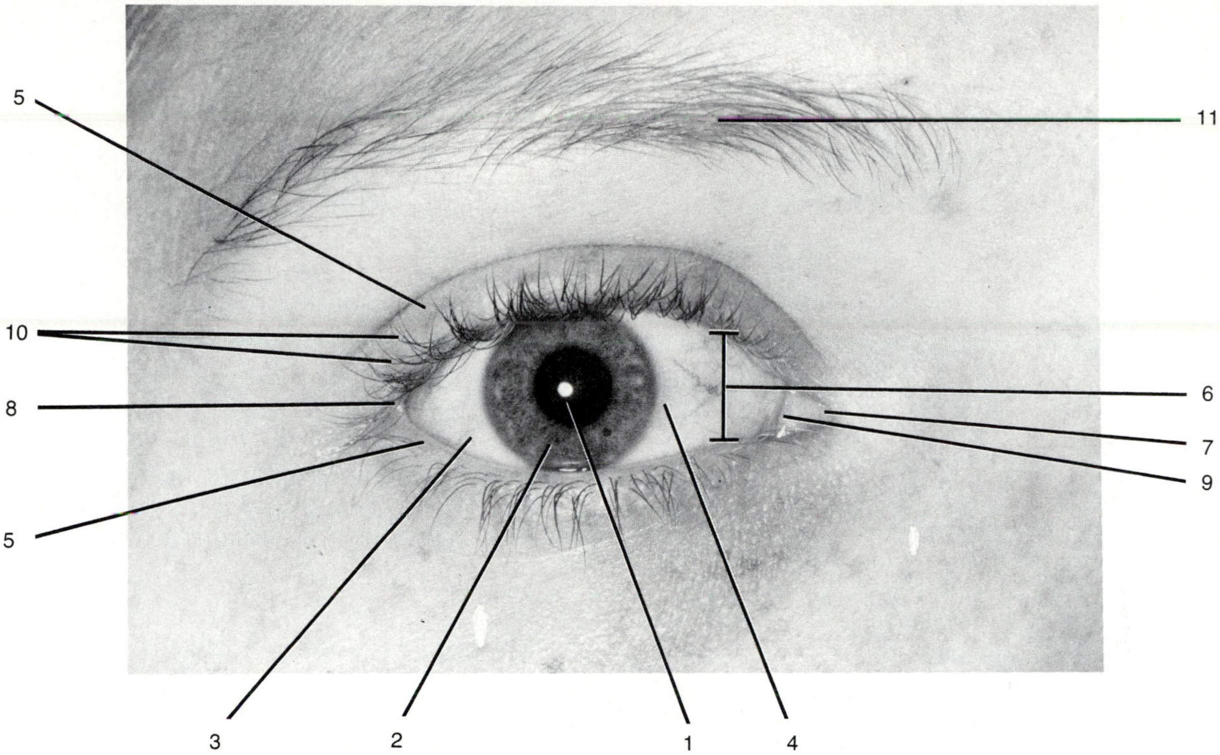

1. Pupil. Opening of center of iris of eyeball for light transmission.
2. Iris. Circular pigmented muscular membrane behind cornea.
3. Sclera. "White" of eye, a coat of fibrous tissue that covers entire eyeball except for cornea.
4. Conjunctiva. Membrane that covers exposed surface of eyeball and lines eyelids.
5. Palpebrae (eyelids). Folds of skin and muscle lined by conjunctiva.
6. Palpebral fissure. Space between eyelids when they are open.
7. Medial commissure. Site of union of upper and lower eyelids near nose.
8. Lateral commissure. Site of union of upper and lower eyelids away from nose.
9. Lacrimal caruncle. Fleshy, yellowish projection of medial commissure that contains modified sweat and sebaceous glands.
10. Eyelashes. Hairs on margins of eyelids, usually arranged in two or three rows.
11. Eyebrows. Several rows of hairs superior to upper eyelids.

FIGURE 14.10 Surface anatomy of eyeball and accessory structures.

7. The ***retina*** is the white inner coat beneath the choroid coat and is easily separated from it.

8. The ***choroid coat*** is a dark, iridescent-colored tissue that gets its iridescence from a special structure called the ***tapetum lucidum***. The tapetum lucidum, which is not present in the human eye, functions to reflect some light back onto the retina.

9. Finally, identify the ***blind spot***, the point at which the retina is attached to the back of the eyeball.

5. Ophthalmoscopic Examination of the Eyeball

Examinations of the eye include inspection of the fundus (interior) of the eyeball. This is accomplished by using an instrument called an ***ophthalmoscope*** (op-THAL-mō-skōp). It consists of a light source, a set of mirrors or a prism arranged to reflect light so that the fundus of the eyeball is illuminated, and a set of lenses arranged on a rotating disc. Ophthalmoscopic examination of the eyeball permits examination of the retina (nervous

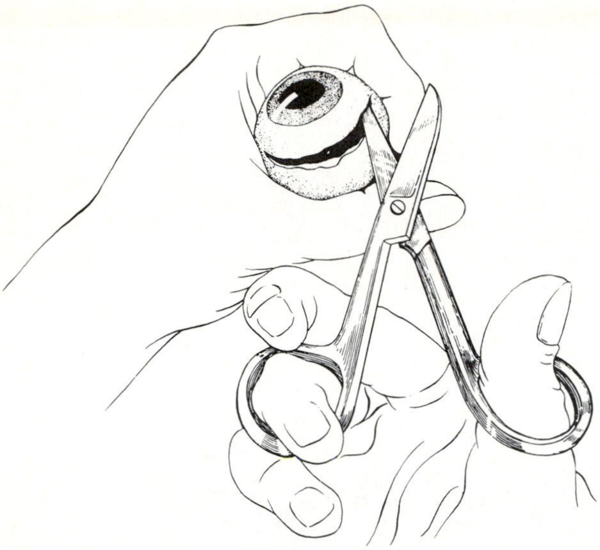

FIGURE 14.11 Procedure for dissecting a vertebrate eye.

tunic), optic disc, macula lutea, and blood vessels and is useful in detecting changes associated with conditions such as diabetes mellitus, atherosclerosis, and cataracts. Without using eye drops to dilate the pupil, a physician can see about 15% of the retina; if the pupil is dilated, about half of the retina can be visualized.

The procedure for using the ophthalmoscope is as follows:

CAUTION! *Please limit your examination time to one minute.*

PROCEDURE

1. In a dimly lit or dark room, seat your partner comfortably and have her or him look straight ahead at an object at eye level.
2. To examine your partner's right eye, hold the ophthalmoscope in you right hand and use your right eye. Reverse hands and eyes when viewing your partner's left eye.
3. With your index finger on the edge of the lens selection disc, rotate the disc so that the "O" is in position.
4. Hold the instrument about 15 cm (6 in.) from the subject's eye.

CAUTION! *Direct light toward the edge of the pupil rather than directly in its center.*

(If you direct light toward the center of the pupil, light will be refracted from the subject's cornea back into your eye.) You should now see a red circular area in the fundus of the eyeball.

5. Now, with the red area in view, move the ophthalmoscope to *within 5 cm (2 in.) of the subject*, and, while still directing the beam of light toward the edge of the pupil, rotate the lens selector until the optic disc is in sharp focus (Figure 14.12). Positive (+) settings (may be printed in black) correct for farsightedness and negative (−) settings (may be printed in red) correct for nearsightedness. Reduce the light, if necessary. Examine the optic disc carefully and note its blood vessels. The disc should have a sharp outline.
6. Examine the macula lutea lateral to the optic disc. It lacks blood vessels and should be darker than the optic disc. Locate the central fovea, a slightly lighter area in the center of the macula lutea.

6. Testing for Visual Acuity

The acuteness of vision may be tested by means of a ***Snellen Chart***. It consists of letters of different sizes which are read at a distance normally designated at 20 ft. If the subject reads to the line that is marked "50," he or she is said to possess 20/50 vision in that eye, meaning that he or she is reading at 20 ft what a person who has normal vision can read at 50 ft. If he or she reads to the line marked "20," he or she has 20/20 vision in that eye. The normal eye can sufficiently refract light rays from an object 20 ft away to focus a clear object on the retina. Therefore, if you have 20/20 vision, your eyes are perfectly normal. The higher the bottom number, the larger the letter must be for you to see it clearly, and of course the worse or weaker your eyes are.

PROCEDURE

1. Have the subject stand 20 ft from the Snellen Chart and cover the right eye with a 3" × 5" card.
2. Instruct the subject to slowly read down the chart until he or she can no longer focus the letters.
3. Record the number of the last line (20/20, 20/30, or whichever) that can be successfully read.
4. Repeat this procedure covering the left eye.
5. Now the subject should read the chart using both eyes.

6. Record your results in Section H.6 of the LABORATORY REPORT RESULTS at the end of the exercise and change places.

7. Testing for Astigmatism

The eye, with normal ability to refract light, is referred to as an ***emmetropic*** (em′-e-TROP-ik) eye. It can sufficiently refract light rays from an object 20 ft away to focus a clear object on the retina. If the lens is normal, objects as far away as the horizon and as close as about 20 ft will form images on the sensitive part of the retina. When objects are closer than 20 ft, however, the lens has to sharpen its focus by using the ciliary muscles. Many individuals, however, have abnormalities related to improper refraction. Among these are ***myopia*** (mī-Ō-pē-a) (nearsightedness), ***hypermetropia*** (hī′-per-mē-TRO-pē-a) (farsightedness), and ***astigmatism*** (a-STIG-ma-tizm) (irregularities in the surface of the lens or cornea).

Why do you think nearsightedness can be corrected with glasses containing biconcave lenses? ______________________________

____

____

How would you correct farsightedness? ______

____

____

PROCEDURE

1. In order to determine the presence of astigmatism, remove any corrective lenses if you are wearing them and look at the center of the following astigmatism test chart, first with one eye, then the other:

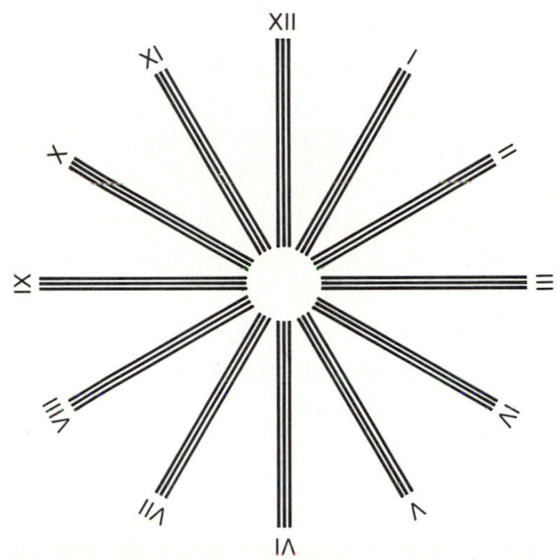

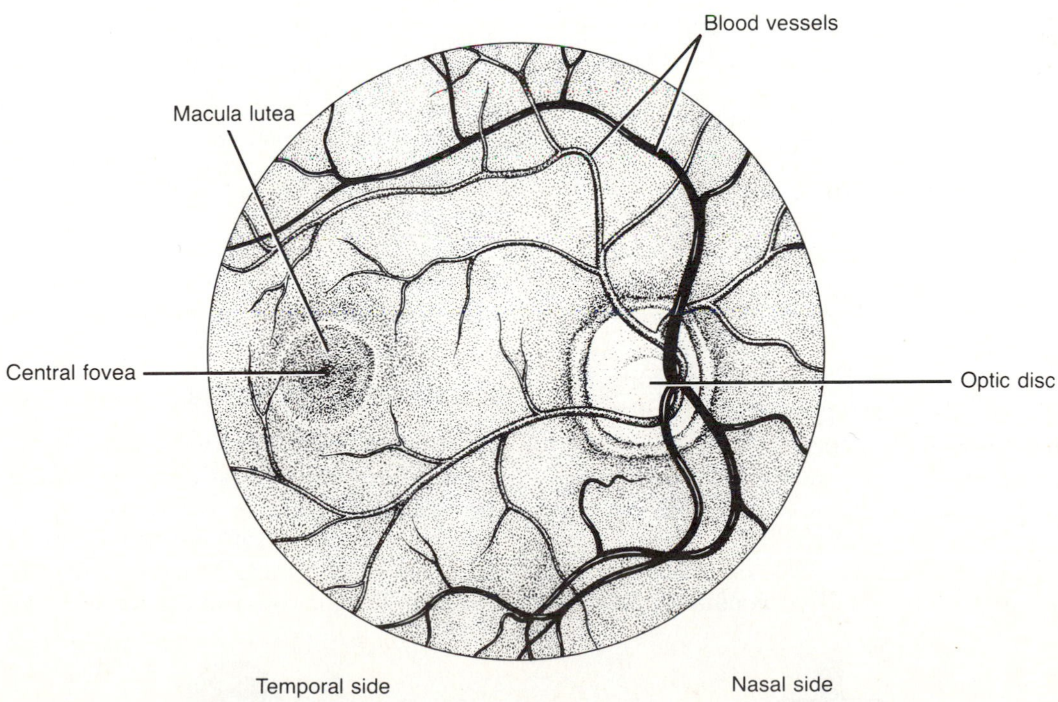

FIGURE 14.12 Diagram of the fundus (back portion of interior) of the right eye as seen by ophthalmoscopy.

2. If all the radiating lines appear equally sharp and equally black, there is no astigmatism.

3. If some of the lines are blurred or less dark than others, astigmatism is present.

4. If you wear corrective lenses, try the test with them on.

8. Testing for the Blind Spot

PROCEDURE

1. Hold this page about 20 in. from your face with the cross in the following diagram directly in front of your right eye. You should be able to see the cross and the circle when you close your left eye.

2. Now, keeping the left eye closed, slowly bring the page closer to your face while fixing the right eye on the cross.

3. At a certain distance the circle will disappear from your field of vision because its image falls on the blind spot.

9. Image Formation

Formation of an image on the retina requires four basic processes, all concerned with focusing light rays. These are: (1) refraction of light rays, (2) accommodation of the lens, (3) constriction of the pupil, and (4) convergence of the eyes.

When light rays traveling through a transparent medium (such as air) pass into a second transparent medium with a different density (such as water), the rays bend at the surface of the two media. This is called ***refraction***. The eye has four such media of refraction—the cornea, aqueous humor, lens, and vitreous humor.

The lens of the eye has the unique ability to change the focusing power of the eye by becoming moderately curved at one moment and greatly curved the next. When the eye is focusing on a close object, the lens curves greatly in order to bend the rays toward the central fovea of the eye. This increase in the curvature of the lens is called ***accommodation***.

a. TESTING FOR NEAR-POINT ACCOMMODATION

The following test determines your ***near-point accommodation***:

PROCEDURE

1. Using any card that has a letter printed on it, close one eye and focus on the letter.

2. Measure the distance of the card from the eye using a ruler or a meter stick.

3. Now *slowly* bring the card as close to your open eye as possible, and stop when you no longer see a clear, detailed letter.

4. Measure and record this distance. This value is your near-point accommodation.

5. Repeat this procedure three times and then test your other eye.

6. Check Table 14.1 to see whether the near point for your eyes corresponds with that recorded for your age group. (***Note:*** Use a letter that is the size of typical newsprint.)

b. TESTING FOR CONSTRICTION OF THE PUPIL

PROCEDURE

1. Place a 3" × 5" card on the side of the nose so that a light shining on one side of the face will not affect the eye on the other side.

2. Shine the light from a lamp or a flashlight on one eye, 6 in. away (approximately 15 cm), for about 5 seconds. Note the change in the size of the pupil of this eye.

3. Remove the light, wait about 3 minutes, and repeat, but this time observe the pupil of the opposite eye.

4. Wait a few minutes, and repeat the test observing the pupils of both eyes.

c. TESTING FOR CONVERGENCE

In humans, both eyes focus on only one set of objects—a characteristic called ***single binocular vision***. The term ***convergence*** refers to a medial movement of the two eyeballs so that they are both directed toward the object being viewed. The nearer the object, the greater the

TABLE 14.1
CORRELATION OF AGE AND NEAR-POINT ACCOMMODATION

Age	Inches	Centimeters
10	2.95	7.5
20	3.54	9.0
30	4.53	11.5
40	6.77	17.2
50	20.67	52.5
60	32.80	83.3

degree of convergence necessary to maintain single binocular vision.

PROCEDURE

1. Hold a pencil or pen about 2 ft from your nose and focus on its point. Now slowly bring the pencil toward your nose.
2. At some moment you should suddenly see two pencil points, or a blurring of the point.
3. Observe your partner's eyes when he or she does this test.

Images are actually focused upside down on the retina. They also undergo mirror reversal. That is, light reflected from the right side of an object hits the left side of the retina and vice versa. Reflected light from the top of the object crosses light from the bottom of the object and strikes the retina below the central fovea. Reflected light from the bottom of the object crosses light from the top of the object and strikes the retina above the central fovea.

The reason why we do not see a topsy-turvy world is that the brain learns early in life to coordinate visual images with the exact location of objects. The brain stores memories of reaching and touching objects and automatically turns visual images right-side up and right-side around.

10. Testing for Binocular Vision, Depth Perception, Diplopia, and Dominance

Humans are endowed with binocular vision. Each eye sees a different view, although there is a large degree of overlap. The cerebral cortex uses these discrepancies to produce ***stereopsis*** or three-dimensional vision, an orientation of the object in space that allows us to perceive its distance from us ***(depth perception)***.

Even though there are slight differences in the views from each eye, we see a single image because the cerebral cortex integrates the images from each eye into a single perception. If this integration is not present, ***diplopia*** (dip-LŌ-pē-a) or double vision results. We do not perceive an "average" or "mean" of both left and right views; rather the view from one of our eyes is ***dominant***, i.e., is the view we always perceive with both eyes open.

PROCEDURE

1. Place a book at arms length on a table. Place your left hand at your side and touch the nearest corner of the book with your right index finger. Close the left eye and repeat; close the right eye and repeat. Note the accuracy of your attempts.

How accurate was your attempt to touch a point with one eye closed? ____________

If the right eye was closed, did you have error to the left or the right? ____________

2. Use a depth perception tester with both eyes open and with each eye closed. Attempt to align the two arrows and measure monocular perception errors using the scale on the base.

Length of depth perception error ____________

3. Focus on a discrete object in the distance and press *very gently* on your left eyelid. Note how many objects you see: ____________ .

4. Stab a pencil through a piece of paper and remove the scrap. Hold the paper at arm's length and focus on a small discrete object which just fills the hole. (You may want to use an X on the blackboard.) *Without moving the paper*, close the left eye and note whether the object is still present. Repeat with right eye closed. Note your dominant eye: ____________ .

11. Testing for Afterimages

The rods and cones of the retina are photoreceptors, that is, they contain light-sensitive pigments which absorb light of different wave lengths. ***Rhodopsin***, the light-sensitive pigment in rods consists of opsin and retinal. When light strikes a molecule of rhodopsin, the retinal changes from a curved to a straight shape and breaks away from the opsin. The energy released as a result of the exchange initiates the nerve impulse that causes us to perceive light.

In bright light, all the rhodopsin is decomposed, so that the rods are nonfunctional. Cones function in bright light, in much the same way as rods, but they absorb light of specific wavelengths (red, blue, or green). The particular color perceived depends on which combinations of cones are stimulated. If all three are stimulated, we see white; if none is stimulated, we see black.

If photoreceptors are stimulated by staring at a bright object for a long period of time, they will continue firing briefly after the stimulus has been removed ***(positive afterimage)***. However, after prolonged firing, these photoreceptors become "bleached" or fatigued, so that they can no longer fire, resulting in a reversal of the image ***(negative afterimage)***.

PROCEDURE

1. Stare at a light source, *but not a bright light*, for 30 seconds. Close the eyes briefly and note the positive afterimage. Open the eyes and focus on a piece of white paper. Note the negative afterimage.

2. Draw a red cross on paper about 1 in. high, with lines ½ in. thick. Stare at the cross without moving the eyes for at least 30 seconds. Close the eyes very briefly; open them and stare at a piece of white paper. Note the positive and negative afterimages. Record your observations in Section H.11 of the LABORATORY REPORT RESULTS at the end of the exercise.

12. Testing for Colorblindness

Colorblindness is an inherited disorder in which certain cones are absent. It is a sex-linked disorder. About 0.5% of all females and about 8% of all males are affected.

PROCEDURE

1. View Ishihara plates in bright light and compare your responses to those in the plate book OR use Holmgren's test for matching colored threads.

2. Note the accuracy of your responses: ______

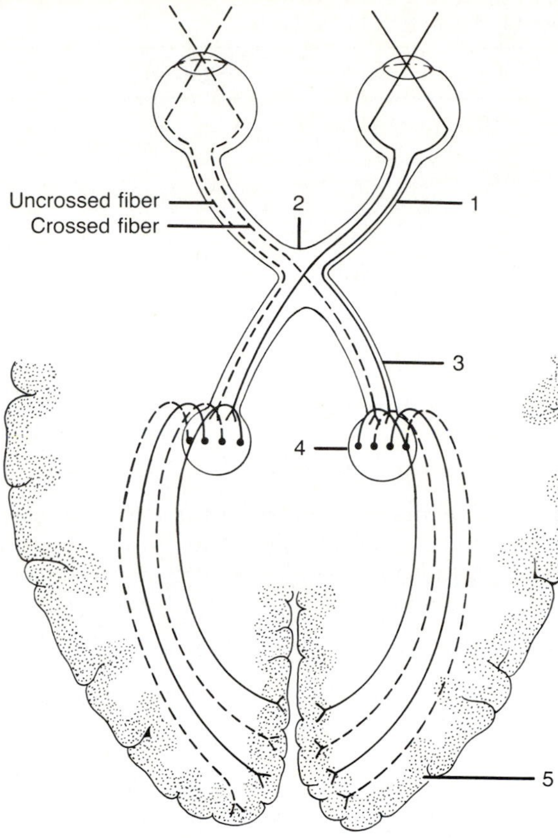

___ Occipital lobe of cerebral cortex
___ Optic chiasma
___ Optic (II) nerve
___ Optic tract
___ Thalamus

FIGURE 14.13 Visual pathway.

13. Visual Pathway

From the rods and cones, impulses are transmitted through bipolar neurons to ganglion cells. The cell bodies of the ganglion cells lie in the retina and their axons leave the eye via the ***optic (II) nerve***. The axons pass through the ***optic chiasma*** (kī-AZ-ma), a crossing point of the optic nerves. Fibers from the medial retina cross to the opposite side. Fibers from the lateral retina remain uncrossed. Upon passing through the optic chiasma, the fibers, now part of the ***optic tract***, enter the brain and terminate in the lateral geniculate nucleus of the thalamus. Here the fibers synapse with the neurons whose axons pass to the visual centers located in the occipital lobes of the cerebral cortex. Label the visual pathway in Figure 14.13.

I. AUDITORY SENSATIONS AND EQUILIBRIUM

In addition to containing receptors for sound waves, the ***ear*** also contains receptors for equilibrium. The ear is subdivided into three principal regions: (1) external or outer ear, (2) middle ear, and (3) internal or inner ear.

1. Structure of Ear

The ***external (outer) ear*** collects sound waves and directs them inward. Its structure consists of the pinna, external auditory canal, and tympanic membrane. The ***pinna (auricle)*** is a trumpet-shaped flap of elastic cartilage covered by thick skin. The rim of the pinna is called the ***helix***, and the inferior portion is referred to as the ***lobule***. The pinna is attached to the head by ligaments and muscles. The ***external auditory canal (meatus)*** is a tube, about 2.5 cm (1 in.) in length that leads from the pinna to the eardrum. The walls of the canal consist of bone lined with cartilage that is continuous with the cartilage of the pinna. Near the exterior opening, the canal contains a few hairs and specialized sebaceous glands called ***ceruminous*** (se-ROO-me-nus) ***glands***, which secrete ***cerumen*** (earwax). The combination of hairs and cerumen prevents foreign objects from entering the ear. The ***tympanic*** (tim-PAN-ik) ***membrane (eardrum)*** is a thin, semitransparent partition of fibrous connective tissue located between the external auditory canal and middle ear.

Examine a model or charts and label the parts of the external ear in Figure 14.14.

The ***middle ear (tympanic cavity)*** is a small, epithelium-lined, air-filled cavity hollowed out of the temporal bone. The area is separated from the external ear by the eardrum and from the internal ear by a very thin bony partition that contains two small membrane-covered openings, called the oval window and the round window. The posterior wall of the middle ear communicates with the mastoid cells of the temporal bone through a chamber called the ***tympanic antrum***. The anterior wall of the middle ear contains an opening that leads into the ***auditory (Eustachian) tube***. The auditory tube connects the middle ear with the nose and nasopharynx. The function of the tube is to equalize air pressure on both sides of the tympanic membrane. Any sudden pressure changes against the eardrum may be equalized by deliberately swallowing.

Extending across the middle ear are three exceedingly small bones called ***auditory ossicles*** (OS-si-kuls). These are known as the malleus, incus, and stapes. Based on their shape, they are commonly named the hammer, anvil, and stirrup, respectively. The "handle" of the ***malleus*** is attached to the internal surface of the tympanic membrane. Its head articulates with the base of the ***incus***, the intermediate bone in the series, which articulates with the stapes. The base of the ***stapes*** fits into a small opening between the middle and inner ear called the ***fenestra vestibuli*** (fe-NES-tra ves-TIB-yoo-lī) (***oval window)***. Directly below the oval window is another opening, the ***fenestra cochlea*** (KOK-lē-a) ***(round window)***. This opening, which separates the middle and inner ears, is enclosed by a membrane called the ***secondary tympanic membrane***.

Examine a model or charts and label the parts of the middle ear in Figures 14.14 and 14.15.

The ***internal (inner) ear*** is also known as the ***labyrinth*** (LAB-i-rinth). Structurally, this consists of two main divisions: (1) an outer bony labyrinth and (2) an inner membranous labyrinth that fits within the bony labyrinth. The ***bony labyrinth*** is a series of cavities within the petrous portion of the temporal bone that can be divided into three portions, named on the basis of shape: vestibule, cochlea, and semicircular canals. The bony labyrinth is lined with periosteum and contains a fluid called the ***perilymph***. This fluid surrounds the ***membranous labyrinth***, a series of sacs and tubes lying inside and having the same general form as the bony labyrinth. Epithelium lines the membranous labyrinth, which is filled with a fluid called the ***endolymph***.

The ***vestibule*** is the oval, central portion of the bony labyrinth (see Figure 14.15). The membranous labyrinth within the vestibule consists of two sacs called the ***utricle*** (YOO-tri-kul) and ***saccule*** (SAK-yool). These sacs are connected to each other by a small duct.

Projecting superiorly and posteriorly from the vestibule are the three bony ***semicircular canals*** (see Figure 14.15). Each is arranged at approximately right angles to the other two. They are called the anterior, posterior, and lateral canals. One end of each canal enlarges into

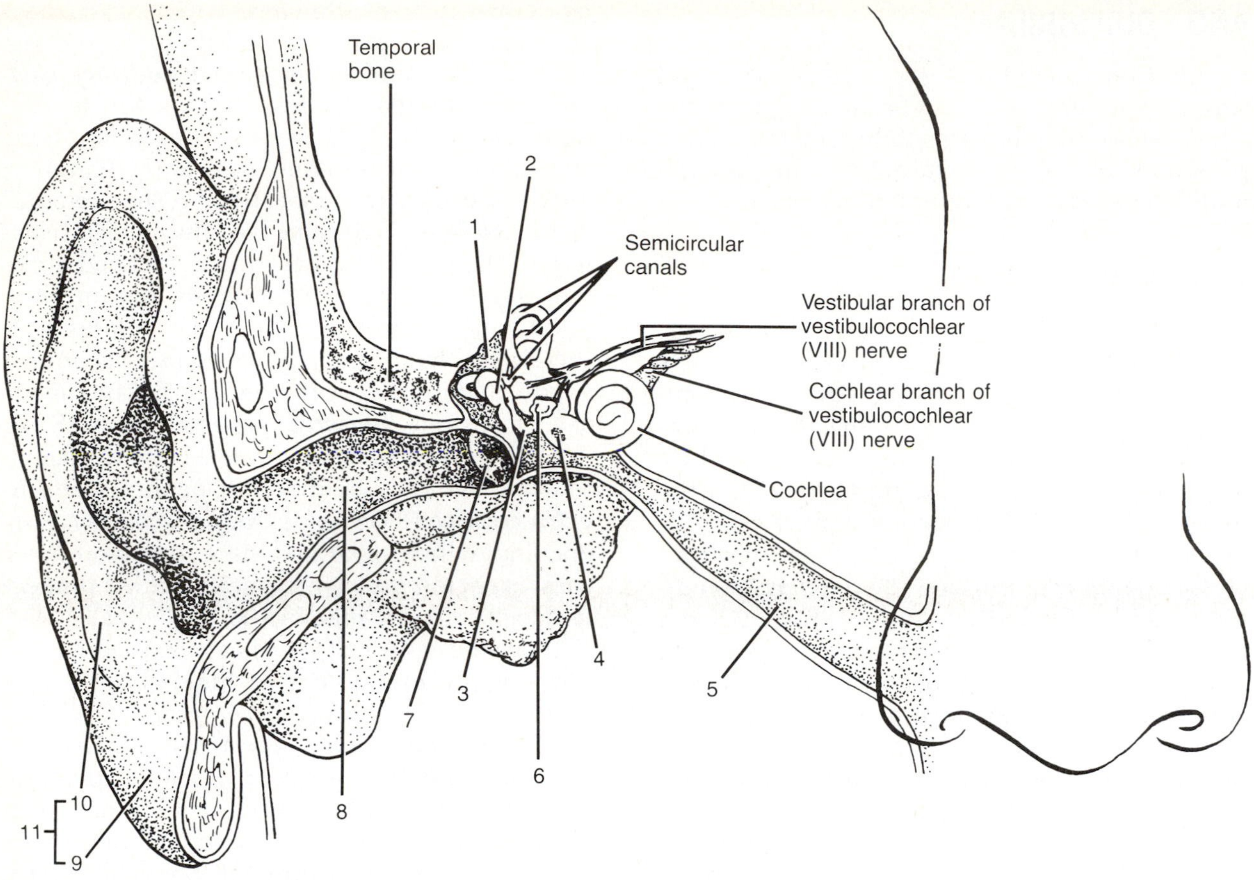

___ Auditory (Eustachian) tube
___ External auditory canal (meatus)
___ Helix
___ Incus
___ Lobule
___ Malleus
___ Pinna (auricle)
___ Round window (fenestra cochlea)
___ Stapes
___ Tympanic antrum
___ Tympanic membrane

FIGURE 14.14 Principal subdivisions of ear.

a swelling called the ***ampulla*** (am-POOL-la). Inside the bony semicircular canals lie portions of the membranous labyrinth, the ***semicircular ducts (membranous semicircular canals)***. These structures communicate with the utricle of the vestibule. Label these structures in Figure 14.16.

Anterior to the vestibule is the ***cochlea*** (KOK-lē-a) (label it in Figure 14.16), which resembles a snail's shell. The cochlea consists of a bony spiral canal that makes about 2¾ turns around a central bony core called the ***modiolus***. A cross section through the cochlea shows that the canal is divided by partitions into three separate channels resembling the letter Y lying on its side. The stem of the Y is a bony shelf that protrudes into the canal. The wings of the Y are composed of the vestibular and basilar membranes. The channel above the partition is called the ***scala vestibuli***. The channel below is known as the ***scala tympani***. The cochlea adjoins the wall of the vestibule, into which the scala vestibuli opens. The scala tympani terminates at the round window. The perilymph of the vestibule is continuous with that of the scala vestibuli. The third channel (between the wings of the Y) is the membranous labyrinth, the ***cochlear duct*** (***scala media***). This duct is separated from the scala vestibuli by the ***vestibular membrane*** and from the scala tym-

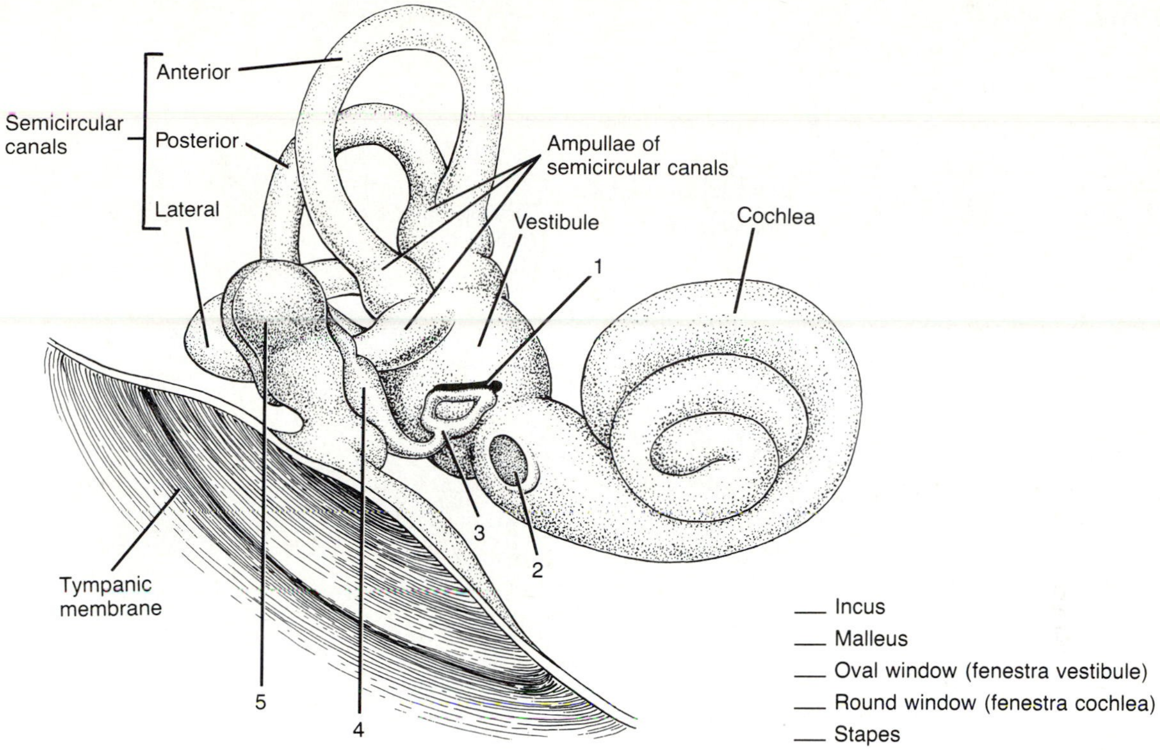

FIGURE 14.15 Ossicles of middle ear.

pani by the ***basilar membrane***. Resting on the basilar membrane is the ***spiral organ (organ of Corti)***, the organ of hearing. Label these structures in Figure 14.17.

The spiral organ (organ of Corti) is a series of epithelial cells on the inner surface of the basilar membrane. This structure is composed of supporting cells and hair cells, which are receptors for auditory sensations. The hair cells have long hairlike processes at their free ends that extend into the endolymph of the cochlear duct. The basal ends of the hair cells are in contact with fibers of the cochlear branch of the vestibulocochlear (VIII) nerve. Projecting over and in contact with the hair cells of the spiral organ is the ***tectorial membrane***, a very delicate and flexible gelatinous membrane. Label the tectorial membrane in Figure 14.17.

Obtain a prepared microscope slide of the spiral organ (organ of Corti) and examine under high power. Now label Figure 14.18.

2. Surface Anatomy

Refer to Figure 14.19 for a summary of several surface anatomy features of the ear.

3. Otoscopy

An ***otoscope*** (Ō-tō-skōp) is an instrument used to examine the tympanic membrane. Otoscopic visualization of the tympanic membrane can detect infection, lesions, and the orientation of the malleus to the tympanic membrane.

PROCEDURE

1. Clean the speculum (earpiece) of the otoscope with alcohol and turn on the light source.
2. Grasp the pinna of the subject's ear and *gently* pull it upward, backward, and a little laterally.
3. *Gently and slowly* insert the speculum into the external auditory canal (meatus) and angle the speculum downward and forward.
4. Observe the canal and tympanic membrane. The tympanic membrane should be pearly white; tympanic blood vessels should be visible.
5. Record your observations. Describe the structures viewed with the otoscope. ______________

____

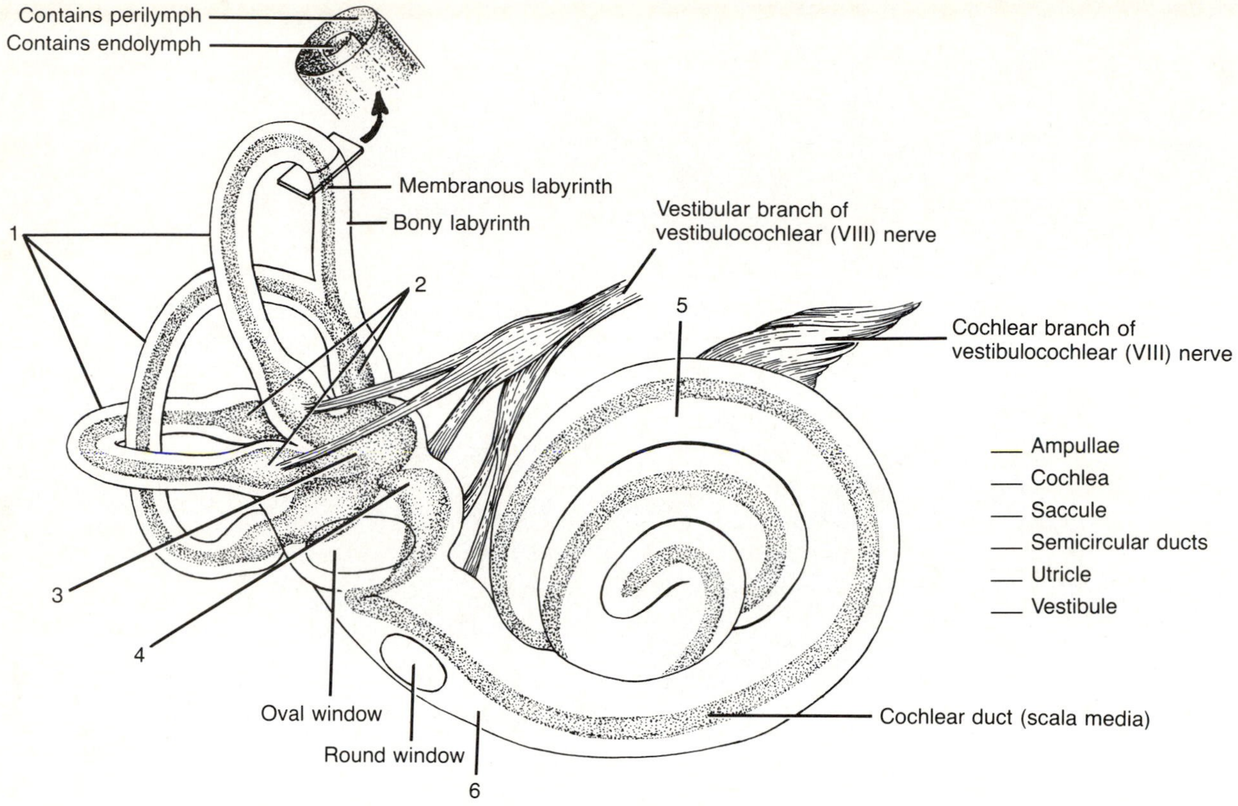

FIGURE 14.16 Details of inner ear.

4. Tests for Auditory Acuity

Sound waves result from the alternate compression and decompression of air. The waves have both a frequency and an amplitude. The ***frequency*** is the distance between crests of a sound wave. It is measured in ***Hertz (Hz)*** or ***cycles per second*** and is perceived as ***pitch***. The higher the frequency of a sound, the higher its pitch. Different frequencies displace different areas of the basilar membrane. The ***amplitude*** of a sound wave is perceived as ***intensity (loudness)***. Intensity is measured in ***decibels (db)***. Population norms have been calculated that measure the intensity of a sound that can just be heard, i.e., the ***threshold*** of sound. These levels are said to have an intensity of 0 db. Each 10 db reflects a ten-fold increase in intensity; thus a sound is 10 times louder than threshold at 10 db, 100 times greater at 20 db, one million times greater at 60 db, and so on.

a. SOUND LOCALIZATION TEST

The cerebral cortex localizes the source of a sound by evaluating the time lag between the entry of sound into each ear. Without turning the head, it is impossible to distinguish between the origin of sounds which arise at 30° ahead of the right ear and 30° behind the right ear.

PROCEDURE

1. In a quiet room, have a subject occlude one ear with a cotton plug and close the eyes.

2. Move a ticking watch to various points 6 inches away from the ear (side, front, top, back) and ask the subject to point to the location of the sound.

3. Note in which positions the sound is best localized. In what regions was sound localization most accurate? ______________________

____

b. WEBER'S TEST

Weber's test is diagnostic for both conduction and sensorineural deafness. ***Conduction deafness*** results from interference with sound

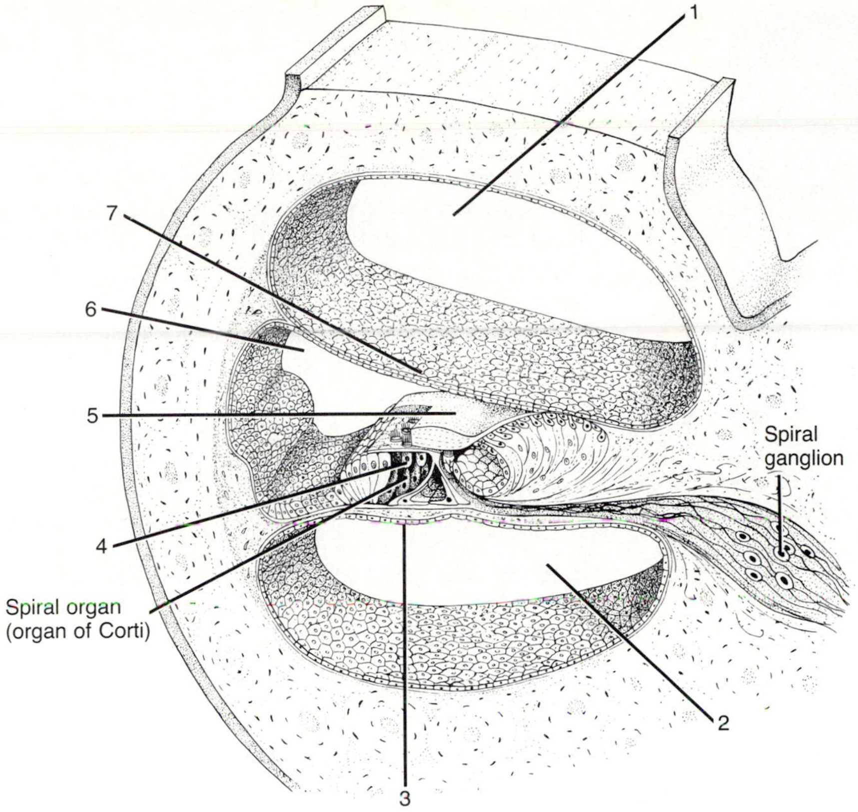

___ Basilar membrane
___ Cochlear duct (scala media)
___ Hair cells of spiral organ (organ of Corti)
___ Scala tympani
___ Scala vestibuli
___ Tectorial membrane
___ Vestibular membrane

FIGURE 14.17 Cross section through cochlea.

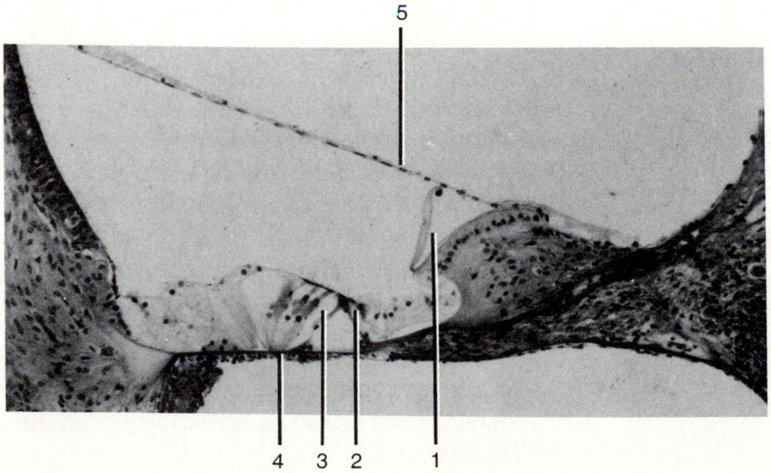

___ Basilar membrane
___ Inner hair cell
___ Outer hair cell
___ Tectorial membrane
___ Vestibular membrane

FIGURE 14.18 Photomicrograph of the spiral organ (organ of Corti). The tectorial membrane is misaligned.

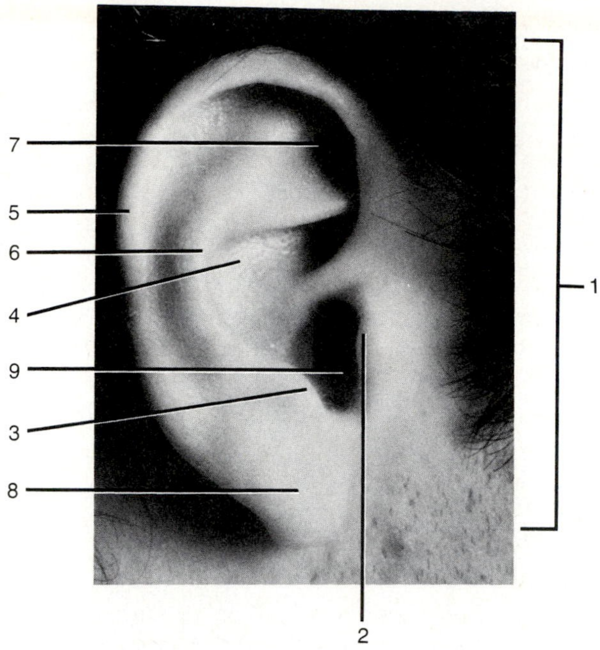

1. Pinna (auricle). Portion of external ear not contained in head, also called trumpet.
2. Tragus. Cartilaginous projection anterior to external opening of ear.
3. Antitragus. Cartilaginous projection opposite tragus.
4. Concha. Hollow of pinna.
5. Helix. Superior and posterior free margin of pinna.
6. Antihelix. Semicircular ridge posterior and superior to concha.
7. Triangular fossa. Depression in superior portion of anti-helix.
8. Lobule. Inferior portion of pinna devoid of cartilage.
9. External auditory canal (meatus). Canal extending from external ear to eardrum.

FIGURE 14.19 Surface anatomy of ear.

waves reaching the inner ear (plugged external auditory canal, fusion of the ossicles, etc.); ***sensorineural deafness*** is caused by damage to the nerve pathway between the cochlea and the brain. This test should not be performed in a quiet room.

PROCEDURE

1. Strike a tuning fork with a mallet and place the tip of the handle on the median line of a subject's forehead.
2. Determine if the tone is equally loud in both ears. ______________________
3. Have the subject occlude the left ear with cotton and repeat the exercise; occlude the right ear and repeat.

Describe your results: ______________________

____

If the sound is equally loud in both unoccluded ears, hearing is normal. If the sound is louder in the occluded ear, hearing is normal. This situation parallels conduction deafness, when the cochlea is not receiving environmental background noise but only the vibrations of the tuning fork. If there is sensorineural deafness, the sound is louder in the normal (or nonoccluded) ear, because sensorineural loss produces a deficit in transmission to the cerebral cortex.

c. RINNE TEST FOR CONDUCTION DEAFNESS

PROCEDURE

1. In a quiet room, strike a tuning fork with a mallet and place the handle of the tuning fork on the right mastoid process of a subject.
2. When the sound is no longer audible to the subject, move the vibrating fork close to the external auditory canal.
3. Ask the subject to note whether sound is still heard.
4. If the subject hears the fork again when moved near the external auditory canal (by air conduction), hearing is *not* impaired.
5. Repeat the test by testing air conduction hearing first by striking a tuning fork and holding it near the external auditory canal. When the sound is no longer audible, place the handle of the tuning fork on the mastoid process.
6. If the subject now hears the tuning fork again (by bone conduction), this indicates some degree of conduction deafness.
7. Repeat 1–6 using the left ear of the subject.
8. Record your results in Section I.4.c of the LABORATORY REPORT RESULTS at the end of the exercise.

With normal hearing ability, air conduction lasts longer than bone conduction, because the sound waves are not absorbed by the tissues. Therefore, the sound is heard outside the external ear. In conduction deafness, bone conduction is better than air conduction because the sound waves can go to the cochlea without interference by damaged structures. Therefore, the sound will not be heard outside the external ear.

d. AUDIOMETRY

An ***audiometer*** (aw′-dē-OM-e-ter) tests auditory acuity by delivering stimuli of different frequencies at varying intensities to determine the hearing threshold for each frequency. The audiometer can use either air conduction or bone conduction. The subject reports whether or not a sound is heard. The threshold for sound detection is measured in decibels (db) of loudness. If a threshold is 30 or more decibels above the population norm for perceiving that sound, the individual is said to have a hearing loss of 30 db for that specific tone. An ***audiogram*** plots hearing loss (or normalcy) for each tested frequency in the auditory spectrum.

Children can perceive sound between 30 and 22,000 Hz. A sensorineural loss accompanies aging. It begins after age 20, and the highest frequencies are affected first. By age 40, the upper limit is about 15,000 Hz; by age 80, it may be as low as 8,000 Hz.

PROCEDURE

1. In a quiet room, have the subject place the earphones with the red earphone on the right ear, blue on the left.
2. With the intensity control set at audiometric zero (the tone which can just be heard by the normal ear), set the frequency control for 125 Hz.
3. Turn the output switch to red (right ear) and push the bar level to send a continuous tone through the right earphone.
4. Instruct the subject to raise a finger when a tone is heard.
5. Increase the sound intensity gradually until it is heard.
6. Record the results in Section I.4.d of the LABORATORY REPORT RESULTS at the end of the exercise.
7. Repeat this procedure for 500, 1000, 2000, 4000, and 8000 Hz. Repeat for the left ear by turning the output switch to blue.
8. Plot the results in Section I.4.d of the LABORATORY REPORT RESULTS at the end of the exercise.

5. Equilibrium Apparatus

The term ***equilibrium*** has two meanings. One kind of equilibrium, called ***static equilibrium***, refers to the orientation of the body (mainly the head) relative to the ground (gravity). The second kind of equilibrium, called ***dynamic equilibrium***, is the maintenance of the position of the body (mainly the head) in response to sudden movements (rotation, acceleration, and deceleration) or to a change in the rate or direction of movement. The receptor organs for equilibrium are the maculae in the saccule and utricle and the cristae in the semicircular ducts.

The ***maculae*** (MAK-yoo-lē) in the walls of the ***utricle*** and ***saccule*** are the receptors concerned with static equilibrium. The maculae are small, flat, plaquelike regions that resemble the spiral organ (organ of Corti) microscopically. Maculae are located in planes perpendicular to each other and contain two kinds of cells: ***hair (receptor) cells*** and ***supporting cells***. The hair cells project ***stereocilia*** (microvilli) and a ***kinocilium*** (conventional cilium). The supporting cells are scattered among the hair cells. Floating over the hair cells is a thick, gelatinous glycoprotein layer, the ***otolithic membrane***. A layer of calcium carbonate crystals, called ***otoliths***, extends over the entire surface of the otolithic membrane. When the head is tilted, the membrane slides over the hair cells in the direction determined by the tilt of the head. This sliding causes the membrane to pull on the stereocilia, thus initiating a nerve impulse that is conveyed via the vestibular branch of the vestibulocochlear (VIII) nerve to the brain (cerebellum). The cerebellum sends continuous nerve impulses to the motor areas of the cerebral cortex in response to input from the maculae in the utricle and saccule, causing the motor system to increase or decrease its nerve impulses to specific skeletal muscles to maintain static equilibrium.

Label the parts of the macula in Figure 14.20a.

Now consider the role of the cristae in the semicircular ducts in maintaining dynamic equilibrium. The three semicircular ducts are positioned at right angles to each other in three planes: frontal (anterior duct); sagittal (posterior duct); and lateral (lateral duct). This positioning permits correction of an imbalance in three planes. In the ampulla, the dilated portion of each duct, is a small elevation called the

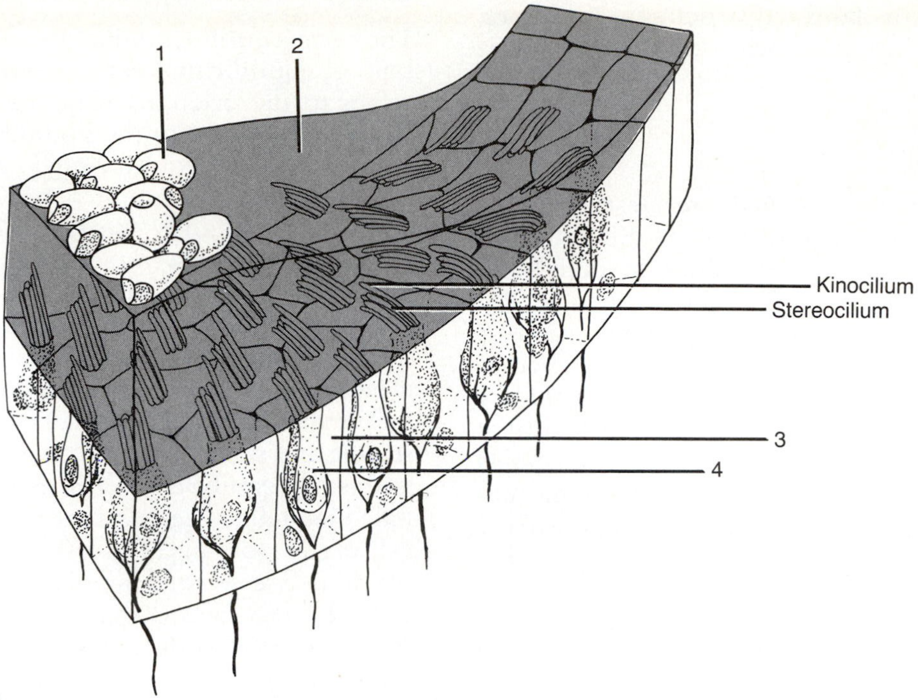

FIGURE 14.20 Equilibrium apparatus. (a) Macula.

crista. Each crista is composed of a group of ***hair (receptor) cells*** and ***supporting cells*** covered by a mass of gelatinous material called the ***cupula***. When the head moves, endolymph in the semicircular ducts flows over the hairs and bends them as water in a stream bends the plant life growing at its bottom. Movement of the hairs stimulates sensory neurons, and nerve impulses pass over the vestibular branch of the vestibulocochlear (VIII) nerve. The nerve impulses follow the same pathway as those involved in static equilibrium and are eventually sent to the muscles that must contract to maintain body balance in the new position.

Label the parts of the crista in Figure 14.20b.

6. Tests for Equilibrium

You can test equilibrium by using a few simple procedures.

a. BALANCE TEST

This test is used to evaluate static equilibrium.

PROCEDURE

1. Instruct a subject to stand perfectly still with the arms at the sides and the eyes closed.
2. Observe any swaying movements. These are easier to detect if the subject stands in front of a blackboard with a light in front of the subject.
3. A mark is made at the edge of the shadow of the subject's shoulders and movement of the shadow is then observed.

If the static equilibrium system is dysfunctional the subject will sway or fall, although other proprioreceptors may compensate for the defect.

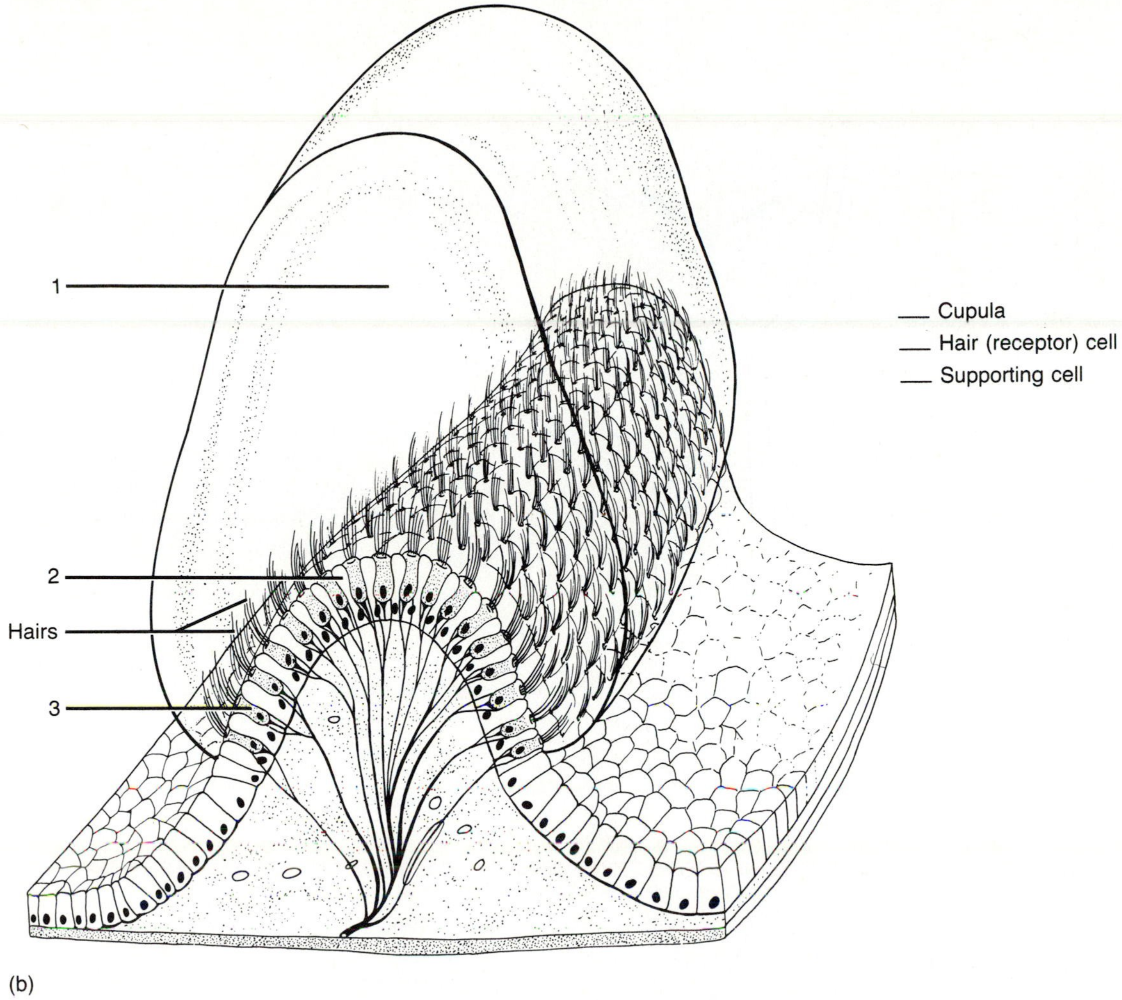

FIGURE 14.20 ***(Continued)*** Equilibrium apparatus. (b) Crista.

b. BARANY TEST

This test evaluates the function of each of the semicircular canals. A subject is rotated, and when rotation is stopped, the momentum causes the endolymph in the lateral (horizontal) canal to continue spinning. The spinning bends the cupula in the direction of rotation. Other sensory input informs us that we are no longer rotating. ***Nystagmus***, the reflex movement of the eyes rapidly and then slowly, attempts to compensate for the loss of balance by visual fixation on an object. When the head is tipped toward the shoulder, the anterior semicircular canals are stimulated, so nystagmus is vertical; when the head is tipped slightly forward, the lateral canals are stimulated by the rotation, so nystagmus occurs laterally; when the head is bent onto the chest, the posterior canal receptors are stimulated, so nystagmus is rotational.

Record your observations in Section I.5.a of the LABORATORY REPORT RESULTS at the end of the exercise.

PROCEDURE

1. Use a person who is not subject to vertigo (dizziness).
2. The subject sits firmly anchored on a stool, legs up on the stool rung, head tilted onto one shoulder and the stool is *very carefully* revolved for a few seconds and suddenly stopped.
3. Observe the subject's eye movements.

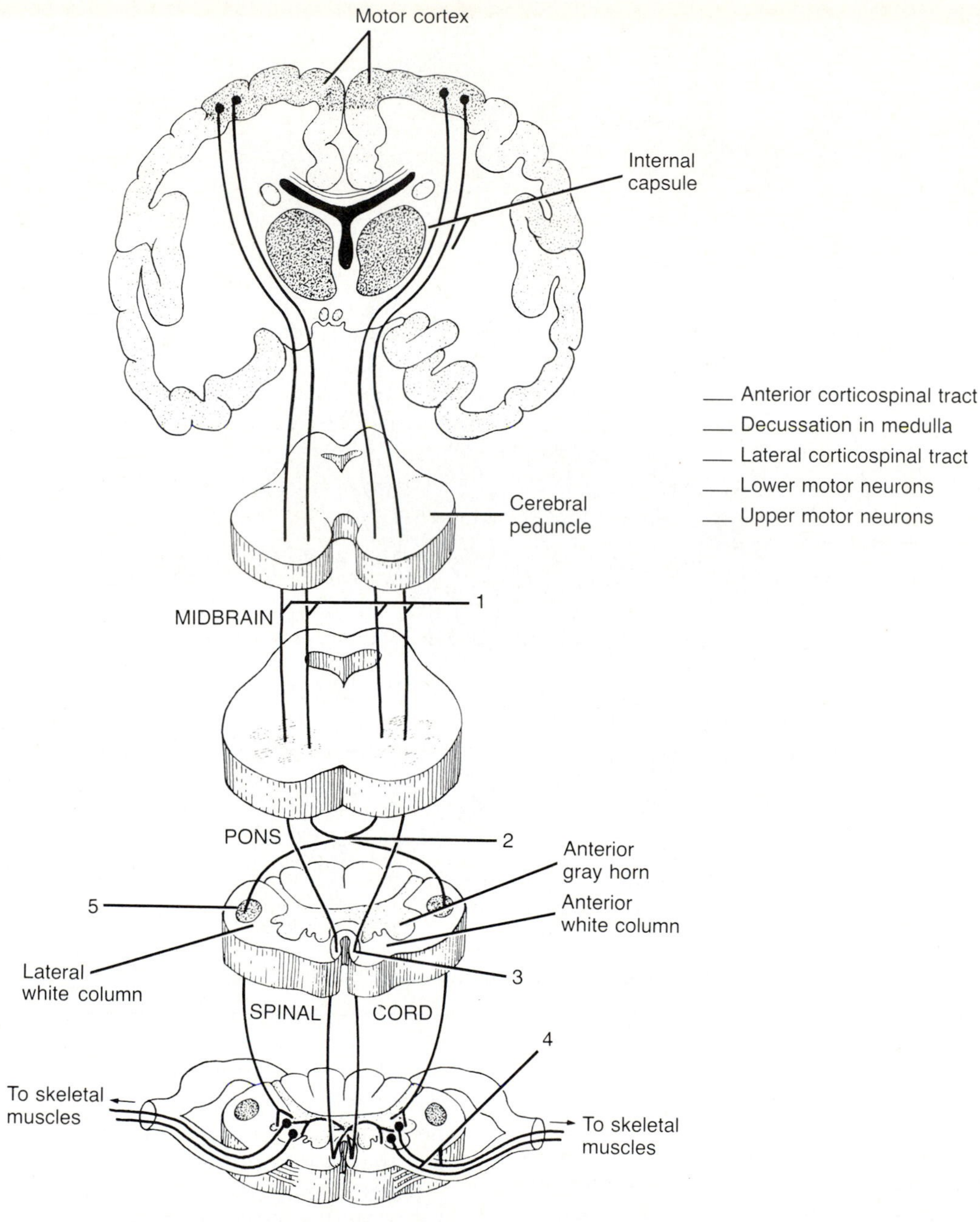

FIGURE 14.21 Pyramidal pathways.

CAUTION! *Watch the subject carefully and be prepared to provide support until the dizziness has passed.*

4. The subject will experience the sensation that the stool is still rotating, which means that the semicircular canals are functioning properly.

5. Repeat on a different subject with the head tipped slightly forward. Use still another subject with the chin touching the chest.

6. Record the direction of nystagmus in all three subjects in Section I.5.b of the LABORATORY REPORT RESULTS at the end of the exercise.

J. SENSORY-MOTOR INTEGRATION

Sensory systems function to keep the central nervous system aware of the external and internal environment. Responses to this information are mediated by motor systems that enable us to move about and change our relationship to the world around us. As sensory information is conveyed into the central nervous system, it becomes part of a large pool of sensory input. We do not respond actively to every bit of input the central nervous system receives. Rather, the incoming information is integrated with other information arriving from all other operating sensory receptors. The integration process occurs not just once but at many stations along the pathways of the central nervous system. It occurs within the spinal cord, brain stem, cerebellum, and cerebral motor cortex. As a result, a motor response to make a muscle contract or a gland secrete can be initiated at any of these levels. The cerebral motor cortex assumes the major role for controlling precise, discrete muscular movements. The basal ganglia largely integrate semivoluntary movements like walking, swimming, and laughing. The cerebellum, although not a control center, assists the motor cortex and basal ganglia by making body movements smooth and coordinated.

When the input reaches the highest center, the process of ***sensory-motor integration*** occurs. This involves not only the utilization of information contained within that center but also the information impinging on that center that is delivered from other centers within the central nervous system. After integration occurs, the output of the center is sent down the spinal cord in two major descending motor pathways: the pyramidal pathways and extrapyramidal pathways.

1. Pyramidal Pathways

Voluntary motor impulses are conveyed from the motor areas of the brain to somatic efferent neurons leading to skeletal muscles via the ***pyramidal*** (pi-RAM-i-dal) ***pathways***. The various tracts of the pyramidal system convey impulses from the cortex that result in precise muscular movements.

Label the various components of the pyramidal pathways in Figure 14.21.

2. Extrapyramidal Pathways

The ***extrapyramidal pathways*** include all descending tracts other than the pyramidal tracts. Generally, these include tracts that begin in the basal ganglia and reticular formation.

ANSWER THE LABORATORY REPORT QUESTIONS AT THE END OF THE EXERCISE

LABORATORY REPORT RESULTS

SENSATIONS AND SENSORY PATHWAYS 14

Student ______________________ Date ______________________

Laboratory Section ______________________ Score/Grade ______________________

SECTION D. TESTS FOR GENERAL SENSES

1. Two-Point Discrimination Test

Part of body	Least distance at which two points can be detected
Tip of tongue	1.4 mm
Tip of finger	________
Side of nose	________
Back of hand	________
Back of neck	36.2 mm

3. Identifying Pressure Receptors

Part of body	Distance between points touched by chalk
Palm of hand	________
Arm	________
Forearm	________
Back of neck	________

6. Identifying Proprioceptors

First trial ________

Second trial ________

Third trial ________

SECTION G. GUSTATORY SENSATIONS

2. Identifying Taste Zones

AREAS OF TONGUE IN WHICH BASIC TASTES ARE DETECTED

	Sweet	Bitter	Salty	Sour
Tip of tongue				
Back of tongue				
Sides of tongue				

4. Taste and Smell

	Sensations when placed on dry tongue	Sensations while chewing (nostrils closed)	Sensations with nostrils opened
Carrot			
Onion			
Potato			
Apple			

SECTION H. VISUAL SENSATIONS

6. Testing for Visual Acuity

Visual acuity, left eye ____________

Visual acuity, right eye ____________

Visual acuity, both eyes ____________

11. Testing for Afterimages

	Appearance of positive afterimage	Appearance of negative afterimage
Bright light		
Red cross		

SECTION I. AUDITORY SENSATIONS AND EQUILIBRIUM

4. Tests for Auditory Acuity

c. RINNE TEST

	Air conduction	Bone conduction
Right ear		
Left ear		

d. AUDIOMETRY

Frequency in Hz	Threshold: left ear	Threshold: right ear
125		
500		
1000		
2000		
4000		
8000		

Audiogram

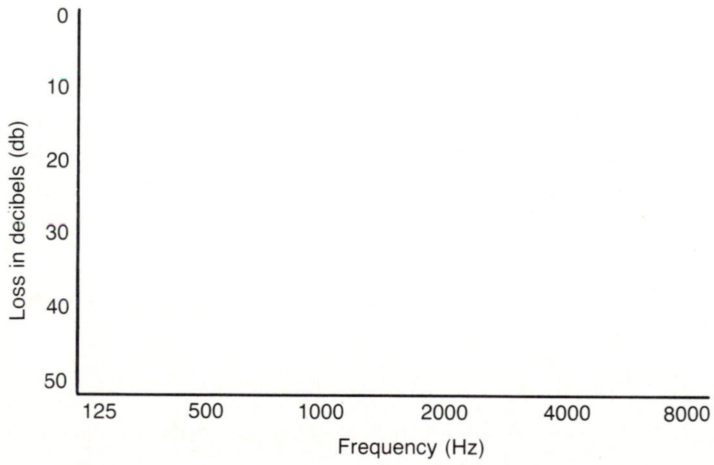

5. Tests For Equilibrium

a. BALANCE TEST

Describe the response of the subject. ______________________________

____

____

b. BARANY TEST

Subject 1 ______________________________

Subject 2 ______________________________

Subject 3 ______________________________

Laboratory Report Questions

SENSATIONS AND SENSORY PATHWAYS 14

Student ______________________ Date ______________________

Laboratory Section ______________________ Score/Grade ______________________

PART I. MULTIPLE CHOICE

_____ 1. The process by which the brain refers sensations to their point of stimulation is referred to as (a) modality (b) projection (c) accommodation (d) convergence

_____ 2. An awareness of the activities of muscles, tendons, and joints is known as (a) referred pain (b) adaptation (c) refraction (d) proprioception

_____ 3. The papillae located in an inverted V-shaped row at the posterior portion of the tongue are the (a) circumvallate (b) filiform (c) fungiform (d) gustatory

_____ 4. The technical name for the "white of the eye" is the (a) cornea (b) conjunctiva (c) choroid (d) sclera

_____ 5. Which is *not* a component of the vascular tunic? (a) choroid (b) macula lutea (c) iris (d) ciliary body

_____ 6. The amount of light entering the eyeball is regulated by the (a) lens (b) iris (c) cornea (d) conjunctiva

_____ 7. Which region of the eye is concerned primarily with image formation? (a) retina (b) choroid (c) lens (d) ciliary body

_____ 8. The densest concentration of cones is found at the (a) blind spot (b) macula lutea (c) central fovea (d) optic disc

_____ 9. Which region of the eyeball contains vitreous humor? (a) anterior chamber (b) posterior cavity (c) posterior chamber (d) conjunctiva

_____ 10. Among the structures found in the middle ear are the (a) vestibule (b) auditory ossicles (c) semicircular canals (d) external auditory canal

_____ 11. The receptors for dynamic equilibrium are the (a) saccules (b) utricles (c) cristae in semicircular ducts (d) spiral organs (organs of Corti)

_____ 12. The auditory ossicles are attached to the tympanic membrane, to each other, and to the (a) semicircular ducts (b) semicircular canals (c) oval window (d) labyrinth

_____ 13. Another name for the internal ear is the (a) labyrinth (b) fenestra (c) cochlea (d) vestibule

_____ 14. Orientation of the body relative to the ground is termed (a) postural reflex (b) tonal reflex (c) dynamic equilibrium (d) static equilibrium

_____ 15. The inability to feel a sensation consciously even though a stimulus is still being applied is called (a) modality (b) projection (c) adaptation (d) afterimage formation

_____ 16. Which sequence best describes the normal flow of tears from the eyes into the nose? (a) lacrimal canals, lacrimal sacs, nasolacrimal ducts (b) lacrimal sacs, lacrimal canals, nasolacrimal ducts (c) nasolacrimal ducts, lacrimal sacs, lacrimal canals (d) lacrimal sacs, nasolacrimal ducts, lacrimal canals

_____ 17. A patient whose lens has lost transparency is suffering from (a) glaucoma (b) conjunctivitis (c) cataract (d) trachoma

_____ 18. The portion of the eyeball that contains aqueous humor is the (a) anterior cavity (b) lens (c) posterior chamber (d) macula lutea

_____ 19. The organ of hearing located within the inner ear is the (a) vestibule (b) oval window (c) modiolus (d) spiral organ (organ of Corti)

_____ 20. The sense organs of static equilibrium are the (a) semicircular ducts (b) membranous labyrinths (c) maculae in the utricle and saccule (d) pinnae

_____ 21. Which receptor does *not* belong with the others? (a) muscle spindle (b) tendon organ (Golgi tendon organ) (c) joint kinesthetic receptor (d) lamellated (Pacinian) corpuscle

_____ 22. The membrane that is reflected from the eyelids onto the eyeball is the (a) retina (b) bulbar conjunctiva (c) sclera (d) choroid

_____ 23. A characteristic of sensations by which one sensation may be distinguished from another is called (a) modality (b) projection (c) adaptation (d) afterimage formation

_____ 24. Which are *not* cutaneous receptors? (a) tactile (Merkel's) discs (b) muscle spindles (c) lamellated (Pacinian) corpuscles (d) corpuscles of touch (Meissner's corpuscles)

_____ 25. Which region of the tongue reacts strongest to bitter tastes? (a) tip (b) center (c) back (d) sides

_____ 26. Which of the following values indicates the best visual acuity? (a) 20/30 (b) 20/40 (c) 20/50 (d) 20/60

_____ 27. Nearsightedness is referred to as (a) emmetropia (b) hypermetropia (c) eumetropia (d) myopia

PART II. COMPLETION

28. Receptors found in blood vessels and viscera are classified as _____________ .

29. Structures that collectively produce and drain tears are referred to as the _____________ .

30. The three zones of the inner nervous layer of the retina (nervous tunic) are the photoreceptor neurons, bipolar neurons, and _____________ neurons.

31. The small area of the retina where no image is formed is referred to as the _____________ .

32. Abnormal elevation of intraocular pressure (IOP) is called _____________ .

33. In the visual pathway, nerve impulses pass from the optic chiasma to the _____________ before passing to the thalamus.

34. The openings between the middle and inner ears are the oval and _____________ windows.

35. The fluid within the bony labyrinth is called _____________ .

36. Tactile sensations include touch, pressure, and _____________ .

37. Receptors for pressure are free (naked) nerve endings, type II cutaneous mechanoreceptors (end organs of Ruffini), and _____________ corpuscles.

38. Proprioceptive receptors that provide information about the degree and rate of angulations of joints are _____________ .

39. The neural pathway for olfaction includes olfactory cells, olfactory bulbs, ____________________ , and cerebral cortex.

40. The posterior wall of the middle ear communicates with the mastoid air cells of the temporal bone through the ____________________ .

41. A thin semitransparent partition of fibrous connective tissue that separates the external auditory meatus from the inner ear is the ____________________ .

42. The fluid in the membranous labyrinth is called ____________________ .

43. The cochlear duct is separated from the scala vestibuli by the ____________________ .

44. The gelatinous glycoprotein layer over the hair cells in the maculae is called the ____________________ .

45. ____________________ is blurred vision caused by an irregular curvature of the surface of the cornea or lens.

46. Image formation requires refraction, accommodation, constriction of the pupil, and ____________________ .

ENDOCRINE SYSTEM 15

You have learned how the nervous system controls the body through nerve impulses that are delivered over neurons. Now you will study the body's other great control system, the ***endocrine system***. The endocrine organs and tissues affect bodily activities by releasing chemical messengers, called ***hormones***, into the bloodstream (the term hormone means "to set in motion"). The nervous and endocrine systems coordinate their activities as an interlocking supersystem. Certain parts of the nervous system stimulate or inhibit the release of hormones. The hormones, in turn, are quite capable of stimulating or inhibiting the flow of particular nerve impulses.

The body contains two different kinds of glands: exocrine and endocrine. ***Exocrine glands*** secrete their products into ducts. The ducts then carry the secretions into body cavities, into the lumens of various organs, or to the external surface of the body. Examples are sudoriferous (sweat), sebaceous (oil), mucous, and digestive glands. ***Endocrine glands***, by contrast, secrete their products (hormones) into the extracellular space around the secretory cells. The secretion passes through the membranes of cells lining blood vessels and into the blood. Because they have no ducts, endocrine glands are also called ***ductless glands***.

A. ENDOCRINE GLANDS

The endocrine glands of the body are the pituitary (hypophysis), thyroid, parathyroids, adrenals (suprarenals), pancreas, testes, ovaries, pineal (epiphysis cerebri), and thymus. Other endocrine tissues include the kidneys, stomach, small intestine, and placenta. The endocrine glands are organs that together form the ***endocrine system***. Locate the endocrine glands on a torso, and, using your textbook or charts for reference, label Figure 15.1.

All hormones maintain homeostasis by changing the physiological activities of cells. A hormone may stimulate changes in the cells of an organ or in groups of organs. Or, the hormone may directly affect the activities of all the cells in the body. The cells that respond to the effects of a hormone are called ***target cells***.

B. PITUITARY GLAND (HYPOPHYSIS)

The hormones of the ***pituitary gland***, also called the ***hypophysis*** (hī-POF-i-sis), regulate so many body activities that the pituitary has been nicknamed the "master gland." This gland lies in the sella turcica of the sphenoid bone and is attached to the hypothalamus of the brain via a stalklike structure, the ***infundibulum***.

The pituitary gland is divided structurally and functionally into an anterior lobe called the ***adenohypophysis*** (ad′-i-nō-hī-POF-i-sis) and a posterior lobe called the ***neurohypophysis*** (noo′-rō-hī-POF-i-sis). The anterior lobe contains many glandular epithelium cells and forms the glandular part of the pituitary. The posterior lobe contains axon terminals of neurons whose cell bodies are in the hypothalamus. These terminals form the neural part of the pituitary. Between the two lobes is a small, relatively avascular rudimentary zone, the ***pars intermedia,*** whose role in humans is obscure.

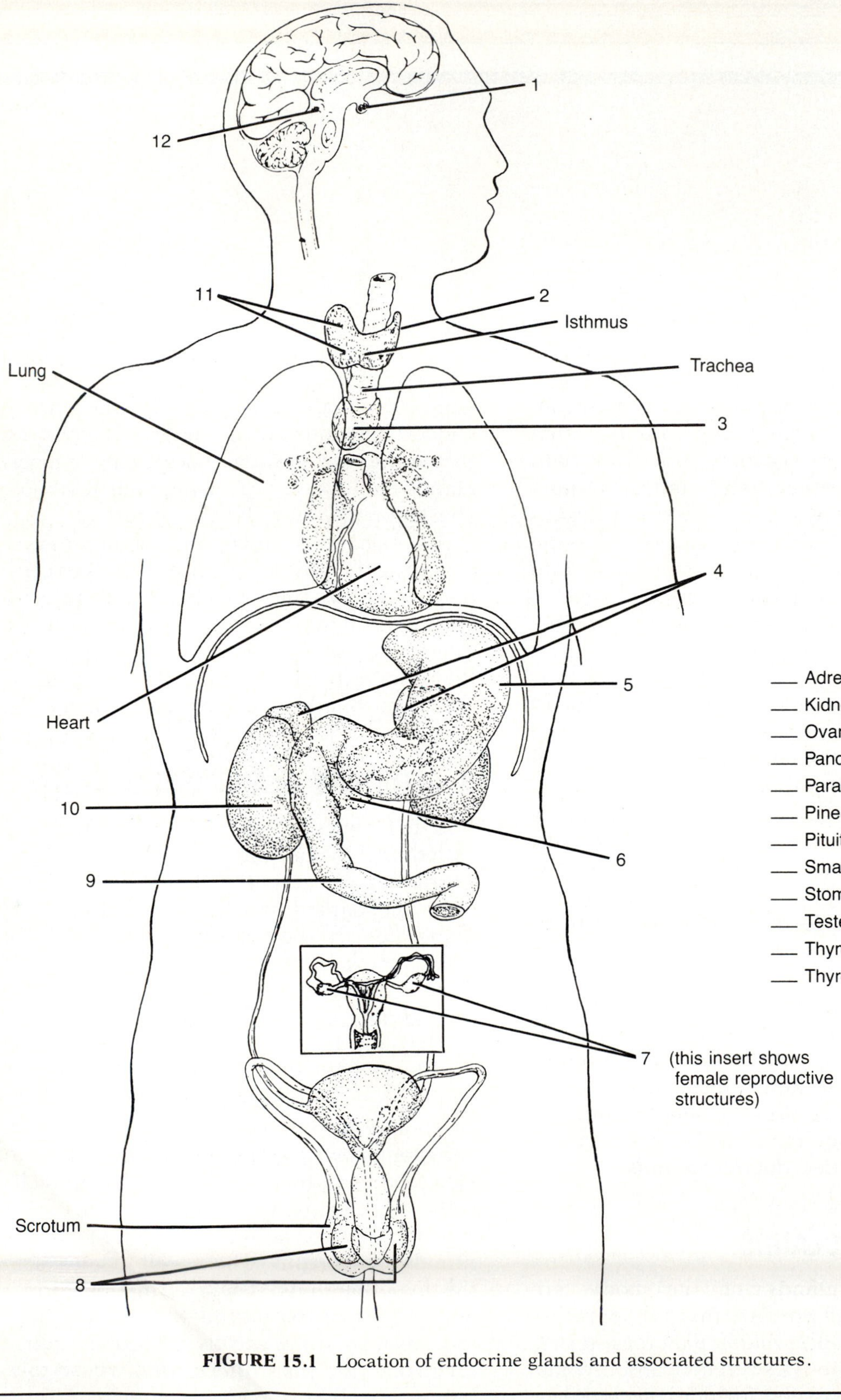

___ Adrenals (suprarenals)
___ Kidney
___ Ovaries
___ Pancreas
___ Parathyroids
___ Pineal (epiphysis cerebri)
___ Pituitary (hypophysis)
___ Small intestine
___ Stomach
___ Testes
___ Thymus
___ Thyroid

FIGURE 15.1 Location of endocrine glands and associated structures.

1. Histology of Pituitary Gland

The adenohypophysis releases hormones that regulate a whole range of body activities, from growth to reproduction. However, the release of these hormones is either stimulated or inhibited by chemical secretions ***(regulating hormones*** or ***factors)*** that are produced by neurosecretory cells in the hypothalamus of the brain. This is one interaction between the nervous system and the endocrine system. The hypothalamic regulating hormones (or factors) reach the adenohypophysis via a network of blood vessels referred to as the ***hypothalamic-hypophyseal*** (hī-pō-FIZ-ē-al) ***portal system***.

When the adenohypophysis receives proper stimulation from the hypothalamus via regulating hormones (or factors), its glandular cells secrete any one of seven hormones. Recently, special staining techniques have established the division of glandular cells into five principal types:

1. ***Somatotroph cells*** produce ***growth hormone (GH)***, which controls general body growth.
2. ***Lactotroph cells*** synthesize ***prolactin (PRL)***, which initiates milk production by the mammary glands.
3. ***Thyrotroph cells*** manufacture ***thyroid-stimulating hormone (TSH)***, which controls the thyroid gland.
4. ***Gonadotroph cells*** produce ***follicle-stimulating hormone (FSH)***, which stimulates the production of eggs and sperm in the ovaries and testes, respectively, and ***luteinizing hormone (LH)***, which stimulates other sexual and reproductive activities.
5. ***Corticolipotroph cells*** synthesize ***adrenocorticotropic*** (ad-rē-nō-kor′-ti-kō-TRŌ-pik) ***hormone (ACTH)***, which stimulates the adrenal cortex to secrete its hormones and ***melanocyte-stimulating hormone (MSH)***, which is related to skin pigmentation.

Except for growth hormone (GH), melanocyte-stimulating hormone (MSH), and prolactin (PRL), all the secretions are referred to as ***tropic hormones***, which means that their target organs are other endocrine glands. Follicle-stimulating hormone (FSH) and luteinizing hormone (LH) are also called ***gonadotropic*** (gō-nad-ō-TRŌ-pik) ***hormones*** because they regulate the functions of the gonads. The gonads (ovaries and testes) are the endocrine glands that produce sex hormones.

Identify somatotroph cells, lactotroph cells, thyrotroph cells, gonadotroph cells, and corticolipotroph cells with the aid of your textbook or a histology textbook.

The posterior lobe (neurohypophysis) of the pituitary is not really an endocrine gland. Instead of synthesizing hormones, the lobe stores hormones synthesized by cells of the hypothalamus. The posterior lobe consists of (1) cells called ***pituicytes*** (pi-TOO-i-sītz), which are similar in appearance to the neuroglia of the nervous system, and (2) axon terminations of secretory nerve cells of the hypothalamus. The cell bodies of the neurons, called ***neurosecretory cells***, originate in nuclei in the hypothalamus. The fibers project from the hypothalamus, form the ***hypothalamic-hypophyseal tract***, and terminate on blood capillaries in the neurohypophysis. The cell bodies of the neurosecretory cells produce the hormones ***oxytocin*** (ok′-sē-TŌ-sin), or ***OT***, and ***anti-diuretic hormone (ADH)***. These hormones are transported in the neuron fibers into the neurohypophysis and are stored in the axon terminals resting on the capillaries. When properly stimulated, the hypothalamus sends impulses over the neurons. The impulses cause release of hormones from the axon terminals into the blood.

Examine a prepared slide of the neurohypophysis under high power. Identify the pituicytes and axon terminations of neurosecretory cells with the aid of your textbook.

2. Hormones of Pituitary Gland

Using your textbook as a reference, give the major functions for the hormones listed below.

a. PITUITARY HORMONES SECRETED BY ADENOHYPOPHYSIS

Growth hormone (GH). Also called ***somatotropin*** and ***somatotropic hormone (STH)*** ______

______

______

______

Thyroid-stimulating hormone (TSH). Also called ***thyrotropin*** ______________________________

____

____

Adrenocorticotropic hormone (ACTH) ______

____

____

____

Follicle-stimulating hormone (FSH)

In female. ______________________________

____

____

In male. ______________________________

____

____

Luteinizing hormone (LH)

In female. ______________________________

____

____

In male. ______________________________

____

____

Prolactin (PRL). Also called ***lactogenic hormone***

____

____

Melanocyte-stimulating hormone (MSH) ____

____

____

____

____

b. PITUITARY HORMONES STORED BY NEUROHYPOPHYSIS

Oxytocin (OT) ______________________________

____

____

____

Antidiuretic hormone (ADH) ______________

____

____

____

C. THYROID GLAND

The ***thyroid gland*** is the endocrine gland located just below the larynx. The ***right and left lateral lobes*** lie on either side of the trachea and are connected by a mass of tissue called an ***isthmus*** (IS-mus) that lies in front of the trachea just below the cricoid cartilage. The ***pyramidal lobe***, when present, extends upward from the isthmus.

1. Histology of Thyroid Gland

Histologically, the thyroid consists of spherical sacs called ***thyroid follicles***. The walls of each follicle consist of epithelial cells that reach the surface of the lumen of the follicle ***(follicular cells)*** and epithelial cells that do not reach the lumen ***(parafollicular*** or ***C cells)***. The follicular cells manufacture the hormones ***thyroxine*** (thī-ROX-sēn) ***(T_4 or tetraiodothyronine)*** and ***triiodothyronine*** (trī-ī-od-ō-THĪ-rō-nēn) ***(T_3)***. Together these hormones are referred to as the ***thyroid hormones***. Thyroxine is considered to be the principal hormone produced by follicular cells. The parafollicular cells produce the hormone ***calcitonin*** (kal-si-TŌ-nin) ***(CT)***. Each thyroid follicle is filled with a glycoprotein called ***thyroglobulin (TBG)*** and stored thyroid hormones, together called ***thyroid colloid***.

Examine a prepared slide of the thyroid gland under high power. Identify the thyroid follicles, epithelial cells forming the follicle, and thyroid colloid. Now label the photomicrograph in Figure 15.2.

2. Hormones of Thyroid Gland

Using your textbook as a reference, give the major functions of the hormones listed below.

Thyroxine (T_4) and ***triiodothyronine (T_3)*** ____

____

____

____

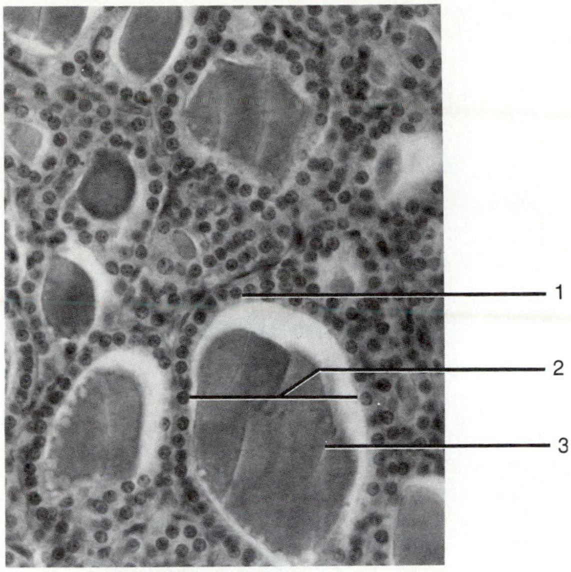

___ Epithelium of follicle
___ Thyroid colloid
___ Thyroid follicle

FIGURE 15.2 Histology of thyroid gland.

Calcitonin (CT) ______________________________

____

____

D. PARATHYROID GLANDS

Embedded on the posterior surfaces of the lateral lobes of the thyroid are small, round masses of tissue called the ***parathyroid glands***. Typically two parathyroids, superior and inferior, are attached to each lateral thyroid lobe.

1. Histology of Parathyroid Glands

Histologically, the parathyroids contain two kinds of epithelial cells. The first, a larger cell called a ***principal (chief) cell***, is believed to be the major synthesizer of ***parathyroid hormone (PTH)***. The second, a smaller cell called an ***oxyphil cell***, may serve as a reserve cell for hormone synthesis.

Examine a prepared slide of the parathyroid glands under high power. Identify the principal and oxyphil cells. Now label the photomicrograph in Figure 15.3.

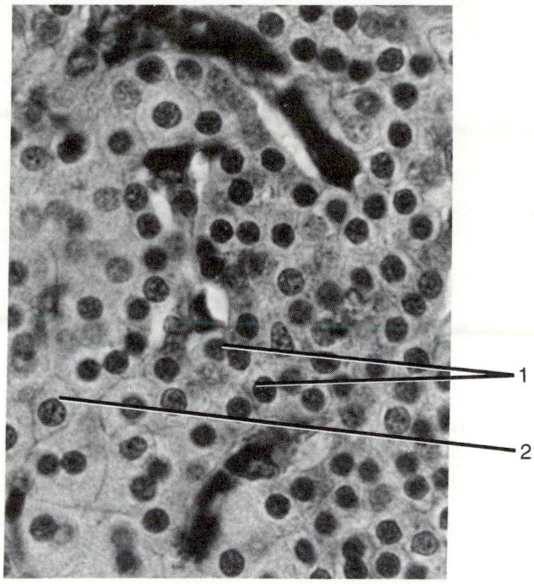

___ Oxyphil cell
___ Principal (chief) cells

FIGURE 15.3 Histology of parathyroids.

2. Hormone of Parathyroid Glands

Using your textbook as a reference, complete the following table by writing the major functions of the hormone listed:

Parathyroid hormone (PTH). Also called ***parathormone***. ______________________________

____

____

E. ADRENAL (SUPRARENAL) GLANDS

The two ***adrenal (suprarenal) glands*** are superior to each kidney and each is structurally and functionally differentiated into two regions: the outer ***adrenal cortex***, which forms the bulk of the gland, and the inner ***adrenal medulla***. Covering the gland are a thick layer of fatty connective tissue and an outer, thin fibrous capsule.

1. Histology of Adrenal Cortex

Histologically, the adrenal cortex is subdivided into three zones, each of which has a different cellular arrangement and secretes different steroid hormones. The outer zone, directly underneath the connective tissue capsule, is called the ***zona glomerulosa***. Its cells, arranged in arched loops or round balls, primarily secrete a group of hormones called ***mineralocorticoids*** (min′-er-al-ō-KŌR-ti-koyds). The middle zone, or ***zona fasciculata***, is the widest of the three zones and consists of cells arranged in long, straight cords. The zona fasciculata secretes mainly ***glucocorticoids*** (gloo′-kō-KŌR-ti-koyds). The inner zone, the ***zona reticularis***, contains cords of cells that branch freely. This zone synthesizes minute amounts of ***gonadocorticoids*** (gō-na-dō-KŌR-ti-koyds) or sex hormones, chiefly male hormones (androgens).

Examine a prepared slide of the andrenal cortex under high power. Identify the capsule, zona glomerulosa, zona fasciculata, and zona reticularis. Now label Figure 15.4.

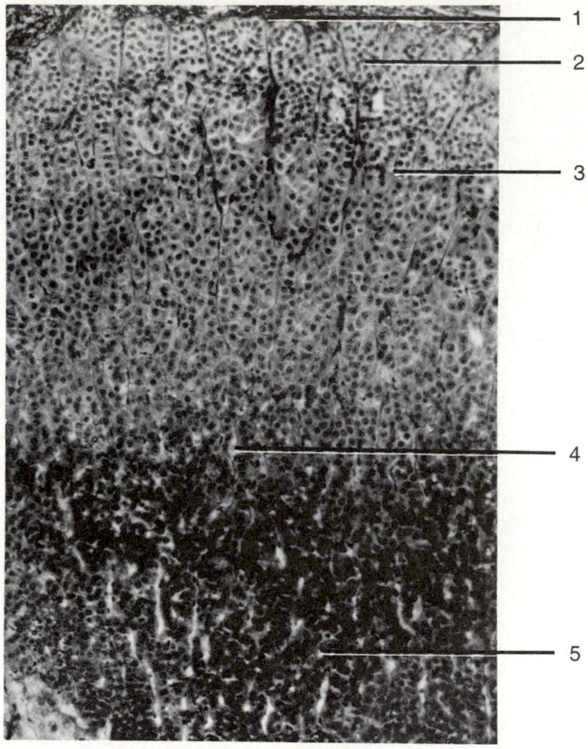

___ Adrenal medulla
___ Capsule
___ Zona fasciculata
___ Zona glomerulosa
___ Zona reticularis

FIGURE 15.4 Histology of adrenal (suprarenal) gland.

2. Hormones of Adrenal Cortex

Using your textbook as a reference, give the major functions of the hormones listed below:

a. MINERALOCORTICOIDS

Aldosterone (only one of three mineralocorticoids, but responsible for 95% of their activity)

____

____

____

____

b. GLUCOCORTICOIDS

Cortisol (hydrocortisone), ***corticosterone***, and ***cortisone*** ______________________________

____

____

____

c. GONADOCORTICOIDS

Estrogens (in female). See Estrogens under Hormones of Ovaries.

Androgens (in male). See Testosterone under Hormones of Testes.

3. Histology of Adrenal Medulla

The adrenal medulla consists of hormone-producing cells called ***chromaffin*** (krō-MAF-in) ***cells***. These cells develop from the same source as the postganglionic cells of the sympathetic division of the nervous system. They are directly innervated by preganglionic cells of the sympathetic division of the autonomic nervous system and may be regarded as postganglionic cells that are specialized to secrete. In all other visceral effectors, preganglionic sympathetic fibers first synapse with postganglionic neurons before innervating the effector. In the adrenal medulla, however, the preganglionic fibers pass directly into the chromaffin cells of the gland. Secretion of hormones from the chromaffin cells is directly controlled by the autonomic nervous system, and innervation by the preganglionic fibers allows the gland to respond extremely rapidly to a stimulus. The adrenal medulla secretes the hormones ***epinephrine*** and ***norepinephrine (NE)***.

Examine a prepared slide of the adrenal medulla under high power. Identify the chromaffin cells. Now locate the cells in Figure 15.4.

4. Hormones of Adrenal Medulla

Using your textbook as a reference, give the major functions of the hormones listed below.

Epinephrine and ***norepinephrine (NE)*** ______

______

______

______

F. PANCREAS

The ***pancreas*** is classified as both an endocrine and an exocrine gland. Thus, it is referred to as a ***heterocrine gland***. We shall discuss only its endocrine functions now. The pancreas is a flattened organ located posterior and slightly inferior to the stomach. The adult pancreas consists of a head, body, and tail.

1. Histology of Pancreas

The endocrine portion of the pancreas consists of clusters of cells called ***pancreatic islets (islets of Langerhans)***. They contain three kinds of cells: (1) ***alpha cells***, which have more distinguishable plasma membranes, are usually peripheral in the islet, and secrete the hormone ***glucagon***; (2) ***beta cells***, which generally lie deeper within the islet and secrete the hormone ***insulin***; and (3) ***delta cells***, which secrete ***somatostatin*** or ***growth hormone-inhibiting factor (GHIF)***, a substance that inhibits the secretion of growth hormone (GH), glucagon, and insulin. The pancreatic islets are surrounded by blood capillaries and by the cells called ***acini*** that form the exocrine part of the gland.

Examine a prepared slide of the pancreas under high power. Identify the alpha cells, beta cells, delta cells, and acini (clusters of cells that secrete digestive enzymes) around the pancreatic islets. Now label Figure 15.5.

2. Hormones of Pancreas

Using your textbook as a reference, give the major functions of the hormones listed below.

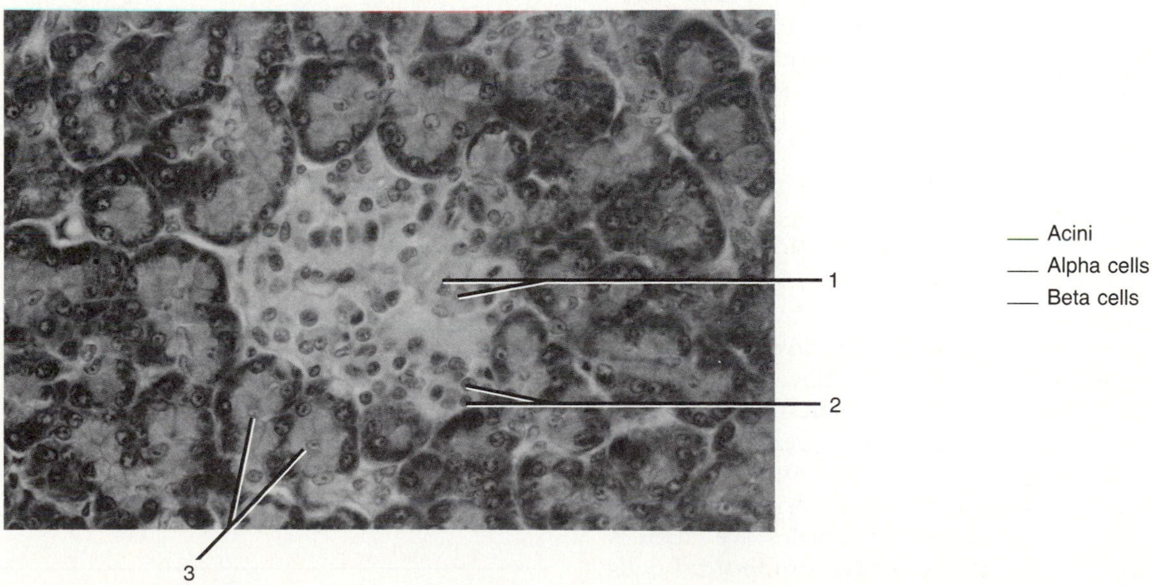

FIGURE 15.5 Histology of pancreas.

Glucagon ______________________________

____

____

Insulin ______________________________

____

____

Somatostatin ______________________________

____

____

G. TESTES

The ***testes*** (male gonads) are paired oval glands enclosed by the scrotum. They are covered by a dense layer of white fibrous tissue, the ***tunica albuginea*** (al′-byoo-JIN-ē-a), which extends inward and divides each testis into a series of internal compartments called ***lobules***. Each lobule contains one to three tightly coiled tubules, the convoluted ***seminiferous tubules***, which produce sperm by a process called ***spermatogenesis*** (sper′-ma-tō-JEN-e-sis).

1. Histology of Testes

A cross section through a seminiferous tubule reveals that it is packed with sperm cells in various stages of development. The most immature cells, called ***spermatogonia*** (sper′-ma-tō-GŌ-nē-a), are located near the basement membrane. Toward the lumen in the center of the tube, layers of progressively more mature cells can be observed. In order of advancing maturity, these are ***primary spermatocytes*** (SPER-ma-tō-sīts), ***secondary spermatocytes***, and ***spermatids***. By the time a ***sperm cell (spermatozoon)*** (sper′-ma-tō-ZŌ-on) has reached full maturity, it is in the lumen of the tubule and begins to be moved through a series of ducts. Between the developing sperm cells in the tubules are ***sustentacular*** (sus′-ten-TAK-yoo-lar) ***cells (Sertoli cells)***. They produce secretions that supply nutrients to the spermatozoa and secrete the hormone ***inhibin***. Between the seminiferous tubules are clusters of ***interstitial endocrinocytes (interstitial cells of Leydig)*** that secrete the primary male hormone ***testosterone***.

Examine a prepared slide of the testes under high power. Identify the basement membrane, spermatogonia, spermatids, spermatozoa, sustentacular cells, interstitial endocrinocytes (interstitial cells of Leydig), and lumen of the seminiferous tubule. Now label Figure 15.6.

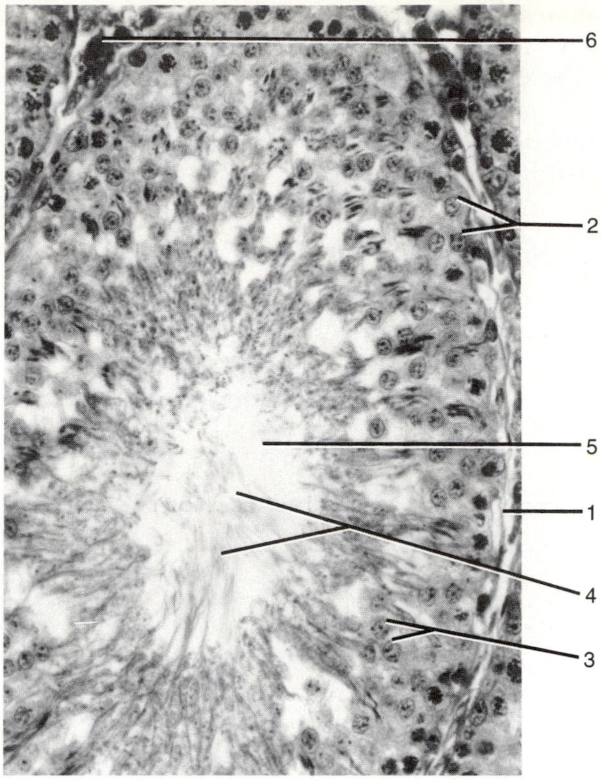

___ Basement membrane
___ Interstitial endocrinocytes (interstitial cells of Leydig)
___ Lumen of seminiferous tubule
___ Spermatids
___ Spermatogonia
___ Spermatozoa

FIGURE 15.6 Histology of testes.

2. Hormones of Testes

Using your textbook as a reference, give the major functions of the hormones listed below.

Testosterone ______________________________

____

____

Inhibin ______________________________

____

____

H. OVARIES

The ***ovaries*** (female gonads) are paired glands resembling unshelled almonds in size and shape. They are positioned in the superior pelvic cavity, one on each side of the uterus, and are maintained in position by a series of ligaments. They are (1) attached to the ***broad ligament*** of the uterus, which is itself part of the parietal peritoneum, by a fold of peritoneum called the ***mesovarium***; (2) anchored to the uterus by the ***ovarian ligament***; and (3) attached to the pelvic wall by the ***suspensory ligament***. Each ovary also contains a ***hilus***, which is the point of entrance for blood vessels and nerves.

1. Histology of Ovaries

Histologically, each ovary consists of the following parts:

1. ***Germinal epithelium***—A layer of simple cuboidal epithelium that covers the free surface of the ovary.
2. ***Tunica albuginea***—A capsule of collagenous connective tissue immediately deep to the germinal epithelium.
3. ***Stroma***—A region of connective tissue deep to the tunica albuginea. This tissue is composed of an outer, dense layer called the ***cortex*** and an inner, loose layer known as the ***medulla***. The cortex contains ovarian follicles.
4. ***Ovarian follicles***—Oocytes (immature ova) and their surrounding tissues in various stages of development. Primary follicles are smaller and secondary follicles are larger.
5. ***Vesicular ovarian (Graafian) follicle***—An endocrine structure composed of a mature ovum and its surrounding tissues. The vesicular ovarian follicle secretes hormones called ***estrogens***.
6. ***Corpus luteum***—A glandular body that develops from a vesicular ovarian (Graafian) follicle after ovulation of a secondary oocyte (potential mature ovum). It produces the hormones ***progesterone (PROG), estrogens***, and ***relaxin***.

Examine a prepared slide of the ovary under high power. Identify the germinal epithelial cells, tunica albuginea, cortex, medulla, primary follicle, secondary follicle, vesicular ovarian (Graafian) follicle, and corpus luteum. Label Figure 15.7.

2. Hormones of Ovaries

Using your textbook as a reference, give the major functions of the hormones listed below.

Estrogens ______________________________

____

____

____

Progesterone (PROG) ______________________________

____

____

____

Relaxin ______________________________

____

____

____

I. PINEAL GLAND (EPIPHYSIS CEREBRI)

The cone-shaped endocrine gland attached to the roof of the third ventricle is known as the ***pineal*** (pī̄n-Ē-al) ***gland (epiphysis cerebri)***.

1. Histology of Pineal Gland

The gland is covered by a capsule formed by the pia mater and consists of masses of neuroglial cells and parenchymal cells called ***pinealocytes*** (pin-ē-AL-ō-sīts). Around the cells are scattered preganglionic sympathetic fibers. The pineal gland starts to calcify at about the time of puberty. Such calcium deposits are called ***brain sand***. Contrary to a once widely held belief, no evidence suggests that the pineal atrophies with age and that the presence of brain sand indicates atrophy. Rather, brain sand may even denote increased secretory activity.

The physiology of the pineal gland is still somewhat obscure. The gland secretes ***melatonin***.

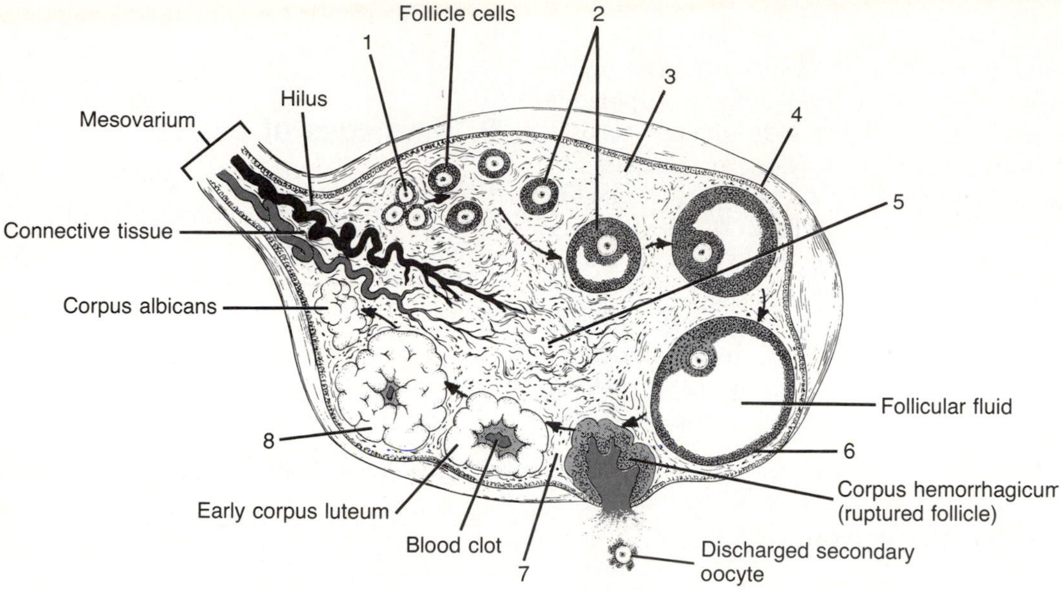

___ Corpus luteum (mature)
___ Cortex of stroma
___ Germinal epithelium
___ Medulla of stroma
___ Primary follicle
___ Secondary follicles
___ Tunica albuginea
___ Vesicular ovarian (Graafian) follicle

FIGURE 15.7 Histology of ovary.

2. Hormone of Pineal Gland

Using your textbook as a reference, give the major functions of the hormone listed below.

Melatonin ______________________________

____

____

J. THYMUS GLAND

The ***thymus gland*** is a bilobed lymphatic gland located in the upper mediastinum posterior to the sternum and between the lungs. The gland is conspicuous in infants and, during puberty, reaches maximum size. After puberty, thymic tissue, which consists primarily of ***lymphocytes***, is replaced by fat. By the time a person reaches maturity, the gland has atrophied.

1. Histology of Thymus Gland

Lymphoid tissue of the body consists primarily of lymphocytes that may be distinguished into two kinds: B cells and T cells. Both are derived originally in the embryo from lymphocytic stem cells in bone marrow. Before migrating to their positions in lymphoid tissue, the descendants of the system cells follow two distinct pathways. About half of them migrate to the thymus gland, where they are processed to become thymus-dependent lymphocytes, or ***T cells***. The thymus gland confers on them the ability to destroy antigens (foreign microbes and substances). The remaining stem cells are processed in some as yet undetermined area of the body, possibly the fetal liver and spleen, and are known as ***B cells***. These cells, perhaps under the influence of hormones produced by the thymus gland—***thymosin, thymic humoral factor (THF), thymic factor (TF),*** and ***thymopoietin***—differentiate into plasma cells. Plasma cells, in turn, produce antibodies against antigens.

2. Hormones of Thymus Gland

Using your textbook as a reference, write the major functions of the hormones listed:

Thymosin, thymic humoral factor (THF), thymic factor (TF), and thymopoietin ________

____

____

K. OTHER ENDOCRINE TISSUES

Body tissues other than endocrine glands also secrete hormones. The gastrointestinal tract synthesizes several hormones that regulate digestion in the stomach and small intestine. Among these hormones are ***stomach gastrin, intestinal gastrin, secretin, cholecystokinin (CCK), enterocrinin***, and ***gastric inhibitory peptide (GIP)***.

The placenta produces ***human chorionic gonadotropin (HCG), estrogens, progesterone (PROG), relaxin***, and ***human chorionic somatomammotroph (HCS)***, all of which are related to pregnancy.

When the kidneys (and liver, to a lesser extent) become hypoxic (subject to below normal levels of oxygen), they release an enzyme called ***renal erythropoietic factor***. This is secreted into the blood to act on a plasma protein produced in the liver to form a hormone called ***erythropoietin*** (ē-rith′-rō-POY-ē-tin).

Vitamin D, produced by the skin in the presence of sunlight, is converted to its active hormonal form in the kidneys and liver.

The atria of the heart secrete a hormone called ***atrial natriuretic factor (ANF)***, released in response to increased blood volume.

Using your textbook as a reference, write the major functions of the hormones listed:

Stomach gastrin _________________________

____

Intestinal gastrin ________________________

____

Secretin ________________________________

____

Cholecystokinin (CCK) _____________________

____

Enterocrinin ____________________________

____

Gastric inhibitory peptide (GIP) ____________

____

Human chorionic gonadotropin (hCG) ________

____

Human chorionic somatomammotropin (hCS)

____

Erythropoietin __________________________

____

Vitamin D ______________________________

____

Atrial natriuretic factor (ANF) ______________

____

L. PHYSIOLOGY OF THE ENDOCRINE SYSTEM

1. Determination of Basal Metabolic Rate in Normal, Hyperthyroid, and Hypothyroid Rats

As was mentioned previously, the thyroid gland produces two hormones related to metabolism, thyroxine (T_4) and triiodothyronine (T_3). This laboratory exercise will help you to determine the effect of a normally functioning, hyperactive, or hypoactive thyroid gland upon a rat's basal metabolic rate (BMR).

a. PREPARATION OF RATS

PROCEDURE

1. Hyperthyroid rat: animals should be injected by your instructor or a laboratory assistant with 25 μg of L-thyroxine interperitoneally every third day for a two-week period.

2. Hypothyroid rat: Rats should be fed by your instructor or a laboratory assistant *either* 0.5% thiouracil in their rat chow *or* 0.02% propylthiouracil *or* 1% potassium perchlorate in their drinking water. (***Note:*** If tetany appears in this group of animals, 1% lactate or gluconate

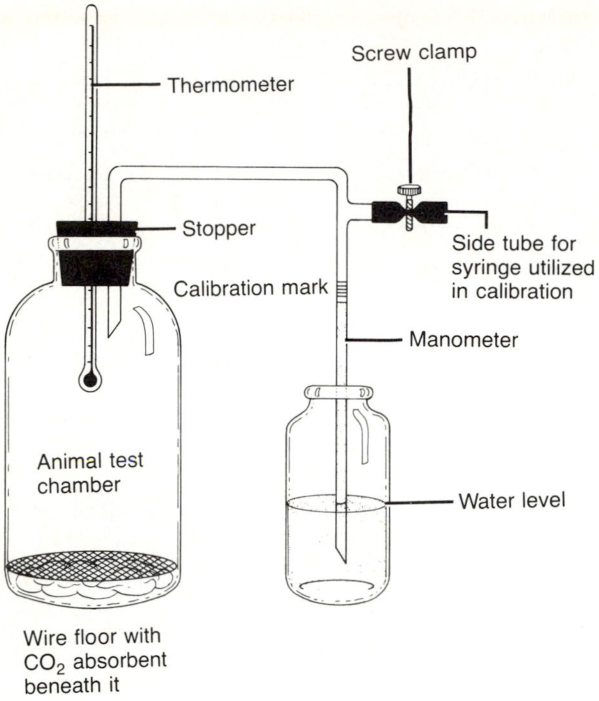

FIGURE 15.8 Experimental setup for measuring basal metabolic rate (BMR) in rats.

should be given in their drinking water.) Maintain this dosage for two weeks.

b. CALIBRATION OF SYSTEM

PROCEDURE

1. Before starting the experiment, the system must be calibrated.

2. Place an object of approximately the same volume as the rat in the system (four or five No. 10 stoppers is approximately the same volume as a 3-month-old rat).

3. Completely seal the chamber, *making sure that the syringe is set at 0.*

4. Withdraw 40 ml of air from the system via the syringe and mark the new water level (Figure 15.8).

c. EXPERIMENTAL PROCEDURE

PROCEDURE

1. Remove the syringe from the rubber tubing and clamp the tube.

2. Have your instructor or a laboratory assistant weigh the rat.

3. Remove the stoppers from the system and have your instructor or a laboratory assistant replace them with an experimental animal.

4. Seal the system and record the time and temperature.

5. As the animal utilizes oxygen, it produces carbon dioxide. The carbon dioxide will be removed by the soda lime, causing the fluid to rise in the tubing as a result of decreased air volume within the chamber.

6. Record the time required for the water to rise to the calibration mark.

7. Also record the temperature within the system at this time.

8. After recording your calculations, repeat the procedure with another experimental animal until all three classes of animals are utilized.

9. Complete calculations as outlined with the example below. Compute average values for each class of animals.

Sample calculations:

Rat weight = 150 g
40 ml oxygen consumed in 10 min

$$\text{Oxygen consumption} = \frac{40 \text{ ml}}{10 \text{ min}} = 4 \text{ ml/min}$$

Reduced to standard conditions:
Temperature = 27°C
Barometric pressure = 731 mm Hg

$$4 \times \frac{731}{760} \times \frac{273}{300} \doteq 3.5 \text{ ml/min}$$

$$(3.5 \text{ ml/min})/0.150 \text{ kg} = 23.33 \text{ ml/kg/min}$$

10. Record all calculations in Section L.1 of the LABORATORY REPORT RESULTS at the end of the exercise.

2. Demonstrations on Effects of Estrogens and Testosterone

Review the physiology of the male and female sex hormones upon the function and structure of the accessory sex organs as outlined in your textbook. These accessory sex organs are very sensitive to the effects of these hormones. The rats used in this demonstration have been prepared by your instructor or a laboratory assistant as follows:

Females	Males
Control	Control
Ovariectomized	Castrated
Ovariectomized and injected with estrogen	Castrated and injected with testosterone
Estrogen injection only	Testosterone injection only

Compare the size of uterus in females to the size of seminal vesicles in males.

Record your observations in Section L.2 of the LABORATORY REPORT RESULTS at the end of the exercise.

3. Hyperinsulinemia in Fish

Insulin causes the level of blood glucose to drop (hypoglycemia) by promoting the transport of glucose through the plasma membranes of cells. Insulin also promotes the conversion of glucose into glycogen and fat and increases the rate of synthesis of proteins from amino acids.

The regulation of insulin secretion is a direct effect of the level of sugar in the blood. When blood sugar levels fall below normal limits, the alpha cells of the pancreas secrete ***glucagon***. This hormone promotes ***glycogenolysis***, the conversion of glycogen to glucose. The release of glucose into the blood counteracts the hypoglycemia through a ***negative feedback mechanism***.

Hypoinsulinism (diabetes mellitis) results in ***hyperglycemia*** and ***glucosuria*** (urinary excretion of glucose). The high osmotic concentration of glucose in the urine causes the excretion of large urinary volumes ***(polyuria)*** and consequent thirst ***(polydipsia)***.

Hypoinsulinism also depletes the supply of glucose available for the energy requirements for cellular metabolism. Consequently, the body shifts to a higher rate of protein and fat metabolism. The breakdown of high quantities of fats releases ketone (acetone) bodies. These byproducts of fat metabolism have a characteristic acetone odor and lower the blood pH, resulting in acidosis and eventually coma.

Hyperinsulinism can result from a tumor of the beta cells in the pancreatic islets (islets of Langerhans) or from administration of excess insulin by a diabetic. High insulin levels cause hypoglycemia because insulin promotes glucose uptake primarily by liver, muscle, and fat cells. As blood sugar levels drop sharply, the brain is deprived of its major energy source. When blood glucose falls below a critical level, ***hypoglycemic (insulin) shock*** causes irritability, convulsions, and eventually coma. It is difficult to distinguish the coma resulting from hypoinsulinism from that of hyperinsulinism, even though the causes are radically different.

The effects of hyperinsulinism in fish are comparable to those in humans. The insulin molecule is small enough to diffuse through the gill membranes into the fish blood.

PROCEDURE

1. Mark two beakers I and G. In beaker I, place 200 ml of water and 10 drops of commercial insulin. Stir the mixture to disperse the insulin. In beaker G, place 200 ml of 5% glucose solution.

2. Place a small (1–2 inch) guppy, sunfish, or goldfish in beaker I.

3. Observe the behavior of the fish and note swimming movements, convulsive movements, and coma. If no changes occur within a minuute, replace the fish in the aquarium and begin the procedure with another fish.

4. When the fish becomes comatose, note the length of time it was subjected to the insulin-containing environment and quickly move it to beaker G.

Time required for the fish to become comatose:

____

5. Note the time required for the fish's behavior to return to normal and then replace the fish in the aquarium.

Time required for recovery: _______________

How did the glucose solution promote recovery? ________________________

____

If the fish were placed in a glucagon-containing solution after exposure to insulin, do you think the recovery would have been slower or faster than that caused by placement in glucose? Why? (The glucagon molecule is smaller than the insulin molecule.) _______________

ANSWER THE LABORATORY REPORT QUESTIONS AT THE END OF THE EXERCISE

ENDOCRINE SYSTEM 15

Student ____________ **Date** ____________

Laboratory Section ____________ **Score/Grade** ____________

SECTION L. PHYSIOLOGY OF THE ENDOCRINE SYSTEM

1. Complete the following table based upon your calculations.

	Control		Hyperthyroid		Hypothyroid	
	Your value	Class average	Your value	Class average	Your value	Class average
Time to consume 40 ml oxygen						
Rat weight						
Oxygen consumption						

2. Record your observations for each of the demonstrations.

Males

Normal

Castrated

Castrated plus injections

Injections only

Females

Ovariectomized

Ovariectomized plus injections

Injections only

LABORATORY REPORT QUESTIONS

ENDOCRINE SYSTEM 15

Student ______________________ Date ______________________

Laboratory Section ______________________ Score/Grade ______________________

PART 1. MULTIPLE CHOICE

_____ 1. Somatotroph cells, gonadotroph cells, and corticolipotroph cells are associated with the (a) thyroid (b) anterior pituitary (c) parathyroids (d) adrenals

_____ 2. The posterior pituitary is *not* an endocrine gland because it (a) has a rich blood supply (b) is not near the brain (c) does not make hormones (d) contains ducts

_____ 3. Which hormone assumes a role in the development and discharge of a secondary oocyte? (a) GH (b) TSH (c) LH (d) PRL

_____ 4. The endocrine gland that is probably malfunctioning if a person has a high metabolic rate is the (a) thymus (b) posterior pituitary (c) anterior pituitary (d) thyroid

_____ 5. The antagonistic hormones that regulate blood calcium level are (a) GH-TSH (b) insulin-glucagon (c) aldosterone-cortisone (d) CT-PTH

_____ 6. The endocrine gland that develops from the sympathetic nervous system is the (a) adrenal medulla (b) pancreas (c) thyroid (d) anterior pituitary

_____ 7. The hormone that aids in sodium conservation and potassium excretion is (a) hydrocortisone (b) CT (c) ADH (d) aldosterone

_____ 8. Which of the following hormones is sympathomimetic? (a) insulin (b) oxytocin (OT) (c) epinephrine (d) testosterone

_____ 9. Which hormone lowers blood sugar level? (a) glucagon (b) melatonin (c) insulin (d) cortisone

_____ 10. The endocrine gland that may assume a role in regulation of the menstrual cycle, release of aldosterone, and normal brain physiology is the (a) pineal (b) thymus (c) thyroid (d) adrenal

PART II. COMPLETION

11. The hypophysis is attached to the hypothalamus by a stalklike structure called the ______________________ .

12. A hormone that acts on another endocrine gland and causes that gland to secrete its own hormones is called a(n) ______________________ hormone.

13. ______________________ hormone is responsible for increasing skin pigmentation.

14. The hormone that helps cause contraction of the smooth muscle of the pregnant uterus is ______________________ .

15. Histologically, the spherical sacs that compose the thyroid gland are called thyroid ______________ .

16. The thyroid hormones associated with metabolism are triiodothyronine and ______________ .

17. Principal and oxyphil cells are associated with the ______________ gland.

18. The zona glomerulosa of the adrenal cortex secretes a group of hormones called ______________ .

19. The hormones that promote normal metabolism, provide resistance to stress, and function as anti-inflammatories are ______________ .

20. The pancreatic hormone that raises blood sugar level is ______________ .

21. In spermatogenesis, the most immature cells near the basement membrane are called ______________ .

22. Cells within the testes that secrete testosterone are known as ______________ .

23. The ovaries are attached to the uterus by means of the ______________ ligament.

24. The female hormones that help cause the development of secondary sex characteristics are called ______________ .

25. The endocrine gland that assumes a direct function in immunity is the ______________ .

26. Any hormone that regulates the functions of the gonads is classified as a(n) ______________ hormone.

27. The hormone that is stored in the neurohypophysis that prevents excessive urine production is ______________ .

28. The region of the adrenal cortex that synthesizes androgens is the ______________ .

29. The hormone-producing cells of the adrenal medulla are called ______________ cells.

30. Together, alpha cells, beta cells, and delta cells constitute the ______________ .

31. Regulating factors from the hypothalamus reach the adenohypophysis by a network of blood vessels called the ______________ portal system.

32. FSH and LH are produced by ______________ cells of the adenohypophysis.

33. The portion of an ovary that produces progesterone (PROG), estrogens, and relaxin is the ______________ .

34. Calcium deposits in the pineal gland are referred to as ______________ .

35. Under the influence of thymic hormones, B cells develop into antibody-producing ______________ cells.

BLOOD 16

The blood, heart, and blood vessels constitute the ***cardiovascular system***. In this exercise you will examine the characteristics of ***blood***, a connective tissue also known as ***vascular tissue***.

A. COMPONENTS OF BLOOD

Blood is composed of two portions: (1) ***plasma***, a liquid that contains dissolved substances, and (2) ***formed elements***, cells and cell-like bodies suspended in the plasma. In clinical practice, the most common classification of the formed elements of the blood is the following:

Erythrocytes (e-RITH-rō-sīts) or ***red blood cells***
Leucocytes (LOO-kō-sīts) or ***white blood cells***
  Granular leucocytes (granulocytes)
    Neutrophils
    Eosinophils
    Basophils
  Agranular leucocytes (agranulocytes)
    Lymphocytes
    Monocytes
Thrombocytes (THROM-bō-sīts) or ***platelets***

The origin and subsequent development of these formed elements can be seen in Figure 16.1. Blood cells are formed by a process called ***hematopoiesis*** (hē-ma-tō-poy-Ē-sis), and the process by which erythrocytes are formed is called ***erythropoiesis*** (e-rith′-rō-poy-Ē-sis). The immature cells that are eventually capable of developing into mature blood cells are called ***hemocytoblasts*** (hē-mō-SĪ-tō-blasts) (see Figure 16.1). Mature blood cells are constantly being replaced, so special cells called ***reticuloendothelial cells*** have the responsibility of clearing away the dead, disintegrating cell bodies so that small blood vessels are not clogged.

The shapes of the nuclei, staining characteristics, and color of cytoplasmic granules are all useful in differentiation and identification of the various white blood cells. Red cells are biconcave discs without nuclei and can be identified easily.

B. PLASMA

When the formed elements are removed from blood, a straw-colored liquid called ***plasma*** is left. This liquid consists of about 92% water and about 8% solutes. Among the solutes are proteins (albumins, globulins, and fibrinogen), nonprotein nitrogen (NPN) substances (urea, uric acid, and creatine), foods (amino acids, glucose, fatty acids, glycerides, and glycerol), regulatory substances (enzymes and hormones), respiratory gases (oxygen and carbon dioxide), and electrolytes (Na^+, K^+, Ca^{2+}, Mg^{2+}, Cl^-, HCO_3^-, SO_4^{2-}, and PO_4^{3-}).

1. Physical Characteristics

CAUTION! *Please reread Section A. "General Safety Precautions and Procedures" and Section B. "Precautions Related to Working With Blood, Blood Products, or Other Body Fluids" on page xi at the beginning of the laboratory manual before you begin any if the following experiments. You should also read the experiments before you perform them to be sure that you understand all the procedures and safety precautions.*

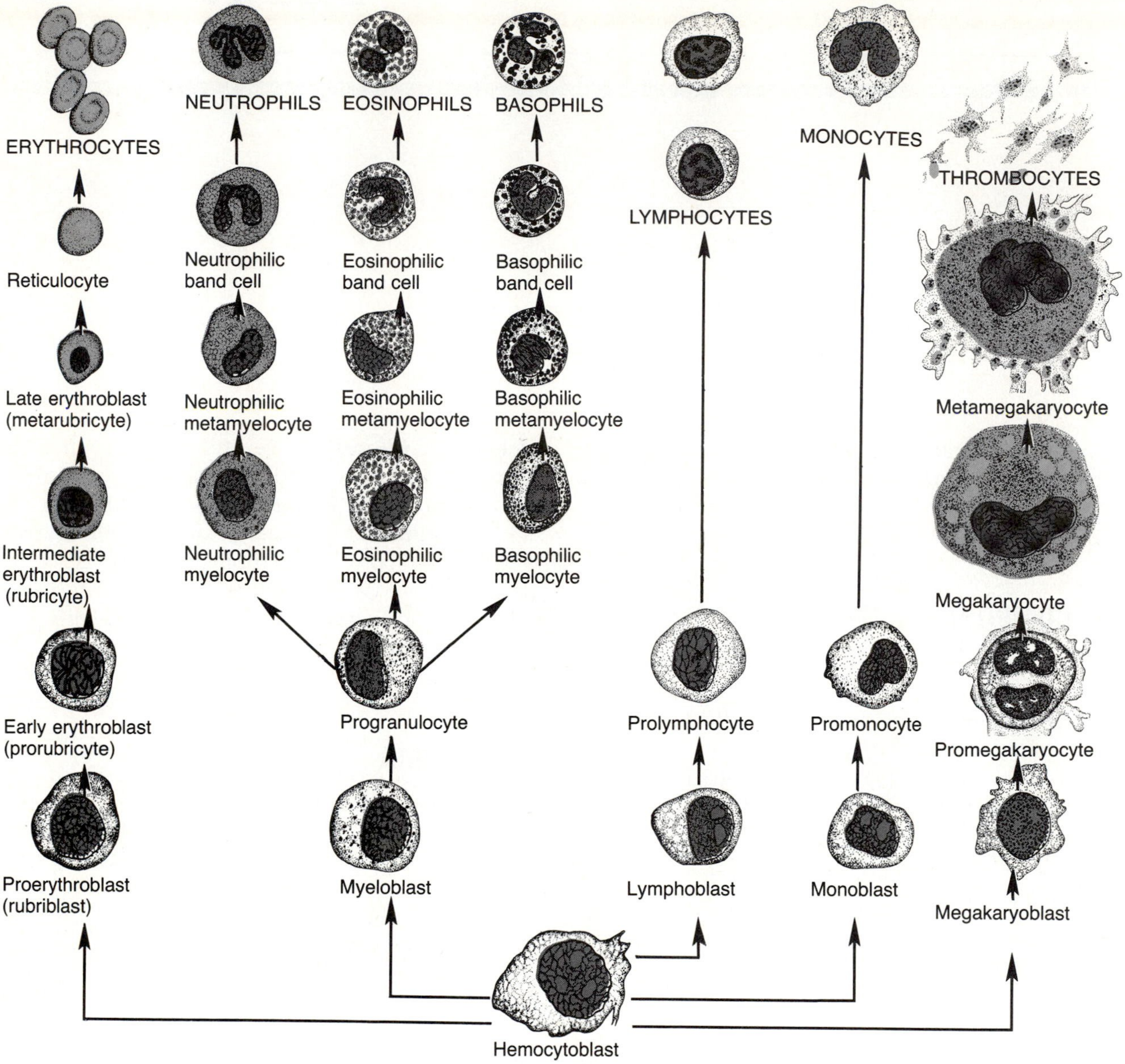

FIGURE 16.1 Origin, development, and structure of blood cells.

PROCEDURE

1. Obtain about 5 ml of plasma, either obtained from a clinical agency or by centrifuging whole blood from a mammal (other than a human).

2. If you are going to centrifuge the whole blood, treat the sample with calcium oxalate to prevent clotting. After centrifugation, the bottom of the test tube will contain calcium oxalate. Right above this will be a layer of red blood cells, followed by a layer of leucocytes and platelets (buffy coat). The top layer is plasma. Visual inspection of these layers will give an indication of the relative proportions of the various components of blood.

3. Using a graduated cylinder, place 5 ml of plasma in a test tube. Note its color. Test the pH of the plasma by dipping the end of a length of pH paper into the plasma and comparing the paper color with the color scale on the pH container. Record the pH in Section B.1 of the LABORATORY REPORT RESULTS at the end of the exercise.

4. Hold the test tube up to a source of natural light and note the color and transparency of

the plasma. Record the color and transparency in Section B.1 of the LABORATORY REPORT RESULTS at the end of the exercise.

5. *While still wearing rubber gloves,* place some plasma between your thumb and forefinger and press them together for a few seconds. Gently pull your fingers apart and note the consistency of the plasma. Record the consistency in Section B.1 of the LABORATORY REPORT RESULTS at the end of the exercise.

6. *Drop the test tube and plasma into a container of fresh bleach solution and dispose of the pH paper as per your instructor's direction.*

2. Chemical Constituents

CAUTION! *Please reread Section A. "General Safety Precautions and Procedures" on page xi and Section C. "Precautions Related to Working With Reagents" on page xii at the beginning of the laboratory manual before you begin any of the following experiments. You should also read the experiments before you perform them to be sure that you understand all the procedures and safety precautions.*

PROCEDURE

1. Add 10 ml of distilled water to 5 ml of plasma in a test tube and boil in a water bath.

CAUTION! *Make sure that the mouth of the test tube is pointed away from you and all other persons in the area.*

2. After the water has boiled, add two drops of 10% acetic acid to the contents of the test tube.

3. Using a test tube holder, remove the test tube from the water bath. Filter the mixture in the test tube and save the contents of the filter paper.

4. Test the filtrate for the presence of glucose by adding 2 ml of plasma and 2 ml of Benedict's solution to a test tube.

5. Place the test tube in a boiling water bath for 5 minutes.

CAUTION! *Make sure that the mouth of the test tube is pointed away from you and all other persons in the area.*

A green, yellow, orange, or red color indicates the presence of glucose. Record your results in Section B.2 of the LABORATORY REPORT RESULTS at the end of the exercise.

6. Test the contents of the filter paper for the presence of protein. If protein is present, it will be coagulated by the acetic acid and filtered out of the plasma. Place the contents of the filter paper in a test tube and add 3 ml of water and 3 ml of Biuret reagent. A violet or purple color indicates the presence of protein. Record your results in Section B.2 of the LABORATORY REPORT RESULTS at the end of the exercise.

C. ERYTHROCYTES

Erythrocytes (red blood cells, RBCs) are biconcave in appearance, have no nucleus, and can neither reproduce nor carry on extensive metabolic activities (Figure 16.1). The interior of the cell contains a red pigment called hemoglobin, which is responsible for the red color of blood. Erythrocytes function to combine with oxygen and, to a lesser extent, carbon dioxide and transport them through the blood vessels. A red blood cell becomes nonfunctional in about 120 days. A healthy male has about 5.4 million red blood cells per cubic millimeter (mm^3) of blood, a healthy female, about 4.8 million. Erythropoiesis and red cell destruction normally proceed at the same pace. A diagnostic test that informs the physician about the rate of erythropoiesis is the ***reticulocyte*** (re-TIK-yoo-lō-sīt) ***count***. The immature red blood cells called ***reticulocytes*** (Figure 16.1) appear in the blood in a certain concentration. An abnormally low or abnormally high concentration of reticulocytes indicates disease.

The different laboratory blood tests performed and evaluated by physicians are invaluable in their search for the causes of bodily malfunction, and are very useful tools in medicine. The following practical routine tests give us a clear insight on important characteristics of blood.

NOTE: Although many blood tests in this exercise are sufficient for laboratory demonstration, they are not necessarily clinically accurate.

D. RED BLOOD CELL TESTS

CAUTION! *Please reread Section A. "General Safety Precautions and Procedures" on page xi and Section B. "Precautions Related to Working With Blood, Blood Products, or Other Body Fluids" on page xi at the beginning of the laboratory manual before you begin any of the following experiments. You should also read the experiments before you perform them to be sure that you understand all the procedures and safety precautions.*

1. Finger Puncture Procedure

Blood for various tests is obtained by finger puncture, which is described shortly. The various tests involving blood require *several punctures* because samples must be taken at different times.

CAUTION! *In order to avoid each student providing blood samples for all the tests, each student should provide blood for only one or two tests from a single puncture. The results of the various tests can be tabulated on the blackboard and shared by the entire class.*

PROCEDURE

1. To enhance your chances for a successful finger puncture the first time, warm your finger if it is cold, shake your finger several times to dry the alcohol and force more blood into it, and gently milk blood toward the puncture site.
2. Obtain 70% alcohol, a sterile lancet, and cotton.
3. *Thoroughly* cleanse the end of the third or fourth finger with alcohol.
4. Open the lancet packet at the blunt end; *do not touch the sharp end.*
5. *Take a sample of YOUR OWN BLOOD as follows.* Remove the rubber glove from the hand from which you will take blood. Holding your finger firmly, take the lancet and puncture the finger using a quick, jabbing motion. The puncture is made toward the side of the finger rather than in the full fleshy part. (Do not begin the motion from a great distance away from the finger. The closer you are and the quicker the motion, the less the discomfort that results.)
6. If the puncture is made properly, the blood should flow freely from the wound. If not, squeeze the finger at its base and move toward the tip.
7. If an additional puncture is required, discard the first lancet and *use a new sterile one.*

CAUTION! *Make sure the used lancets are discarded immediately and properly in a container of fresh bleach solution to prevent subsequent injury, and to prevent passage of diseases such as hepatitis and AIDS from one person to another.*

8. Do not use the first drop, which usually coagulates fast.
9. Finally, place a piece of cotton between the finger and the thumb and press them together firmly.
10. *Use an adhesive bandage to cover the puncture site.*

2. Filling of Hemocytometer (Counting Chamber)

The procedure for filling the hemocytometer is as follows. Please read it carefully *before* doing 3. Red Blood Cell Count.

PROCEDURE

1. Obtain a hemocytometer and cover slip (Figure 16.2).
2. Clean the hemocytometer thoroughly and carefully with alcohol.
3. Place the cover slip on the hemocytometer.
4. Using a Unopette Reservoir System® pipette filled with the proper dilution of blood, place the tip of the pipette on the polished surface of the hemocytometer next to the edge of the cover slip. Although we recommend use of this system, your instructor might wish to use an alternate procedure. *If a different system is used for pipetting and diluting, use rubber suction bulbs or pipette pumps. Do not pipette by mouth.*
5. Deposit a small drop of diluted blood by squeezing the sides of the reservoir, but do not leave the tip of the pipette in contact with the hemocytometer for more than an instant because this will cause the chamber to overfill. The diluted blood must not overflow the moat; overfilling the moat results in an inaccurate cell count. A properly filled hemocytometer has a blood specimen only within the space between the cover glass and counting area.

3. Red Blood Cell Count

The purpose of a red blood cell count is to determine the number of circulating red blood

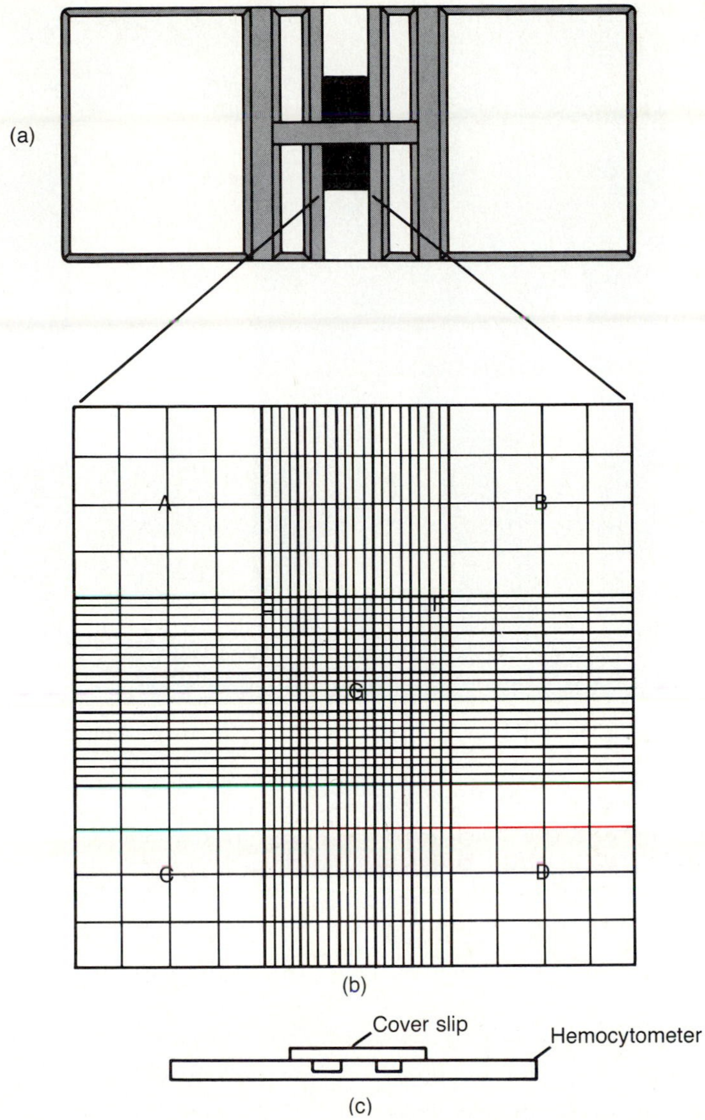

FIGURE 16.2 Various parts of a hemocytometer (counting chamber). (a) Top view. (b) Enlargement of grid on the glass surface of the hemocytometer. The areas used for red cell counts are labeled E, F, G, H, and I, and those for white cell counts are labeled A, B, C, D. (c) Side view with cover slip.

cells per cubic millimeter (mm^3) of blood. Red blood cells carry oxygen to all tissues; thus a drastic reduction in the red cell count will cause immediate reduction in available oxygen.

A decrease in red blood cells can result from a variety of conditions, including impaired cell production, increased cell destruction, and acute blood loss. When the red cell count is increased above normal limits, the condition is called ***polycythemia*** (pol′-ē-sī-THĒ-mē-a).

The procedure we will use for determining the number of red blood cells per cubic millimeter (mm^3) of blood is as follows:

PROCEDURE

1. Obtain a Unopette Reservoir System® for red blood cell determination (the color of the reservoir's bottom surface is red). Identify the reservoir chamber, diluent fluid inside the chamber, pipette, and protective shield for the pipette (Figure 16.3). The diluent fluid contains

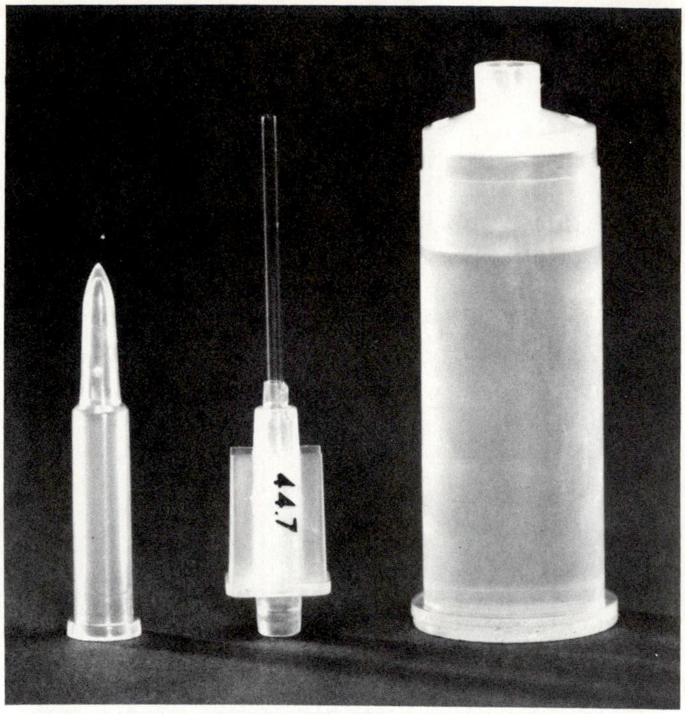

Pipette shield Pipette Reservoir chamber containing diluent fluid

FIGURE 16.3 Photograph of the Unopette Reservoir System®.

isotonic saline and sodium azide, an antibacterial agent.

CAUTION! *Do not ingest this fluid; it is for in vitro diagnostic use only.*

2. Hold the reservoir on a flat surface in one hand and grasp the pipette assembly in the other hand.

3. Push the tip of the pipette shield firmly through the diaphragm in the neck of the reservoir (Figure 16.4a).

4. Pull out the assembly unit and with a twist remove the protective shield from the pipette assembly.

5. Follow the finger puncture procedure outlined in D.1, and *immediately dispose of the lancet in a container of fresh bleach solution.*

6. Hold the pipette *almost* horizontally and touch the tip of it to the forming drop of blood (Figure 16.4b). The pipette will fill by capillary action and, when the blood reaches the end of the capillary bore in the neck of the pipette, the filling action will stop.

7. Carefully wipe any excess blood from the tip of the pipette using a Kimwipe or similar wiping tissue, *making certain that no sample is removed from the capillary bore.*

8. Squeeze the reservoir *slightly* to expel a small amount of air.

CAUTION! *Do not expel any liquid. If the reservoir is squeezed too hard, the specimen may be expelled through the overflow chamber, resulting in contamination of the fingers. Reagent contains sodium azide, which is extremely toxic and yields explosive products in metal sinks or pipes. Azide compounds should be diluted with running water before being discarded and disposed of per the directions of your instructor.*

While still maintaining pressure on the reservoir, cover the opening of the overflow chamber of the pipette with your index finger and push the pipette securely into the reservoir neck (Figure 16.4c).

9. Release the pressure on the reservoir and remove your finger from the pipette opening. This will draw the blood into the diluent fluid.

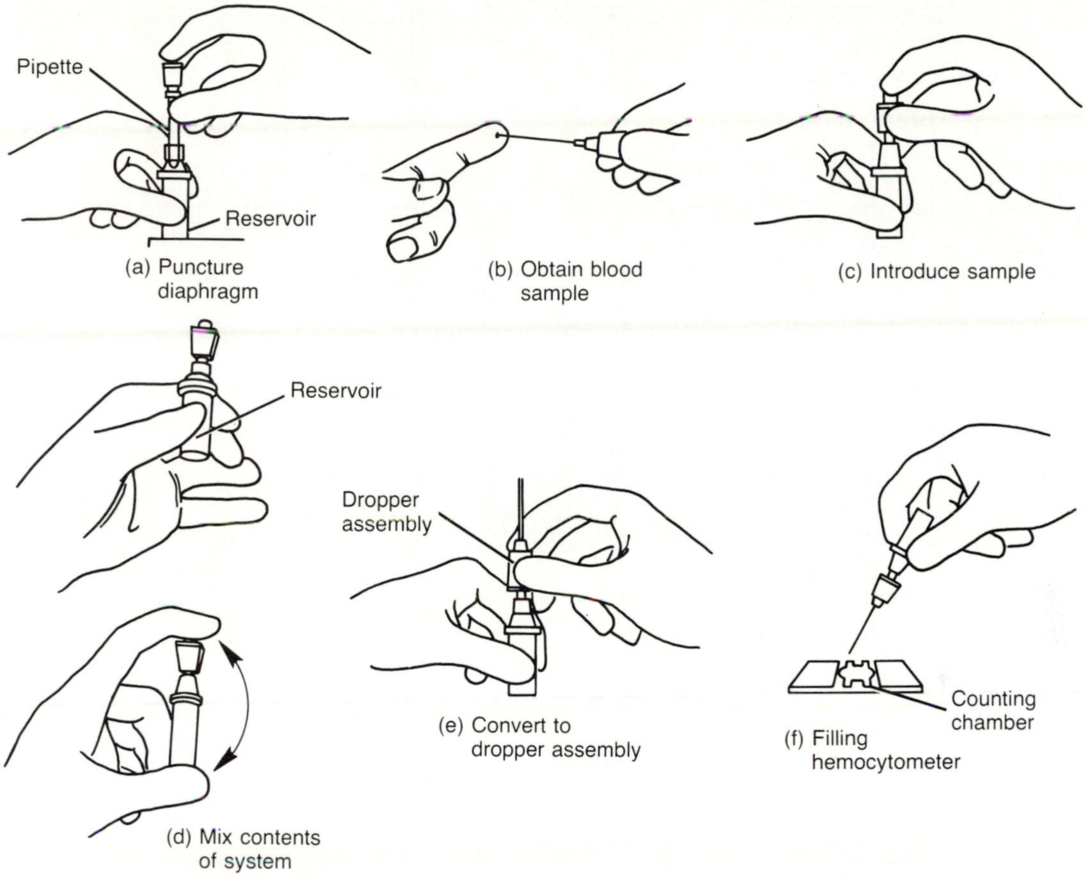

FIGURE 16.4 Preparation of the Unopette Reservoir System®. (Modified from *RBC Determination for Manual Methods* and *WBC Determination for Manual Methods*, Becton-Dickinson, Division of Becton, Dickinson and Company.)

10. Mix the contents of the reservoir chamber by squeezing the reservoir gently several times. It is important to *squeeze gently* so that the diluent fluid is not forced out of the chamber. In addition to squeezing the reservoir chamber, invert it *gently* several times (Figure 16.4d).

11. Remove the pipette from the reservoir chamber, reverse its position, and replace it on the reservoir chamber (Figure 16.4e). This converts the apparatus into a dropper assembly (Figure 16.4f).

12. Squeeze a few drops out of the reservoir chamber into a disposal container or wipe with Kimwipe or similar wiping tissue. You are now ready to fill the hemocytometer.

13. Follow the procedure outlined in D.2, filling the hemocytometer.

14. Place the hemocytometer on the microscope stage and allow the red blood cells to settle in the hemocytometer (approximately 1–2 minutes).

15. Using the high power lens, count the cells in each of the five squares (E, F, G, H, I) as shown in Figure 16.5. It is suggested that you use a hand counter. As you count, move up and down in a systematic manner. ***Note:*** To avoid overcounting cells at the boundaries, cells that touch the lines on the left and top sides of the hemocytometer should be counted, but not the ones that touch the boundary lines on the right and bottom sides.

16. Multiply the total number of red blood cells counted in the five squares by 10,000 to obtain the number of red blood cells in 1 mm^3 of blood. Record your value in Section D.1 of the LABORATORY REPORT RESULTS at the end of the exercise.

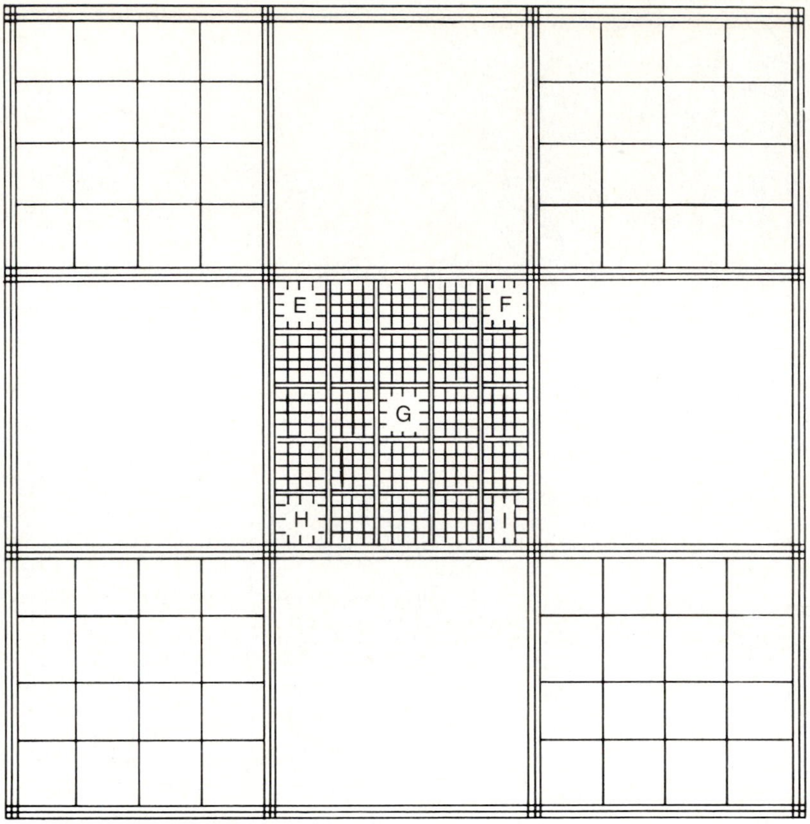

FIGURE 16.5 Hemocytometer (counting chamber). Areas E, F, G, H, and I are all counted in red blood cell count. The large square containing the smaller squares E, F, G, H, and I is seen in microscopic field under 10× magnification. To count the smaller squares (E, F, G, H, and I), the 45× lens is used and therefore E, F, G, H, and I each encompass the entire microscopic field.

17. CAUTION! *Place the hemocytometer and cover slip in a container of fresh bleach solution and dispose of all other materials as directed by your instructor.* The hemocytometer should be saved, not discarded.

4. Red Blood Cell Volume (Hematocrit)

The percentage of blood volume occupied by the red blood cells is called the ***hematocrit (Hct)*** or ***packed cell volume (PCV)***. Its measurement is a good routine test for anemia. When a tube of blood is centrifuged, the erythrocytes pack into the bottom part of the tube with the plasma on top. The white blood cells and platelets are found in a thin area, the buffy layer, above the red blood cells.

The Readacrit® centrifuge (Figure 16.6) incorporates a built-in hematocrit scale and tube-holding compartments which, when used with special precalibrated capillary tubes, permit direct reading of the hematocrit value by measuring the length of the packed red cell column. Readacrit® centrifuges present the final hematocrit value without requiring computation by the operator. If the Readacrit® centrifuge or tube reader is not used for direct reading, measure the total length of blood volume, divide this quantity into length of packed red cells, and multiply by 100 for percentage.

The calculation is as follows:

$$\frac{\text{length of packed red cells}}{\text{total length of blood volume}} \times 100 = \text{\% volume of whole blood occupied by red cells (hematocrit)}$$

In males the normal range is between 40% and 54%, with an average of 47%. In females the normal range is between 38% and 47%,

FIGURE 16.6 Centrifuge used for spinning blood to determine hematocrit.

with an average of 42%. Anemic blood may have a hematocrit of 15%; polycythemic blood may have a hematocrit of 65%.

The following procedure for testing red blood cell volume is a micromethod requiring only a drop of blood:

PROCEDURE

1. Produce a free flow of blood from the tip of your own finger by following the finger puncture procedure outlined in D.1.

CAUTION! *Immediately dispose of the lancet in a container of fresh bleach solution.*

2. Place the unmarked (clear) end of the capillary tube into the drop of blood (Figure 16.7). Hold the tube slightly below the level of the blood and do not move the tip from the blood or an air bubble will enter the column.

CAUTION! *If an air bubble is present, immediately dispose of the capillary tube in a fresh bleach solution and begin again.*

3. Allow blood to flow about two-thirds of the way into the tube. The tube will fill easily if you hold the open end below the blood source.

4. Place your finger over the marked (red) end of the tube and keep it in that position as you seal the blood end of the tube with Seal-ease® or clay or other similar sealing material.

5. According to the directions of your laboratory instructor, place the tube into the centrifuge, making sure that the sealed end is against the rubber ring at the circumference. Have your laboratory instructor make sure that the tubes are properly balanced. Write the number that indicates the placement of your tube. ____

6. When all the tubes from your laboratory section are in place, secure the inside cover of the centrifuge and close the outside cover. Centrifuge on high speed for four minutes.

7. Remove your tube and determine the hematocrit value by reading the length of packed red cells directly on the centrifuge scale, or by placing the tube in the tube reader (Figure 16.8) and following instructions on the reader, or by calculation using the equation mentioned previously (page 346). When using the tube reader, place the base of the Seal-ease® or clay on the base line and move tube until the top of the plasma is at the line marked 100.

CAUTION! *Immediately dispose of your capillary tube in a fresh bleach solution.*

Record your results in Section D.2 of the LABORATORY REPORT RESULTS at the end of the exercise.

What effect would dehydration have on hematocrit (Hct)? ______________________

____

5. Sedimentation Rate

If citrated blood is allowed to stand vertically in a tube, the red blood cells fall to the bottom of the tube and leave clear plasma above them. (Citrate binds calcium and thus prevents coagulation). The distance that the cells fall in 1 hour can be measured and is called the ***sedimentation rate***. It is a function of the amount of fibrinogen and gamma globulin present in plasma and the tendency of the red blood cells to adhere to one another. This rate is greater than normal during menstruation, pregnancy, malignancy, and most infections. Sedimentation rate may

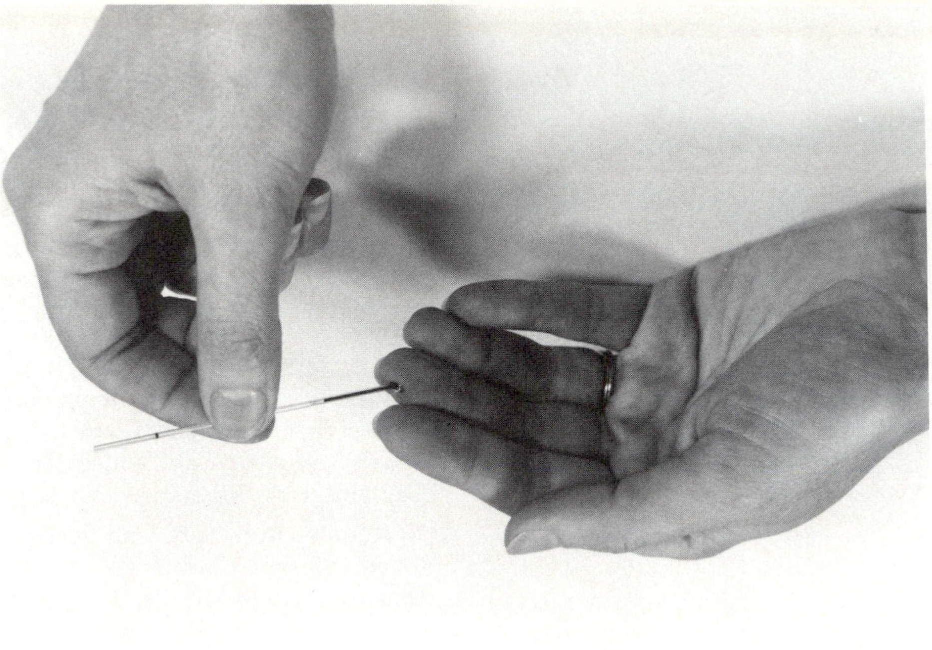

FIGURE 16.7 A heparinized capillary tube is filled with blood from freely flowing finger puncture. The tube is used for hematocrit determination.

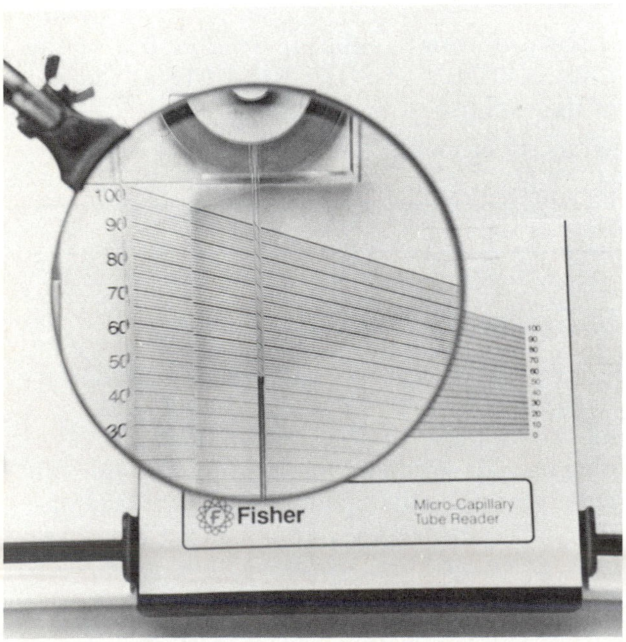

FIGURE 16.8 Microhematocrit tube reader.

be decreased in liver disease. A high rate may indicate tissue destruction in some part of the body; therefore, sedimentation rate is considered a valuable nonspecific diagnostic tool. The normal rate for adults is 0 to 6 mm per hour, for children 0 to 8 mm per hour.

The following method for determining sedimentation rate requires only one drop of blood and is called the Landau micromethod.

PROCEDURE

1. Obtain a drop of your own blood by following the finger puncture procedure outlined in D.1.

CAUTION! *Immediately dispose of the lancet in a container of fresh bleach solution.*

2. Using the mechanical suction device, draw the sodium citrate up to the first line that encircles the pipette.
3. Draw free-flowing blood up into the pipette to the second line. Do not remove the tip of the pipette from the blood or air bubbles will enter the column.

CAUTION! *If air bubbles are drawn into the blood, carefully expel the mixture onto a clean microscope slide, dispose of the slide in a container of fresh bleach solution, place the pipette into a container of fresh bleach solution, and begin again.*

4. Draw the mixture of fluids up into the bulb and mix by expelling it into the lumen of the tube. Draw and mix the fluids six times. If any air bubbles appear, use the procedure described in step 3.

5. Adjust the level of the blood as close to zero as possible. (Exactly zero is very difficult to get.)

6. Remove the suction device by placing the lower end of the pipette on the *gloved* index finger of the left hand before removing the device from the other end. The blood will leave the pipette if the end is not held closed.

7. Place the lower end of the pipette on the base of the pipette rack (Figure 16.9) and the opposite end at the top of the rack. The tube must be exactly perpendicular. Record the time at which it is put in the rack. Time ______ .

8. One hour later, measure the distance from the top of the plasma to the top of the red blood cell layer with an accurate millimeter scale, and record your results in Section D.3 of the LABORATORY REPORT RESULTS at the end of the exercise.

9. CAUTION! *Dispose of the pipette in the container of fresh bleach solution. Using a medicine dropper, rinse the base of the pipette rack with fresh bleach solution.*

6. Hemoglobin Determination

It is possible for a person to have anemia even if their red blood cell count is normal. For example, the red blood cells present may be deficient in hemoglobin or they may be smaller than normal. Thus, the amount of hemoglobin per unit volume of blood, and not necessarily the number of red blood cells, is the determining factor for anemia. The hemoglobin content of blood is expressed in grams (g) per 100 ml of blood and the normal ranges vary with the technique used. An average range is 12–15 g/100 ml in females and 13–16 g/100 ml in males.

One procedure that we will use for determining hemoglobin uses an apparatus called a ***hemoglobinometer*** (Figure 16.10a). This instrument compares the absorption of light by

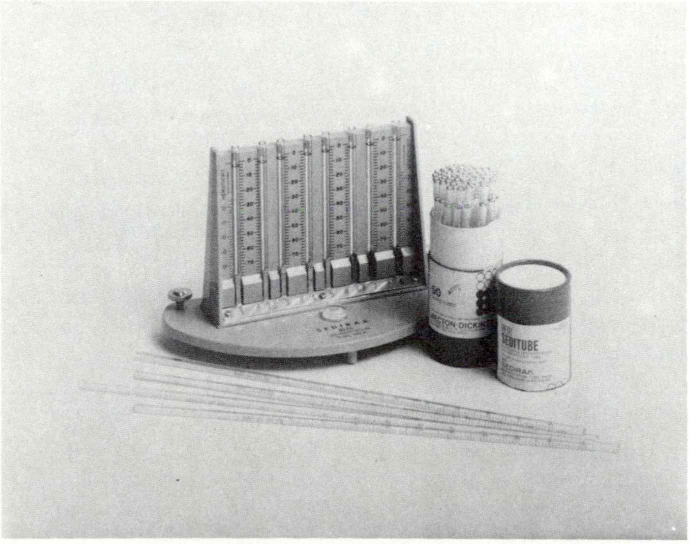

FIGURE 16.9 One type of sedimentation tube rack.

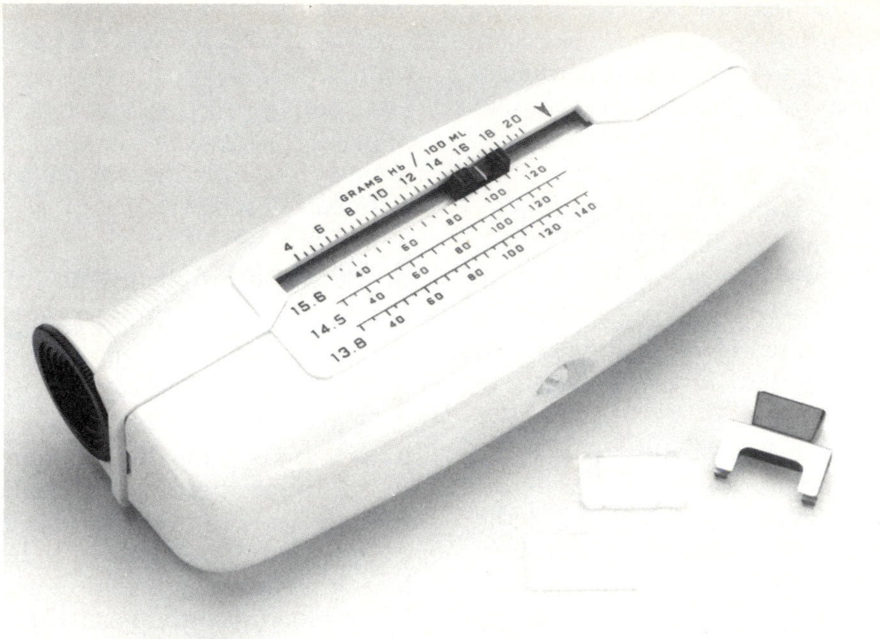

(a)

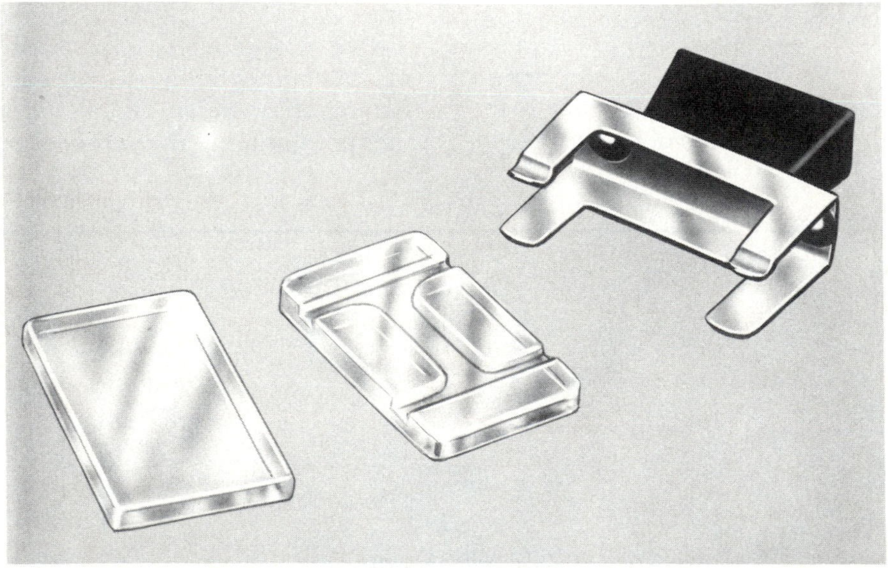

(b)

FIGURE 16.10 Hemoglobinometer. (a) Cambridge Instruments Hemoglobin-Meter. (b) Specimen chamber.

hemoglobin in a blood sample of known depth to that of a standardized glass plate.[1]

[1]If UNOPETTE tests are available, follow the procedure outlined in UNOPETTE test no. 5857, "Cyanmethemoglobin Determination for Manual Methods," for hemoglobin estimation.

PROCEDURE

1. Obtain a hemoglobinometer and examine its parts (Figure 16.10a). Note the four scales on the side of the instrument. The top scale gives hemoglobin content in g/100 ml of blood; the other three scales give comparative readings, that is, the amount of hemoglobin expressed

as a percent of 15.6, 14.5, or 13.8 g/100 ml of blood.

2. Remove the two pieces of glass from the blood chamber assembly (Figure 16.10b) and clean them with alcohol and wipe them with lens paper. The piece of glass that contains the moat will receive the blood sample; the other piece of glass serves as a cover glass.

3. Replace the glass pieces half-way onto the clip with the moat plate on the bottom.

4. Obtain a sample of your own blood via a finger puncture according to the procedure outlined in D.1.

CAUTION! *Dispose of the lancet immediately in the container of fresh bleach solution.*

5. Touch the finger to the moat and completely cover the raised surface of the plate with blood (Figure 16.11a).

6. Hemolyze the blood by agitating it with the tip of a hemolysis applicator for about 30 to 45 seconds. Hemolysis is complete when the appearance of the blood changes from cloudy to transparent. This change reflects the lysis of the red blood cell membranes and the liberation of hemoglobin.

7. Push the specimen chamber into the clip and push the clip into the hemocytometer (Figure 16.11b).

8. Holding the hemocytometer in your left hand, look through the eyepiece while depressing the light switch button (Figure 16.11c). You will see a split field.

9. Move the slide button back and forth with the right index finger until the two sides of the field match in color and shading.

10. Read the value indicated on the scale marked 15.6. Determine the grams of hemoglobin per 100 ml of the blood sample by reading the number above the index mark on the hemocytometer. Record your results in Section D.4 of the LABORATORY REPORT RESULTS at the end of the exercise.

11. CAUTION! Remove the glass pieces from the chamber and *wash in fresh bleach solution*. Dry the glass pieces with lens paper.

What type of anemia would reveal a normal or slightly below normal hematocrit, yet a significantly reduced hemoglobin level?

ALTERNATE PROCEDURE

The ***Tallquist measurement*** of hemoglobin is an old and somewhat inaccurate technique, although it is quick and inexpensive. The hemoglobinometer gives more accurate (±5%) results.

1. Obtain a sample of your own blood using the finger puncture technique outlined in D.1.

CAUTION! *Dispose of the used lancet immediately in the container of fresh bleach solution.*

2. Place one drop of blood on the Tallquist® paper.

3. As soon as the blood no longer appears shiny, match its color with the scale provided. Record your observations in Section D.4 of the LABORATORY REPORT RESULTS at the end of the exercise.

7. Oxyhemoglobin Saturation

The ability of blood to carry oxygen is determined by proper respiratory system function, hematocrit (Hct), and the ability of hemoglobin

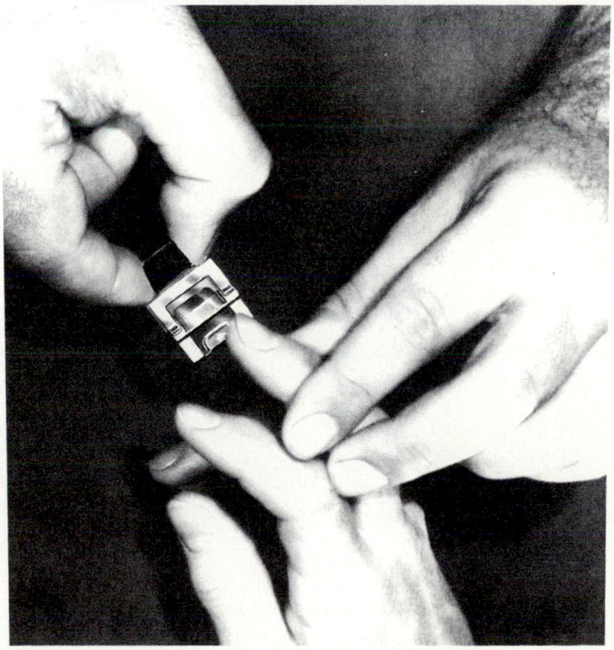

(a)

FIGURE 16.11 Use of the Hemoglobin-Meter. (a) Obtaining a specimen.

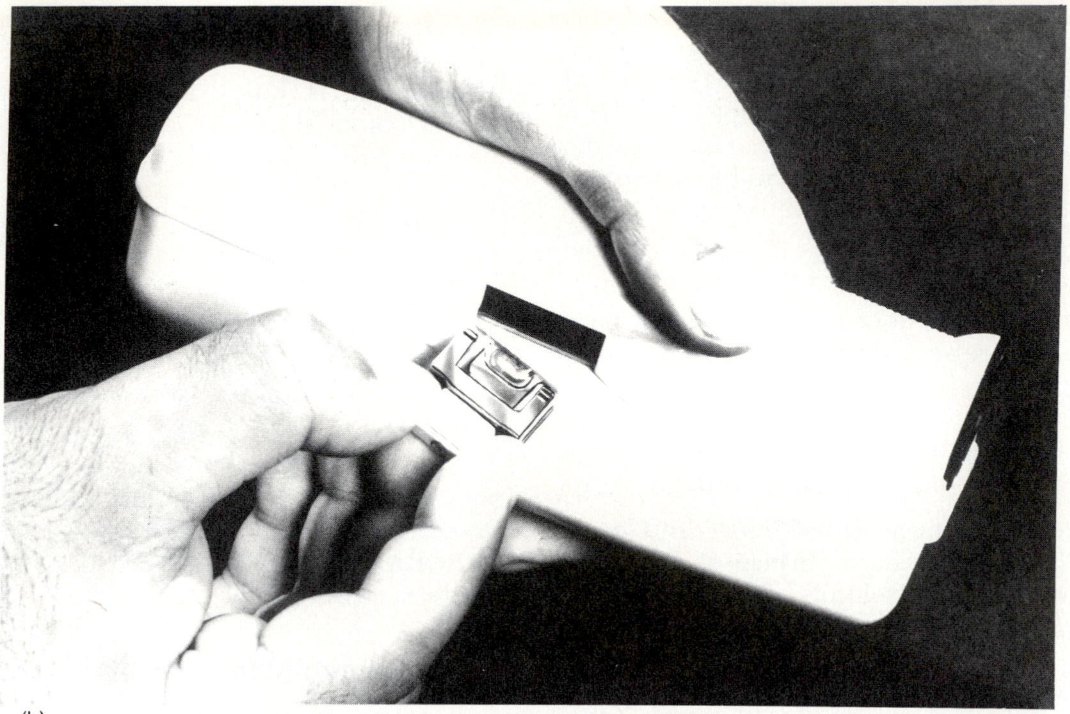

(b)

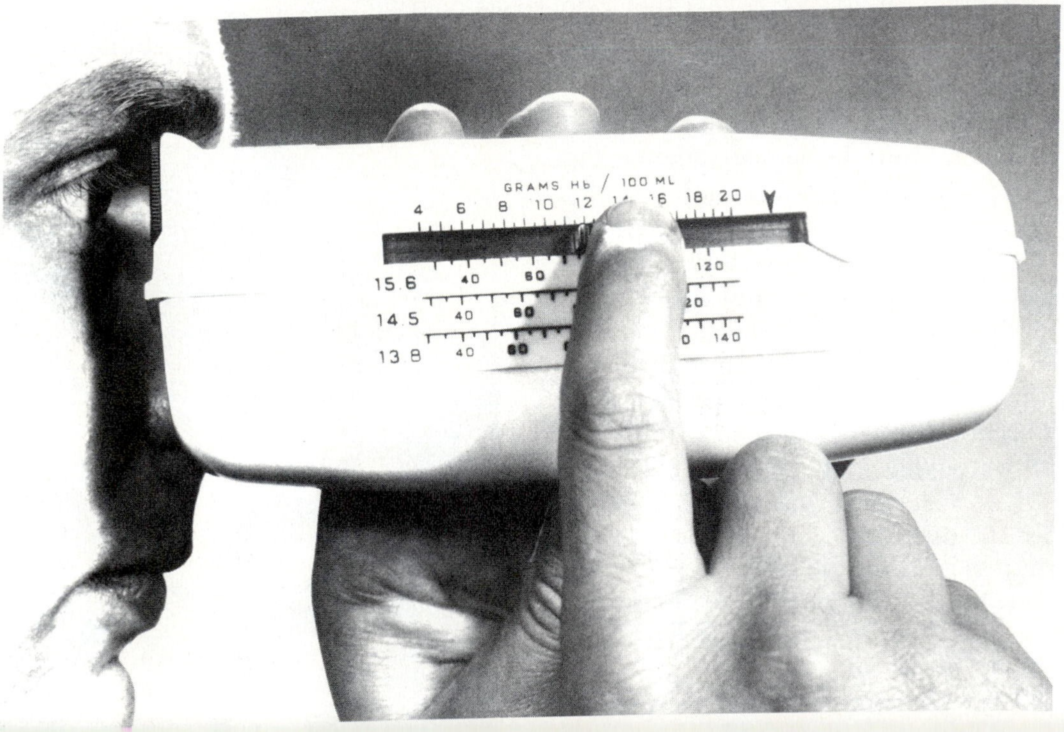

(c)

FIGURE 16.11 ***(Continued)*** (b) Inserting the sample. (c) Examining the specimen.

to bind reversibly with oxygen within the alveoli of the lungs and release oxygen at the tissue level. Hemoglobin's interaction with oxygen is measured by an ***oxygen-hemoglobin dissociation curve***. Such a curve demonstrates that the number of available oxygen-binding sites on hemoglobin that actually bind to oxygen is proportional to the pressure of oxygen ($P0_2$). Arterial blood usually has a $P0_2$ of 95, causing 97% of the available hemoglobin to be saturated with oxygen. Venous blood, however, has a $P0_2$ of 40 so only 75% of the hemoglobin is satuated with oxygen; 25% is given off to tissues. The determination of percent saturation for hemoglobin is a very sensitive test of pulmonary function. However, such a test needs to be interpreted carefully, as an individual with a normally functioning pulmonary system may exhibit abnormally low oxyhemoglobin saturation due to a variety of possible causes, such as ***methemoglobinemia*** (transformation of normal oxyhemoglobin due to the reduction of normal Fe^{2+} to Fe^{3+} within the hemoglobin molecule) or ***carbon monoxide (CO) poisoning***. In order to determine the percent hemoglobin saturation in an unknown sample of blood one would need to compare the absorption spectrum of the blood sample to that obtained from a pure sample of ***oxyhemoglobin (HbO_2), carboxyhemoglobin (Hb-CO)***, and ***reduced hemoglobin (Hb)*** (carboxyhemoglobin is included due to the normal presence of carbon monoxide in today's polluted atmosphere). This is shown in Figure 16.12. The spectrum obtained with the unknown sample of blood would be some combination of these three, since it would contain a certain percentage of each form of hemoglobin. Such a test is possible due to the fact that all three of these types of hemoglobin compounds are different colors, and would therefore absorb different portions of the light spectrum. Through a complex determination of the relative contribution of each type of hemoglobin to the resulting spectrum for the unknown sample, the percentage of each hemoglobin form may be determined.

8. Spectrum of Oxyhemoglobin and Reduced Hemoglobin

The following procedure details a simplified, although less accurate method for determining the absorption spectra for fully oxygenated and deoxygenated hemoglobin. Exposing blood to air saturates it with oxygen, while sodium dithionate reduces it by removing oxygen.

PROCEDURE

1. Turn on the spectrophotometer and let it warm up for several minutes.
2. After the instrument has warmed up, set the wavelength at 500 nm and adjust the meter needle to 0% transmittance (absorbance at infinity) by using the zero control knob.
3. Place a cuvette containing distilled water into the cuvette holder and adjust the needle to 100% transmittance by using the light control knob.
4. Place 8 ml of distilled water in a clean test tube.
5. Produce a free flow of blood from your own finger by following the finger puncture procedure outlined in D.1.

CAUTION! *Dispose of the lancet immediately in a fresh bleach solution.*

6. Mix the distilled water with the blood by placing a stopper in the mouth of the test tube and then inverting the test tube, thereby adding the blood to the distilled water.
7. Transfer half of the test tube contents (4 ml) into a second clean test tube.
8. Add 0.2 ml of 1% sodium dithionite solution to the second test tube and mix thoroughly. (***Note:*** In order for the procedure to work properly, the sodium dithionite solution must be fresh and made up just prior to use, and a spectral determination must be completed within 5 minutes of addition of sodium dithionite to the second tube.)
9. Record the absorbances of solutions 1 (oxyhemoglobin) and 2 (reduced hemoglobin) at 500 nm wavelength.
10. Standardize the spectrophotometer at 510 nm utilizing the cuvette containing only distilled water as outlined in steps 2 and 3 above.
11. Determine the absorbances for solutions 1 and 2 at 510 nm.
12. Repeat procedures 2, 3, and 9 at each of the following wavelengths: 520, 530, 540, 560, 570, 580, 590, 600 nm.
13. Record and graph your results in Section D.5 of the LABORATORY REPORT RESULTS

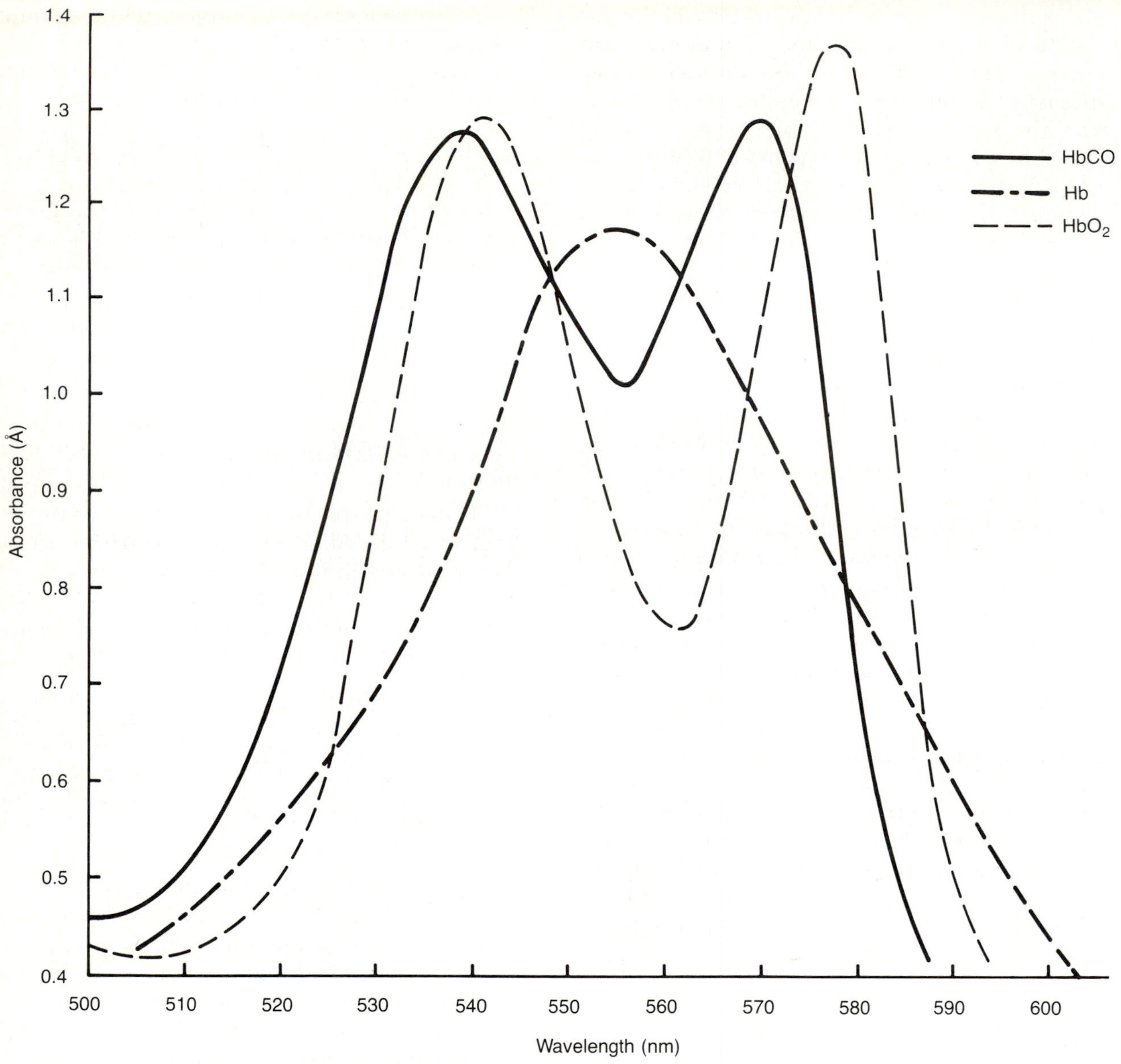

FIGURE 16.12 Absorption spectra of carboxyhemoglobin (HbCO), reduced hemoglobin (Hb), and oxygemoglobin (HbO_2).

at the end of the exercise. Be sure to compare and give physiological reasons for the observed differences in the two spectra.

E. LEUCOCYTES

Leucocytes (white blood cells, WBCs) are different from red blood cells in that they have nuclei and do not contain hemoglobin (see Figure 16.1). They are less numerous than red blood cells, averaging from 5000 to 9000 cells per cubic millimeter (mm^3) of blood. The ratio, therefore, of red blood cells to white blood cells is about 700:1.

As Figure 16.1 shows, leucocytes can be differentiated by their appearance. They are divided into two major groups, granular leucocytes and agranular leucocytes. ***Granular leucocytes***, which are formed from red bone marrow, have large granules in the cytoplasm that can be seen under a microscope and possess lobed nuclei. The three types of granular leucocytes are ***neutrophils, eosinophils,*** and ***basophils***. ***Agranular leucocytes,*** which are formed from lymphoid and myeloid tissue, contain small granules that cannot be seen under a light microscope, and usually have spherical nuclei. The two types of agranular leucocytes are ***lymphocytes*** and ***monocytes***.

Leucocytes as a group function in phagocytosis, producing antibodies, and combating allergies. The life span of a leucocyte usually ranges from a few hours to a few months. Some lymphocytes, called T and B memory cells, can live throughout one's life once they are formed.

F. WHITE BLOOD CELL TESTS

CAUTION! *Please reread Section A. "General Safety Precautions and Procedures" on page xi and Section B. "Precautions Related to Working With Blood, Blood Products, or Other Body Fluids" on page xi at the beginning of the laboratory manual before you begin any of the following experiments. You should also read the experiments before you perform them to be sure that you understand all the procedures and safety precautions.*

1. White Blood Cell Count

This procedure determines the number of circulating white blood cells in the body. Because white blood cells are a vital part of the body's immune defense system, any abnormalities in the white cell count must be carefully noted.

An increase in number ***(leucocytosis)*** may result from such conditions as bacterial or viral infection, metabolic disorders, chemical and drug poisoning, and acute hemorrhage. A decrease in number ***(leucopenia)*** may result from typhoid infection, measles, infectious hepatitis, tuberculosis, or cirrhosis of the liver.

The procedure for determining the number of white blood cells per mm^3 of blood is as follows:

PROCEDURE

1. Follow steps 1–14 of D.3, Red Blood Cell Count, but use the Unopette Reservoir System® for white blood cell determination (the color of the reservoir's bottom surface is white).

2. Using the 10× lens, count the cells in each of the four corner squares (A, B, C, D in Figure 16.13) of the hemocytometer. The direction to follow when counting white blood cells is indicated in square A in Figure 16.11. ***Note:*** To avoid overcounting of cells at the boundaries, the cells that touch the lines on the left and top sides of the hemocytometer should be counted, but not the ones that touch the boundary lines on the right and bottom sides. It is suggested that you use a hand counter.

3. Multiply the results by 50 to obtain the amount of circulating white blood cells per mm^3 of blood and record your results in Section F.6 of the LABORATORY REPORT RESULTS at the end of the exercise.

4. The factor of 50 is the dilution factor, or $2.5 \times 20 = 50$. The volume correction factor of 2.5 is arrived at in this manner: each of the corner areas (A, B, C, and D) is exactly 1 mm^3 by 0.1 mm deep (Figure 16.13). Therefore, the volume of each of these corner areas is 0.1 mm^3. Because four of them are counted, the total volume of diluted blood examined is 0.4 mm^3. However, because we want to know the number of cells in 1 mm^3 instead of 0.4 mm^3, we must multiply our count by 2.5 ($0.4 \times 2.5 = 1.0$).

5. CAUTION! *Place the hemocytometer and cover slip in a container of fresh bleach solution and dispose of all other materials as directed by your instructor*. The hemocytometer should be saved, not discarded.

2. Differential White Cell Count

The purpose of a differential white cell count is to determine the *percentages* of each of the five normally circulating types of white cells in a total count of 100 white blood cells. A normal differential count might appear as follows:

	Normal
Neutrophils	60% to 70%
Eosinophils	2% to 4%
Basophils	0.5% to 1%
Lymphocytes	20% to 25%
Monocytes	3% to 8%

Significant elevations of different types of white blood cells usually indicate specific pathological conditions. For example, a high neutrophil count indicates response

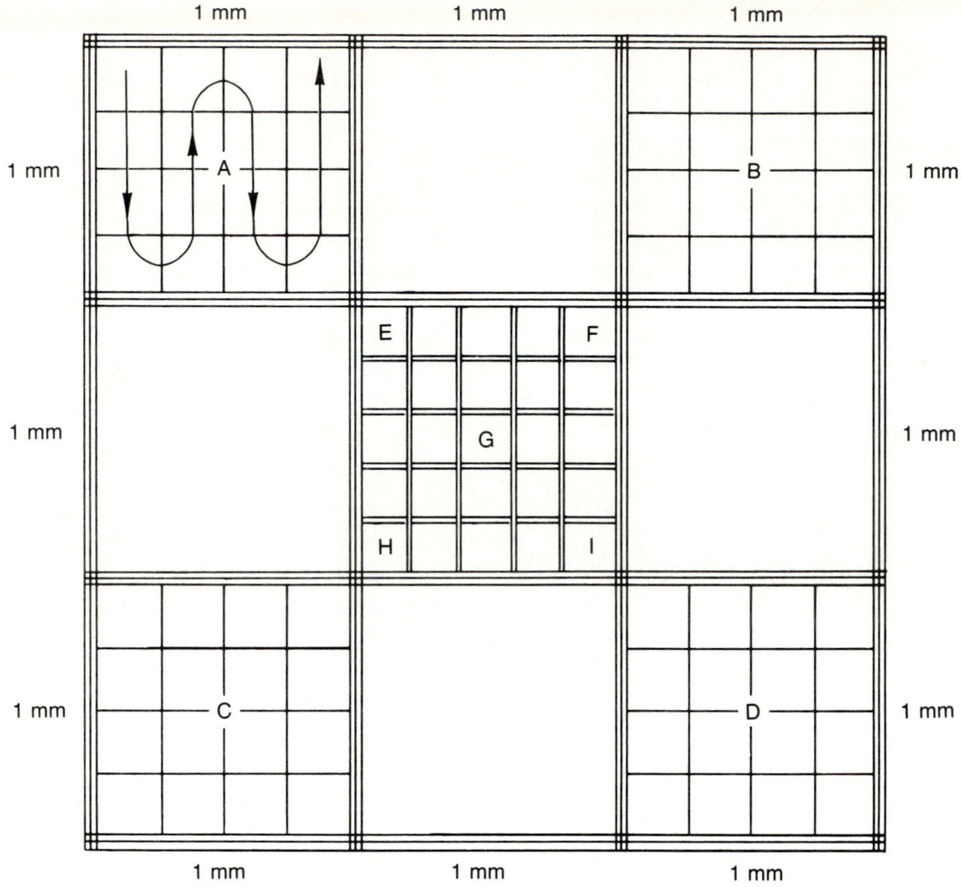

FIGURE 16.13 Hemocytometer (counting chamber) for white blood cell count (as seen though scanning lens). A, B, C, and D are areas counted in white blood cell count (when viewed through 10× lens, 1 sq mm is seen in microscopic field). Areas A, B, C, and D each equal 1 sq mm; therefore, a total of 4 sq mm is counted in white cell count.

to tissue destruction by invading bacteria. Lymphocytes predominate in antigen-antibody reactions, specific leukemias, and in infectious mononucleosis. An increase in eosinophils may be seen in allergic reactions. An elevated percentage of monocytes may result from chronic infections. An increase in basophils is rare and denotes a specific type of leukemia and allergic reactions.

The procedure for making a differential white blood cell count is as follows:

PROCEDURE

1. Obtain a drop of your own blood via a finger puncture according to the procedure outlined in D.1 and place one drop of blood on a glass slide.

CAUTION! *Immediately dispose of the lancet in a container of fresh bleach solution.*

2. Use a second slide as a spreader (Figure 16.14).
3. Draw the spreader toward the drop of blood (in this direction →) until it touches the drop. The blood should fan out to the edges of the spreader slide.
4. Keeping the spreader at a 25° angle, press the edge of the spreader firmly against the slide and push the spreader rapidly over the entire length of the slide (in this direction ←). The drop of blood will thin out toward the end of the slide.
5. Let the smear dry.
6. To stain the slide follow this procedure:

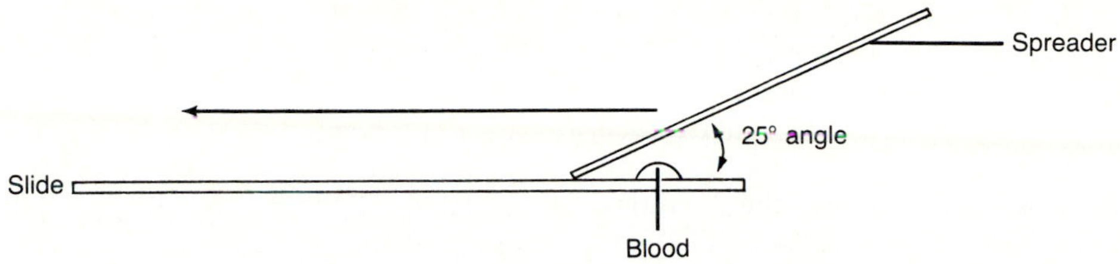

FIGURE 16.14 Procedure to prepare a blood smear.

a. Place the slide on staining rack.

b. Cover the entire slide with Wright's stain.

c. Let stain stand for 1 minute.

d. Add 25 drops of buffer solution and mix completely with Wright's stain.

e. Let mixture stand for 8 minutes.

f. Wash mixture off completely with distilled water.

7. Let the slide dry completely before counting.

8. To proceed with counting the cells, use an area of slide where blood is thinnest. Count cells under an oil-immersion lens. Count a total of 100 white cells.

9. The actual counting may be done by moving the slide either up and down or from side to side as follows:

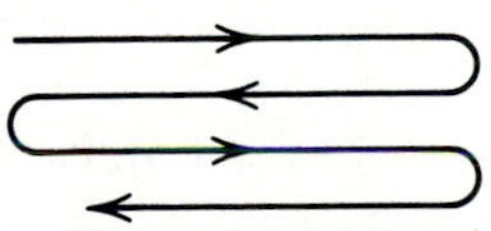

10. Record each white blood cell you observe by making a mark in the chart below until you have recorded 100 cells.

ALTERNATE PROCEDURE

1. Obtain a prepared slide of stained blood.

2. Repeat steps 9 and 10 in the preceding procedure.

G. THROMBOCYTES

Thrombocytes (platelets) are formed from fragments of the cytoplasm of megakaryocytes (see Figure 16.1). The fragments become enclosed in pieces of cell membrane from the megakaryocytes and develop into platelets. Platelets are very small, disc-shaped cell fragments without nuclei. Between 250,000 and 400,000 are found in each cubic millimeter of blood. They function to prevent fluid loss by starting a chain of reactions that results in blood clotting. They have a short life span, probably only one week, because they are expended in clotting and are just too simple to carry on extensive metabolic activity.

Examine a prepared slide of a stained smear of blood cells using the oil immersion objective. With the aid of your textbook, identify red

Type of leucocyte	**Number observed**	**Percentage**
Neutrophils		
Eosinophils		
Basophils		
Lymphocytes		
Monocytes		

blood cells, neutrophils, basophils, eosinophils, lymphocytes, monocytes, and thrombocytes. Using colored crayons or pencil crayons, draw and label all of the different blood cells in Section F.7 of the LABORATORY REPORT RESULTS at the end of the exercise. Be very accurate in drawing and coloring the granules and the nuclear shapes, because both of these are used to identify the various types of cells.

H. CLOTTING TESTS

Blood normally remains in a liquid state in the vessels but thickens and forms a jelly, or gel, when drawn from the body. The gel eventually separates from the liquid portion and produces the ***clot***, which consists of insoluble fibers in which the formed elements of the blood are trapped. The straw-colored liquid portion, called ***serum***, is plasma without its clotting proteins.

The process of clotting is called ***coagulation***. Its function is to prevent blood loss when a blood vessel is ruptured. Clotting is a complex process requiring many coagulation factors and pathways, but basically occurs in three stages. In stage 1, prothrombin activator forms. In stage 2, prothrombin, a plasma protein, is converted into thrombin, an enzyme. In stage 3, thrombin catalyzes the conversion of fibrinogen, another plasma protein, into fibrin.

The time required for blood to coagulate when exposed to air, usually from 2 to 6 minutes, is known as ***clotting time***. This time is used as an index of a person's blood-clotting properties.

1. Clotting Time (Slide Method)

The procedure for testing clotting time by the slide method is as follows:

PROCEDURE

1. Perform a finger puncture to obtain a sample of your own blood, following the procedure outlined in D.1.

CAUTION! *Immediately dispose of the lancet in a container of fresh bleach solution.*

2. Discard the first drop of blood since it will clot faster than the second drop.
3. Begin timing at the appearance of the second drop of blood.
4. Put a glass slide under several thicknesses of paper toweling and place 3–4 drops of blood on the slide.
5. At four minutes and at 60 second intervals after that time, *slowly* pass an applicator stick or dissecting pin through the blood and raise it *slowly* about 5 mm above the blood until a fibrin strand appears at the end of the stick or needle or dissecting pin. This represents coagulation.
6. Stop the watch when the strand appears.
7. Record your results in Section H of the LABORATOY REPORT RESULTS at the end of the exercise.
8. CAUTION! *Promptly dispose of the applicator stick or dissecting pin in the container of fresh bleach. Also, dispose of the paper towels in the plastic bag provided.*

I. BLOOD GROUPING (TYPING)

The plasma membranes of red blood cells contain genetically determined antigens called ***agglutinogens*** (ag′-loo-TIN-ō-jens). The plasma of blood contains genetically determined antibodies called ***agglutinins*** (a-GLOO-ti-nins). The antibodies cause the ***agglutination (clumping)*** of the red blood cells carrying the corresponding antigen.

These proteins, agglutinogens and agglutinins, are responsible for the two major classifications of blood groups: the ABO group and the Rh system. In addition to the ABO group and the Rh system, other human blood groups include: MNSs, P, Lutheran, Kell, Lewis, Duffy, Kidd, Diego, and Sutter. Fortunately these different antigenic factors do not exhibit extreme degrees of antigenicity and, therefore, usually cause very weak transfusion reactions or no reaction at all.

1. ABO Group

This major blood grouping is based on two agglutinogens symbolized as *A* and *B* (Figure 16.15). Individuals whose red blood cells produce only agglutinogen *A* have blood type A. Individuals who produce only agglutinogen *B* have blood type B. If both *A* and *B* agglutinogens are produced, the result is type AB, whereas the absence of both *A* and *B* agglutinogens results in the so-called type O.

The agglutinins in blood plasma are antibody *a* (anti-A), which attacks agglutinogen *A*, and antibody *b* (anti-B), which attacks agglutinogen *B*. The agglutinogens and agglutinins formed by each of the four blood types and the various agglutination reactions that occur when whole blood samples are mixed with serum are shown in Figure 16.15. You do not have agglutinins that attack the agglutinogens of your own erythrocytes. For example, a type A person has agglutinogen *A* but not agglutinin *a* (anti-A).

Agglutinogens and agglutinins are of critical importance in blood transfusions. In an incompatible blood transfusion, the donated erythrocytes are attacked by the recipient agglutinins, causing the blood cells to agglutinate. Agglutinated cells become lodged in small capillaries throughout the body and, over a period of hours, the cells swell, rupture, and release hemoglobin into the blood. Such a reaction as it relates to erythrocytes is called ***hemolysis*** (*lysis* = dissolve). The degree of agglutination depends on the titer (strength or amount) of agglutinin in the blood. Agglutinated cells can block blood vessels and may lead to kidney or brain damage and death, and the liberated hemoglobin may also cause kidney damage.

Because cells of type O blood contain neither of the two antigens (*A* or *B*), moderate amounts of this blood can be transfused into a recipient of any ABO blood type without

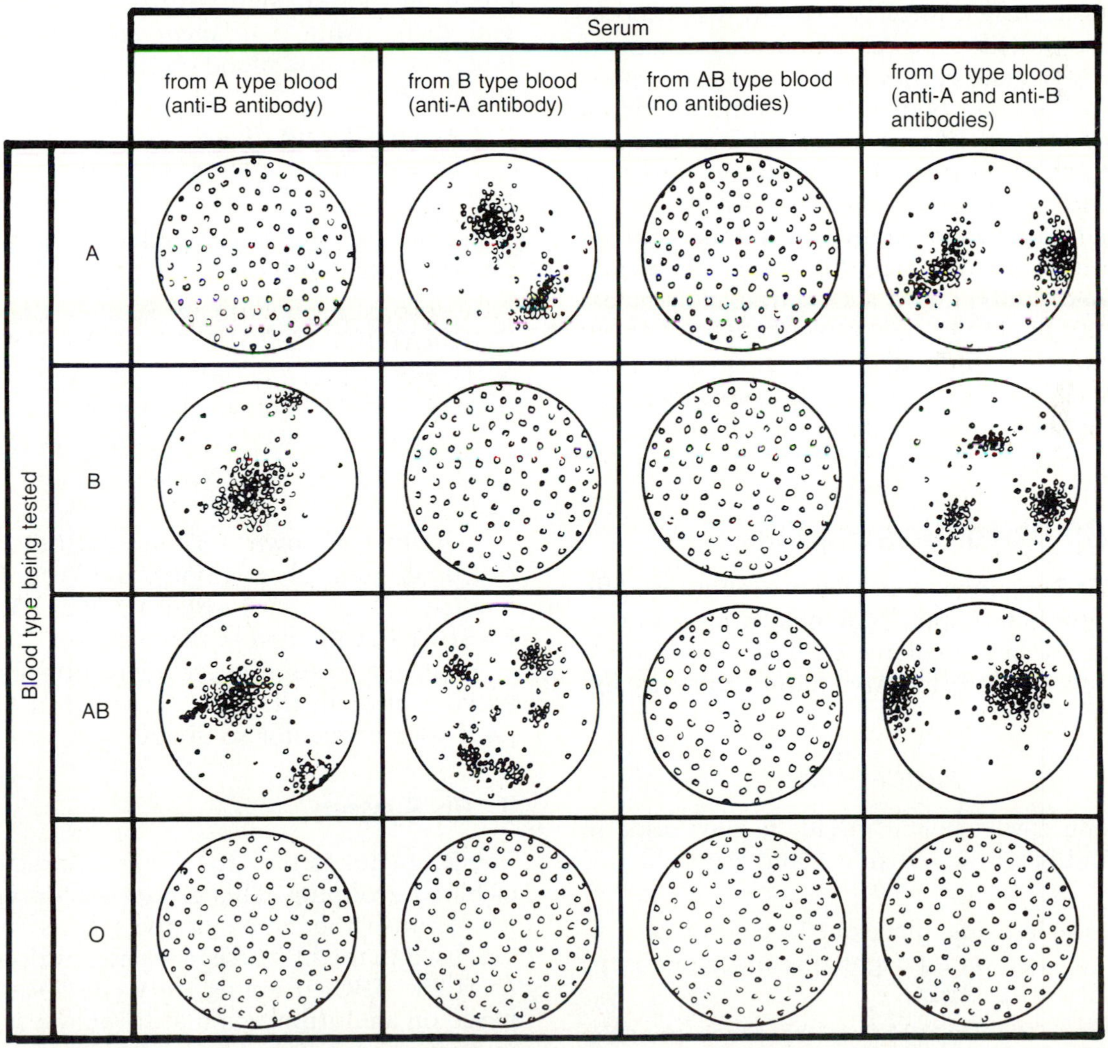

FIGURE 16.15 Chart to show agglutination reactions.

TABLE 16.1
SUMMARY OF ABO SYSTEM INTERACTIONS

Blood type	Agglutinogen (antigen)	Agglutinin (antibody)	Plasma causes agglutination of	Cells agglutinate by plasma of
A	*A*	*b*	B, AB	B, O
B	*B*	*a*	A, AB	A, O
AB	*A, B*	Neither *A* nor *B*	—	A, B, O
O	Neither *A* nor *B*	*a, b*	A, B, AB	—

immediate agglutination. For this reason, type O blood is referred to as the ***universal donor***. However, transfusing large amounts of type O blood into a recipient of Type A, B, or AB can cause *delayed* agglutination of the recipient's erythrocytes, because the transfused antibodies (type O blood contains *a* and *b* antibodies) are not sufficiently diluted to prevent the reaction.

Persons with type AB blood are sometimes called ***universal recipients*** because their blood contains no antibodies to agglutinate donor erythrocytes. *Small quantities* of blood from all other ABO types can be transfused into type AB blood without adverse reaction. However, again, if *large quantities* of type A, B, or O blood are transfused into type AB, the antibodies in the donor blood might accumulate in sufficient quantity to clump the recipient's RBCs.

Table 16.1 summarizes the various interreactions of the four blood types. Table 16.2 lists the incidence of these different blood types in the United States, comparing some races.

2. ABO Blood Grouping Test

NOTE: Remember that this procedure is sufficient for laboratory demonstration, but it should not be considered clinically accurate.

The procedure for ABO sampling is as follows:

PROCEDURE

1. Using a wax pencil, divide a glass slide in half and label the left side A and the right side B.
2. On the left side, place one large drop of anti-A serum, and on the right side place one large drop of anti-B serum.
3. Next to the drops of antisera, place one drop of your own blood obtained by finger puncture according to the procedure outlined in D.1, being careful not to mix samples on the left and right sides.

CAUTION! *Immediately dispose of the lancet in a container of fresh bleach solution.*

4. Using a mixing stick or a toothpick, mix the blood on the left side with the anti-A serum, and then, using a *different stick or different toothpick*, mix the blood on the right side with the anti-B.
5. CAUTION! *Immediately dispose of the stick or toothpick in a container of fresh bleach solution.*
6. *Gently* tilt the slide back and forth and observe it for 1 minute.
7. Record your results in Section I.1 of the LABORATORY REPORT RESULTS at the end of the exercise, using "+" for clumping (agglutination) and "−" for no clumping.
8. CAUTION! *Dispose of your slide in the container of fresh bleach solution.*

Identify your own ABO system blood type based on your observations and record it in Section I.2 of the LABORATORY REPORT RESULTS at the end of the exercise.

You can also summarize the results obtained by your class in the table provided in Section I.4 if your instructor so directs.

3. Rh System

The Rh factor is the other major classification of blood grouping. This group was designated Rh because the blood of the rhesus monkey was used in the first research and development. As is the ABO grouping, this classification is based on agglutinogens that lie on the surfaces of erythrocytes. The designation Rh^+ (positive) is given to those that have the agglutinogen, and Rh^- (negative) is for those that lack the

TABLE 16.2
INCIDENCE OF HUMAN BLOOD GROUPS IN THE UNITED STATES

	Blood groups (percentages)				
	O	A	B	AB	Rh^+
Caucasians	45	41	10	4	85
Blacks	48	27	21	4	88
Japanese	31	38	22	9	about 100
Chinese	36	28	23	13	about 100
American Indians	23	76	0	1	about 100
Hawaiians	37	61	1.5	0.5	about 100

agglutinogen. The estimation is that 85% of whites and 88% of blacks in the United States are Rh^+, whereas 15% of whites and 12% of blacks are Rh^-. (See Table 16.2).

Rh factor is extremely important in pregnancy and childbirth. Under normal circumstances, human plasma does not contain anti-Rh antibodies. If however, a woman who is Rh^- becomes pregnant with an Rh^+ child, her blood may produce antibodies that will react with the blood of a subsequent child.[2] The first child is unaffected because the mother's body has not yet produced these antibodies. This is a serious reaction and hemolysis may occur in the fetal blood.

The hemolysis produced by this fetal-maternal incompatibility is called ***hemolytic disease of newborn*** or ***HND (erythroblastosis fetalis)*** and could be fatal for the newborn infant. A drug called Rho-GAM, given to RH^- mothers immediately after delivery or abortion, prevents the production of antibodies by the mother so that the fetus of the next pregnancy is protected.

4. Rh Blood Grouping Test

NOTE: Remember that this procedure is sufficient for laboratory demonstration, but it should not be considered clinically accurate, since the anti-Rh (anti-D) serum deteriorates rapidly at room temperature and the anti-Rh agglutinins are far less potent in their agglutinizing capability than the anti-A or anti-B antibodies. This test is less accurate than the ABO determination.

The procedure for Rh grouping is as follows:

PROCEDURE

1. Place one large drop of Anti-Rh (anti-D) serum on a glass slide.[3]
2. Following the finger puncture procedure outlined in D.1, add one drop of your own blood and mix using a mixing stick or toothpick.

CAUTION! *Immediately dispose of the lancet and stick or toothpick in a container of fresh bleach solution.*

3. Place the slide on a preheated warming box, and gently rock the box back and forth for 2 minutes. (Unlike ABO typing, Rh typing is better done on a heated warming box.)
4. Record your results, using "+" for clumping and "−" for no clumping. Record whether you are Rh^+ or Rh^- in Section I.5 of the LABORATORY REPORT RESULTS at the end of the exercise. You can also summarize your class's results in the table provided in Section I.6 of the LABORATORY REPORT RESULTS if your instructor so directs.
5. CAUTION! *Dispose of your slide in the container of fresh bleach solution.*

ANSWER THE LABORATORY REPORT QUESTIONS AT THE END OF THE EXERCISE

[2]Be sure to note the important difference in the produciton of antibodies in the ABO and Rh systems. An Rh^- person can *produce antibodies in response to the stimulus of invading Rh antigen;* by contrast, any antibodies of the ABO system that exist in the blood of a person *occur naturally and are present regardless of whether or not ABO antigens are introduced.*

[3]The Rh antigen is more specifically termed the D antigen after the Fisher-Race nomenclature, which is based on genetic concepts or theories of inheritance.

LABORATORY REPORT RESULTS

BLOOD 16

Student ______________________ Date ______________________

Laboratory Section ______________________ Score/Grade ______________________

SECTION B. PLASMA

1. ***Physical Characteristics***

   pH ______________

   Color ______________

   Transparency (clear, translucent, opaque) ______________

   Consistency (slippery, watery, granular, sticky) ______________

2. ***Chemical Constituents***

   Is glucose present? ______________

   Is protein present? ______________

SECTION D. RED BLOOD CELL TESTS

1. Red blood cell count results: ______________ RBCs per mm^3.
2. Red blood cell volume (hematocrit) results: ______________ %.
3. Sedimentation rate results: ______________ mm per hr.
4. Hemoglobin determination results: hemoglobinometer, ______________ g per 100 ml.

   Tallqvist® paper, ______________ g per 100 ml.

5. Complete the following table:

Wavelength (nm)

	500	510	520	530	540	550	560	570	580	590	600
Solution 1											
Solution 2											

Graph your results here.

SECTION F. WHITE BLOOD CELL TESTS

6. White blood cell count results: ______________ WBCs per mm^3.
7. Drawings of various blood cells.

SECTION H. CLOTTING TESTS

Clotting time (slide method) results: ______________ min., ______________ sec.

SECTION I. BLOOD GROUPING (TYPING)

1. In determining your ABO blood grouping, did you observe clumping when your blood was mixed with

__________ anti-A serum only

__________ anti-B serum only

__________ both anti-A and anti-B serums

__________ neither anti-A nor anti-B serum

2. Based on your observations, what is your ABO grouping?

_______ A _______ B _______ AB _______ O

3. Based on your observations, briefly explain why you identify your ABO blood grouping as you do. ____

____

____

4. Record the results of the ABO blood grouping test done by your class.

Type	Anti-A (present or absent)	Anti-B (present or absent)	Number of individuals	Class percentage
A				
B				
AB				
O				

5. Based on your observations, are you Rh^+ or Rh^-?

_______ Rh^+ _______ Rh^-

6. Record the results of the Rh test done by your class.

Type	Number of individuals	Class percentage
Rh^+		
Rh^-		

LABORATORY REPORT QUESTIONS

BLOOD 16

Student ______________________ Date ______________________

Laboratory Section ______________________ Score/Grade ______________________

PART I. MULTIPLE CHOICE

_____ 1. The process by which all blood cells are formed is called (a) hemocytoblastosis (b) erythropoiesis (c) hematopoiesis (d) leucocytosis

_____ 2. An inability of body cells to receive adequate amounts of oxygen may indicate a malfunction of (a) neutrophils (b) leucocytes (c) lymphocytes (d) erythrocytes

_____ 3. Special cells of the body that have the responsibility of clearing away dead, disintegrating bodies of red and white blood cells are called (a) agranular leucocytes (b) reticuloendothelial cells (c) erythrocytes (d) thrombocytes

_____ 4. The name of the test procedure that informs the physician about the rate of erythropoiesis is called the (a) reticulocyte count (b) sedimentation rate (c) hemoglobin count (d) differential count

_____ 5. The normal red blood cell count per cubic millimeter in a male is about (a) 5.4 million (b) 7 million (c) 4 million (d) more than 9 million

_____ 6. The normal number of leucocytes per cubic millimeter is (a) 5000 to 9000 (b) 8000 to 12,000 (c) 2000 to 4000 (d) over 15,000

_____ 7. Under the microscope, red blood cells appear as (a) circular discs with centrally located nuclei (b) circular discs with lobed nuclei (c) oval discs with many nuclei (d) biconcave discs without nuclei

_____ 8. An increase in the number of white blood cells is called (a) leucopenia (b) hematocrit (c) polycythemia (d) leucocytosis

_____ 9. Thrombocytes are formed from a special large cell that breaks up into small fragments. This cell is called a(n) (a) eosinophil (b) hemocytoblast (c) megakaryocyte (d) platelet

_____ 10. The blood type showing the highest incidence in Caucasians in the United States is (a) A (b) O (c) AB (d) B

PART II. COMPLETION

11. Another name for red blood cells is ______________________ .

12. Blood gets its red color from the presence of ______________________ .

13. The life span of a red blood cell is approximately ______________________ .

14. A good method for routine testing for anemia is ______________________ .

15. The normal sedimentation rate value for adults is ______________________ .

16. The normal ratio of red blood cells to white blood cells is about ______________________ .

17. The granular leucocytes are formed from ____________ tissue.
18. The number of thrombocytes per cubic millimeter found normally in blood is ____________ .
19. The function of thrombocytes is to prevent blood loss by starting a chain of reactions resulting in ____________ .
20. In blood groupings (typing), the antigens are also called ____________ .
21. The hemolysis produced by fetal-maternal incompatability of blood cells is called ____________ .
22. The part of the cell where agglutinogens are located is ____________ .

PART III. MATCHING

____ 23. Hemocytoblast
____ 24. Polycythemia
____ 25. A high neutrophil count
____ 26. A high monocyte count
____ 27. Leucopenia
____ 28. A high lymphocyte count
____ 29. Plasma
____ 30. A high eosinophil count
____ 31. Serum

A. Response to tissue destruction by invading bacteria
B. An increase in the normal red blood cell count
C. Leukemia and infectious mononucleosis
D. Immature cells that develop into mature blood cells
E. Chronic infections
F. Liquid portion of blood without the formed elements and clotting substances
G. An allergic reaction
H. A decrease in the normal white blood cell count
I. Liquid portion of blood without the formed elements

HEART 17

The ***heart*** is a hollow, muscular organ that pumps blood through miles and miles of blood vessels. This organ is located in the mediastinum, between the lungs, with two-thirds of its mass lying to the left of the body's midline. Its pointed end, the ***apex***, projects downward to the left, and its broad end, the ***base***, projects upward to the right. The main parts of the heart and associated structures to be discussed here are the pericardium, wall, chambers, great vessels, and valves.

A. PERICARDIUM

A loose-fitting serous membrane called the ***pericardium (pericardial sac)*** encloses the heart (Figure 17.1). The membrane is composed of two layers, the fibrous pericardium and the serous pericardium. The ***fibrous pericardium*** forming the outer layer is tough connective tissue that adheres to the parietal pleura and anchors the heart in the mediastinum. The inner layer, the ***serous pericardium***, is a delicate membrane that is a double layered structure. The outer ***parietal layer*** of the serous pericardium is directly beneath the fibrous pericardium. The inner ***visceral layer*** of the serous pericardium is also called the ***epicardium***. Between the parietal and visceral layers of the serous pericardium is a potential space, the ***pericardial cavity***, which contains pericardial fluid and functions to prevent friction between the layers as the heart beats. Identify these structures using a specimen, model, or chart of a heart and label Figure 17.1.

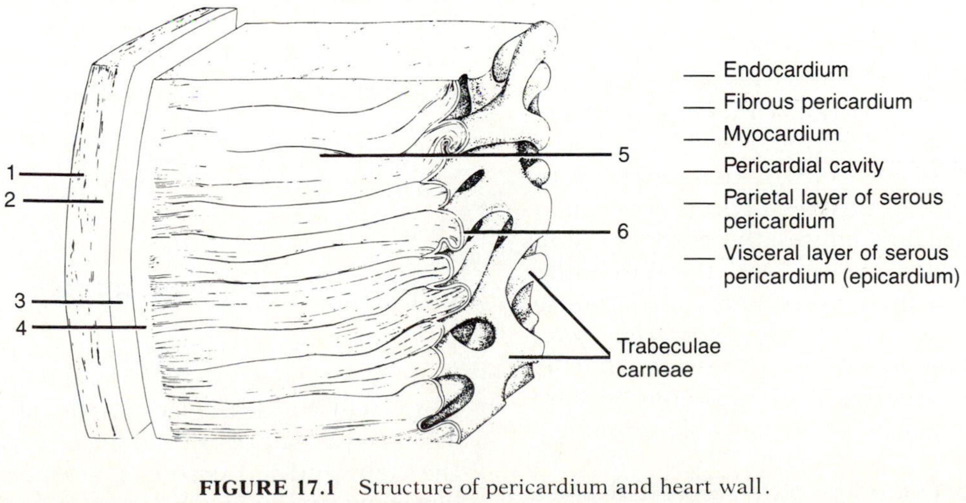

FIGURE 17.1 Structure of pericardium and heart wall.

B. HEART WALL

Three layers of tissue compose the heart: the visceral layer of the serous pericardium (external layer), the myocardium (middle layer), and the endocardium (inner layer). The ***visceral layer of the serous pericardium (epicardium)*** is the thin transparent outer layer of the heart wall. The ***myocardium***, which is composed of cardiac muscle tissue, forms the bulk of the heart and is responsible for contraction. The ***endocardium*** is a thin layer of endothelium that lines the inside of the myocardium and covers the heart valves and the tendons that hold them open. Label the layers of the heart in Figure 17.1.

C. CHAMBERS OF HEART

The interior of the heart is divided into four cavities called ***chambers***, which receive circulating blood. The two superior chambers are known as ***right*** and ***left atria*** and are separated internally by a partition called the ***interatrial septum***. A prominent feature of this septum is an oval depression, the ***fossa ovalis***, which corresponds to the site of the ***foramen ovale***, an opening in the interatrial septum of the fetal heart that helps blood bypass the nonfunctioning lungs. Each atrium has an appendage called an ***auricle*** (OR-i-kul), so named because its shape resembles a dog's ear. The auricles increase the surface area of the atria. The lining of the atria is smooth, except for the anterior walls and linings of the auricles, which contain projecting muscle bundles called ***pectinate*** (PEK-ti-nāt) ***muscles***.

The two inferior and larger chambers, called the ***right*** and ***left ventricles***, are separated internally by a partition called the ***interventricular septum***. Externally, a groove known as the ***coronary sulcus*** (SUL-kus) separates the atria from the ventricles. The groove encircles the heart and houses the coronary sinus (a large cardiac vein) and the circumflex branch of the left coronary artery. The ***anterior interventricular sulcus*** and ***posterior interventricular sulcus*** separate the right and left ventricles externally. They also contain coronary blood vessels and a variable amount of fat.

Label the chambers and associated structures of the heart in Figures 17.2 and 17.3.

D. GREAT VESSELS AND VALVES OF HEART

The right atrium receives blood from every area of the body except the lungs. The blood enters through three veins: the ***superior vena cava (SVC)*** bringing blood from the upper body, the ***inferior vena cava (IVC)*** bringing blood from the lower body, and the ***coronary sinus*** bringing blood from most of the vessels supplying the heart wall.

The blood is passed from the right atrium into the right ventricle through the atrioventricular valve called the ***tricuspid valve***, which consists of three cusps (flaps). The right ventricle then pumps the blood through the ***pulmonary semilunar valve*** into the ***pulmonary trunk***. The pulmonary trunk divides into a ***right*** and ***left pulmonary artery***, each of which carries blood to the lungs where the blood releases its carbon dioxide and takes on oxygen. It returns to the heart via four ***pulmonary veins*** that empty freshly oxygenated blood into the left atrium. The blood is then passed into the left ventricle through another atrioventricular valve called the ***bicuspid (mitral) valve***, which consists of two cusps. The cusps of the tricuspid and bicuspid valves are connected to cords called ***chordae tendineae*** (KOR-dē TEN-din-ē-ē), which in turn attach to projections in the ventricular walls called ***papillary muscles***. The left ventricle pumps the blood through the ***aortic semilunar valve*** into the ***ascending aorta***. From this vessel, aortic blood is passed into the ***coronary arteries, arch of the aorta, thoracic aorta***, and ***abdominal aorta***. These blood vessels transport the blood to all areas of the body except the lungs. The function of the heart valves is to permit the blood to flow in only one direction.

Label the great vessels and valves of the heart in Figures 17.2 and 17.3a and b.

E. BLOOD SUPPLY OF HEART

Because of its importance in myocardial infarction (heart attack), the blood supply of the heart will be described briefly at this point. The arterial supply of the heart is provided by the right and left coronary arteries. The ***right coronary artery*** originates as a branch of the ascending aorta, descends in the coronary sul-

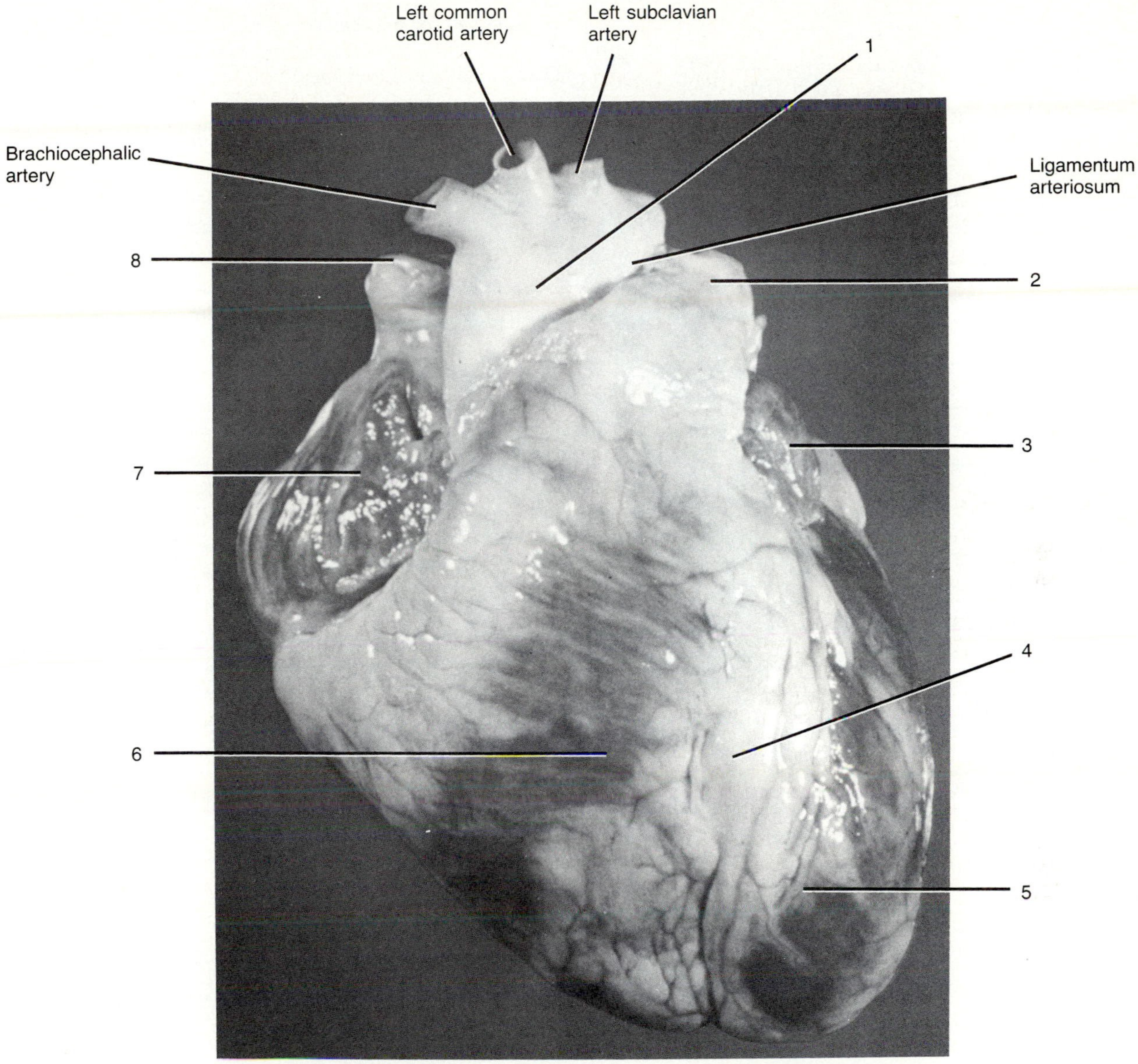

___ Anterior interventricular sulcus
___ Arch of aorta
___ Left auricle
___ Left ventricle
___ Pulmonary trunk
___ Right atrium
___ Right ventricle
___ Superior vena cava (SVC)

FIGURE 17.2 Photograph of external view of human heart.

cus, and gives off a ***marginal branch*** that supplies the right ventricle. The right coronary artery continues around the posterior surface of the heart in the posterior interventricular sulcus. This portion of the artery is known as the ***posterior interventricular branch*** and supplies the right and left ventricles. The ***left coronary artery*** also originates as a branch of the ascending aorta. Between the pulmonary trunk and left auricle, the left coronary artery divides into two branches: anterior interventricular and circumflex. The ***anterior interventricular branch*** passes in the anterior interventricular sulcus and supplies the right and left ventricles. The ***circumflex branch*** circles toward the posterior surface of the heart in the coronary sul-

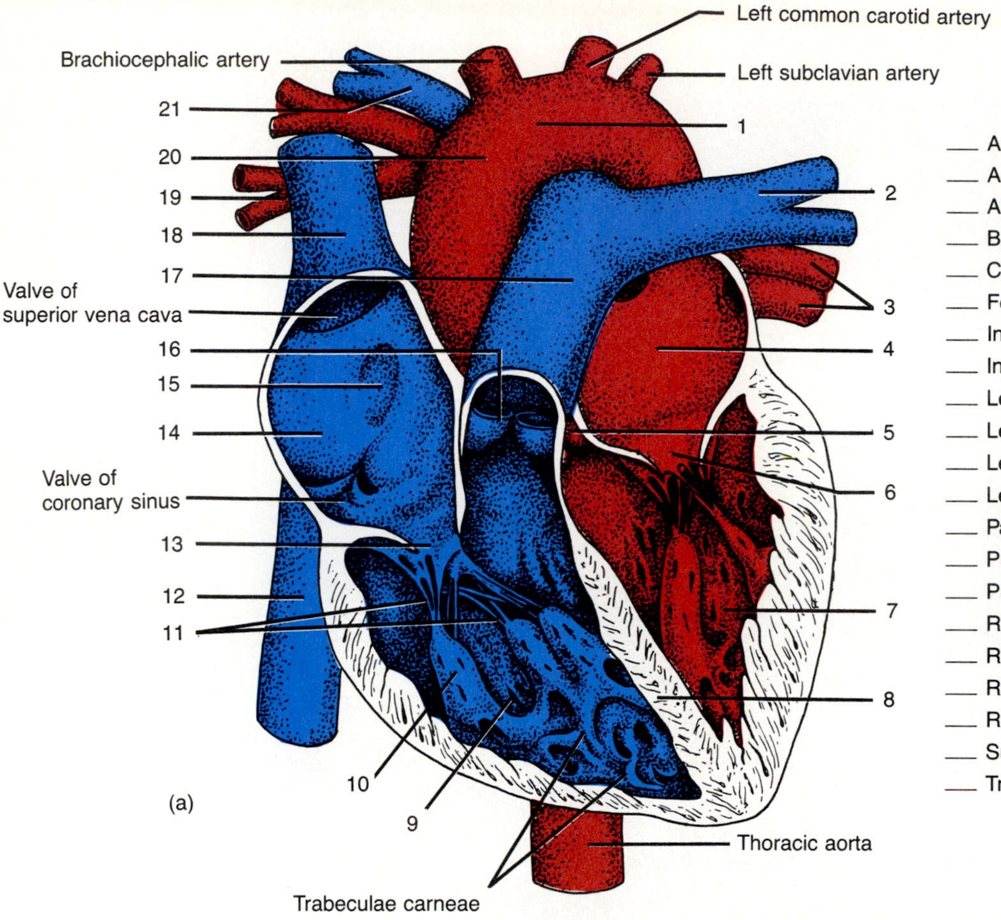

— Arch of aorta
— Aortic semilunar valve
— Ascending aorta
— Bicuspid valve
— Chordae tendineae
— Fossa ovalis
— Inferior vena cava (IVC)
— Interventricular septum
— Left atrium
— Left pulmonary artery
— Left pulmonary veins
— Left ventricle
— Papillary muscle
— Pulmonary semilunar valve
— Pulmonary trunk
— Right atrium
— Right pulmonary artery
— Right pulmonary veins
— Right ventricle
— Superior vena cava (SVC)
— Tricuspid valve

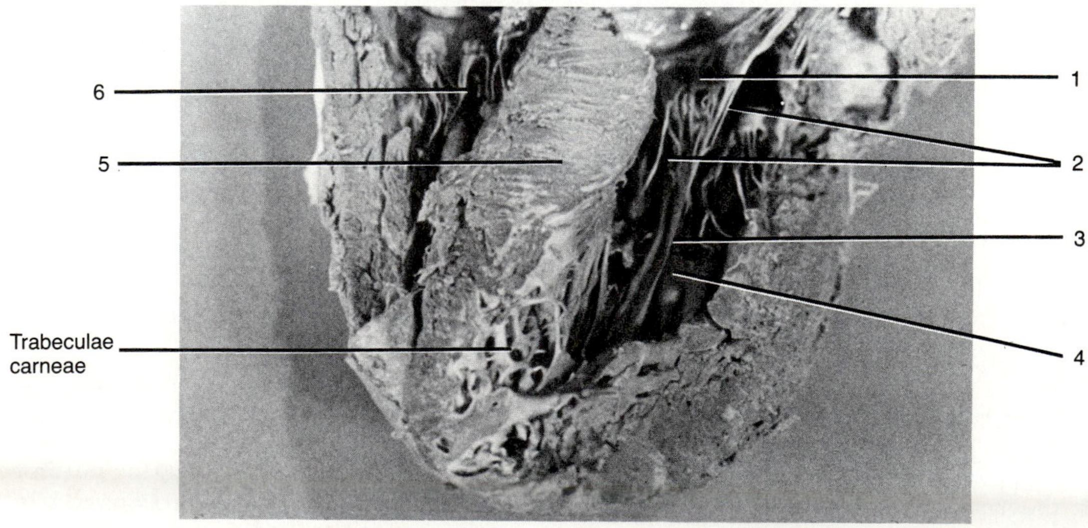

(b)

— Bicuspid valve
— Chordae tendineae
— Interventricular septum
— Left ventricle
— Papillary muscle
— Right ventricle

FIGURE 17.3 Structure of human heart. Red-colored vessels carry oxygenated blood; blue-colored vessels carry deoxygenated blood. (a) Diagram of internal view. (b) Photograph of internal view.

cus and distributes blood to the left ventricle and left atrium.

Label these arteries in Figure 17.4a.

Most blood from the heart drains into the ***coronary sinus***, a venous channel in the posterior portion of the coronary sulcus between the left atrium and left ventricle. The principal tributaries of the coronary sinus are the ***great cardiac vein***, which drains the anterior aspect of the heart, and the ***middle cardiac vein***, which drains the posterior aspect of the heart.

Label these veins in Figure 17.4b.

F. DISSECTION OF CAT HEART

CAUTION! *Please reread Section D. "Precautions Related to Dissection" at the beginning of the laboratory manual on page xii before you begin your dissection.*

PROCEDURE

1. Before examining the cat's heart and major blood vessels, you must open the thoracic cavity.

2. On the ventral surface of the animal, pal-

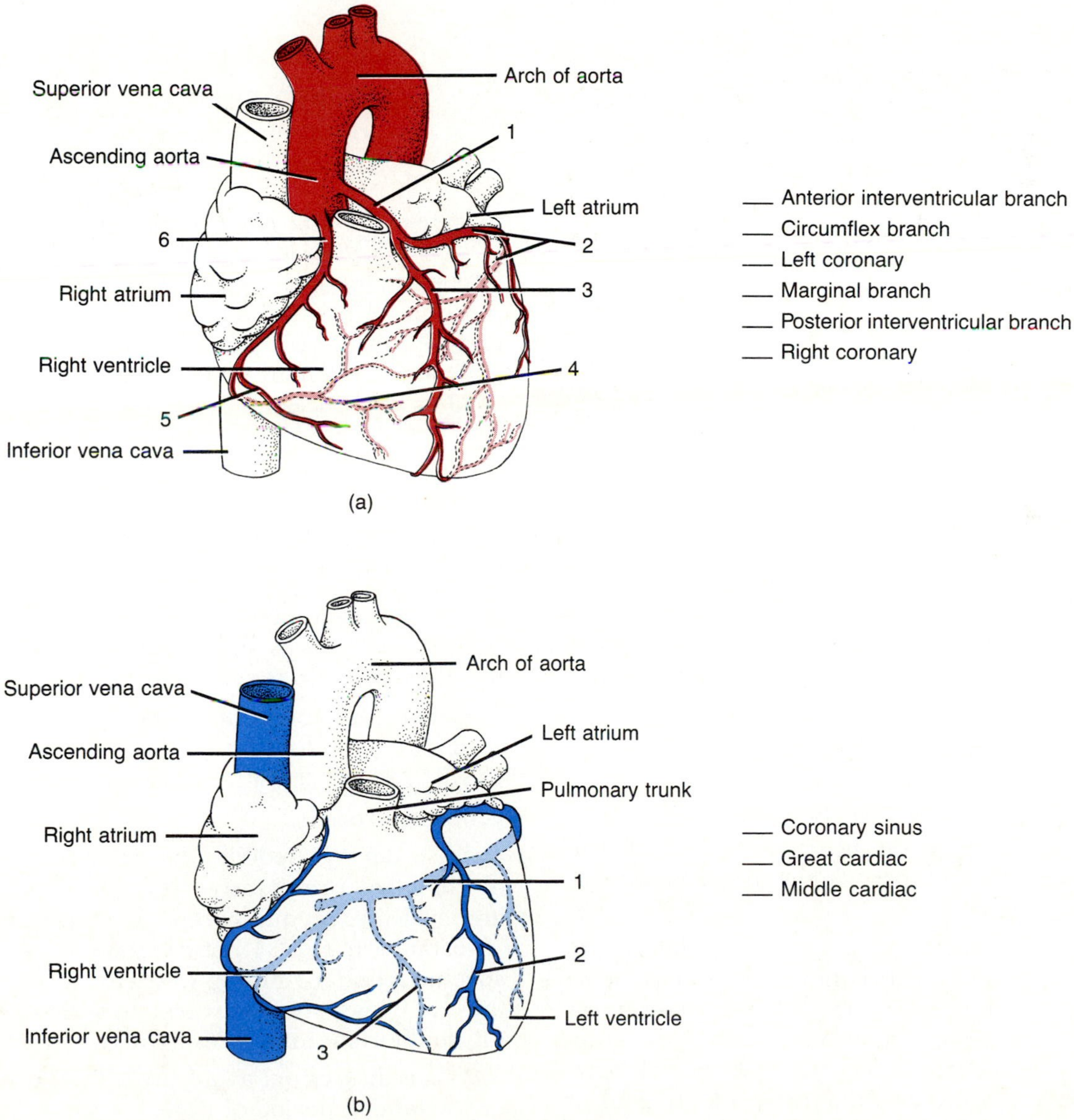

FIGURE 17.4 Coronary (cardiac) circulation. (a) Anterior view of arterial distribution. (b) Anterior view of venous drainage.

pate the bottom edge of the rib cage. Using heavy-duty scissors, penetrate the chest cavity immediately below the lowest rib and make a longitudinal incision about 1/2 in. to the right of the sternum.

3. Cut through the ribs, extending the incision anteriorly to the apex of the thorax. Be careful not to cut into any internal organs.

4. Spread apart the walls of the thorax and locate the diaphragm at the caudal end of the incision. *Be careful not to break any of the ribs*, because the sharp edge of a broken rib could interfere with your examination of the viscera.

5. Make two additional cuts laterally and dorsally toward the spine on both sides, keeping the incisions just anterior to the diaphragm.

6. Note the ***right*** and ***left atria*** and the ***right*** and ***left ventricles*** (Figure 17.5). Each atrium has an earlike flap on its ventral surface called an ***auricle***. Identify the ***coronary sulcus*** between the right atrium and right ventricle and the ***anterior longitudinal sulcus*** between the right ventricle and left ventricle.

7. Carefully remove the thymus gland and any fat so that you can clearly see the ventral surface of the heart and its attached vessels (Figure 17.5). Be careful not to damage any nerves in this area.

8. The ***pericardium***, the membrane surrounding the heart, consists of two layers. The outer ***parietal pericardium*** is a tough membrane that surrounds the heart and blood vessels that join the heart. This layer can be removed. The inner ***visceral pericardium*** invests the heart muscle (myocardium) closely and is very difficult to remove.

9. The ***apex*** of the heart is directed caudally and to the left and its ***base*** is directed cranially and to the right.

10. The cat's heart can be examined without removal. Leaving the heart in place will help you examine and identify the major blood vessels later on.

11. If a detailed dissection of the cat heart is to be done, follow the directions of the sheep heart dissection and identify the following structures of the cat heart, which correlate with structures of the sheep heart. Because the cat heart is small, it may be difficult to identify all of them.

1. Opening of the superior vena cava
2. Opening of the inferior vena cava
3. Fossa ovalis
4. Opening of the coronary sinus
5. Pectinate muscle
6. Interatrial septum
7. Right ventricle
    a. tricuspid valve
    b. chordae tendineae
    c. papillary muscles
    d. interventricular septum
    e. pulmonary semilunar valve
8. Left atrium: openings of pulmonary veins
9. Left ventricle
    a. bicuspid (mitral) valve
    b. chordae tendineae
    c. papillary muscles
    d. aortic semilunar valve

G. DISSECTION OF SHEEP HEART

The anatomy of the sheep heart closely resembles that of the human heart. Use Figures 17.5 and 17.6 as references for this dissection. In addition, models of human hearts can also be used as references.

CAUTION! *Please reread Section D. "Precautions Related to Dissection" at the beginning of the laboratory manual on page xii before you begin your dissection.*

1. In order to minimize the potential for infection, rubber gloves should be worn when performing dissections.

2. Laboratory coats should be worn to protect your skin and clothing.

3. Dissections should be performed in well-ventilated rooms.

4. In order to reduce the irritating effects of chemical preservatives to your skin, eyes, and nose, soak and wrap your specimen in "Biostat." If this is not available, hold your specimen under running water for several minutes to wash away excess preservative and dilute the remaining preservative.

5. When dissecting, avoid breathing fumes for any extended period of time.

6. When dissecting, there is always the possibility of skin cuts or punctures from dissecting equipment or the specimens themselves, such

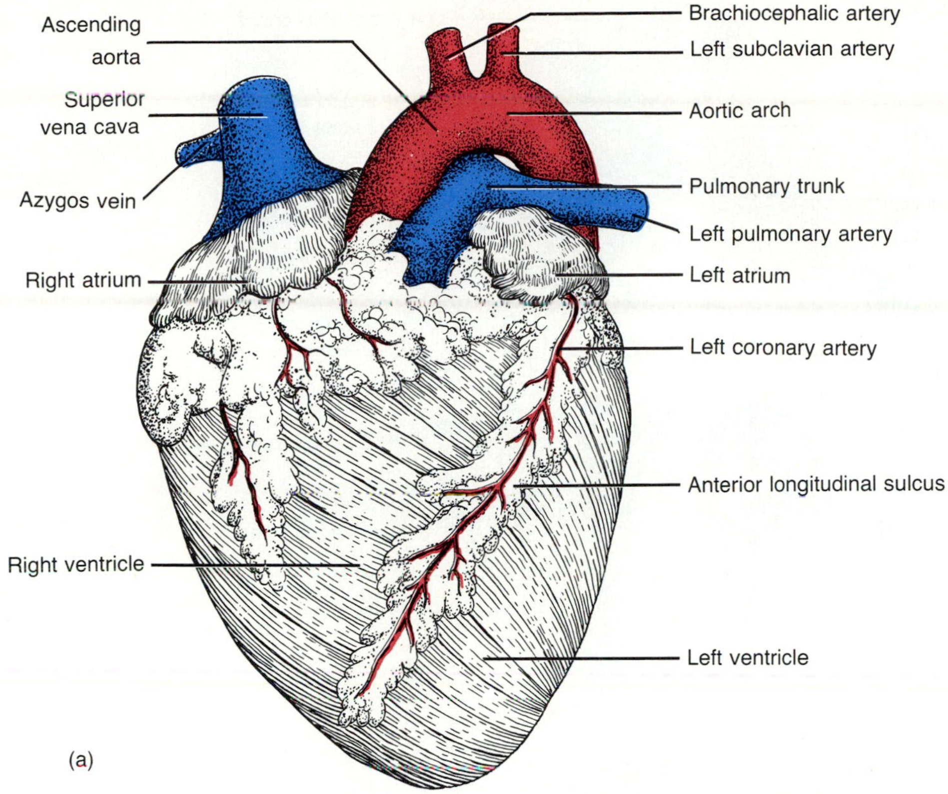

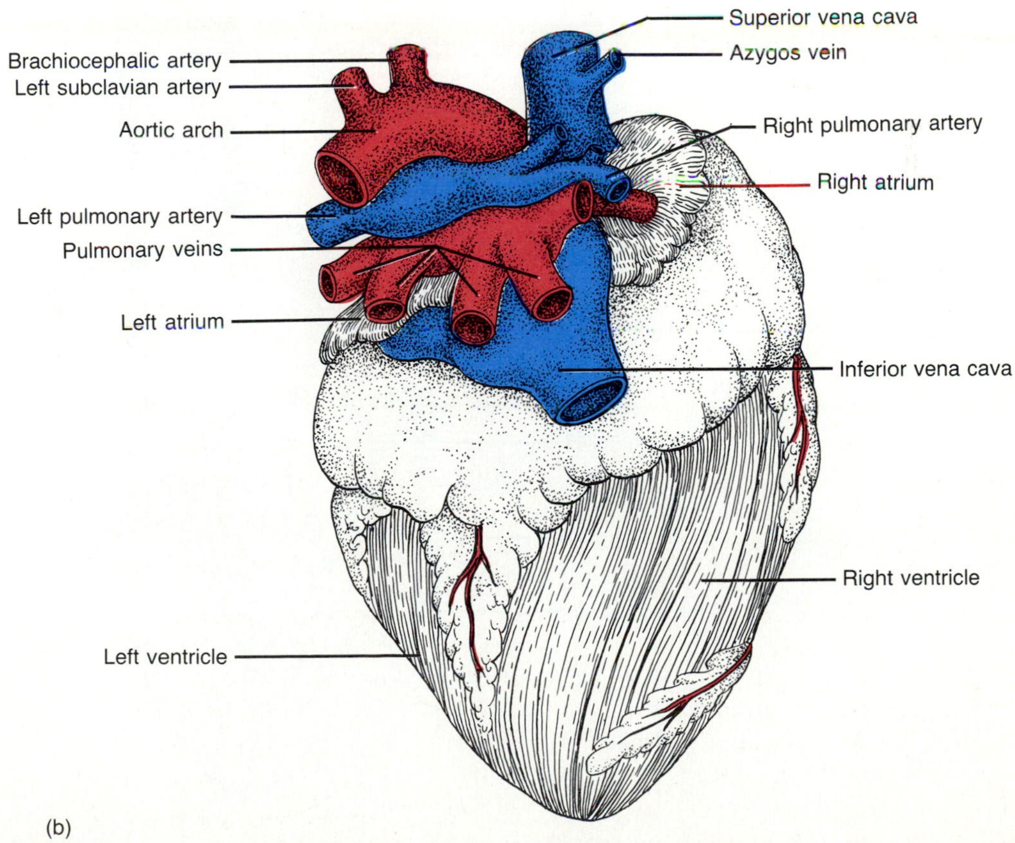

FIGURE 17.5 External structure of a cat or sheep heart. (a) Ventral view. (b) Dorsal view.

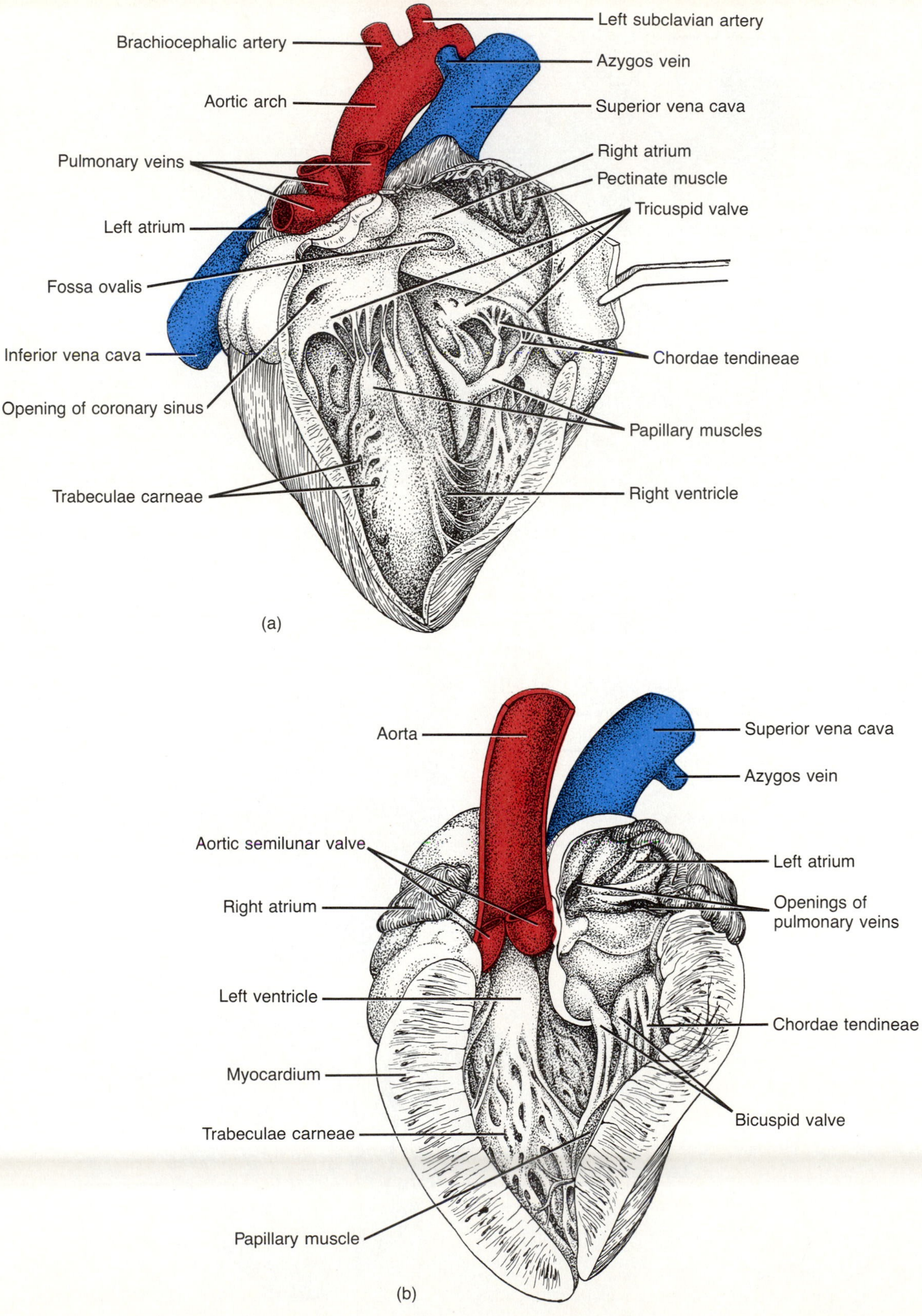

FIGURE 17.6 Internal structure of a cat or sheep heart. (a) View of right atrium and right ventricle. (b) View of left atrium and left ventricle.

as the teeth or claws of animals. Should you sustain a cut or puncture in this manner, wash your hands with a disinfectant soap, notify your instructor, and seek immediate medical attention to decrease the possibility of infection.

First examine the ***pericardium***, a fibroserous membrane that encloses the heart, which may have already been removed in preparing the sheep heart for preservation. The ***myocardium*** is the middle layer and constitutes the main muscle portion of the heart. The ***endocardium*** (the third layer) is the inner lining of the heart. Use the figures to determine which is the ventral surface of the heart and then identify the ***pulmonary trunk*** emerging from the anterior ventral surface, near the midline, and medial to the ***left auricle***. A longitudinal depression on the ventral surface, called the ***anterior longitudinal sulcus***, separates the right ventricle from the left ventricle. Locate the ***coronary blood vessels*** lying in this sulcus.

PROCEDURE

1. Remove any fat or pulmonary tissue that is present.

2. In cutting the sheep heart open to examine the chambers, valves, and vessels, the anterior longitudinal sulcus is used as a guide.

3. Carefully make a shallow incision through the ventral wall of the pulmonary trunk and the right ventricle, trying not to cut the dorsal surface of either structure.

4. The incision is best made *less than an inch to the right of, and parallel to,* the previously mentioned anterior longitudinal sulcus.

5. If necessary, the incision can be continued to where the pulmonary trunk branches into a ***right pulmonary artery***, which goes to the right lung, and a ***left pulmonary artery***, which goes to the left lung. The ***pulmonary semilunar valve*** of the pulmonary artery can be clearly seen upon opening it. In any of these internal dissections of the heart, any coagulated blood or latex should be immediately removed so that all important structures can be located and identified.

6. Keeping the cut still parallel to the sulcus, extend the incision around and through the dorsal ventricular wall until you reach the ***interventricular septum.***

7. Now examine the dorsal surface of the heart and locate the thin-walled ***superior vena cava*** directly above the ***right auricle***. This vein proceeds posteriorly straight into the right atrium.

8. Make a second longitudinal cut, this time through the superior vena cava (dorsal wall).

9. Extend the cut posteriorly through the right atrium on the left of the right auricle. Proceed posteriorly to the dorsal right ventricle wall and join your first incision.

10. The entire internal right side of the heart should now be clearly seen when carefully spread apart. The interior of the superior vena cava, right atrium, and right ventricle will now be examined. Start with the right auricle and locate the ***pectinate muscle***, the large opening of the ***inferior vena cava*** on the left side of the right atrium, and the opening of the ***coronary sinus*** just below the opening of the inferior vena cava. By using a dull probe and gentle pressure, most of the vessels can be traced to the dorsal surface of the heart.

11. Now find the wall that separates the two atria, the ***interatrial septum***. Also find the ***fossa ovalis***, an oval-shaped depression ventral to the entrance of the inferior vena cava.

12. The ***tricuspid valve*** between the right atrium and the right ventricle should be examined to locate the three cusps as its name indicates. From the cusps of the valve itself, and tracing posteriorly, the ***chordae tendineae***, which hold the valve in place, should be identified. Still tracing posteriorly, the chordae are seen to originate from the ***papillary muscles***, which themselves originate from the wall of the right ventricle itself.

13. Look carefully again at the dorsal surface of the left atrium and locate as many ***pulmonary veins*** (normally, four) as possible.

14. Make your third longitudinal cut through the most lateral of the pulmonary veins that you have located.

15. Continue posteriorly through the left atrial wall and the left ventricle to the ***apex*** of the heart.

16. Compare the difference in the thickness of the wall between the right and left ventricles. Explain your answer.

17. Examine the ***bicuspid (mitral) valve***, again counting the cusps. Determine if the left side of

the heart has basically the same structures as studied on the right side.

18. Probe from the left ventricle to the ***aorta*** as it emerges from the heart, examining the ***aortic semilunar valve***. Find the openings of the right and left main coronary arteries.

19. Locate now the ***brachiocephalic artery***, which is one of the first branches from the arch of the aorta. This artery continues branching and terminates by supplying the arms and head as its name indicates.

20. Connecting the aorta with the pulmonary artery is the remnant of the ***ductus arteriosus***, called the ***ligamentum arteriosum***. It may not be present in your sheep heart.

ANSWER THE LABORATORY REPORT QUESTIONS AT THE END OF THE EXERCISE

LABORATORY REPORT QUESTIONS

HEART 17

Student ______________________ Date ______________________

Laboratory Section ______________________ Score/Grade ______________________

PART I. MULTIPLE CHOICE

_____ 1. Which of the following veins drains the blood from most of the vessels supplying the heart wall? (a) vasa vasorum (b) superior vena cava (c) coronary sinus (d) inferior vena cava

_____ 2. The atrioventricular valve on the same side of the heart as the origin of the aorta is the (a) aortic semilunar (b) tricuspid (c) bicuspid (d) pulmonary semilunar

_____ 3. Which valve does the blood go through just before entering the pulmonary trunk on the way to the lungs? (a) tricuspid (b) pulmonary semilunar (c) aortic semilunar (d) bicuspid

_____ 4. The pointed end of the heart that projects downward and to the left is the (a) costal surface (b) base (c) apex (d) coronary sulcus

_____ 5. Which of these structures is more internal? (a) fibrous pericardium (b) visceral layer of serous pericardium (c) parietal layer of serous pericardium (d) myocardium

_____ 6. The musculature of the heart is referred to as the (a) endocardium (b) myocardium (c) epicardium (d) pericardium

_____ 7. The depression in the interatrial septum corresponding to the foramen ovale of fetal circulation is the (a) interventricular sulcus (b) pectinate muscle (c) chordae tendineae (d) fossa ovalis

PART II. COMPLETION

8. Malfunction of the ______________________ valve would interfere with the flow of blood from the right atrium to the right ventricle.

9. Deoxygenated blood is sent to the lungs through the ______________________ .

10. The loose-fitting serous membrane that encloses the heart is called the ______________________ .

11. The two inferior chambers of the heart are separated by the ______________________ .

12. The earlike flap of tissue on each atrium is called a(n) ______________________ .

13. The large vein that drains blood from superior parts of the body and empties into the right atrium is the ______________________ .

14. The cusps of atrioventricular valves are prevented from inverting by the presence of cords called ______________________ , which are attached to papillary muscle.

15. A groove on the surface of the heart that houses blood vessels and a variable amount of fat is called a(n) ______________________ .

16. The branch of the left coronary artery that distributes blood to the left atrium and left ventricles is the ______________________ .

PART III. SPECIAL EXERCISE

Draw a model of the heart and carefully label the four chambers, the four valves in their proper places, and the major blood vessels entering and exiting from the heart.

BLOOD VESSELS 18

Blood vessels are networks of tubes that carry blood throughout the body. Blood vessels are called arteries, arterioles, capillaries, venules, or veins. In this exercise, you will study the histology of blood vessels and identify the principal arteries and veins of the human cardiovascular system.

A. ARTERIES AND ARTERIOLES

Arteries are blood vessels that carry blood *away* from the heart to body tissues. Arteries are constructed of three coats of tissue called ***tunics*** and a hollow core, called a ***lumen***, through which blood flows (Figure 18.1). The inner coat is called the ***tunica interna*** and consists of a lining of endothelium in contact with the blood and a layer of elastic tissue called the internal elastic membrane. The middle coat, or ***tunica media***, is usually the thickest layer and consists of elastic fibers and smooth muscle. This tunic is responsible for two major properties of arteries: ***elasticity*** and ***contractility***. The outer coat, or ***tunica externa***, is composed principally of elastic and collagenous fibers. An external elastic membrane may separate the tunica externa from the tunica media.

Obtain a prepared slide of a cross section of an artery and identify the tunics using Figure 18.1 as a guide.

As arteries approach various tissues of the body, they become smaller and are known as ***arterioles***. Arterioles play a key role in regulating blood flow from arteries into capillaries. When arterioles enter a tissue, they branch into countless microscopic blood vessels called capillaries.

B. CAPILLARIES

Capillaries are microscopic blood vessels that connect arterioles and venules. Their function is to permit the exchange of nutrients and wastes between blood and body tissues. This function is related to the fact that capillaries consist of only a single layer of endothelium.

C. VENULES AND VEINS

When several capillaries unite, they form small veins called ***venules***. They collect blood from capillaries and drain it into veins.

Veins are composed of the same three tunics as arteries, but there are variations in their relative thicknesses. The tunica interna of veins is extremely thin compared to that of their accompanying arteries. In addition, the tunica media of veins is much thinner than that of accompanying arteries, and the tunica externa is thicker in veins (Figure 18.1). Functionally, veins return blood from tissues *to* the heart.

Obtain a prepared slide of a cross section of an artery and its accompanying vein and compare them, using Figure 18.1 as a guide.

D. CIRCULATORY ROUTES

The two basic circulatory routes, are systemic and pulmonary (Figure 18.2). Other circulatory routes, which are all subdivisions of systematic circulation, include hepatic portal, coronary (cardiac) circulation, fetal circulation, and cerebral arterial circle (circle of Willis). The latter is found at the base of the brain (see Table 18.2).

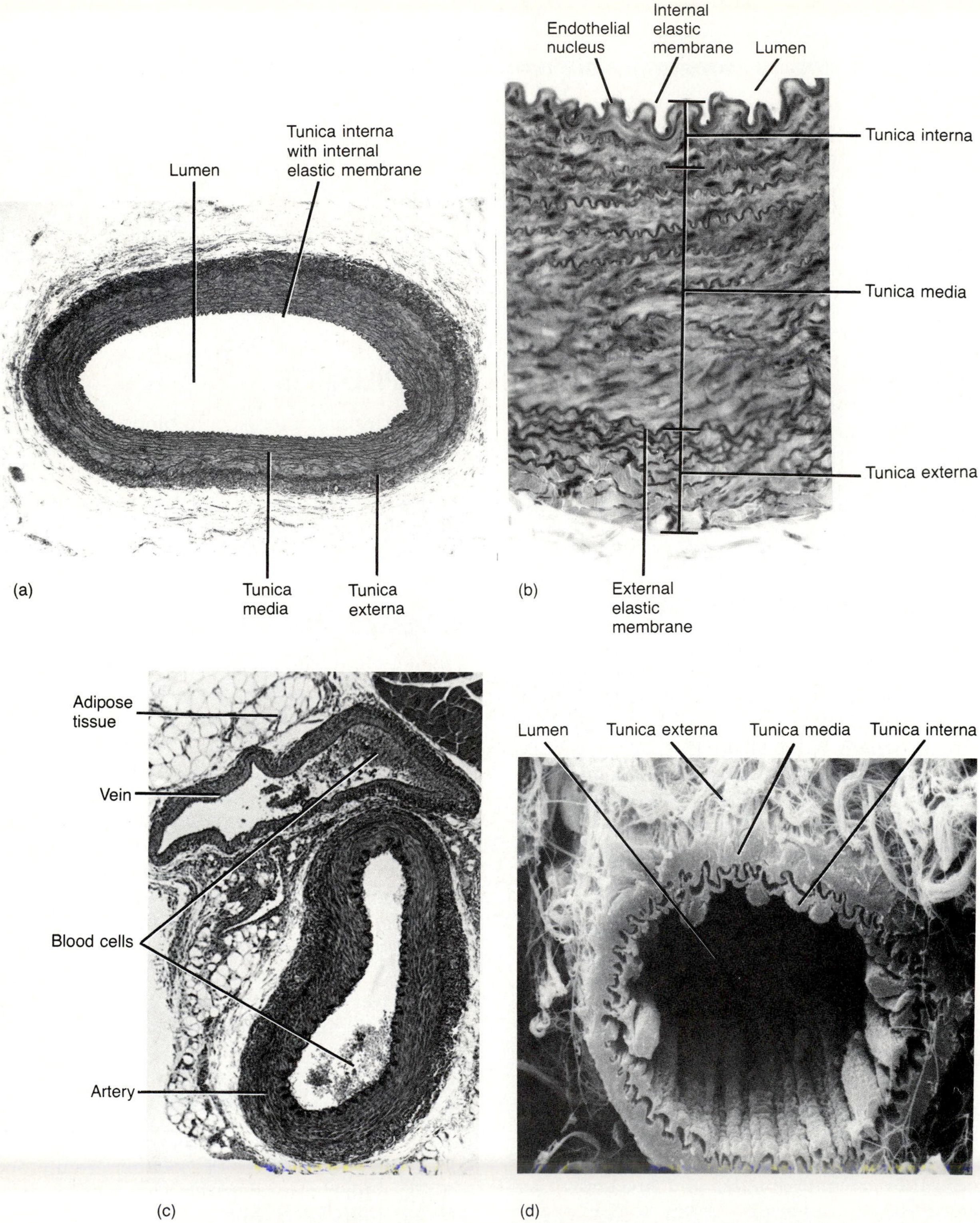

FIGURE 18.1 Histology of blood vessels. (a) Cross section of an artery. (b) Enlarged aspect of an arterial wall. (c) Comparison of structure of an artery and its accompanying vein. (d) Scanning electron micrograph of an artery at a magnification of 5170×. ((d) from *Tissues and Organs: A Text-Atlas of Scanning Electron Microscopy* by Richard G. Kessel and Randy H. Kardon. W. H. Freeman and Company. Copyright © 1979.)

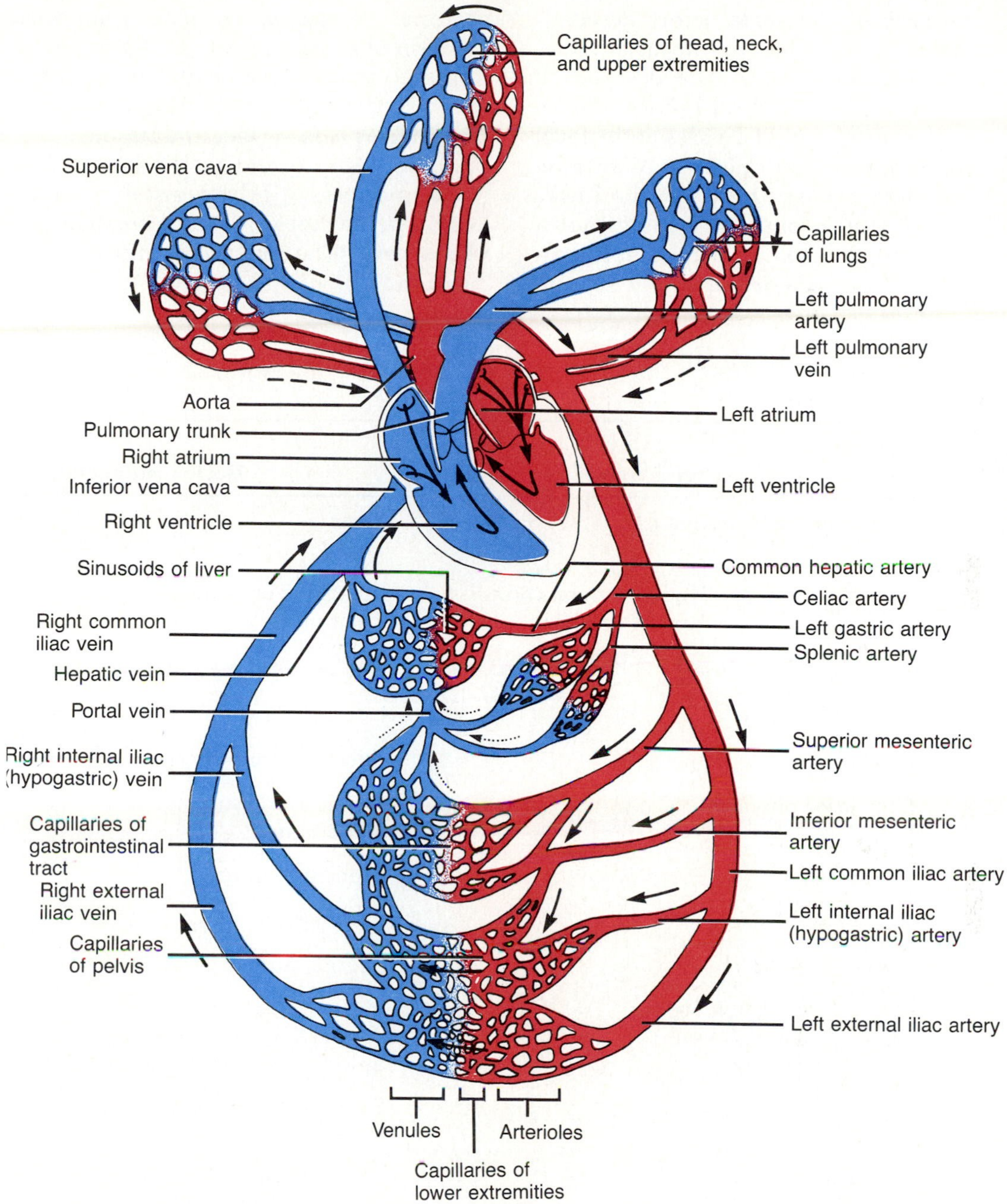

FIGURE 18.2 Circulatory routes. Systemic circulation is indicated by solid arrows; pulmonary circulation by broken arrows; and hepatic portal circulation by dotted arrows.

1. Systemic Circulation

The largest route is the ***systemic circulation*** (see Figures 18.3 through 18.11 and Tables 18.1 through 18.11). This route includes the flow of blood from the left ventricle to all parts of the body. The function of the systemic circulation is to carry oxygen and nutrients to all body tissues and to remove carbon dioxide and other wastes from the tissues. All systemic arteries branch from the ***aorta***. As the aorta emerges from the left ventricle, it passes upward and deep to the pulmonary trunk. At this point, it is called the ***ascending aorta***. The ascending aorta gives off two coronary branches to the heart muscle. Then it turns to the left, forming the ***arch of the aorta*** before descending to the level of the fourth thoracic vertebra as the ***descending aorta***. The descending aorta lies close to the vertebral bodies, passes through the diaphragm, and divides at the level of the fourth lumbar vertebra into two ***common iliac arteries***, which carry blood to the lower extremities. The section of the descending aorta between the arch of aorta and the diaphragm is referred to as the ***thoracic aorta***. The section between the diaphragm and the common iliac arteries is termed the ***abdominal aorta***. Each section of the aorta gives off arteries that continue to branch into distributing arteries leading to organs and finally into the arterioles and capillaries that service the tissues.

Blood is returned to the heart through the systemic veins. All the veins of the systemic circulation flow into either the ***superior*** or ***inferior vena cava***. They in turn empty into the right atrium.

TABLE 18.1
AORTA AND ITS BRANCHES

Division of aorta	Arterial branch		Region supplied
Ascending aorta	Right and left coronary		Heart
Arch of aorta	Brachiocephalic	Right common carotid	Right side of head and neck
		Right subclavian	Right upper extremity
	Left common carotid		Left side of head and neck
	Left subclavian		Left upper extremity
	Intercostals		Intercostal and chest muscles, pleurae
Thoracic aorta	Superior phrenics		Posterior and superior surfaces of diaphragm
	Bronchials		Bronchi of lungs
	Esophageals		Esophagus
Abdominal aorta	Inferior phrenics		Inferior surface of diaphragm
	Celiac	Common hepatic	Liver
		Left gastric	Stomach and esophagus
		Splenic	Spleen, pancreas, stomach
	Superior mesenteric		Small intestine, cecum, ascending and transverse colons
	Suprarenals		Adrenal (suprarenal) glands
	Renals		Kidneys
	Gonadals	Testiculars or	Testes
		Ovarians	Ovaries
	Inferior mesenteric		Transverse, descending, sigmoid colons; rectum
	Common iliacs	External iliacs	Lower extremities
		Internal iliacs (hypogastrics)	Uterus, prostate gland, muscles of buttocks, urinary bladder

TABLE 18.2
ARCH OF AORTA (Figures 18.3, 18.4, and 18.5)

Branch	Description and region supplied
Brachiocephalic	***Brachiocephalic artery*** is the first branch off arch of aorta. Divides to form right subclavian artery and right common carotid artery. ***Right subclavian artery*** extends from brachiocephalic to first rib, passes into armpit (axilla) and supplies arm, forearm, and hand. This artery is a good example of the same vessel having different names as it passes through different regions. Continuation of right subclavian into axilla is called ***axillary artery.*** From here, it continues into arm as ***brachial artery.*** At bend of elbow, brachial artery divides into medial ***ulnar*** and lateral ***radial arteries.*** These vessels pass down to palm, one on each side of forearm. In palm, branches of two arteries anastomose to form two palmar arches—***superficial palmar arch*** and ***deep palmar arch.*** From these arches arise ***digital arteries,*** which supply fingers and thumb (Figure 18.3). Before passing into axilla, right subclavian gives off major branch to brain called ***vertebral artery.*** Right vertebral artery passes through foramina of transverse processes of cervical vertebrae and enters skull through foramen magnum to reach undersurface of brain. Here it unites with left vertebral artery to form ***basilar artery*** (Figures 18.4 and 18.5.) ***Right common carotid artery*** passes upward in neck. At upper level of larynx, it divides into ***right external*** and ***right internal carotid arteries.*** External carotid supplies right side of thyroid gland, tongue, face, ear, scalp, and dura mater. Internal carotid supplies brain, right eye, and right sides of forehead and nose (Figure 18.4). Anastomoses of left and right internal carotids along with basilar artery form a somewhat hexagonal arrangement of blood vessels at the base of the brain near the sella turcica called the ***cerebral arterial circle (circle of Willis).*** From this anastomosis arise arteries supplying brain. Essentially, cerebral arterial circle is formed by union of ***anterior cerebral arteries*** (branches of internal carotids) and ***posterior cerebral arteries*** (branches of basilar artery). Posterior cerebral arteries are connected with internal carotids by ***posterior communicating arteries.*** Anterior cerebral arteries are connected by ***anterior communicating arteries.*** Cerebral arterial circle equalizes blood pressure to brain and provides alternate routes for blood to brain should arteries become damaged.
Left common carotid	***Left common carotid*** is the second branch off the arch of aorta. Corresponding to right common carotid, it divides into basically same branches with same names—except that arteries are now labeled "left" instead of "right."
Left subclavian	***Left subclavian artery*** is third branch off arch of aorta. It distributes blood to left vertebral artery and vessels of left upper extremity. Arteries branching from left subclavian are named like those of right subclavian.

Label Figures 18.3, 18.4, and 18.5.

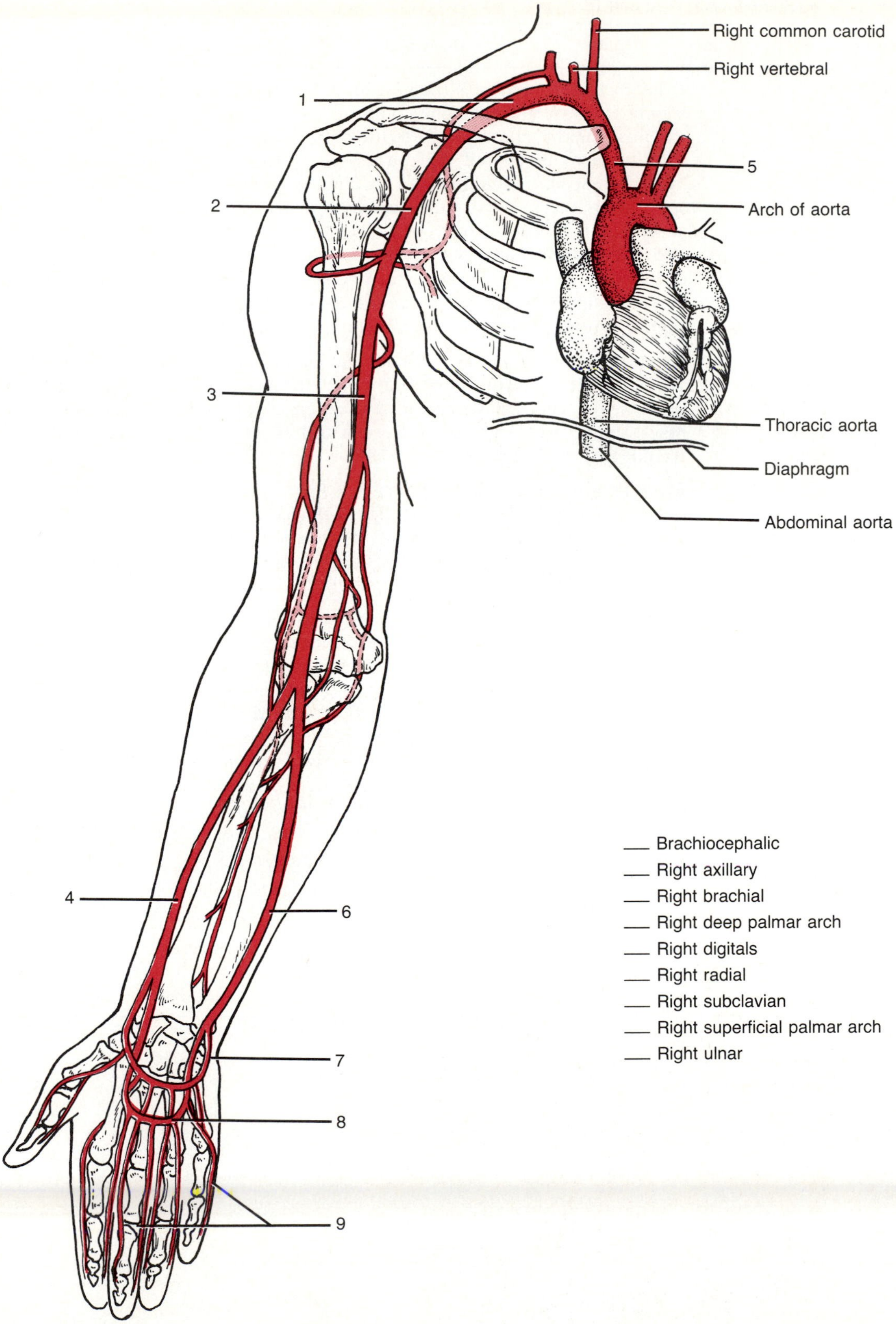

FIGURE 18.3 Anterior view of arteries of right upper extremity.

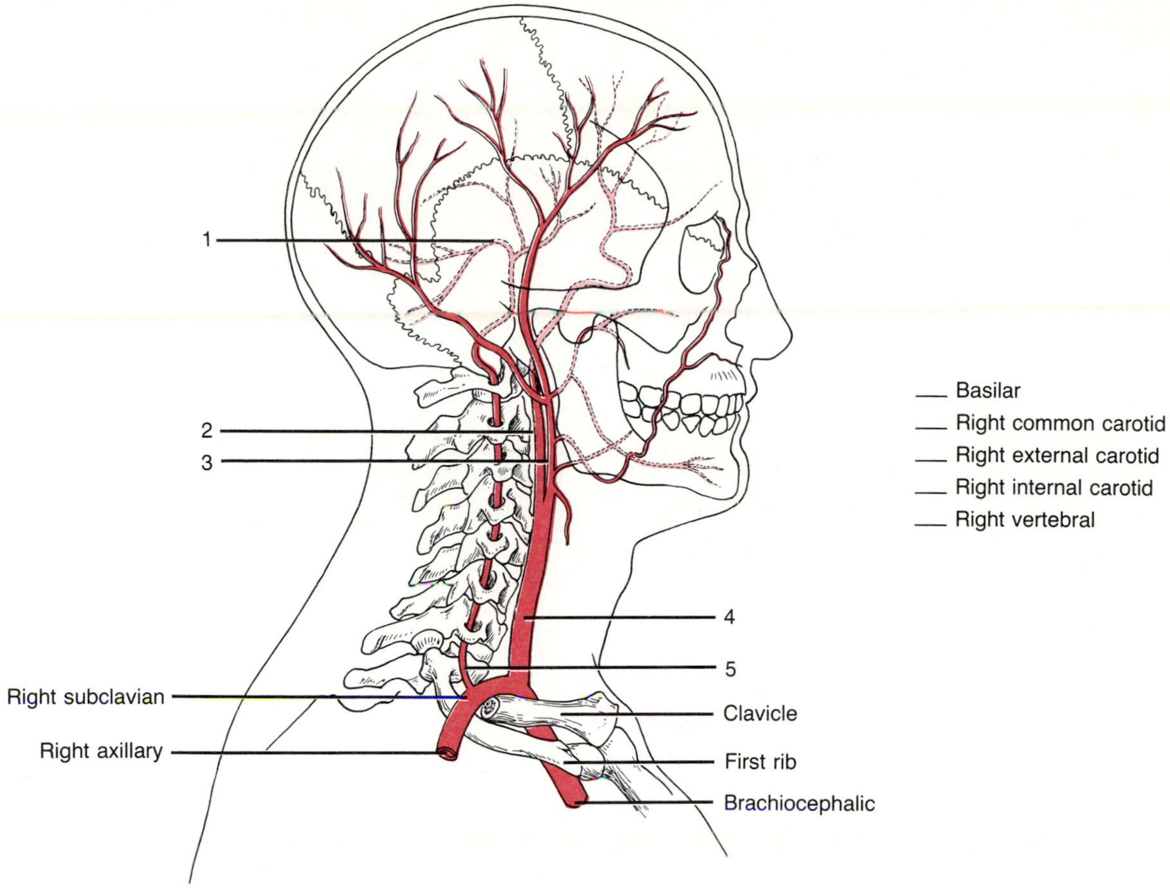

FIGURE 18.4 Right lateral view of arteries of neck and head.

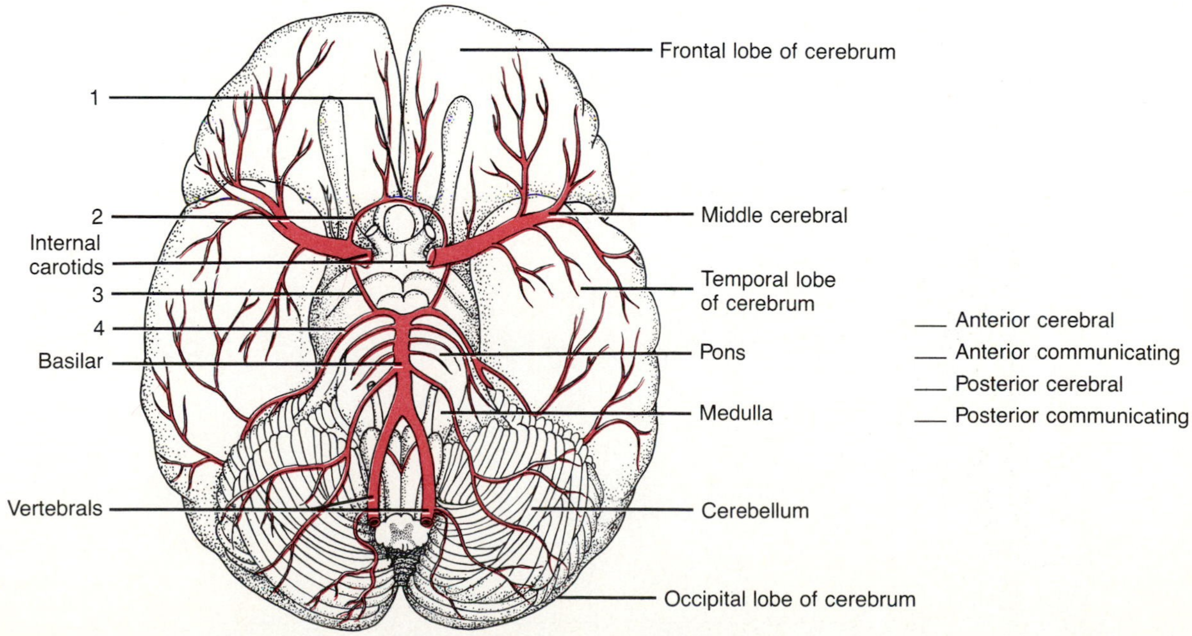

FIGURE 18.5 Arteries of base of brain.

Write the names of the missing veins in the following scheme of circulation. Be sure to specify left or right where applicable.

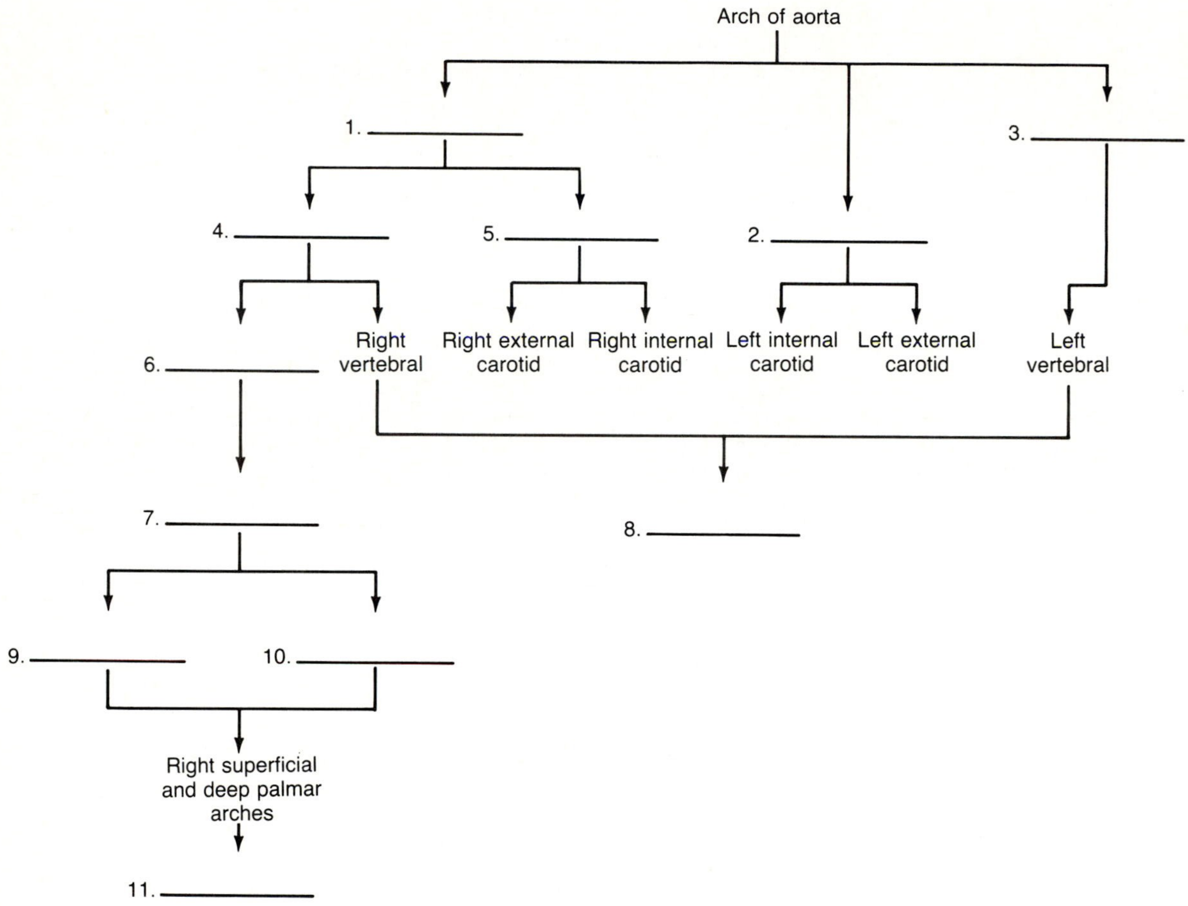

TABLE 18.3
THORACIC AORTA

Branch	Description and region supplied
	Thoracic aorta runs from fourth to twelfth thoracic vertebrae, sending off numerous small arteries to viscera and skeletal muscles of the chest. Branches of an artery that supply viscera are called ***visceral branches.*** Those that supply body wall structures are ***parietal branches.***
VISCERAL	
Pericardial	Several minute ***pericardial arteries*** supply blood to dorsal aspect of pericardium.
Bronchial	One ***right*** and two ***left bronchial arteries*** supply the bronchial tubes, visceral pleurae, bronchial lymph nodes, and esophagus.
Esophageal	Four or five ***esophageal arteries*** supply the esophagus.
Mediastinal	Numerous small ***mediastinal arteries*** supply blood to structures in the posterior mediastinum.
PARIETAL	
Posterior intercostal	Nine pairs of ***posterior intercostal arteries*** supply (1) the intercostal, pectoral, and abdominal muscles; (2) overlying subcutaneous tissue and skin; (3) mammary glands; and (4) vertebral canal and its contents.
Subcostal	The ***left*** and ***right subcostal arteries*** have a distribution similar to that of the posterior intercostals.
Superior phrenic	Small ***superior phrenic arteries*** supply the superior surface of the diaphragm.

TABLE 18.4
ABDOMINAL AORTA (Figure 18.6)

Branch	Description and region supplied
VISCERAL	Supply viscera.
Celiac	***Celiac artery (trunk)*** is first visceral aortic branch below diaphragm. The artery has three branches: (1) ***common hepatic artery,*** (2) ***left gastric artery,*** and (3) ***splenic artery.*** The common hepatic artery has three main branches: (1) ***hepatic artery proper,*** a continuation of the common hepatic artery, that supplies the liver and gallbladder; (2) ***right gastric artery,*** that supplies the stomach and duodenum; and (3) ***gastroduodenal artery,*** that supplies the stomach, duodenum, and pancreas. The left gastic artery supplies the stomach and its ***esophageal branch*** supplies the esophagus. The splenic artery supplies the spleen and has three main branches: (1) ***pancreatic arteries,*** that supply the pancreas; (2) ***left gastroepiploic artery,*** that supplies the stomach; and (3) ***short gastric arteries,*** that supply the stomach.
Superior mesenteric	The ***superior mesenteric artery*** has several principal branches: (1) ***inferior pancreaticoduodenal artery,*** that supplies the pancreas and duodenum; (2) ***jejunal*** and ***ileal arteries,*** that supply the jejunum and ileum; (3) ***ileocolic artery,*** that supplies the ileum and ascending colon; (4) ***right colic artery,*** that supplies the ascending colon; and (5) ***middle colic artery,*** that supplies the transverse colon.
Suprarenals	Right and left ***suprarenal arteries*** supply blood to adrenal (suprarenal) glands. The glands are also supplied by branches of the renal and inferior phrenic arteries.
Renals	Right and left ***renal arteries*** carry blood to kidneys and adrenal (suprarenal) glands.
Gonadals (testiculars or ovarians)	Right and left ***testicular arteries*** extend into scrotum and terminate in testes; right and left ***ovarian arteries*** are distributed to ovaries.
Inferior mesenteric	The principal branches of the ***inferior mesenteric artery*** are the: (1) ***left colic artery,*** that supplies the transverse and descending colons; (2) ***sigmoid arteries,*** that supply the descending and sigmoid colons; and (3) ***superior rectal artery,*** that supplies the rectum.
PARIETAL	Supply structures of body wall.
Inferior phrenics	***Inferior phrenic arteries*** are distributed to undersurface of diaphragm and adrenal (suprarenal) glands.
Lumbars	***Lumbar arteries*** supply spinal cord and its meninges and muscles and skin of lumbar region of back.
Middle sacral	***Middle sacral artery*** supplies sacrum, coccyx, and rectum.

Label Figure 18.6.

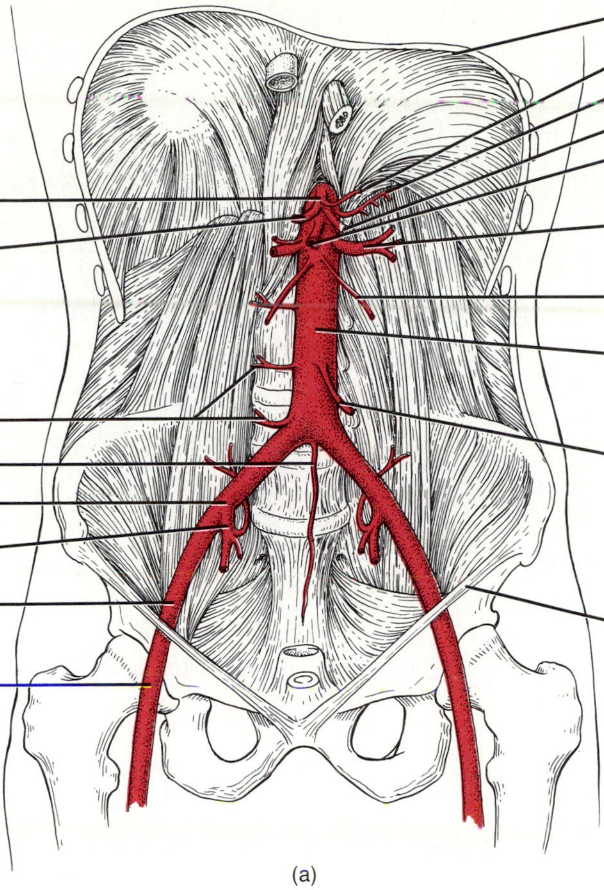

(a)

___ Celiac
___ Common hepatic
___ Inferior mesenteric
___ Left adrenal (suprarenal)
___ Left gastric
___ Left gonadal (testicular or ovarian)
___ Left renal
___ Middle sacral
___ Right lumbars
___ Splenic
___ Superior mesenteric

FIGURE 18.6 Abdominal arteries. (a) Abdominal aorta and its principal branches in anterior view.

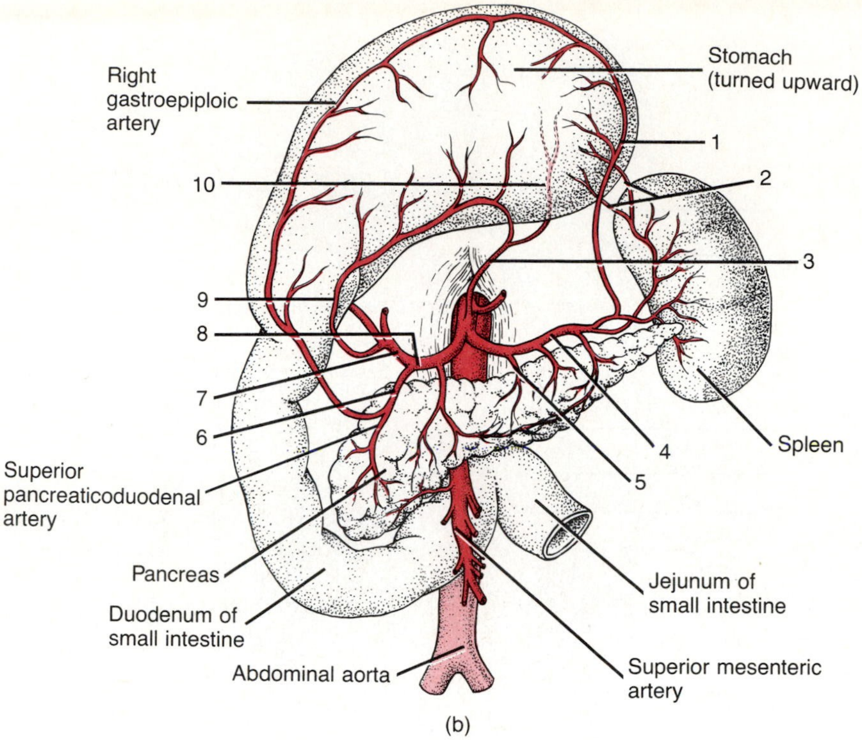

___ Common hepatic
___ Esophageal branch of left gastric
___ Gastroduodenal
___ Hepatic artery proper
___ Left gastric
___ Left gastroepiploic
___ Pancreatic
___ Right gastric
___ Short gastric
___ Splenic

FIGURE 18.6 ***(Continued)*** Abdominal arteries. (b) Branches of common hepatic, left gastric, and splenic arteries in anterior view.

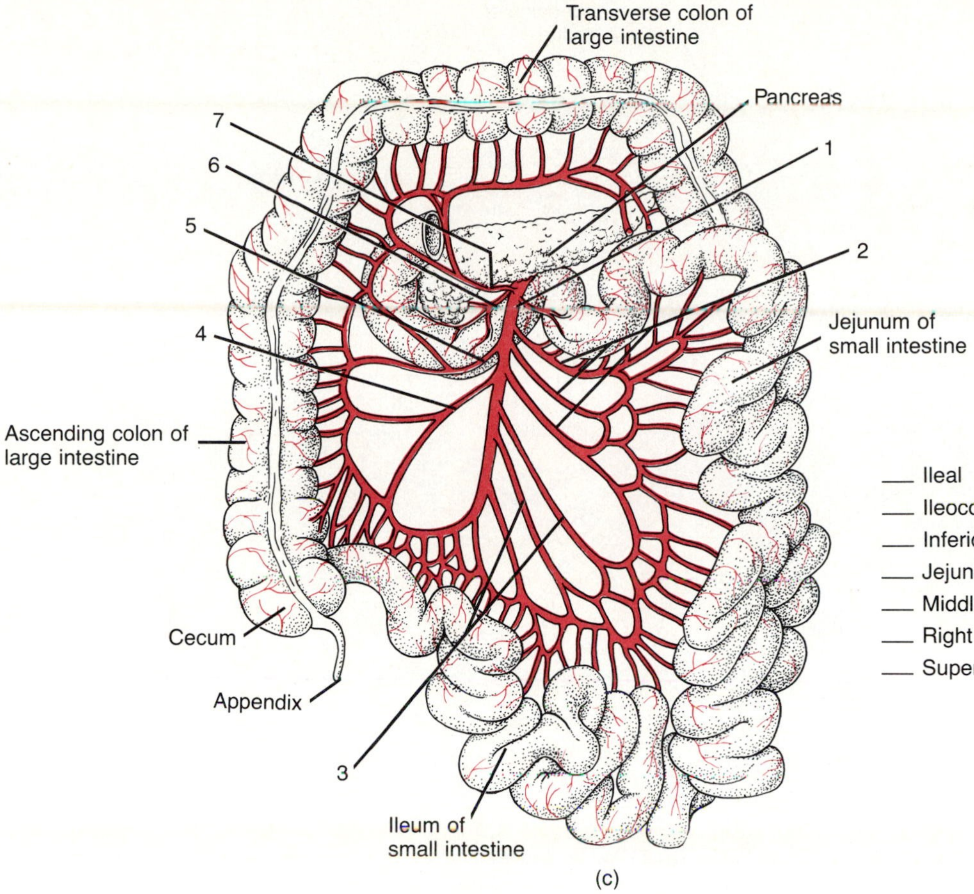

___ Ileal
___ Ileocolic
___ Inferior pancreaticoduodenal
___ Jejunal
___ Middle colic
___ Right colic
___ Superior mesenteric

FIGURE 18.6 ***(Continued)*** Abdominal arteries. (c) Branches of superior mesenteric artery in anterior view.

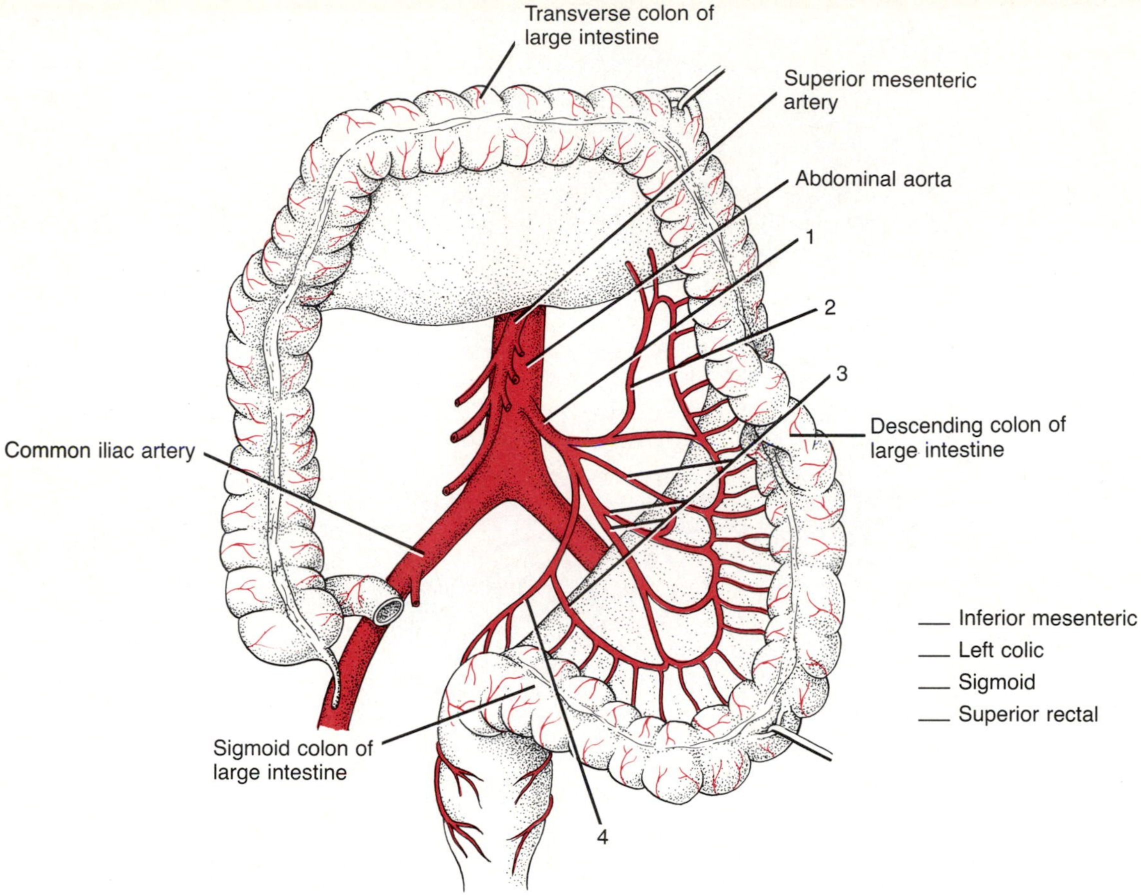

(d)

FIGURE 18.6 ***(Continued)*** Abdominal arteries. (d) Branches of inferior mesenteric artery in anterior view.

TABLE 18.5
ARTERIES OF PELVIS AND LOWER EXTREMITIES (Figure 18.7)

Branch	Description and region supplied
Common iliacs	At about level of fourth lumbar vertebra, abdominal aorta divides into right and left ***common iliac arteries.*** Each passes downward about 5 cm (2 in.) and gives rise to two branches: internal iliac and external iliac.
Internal iliacs	***Internal iliac (hypogastric) arteries*** form branches that supply psoas major, quadratus lumborum, medial side of each thigh, urinary bladder, rectum, prostate gland, uterus, ductus (vas) deferens, and vagina.
External iliacs	***External iliac arteries*** diverge through greater (false) pelvis, enter thighs, and become right and left ***femoral arteries.*** Both femorals send branches back up to genitals and wall of abdomen. Other branches run to muscles of thigh. Femoral continues down medial and posterior side of thigh at back of knee joint, where it becomes ***popliteal artery.*** Between knee and ankle, popliteal runs down back of leg and is called ***posterior tibial artery.*** Below knee, ***peroneal artery*** branches off posterior tibial to supply structures on medial side of fibula and calcaneus. In calf, ***anterior tibial artery*** branches off popliteal and runs along front of leg. At ankle, it becomes ***dorsalis pedis artery.*** At ankle, posterior tibial divides into ***medial*** and ***lateral plantar arteries.***

Label Figure 18.7.

Write the names of the missing arteries in the following scheme of circulation. Be sure to specify left or right, where applicable.

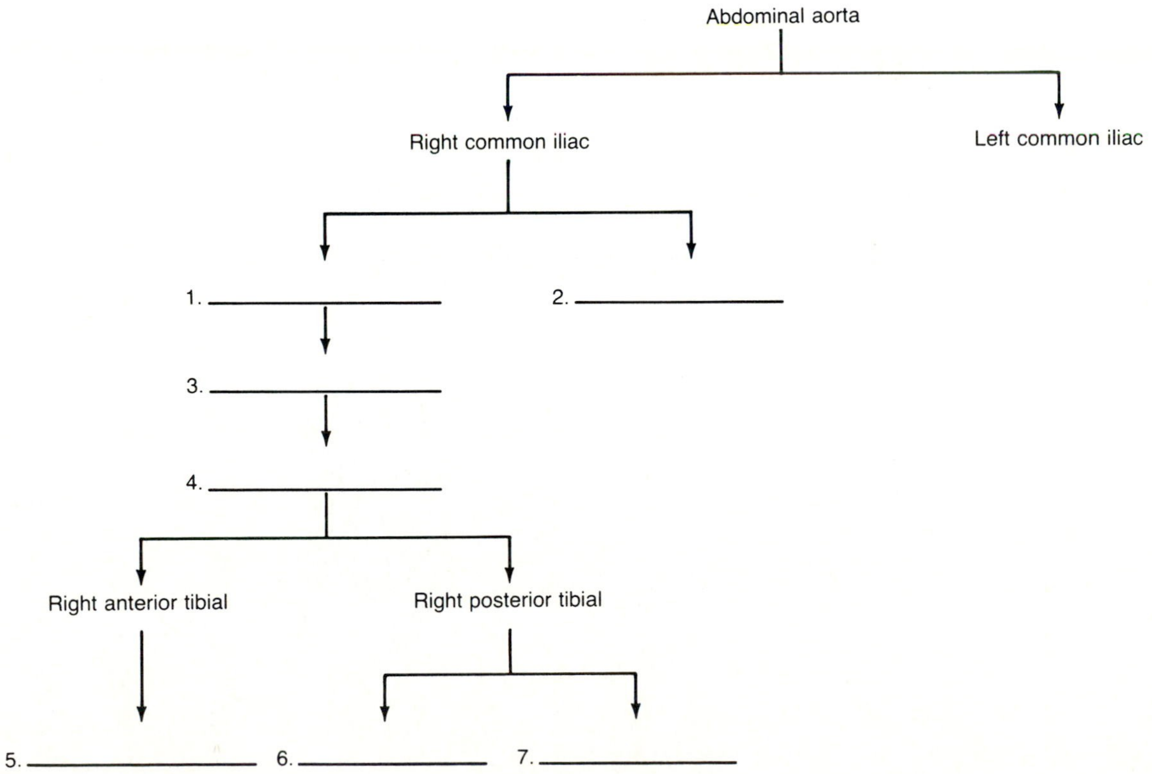

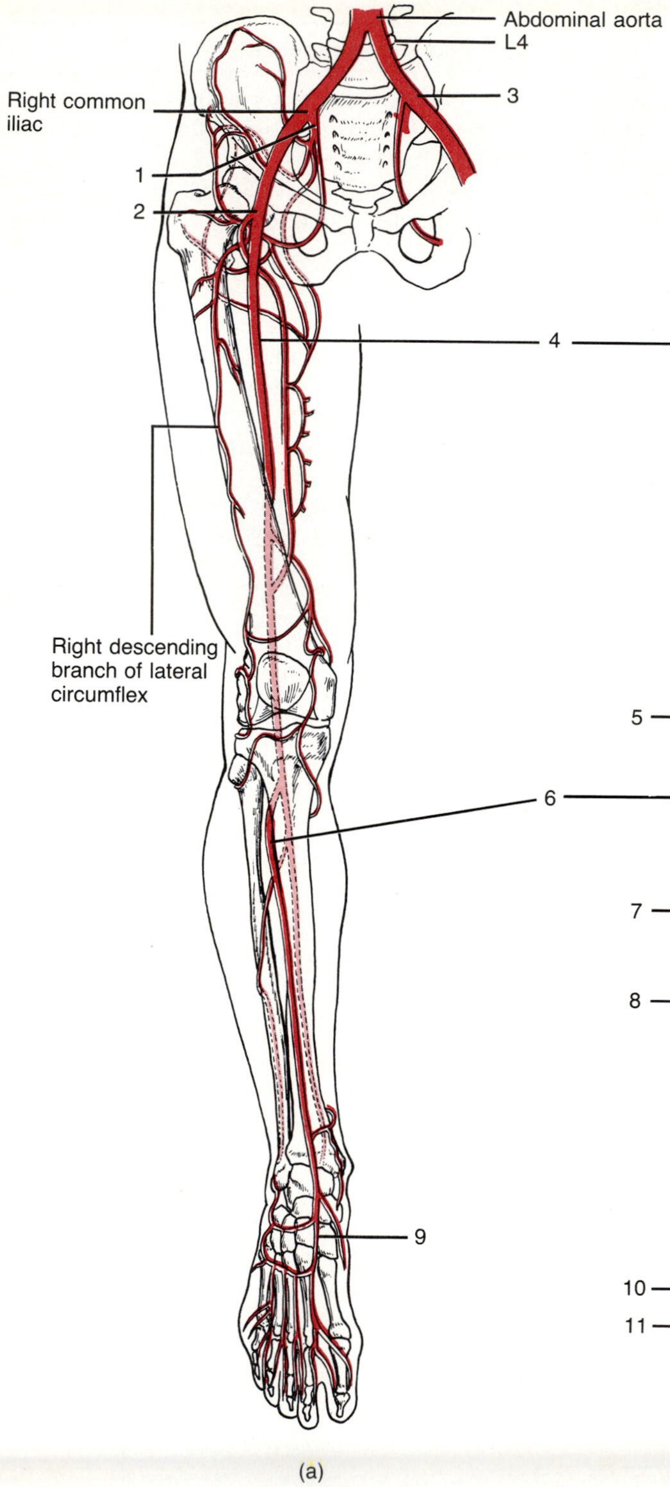

___ Left common iliac
___ Right anterior tibial
___ Right dorsalis pedis
___ Right external iliac
___ Right femoral
___ Right internal iliac (hypogastric)
___ Right lateral plantar
___ Right medial plantar
___ Right peroneal
___ Right popliteal
___ Right posterior tibial

FIGURE 18.7 Arteries of pelvis and right lower extremity. (a) Anterior view. (b) Posterior view.

TABLE 18.6
VEINS OF SYSTEMIC CIRCULATION (See Figure 18.9)

Vein	Description and region drained
	All systemic and cardiac veins return blood to the right atrium of the heart through one of three large vessels. Return flow in coronary (cardiac) circulation is taken up by ***cardiac veins,*** which empty into the large vein of the heart, the ***coronary sinus.*** From here, the blood empties into the right atrium of the heart. Return flow in systemic and coronary circulation empties into the superior vena cava, inferior vena cava, or coronary sinus.
Superior vena cava (SVC)	Veins of the head and neck, upper extremities, and some from thorax empty into the ***superior vena cava (SVC).***
Inferior vena cava (IVC)	Veins of the abdomen, pelvis, and lower extremities and some from thorax empty into the ***inferior vena cava (IVC).***

Write the names of the missing veins in the following scheme of circulation. Be sure to specify left or right, where applicable.

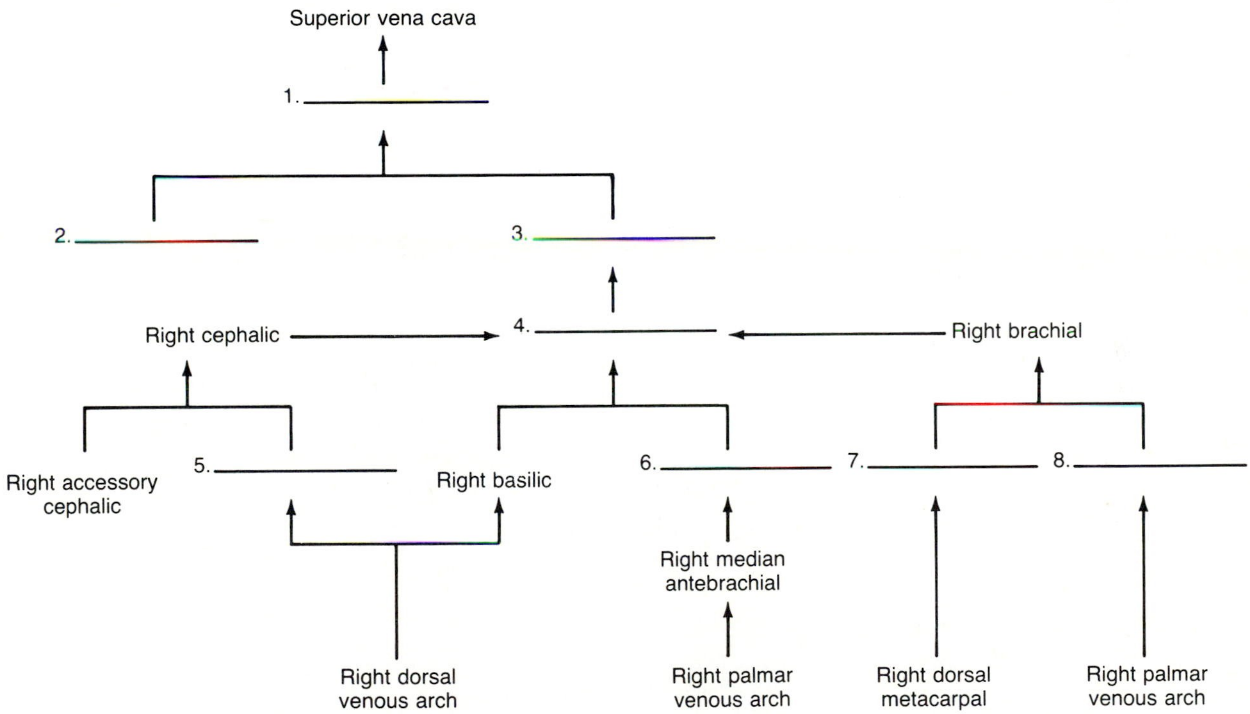

TABLE 18.7
VEINS OF HEAD AND NECK (Figure 18.8)

Vein	Description and region drained
Internal jugulars	Right and left ***internal jugular veins*** arise as continuation of ***sigmoid sinuses*** at base of skull. Intracranial vascular sinuses are located between layers of dura mater and receive blood from brain. Other sinuses that drain into internal jugular include ***superior sagittal sinus, inferior sagittal sinus, straight sinus,*** and ***transverse (lateral) sinuses.*** Internal jugulars descend on either side of neck. They receive blood from face and neck and pass behind clavicles, where they join with right and left ***subclavian veins.*** Unions of internal jugulars and subclavians form right and left ***brachiocephalic veins.*** Form here blood flows into ***superior vena cava.***
External jugulars	Right and left ***external jugular veins*** run down neck along outside of internal jugulars. They drain blood from parotid (salivary) glands, facial muscles, scalp, and other superficial structures into ***subclavian veins.***

Label Figure 18.8.

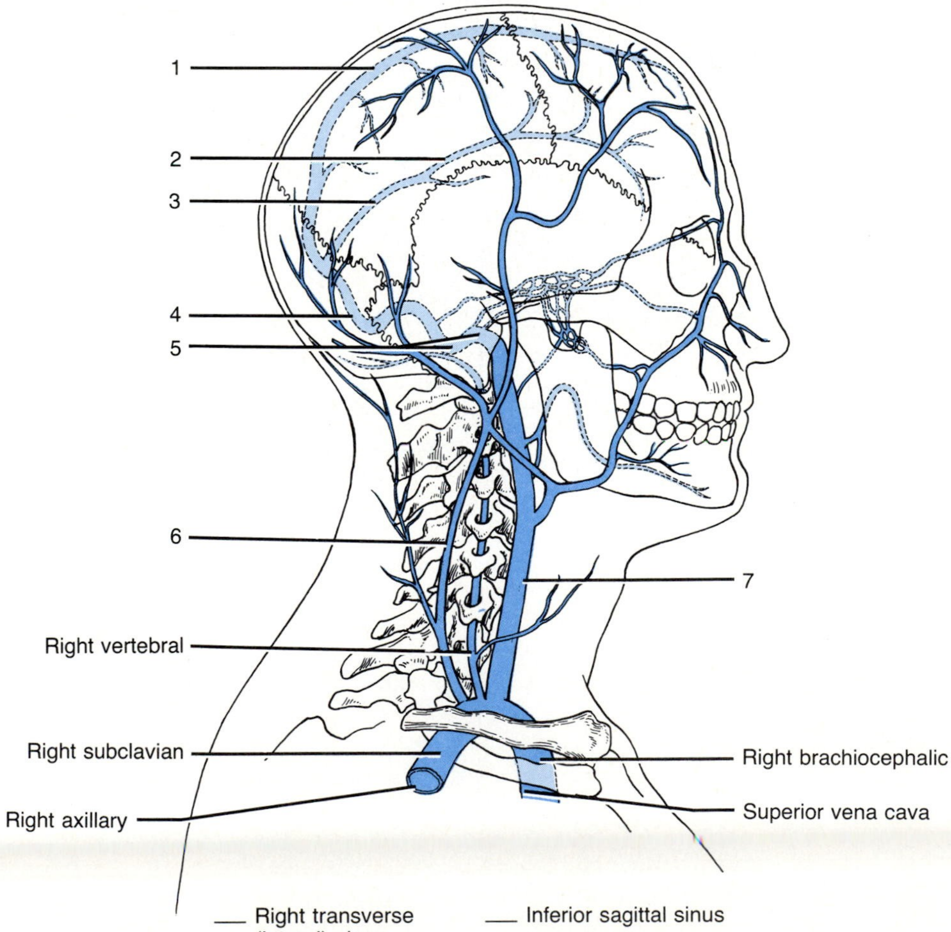

___ Right transverse (lateral) sinus
___ Straight sinus
___ Superior sagittal sinus
___ Inferior sagittal sinus
___ Right external jugular
___ Right internal jugular
___ Right sigmoid sinus

FIGURE 18.8 Veins of neck and head in right lateral view.

TABLE 18.8
VEINS OF UPPER EXTREMITIES (Figure 18.9)

Vein	Description and region drained
	Blood from each upper extremity is returned to heart by deep and superficial veins. Both sets of veins contain valves.
SUPERFICIAL	Superficial veins are just below skin and anastomose extensively with each other and with deep veins.
Cephalics	***Cephalic vein*** of each upper extremity begins in the medial part of ***dorsal arch*** of hand and winds upward around radial border of forearm. Just below elbow, vein unites with ***accessory cephalic vein*** to form cephalic vein of upper extremity and eventually empties into ***axillary vein.***
Basilics	***Basilic vein*** of each upper extremity originates in ulnar part of dorsal arch. It extends along posterior surface of ulna to point below elbow where it receives ***median cubital vein.*** If a vein must be punctured for an injection, transfusion, or removal of a blood sample, median cubitals are preferred. The median cubital joins the basilic vein to form the ***axillary vein.***
Median antebrachials	***Median antebrachial veins*** drain ***palmar venous arch,*** ascend on ulnar side of anterior forearm, and end in median cubital veins.
DEEP	Located deep in the body, usually accompany arteries, and bear same names as corresponding arteries.
Radials	***Radial veins*** receive dorsal metacarpal veins.
Ulnars	***Ulnar veins*** receive tributaries from ***palmar venous arch.*** Radial and ulnar veins unite in bend of elbow to form brachial veins.
Brachials	Located on either side of brachial artery, ***brachial veins*** join into axillary veins.
Axillaries	***Axillary veins*** are a continuation of brachials and basilics. Axillaries end at first rib, where they become subclavians.
Subclavians	***Right*** and ***left subclavian veins*** unite with internal jugulars to form ***brachiocephalic veins.*** Thoracic duct of lymphatic system delivers lymph into left subclavian vein at junction with internal jugular. Right lymphatic duct delivers lymph into right subclavian vein at corresponding junction.

Label Figure 18.9.

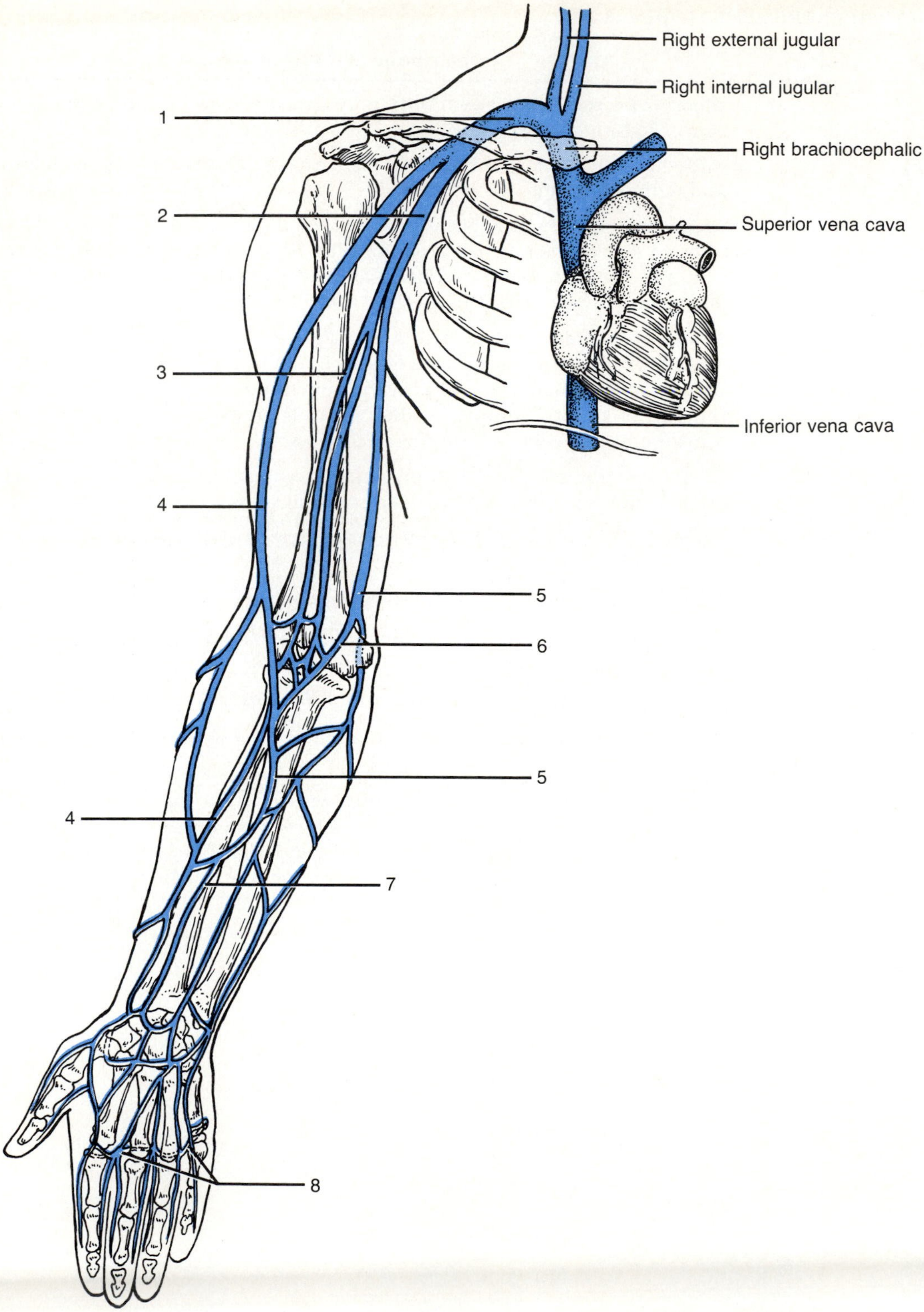

___ Right axillary
___ Right basilic
___ Right brachial
___ Right cephalic
___ Right median antebrachial
___ Right median cubital
___ Right palmar venous arch
___ Right subclavian

FIGURE 18.9 Veins of right upper extremity in anterior view.

TABLE 18.9
VEINS OF THORAX (Figure 18.10)

Vein	Description and region drained
	Principal thoracic vessels that empty into superior vena cava are brachiocephalic and azygos veins.
Brachiocephalic	Right and left ***brachiocephalic veins,*** formed by union of subclavians and internal jugulars, drain blood from head, neck, upper extremities, mammary glands, and upper thorax. Brachiocephalics unite to form ***superior vena cava.***
Azygos	***Azygos veins,*** besides collecting blood from thorax, serve as bypass for inferior vena cava that drains blood from lower body. Several small veins directly link azygos veins with inferior vena cava. Large veins that drain lower extremities and abdomen may drain blood into azygos. If inferior vena cava or hepatic portal vein becomes obstructed, azygos veins can return blood from lower body to superior vena cava.
Azygos	***Azygos vein*** lies in front of vertebral column, slightly right of midline. The vein begins as continuation of right ascending lumbar vein and connects with inferior vena cava, right common iliac, and lumbar veins. Azygos receives blood from (1) right ***intercostal veins*** that drain chest muscles; (2) hemiazygos and accessory hemiazygos veins; (3) several ***esophageal, mediastinal,*** and ***pericardial veins;*** and (4) right ***bronchial vein.*** Vein ascends to fourth thoracic vertebra, arches over right lung, and empties into superior vena cava.
Hemiazygos	***Hemiazygos vein*** is in front of vertebral column and slightly left of midline, beginning as continuation of left ascending lumbar vein. The vein receives blood from lower four or five ***intercostal veins*** and some ***esophageal*** and ***mediastinal veins.*** At level of ninth thoracic vertebra, it joins azygos vein.
Accessory hemiazygos	***Accessory hemiazygos vein*** is also in front and to left of vertebral column. It receives blood from three or four ***intercostal veins*** and left ***bronchial vein*** and joins azygos at level of eighth thoracic vertebra.

Label Figure 18.10, using Figure 18.6 for reference, if necessary.

TABLE 18.10
VEINS OF ABDOMEN AND PELVIS (Figure 18.10)

Vein	Description and region drained
Inferior vena cava	***Inferior vena cava,*** the largest vein of the body, is formed by union of two common iliac veins that drain lower extremities and abdomen. Inferior vena cava extends upward through abdomen and thorax to right atrium. Numerous small veins enter the inferior vena cava. Most carry return flow from branches of abdominal aorta and names correspond to names of arteries.
Common iliacs	***Common iliac veins*** are formed by union of internal (hypogastric) and external iliac veins and represent distal continuation of inferior vena cava at its bifurcation.
Internal iliacs (hypogastrics)	Tributaries of ***internal iliac (hypogastric) veins*** basically correspond with branches of external iliac arteries. Internal iliacs drain gluteal muscles, medial side of thigh, urinary bladder, rectum, prostate gland, ductus (vas) deferens, uterus, and vagina.
External iliacs	***External iliac veins*** are continuation of femoral veins and receive blood from lower extremities and inferior part of anterior abdominal wall.
Renals	***Renal veins*** drain kidneys.
Gonadals (testiculars or ovarians)	***Testicular veins*** drain testes (left testicular vein empties into left renal vein) and ***ovarian veins*** drain ovaries (left ovarian vein empties into left renal vein).
Suprarenals	***Suprarenal veins*** drain adrenal (suprarenal) glands (left suprarenal vein empties into left renal vein).
Inferior phrenics	***Inferior phrenic veins*** drain diaphragm (left inferior phrenic vein sends tributary to left renal vein).
Hepatics	***Hepatic veins*** drain liver.
Lumbars	A series of parallel ***lumbar veins*** drain blood from both sides of posterior abdominal wall. Lumbars connect at right angles with right and left ***ascending lumbar veins,*** which form origin of corresponding azygos or hemiazygos vein. Lumbars drain blood into ascending lumbars and then run to inferior vena cava, where they release remainder of flow.

Label Figure 18.10.

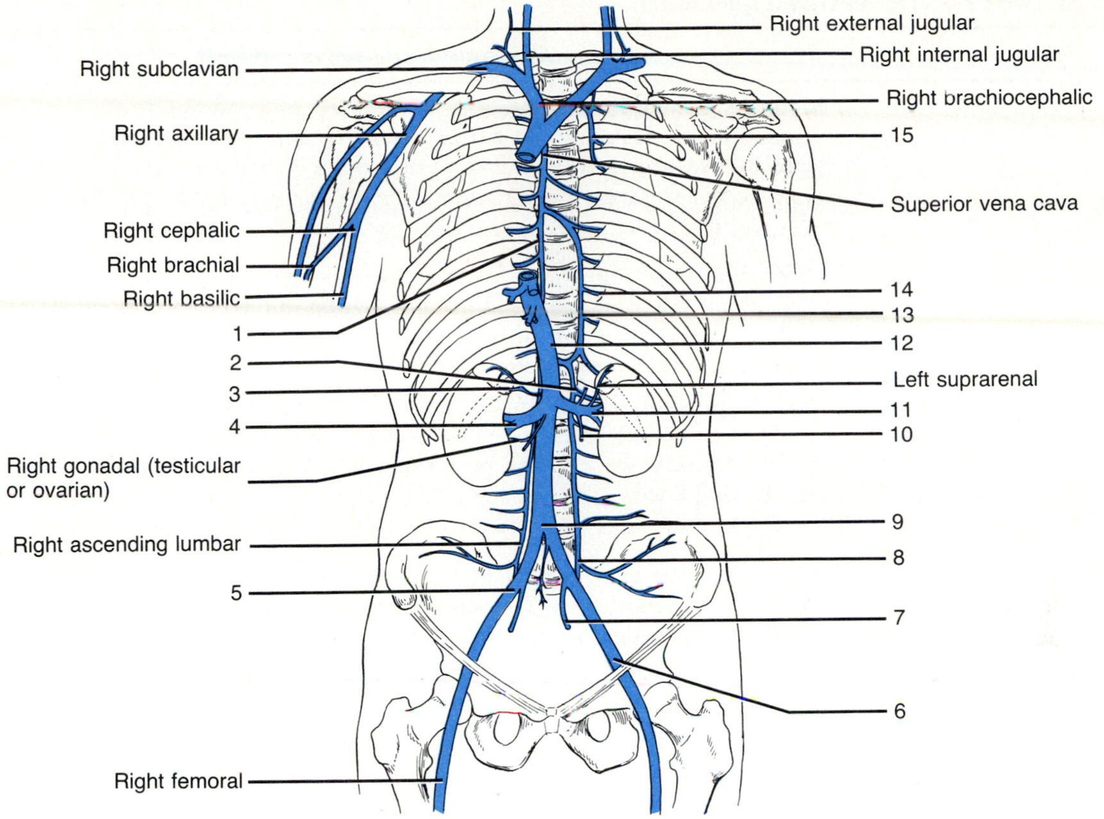

___ Accessory hemiazygos
___ Azygos
___ Hemiazygos
___ Hepatic
___ Inferior vena cava
___ Left ascending lumbar
___ Left external iliac
___ Left gonadal (testicular or ovarian)
___ Left inferior phrenic
___ Left intercostals
___ Left internal iliac (hypogastric)
___ Left renal
___ Right adrenal (suprarenal)
___ Right common iliac
___ Right renal

FIGURE 18.10 Veins of thorax, abdomen, and pelvis in anterior view.

TABLE 18.11
VEINS OF LOWER EXTREMITIES (Figure 18.11)

Vein	Description and region drained
	Blood from each lower extremity is returned by superficial set and deep set of veins. Superficials are formed from extensive anastomoses close to surface. Deep veins follow large arterial trunks. Both sets have valves.
SUPERFICIAL	Main superficial veins are great saphenous and small saphenous. Both, especially great saphenous, may become varicosed.
Great saphenous	***Great saphenous vein***, longest vein in body, begins at medial end of ***dorsal venous arch*** of foot. The vein passes in front of medial malleolus and then upward along medial aspect of leg and thigh. It receives tributaries from superficial tissues, connects with deep veins, and empties into femoral vein in groin.
Small saphenous	***Small saphenous vein*** begins at lateral end of dorsal venous arch of foot. The vein passes behind lateral malleolus and ascends under skin of back of leg. It receives blood from foot and posterior portion of leg and empties into popliteal vein behind knee.
DEEP	
Posterior tibial	***Posterior tibial vein*** is formed by union of ***medial*** and ***lateral plantar veins*** behind medial malleolus. The vein ascends deep in muscle at back of leg, receives blood from ***peroneal vein,*** and unites with anterior tibial vein just below knee. These veins begin in the ***plantar arch*** under the bones of the foot.
Anterior tibial	***Anterior tibial vein*** is upward continuation of ***dorsalis pedis*** veins in foot. The vein runs between tibia and fibula and unites with posterior tibial to form popliteal vein.
Popliteal	***Popliteal vein,*** just behind knee, receives blood from anterior and posterior tibials and small saphenous vein.
Femoral	***Femoral vein*** is upward continuation of popliteal just above knee. Femorals run up posterior of thighs and drain deep structures of thighs. After receiving great saphenous veins in groin, they continue as right and left ***external iliac veins.*** Right and left ***internal iliac veins*** receive blood from pelvic wall and viscera, external genitals, buttocks, and medial aspect of thigh. Right and left ***common iliac veins*** are formed by union of internal and external iliacs. Common iliacs unite to form inferior vena cava.

Label Figure 18.11.

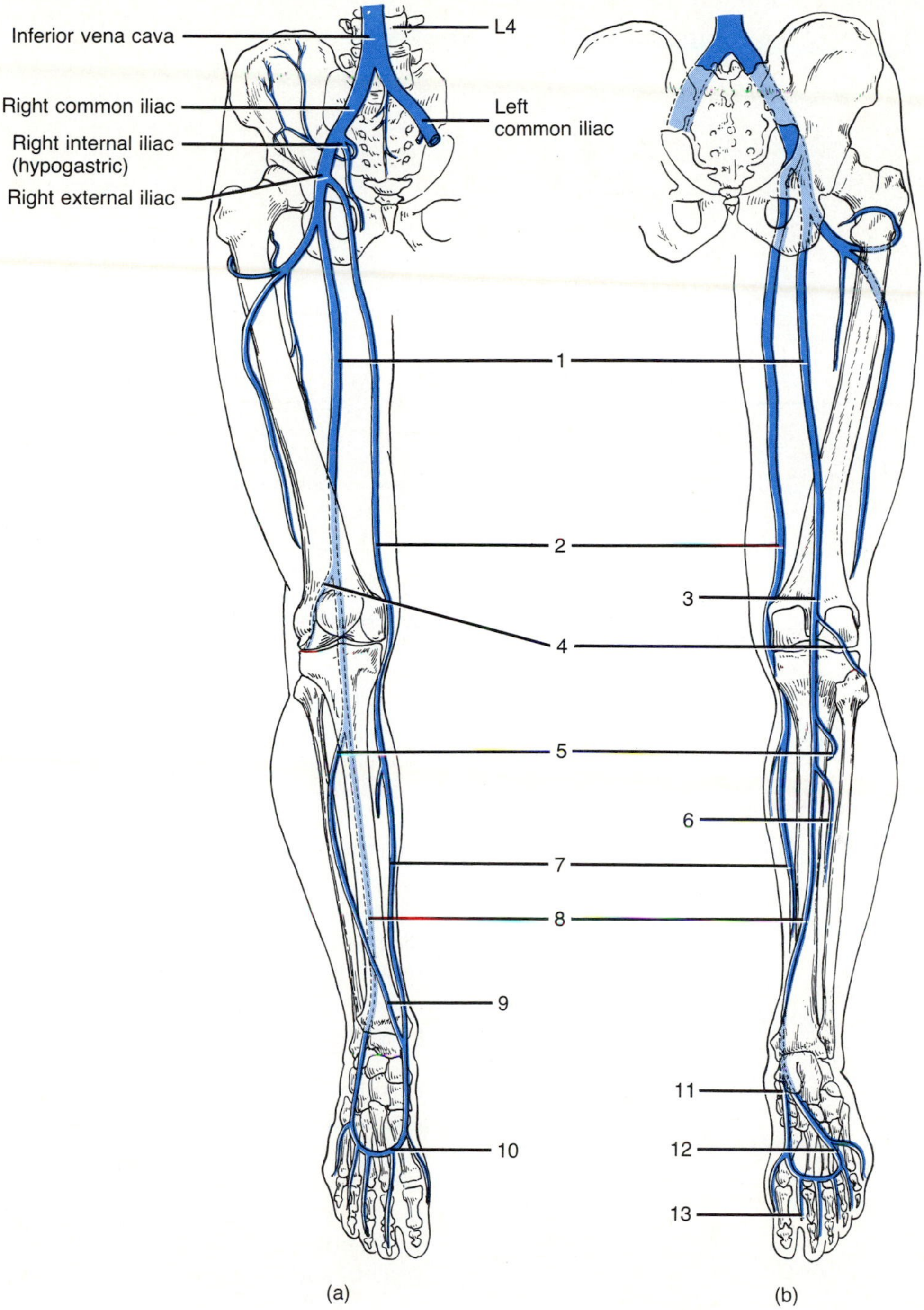

___ Plantar arch
___ Right anterior tibial
___ Right dorsalis pedis
___ Right dorsal venous arch
___ Right femoral
___ Right great saphenous
___ Right great saphenous (Note that two numbers are given to this vein)
___ Right lateral plantar
___ Right medial plantar
___ Right peroneal
___ Right popliteal
___ Right posterior tibial
___ Right small saphenous

FIGURE 18.11 Veins of pelvis and right lower extremity. (a) Anterior view. (b) Posterior view.

Write the names of the missing veins in the following scheme of circulation. Be sure to indicate left or right where applicable.

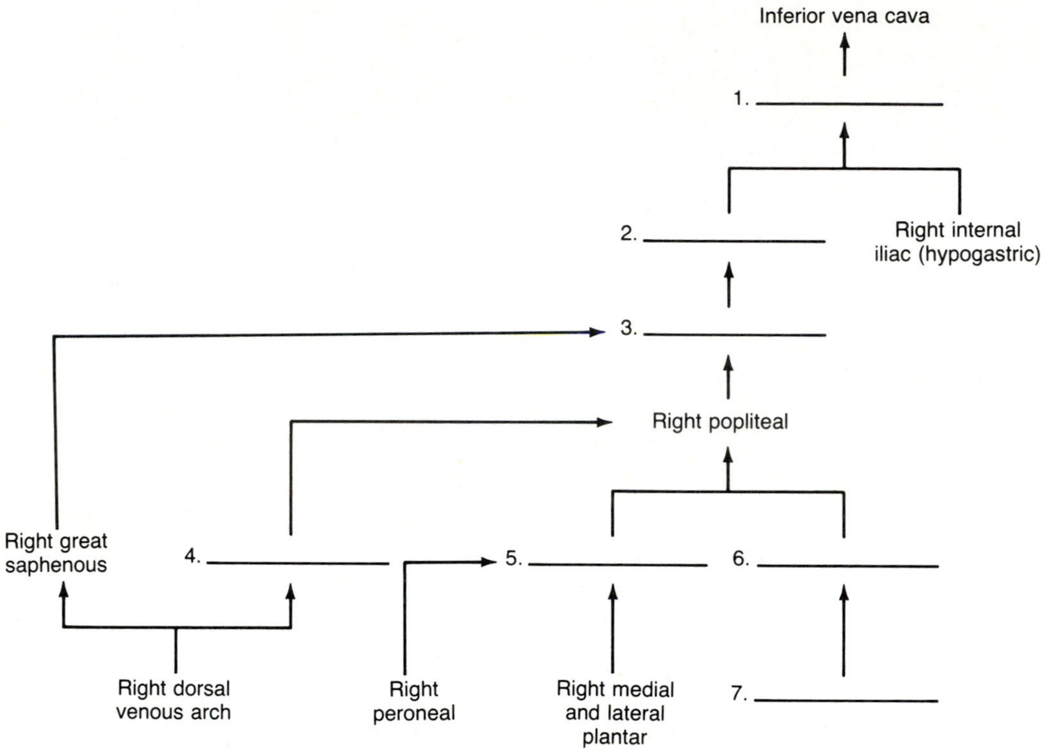

2. Hepatic Portal Circulation

Hepatic portal circulation refers to the flow of venous blood from the organs of the gastrointestinal tract to the liver before it is returned to the heart (Figure 18.12). The blood vessels contained in this circulatory route include the ***hepatic portal vein, superior mesenteric vein, splenic vein, gastric vein, pyloric vein, gastroepiploic vein, pancreatic veins, inferior mesenteric veins***, and ***cystic vein***. Ultimately, blood leaves the liver through the hepatic veins, which enter the inferior vena cava.

Using your textbook, charts, or models for reference, label Figure 18.12.

3. Pulmonary Circulation

Pulmonary circulation refers to the flow of deoxygenated blood from the right ventricle to the lungs and the return of oxygenated blood from the lungs to the left atrium (Figure 18.13). The ***right*** and ***left pulmonary arteries*** are the only postnatal (after birth) arteries that carry deoxygenated blood, and the ***pulmonary veins*** are the only postnatal veins that carry oxygenated blood.

Study a chart or model of the pulmonary circulation, trace the path of blood through it, and label Figure 18.13.

4. Fetal Circulation

The developing fetus has nonfunctional lungs and a nonfunctional gastrointestinal tract, and this difference from an adult's circulatory system produces the ***fetal circulation*** (Figure 18.14). The lungs of the fetus are not active in respiration because the fetus derives its oxygen and nutrients and eliminates its carbon dioxide and wastes through the maternal blood. An opening, the ***foramen ovale*** (fo-RĀ-men ō-VAL-ē), is found between the right and left atria. Most blood leaving the right atrium passes through this opening into the left atrium rather than into the right ventricle. Blood that does enter the right ventricle is pumped into the pulmonary trunk, but little of this blood reaches the lungs. Most is conveyed from the pulmonary trunk to the aorta via a small ves-

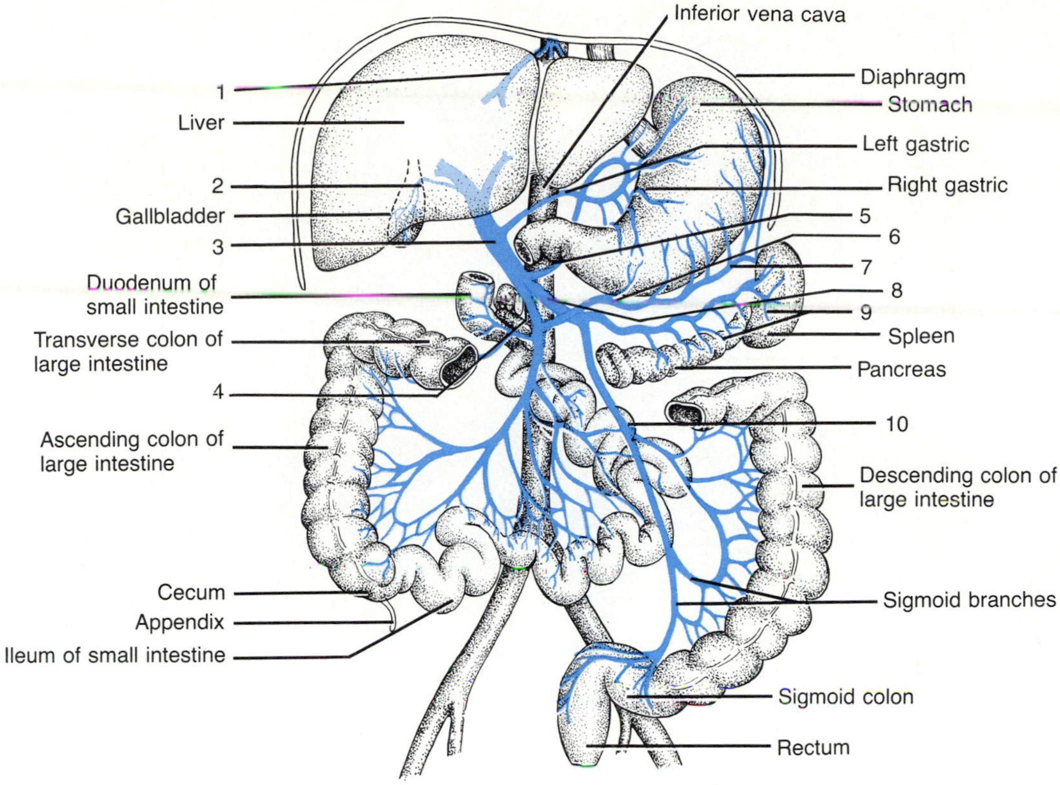

__ Cystic
__ Hepatic
__ Hepatic portal
__ Inferior mesenteric
__ Left gastroepiploic
__ Pancreatic
__ Pyloric
__ Right gastroepiploic
__ Splenic
__ Superior mesenteric

FIGURE 18.12 Hepatic portal circulation.

sel, the ***ductus arteriosus*** (DUK-tus ar-tē-rē-Ō-sus), which closes shortly after birth.

A highly specialized structure called the ***placenta*** (pla-SEN-ta) accomplishes the exchange of substances between maternal and fetal circulation. Other special blood vessels include two ***umbilical arteries*** and an ***umbilical vein***, which extend through the ***umbilical cord***, and the ***ductus venosus*** (DUK-tus ve-NŌ-sus) in the liver.

Label Figure 18.14.

E. BLOOD VESSEL EXERCISE

For each vessel listed, indicate the region supplied (if an artery) or the region drained (if a vein):

1. ***Coronary artery*** ____________________
2. ***Internal iliac veins*** ____________________
3. ***Lumbar arteries*** ____________________
4. ***Renal artery*** ____________________
5. ***Left gastric artery*** ____________________
6. ***External jugular vein*** ____________________

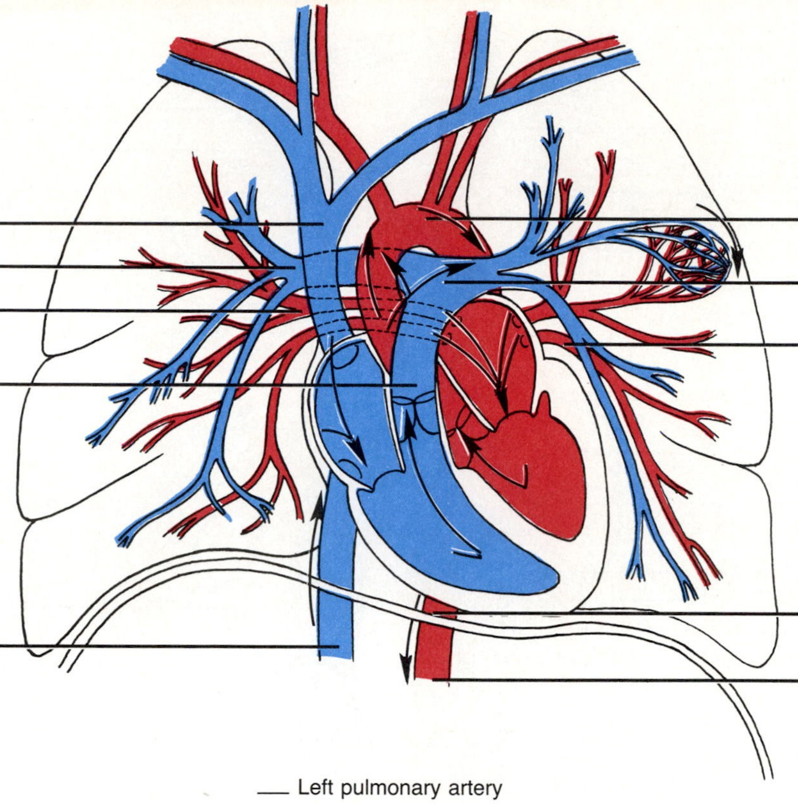

___ Left pulmonary artery
___ Left pulmonary vein
___ Pulmonary trunk
___ Right pulmonary artery
___ Right pulmonary vein

FIGURE 18.13 Pulmonary circulation.

7. ***Left subclavian artery*** ________________

8. ***Axillary vein*** ________________

9. ***Brachiocephalic veins*** ________________

10. ***Transverse sinuses*** ________________

11. ***Hepatic artery*** ________________

12. ***Inferior mesenteric artery*** ________________

13. ***Suprarenal artery*** ________________

14. ***Inferior phrenic artery*** ________________

15. ***Great saphenous vein*** ________________

16. ***Popliteal vein*** ________________

17. ***Azygos vein*** ________________

18. ***Internal iliac (hypogastric) artery*** ________________

19. ***Internal carotid artery*** ________________

20. ***Cephalic vein*** ________________

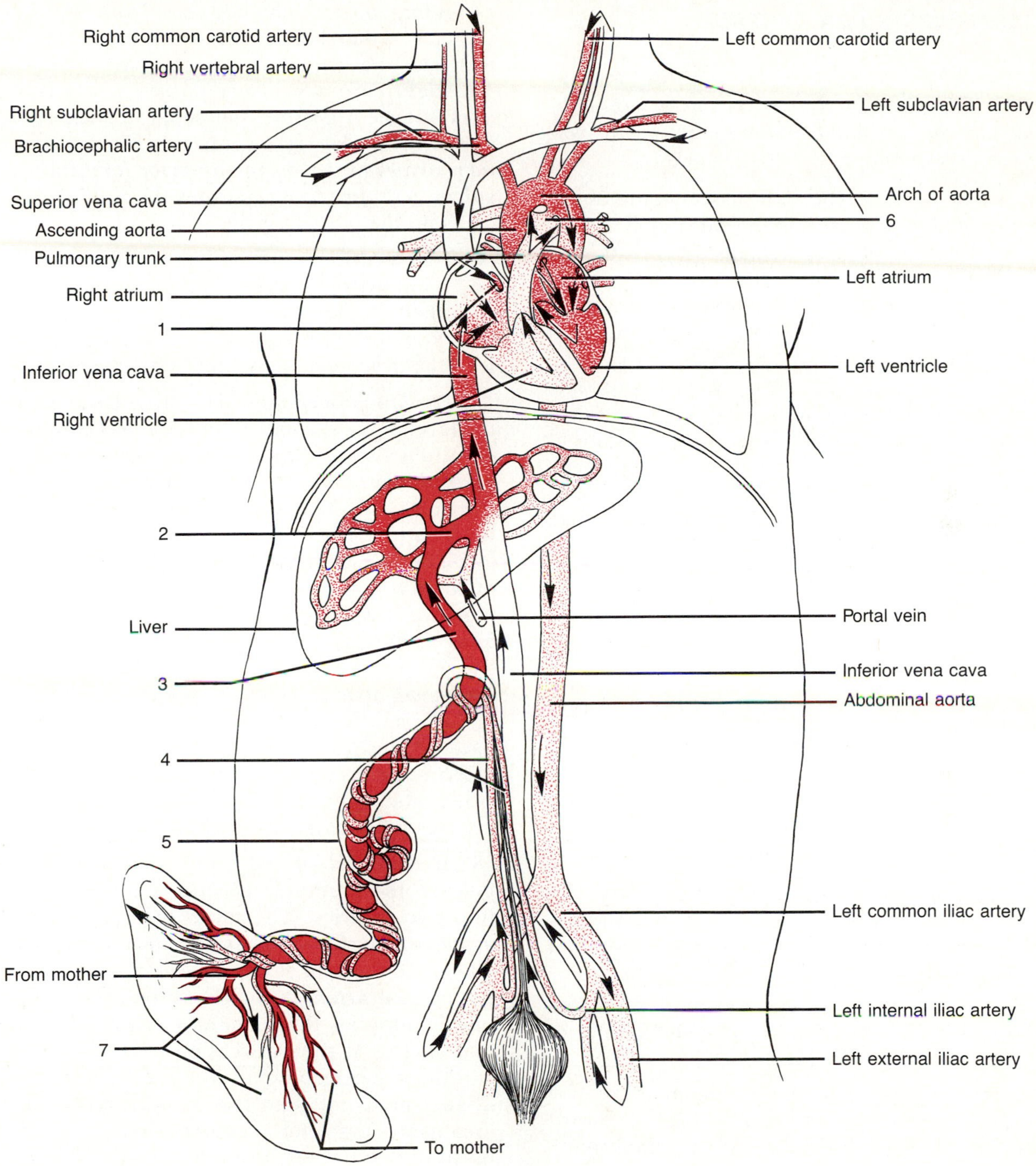

___ Ductus arteriosus
___ Ductus venosus
___ Foramen ovale
___ Placenta
___ Umbilical arteries
___ Umbilical cord
___ Umbilical vein

FIGURE 18.14 Fetal circulation.

F. DISSECTION OF CAT CARDIOVASCULAR SYSTEM

CAUTION! *Please reread Section D. "Precautions Related to Dissection" at the beginning of the laboratory manual on page xii before you begin dissection.*

In this section, you will dissect some of the principal blood vessels in the cat. Blood vessels should be studied in specimens that are at least doubly injected, meaning that the arteries have been injected with red latex and the veins have been injected with blue latex. Triply injected specimens are necessary to observe all of the hepatic portal system. This system is usually injected with yellow latex.

As you dissect blood vessels, bear in mind that they vary somewhat in position from one specimen to the next. Thus, the pattern of blood vessels in your specimen may not be exactly like those shown in the illustrations. Also remember that when you dissect blood vessels, you must free them from surrounding tissues so that they are clearly visible. When doing so, use a forceps or a blunt probe, not a scalpel. Finally, you obtain best results if you dissect blood vessels on the side of the body opposite the side on which you dissected muscles.

1. Arteries

PROCEDURE

1. To examine the major blood vessels of the cat, extend the original thoracic longitudinal incisions posteriorly through the abdominal wall.

2. Using a scissors, cut the wall from the diaphragm to the pubic region. Be careful not to damage viscera below.

3. On both sides of the abdomen make two additional cuts, laterally and dorsally, toward the spine about 4 in. below the diaphragm. Reflect the flaps of the abdominal wall.

4. Locate the ***pulmonary trunk*** exiting from the right ventricle (see Figure 17.5). The trunk branches into a right and a left pulmonary artery. The ***left pulmonary artery*** passes ventral to the aorta to reach the left lung whereas the ***right pulmonary artery*** passes between the arch of the aorta and the heart to reach the right lung.

5. Dorsal to the pulmonary trunk is the ***ascending aorta***, which arises from the left ventricle. At the base of the ascending aorta, try to locate the origins of the coronary arteries. The ***right coronary artery*** passes in the ***coronary sulcus***, a groove that separates the right atrium from the right ventricle. The ***left coronary artery*** passes in the ***anterior longitudinal sulcus***, a groove that separates the right ventricle from the left ventricle.

6. The ascending aorta continues cranially and then turns to the left. This part of the aorta is called the ***arch of the aorta.*** Originating from the arch of the aorta are two large arteries—the brachiocephalic and left subclavian.

7. The ***brachiocephalic artery*** divides at about the level of the second rib into the ***left common carotid artery*** and then into the ***right common carotid artery*** and the ***right subclavian artery*** (Figure 18.15). The left common carotid artery ascends in the neck between the internal jugular vein and the trachea. Among the branches of the left common carotid artery are (1) the ***superior thyroid artery***, which originates at the level of the thyroid cartilage and supplies the thyroid gland and muscles of the neck; (2) the ***laryngeal artery***, which supplies the larynx; and (3) the ***occipital artery***, which supplies the back of the neck. The right common carotid artery supplies the same structures on the right side of the body. The right and left subclavian arteries will be discussed shortly.

8. At the level of the cranial border of the larynx, the common carotid arteries divide into two arteries. The large ***external carotid artery*** primarily supplies the head structures outside the cranial cavity. The very small ***internal carotid artery***, which may be absent, primarily supplies the structures within the cranial cavity. At the bifurcation of the common carotid into external and internal carotids, the dilated area is called the carotid sinus. This area contains stretch receptors that inform the brain of fluctuations in blood pressure so that necessary adjustments can be made. The aortic sinus in the aorta has a similar function.

9. The right and left subclavian arteries supply the head, neck, thoracic wall, and forelimbs. Among their important branches are (1) the ***vertebral arteries***, which supply the brain; (2) the ***internal mammary arteries***, which primarily supply the ventral thoracic

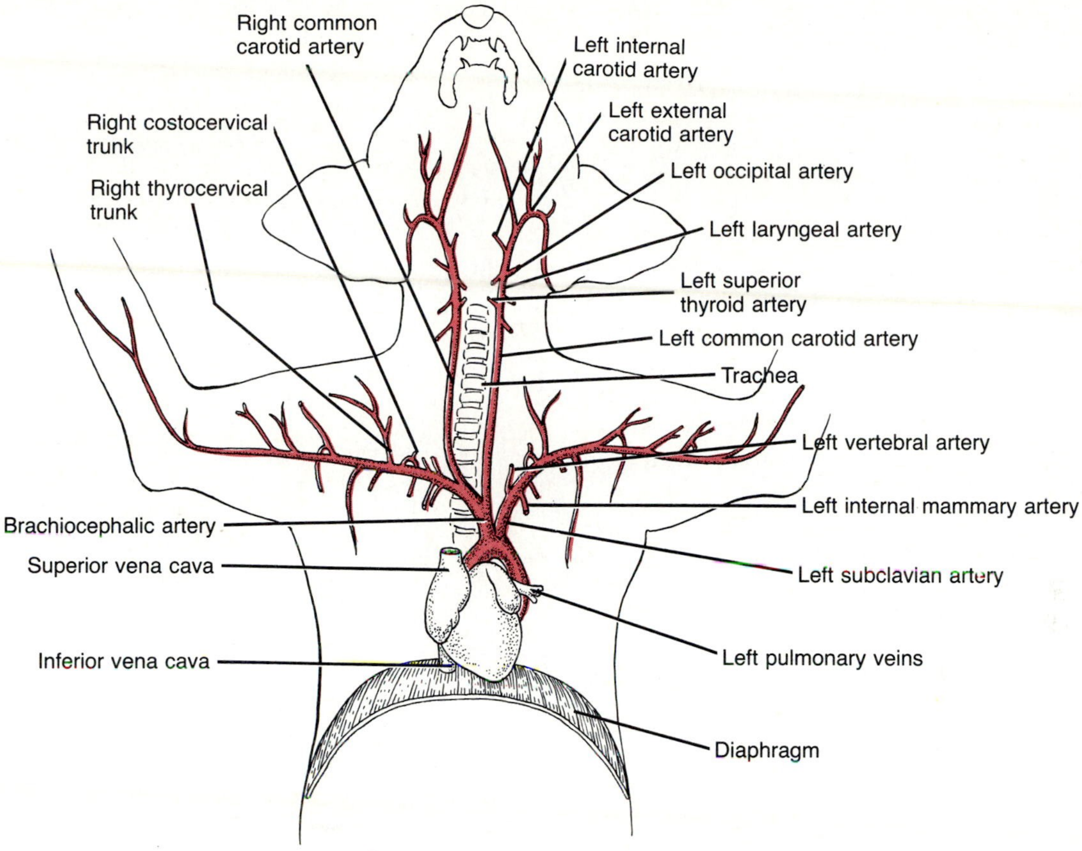

FIGURE 18.15 Arteries of forelimbs, neck, and head in ventral view.

wall; (3) the ***costocervical trunk***, which supplies the neck, back, and intercostal muscles; and (4) the ***thyrocervical trunk***, which supplies some neck and back muscles.

10. The continuation of the subclavian artery out of the thoracic cavity across the axillary space is called the ***axillary artery*** (Figure 18.16). Branches of the axillary artery include (1) the ***ventral thoracic artery***, which supplies the pectoral muscles; (2) the ***long thoracic artery***, which supplies the latissimus dorsi muscle; and (3) the ***subscapular artery***, which supplies the shoulder muscles, dorsal muscles of the arm, and muscles of the scapula.

11. The continuation of the axillary artery into the arm is called the ***brachial artery***. A branch of the brachial artery, the ***anterior humeral circumflex artery***, arises just distal to the subscapular artery and supplies the biceps brachii muscle. Another branch of the brachial artery, the ***deep brachial artery***, is distal to the anterior humeral circumflex. It supplies the triceps brachii, epitrochlearis, and latissimus dorsi muscles. Distal to the elbow, the brachial artery continues as the ***radial artery.*** In the forearm, the radial artery gives off a small branch, the ***ulnar artery.*** The radial and ulnar arteries supply the musculature of the forearm.

12. Continue your dissection of arteries by returning to the arch of the aorta. Follow it caudally as it passes through the thoracic cavity. At this point the aorta is known as the ***thoracic aorta.*** Arising from the thoracic aorta are (1) the ***intercostal arteries*** (10 pairs), which supply the intercostal muscles; (2) the ***bronchial arteries***, which supply the bronchi and lungs; and (3) the ***esophageal arteries***, which supply the esophagus.

13. The aorta passes through the diaphragm at the level of the second lumbar vertebra. The portion of the aorta between the diaphragm and pelvis is called the ***abdominal aorta.***

14. The first branch of the abdominal aorta immediately caudal to the diaphragm is

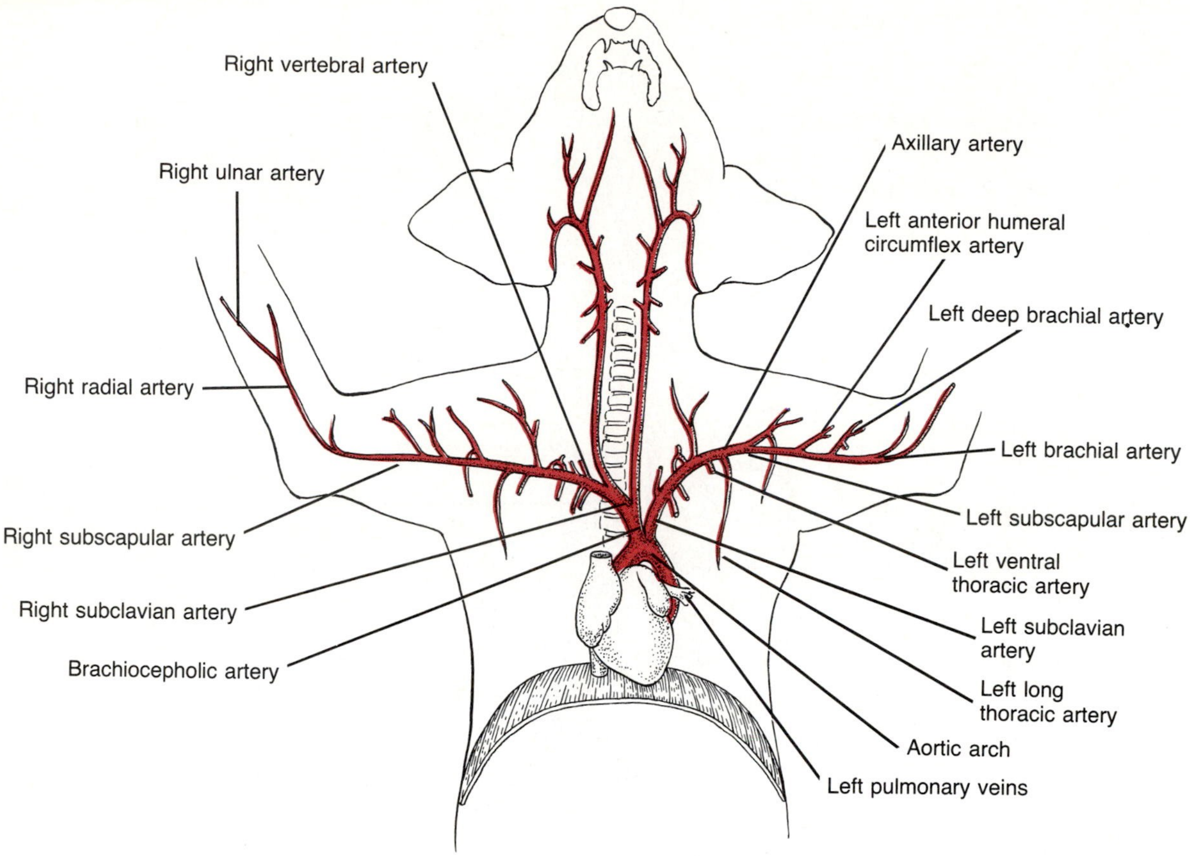

FIGURE 18.16 Arteries of forelimbs in ventral view.

the ***celiac trunk*** (Figure 18.17). It consists of three branches. The ***hepatic artery*** supplies the liver, the ***left gastric artery*** along the lesser curvature of the stomach supplies the stomach, and the ***splenic artery*** supplies the spleen. See if you can also locate the ***cystic artery***, a branch of the hepatic artery which supplies the gallbladder.

15. Caudal to the celiac trunk is the ***superior mesenteric artery.*** Its branches are the ***inferior pancreatoduodenal artery*** to the pancreas and duodenum, the ***ileocolic artery*** to the ileum and cecum, the ***right colic artery*** to the ascending colon, the ***middle colic artery*** to the transverse colon, and the ***intestinal arteries*** to the jejunum and ilium.

16. Just caudal to the superior mesenteric artery, the abdominal aorta gives off a pair of ***adrenolumbar arteries.*** Each gives off an ***inferior phrenic artery*** to the diaphragm, an ***adrenal (suprarenal) artery*** to the adrenal gland, and a ***parietal branch*** to the muscles of the back.

17. The paired ***renal arteries***, caudal to the adrenolumbar arteries, supply the kidneys.

18. In the male, the paired ***internal spermatic arteries*** or ***testicular arteries*** arise from the abdominal aorta caudal to the renal arteries. The internal spermatic arteries pass caudally to the internal inguinal ring and through the inguinal canal where they accompany the ductus deferens of each testis.

19. In the female, the paired ***ovarian arteries*** arise from the abdominal aorta caudal to the renal arteries. They pass laterally in the broad ligament and supply the ovaries.

20. The unpaired ***inferior mesenteric artery*** arises caudal to the internal spermatic or ovarian arteries. Its branches include a ***left colic artery*** to the descending colon and a ***superior hemorrhoidal artery*** to the rectum.

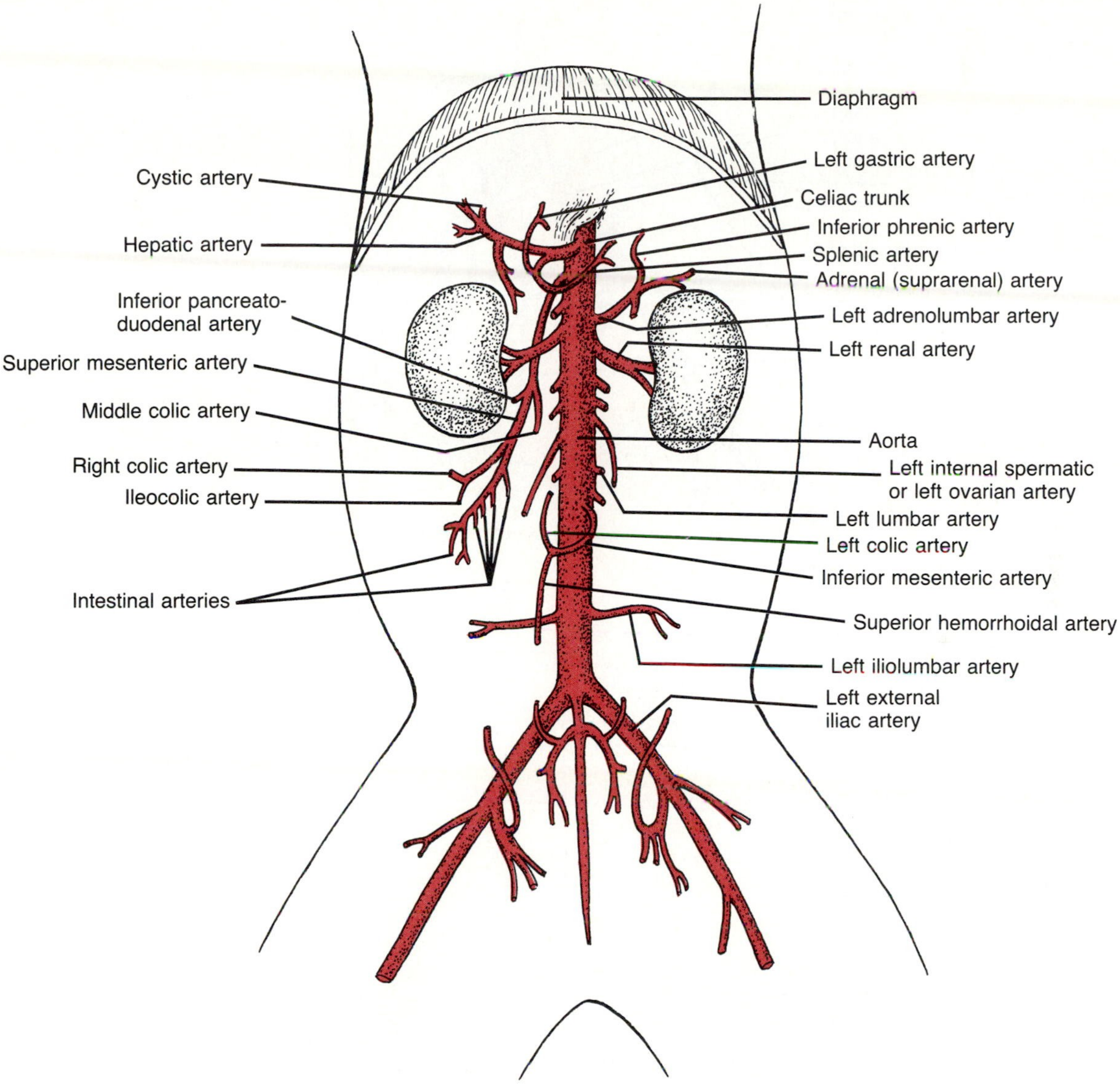

FIGURE 18.17 Arteries of abdomen in ventral view.

21. Seven pairs of ***lumbar arteries*** arise from the dorsal surface of the abdominal aorta and supply the muscles of the dorsal abdominal wall. A pair of ***iliolumbar arteries*** arises near the inferior mesenteric artery and also supplies the muscles of the dorsal abdominal wall.

22. Near the sacrum, the abdominal aorta divides into right and left ***external iliac arteries*** (Figure 18.18). Also from this point of divison, a short common artery extends caudally and gives rise to the paired ***internal iliac (hypogastric) arteries*** and unpaired ***median sacral artery*** that passes along the ventral aspect of the tail. The first branch of the internal iliac artery, the ***umbilical artery***, supplies the urinary bladder. Another branch of the internal iliac artery, the ***superior gluteal artery***, supplies the gluteal muscles and other muscles of the thigh. Still another branch of the internal iliac artery, the ***middle hemorrhoidal artery***, supplies the rectum, anus, urethra, and adjacent structures. The terminal portion of the internal iliac artery, the ***inferior gluteal artery***, supplies thigh muscles.

23. Now we can return to the external iliac arteries. Just before leaving the abdominal cav-

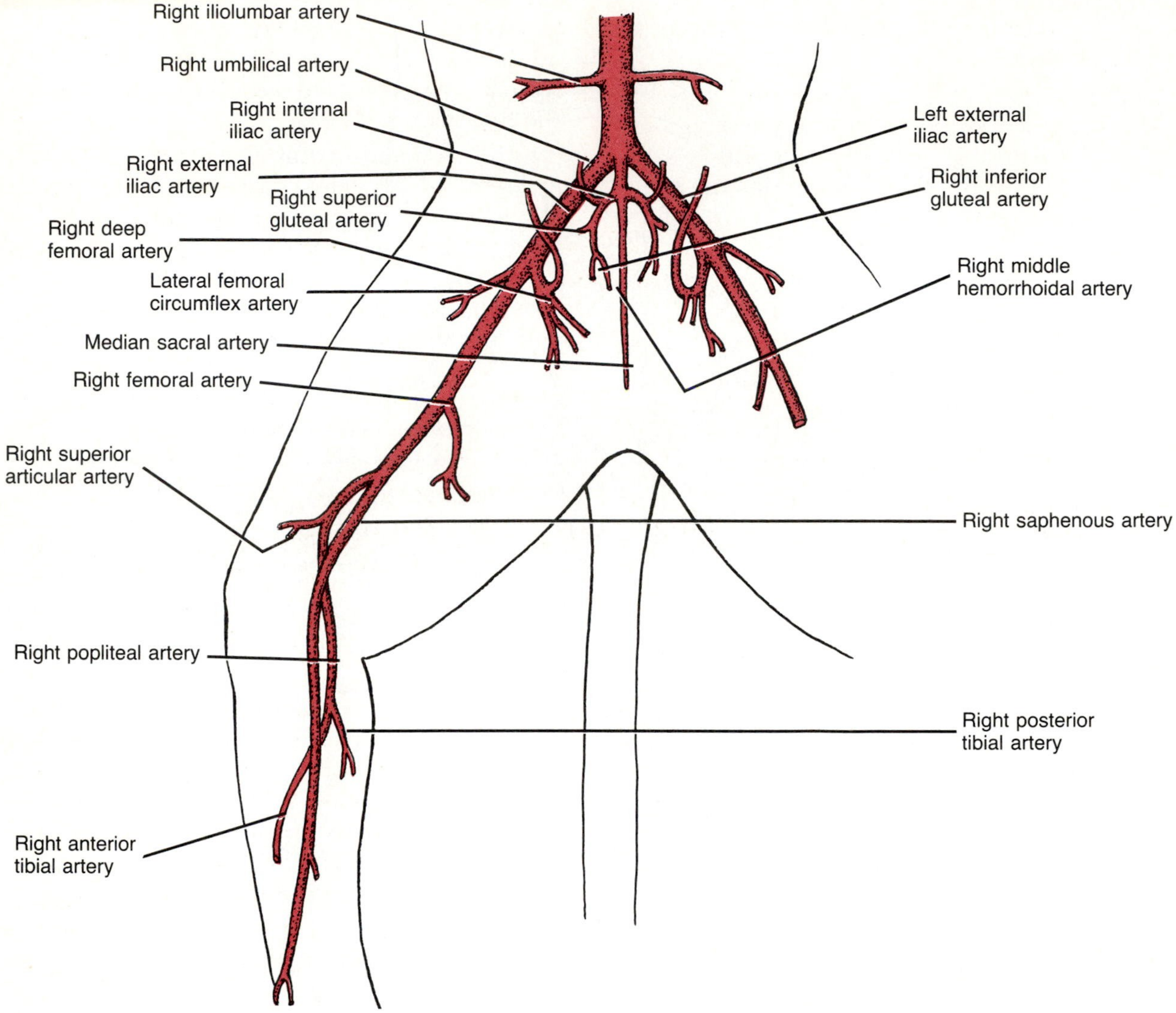

FIGURE 18.18 Arteries of right hindlimb in ventral view.

ity, each external iliac artery gives off a branch, the ***deep femoral artery***, which supplies the muscles of the thigh. Branches of the deep femoral artery supply the rectus abdominis muscle, the urinary bladder, and muscles on the medial aspect of the thigh.

24. The external iliac artery continues outside the abdominal cavity and into the thigh as the ***femoral artery.*** Passing through the thigh, it gives off one branch as the ***lateral femoral circumflex artery*** to the rectus femoris and vastus medialis. Just above the knee, the femoral artery gives off a branch as the ***superior articular artery*** and another as the ***saphenous artery.*** Both supply thigh muscles.

25. In the region of the knee, the femoral artery is known as the ***popliteal artery.*** It divides into an ***anterior tibial artery*** and a ***posterior tibial artery***, which supply the leg and foot.

2. Veins

PROCEDURE

1. In learning the arteries of the cat, we began our study with the heart and then worked outward from the heart to various regions of the body. In learning the veins of the cat, we will begin our study with the various regions of the body and work toward the heart. Whereas

arteries supply various parts of the body with blood, veins drain various parts of the body. That is, veins remove blood and take it back to the heart.

2. The veins of the pelvis and thigh generally correspond to the arteries in those regions. Whereas the artery supplies the region with blood, the corresponding vein drains the same region. An important difference between the arteries and veins of the pelvis is that the internal iliac vein joins the external iliac vein directly to form the common iliac vein. The cat has no common iliac artery.

3. The deep veins of the leg are the ***anterior tibial vein*** and the ***posterior tibial vein*** (Figure 18.19). At the knee the two unite to form the ***popliteal vein***, another deep vein. In the thigh, the politeal vein is known as the ***femoral vein.*** The superficial veins of the lower extremity are the ***greater saphenous vein***, which runs along the medial surface and joins the femoral vein, and the ***lesser saphenous vein***, which runs along the dorsal surface and joins a tributary of the internal iliac vein.

4. The ***external iliac vein*** is a continuation of the femoral vein. It has tributaries corresponding to the arteries and returns blood distributed by these arteries. The ***internal iliac (hypogastric) vein***, with its tributaries, returns blood distributed by the internal iliac artery and its branches. The ***common iliac vein*** is formed by the union of the external and internal iliac veins.

5. The ***inferior vena cava (postcava)***, formed by the union of the common iliac veins, enters the caudal portion of the right atrium of the heart. The tributaries of the inferior vena cava, exclusive of the hepatic portal system, include the ***iliolumbar, lumbar, renal, adrenolumbar***, and ***internal spermatic*** or ***ovarian veins.*** These correspond to the paired branches of the abdominal aorta. Cranial to the diaphragm, the inferior vena cava has no tributaries. After piercing the diaphragm, the inferior vena cava ascends and enters the right atrium.

6. The ***hepatic portal system*** consists of a series of veins that drain blood from the small and large intestines, stomach, spleen, and pancreas and convey the blood to the liver. The blood is delivered to the liver by the hepatic portal vein. Within the liver, the hepatic portal vein gives branches to each lobe. These branches terminate in capillaries called ***liver sinusoids.*** The sinusoids are drained by the hepatic veins, which leave the liver and join the inferior vena cava. Blood delivered to the liver by the hepatic artery also passes through the liver sinusoids.

7. Unless your specimen is triply injected, locating all the veins of the hepatic portal system will be very difficult. However, using Figure 18.20 as a guide, identify as many of the vessels as you can. Notice that the hepatic portal vein receives blood from the spleen and from all digestive organs except the liver. Blood is delivered to the hepatic portal vein by veins corresponding to the arteries that distribute blood to these viscera. In the cat, the ***hepatic portal vein*** is formed by the union of the ***gastrosplenic vein*** and the ***superior mesenteric vein.*** The superior mesenteric vein is formed by the ***inferior mesenteric vein, intestinal veins***, and ***ileocolic vein***, all of which drain the small intestine, and by the ***anterior pancreatoduodenal vein*** from the pancreas and small intestine (Figure 18.20).

8. In the forearm identify the ulnar and radial veins. The ***ulnar vein*** joins the ***radial vein*** (Figure 18.21). Also find the ***cephalic vein***, which joins the ***transverse scapular vein***, and the ***median cubital vein***, a communicating vein between the cephalic and brachial veins.

9. In the arm, identify the ***brachial vein*** and its tributaries. In general, this vein parallels the brachial artery and its branches.

10. The ***axillary vein*** is a continuation of the brachial vein in the axilla. The tributaries to the axillary vein generally correspond to the branches of the axillary artery.

11. The axillary vein is joined by the ***subscapular vein*** in the axillary region. In the shoulder, the axillary vein is known as the ***subclavian vein.*** Follow the subclavian medially and locate the superior vena cava.

12. The ***superior vena cava (precava)*** enters the anterior part of the right atrium and returns blood to the heart from all areas cranial to the diaphragm, except the heart itself. The superior vena cava is formed by the union of the two ***brachiocephalic veins.***

13. Entering the right side of the superior vena cava immediately cranial to the heart is the single ***azygos vein.*** This vein collects blood from

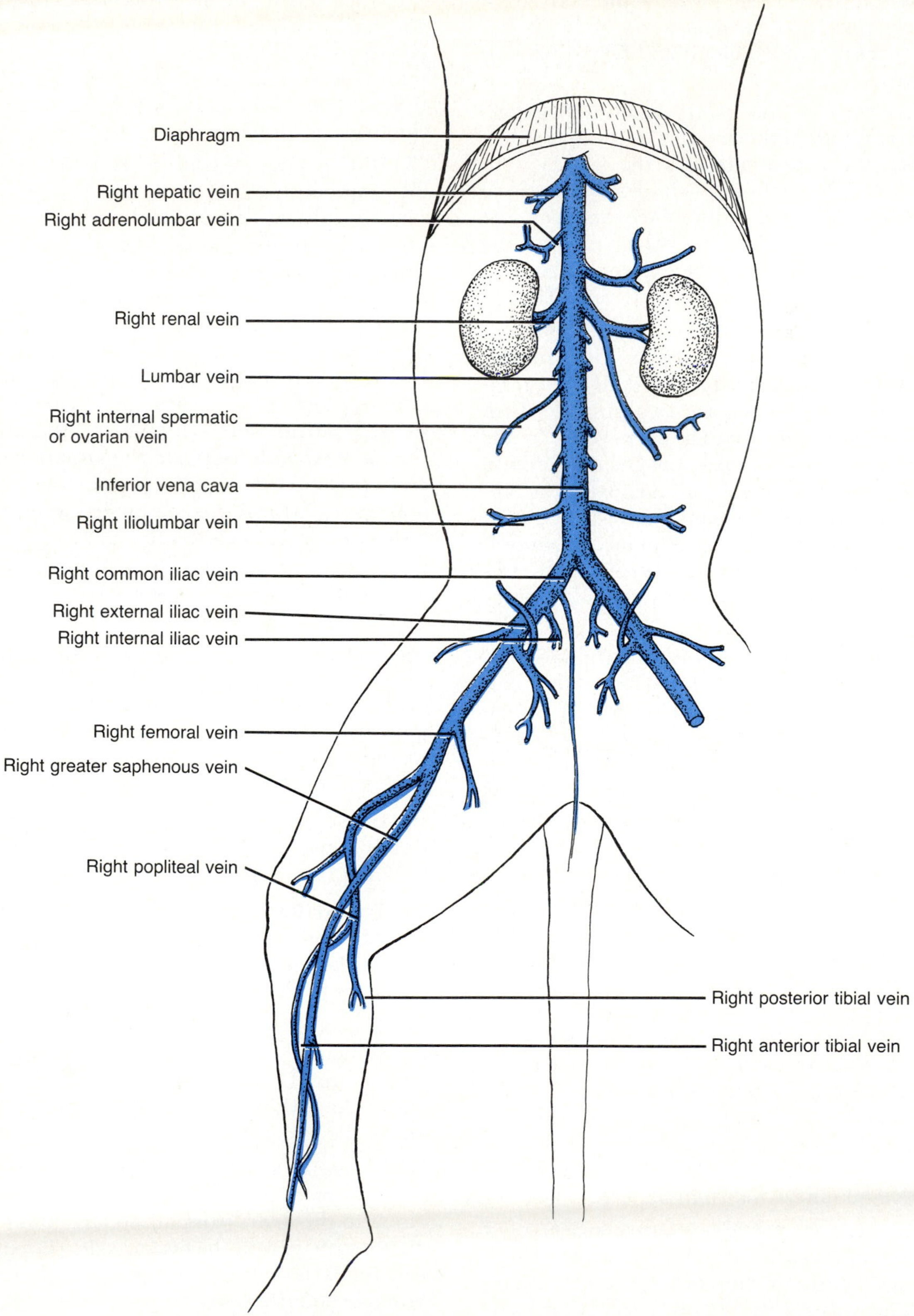

FIGURE 18.19 Veins of right hindlimb and abdomen (exclusive of hepatic portal system) in ventral view.

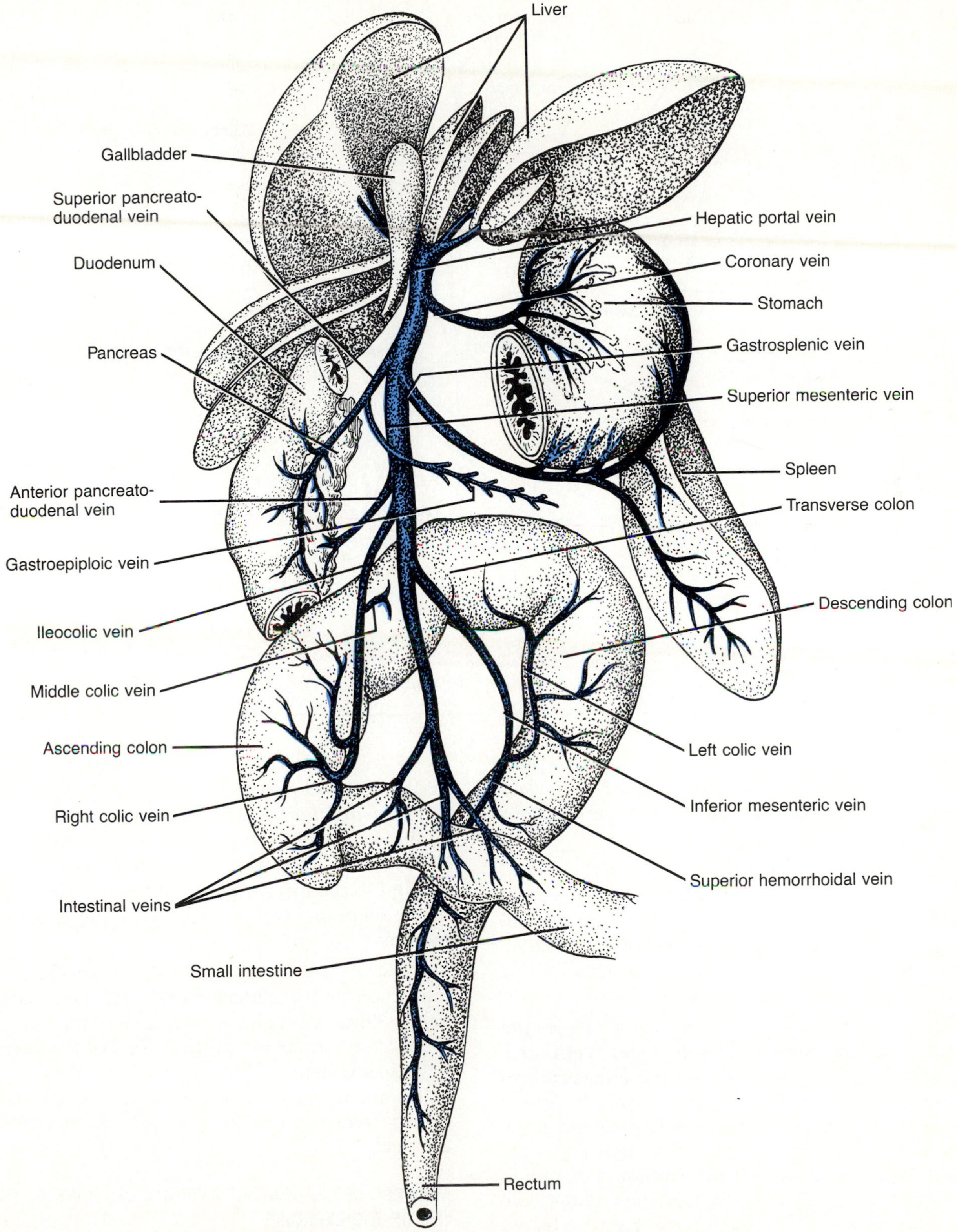

FIGURE 18.20 Hepatic portal system in ventral view.

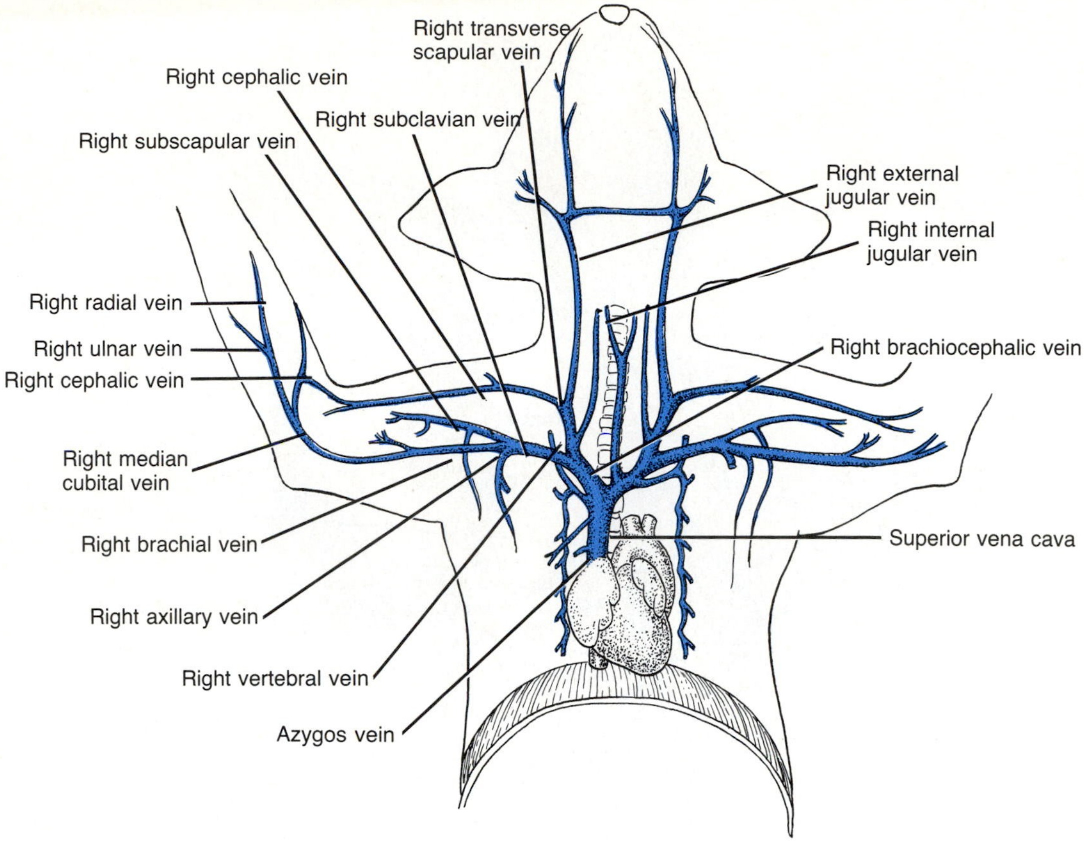

FIGURE 18.21 Veins of right forelimb in ventral view.

intercostal veins, from the body wall, from the diaphragm, from the ***esophageal veins***, and from the ***bronchial veins.*** Identify the azygos vein (Figure 18.21).

14. The superior vena cava also receives blood from the chest wall from the ***internal mammary veins.*** These veins unite as the ***sternal vein*** before entering the superior vena cava. The ***right vertebral vein*** may join the superior vena cava or the brachiocephalic vein.

15. From the superior vena cava, trace the ***left brachiocephalic (innominate) vein*** cranially (Figure 18.22). Identify the ***left vertebral vein.*** Note that the left brachiocephalic vein is formed by the union of the ***left external jugular vein***, which drains the head outside the cranial cavity, and the ***left subclavian vein***, which drains the arm. The ***right brachiocephalic vein*** is formed by the union of the right external jugular and right subclavian veins.

16. If you trace the external jugular vein cranially, you will note that it receives the smaller ***internal jugular vein*** above the point of union of the external jugular and subclavian veins. The internal jugular vein drains the brain. Cranial to the internal jugular vein is the ***transverse scapular vein*** which joins the external jugular.

17. The ***external jugular vein*** is formed by the union of the ***anterior*** and ***posterior facial veins.*** These veins have tributaries that return blood distributed by the branches of the external carotid artery. The external jugular veins communicate near their points of formation by a large ***transverse vein*** that passes ventral to the neck.

ANSWER THE LABORATORY REPORT QUESTIONS AT THE END OF THE EXERCISE

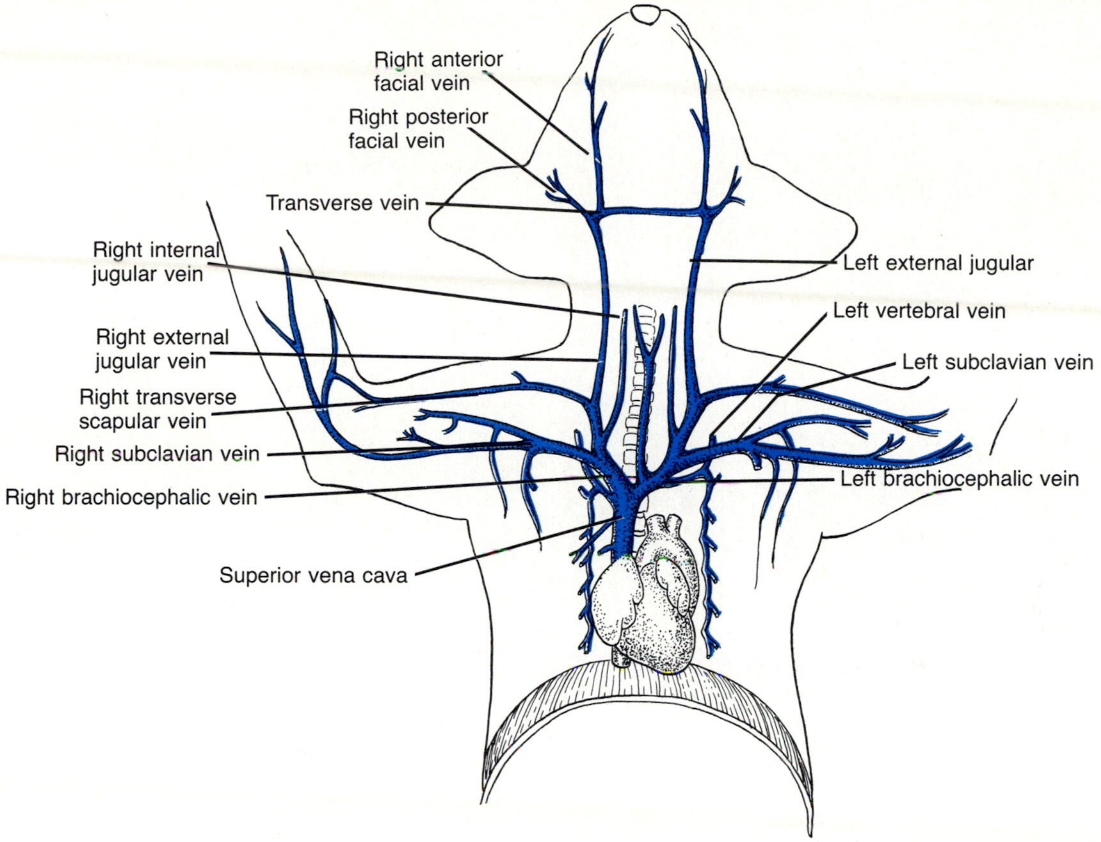

FIGURE 18.22 Veins of neck and head in ventral view.

LABORATORY REPORT QUESTIONS

BLOOD VESSELS 18

Student ______________________ **Date** ______________________

Laboratory Section ______________________ **Score/Grade** ______________________

PART I. MULTIPLE CHOICE

_____ 1. The largest of the circulatory routes is (a) systemic (b) pulmonary (c) coronary (d) hepatic portal

_____ 2. All arteries of systemic circulation branch from the (a) superior vena cava (b) aorta (c) pulmonary artery (d) coronary artery

_____ 3. The arterial system that supplies the brain with blood is the (a) hepatic portal system (b) pulmonary system (c) cerebral arterial circle (circle of Willis) (d) carotid system

_____ 4. An obstruction in the inferior vena cava would hamper the return of blood from the (a) head and neck (b) upper extremities (c) thorax (d) abdomen and pelvis

_____ 5. Which statement best describes arteries? (a) all carry oxygenated blood to the heart (b) all contain valves to prevent the backflow of blood (c) all carry blood away from the heart (d) only large arteries are lined with endothelium

_____ 6. Which statement is *not* true of veins? (a) they have less elastic tissue and smooth muscle than arteries (b) they contain more fibrous tissue than arteries (c) most veins in the extremities have valves (d) they always carry deoxygenated blood

_____ 7. A thrombus in the first branch of the arch of the aorta would affect the flow of blood to the (a) left side of the head and neck (b) myocardium of the heart (c) right side of the head and neck and right upper extremity (d) left upper extremity

_____ 8. If a vein must be punctured for an injection, transfusion, or removal of a blood sample, the likely site would be the (a) median cubital (b) subclavian (c) hemiazygous (d) anterior tibial

_____ 9. In hepatic portal circulation, blood is eventually returned to the inferior vena cava through the (a) superior mesenteric vein (b) hepatic portal vein (c) hepatic artery (d) hepatic veins

_____ 10. Which of the following are involved in pulmonary circulation? (a) superior vena cava, right atrium, and left ventricle (b) inferior vena cava, right atrium, and left ventricle (c) right ventricle, pulmonary artery, and left atrium (d) left ventricle, aorta, and inferior vena cava

_____ 11. If a thrombus in the left common iliac vein dislodged, into which arteriole system would it first find its way? (a) brain (b) kidneys (c) lungs (d) left arm

_____ 12. In fetal circulation, the blood containing the highest amount of oxygen is found in the (a) umbilical arteries (b) ductus venosus (c) aorta (d) umbilical vein

_____ 13. The greatest amount of elastic tissue found in the arteries is located in which coat? (a) tunica interna (b) tunica media (c) tunica externa (d) tunica adventitia

_____ 14. Which coat of an artery contains endothelium? (a) tunica interna (b) tunica media (c) tunica externa (d) tunica adventitia

_____ 15. Permitting the exchange of nutrients and gases between the blood and tissue cells is the primary function of (a) capillaries (b) arteries (c) veins (d) arterioles

_____ 16. The circulatory route that runs from the gastrointestinal tract to the liver is called (a) coronary circulation (b) pulmonary circulation (c) hepatic portal circulation (d) cerebral circulation

_____ 17. Which of the following statements about systemic circulation is *not* correct? (a) its purpose is to carry oxygen and nutrients to body tissues and to remove carbon dioxide (b) all systemic arteries branch from the aorta (c) it involves the flow of blood from the left ventricle to all parts of the body except the lungs (d) it involves the flow of blood from the body to the left atrium

_____ 18. The opening in the septum between the right and left atria of a fetus is called the (a) foramen ovale (b) ductus venosus (c) foramen rotundum (d) foramen spinosum

_____ 19. The branch of the umbilical vein in the fetus that connects with the inferior vena cava, bypassing the liver, is the (a) foramen ovale (b) ductus venosus (c) ductus arteriosus (d) patent ductus

_____ 20. Which of the vessels does *not* belong with the others? (a) brachiocephalic artery (b) left common carotid artery (c) celiac artery (d) left subclavian artery

PART II. MATCHING

_____ 21. aortic branch that supplies the head and associated structures

_____ 22. artery that distributes blood to the small intestine and part of the large intestine

_____ 23. vessel into which veins of the head and neck, upper extremities, and thorax enter

_____ 24. vessel into which veins of the abdomen, pelvis, and lower extremities enter

_____ 25. vein that drains the head and associated structures

_____ 26. longest vein in the body

_____ 27. vein just behind the knee

_____ 28. artery that supplies a major part of the large intestine and rectum

_____ 29. first branch off of the arch of the aorta

_____ 30. arteries supplying the heart

A. Inferior vena cava
B. Superior vena cava
C. Superior mesenteric
D. Common carotid
E. Jugular
F. Brachiocephalic
G. Coronary
H. Inferior mesenteric
I. Popliteal
J. Great saphenous

CARDIOVASCULAR PHYSIOLOGY 19

A. CARDIAC CONDUCTION SYSTEM AND ELECTROCARDIOGRAM (ECG OR EKG)

The heart is innervated by the autonomic nervous system (ANS) which modulates, but does not initiate, the cardiac cycle (heartbeat). The heart can continue to contract if separated from the ANS. This is possible because the heart has an intrinsic pacemaker termed the ***sinoatrial (SA) node***. The sinoatrial node is connected to a series of specialized muscle fibers (cells) that comprise the ***cardiac conduction system***. This system distributes the electrical impulses that stimulate the cardiac muscle fibers to contract.

The sinoatrial node is located in the right atrial wall, inferior to the opening of the superior vena cava. Once an action potential is initiated by this node, the impulse spreads out over both atria, causing them to contract and, at the same time, the impulse spreads to the ***atrioventricular (AV) node*** near the inferior portion of the interatrial septum via three specialized internodal pathways, ultimately causing the AV node to depolarize. From the atrioventricular node, a tract of conducting fibers called the ***atrioventricular (AV) bundle (bundle of His)*** extends to the top of the interventricular septum and continues down both sides of the septum as the ***right*** and ***left bundle branches***. The AV bundle distributes the electrical impulses over the medial surfaces of the ventricles via ***conduction myofibers (Purkinje fibers)*** that emerge from the bundle branches and pass into the cells of the ventricular myocardium. As the impulse reaches the ventricular fibers, they are depolarized, initiating contraction.

Label the components of the cardiac conduction system in Figure 19.1.

These impulses generate electrical currents that may be detected on the surface of the body. An ***electrocardiogram (ECG*** or ***EKG)*** is a recording of the electrical changes accompanying each cardiac cycle. An ***electrocardiograph*** is the instrument used to record these changes. As the electrical impulses are transmitted throughout the cardiac conduction system and myocardial cells, a different electrical impulse is generated. These impulses are transmitted from the electrodes to a recording needle that graphs the impulses as a series of up-and-down waves called ***deflection waves*** (Figure 19.2).

1. Electrocardiogram Readings

In a typical record (Figure 19.2), three clearly recognizable waves accompany each cardiac cycle. The electrocardiogram recording represents only the electrical events of the cardiac cycle, *not the mechanical events*. The first wave, the ***P wave***, indicates depolarization of the atria, that is, the spread of an electical impulse from the SA node throught the atria. A fraction of a second after the P wave begins, the atria contract. A deflection wave, called the ***QRS wave (complex)***, represents ventricular depolarization, that is, the spread of an electrical impulse through the ventricles. The third deflection is the ***T wave***, indicating ventricular repolarization. No deflection appears for atrial repolarization because depolarization of the ventricles masks this event.

The ***P-R interval*** lasts from the beginning of the P wave to the beginning of the QRS wave

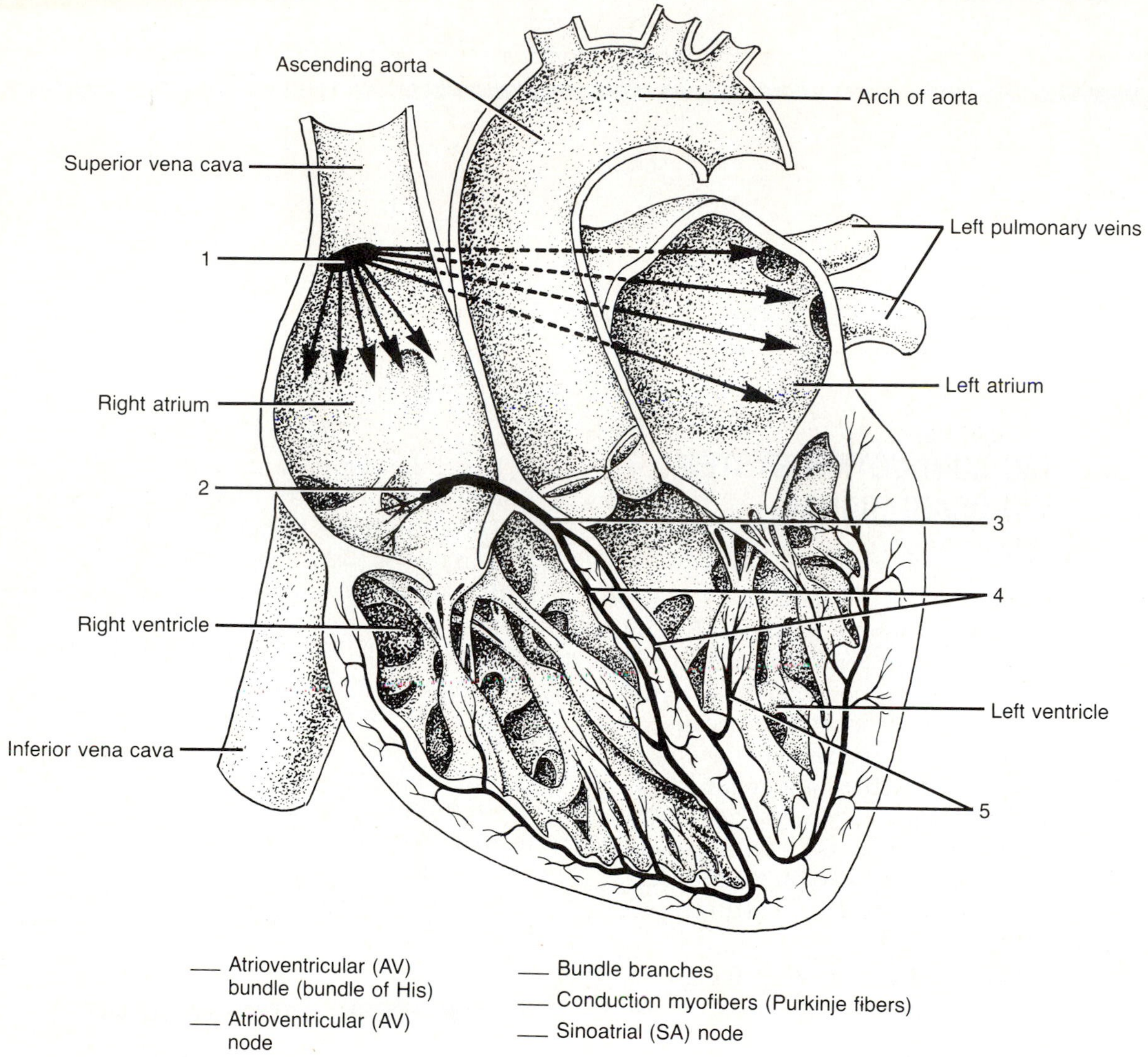

FIGURE 19.1 Conduction system of heart.

and represents the time needed for impulses to pass from the sinoatrial (SA) node to the ventricles. It is lengthened in conditions such as atherosclerotic heart disease and rheumatic fever that cause damage to the conduction pathways to the ventricles.

The ***P-R segment*** from the end of the P wave to the beginning of the QRS complex is normally at the baseline, indicating no change in the electrical potential at the surface of the heart. During this time, impulses are traveling within the myocardium, from the atrioventricular (AV) node to the conduction myofibers (Purkinje fibers).

The ***S-T segment*** begins at the end of the S wave and terminates at the beginning of the T wave and represents the time between the end of ventricular depolarization and ventricular repolarization. During this time, there is no electrical change, so the record is at the baseline. The configuration of the S-T segment is altered when ventricular muscle is damaged by events such as myocardial infarction (MI) or myocardial ischemia.

In reading and interpreting an electrocardiogram, you must note the *lead* type or configuration from which the electrocardiogram was recorded, the size of the deflection waves, and certain time intervals. An attempt will not be made to interpret an electrocardiogram, but only to become familiar with a normal ECG. The ECG is invaluable in diagnosing abnor-

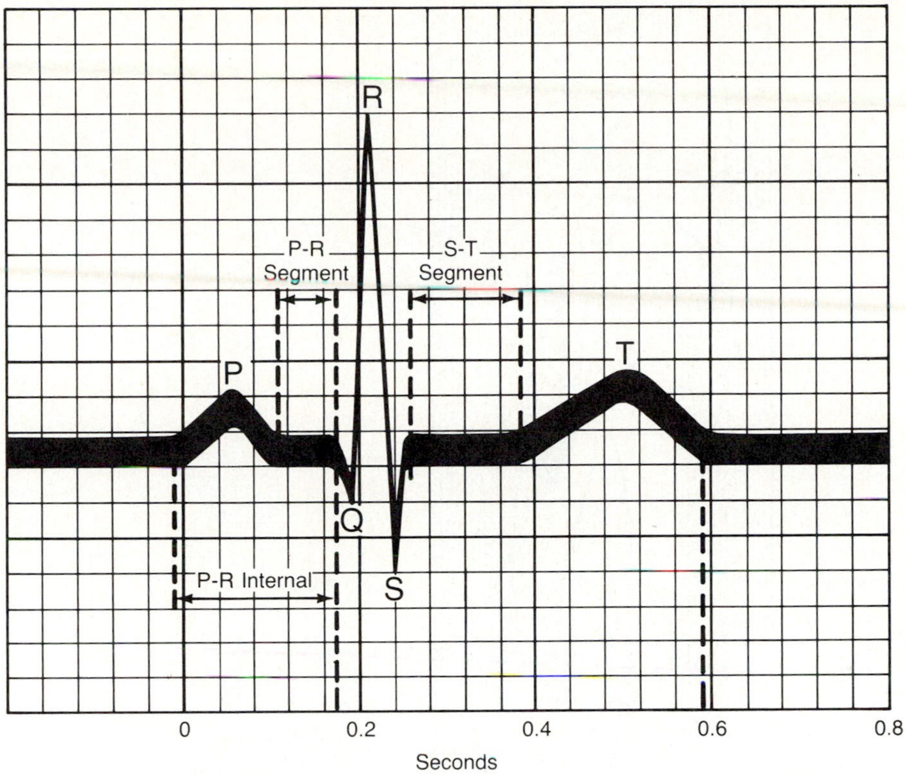

FIGURE 19.2 Recordings of a normal electrocardiogram (ECG), lead II.

mal cardiac rhythms and conducting patterns, detecting the presence of fetal life, determining the presence of more than one fetus, and following the course of recovery from a heart attack.

In the following procedure a physiograph-type polygraph is used, but an electrocardiograph or oscilloscope may also be used (Figure 19.3a).

2. Use of Polygraph

In recording the electrocardiogram, we will utilize only the standard *three limb leads*. A ***lead*** is defined as two electrodes working in pairs. ***Electrodes*** are sensing devices that are made of sensitive metal plates or small rods that can detect electrophysiological phenomena such as changes in electrical potential of the skin or nerves (see Exercise 9). Usually only two electrodes will be in actual use at any one time. The third electrode will be automatically switched off by the ***lead-selector switch*** of the polygraph. The three standard limb leads most commonly used in electrocardiography are:

Lead	Electrodes Used
I	Right wrist and left wrist
II	Right wrist and left ankle
III	Left wrist and left ankle

For most general purposes three leads are usually used. However, elecrocardiologists use additional leads, including several chest wall electrodes (Figure 19.3b). These leads are positioned around the chest, and encircle the heart so that there are six intersecting lines on a horizontal plane through the atrioventricular node.

The procedure for recording electrocardiograms using the physiograph is a follows:

PROCEDURE

1. Either you or your laboratory partner should lie on a table or cot, roll down long stockings or socks to the ankles, and *remove all metal jewelry*.

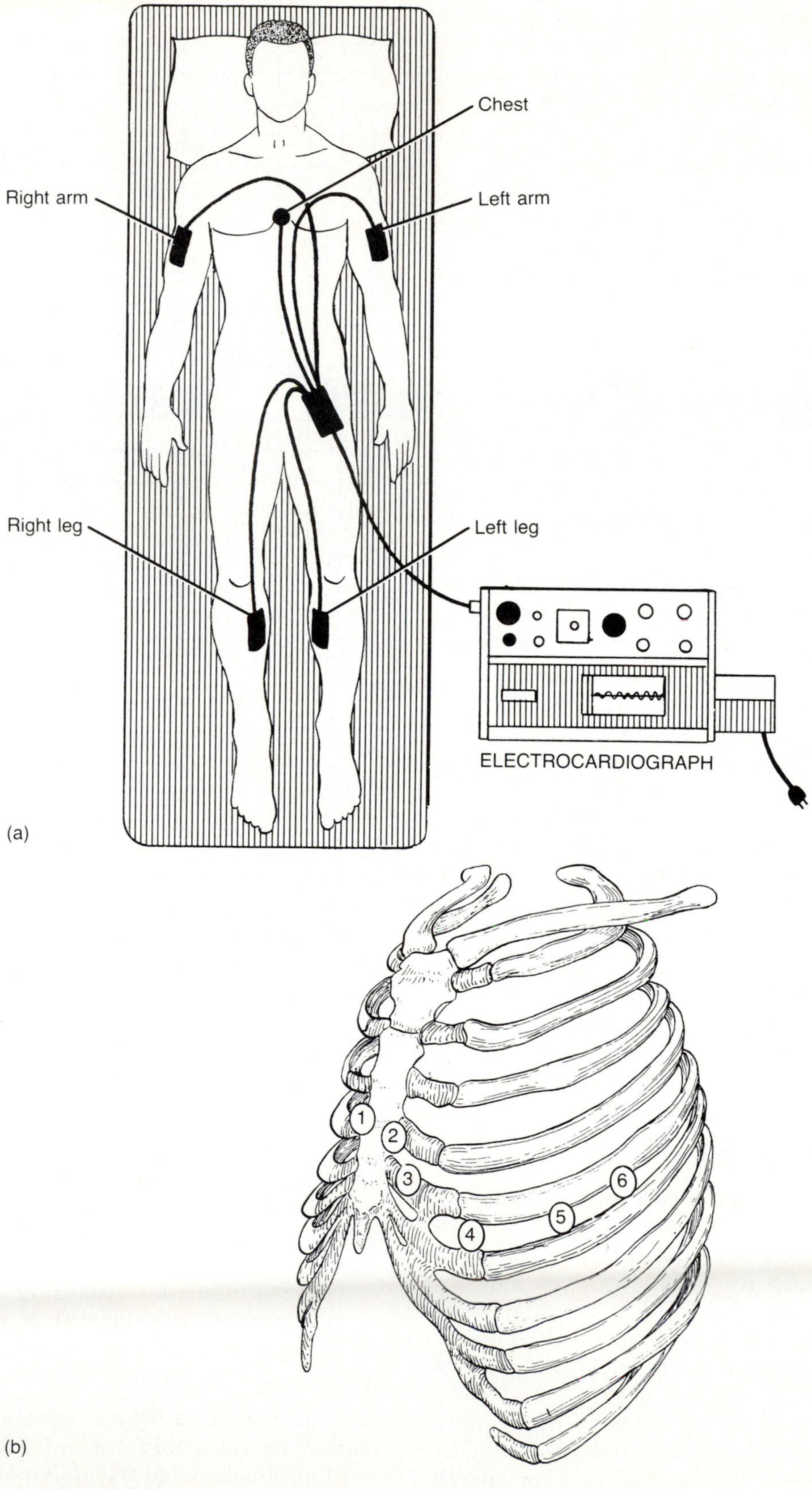

FIGURE 19.3 Electrocardiograph. (a) Connections. (b) Normal chest wall leads.

2. Swab the skin with alcohol where electrodes will be applied. Electrode cream, jelly, or saline paste is applied to the skin only where the electrode will make contact, such as to the inner forearm just above the wrists and to the inside of the legs just above the ankles.

3. The cream is then applied on the concave surface of the electrode plates, which are then fastened securely to the limb area surfaces with rubber straps or adhesive tape (rings). Adjust the straps to fit snugly, but *not so tight as to impede blood flow*.

4. Firmly connect the proper end of the electrode cables to the electrode plates, using the standard limb leads previously described, and connect the other end to a cardiac preamplifier or to a self-contained electrocardiograph.

5. After you apply the electrodes and connect the instrument to the subject, ascertain the following before actually recording:

a. Recording power switch is on.

b. Paper is sufficient for the entire recording, and the pen is centered properly on the paper with ink flowing freely.

c. Preamplifier sensitivity has been set at 10 mm per millivolt.

d. Subject is lying quietly and is relaxed.

Note: Usually five or six ECG complex recordings from each lead are sufficient. All electrocardiographers have chosen a standard paper speed of 25 mm per second.

6. Turn the instrument on and record for 30 seconds from each lead. Obtain directions from your laboratory instructor on the procedure to be followed when switching from lead to lead. When the recording leads are switched, note the change by marking it on the electrocardiogram. The subject must remain very still and relax completely during the recording period.

7. When the recording is finished, turn off the reading instrument and disconnect the leads from the subject and remove the electrodes, cleaning off the excess cream or paste. The skin can be washed with water to remove electrode cream.

8. Identify and letter the P, QRS, and T waves, and compare the waves with those shown in Figure 19.2.

9. Calculate the duration of the waves and the P-R interval. Attach this recording to Section A of the LABORATORY REPORT RESULTS at the end of the exercise.

3. Determination of Mean Electrical Axis

As was previously outlined, depolarization of the entire ventricular myocardium does not occur simultaneously. The depolarization starts within the interventricular septum and then spreads to the ventricular walls. While the septum is depolarized and the ventricular walls are not, an electrical difference (voltage) exists between these two locations, thereby producing the R wave of the QRS complex. When both locations are depolarized, no electrical difference occurs, thereby producing the S wave of the QRS complex and completing that phase of the ECG. The spread of the action potential through the ventricular myocardium depends upon several factors. These include the size of the ventricles, orientation of the heart during that particular instant of the cardiac cycle, and the ease with which action potentials spread throughout the various parts of the ventricular conducting system and myocardium. Because of this, it is clinically useful to determine the ***mean electrical axis*** (average direction of depolarization) of the cardiac cycle. This is done by observing the voltages of the QRS complex of the ECG from two different orientations using Leads I and III of the ECG. Lead I crosses the heart from left arm to right arm, thereby giving a horizontal axis of observation, while lead III approximately parallels the heart, going from the left arm to the left leg, thereby providing a 120° axis of observation. When these two observations are taken together the average heart has a mean electrical axis of 59° (Figure 19.4).

PROCEDURE

The procedure for determining the mean electrical axis from the electrocardiogram is as follows:

1. Record an ECG utilizing Lead I (with sensitivity setting at 1) from a subject that is reclining quietly.

2. Find a QRS complex and count the number of millimeters (small boxes on the paper) that it projects above the top of the ***isoelectric point*** (baseline) of the recording. Do this for several

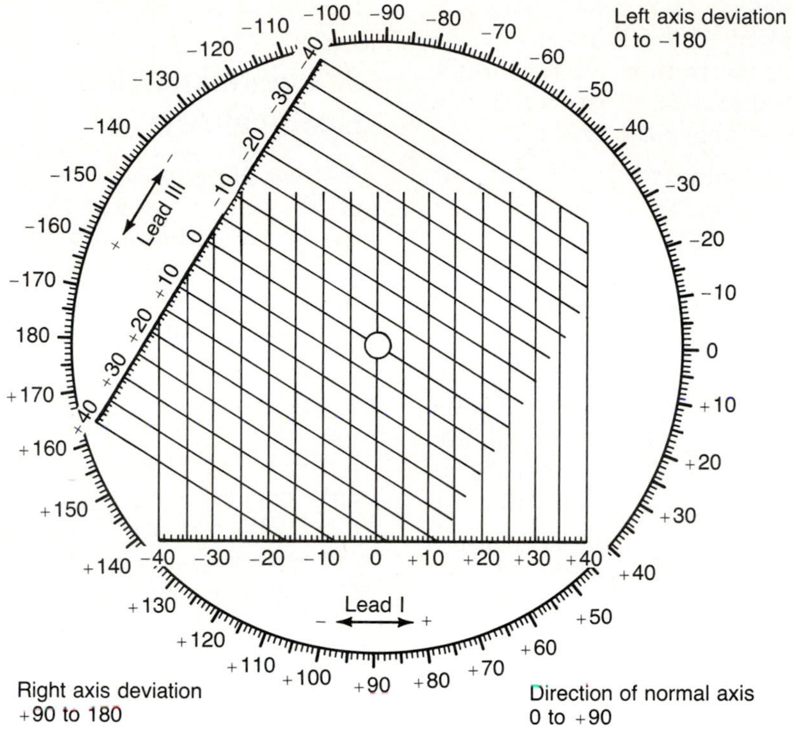

FIGURE 19.4 Grid chart showing the direction of the electrical axis of the heart as determined by a human electrocardiogram.

QRS complexes and record the average value in the space below.

\+ __________

3. *Utilizing the same QRS complexes that were utilized for number 2 above* count the number of millimeters that the Q and S (or if the QRS complex is inverted utilize the R wave) waves project below the isoelectric point. Do this for several QRS complexes and record the average value in the space below.

− __________

4. Algebraically add the two values obtained above as demonstrated below and enter the value in the space, keeping the (−) sign if the sum is negative or the (+) sign if the sum is positive.

$$(+59) + (-65) = (-6)$$

Sum = __________

5. Record an ECG utilizing Lead III (with the sensitivity at the same setting) from the same subject while they are reclining quietly.

6. Find several QRS complexes and repeat steps 3 through 5 above, recording the values in the spaces below.

\+ __________

− __________

Sum __________

7. Utilizing a ruler or other straight edge mark a line on the grid chart in Section A of the LABORATORY REPORT RESULTS on the axis corresponding to the algebraic sum obtained from Lead I.

8. Using a ruler or other straight edge mark a line on graph paper corresponding to the algebraic sum obtained from Lead III.

9. Connect the two lines on the graph paper *making sure to pass through the center of the grid chart.* Record the mean electrical axis.

Mean electrical axis __________

B. CARDIAC CYCLE

In a normal heartbeat, the two atria contract while the two ventricles relax. Then, when the two ventricles contract, the two atria relax. The term ***systole*** (SIS-tō-lē) refers to the phase of contraction; ***diastole*** (dī-AS-tō-lē) is the phase of relaxation. A ***cardiac cycle***, or complete heartbeat, consists of a systole and diastole of both atria plus the systole and diastole of both ventricles. In a complete cardiac cycle, the atria are in systole 0.1 second and in diastole 0.7 second. By contrast, the ventricles are in systole 0.3 second and in diastole 0.5 second. If we assume an average heartbeat of 75 times per minute, each cardiac cycle requires about 0.8 seconds (Figure 19.5).

C. ISOLATED TURTLE (OR FROG) HEART EXPERIMENTS

Before beginning these experiments, review the procedure for pithing in Excerise 9, Section G.1 and proper use of the polygraph and stimulator in Exercise 9, Section G.2.

1. Observation of Cardiac Cycle

CAUTION! *Please reread Section D. "Precautions Related to Dissection" at the beginning of the laboratory manual on page xii before you begin your dissection.*

PROCEDURE

NOTE: *It is strongly suggested that either your instructor or a laboratory assistant perform the procedure for pithing a frog or turtle. The procedure requires some practice and may prove difficult for inexperienced individuals.*

1. Obtain a double pithed turtle (or frog) from your instructor. The turtle heart is shown in Figure 19.6. Both the turtle and frog hearts differ from the mammalian heart anatomically in that they have only a single ventricle from which all blood exits. The large sinus venosus on the dorsal surface serves as the pacemaker.

2. Fasten the specimen to a board ventral side up, and expose the heart by cutting through the body wall of the thoracic region.

Turtle	**Frog**
Using a saw and scalpel, remove the ventral shell and associated musculature. Carefully remove the mediastinum and pericardial sac.	Make a midline incision from the abdomen to the jaw and extend the incision laterally. Pin the body flaps to the board.

3. *Throughout this exercise keep the heart moist with Krebs-Ringer's solution.*

4. Observe the sequence of contraction of the atria and the ventricle and determine the rate of contraction for 30 seconds. Calculate the heart rate per minute and record it.

Heart rate = __________ beats/minute.

Estimate the relationship of atrial and ventricular cycles by shading in a portion of the bars below to indicate systole and leaving the portion representing diastole unshaded.

Atrial	
Ventricular	

2. Recording The Cardiac Cycle

PROCEDURE

1. **Turtle**	**Frog**
Tie a thread ligature tightly around the cardiac frenulum at the ventricular apex (Figure 19.5) and cut the frenulum *distal* to the thread.	Tie a thread to a small metal hook and carefully insert the hook into the tip of the ventricular myocardium. *Do not* penetrate the chamber.

2. Connect the thread to a force transducer.

3. Keep just enough tension on the thread to have a force tracing on the polygraph.

4. Cut the frenulum *distal to* the thread.

5. Connect an atrium to a second force transducer by either inserting a small metal hook or bent pin through the atrium and connecting it to the transducer with a thread, or by tying a thread ligature to the atrium and connecting it to the transducer. Again regulate the tension in a manner similar to that for the ventricle.

6. Record several cardiac cycles and paste a strip of the recording paper in the space provided in Section C.2 of the LABORATORY REPORT RESULTS at the end of the excerise. Label the atrial (smaller) and ventricular (larger) systolic waves. Calculate and record the heart rate per minute.

Heart rate = __________ beats/minute

Why do the ventricular waves have greater amplitude than the atrial waves? __________

3. Intrinsic Control of the Cardiac Cycle

Intrinsic control of the cardiac cycle refers to control mechanisms arising totally within the

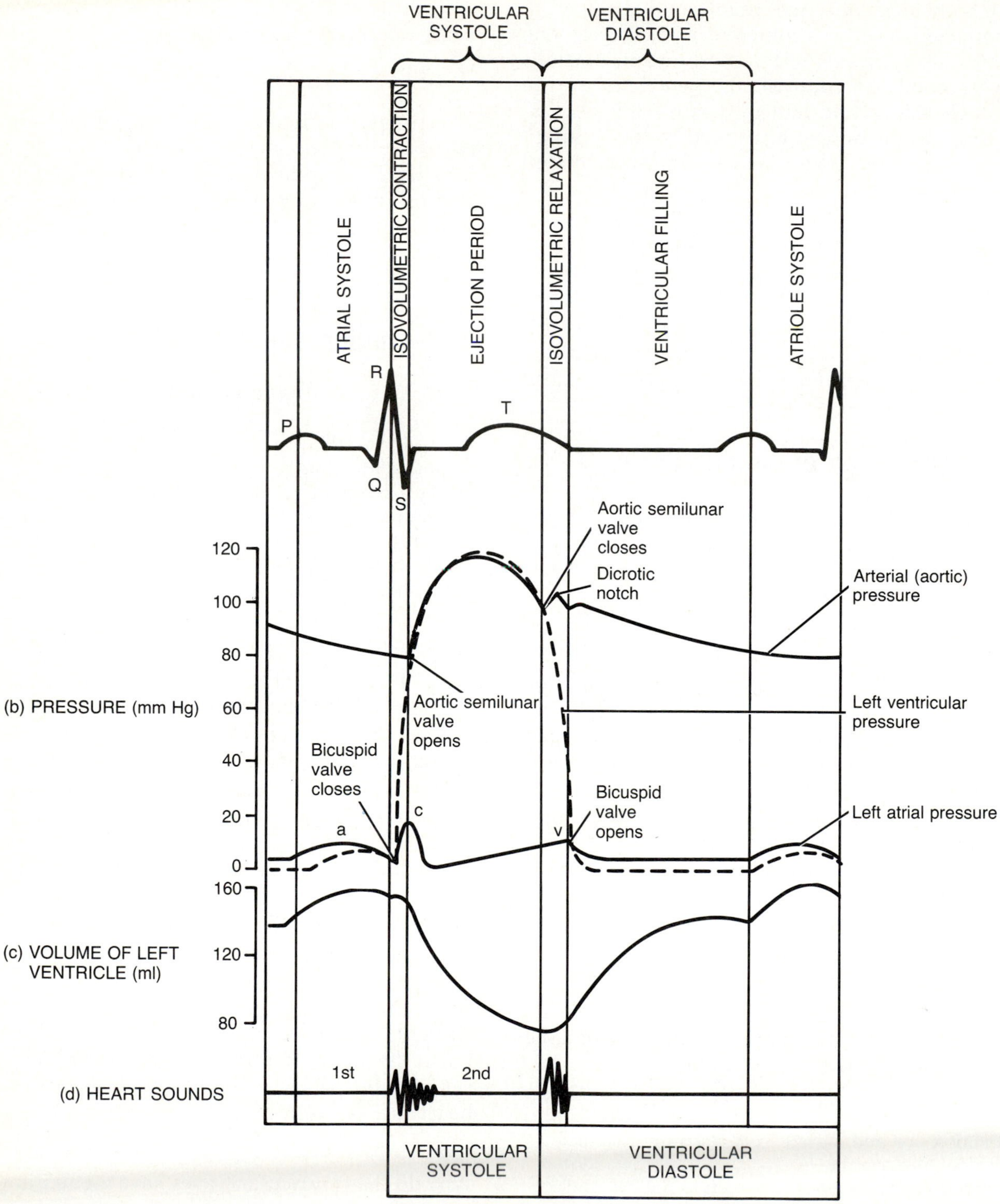

FIGURE 19.5 Cardiac cycle. (a) ECG related to the cardiac cycle. (b) Left atrial, left ventricular, and arterial (aortic) pressure changes along with the opening and closing of valves during the cardiac cycle. The letters a, c, and v represent the elevation waves of the left atrial pressure curve. (c) Left ventricular volume during the cardiac cycle. (d) Heart sounds related to the cardiac cycle.

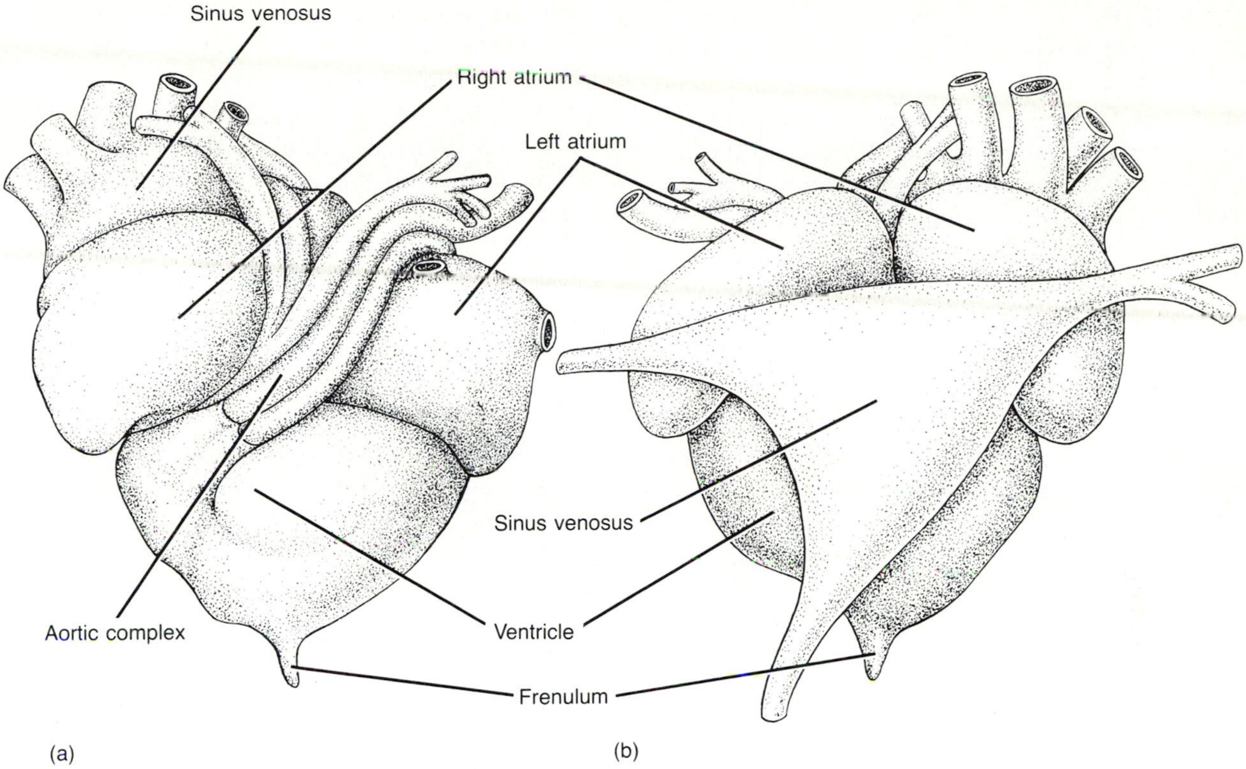

FIGURE 19.6 Turtle heart. (a) Ventral view. (b) Dorsal view.

heart, devoid of any external influence from nerves or hormones, or both. The major intrinsic control mechanism is ***Starling's Law of the Heart***, and is attributed to a length-tension interaction between tha actin and myosin fibers of the myocardium. Simply stated, Starling's Law says that the heart will pump out all of the blood it receives, within physiological limits.

PROCEDURE

1. *Throughout this exercise keep the heart moist with Krebs-Ringer's solution.*

2. Utilizing the same preparation as that from the previous procedures, place minimal tention on the ventricle by regulating the tension of the ligature tied to the cardiac frenulum.

3. While continuously recording, increase the tension of the ligature tied to the cardiac frenulum in step-wise increments by raising the height of the transducer on its stand. Allow several beats at each tension setting.

4. Graph the relationship between resting tension and developed tension in Section C.3 of the LABORATORY REPORT RESULTS at the end of the exercise and explain your observations.

What normal physiological event is mimicked by artificially increasing tension on the heart?

3. Extrinsic Control of the Cardiac Cycle

Extrinisic control of the cardiac cycle is exerted through the influences of hormones of the adrenal medulla and autonomic nervous system. The parasympathetic nervous system innervates the heart through the vagus (x) nerves, while the sympathetic nervous system innervates the heart through branches of the stellate and cervical ganglia.

PROCEDURE

1. Utilizing the same preparation as that from the previous procedures, *keep the heart moist with Krebs-Ringer's solution.*

2. Use a medicine dropper and apply a 1:1000 solution of epinephrine to the external surface

of the heart. Note the changes, if any, in the cardiac cycle.

3. Monitor any changes in the cardiac cycle for 2 to 3 minutes.

4. Using a medicine dropper, rinse with Krebs-Ringer's solution and allow several minutes for recovery before proceeding to the next section. Using a medicine dropper, *gently* aspirate any accumulated fluid.

5. Attach the tracing in the space provided in Section C.4.a of LABORATORY REPORT RESULTS at the end of the exercise and explain your observations.

6. Expose the vagus (x) nerve in the neck and pass a ligature beneath it but *do not tie the ligature*.

Turtle	Frog
The vagus (x) nerve lies deep in the neck muscles beside the carotid artery. Use dissection scissors to expose the muscles of the neck and blunt dissection with a glass probe to isolate the left or right nerve.	The vagus (x) nerve lies beside the external jugular vein. Use blunt dissection with a glass probe to separate the left or right nerve.

7. Because different preparations will have varying thresholds, set the electrical stimulator at a frequency of 25 per second and a duration of 2 milliseconds (msec) and determine what is the threshold voltage necessary to affect a noticeable change in the heart rate or contraction strength while continuously recording.

8. Attach the tracing in the space provided in Section C.4.b of the LABORATORY REPORT RESULTS at the end of the exercise and explain your observations.

9. After allowing several minutes for recovery, use a medicine dropper and apply a 1:1000 solution of acetylcholine to the surface of the heart.

10. Monitor the changes, if any, in the cardiac cycle for 2 to 3 minutes and then, using a medicine dropper, rinse with Krebs-Ringer's solution. Allow several minutes for recovery. Using a medicine dropper, *gently* aspirate any accumulated fluid.

11. Attach the tracing in the space provided in Section C.4.c of the LABORATORY REPORT RESULTS at the end of the exercise and explain your observations.

12. After several minutes for recovery, use a medicine dropper and apply a 1:1000 solution of atropine sulfate to the surface of the heart. Note any changes in the cardiac cycle.

13. Monitor any changes in the cardiac cycle for 2 to 3 minutes and then, using a medicine dropper, rinse with Krebs-Ringer's solution. Allow several minutes for recovery. Using a medicine dropper, *gently* aspirate any accumulated fluid.

14. Attach the tracing in the space provided in Section C.4.d of the LABORATORY REPORT RESULTS at the end of the exercise and explain your observations.

15. Using medicine droppers, reapply atropine sulfate and immediately follow it with an external application of 1:1000 acetylcholine solution.

16. Note the effects, if any, of acetylcholine after application of atropine.

17. Attach the tracing in the space provided in Section C.4.e of the LABORATORY REPORT RESULTS at the end of the exercise and explain your observations.

18. After another application of atropine sulfate with a medicine dropper, restimulate the right and left vagi, as previously outlined. Note the changes, if any.

19. Using a medicine dropper, rinse the heart with Krebs-Ringer's solution.

20. Attach the tracing in the space provided in Section C.4.f of the LABORATORY REPORT RESULTS at the end of the exercise and explain your observations.

Which of the solution(s) mimics the effect of the sympathetic stimulation? ______________

Of parasympathetic stimulation? ______________

5. Demonstration of Refractory Period

The concept of refractory period in cardiac muscle is very similar to that discussed in conjunction with nerve action potential propagation and skeletal muscle contraction. Compared to skeletal muscle, cardiac muscle has a long duration of contraction and a long refractory period. Normally, the fibers (cells) cannot be stimulated again until they have relaxed from their first contraction. Thus, summation and tetany are not possible, and the heart pumps rhythmically. Stimulation of cardiac muscle at some points in the cycle may produce an ***extrasystole*** (extra beat) followed by a compensatory pause.

PROCEDURE

1. Use the frog or turtle heart preparation from the preceding exercises, and *remember to keep the heart moist with Krebs-Ringer's solution*.

2. Set the paper drive on the polygraph to its highest speeed.

3. While continuously recording, stimulate the ventricle with a single stimulus of about 50 volts for 2 milliseconds (msec). Make sure that both electrodes touch the ventricle.

4. Allowing several contractions between stimuli, stimulate the ventricle at the following five points of ventricular systole.

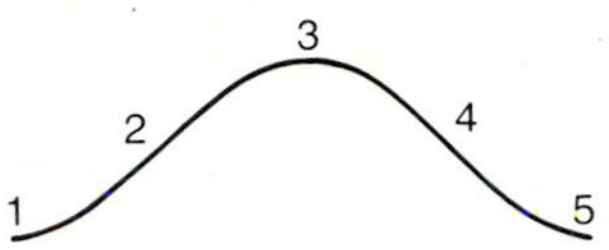

5. Note whether an extrasystole occurs, at any or all of the five points and, if so, whether it is followed by a compensatory pause.

6. Continuously stimulate the ventricle for 10 seconds to see if you can elicit tetany.

7. Attach the tracing in the space provided in Section C.5 of the LABORATORY REPORT RESULTS at the end of the exercise and explain your observations.

D. HEART SOUNDS

The beating of a human heart consists of two well-defined sounds during each cardiac cycle. These two sounds can be easily heard by using an ordinary ***stethoscope***. The first sound is created by the closure of the atrioventricular valves soon after the ventricular systole begins (see Figure 19.5.d). The second sound is created as the semilunar valves close near the end of ventricular systole.

Heart sounds provide valuable information about the valves. Peculiar sounds may be called ***murmurs***. Some murmurs are caused by the noise made by a little blood flowing back in an atrium because of improper closure of an atrioventricular valve. Murmurs do not always indicate that the valves are not functioning properly, and many have no clinical significance.

Listening to sounds of the body is called ***auscultation*** (aws-kul-TĀ-shun).

1. Use of Stethoscope

PROCEDURE

1. The stethoscope should be used in a quiet room.

2. The earpieces of the stethoscope *should be cleaned with alcohol* just before using and should also be pointed slightly forward when placed in the ears. They will be more comfortable in this position, and it will be easier to hear through them.

3. Listen to the heart sounds of your laboratory partner by placing the diaphragm of the stethoscope next to the skin at several positions on the chest wall illustrated in Figure 19.7.

4. The first sound is best heard at the apex of the heart, which is located approximately at the fifth intercostal space at the midline of the clavicles (Figure 19.7).

5. The second sound is best heard over the area between the second right costal cartilage and the second intercostal space on the left side (Figure 19.7).

6. CAUTION! *Assuming that your partner has no known or apparent cardiac or other health problems and is capable of such an activity*, ask her/him to run in place about 25 steps. Listen to the heart sounds again.

7. Answer all questions pertaining to Section D of the LABORATORY REPORT RESULTS at the end of the exercise.

E. PULSE RATE

During ventricular systole, a wave of pressure, called ***pulse***, is produced in the arteries due to ventricular contraction. Pulse rate and heart rate are essentially the same. Pulse can be felt readily where an artery is near the surface of the skin and over the surface of a bone. Pulse rates vary considerably in individuals because of time of day, temperature, emotions, stress, and other factors. The normal adult pulse rate of the heart at rest is within a range of 72 to 80 beats per minute. With practice you can learn to take accurate pulse rates by counting the beats per 15 seconds and multiplying by 4 to obtain beats per minute. The term ***tachycardia*** (tak′-e-KAR-dē-a) is applied

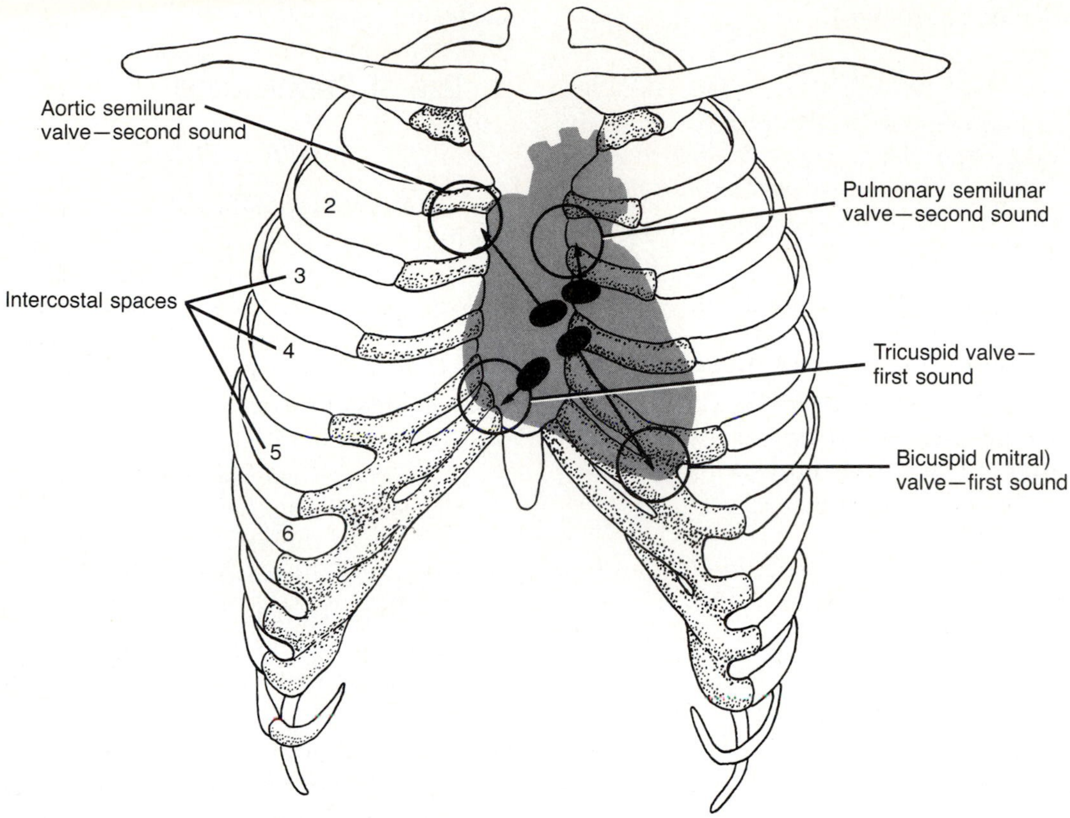

FIGURE 19.7 Surface areas where heart valve sounds are best heard.

to a rapid heart rate or pulse rate (over 100 beats/minute). ***Bradycardia*** (brād-e-KAR-dē-a) indicates a slow heart rate or pulse rate (below 50 beats/minute).

Although the pulse may be detected in most surface arteries, the pulse rate is usually determined on the ***radial artery*** of the wrist.

1. Radial Pulse

PROCEDURE

1. Using your index and middle finger palpate your laboratory partner's radial artery.
2. The thumb should never be used because it has its own prominent pulse.
3. Palpate the area behind your partner's thumb just inside the bony prominence on the lateral aspect of the wrist.
4. CAUTION! *Do not apply too much pressure.*
5. Count the pulse, change positions, and record your results in Section E of the LABORATORY REPORT RESULTS at the end of the exercise.
6. Calculate the ***median pulse rate*** of the entire class.

2. Carotid Pulse

PROCEDURE

1. Using the same fingers that you used for the radial pulse, place them on either side of your partner's larynx.
2. *Gently* press downward and toward the back until you feel the pulse. You must feel the pulse clearly with at least two fingers, so adjust your hand accordingly.
3. The radial and carotid pulse can be compared under the following conditions: (a) sitting quietly, (b) standing quietly, (c) right after walking 60 steps, (d) right after running in place 60 steps, *assuming that your partner has no known or apparent cardiac or other health problems and is capable of such an activity*. Notice how long it takes the pulse to return to normal after the walking and running exercises.

4. Record your results in Section E of the LABORATORY REPORT RESULTS at the end of the exercise.

5. Compare your radial and carotid pulse in the table provided.

F. BLOOD PRESSURE (AUSCULTATION METHOD)

Blood pressure, clinically, refers to the pressure only in the large arteries even though the term blood pressure may be defined as the pressure exerted by the blood on the walls of any blood vessel. Blood pressure is recorded in millimeters of mercury (mm Hg) and is normally taken in the brachial artery. It is measured by an instrument called a ***sphygmomanometer*** (sfig′-mō-ma-NOM-e-ter) (*sphygmo* = pulse).

A commonly used sphygmomanometer (Figure 19.8) consists of an inflatable rubber cuff attached by a rubber tube to a compressible hand pump or bulb. Another tube attaches to a cuff and to a mercury column marked off in millimeters or an anaeroid gauge that measures the pressure in millimeters of mercury (mm Hg).

Stethoscopes are used in conjunction with sphygmomanometers to locate the pulse. Pulse sounds are difficult to hear if the room is noisy.

FIGURE 19.8 Use of sphygmomanometer.

The highest pressure in the artery, occurring during ventricular systole, is termed ***systolic blood pressure***. The lowest pressure, occurring during ventricular diastole, is termed ***diastolic blood pressure***.

Arterial pressure can be measured in a variety of ways. This exercise employs the ausculatory method, in which the sounds of blood flow are heard with a stethoscope. Blood flow in an artery is impeded with a sphygmomanometer. When the cuff of the sphygmomanometer applies sufficient pressure to completely occlude blood flow, no sounds can be heard distal to the cuff because no blood can flow through the artery. When cuff pressure drops just below the maximal (systolic) pressure in the artery, blood is heard passing through the vessel. When cuff pressure drops below the lowest (diastolic) pressure in the vessel, the sound becomes muffled and usually disappears.

Blood pressure can be taken in any artery which can be occluded easily. The brachial artery has the advantage of being at approximately the same level as the heart, so brachial pressure closely reflects aortic pressure.

Blood pressures are usually expressed as a ratio of systolic/diastolic. The average blood pressure of a young adult is about 120 mm Hg systolic and 80 mm Hg diastolic, abbreviated to 120/80. The difference betweeen systolic and diastolic pressure is called ***pulse pressure***, which averages 40 mm Hg and indicates whether or not the arteries are functioning properly. The normal ratio of systolic pressure to diastolic pressure to pulse pressure is 3:2:1.

The procedure for determining blood pressure using the sphygmomanometer is as follows:

PROCEDURE

1. Either you or your laboratory partner should be comfortably seated, at ease, with your arm bared, slightly flexed, abducted, and perfectly relaxed.

2. Wrap the deflated cuff of the sphygmomanometer around the arm with the lower edge about 2.54 cm (1 in.) above the antecubital space. Close the valve on the neck of the rubber bulb.

3. Clean the earpieces of the stethoscope with alcohol before using it. Using the diaphragm of the stethoscope, find the pulse in the brachial

artery just above the bend of the elbow, on the inner margin of the biceps brachii muscle.

4. Inflate the cuff by squeezing the bulb until the air pressure within it just exceeds 170 mm Hg. At this point the wall of the brachial artery is compressed tightly, and no blood can flow through.

5. Place the diaphragm of the stethoscope firmly over the brachial artery and, while watching the pressure gauge, slowly turn the valve, releasing air from the cuff. Listen carefully as you watch the pressure fall and note the pressure on the gauge when you hear a loud, rapping noise. This is the ***systolic pressure*** reading. These rapping or thumping sounds that are heard are clinically called ***Korotkoff*** (kō-ROT-kof) ***sounds***.

6. Continue listening as the pressure falls. The pressure recorded on the mercury column when the sounds become faint or disappear is the ***diastolic pressure*** reading. It measures the force of blood in arteries during ventricular relaxation and specifically reflects the peripheral resistance of the arteries.

7. Repeat this procedure for both readings two or three times to see if you get consistent results. *Allow a few minutes between readings*. Record all results in Section F of the LABORATORY REPORT RESULTS at the end of the exercise.

8. Have your partner stand and record his or her blood pressure several times for each arm. Record all results in the table provided in Section F of the LABORATORY REPORT RESULTS at the end of the exercise.

9. Now, *assuming that your partner has no known or apparent cardiac or other health problems, and is capable of such an activity* have your partner do some exercise, such as running in place 40 or 50 steps, and measure the blood pressure again. Have your partner run in place again and measure his or her blood pressure using the other arm. Record all results in the table provided in Section F of the LABORATORY REPORT RESULTS at the end of the exercise.

Record the pulse pressure in the table provided in Section F of the LABORATORY REPORT RESULTS at the end of the exercise.

Would the *measured* blood pressure in an obese individual be abnormally high or low due to artifacts in measurement? ______

G. CONTROL OF PERIPHERAL BLOOD FLOW

Peripheral blood flow is controlled by a variety of factors, one of which is the ***autonomic nervous system (ANS).*** Such control is mainly accomplished by the ***sympathetic*** branch of the ANS. A peripheral vascular bed is composed of ***arterioles***, ***metarterioles***, ***capillaries***, and ***venules***. Metarterioles contain ***precapillary smooth muscle sphincters*** that assist in limiting blood flow to capillary beds by controlled contraction and relaxation. The sympathetic nervous system innervates all components of a peripheral vascular bed, with the exception of the capillaries. Sympathetic outflow to the peripheral vasculature is controlled by many factors, some of which are arterial blood pressure, as sensed by the baroreceptors, and PCO_2, PO_2, and pH, as sensed by central and peripheral chemoreceptors. With increased sympathetic nervous system output to vascular smooth muscle, contraction will occur and vessel diameter will decrease, thereby increasing resistance and decreasing peripheral blood flow through the capillary bed.

Peripheral blood flow is also controlled by a series of local factors that alter smooth muscle contraction within the vessels. Changes in pH, PO_2, PCO_2, ***histamine***, ***bradykinin***, or ***prostaglandin*** concentrations will cause either contraction or relaxation of smooth muscle within the peripheral vasculature, thereby altering capillary bed blood flow. Various factors, such as inflammation or exercise will alter the local concentration of one or more of these compounds, and therefore alter smooth muscle contraction.

H. PERIPHERAL BLOOD FLOW IN THE FROG

This experiment may be performed on either an intact or pithed frog. Review the proper use of a microscope in Exercise 1, Sections A.1 through A.4, and if the experiment is to be performed on a pithed frog review the procedures for pithing in Exercise 9, Section G.1.

NOTE: *It is strongly suggested that either your instructor or a laboratory assistant perform the procedure for pithing a frog. The procedure requires some practice and may prove difficult for inexperienced individuals.*

PROCEDURE

1. Strap a frog to a frog board in the prone position by wrapping a piece of moist cloth around the body, head, and legs and securing with several rubber bands, being sure not to secure the animal too tightly and thereby causing internal injury.

2. Pin the webbing of one foot over the hole in the frog board and secure the frog board to the stage of the microscope with rubberbands in such a way that the hole in the board is located directly over the hole in the microscope stage.

3. Continuously moisten the restraining cloth and foot webbing with water throughout the entire experiment.

1. Control Observations

PROCEDURE

1. Using an adequate amount of light, observe the blood flow through the peripheral vasculature of the frog's foot with either the 10× or 20× objective.

2. Identify an arteriole by its characteristic pulsating, rapid flow of blood.

3. A venule can be characterized by a slower, more constant blood flow, while a capillary may be recognized by its small diameter and slow blood flow. Red blood cells usually pass through a capillary in single file, thereby aiding in the vessel's identification.

4. Attempt to locate the junction of an arteriole and capillary (metarterioles cannot be distinguished readily in such a preparation) in the vascular bed. Although the precapillary sphincter probably will not be visible, one should be able to observe the irregular blood flow from the arteriole into the capillary due to the activity of the precapillary sphincter.

5. Make a drawing of the peripheral vascular bed visible in you microscope and label the various vessels in the space provided in Section H.1 of the LABORATORY REPORT RESULTS at the end of the exercise.

2. Effect of Histamine

PROCEDURE

1. Dry the webbing of the foot by gently blotting with paper towel.

2. *While observing the peripheral blood flow through the microscope,* use a medicine dropper to add several drops of 1:10,000 histamine solution and note the changes, if any, in peripheral blood flow.

3. Record your observations in Section H.2 of the LABORATORY REPORT RESULTS at the end of the excercise.

3. Effect of Antihistamines

PROCEDURE

1. After rinsing the foot several times with plain water, gently blot dry and add several drops of antihistamine solution to the webbing of the foot *while observing the peripheral blood flow through the microscope.*

2. *Without rinsing the foot,* use a medicine dropper and add several drops of 1:10,000 histamine solution and note the changes, if any, in peripheral blood flow.

3. Record and explain your observations in Section H.3 of the LABORATORY REPORT RESULTS at the end of the exercise.

4. Effect the Epinephrine

PROCEDURE

1. After rinsing the foot several times with plain water, gently blot dry and use a medicine dropper to add several drops of 1:1,000 epinephrine solution to the webbing of the foot *while observing the peripheral blood flow through the microscope.*

2. Record and explain your observations in Section H.4 of the LABORATORY REPORT RESULTS at the end of the exercise.

ANSWER THE LABORATORY REPORT QUESTIONS AT THE END OF THE EXERCISE

LABORATORY REPORT RESULTS

CARDIOVASCULAR PHYSIOLOGY 19

Student ______________________ **Date** ______________________

Laboratory Section ______________________ **Score/Grade** ______________________

SECTION A. CARDIAC CONDUCTION SYSTEM AND ELECTROCARDIOGRAM (ECG OR EKG)

Attach examples of the electrocardiogram strips you obtained.

LEAD I

Electrocardiogram Strip

LEAD II

Electrocardiogram Strip

LEAD III

Electrocardiogram Strip

SECTION C. ISOLATED TURTLE (OR FROG) HEART EXPERIMENTS

2. Recording the cardiac cycle:

3. Intrinsic control of the cardiac cycle:

4. Extrinsic control of the cardiac cycle:
   a. Application of epinephrine:

   b. Vagal stimulation:

   c. Application of acetylcholine:

d. Application of atropine sulfate:

e. Application of atropine sulfate immediately followed by acetylcholine:

f. Application of atropine sulfate and vagal stimulation:

5. Refractory period:

SECTION D. HEART SOUNDS

1. Which heart sound is the loudest? __________
2. Did you hear a third sound? __________
3. Where does the first sound originate? __________
4. Where does the second sound originate? __________
5. After you exercised, how did the heart sounds differ from before? __________
6. Did they differ in rate and intensity? __________
7. Did the first or second increase in loudness? __________

SECTION E. PULSE RATE

1. Radial pulse rate count results: __________ pulses per minute
2. Have all radial pulse rates put on the blackboard, arranging them from the highest to the lowest. The median pulse rate is found exactly halfway down from the top.

What is the median radial pulse rate of the class? __________

What was the highest rate? __________

What was the lowest rate? __________

3. Compare your radial and carotid pulse rates by filling in the following table.

	Radial pulse rate	Carotid pulse rate
Sitting quietly		
Standing quietly		
After walking		
After running in place		

SECTION F. BLOOD PRESSURE (AUSCULTATION METHOD)

Record your systolic and diastolic blood pressures in the following table.

	Systolic pressure		Diastolic pressure		Pulse pressure
	Left arm	Right arm	Left arm	Right arm	
Sitting					
Standing					
After running					

SECTION H. PERIPHERAL BLOOD FLOW IN THE FROG

1. Draw a diagram indicating the vascular bed observed during your control observations.

2. Indicate whether blood flow increased, decreased, or remained unchanged following addition of histamine. Explain the physiological mechanism for your observations. ______

______

3. a. Indicate how blood flow changed, if at all, following addition of antihistamine solution to the webbing of the foot. Explain the physiological mechanism for your observation. ______

______

b. Indicate how blood flow changed following the addition of the histamine solution while the antihistamine solution was still in place and indicate a possible physiological mechanism for antihistamine ______

______

4. Indicate how blood flow changed, if at all, following the addition to epinephrine solution to the webbing of the frog's foot and give a physiological mechanism for your observation. ______

______

Also indicate how the autonomic nervous system could give rise to a physiological condition opposite to that observed in your experiment. ______

______

LABORATORY REPORT QUESTIONS

CARDIOVASCULAR PHYSIOLOGY 19

Student ______________________ Date ______________________

Laboratory Section ______________________ Score/Grade ______________________

PART I. MULTIPLE CHOICE

_____ 1. During atrial systole, all of the following occur *except* (a) deoxygenated blood passes into the right ventricle (b) oxygenated blood passes into the left ventricle (c) the ventricles are in diastole (d) the semilunar valves are open

_____ 2. The two distinct heart sounds, described phonetically as lubb and dupp, represent (a) contraction of the ventricles and relaxation of the atria (b) contraction of the atria and relaxation of the ventricles (c) closing of the atrioventricular and semilunar valves (d) surging of blood into the pulmonary artery and aorta

_____ 3. When the semilunar valves are open during a cardiac cycle, which of the following occur? I—atrioventricular valves are closed; II—ventricles are in systole; III—ventricles are in diastole; IV—blood enters the aorta; V—blood enters the pulmonary artery; VI—atria contract. (a) I, II, IV and V (b) I, II, and VI (c) II, IV, and V (d) I, III, IV and VI

_____ 4. A pulse rate of 100 times per minute indicates (a) bradychardia (b) myocardia (c) tachycardia (d) endocardia

_____ 5. Pulse pressure provides clinical information concerning (a) condition of the arteries (b) ventricular contractions (c) state of the atrioventricular valves (d) state of the semilunar valves

PART II. COMPLETION

6. Systole and diastole of both atria plus systole and diastole of both ventricles is called ______________________ .

7. Blood flow through the heart is controlled by speed of the cardiac cycle, venous return to the heart, opening and closing of the valves, and ______________________ .

8. Heart sounds provide valuable information about the ______________________ .

9. Abnormal or peculiar heart sounds are called ______________________ .

10. Although the pulse may be detected in most surface arteries, pulse rate is usually determined on the ______________________ .

11. The heart has an intrinsic regulating system called the cardiac ______________________ system.

12. Electrical impulses accompanying the cardiac cycle are recorded by the ______________________ .

13. The typical ECG produces three clearly recognizable waves. The first wave, which indicates depolarization of the atria, is called the ______________________ .

14. Various up-and-down impulses produced by an ECG are called ____________________ .
15. The instrument normally used to measure blood pressure is called a(n) ____________________ .
16. The artery that is normally used to evaluate blood pressure is the ____________________ .
17. Rapping or thumping sounds heard clinically when blood pressure is being taken are called ____________________ sounds.
18. The difference between systolic and diastolic pressure is called ____________________ .
19. An average blood pressure value for an adult is ____________________ .
20. An average pulse pressure is ____________________ .

LYMPHATIC SYSTEM 20

The ***lymphatic*** (lim-FAT-ik) ***system*** is composd of a pale yellow fluid called lymph, vessels that transport lymph called lymphatics, and a number of structures and organs that contain lymphatic (lymphoid) tissue. Lymphatic tissue is a specialized form of reticular connective tissue that contains large numbers of lymphocytes and is found in lymph nodes, tonsils, thymus gland, spleen, and bone marrow (see Figure 20.3a).

Lymphangiography (lim-fan′-jē-OG-ra-fē) is the x-ray examination of lymphatic vessels and lymph organs after they are filled with a radiopaque substance. Such an x-ray is called a ***lymphangiogram*** (lim-FAN-jē-ō-gram). Lymphangiograms are useful in detecting edema and carcinomas, and in locating lymph nodes for surgical and radiotherapeutic treatment.

A. LYMPHATIC VESSELS

Lymphatic vessels originate as microscopic vessels in spaces between cells called ***lymph capillaries***. They originate throughout the body, but not in avascular tissue, the central nervous system, splenic pulp, and bone marrow. They are slightly larger and more permeable in only one direction than blood capillaries. Lymph capillaries also differ from blood capillaries in that they end blindly; blood capillaries have an arterial and a venous end. In addition, lymph capillaries are structurally adapted to ensure the return of proteins to the cardiovascular system when they leak out of blood capillaries.

Just as blood capillaries converge to form venules and veins, lymph capillaries unite to form larger and larger lymph vessels called ***lymphatics*** (Figure 20.1). Lymphatics resemble veins in structure, but have thinner walls and more valves, and contain lymph nodes at various intervals along their length (Figure 20.1). Ultimately, lymphatics deliver lymph into two main channels—the thoracic duct and the right lymphatic duct. These will be described shortly.

B. LYMPHATIC TISSUE

1. Lymph Nodes

The oval or bean-shaped structures located along the length of lymphatics are called ***lymph nodes***. A lymph node contains a slight depression on one side called a ***hilus*** (HĪ-lus) or ***hilum***, where blood vessels and efferent lymphatic vessels leave the node. Each node is covered by a ***capsule*** of dense connective tissue that extends into the node. The capsular extensions 'are called ***trabeculae*** (tra-BEK-yoo-lē). Internal to the capsule is a supporting network of reticular fibers and reticular cells (fibroblasts and macrophages). The capsule, trabeculae, and reticular fibers and cells constitute the stroma (framework) of a lymph node. The parenchyma of a lymph node is specialized into two regions: cortex and medulla. The outer ***cortex*** contains densely packed lymphocytes arranged in masses called ***lymphatic nodules***. The nodules often contain lighter-staining central areas, the ***germinal centers***, where lymphocytes are produced. The inner region of a lymph node is called the ***medulla***. In the medulla, the lymphocytes are arranged in strands called ***medullary cords***. These cords also contain macrophages and plasma cells.

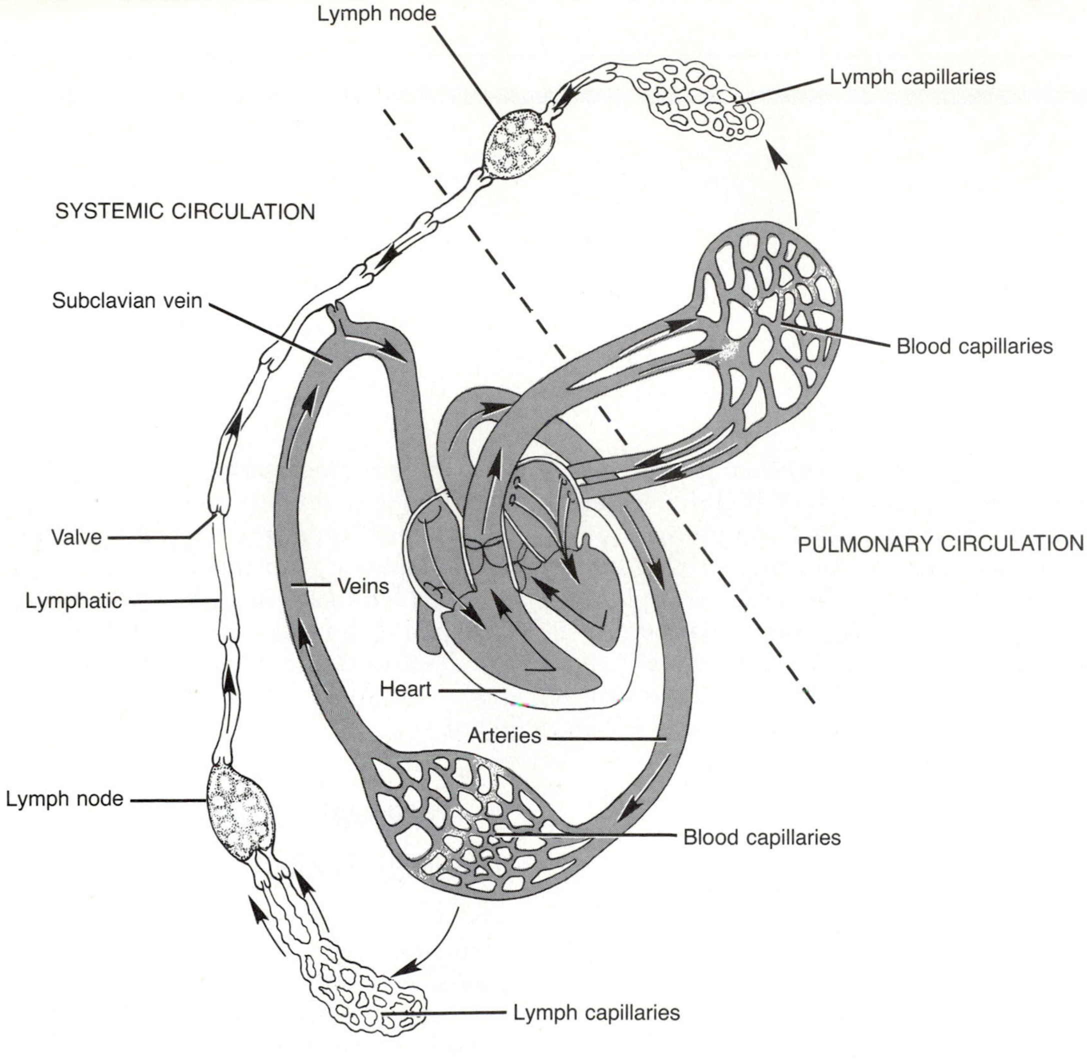

FIGURE 20.1 Relationship of lymphatic system to cardiovascular system.

The circulation of lymph through a node involves afferent (to convey toward a center) lymphatic vessels, sinuses in the node, and efferent (to convey away from a center) lymphatic vessels. ***Afferent lymphatic vessels*** enter the convex surface of the node at several points. They contain valves that open toward the node so that the lymph is directed *inward*. Once inside the node, the lymph enters the sinuses, which are a series of irregular channels. Lymph from the afferent lymphatic vessels enters the ***cortical sinuses*** just inside the capsule. From here it circulates to the ***medullary sinuses*** between the medullary cords. From these sinuses the lymph usually circulates into one or two ***efferent lymphatic vessels***, located at the hilus of the lymph node. Efferent lymphatic vessels are wider than the afferent vessels and contain valves that open away from the node to convey lymph *out* of the node.

Lymph passing from tissue spaces through lymphatics on its way back to the cardiovascular system is filtered through lymph nodes. As lymph passes through the nodes it is filtered of foreign substances. These substances are trapped by the reticular fibers within the node. Then, macrophages destroy the foreign substances by phagocytosis, T cells may destroy them by releasing various products, and/or B cells may develop into plasma cells that produce antibodies that destroy them. Lymph nodes also produce lymphocytes, some of which can circulate to other parts of the body.

Label the parts of a lymph node in Figure 20.2 and the various groups of lymph nodes in Figure 20.3.

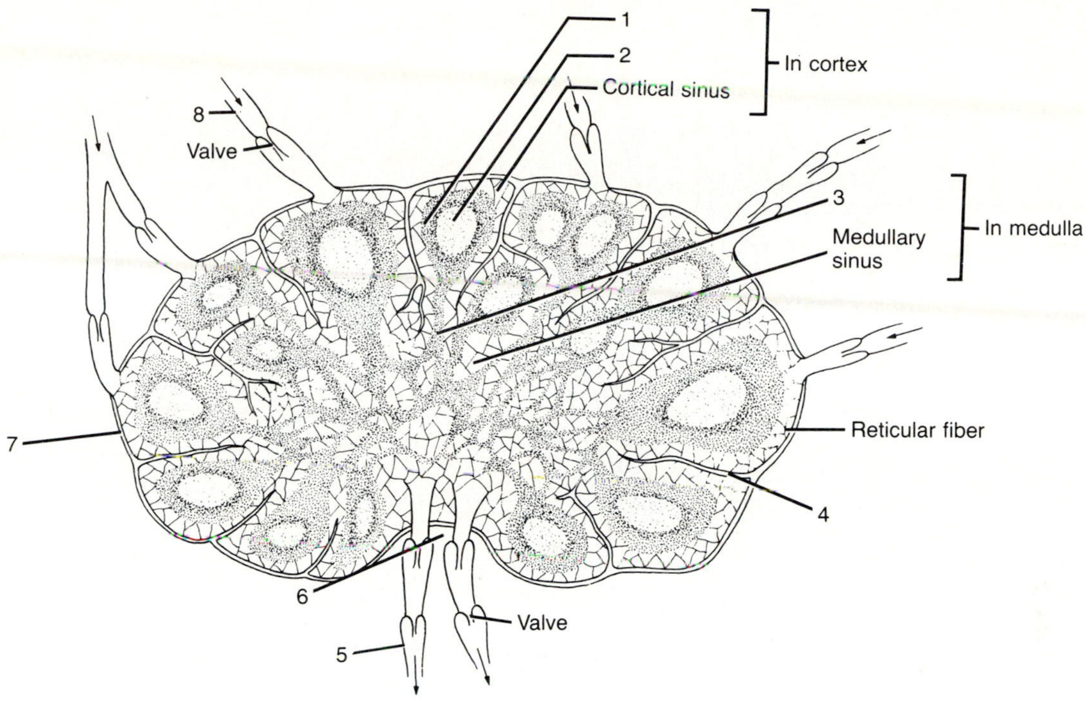

___ Afferent lymphatic vessel
___ Capsule
___ Efferent lymphatic vessel
___ Germinal center
___ Hilus
___ Lymphatic nodule
___ Medullary cord
___ Trabecula

FIGURE 20.2 Structure of a lymph node.

2. Tonsils

Tonsils are multiple aggregations of large lymphatic nodules embedded in a mucous membrane. The tonsils are arranged in a ring at the junction of the oral cavity and pharynx. The tonsils are situated strategically to protect against invasion of foreign substances (Figure 20.3a). Functionally, the tonsils produce lymphocytes and antibodies.

3. Spleen

The oval ***spleen*** is the largest mass of lymphatic tissue in the body. It is situated in the left hypochondriac region between the fundus of the stomach and diaphragm (see Figure 20.3a).

The splenic artery and vein and the efferent lymphatics pass through the hilus. Since the spleen has no afferent lymphatic vessels or lymph sinuses, it does not filter lymph. One key splenic function related to immunity is the production of B cells, which develop into antibody-producing plasma cells. The spleen also phagocytizes bacteria and worn-out and damaged red blood cells and platelets. In addition, the spleen stores and releases blood in case of demand, such as during hemorrhage. Sympathetic impulses cause the smooth muscle of the capsule of the spleen to contract. During early fetal development, the spleen participates in blood cell formation.

4. Thymus Gland

Usually a bilobed lymphatic organ, the ***thymus gland*** is located in the superior mediastinum, posterior to the sternum and between the lungs (Figure 20.3a). Its role in immunity is to help produce T cells that destroy invading microbes, including the AIDS virus, directly or indirectly by producing various substances.

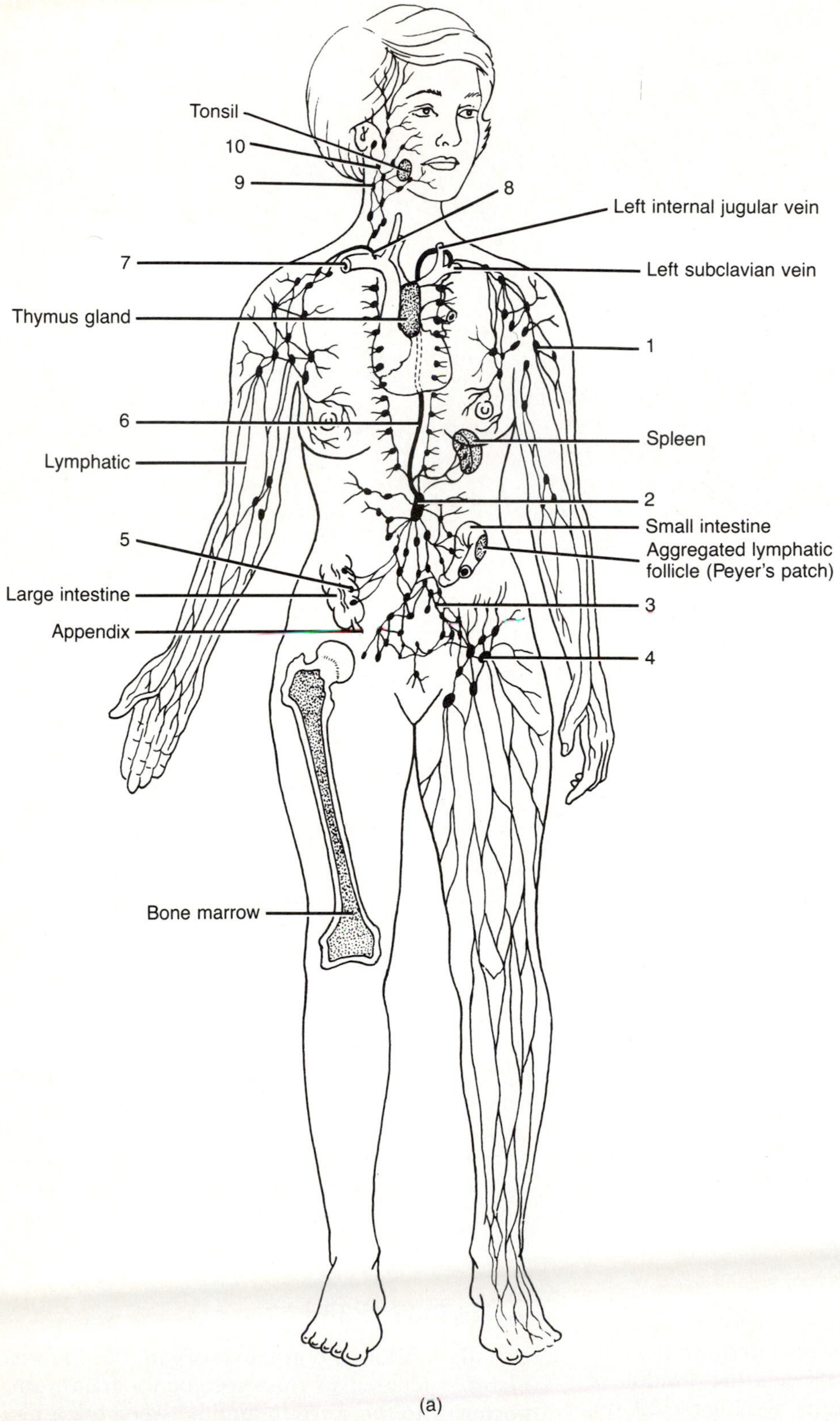

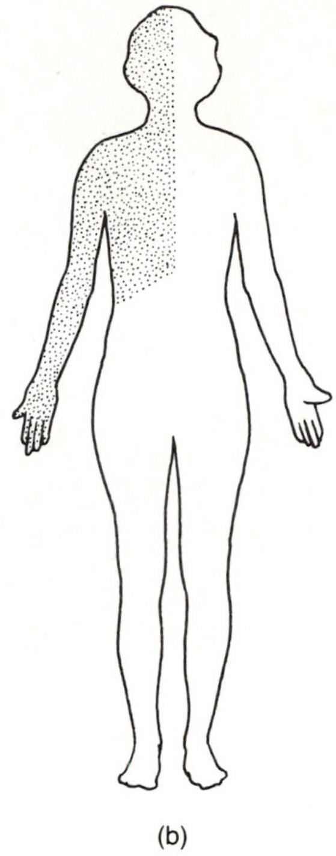

___ Axillary node
___ Cervical nodes
___ Cisterna chyli
___ Iliac nodes
___ Inguinal nodes
___ Intestinal nodes
___ Right lymphatic duct
___ Right subclavian vein
___ Submandibular nodes
___ Thoracic duct

FIGURE 20.3 Lymphatic system. (a) Location of the principal components of the lymphatic system. (b) The light gray area indicates those portions of the body drained by the right lymphatic duct. All other areas of the body are drained by the thoracic duct.

C. LYMPH CIRCULATION

When plasma is filtered by blood capillaries, it passes into the interstitial spaces; it is then known as interstitial fluid. When this fluid passes from interstitial spaces into lymph capillaries, it is called lymph. Lymph from lymph capillaries flows into lymphatics that run toward lymph nodes. At the nodes, afferent vessels penetrate the capsules at numerous points, and the lymph passes through the sinuses of the nodes. Efferent vessels from the nodes unite to form ***lymph trunks***.

The principal trunks pass their lymph into two main channels, the thoracic duct and the right lymphatic duct. The ***thoracic (left lymphatic) duct*** begins as a dilation in front of the second lumbar vertebra called the ***cisterna chyli*** (sis-TER-na KĪ-lē). The thoracic duct is the main collecting duct of the lymphatic system and receives lymph from the left side of the head, neck, and chest, the left upper extremity, and the entire body below the ribs.

The ***right lymphatic duct*** drains lymph from the upper right side of the body. Ultimately, the thoracic duct empties all of its lymph into the junction of the left internal jugular vein and left subclavian vein, and the right lymphatic duct empties all of its lymph into the junction of the right internal jugular vein and right subclavian vein. Thus, lymph is drained back into the blood and the cycle repeats itself continuously.

Edema, an excessive accumulation of interstitial fluid in tissue spaces, may be caused by an obstruction, such as an infected node or a blockage of vessels, in the pathway between the lymphatic capillaries and the subclavian veins. Another cause is excessive lymph formation and increased permeability of blood capillary walls. A rise in capillary blood pressure, in which interstitial fluid is formed faster than it is passed into lymphatics, also may result in edema.

D. DISSECTION OF CAT LYMPHATIC SYSTEM

CAUTION! *Please reread Section D. "Precautions Related to Dissection" at the beginning of the laboratory manual on page xii before you begin your dissection.*

Even if your cat has not been injected for examination of the lymphatic system, certain parts can be observed.

1. ***Lymphatic vessels***—The major lymphatic vessel is the ***thoracic duct***, a brownish vessel in the left pleural cavity dorsal to the aorta. It passes deep to most of the blood vessels at the base of the neck and enters the left external jugular vein near the entrance of the internal jugular vein (Figure 20.4). Using the diagram as a guide, see if you can find the ***right lymphatic duct***.
2. ***Lymph nodes***—These are small ovoid or round masses of lymphatic tissue connected by lymph vessels. You may be able to identify lymph nodes in the axillary and inguinal regions, at the base of the neck, at the angle of the jaws, and in the mesentery of the small intestine (Figure 20.4).
3. ***Lymphatic organs***—Lymphatic organs include the ***thymus gland***, ***tonsils***, and ***spleen***.

ANSWER THE LABORATORY REPORT QUESTIONS AT THE END OF THE EXERCISE

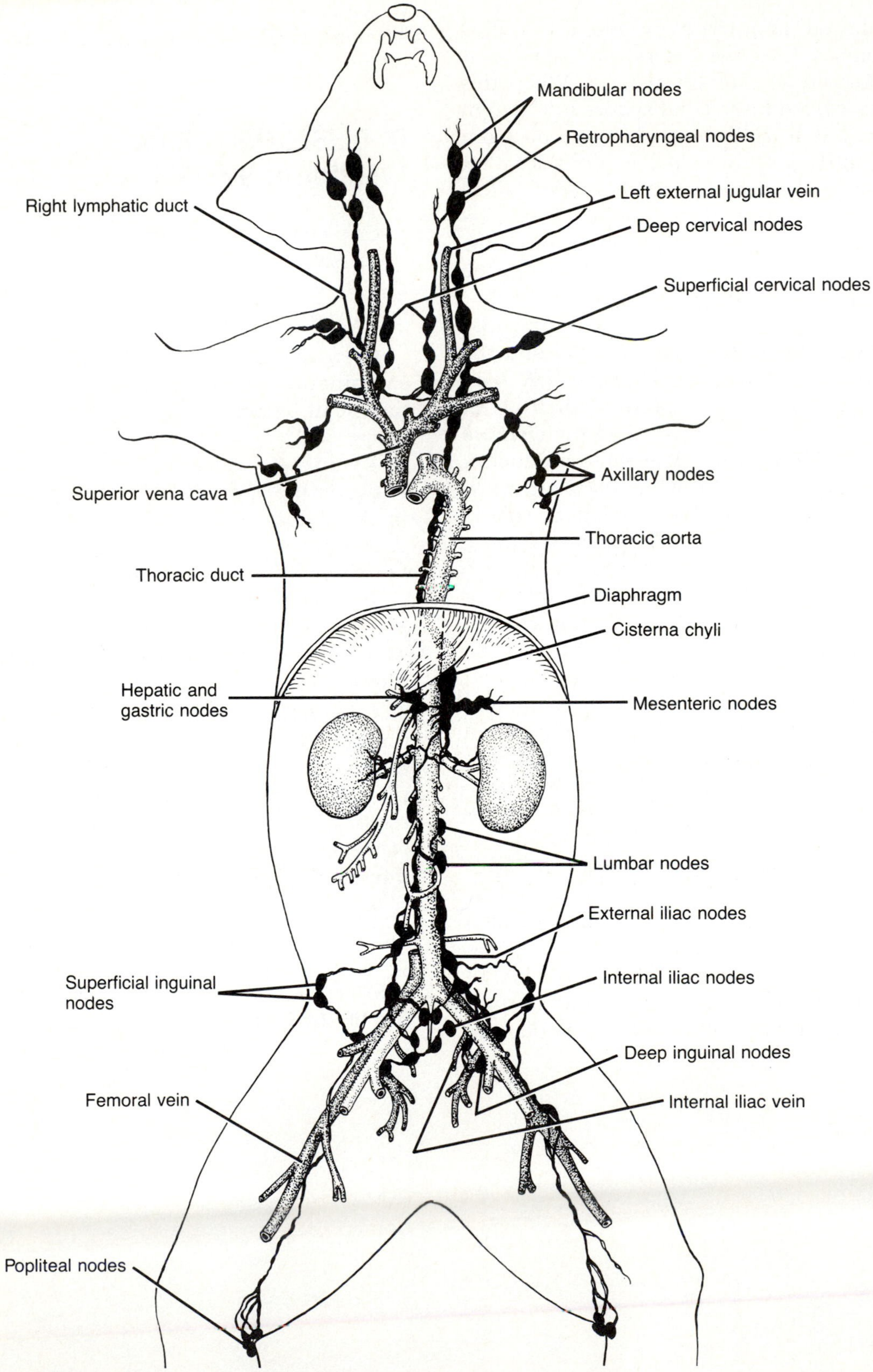

FIGURE 20.4 Lymphatic system of cat.

LABORATORY REPORT QUESTIONS

LYMPHATIC SYSTEM 20

Student ______________________ **Date** ______________________

Laboratory Section ______________________ **Score/Grade** ______________________

PART I. COMPLETION

1. Small masses of lymphatic tissue located along the length of the lymphatics are called ______________________ .
2. Lymphatics have thinner walls than veins, but resemble veins in that they also have ______________________ .
3. All lymphatics converge, get larger, and eventually merge into two main channels, the thoracic duct and the ______________________ .
4. Cells in the lymph nodes that carry on phagocytosis are ______________________ .
5. Lymph is conveyed out of lymph nodes in ______________________ vessels.
6. B cells in lymph nodes produce certain cells that are responsible for the production of antibodies. These cells are called ______________________ .
7. The x-ray examination of lymphatic vessels and lymph organs after they are filled with a radiopaque substance is called ______________________ .
8. This x-ray examination is useful in detecting edema and ______________________ .
9. The largest mass of lymphatic tissue is the ______________________ .
10. The main collecting duct of the lymphatic system is the ______________________ duct.

RESPIRATORY SYSTEM 21

Cells need a continuous supply of oxygen (O_2) for various metabolic reactions that produce energy, some of which is stored in ATP for cellular use. As a result of these reactions, cells also release quantities of carbon dioxide (CO_2). Since an excessive amount of carbon dioxide produces acid conditions that are poisonous to cells, the gas must be eliminated quickly and efficiently. The two systems that supply oxygen and eliminate carbon dioxide are the cardiovascular system and the respiratory system. The ***respiratory system*** consists of the nose, pharynx, larynx, trachea, bronchi, and lungs (Figure 21.1). The cardiovascular system transports the gases in the blood between the lungs and the cells. The term ***upper respiratory system*** refers to the nose, throat, and associated structures. The ***lower respiratory system*** referes to the remainder of the system.

The overall exchange of gases between the atmosphere, blood, and cells is called ***respiration***. Three basic processes are involved. The first process, ***pulmonary ventilation***, or breathing, is the inspiration (inflow) and expiration (outflow) of air between the atmosphere and the lungs. The second and third processes involve the exchange of gases within the body. ***External respiration*** is the exchange of gases between the lungs and blood. ***Internal respiration*** is the exchange of gases between the blood and the cells.

Using your textbook, charts, or models for reference, label Figure 21.1.

A. ORGANS OF RESPIRATORY SYSTEM

1. Nose

The ***nose*** has an external portion and an internal portion found inside the skull. Both the external and internal nose are divided internally by a vertical partition called the ***nasal septum***. The undersurface of the external nose contains two openings called the ***external nares*** (NA-rēz) or ***nostrils***. The internal nose is formed by the ethmoid, maxillae, inferior conchae, and palatine bones, and communicates with the paranasal sinuses. Sometimes the palatine and maxillary bones fail to fuse completely during embryonic life and the condition called ***cleft palate*** results. The interior of the nose is the ***nasal cavity***, which communicates with the throat through two openings, the ***internal nares (choanae)***. The nasal cavity is separated from the ***oral cavity*** below it by the palate, which is composed of the anterior, bony ***hard palate*** and the posterior muscular ***soft palate***. The anterior portion of the nasal cavity, just inside the nostrils, is called the ***vestibule***, which is lined with coarse hairs. Three shelves formed by projections of the ***superior***, ***middle***, and ***inferior nasal conchae*** (KON-kē) extend out of the lateral wall of each nasal cavity. The conchae almost reach the nasal septum and subdivide each nasal cavity into a series of groovelike passageways called the ***superior***, ***middle***, and ***inferior meatuses***.

The interior structures of the nose are specialized for three functions: (1) incoming air is warmed, moistened, and filtered; (2) olfactory stimuli are received; and (3) large hollow resonating chambers are provided for speech sounds. In addition, mucous membranes trap dust particles and, with the help of cilia, move unwanted particles to the throat for elimination.

Label the hard palate, inferior meatus, inferior nasal concha, internal naris, middle meatus, middle nasal concha, nasal cavity, oral cavity, soft palate, superior meatus, superior nasal concha, and vestibule in Figure 21.2.

The surface anatomy of the nose is shown in Figure 21.3.

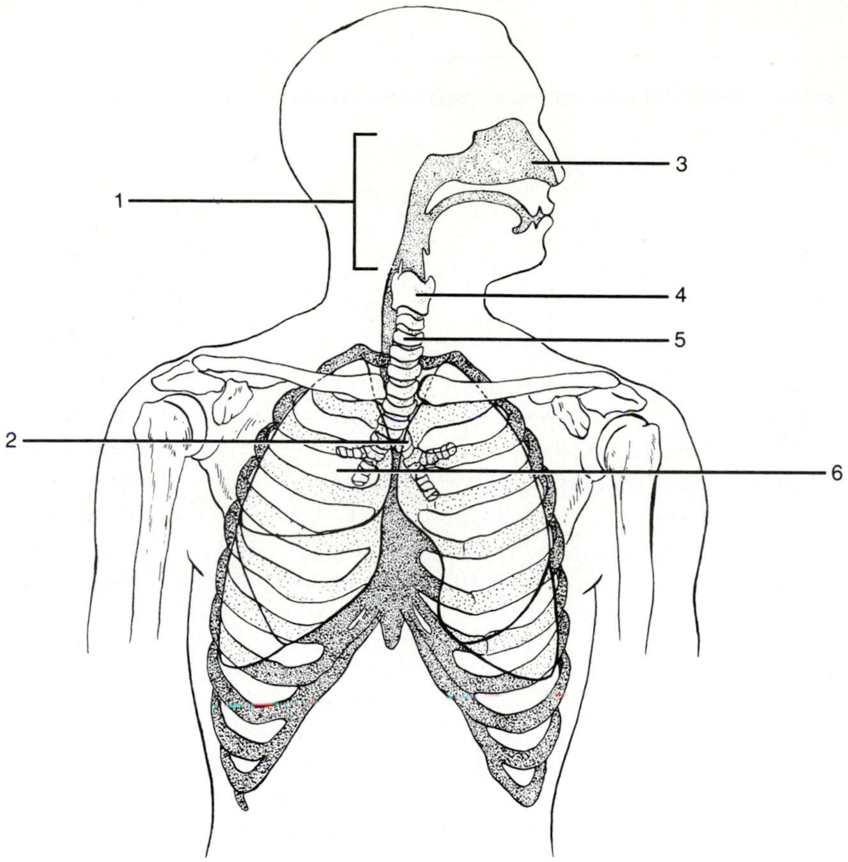

FIGURE 21.1 Organs of respiratory system.

2. Pharynx

The ***pharynx*** (FAR-inks) (throat) is a somewhat funnel-shaped tube about 13 cm (5 in.) long. Lying in back of the nasal and oral cavities and just in front of the cervical vertebrae, the pharynx is a passageway for air and food and a resonating chamber for speech sounds.

The pharynx is composed of a superior portion, called the ***nasopharynx***, a middle portion, the ***oropharynx***, and an inferior portion, the ***laryngopharynx*** (la-rin′-gō-FAR-inks). The nasopharynx consists of ***pseudostratified epithelium*** and has four openings in its wall: two ***internal nares*** plus two openings into the ***auditory (Eustachian) tubes***. The nasopharynx also contains the ***pharyngeal tonsil (adenoid)***. The oropharynx is lined by ***stratified squamous epithelium*** and receives one opening: the ***fauces*** (FAW-sēz). The oropharynx contains the ***palatine*** and ***lingual tonsils***. The laryngopharynx is also lined by ***stratified squamous epithelium*** and becomes continuous with the esophagus posteriorly and the larynx anteriorly.

Label the laryngopharynx, nasopharynx, oropharynx, palatine tonsil, and pharyngeal tonsil in Figure 21.2.

3. Larynx

The ***larynx*** (voice box) is a short passageway connecting the laryngopharynx with the trachea. Its wall is composed of nine pieces of cartilage.

a. ***Thyroid cartilage (Adam's apple)***—Large anterior piece that gives larynx its triangular shape.

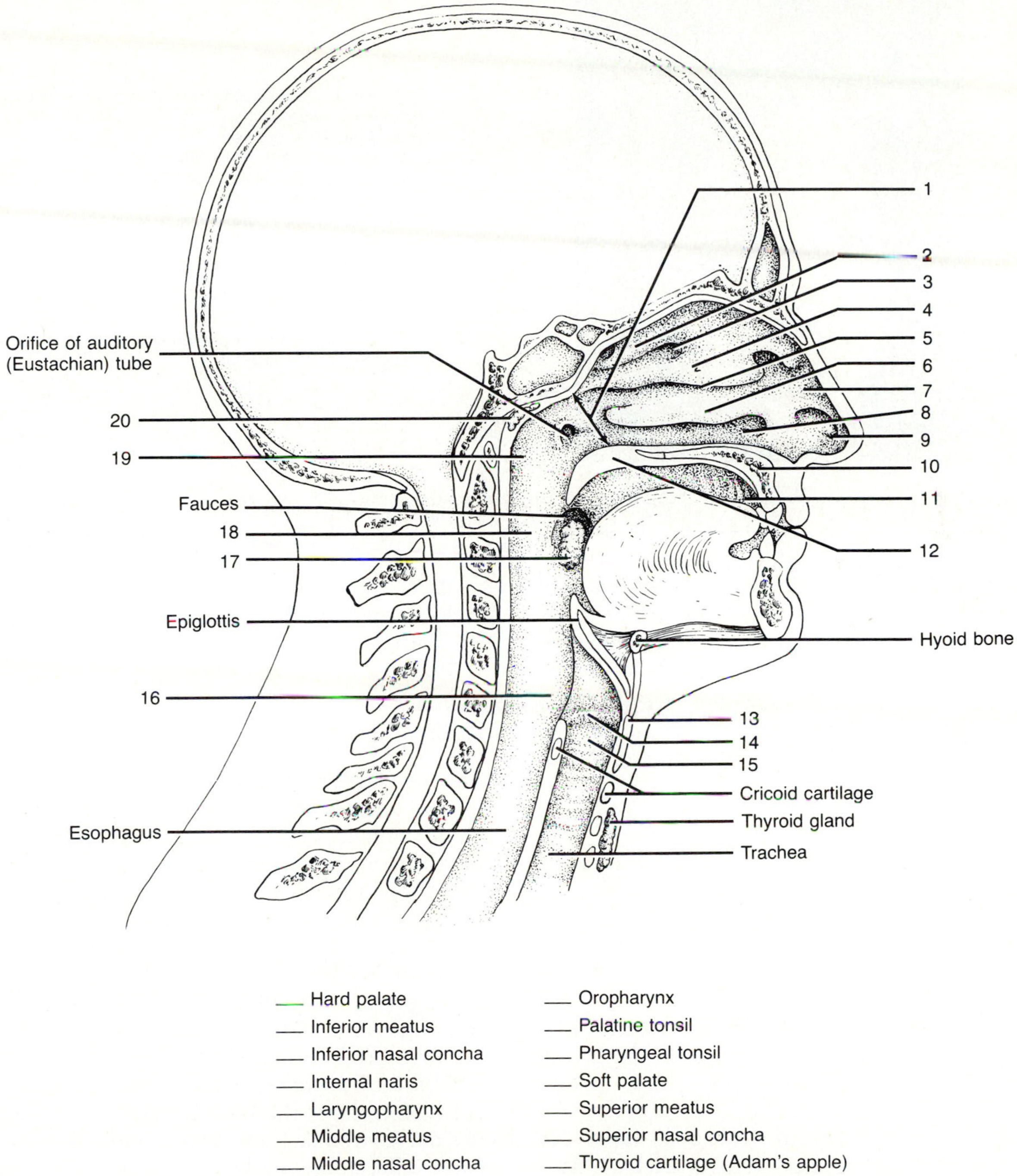

___ Hard palate	___ Oropharynx
___ Inferior meatus	___ Palatine tonsil
___ Inferior nasal concha	___ Pharyngeal tonsil
___ Internal naris	___ Soft palate
___ Laryngopharynx	___ Superior meatus
___ Middle meatus	___ Superior nasal concha
___ Middle nasal concha	___ Thyroid cartilage (Adam's apple)
___ Nasal cavity	___ Ventricular fold (false vocal cord)
___ Nasopharynx	___ Vestibule
___ Oral cavity	___ Vocal fold (true vocal cord)

FIGURE 21.2 Upper respiratory system as seen in midsagittal section.

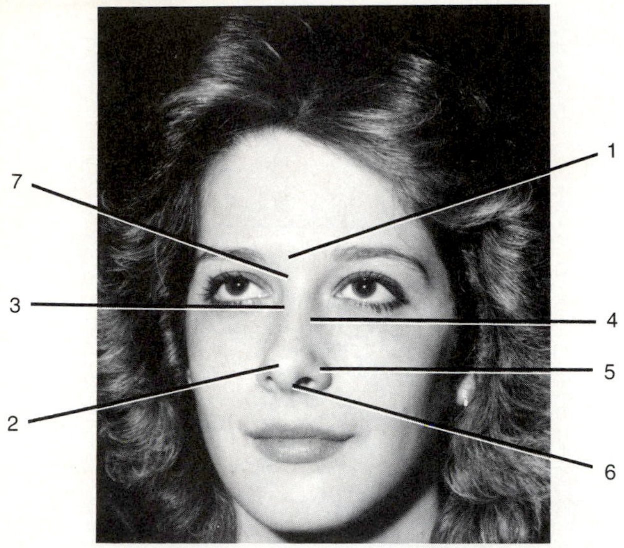

1. **Root.** Superior attachment of nose at forehead located between eyes.
2. **Apex.** Tip of nose.
3. **Dorsum nasi.** Rounded anterior border connecting root and apex; in profile, may be straight, convex, concave, or wavy.
4. **Nasofacial angle.** Point at which side of nose blends with tissues of face.
5. **Ala.** Convex flared portion of inferior lateral surface; unites with upper lip.
6. **External nares.** External openings into nose (nostril).
7. **Bridge.** Superior portion of dorsum nasi, superficial to nasal bones.

FIGURE 21.3 Surface anatomy of nose.

b. ***Epiglottis***—Leaf-shaped cartilage on top of larynx that closes off the larynx so that foods and liquids are routed into the esophagus and kept out of the respiratory system.

c. ***Cricoid*** (KRĪ-koyd) ***cartilage***—Ring of cartilage forming the inferior portion of the larynx that is attached to the first ring of tracheal cartilage.

d. ***Arytenoid*** (ar-i-TĒ-noyd) ***cartilage***—Paired, pyramid-shaped cartilages at superior border of cricoid cartilage that attach vocal folds to the intrinsic pharyngeal muscles.

e. ***Corniculate*** (kor-NIK-yoo-lāt) ***cartilages***—Paired, cone-shaped cartilages at apex of arytenoid cartilages.

f. ***Cuneiform*** (kyoo-NĒ-i-form) ***cartilages***—Paired, club-shaped cartilages anterior to the corniculate cartilages.

With the aid of your textbook, label the laryngeal cartilages shown in Figure 21.4. Also label the thyroid cartilage in Figure 21.2.

The mucous membrane of the larynx is arranged into two pairs of folds, an upper pair called the ***ventricular folds (false vocal cords)*** and a lower pair called the ***vocal folds (true vocal cords)***. The space between the vocal folds when they are apart is called the ***glottis***. Movement of the vocal folds produces sounds; variations in pitch result from (1) varying degrees of tension and (2) varying lengths in males and females.

With the aid of your textbook, label the ventricular folds, vocal folds, and glottis in Figure 21.5. Also label the ventricular and vocal folds in Figure 21.2.

4. Trachea

The ***trachea*** (TRĀ-kē-a) (windpipe) is a tubular air passageway about 12 cm (4½ in.) in length and 2.5 cm (1 in.) in diameter. It lies in front of the esophagus and, at its inferior end (T5), divides into right and left primary bronchi (Figure 21.6). The epithelium of the trachea consists of ***pseudostratified epithelium***. This epithelium consists of ciliated columnar cells, goblet cells, and basal cells. The epithelium offers the same protection against dust as the membrane lining the larynx.

Obtain a prepared slide of pseudostratified epithelium from the trachea, and, with the aid of your textbook, label the ciliated columnar cells, cilia, goblet cells, and basal cells in Figure 21.7.

The trachea consists of smooth muscle, elastic connective tissue, and incomplete ***rings of cartilage*** shaped like a series of letter Cs. The open ends of the Cs are held together by the ***trachealis muscle***. The cartilage provides a rigid support so that the tracheal wall does not collapse inward and obstruct the air passageway, and, because the open parts of the Cs face the esophagus, the latter can expand into the trachea dur-

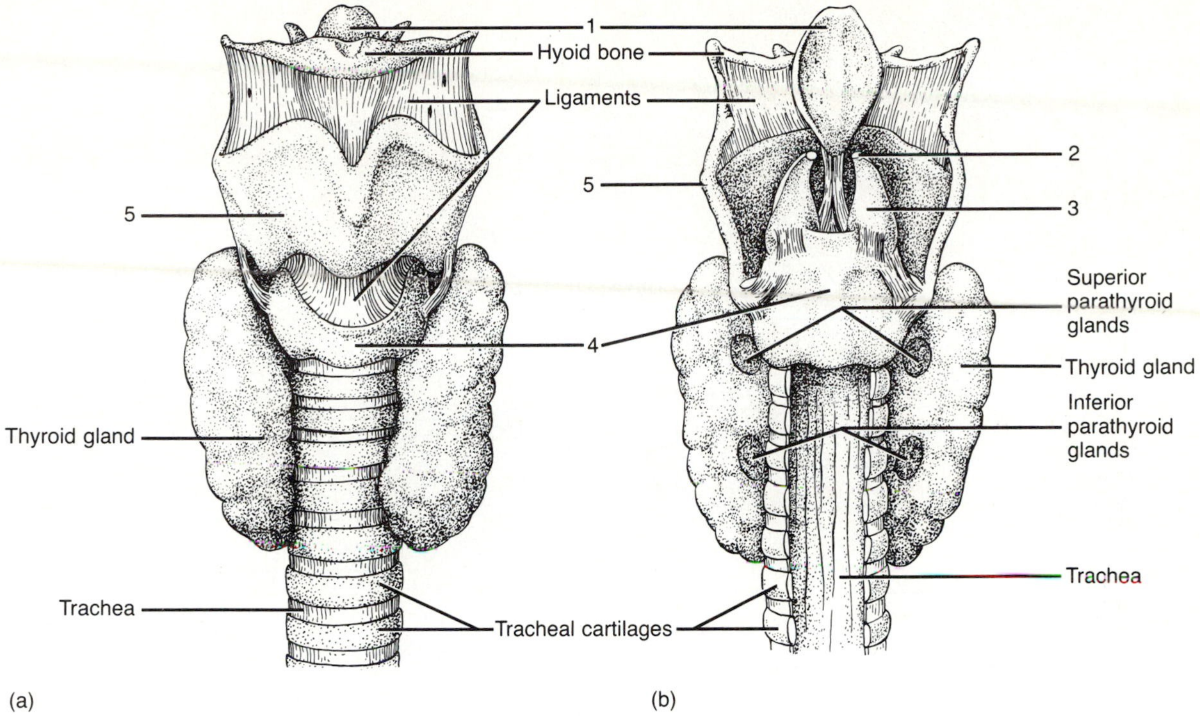

___ Arytenoid cartilage
___ Corniculate cartilage
___ Cricoid cartilage
___ Epiglottis
___ Thyroid cartilage (Adam's apple)

FIGURE 21.4 Larynx. (a) Anterior view. (b) Posterior view.

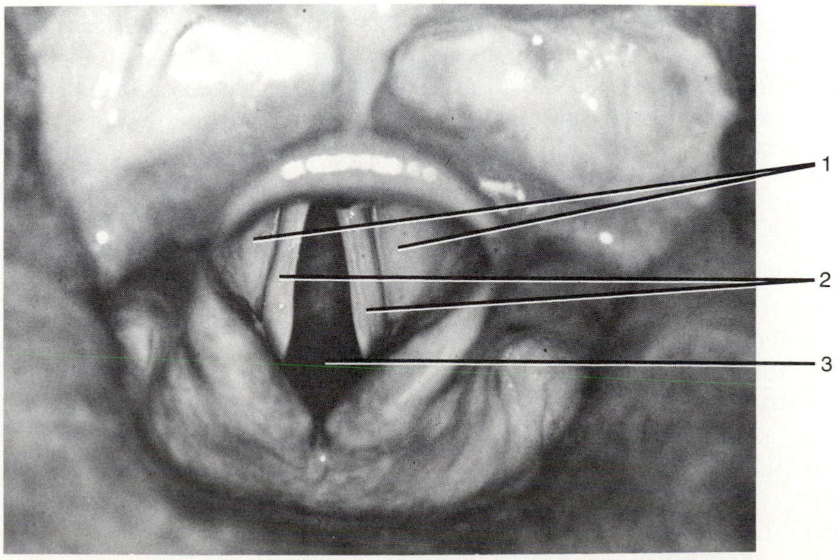

___ Glottis
___ Ventricular folds (false vocal cords)
___ Vocal folds (true vocal cords)

FIGURE 21.5 Photograph of superior view of larnyx.

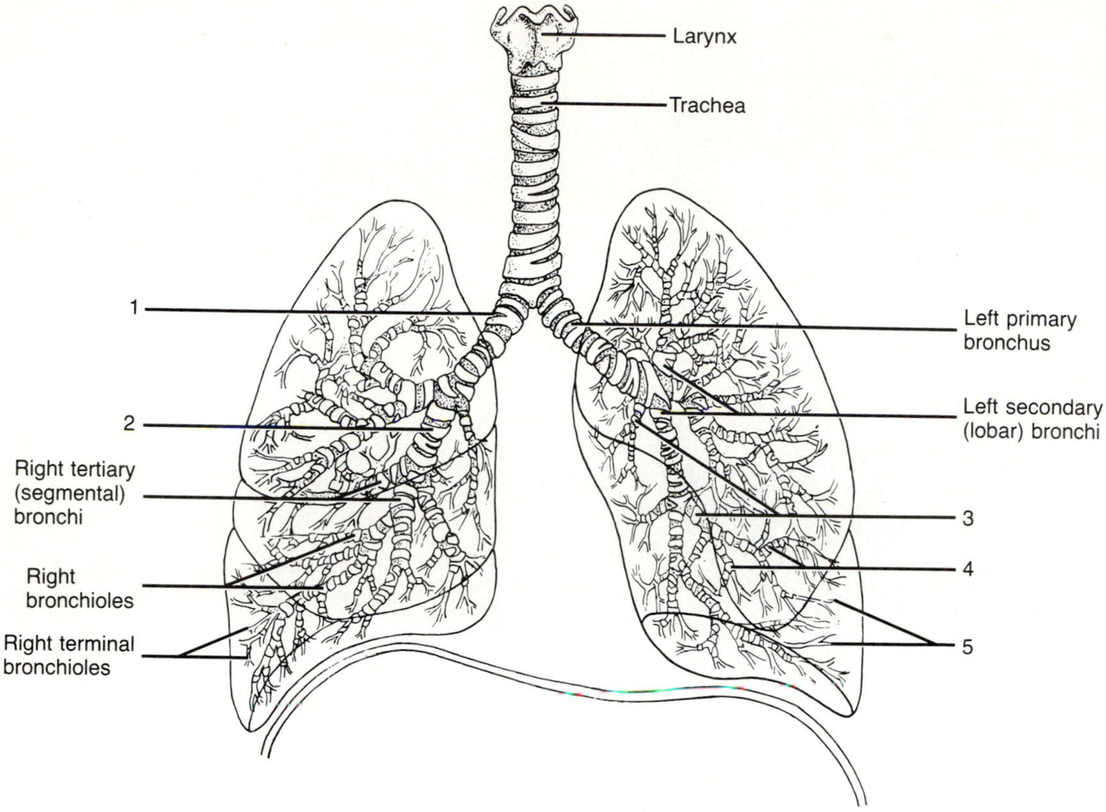

___ Left bronchioles
___ Left terminal bronchioles
___ Left tertiary (segmental) bronchi
___ Right primary bronchus
___ Right secondary (lobar) bronchi

FIGURE 21.6 Air passageways to the lungs. Shown is the bronchial tree in relationship to lungs.

ing swallowing. If the trachea should become obstructed, a ***tracheostomy*** (trā-kē-OS-tō-mē) may be performed. Another method of opening the air passageway is called ***intubation***, in which a tube is passed into the mouth and down through the larynx and the trachea.

5. Bronchi

The trachea terminates by dividing into a ***right primary bronchus*** (BRON-kus), going to the right lung, and a ***left primary bronchus***, going to the left lung. They continue dividing in the lungs into smaller bronchi, the ***secondary (lobar) bronchi*** (BRON-kē), one for each lobe of the lung. These bronchi, in turn, continue dividing into still smaller bronchi called ***tertiary (segmental) bronchi***, which divide into ***bronchioles***. The next division is into even smaller tubes called ***terminal bronchioles***. This entire branching structure of the trachea is commonly referred to as the ***bronchial tree***.

Label Figure 21.6.

Bronchography (brong-KOG-ra-fē) is a technique for examining the bronchial tree. With this procedure, an intratracheal catheter is passed transorally or transnasally through the glottis into the trachea. Then an opaque iodinated medium is introduced, by means of gravity, into the trachea and distributed through the bronchial branches. Roentgenograms of the chest in various positions are taken and the developed film, called ***bronchogram*** (BRONG-kō-gram), provides a picture of the bronchial tree.

6. Lungs

The ***lungs*** are paired, cone-shaped organs lying in the thoracic cavity (see Figures 21.1). The

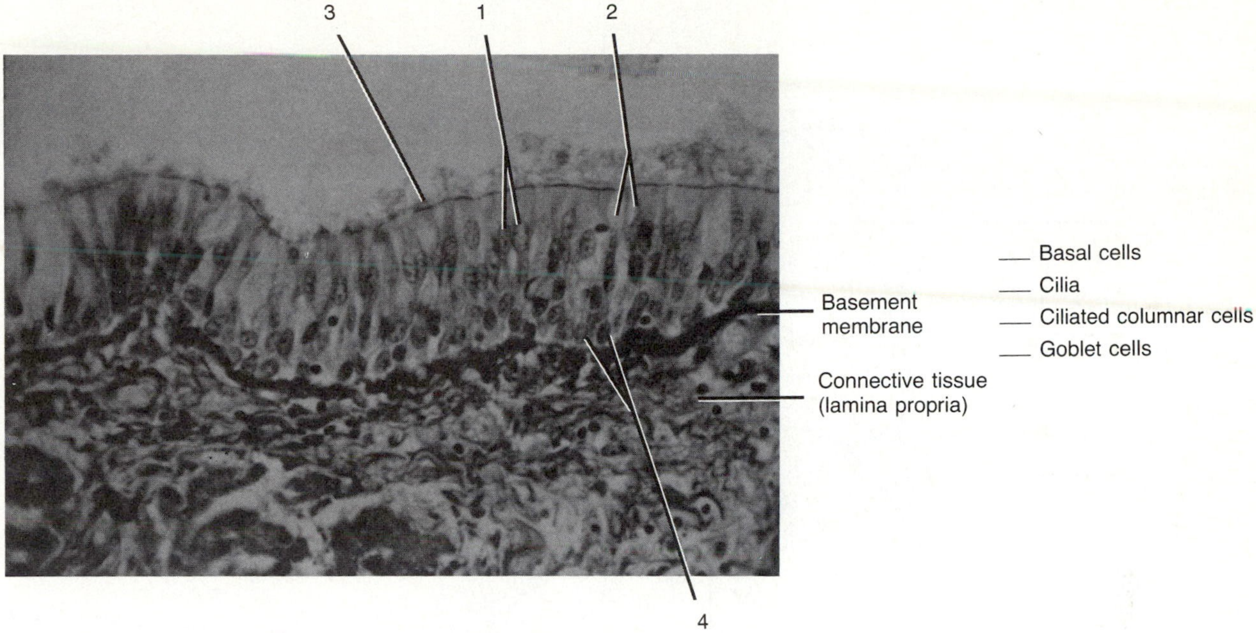

FIGURE 21.7 Histology of trachea.

pleural membrane encloses and protects each lung. Whereas the ***parietal pleura*** lines the wall of the thoracic cavity, the ***visceral pleura*** covers the lungs; the potential space between parietal and visceral pleurae, the ***pleural cavity***, contains a lubricating fluid to reduce friction as the lungs expand and recoil.

Major surface features of the lungs include:

a. ***Base***—Broad inferior portion resting on diaphragm.

b. ***Apex***—Narrow superior portion just above clavicles.

c. ***Costal surface***—Surface lying against ribs.

d. ***Mediastinal surface***—Medial surface.

e. ***Hilus***—Region in mediastinal surface through which bronchial tubes, blood vessels, lymphatics, and nerves enter and exit the lung.

f. ***Cardiac notch***—Medial concavity in left lung in which heart lies.

Each lung is divided into ***lobes*** by one or more ***fissures***. The right lung has three lobes, ***superior***, ***middle***, and ***inferior***; the left lung has two lobes, ***superior*** and ***inferior***. The ***horizontal fissure*** separates the superior lobe from the middle lobe in the right lung; an ***oblique fissure*** separates the middle lobe from the inferior lobe in the right lung and the superior lobe from the inferior lobe in the left lung.

Using your textbook as a reference, label Figure 21.8a.

Each lobe of a lung is divided into many small compartments called ***lobules***. Each lobule is wrapped in elastic connective tissue and contains a lymphatic, arteriole, venule, and branch from a terminal bronchiole. Terminal bronchioles divide into ***respiratory bronchioles***, which, in turn, divide into several ***alveolar*** (al-VĒ-ō-lar) ***ducts***. Around the circumference of alveolar ducts are numerous alveoli and alveolar sacs. ***Alveoli*** (al-VĒ-ō-lī) are cup-shaped outpouchings lined by epithelium and supported by a thin elastic membrane. ***Alveolar sacs*** are two or more alveoli that share a common opening. Over the alveoli, an arteriole and venule disperse into a network of capillaries. Gas is exchanged between the lungs and blood by diffusion across the alveolar and the capillary walls.

Using your textbook as a reference, label Figure 21.8b.

The alveolar wall (Figure 21.9) consists of:

a. ***Squamous pulmonary epithelial cells***—Large cells that form a continuous lining of the alveolar wall, except for occasional septal cells.

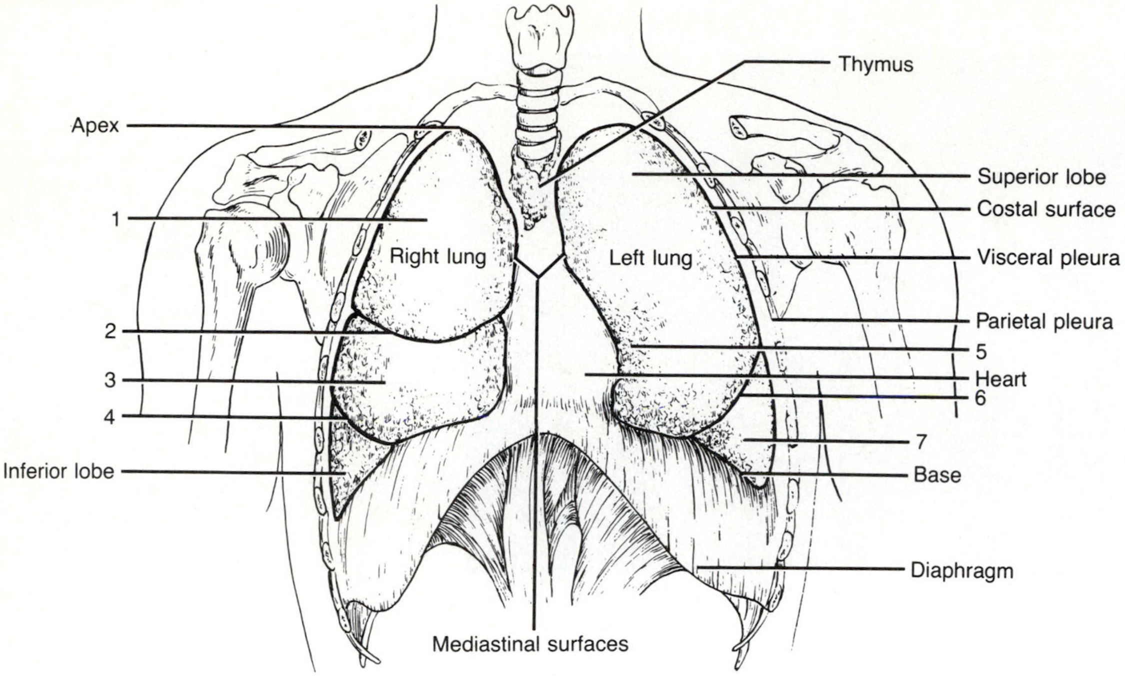

(a)

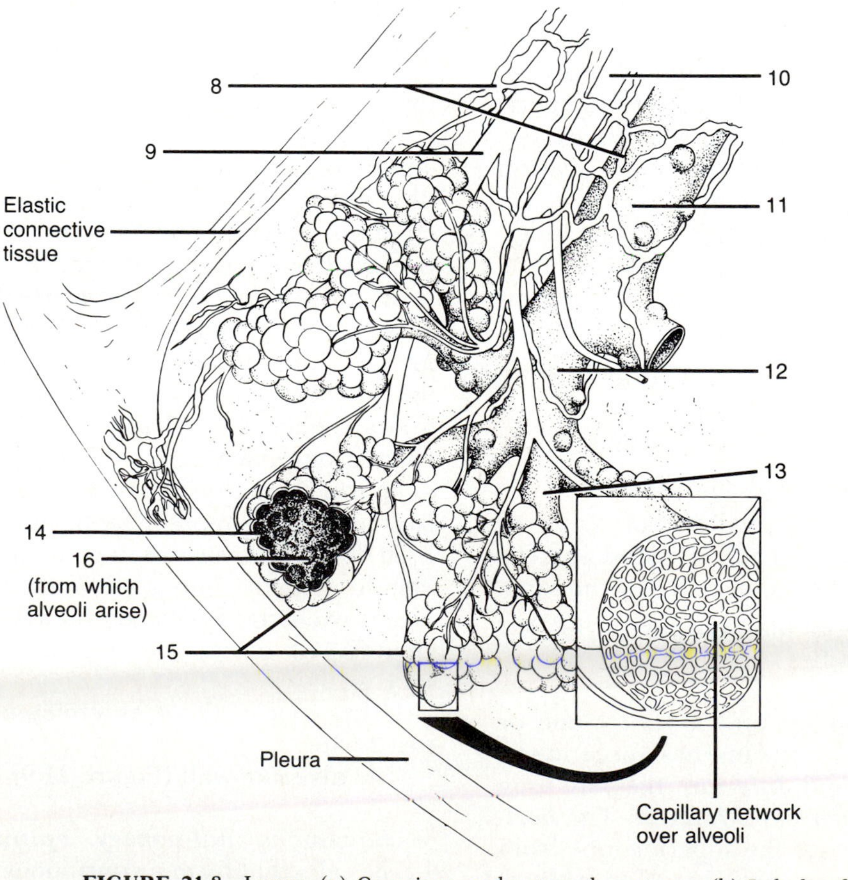

___ Alveolar duct
___ Alveolar sac
___ Alveoli
___ Bronchiole
___ Cardiac notch
___ Horizontal fissure
___ Inferior lobe
___ Lymphatics
___ Middle lobe
___ Oblique fissure of right lung
___ Oblique fissure of left lung
___ Pulmonary arteriole (stippled)
___ Pulmonary venule (clear)
___ Respiratory bronchiole
___ Superior lobe
___ Terminal bronchiole

FIGURE 21.8 Lungs. (a) Coverings and external anatomy. (b) Lobule of lung.

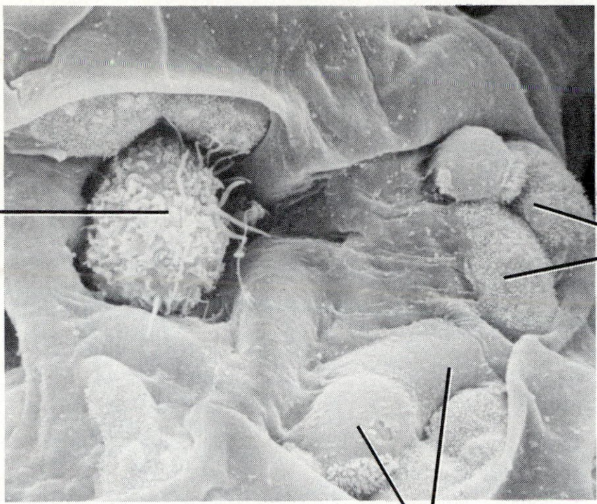

FIGURE 21.9 Scanning electron micrograph of alveolar wall at magnification of 3430×. (From *Tissues and Organs: A Text-Atlas of Scanning Electron Microscopy* by Richard G. Kessel and Randy H. Kardon. W. H. Freeman and Company. Copyright © 1979.)

b. ***Septal cells***—Cuboidal cells dispersed amoung squamous pulmonary epithelial cells that secrete a phospholipid substance called ***surfactant*** (sur-FAK-tant), a surface tension-lowering agent.

c. ***Alveolar macrophages (dust cells)***—Phagocytic cells that remove dust particles and other debris from the alveolar spaces.

Obtain a slide of normal lung tissue and examine it under high power. Using your textbook as a reference, see if you can identify a terminal bronchiole, respiratory bronchiole, alveolar duct, alveolar sac, and alveoli.

If available, examine several pathological slides of lung tissue, such as slides that show emphysema and lung cancer. Compare your observations to the normal lung tissue.

The exchange of respiratory gases between the lungs and blood takes place by diffusion across the alveolar and capillary walls. This membrane, through which the respiratory gases move, is collectively known as the ***alveolar-capillary (respiratory) membrane*** (Figure 21.10). It consists of:

1. A layer of squamous pulmonary epithelial cells with septal cells and free alveolar macrophages (dust cells) that constitute the alveolar (epithelial) wall.
2. An epithelial basement membrane underneath the alveolar wall.
3. A capillary basement membrane that is often fused to the epithelial basement membrane.
4. The endothelial cells of the capillary.

Label the components of the alveolar-capillary (respiratory) membrane in Figure 21.10.

B. DISSECTION OF CAT RESPIRATORY SYSTEM

CAUTION! *Please reread Section D. "Precautions Related to Dissection" at the beginning of the laboratory manual on page xii before you begin your dissection.*

Air enters the respiratory system by one of two routes, the nose or the mouth. Air entering the external nares (nostrils) passes into the nasal cavity, through the internal nares (choanae), and into the nasopharynx, oropharynx, and laryngopharynx. From here it enters

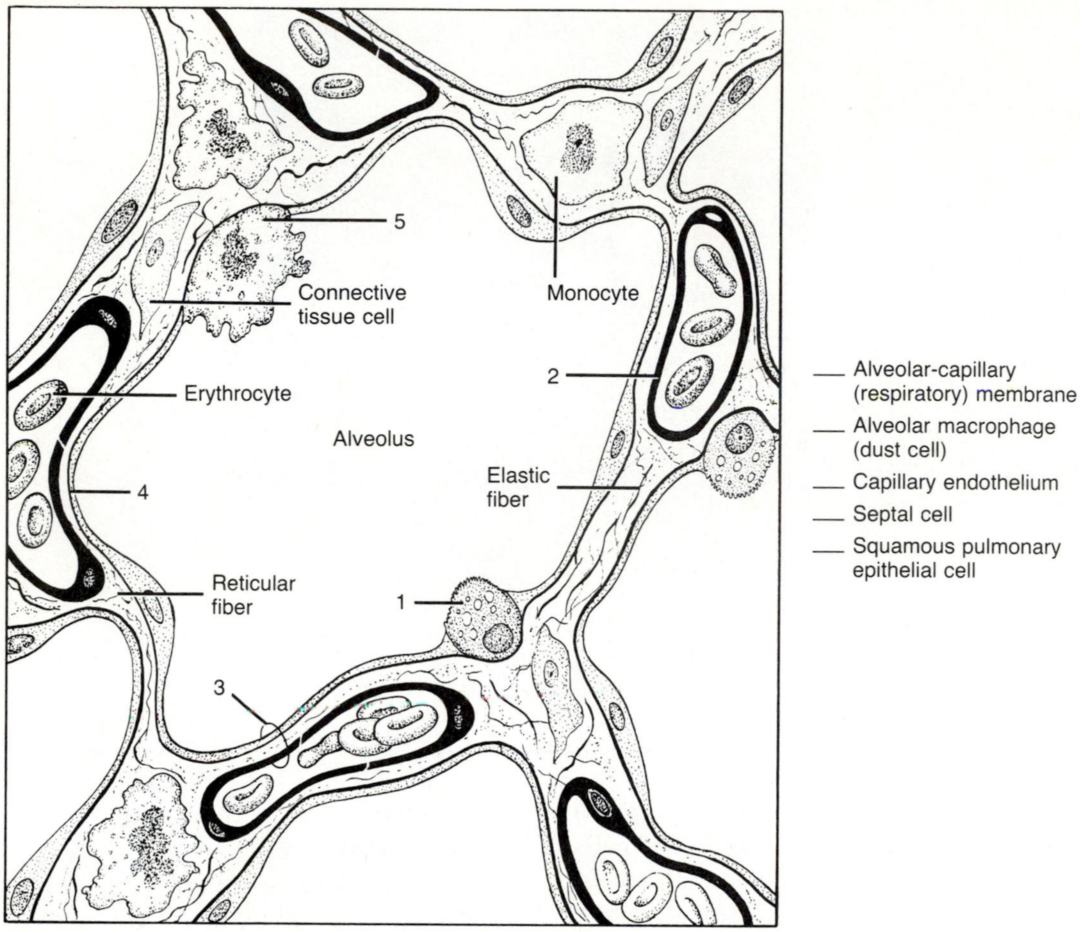

FIGURE 21.10 Alveolar-capillary (respiration) membrane.

the larynx. Air entering the mouth passes through the vestibule, oral cavity, fauces, subdivisions of the pharynx, and into the larynx. The mouth and pharynx of the cat will be studied in detail in Exercise 22, Digestive System.

PROCEDURE

1. Larynx

1. The ***larynx*** (voice box) consists of five cartilages (Figure 21.11).
2. Expose the laryngeal cartilages by stripping off the surrounding tissues on the dorsal, ventral, and left lateral surfaces.
3. Identify the following parts of the larynx:
   a. ***Thyroid cartilage***—Forms most of the ventral and lateral walls.
   b. ***Cricoid cartilage***—Caudal to the thyroid cartilage; its expanded dorsal portion forms most of the dorsal wall of the larynx.
   c. ***Arytenoid cartilages***—Paired, small, triangular cartilages cranial to dorsal part of cricoid cartilage. These are best seen if the larynx is separated from the esophagus.
   d. ***Epiglottis***—Projects cranially from the thyroid cartilage to which it is attached; guards the opening into the larynx. To examine the cranial portion of the larynx, make a midventral incision through the larynx and cranial part of the trachea. Be careful not to damage the blood vessels on either side of the trachea. Spread the cut edges back and identify the following structures.
   e. ***False vocal cords***—Cranial pair of mucous membranes extending across the larynx from the arytenoid cartilages to the base of the epiglottis.

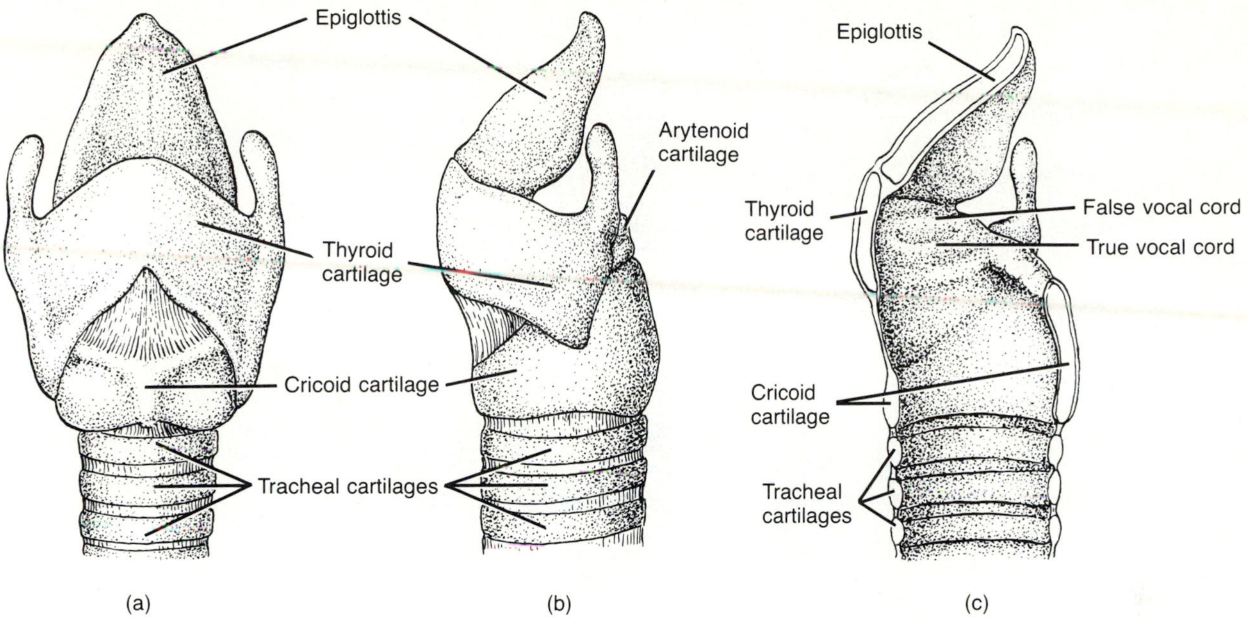

FIGURE 21.11 Larynx of cat. (a) Ventral view. (b) Lateral view. (c) Longitudinal section.

f. ***True vocal cords***—Caudal, larger pair of mucous membranes extending across the larnx from the arytenoid cartilages to the thyroid cartilage; important in sound production.

g. ***Glottis***—Space between the free margins of the true vocal cords.

h. ***Tracheal cartilages***—C-shaped, dorsally incomplete rings of cartilage that support the trachea and prevent it from collapsing.

2. Trachea

1. The ***trachea*** (windpipe) is a tube that extends from the larynx and bifurcates into the bronchial tubes (Figure 21.12a). Note the tracheal cartilages.

2. On either side of the trachea near the larynx you can locate the dark lobes of the ***thyroid gland***. This gland belongs to the endocrine system. Its lobes are connected by a ventral strip of thyroid tissue called the ***isthmus***.

3. Also on either side of the trachea you can identify the large ***common carotid arteries***, the smaller ***internal jugular veins***, and the white, threadlike ***vagus nerve*** adjacent to the common carotid artery.

3. Bronchi

1. The ***bronchi*** are two short tubes formed by the bifurcation of the trachea at the level of the sixth rib. One bronchus enters each lung. The bronchi, like the trachea, consists of C-shaped cartilages.

2. Within the lungs, the bronchi branch repeatedly into ***secondary bronchi***. These branch into ***tertiary bronchi*** which, in turn, branch into ***bronchioles***.

3. Ultimately, the bronchioles terminate in alveoli, or air sacs, where gases are exchanged.

4. After you examine the parts of the lungs, you can trace the bronchial tree into the substance of the lungs. Use Figure 21.6 for reference.

4. Lungs

1. The ***lungs*** are paired, lobed, spongy organs in which respiratory gases are exchanged (Figure 21.12).

2. Each lung is covered by ***visceral pleura***, whereas ***parietal pleura*** lines the thoracic wall.

3. The potential space between the visceral pleurae of the lungs is called the ***mediastinum***. This contains the thymus gland of the endocrine system, heart, trachea, esophagus,

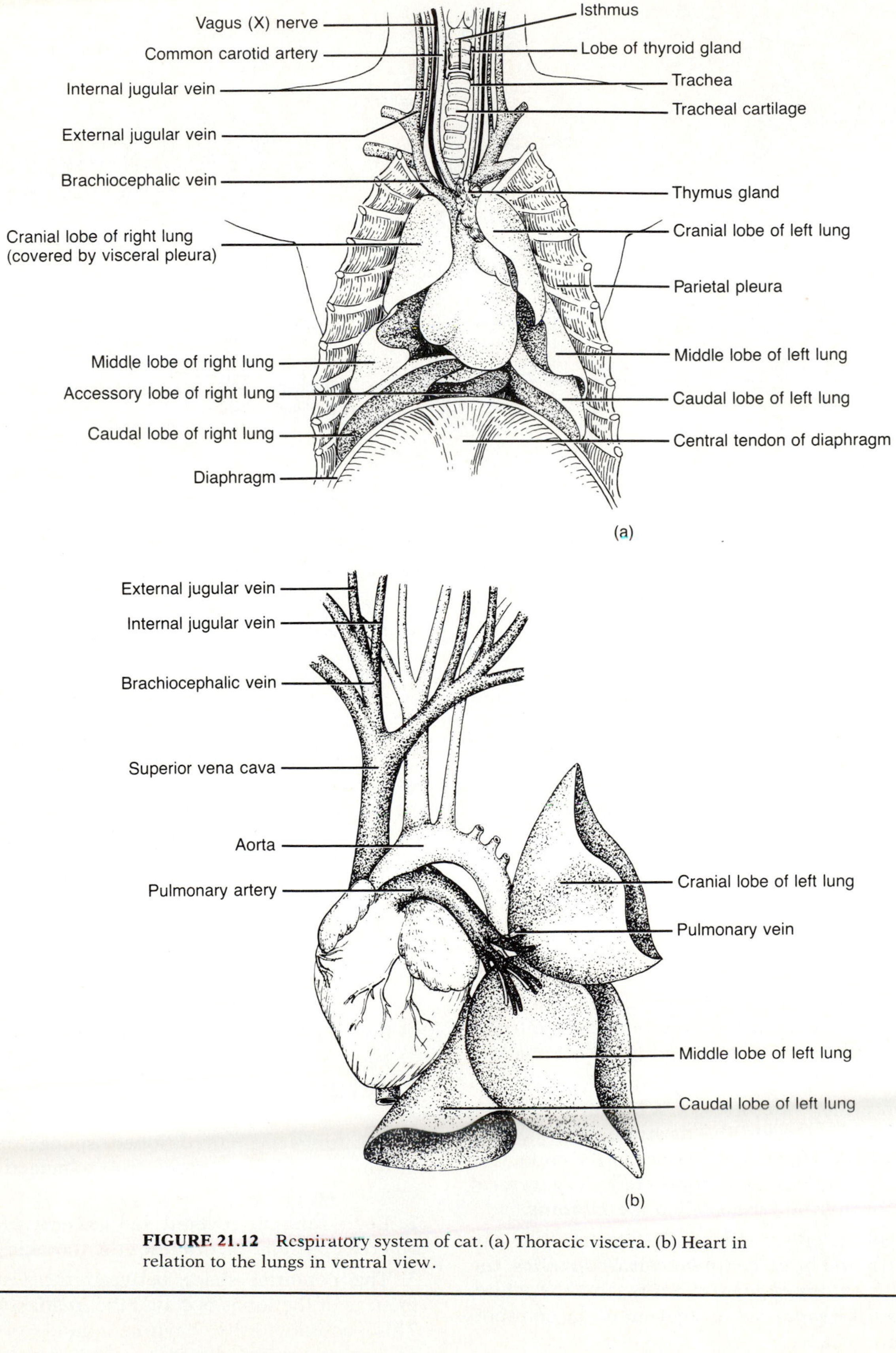

FIGURE 21.12 Respiratory system of cat. (a) Thoracic viscera. (b) Heart in relation to the lungs in ventral view.

nerves, and blood vessels. In the midventral thorax, visceral and parietal pleurae meet to form a ventral partition called the ***mediastinal septum***. This structure is caudal to the heart.

4. Note that the right lung has four lobes and the left lung has three lobes. Try to identify the ***pulmonary arteries*** entering the lungs. They are usually injected with blue latex.

5. Now try to find the ***pulmonary veins*** leaving the lungs. They are usually injected with red latex.

6. If you cut through a lobe of the lung, you should be able to identify branches of the pulmonary arteries, pulmonary veins, and bronchi. The lobes of the lung are labeled in Figure 21.12a.

C. DISSECTION OF SHEEP PLUCK

CAUTION! *Please reread Section D. "Precautions Related to Dissection" at the beginning of the laboratory manual on page xii before you begin your dissection.*

PROCEDURE

1. Preserved sheep pluck may or may not be available for dissection.

2. Pluck consists mainly of a sheep ***trachea***, ***bronchi***, ***lungs***, ***heart***, and ***great vessels***, and a small portion of the ***diaphragm***. It is a good demonstration because it is large and shows the close anatomical correlation between these structures and the systems to which they belong, namely, the respiratory and the cardiovascular system.

3. The heart and its great blood vessels have been described in detail in Exercise 17.

4. Pluck can also be used to examine in great detail the trachea and its relationship to the development of the bronchi until they branch into each lung. In addition, this specimen is sufficiently large that the bronchial tree can be exposed by careful dissection.

5. This dissection is done by starting at the primary and secondary bronchi and slowly and carefully removing lung tissue as the trees forms even smaller branches into lungs.

6. These specimens should be used primarily by the instructor for demonstration, but you can help expose the bronchial tree.

D. LABORATORY TESTS ON RESPIRATION

1. Observation of Living Lung Tissue

CAUTION! *Please reread Section D. "Precautions Related to Dissection" at the beginning of the laboratory manual on page xii before you begin you dissection.*

PROCEDURE

1. ***NOTE***. *It is strongly suggested that either your instructor or a laboratory assistant perform the procedure for pithing a frog. The procedure requires some practice and may prove difficult for inexperienced individuals.*

2. Pin the freshly pithed frog to the dissecting tray so the ventral surface is upward.

3. As illustrated in Figure 2.5 make a midsagittal cut from the abdomen to the level of the forelimbs. Make two lateral cuts to create flaps of skin which can be reflected and pinned to the tray.

4. Carefully remove any liver lobes which obscure your view of the lungs.

5. Wash the blood and any tissue fragments off the lungs with Ringer's solution that has been maintained at room temperature.

6. Place the tray under the dissecting microscope and locate the capillary network that surrounds the alveoli. Observe the blood flow through the capillaries of the lungs.

Describe your observation. ____________________

____

____

2. Mechanics of Pulmonary Ventilation (Breathing)

Pulmonary ventilation (breathing) is the process by which gases are exchanged between the atmosphere and lung alveoli. Air flows between the atmosphere and lungs for the same reason that blood flows through the body—a pressure gradient exists. We breathe in when the pressure inside the lungs is less than the air pressure in the atmosphere. We breathe out when the pressure inside the lungs is greater than the pressure in the atmosphere.

a. INSPIRATION

Breathing in is called ***inspiration (inhalation)***. Just before each inspiration, the air pressure inside the lungs equals the pressure of the atmosphere, which is about 760 mm Hg or 1 atmosphere (atm), at sea level. For air to flow into the lungs, the pressure inside the lungs must become lower than the pressure in the atmosphere. This condition is achieved by increasing the volume of the lungs.

In order for inspiration to occur, the lungs must be expanded. This increases lung volume and thus decreases the pressure in the lungs. The first step toward increasing lung volume involves contraction of the principal inspiratory muscles—the diaphragm and external intercostals.

Contraction of the diaphragm causes it to flatten, lowering its dome. This increases the vertical diameter of the thoracic cavity and accounts for the movement of more than two-thirds of the air that enters the lungs during inspiration. At the same time the diaphragm contracts, the external intercostals contract. As a result, the ribs are pulled upward and the sternum is pushed forward. This increases the anterior-posterior diameter of the thoracic cavity.

During normal breathing, the pressure between the two pleural layers, called ***intrapleural (intrathoracic) pressure***, is always subatmospheric. (It may become temporarily positive only during modified respiratory movements such as coughing or straining during defecation.) Just before inspiration, it is about 756 mm Hg. The overall increase in the size of the thoracic cavity causes intrapleural pressure to fall to about 754 mm Hg. Consequently, the walls of the lungs are sucked outward by the partial vacuum. Expansion of the lungs is further aided by movement of the pleura. The parietal and visceral pleurae are normally strongly attached to each other due to surface tension created by their moist adjoining surfaces. As the thoracic cavity expands, the parietal pleura lining the cavity is pulled in all directions, and the visceral pleura is pulled along with it. When the volume of the lungs increases, the pressure inside the lungs, called the ***intrapulmonic (intraalveolar) pressure***, drops from 760 to 758 mm Hg. A pressure gradient is thus established between the atmosphere and the alveoli. Air rushes from the atmosphere into the lungs, and inspiration takes place. Air continues to move into the lungs until intrapulmonic pressure equals atmospheric pressure.

b. EXPIRATION

Breathing out, called ***expiration (exhalation)***, is also achieved by a pressure gradient, but in this case the gradient is reversed so that the pressure in the lungs is greater than the pressure of the atmosphere. Normal expiration, unlike inspiration, is a passive process since no muscular contractions are involved. Expiration starts when the inspiratory muscles relax. As the external intercostals relax, the ribs move downward, and as the diaphragm relaxes, its dome moves upward. These movements decrease the vertical and anterior-posterior diameters of the thoracic cavity, and it returns to its resting size.

As intrapleural pressure returns to its preinspiration level (756 mm Hg), the walls of the lungs are no longer sucked out. The elastic basement membranes of the alveoli and elastic fibers in bronchioles and alveolar ducts recoil into their relaxed shape, and lung volume decreases. Intrapulmonic pressure increases to 763 mm Hg, and air moves from the area of higher pressure in the alveoli to the area of lower pressure in the atmosphere.

In order to demonstrate pulmonary ventilation, a model lung (bell jar demonstrator) will be used. This apparatus is basically an artificial thorax that mimics the organs of the respiratory system and allows the pressure/volume relationships to be manipulated. A diagram of such a device is shown in Figure 21.13. Some models will not have a stopcock ("glottis"); some will not have a tube opening into the intrapleural space.

PROCEDURE

1. Label the diagram in Figure 21.13 by writing the name of the anatomical structure that corresponds to the following part of the apparatus: glottis, trachea, primary bronchus, lungs, intrapulmonic space, intrapleural space, and diaphragm.
2. Pull on the rubber membrane (diaphragm) to simulate inspiration. What changes occur in

intrapleural pressure? ______________________

____

In intrapulmonic pressure? ____________________

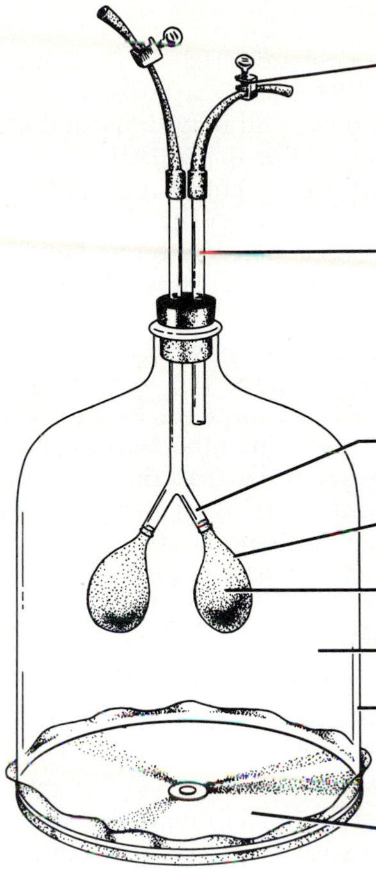

FIGURE 21.13 Model lung.

3. Push the rubber membrane (diaphragm) upward to simulate expiration. What changes occur in intrapleural pressure? ______

______

In intrapulmonic pressure? ______

4. Open the clamp on the tube leading to the intrapleural space. This simulates pneumothorax (air in the pleural cavity). Try to cause inspiration and expiration.

Explain your results. ______

______

______

5. If present, close the stopcock ("glottis") and simulate expiration. This procedure mimics the Valsalva maneuver (forced expiration against a closed glottis). What changes occur in intrathoracic and intrapulmonic pressure? ______

______

3. Measurement of Chest and Abdomen in Respiration

When the diaphragm contacts, the dome shape of this muscle flattens; the flattening pushes against the abdominal viscera and increases thoracic volume which, in turn, decreases intrapulmonic pressure. The volume changes in the abdomen and chest can easily be measured.

PROCEDURE

1. Place a tape measure at the level of the fifth rib (about armpit level) and determine the size of the chest immediately after a

a. normal inspiration
b. normal expiration
c. maximal inspiration
d. maximal expriation

2. Repeat step 1 with the tape measure at the level of the waist to determine abdominal size.

3. Record your results in Section D.1 of the LABORATORY REPORT RESULTS at the end of the exercise.

Which measurement increased during inspiration? ____________________

Which increased during expiration? ____________________

4. Respiratory Sounds

Air flowing through the respiratory tract creates characteristic sounds that can be detected through the use of a stethoscope. Normal breathing sounds include a soft-breezy inspiration as air fills the lungs, followed by a shorter, lower-pitched expiration. Perform the following exercises.

PROCEDURE

1. *Clean the ear plugs of a stethoscope with alcohol.*

2. Place the diaphragm of the stethoscope just below the larynx and listen for bronchial sounds during both inspiration and expiration.

3. Move the stethoscope slowly downward toward the bronchial tubes until the sounds are no longer heard.

4. Place the stethoscope under the scapula, under the clavicle, over different intercostal spaces (spaces between ribs) on the chest, and listen for any sound during inspiration and expiration.

5. Use of Pneumograph

The ***pneumograph*** is an instrument that measures variations in breathing patterns caused by various physical or chemical factors. The chest pneumograph, which is attached to a polygraph recorder via electrical leads, consists of a rubber bellows that fits around the chest just below the rib cage. As the subject breathes, chest movements cause changes in the air pressure within the pneumograph that are transmitted to the recorder. Normal inpiration and expiration can thus be recorded, and the effects of a wide range of physical and chemical factors on these movements can be studied.

PROCEDURE

Perform the following exercises and record you values in Section D.2 of the LABORATORY REPORT RESULTS at the end of the exercise. Label and save all recordings and attach them to Section D.2 of the LABORATORY REPORT RESULTS at the end of the exercise. Also for each recording label the inspiratory and expiratory phases, determine the duration of each, and calculate the respiratory rate.

1. Place the rubber bellows hose around the chest at the level of the sixth rib and attach it at the back. Connect the electrical lead to the recorder and adjust the instrument's centering and sensitivity so that the needle deflects as the subject breathes. If it does not, adjust the pneumograph bellows up or down on the chest, or loosen the degree of the tightness around the subject's chest.

2. Seat the subject so that the pneumograph recording is not visible to the subject.

3. Set the pneumograph at speed one and record normal, quiet breathing (eupnea) for about 30 seconds.

4. Have the subject inhale deeply and hold his or her breath for as long as possible. Record the breathing pattern for 30 seconds after this procedure.

5. Record the respiratory movements of a subject in Section D.2 of the LABORATORY REPORT RESULTS at the end of the exercise during the following activities:

Reading
Swallowing water
Laughing
Yawning
Coughing
Sniffing

6. Measurement of Respiratory Volumes

The word ***respiration*** means one inspiration plus one expiration. A normal adult has 14 to 18 respirations in a minute, during which the lungs exhange specific volumes of air with the atmosphere.

As the following repiratory volumes and capacities are discussed, keep in mind that the values given vary with age, height, and sex.

Each inspiration of normal, quiet breathing pulls about 500 ml (cc) of air into the respiratory passageways. The same amount moves out with each expiration, and this volume of air inspired (or expired) is called ***tidal volume*** (Figure 21.14). Only about 350 ml of this tidal volume reaches the alveoli. The other 150 ml is called ***dead air volume*** because it remains in the dead spaces of the nose, pharynx, larynx, trachea, and bronchi where it is not available for exchange.

If we take a very deep breath, we can inspire much more than 500 ml. The additional inhaled air, called the ***inspiratory reserve volume***, averages 3100 ml above the 500 ml of tidal volume. Thus, our respiratory system can pull in as much as 3600 ml of air. If we inspire normally and then expire as forcibly as possible, we can push out 1200 ml of air in addition to the 500 ml tidal volume. This extra 1200 ml is called ***expiratory reserve volume***. Even after the expiratory reserve volume is expelled, a considerable amount of air still remains in the lungs because the lower intrapleural pressure keeps the alveoli slightly inflated. This air, the ***residual volume***, amounts to about 1200 ml. Even if pneumothorax (air in the pleural cavity) results from opening the chest cavity and atmospheric pressure forces most of the residual volume out of the lungs, a small amount called ***minimal volume*** still remains in the lungs. This air can be demonstrated by placing a piece of lung in water and watching it float.

Lung capacity can be calculated by combining various lung volumes. ***Inspiratory capacity***, the total inspriatory ability of the lungs, is the sum of tidal volume plus inspiratory reserve volume (3600 ml). ***Functional residual capacity*** is the sum of residual volume plus expiratory reserve volume (2400 ml). ***Vital capacity*** is the sum of inspiratory reserve volume, tidal volume, and expiratory reserve volume (4800 ml). Finally, ***total lung capacity*** is the sum of all volumes (6000 ml).

a. USE OF VITAL CAPACITY APPARATUS

Using the vital capacity apparatus (Figure 21.15), determine your own particular vital capacity value.

The procedure for determining forced vital capacity is as follows:

PROCEDURE

1. Place a *sterile* air shield over the mouthpiece.
2. Push black pointer to red pointer at zero.
3. Stand, hold apparatus horizontally at eye level, and completely fill your lungs with air.
4. Apply mouth to air shield and blow air into apparatus as forcefully as possible, completely emptying lungs.

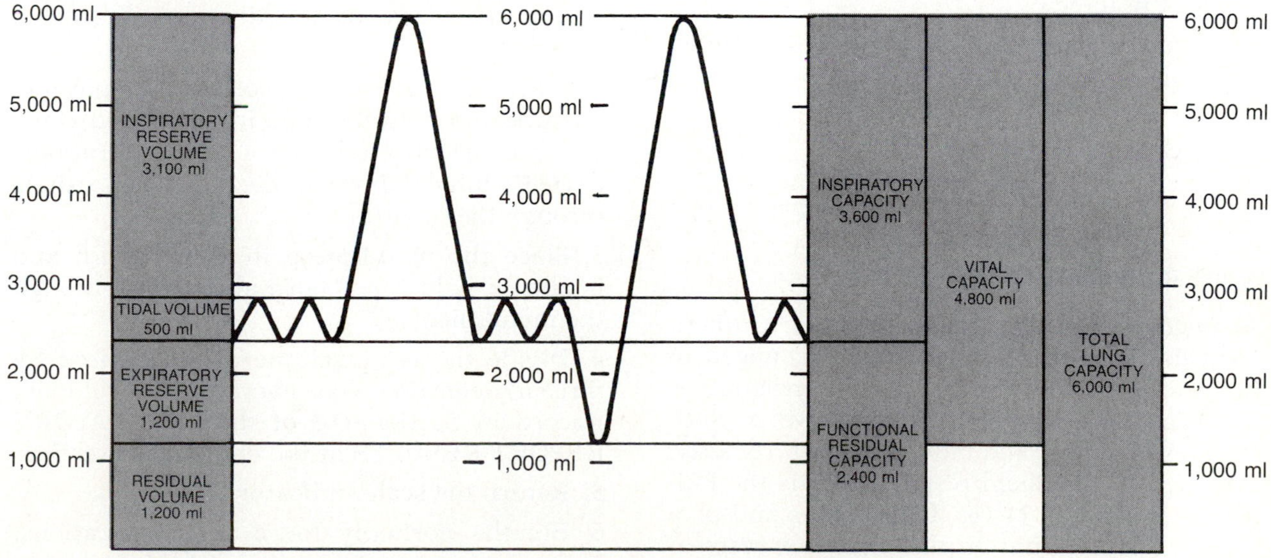

FIGURE 21.14 Spirogram of lung volumes and capacities.

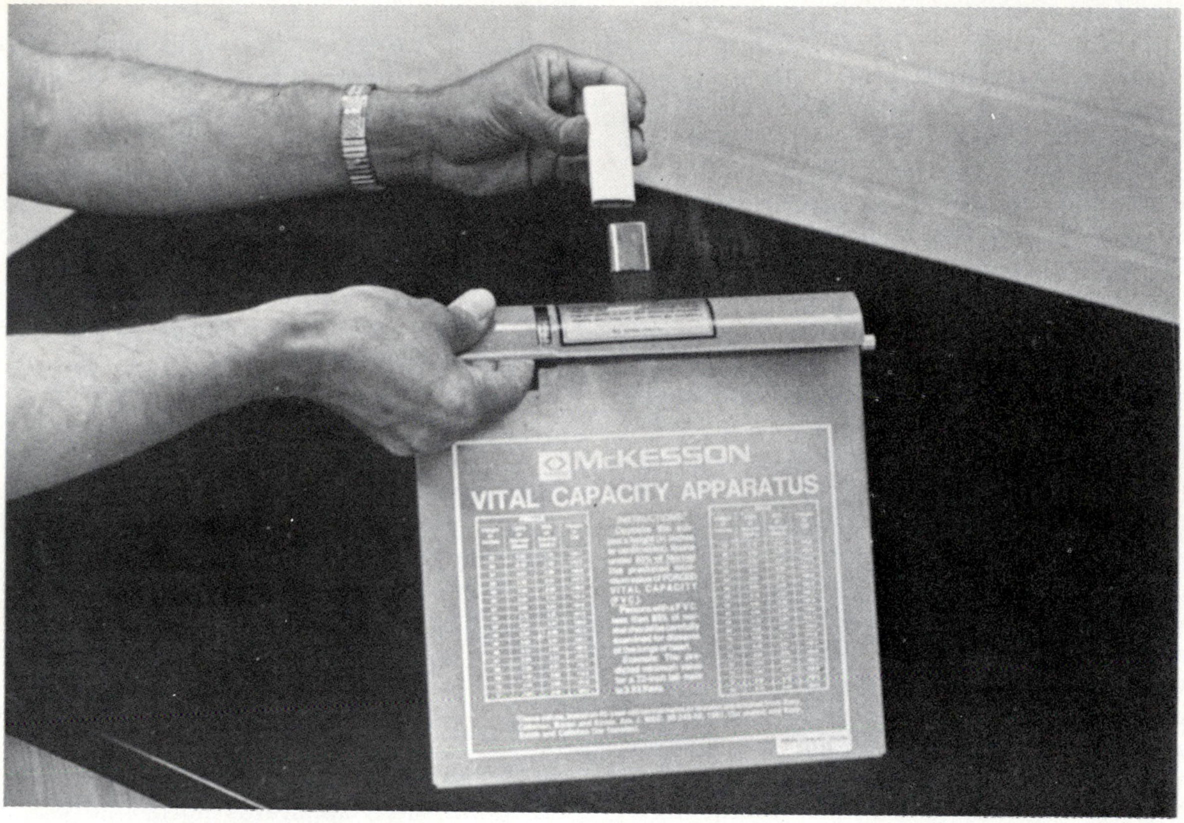

FIGURE 21.15 Placing a disposable air shield on a vital capacity apparatus.

5. Read vital capacity, in liters, at upper edge of black pointer.

6. Compare value with printed tables on apparatus using your height in centimeters as a guide.

7. Discard the air shield.

8. Record your vital capacity: ____________

b. USE OF HANDHELD RESPIROMETER

A ***respirometer (spirometer)*** is an instrument used to measure volumes of air exchanged in breathing. Of the several different respirometers available, we will make use of two: a handheld respirometer and the Collins respirometer. One type of handheld respirometer is the Pulmometer (Figure 21.16). It measures and provides a direct digital display of certain respiratory volumes and capacities.

PROCEDURE

1. Set the needle or the digital indicator to zero.

2. Place a *clean* mouthpiece in the spirometer tube and hold the tube in your hand while you breathe normally for a few respirations. (***NOTE:*** Inhale through the nose and exhale through the mouth.)

3. Place the mouthpiece in your mouth and inhale; exhale three normal breaths through the mouthpiece.

4. Divide the total volume of air expired by three to determine your average tidal volume. Record in Section D.3 of the LABORATORY REPORT RESULTS at the end of the exercise.

5. Return the scale indicator to zero.

6. Breathe normally for a few respirations. Following a normal expiration of tidal volume,

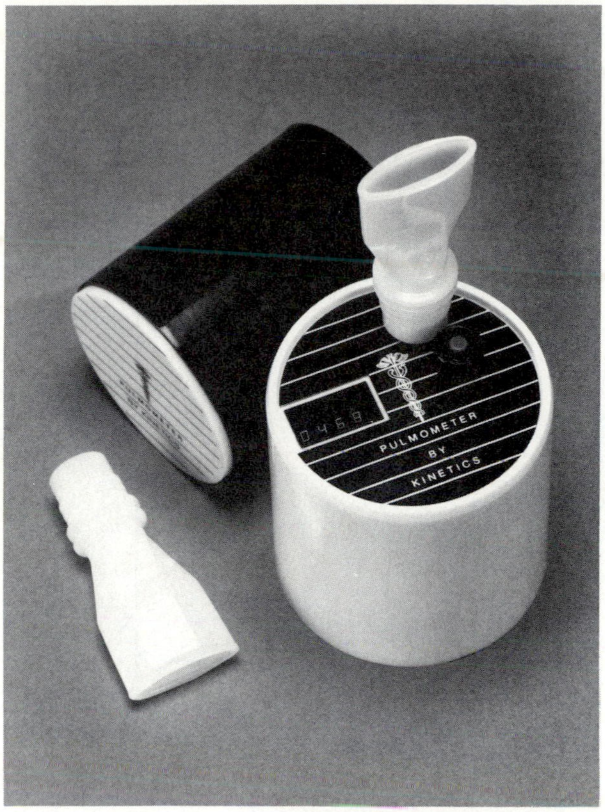

FIGURE 21.16 Handheld respirometer.

forcibly exhale as much air as possible into the mouthpiece.

7. Repeat step 6 twice. Divide the total amount of air expired by three to determine your average expiratory reserve volume. Record this value in Section D.3 of the LABORATORY REPORT RESULTS at the end of the exercise.

8. To determine vital capacity, breathe deeply a few times and inhale as much air as possible. Exhale as fully and as steadily as possible into the spirometer tube.

9. Repeat step 8 twice. Divide the total volume of expired air by three to determine your average vital capacity. Record this value in Section D.3 of the LABORATORY REPORT RESULTS at the end of the exercise.

10. Calculate your inspiratory reserve volume by substituting the known volumes in this equation: Vital capacity = Inspiratory reserve volume + tidal volume + expiratory reserve volume. Record this volume in Section D.3 of the LABORATORY REPORT RESULTS at the end of the exercise.

12. Compare your vital capacity with the normal values shown in Tables 21.1 and 21.2.

13. Sit quietly and count your respirations per minute. Calculate your minute volume of respirations (MVR) by multiplying the breaths per minute by your tidal volume.

$$\frac{\#\text{breaths}}{\text{minute}} \times \frac{\#\text{ml}}{\text{breath}} = \frac{\#\text{ml}}{\text{minute}}$$

Record this value in Section D.3 of the LABORATORY REPORT RESULTS at the end of the exercise.

C. USE OF COLLINS RESPIROMETER

The Collins respirometer, shown in Figure 21.17, is a closed system which allows the measurement of volumes of air inhaled and exhaled. The instrument consists of a weighted drum, containing air, inverted over a chamber of water. The air-filled chamber is connected to the subject's mouth by a tube. When the subject inspires, air is removed from the chamber, causing the drum to sink and producing an upward deflection. This deflection is recorded by the stylus on the graph paper on the kymograph (rotating drum). When the subject expires, air is added, causing the drum to rise and producing a downward deflection. These deflections are recorded as a ***spirogram*** (see Figure 21.14). The horizontal (x) axis of a spirogram is graduated in millimeters (mm) and records elapsed time; the vertical (y) axis is graduated in milliliters (ml) and records air volumes. Spirometric studies measure lung capacities and rates and depths of ventilation for diagnostic purposes. Spirometry is indicated for individuals with labored breathing, and is used to diagnose respiratory disorders such as bronchial asthma and emphysema.

The air in the body is at a different temperature than the air contained in the respirometer; body air is also saturated with water vapor. To make the correction for these differences, one measures the temperature of the respirometer and uses Table 21.3 to obtain a conversion factor (***BTPS*** factor: ***Body Temperature***, ***atmospheric Pressure***, ***Saturated*** with water vapor). All measured volumes are then multiplied by this factor.

TABLE 21.1
PREDICTED VITAL CAPACITIES FOR FEMALES

	Height in centimeters and inches																		
Age	cm 152 in. 59.8	154 60.6	156 61.4	158 62.2	160 63.0	162 63.7	164 64.6	166 65.4	168 66.1	170 66.9	172 67.7	174 68.5	176 69.3	178 70.1	180 70.9	182 71.7	184 72.4	186 73.2	188 74.0
16	3,070	3,110	3,150	3,190	3,230	3,270	3,310	3,350	3,390	3,430	3,470	3,510	3,550	3,590	3,630	3,670	3,715	3,755	3,800
17	3,055	3,095	3,135	3,175	3,215	3,255	3,295	3,335	3,375	3,415	3,455	3,495	3,535	3,575	3,615	3,655	3,695	3,740	3,780
18	3,040	3,080	3,120	3,160	3,200	3,240	3,280	3,320	3,360	3,400	3,440	3,480	3,520	3,560	3,600	3,640	3,680	3,720	3,760
20	3,010	3,050	3,090	3,130	3,170	3,210	3,250	3,290	3,330	3,370	3,410	3,450	3,490	3,525	3,565	3,605	3,645	3,695	3,720
22	2,980	3,020	3,060	3,095	3,135	3,175	3,215	3,255	3,290	3,330	3,370	3,410	3,450	3,490	3,530	3,570	3,610	3,650	3,685
24	2,950	2,985	3,025	3,065	3,100	3,140	3,180	3,220	3,260	3,300	3,335	3,375	3,415	3,455	3,490	3,530	3,570	3,610	3,650
26	2,920	2,960	3,000	3,035	3,070	3,110	3,150	3,190	3,230	3,265	3,300	3,340	3,380	3,420	3,455	3,495	3,530	3,570	3,610
28	2,890	2,930	2,965	3,000	3,040	3,070	3,115	3,155	3,190	3,230	3,270	3,305	3,345	3,380	3,420	3,460	3,495	3,535	3,570
30	2,860	2,895	2,935	2,970	3,010	3,045	3,085	3,120	3,160	3,195	3,235	3,270	3,310	3,345	3,385	3,420	3,460	3,495	3,535
32	2,825	2,865	2,900	2,940	2,975	3,015	3,050	3,090	3,125	3,160	3,200	3,235	3,275	3,310	3,350	3,385	3,425	3,460	3,495
34	2,795	2,835	2,870	2,910	2,945	2,980	3,020	3,055	3,090	3,130	3,165	3,200	3,240	3,275	3,310	3,350	3,385	3,425	3,460
36	2,765	2,805	2,840	2,875	2,910	2,950	2,985	3,020	3,060	3,095	3,130	3,165	3,205	3,240	3,275	3,310	3,350	3,385	3,420
38	2,735	2,770	2,810	2,845	2,880	2,915	2,950	2,990	3,025	3,060	3,095	3,130	3,170	3,205	3,240	3,275	3,310	3,350	3,385
40	2,705	2,740	2,775	2,810	2,850	2,885	2,920	2,955	2,990	3,025	3,060	3,095	3,135	3,170	3,205	3,240	3,275	3,310	3,345
42	2,675	2,710	2,745	2,780	2,815	2,850	2,885	2,920	2,955	2,990	3,025	3,060	3,100	3,135	3,170	3,205	3,240	3,275	3,310
44	2,645	2,680	2,715	2,750	2,785	2,820	2,855	2,890	2,925	2,960	2,995	3,030	3,060	3,095	3,130	3,165	3,200	3,235	3,270
46	2,615	2,650	2,685	2,715	2,750	2,785	2,820	2,855	2,890	2,925	2,960	2,995	3,030	3,060	3,095	3,130	3,165	3,200	3,235
48	2,585	2,620	2,650	2,685	2,715	2,750	2,785	2,820	2,855	2,890	2,925	2,960	2,995	3,030	3,060	3,095	3,130	3,160	3,195
50	2,555	2,590	2,625	2,655	2,690	2,720	2,755	2,785	2,820	2,855	2,890	2,925	2,955	2,990	3,025	3,060	3,090	3,125	3,155
52	2,525	2,555	2,590	2,625	2,655	2,690	2,720	2,755	2,790	2,820	2,855	2,890	2,925	2,955	2,990	3,020	3,055	3,090	3,125
54	2,495	2,530	2,560	2,590	2,625	2,655	2,690	2,720	2,755	2,790	2,820	2,855	2,885	2,920	2,950	2,985	3,020	3,050	3,085
56	2,460	2,495	2,525	2,560	2,590	2,625	2,655	2,690	2,720	2,755	2,790	2,820	2,855	2,885	2,920	2,950	2,980	3,015	3,045
58	2,430	2,460	2,495	2,525	2,560	2,590	2,625	2,655	2,690	2,720	2,750	2,785	2,815	2,850	2,880	2,920	2,945	2,975	3,010
60	2,400	2,430	2,460	2,495	2,525	2,560	2,590	2,625	2,655	2,685	2,720	2,750	2,780	2,810	2,845	2,875	2,915	2,940	2,970
62	2,370	2,405	2,435	2,465	2,495	2,525	2,560	2,590	2,620	2,655	2,685	2,715	2,745	2,775	2,810	2,840	2,870	2,900	2,935
64	2,340	2,370	2,400	2,430	2,465	2,495	2,525	2,555	2,585	2,620	2,650	2,680	2,710	2,740	2,770	2,805	2,835	2,865	2,895
66	2,310	2,340	2,370	2,400	2,430	2,460	2,495	2,525	2,555	2,585	2,615	2,645	2,675	2,705	2,735	2,765	2,800	2,825	2,860
68	2,280	2,310	2,340	2,370	2,400	2,430	2,460	2,490	2,520	2,550	2,580	2,610	2,640	2,670	2,700	2,730	2,760	2,795	2,820
70	2,250	2,280	2,310	2,340	2,370	2,400	2,425	2,455	2,485	2,515	2,545	2,575	2,605	2,635	2,665	2,695	2,725	2,755	2,780
72	2,220	2,250	2,280	2,310	2,335	2,365	2,395	2,425	2,455	2,480	2,510	2,540	2,570	2,600	2,630	2,660	2,685	2,715	2,745
74	2,190	2,220	2,245	2,275	2,305	2,335	2,360	2,390	2,420	2,450	2,475	2,505	2,535	2,565	2,590	2,620	2,650	2,680	2,710

From: *Archives of Environmental Health,* February, 1966, Vol. 12, pp. 146–189, E. A. Gaensler, MD and G. W. Wright, MD

TABLE 21.2
PREDICTED VITAL CAPACITIES FOR MALES

Age	Height in centimeters and inches																		
	cm	152	154	156	158	160	162	164	166	168	170	172	174	176	178	180	182	184	186	188
	in.	59.8	60.6	61.4	62.2	63.0	63.7	64.6	65.4	66.1	66.9	67.7	68.5	69.3	70.1	70.9	71.7	72.4	73.2	74.0
16		3,920	3,975	4,025	4,075	4,130	4,180	4,230	4,285	4,335	4,385	4,440	4,490	4,540	4,590	4,645	4,695	4,745	4,800	4,850
18		3,890	3,940	3,995	4,045	4,095	4,145	4,200	4,250	4,300	4,350	4,405	4,455	4,505	4,555	4,610	4,660	4,710	4,760	4,815
20		3,860	3,910	3,960	4,015	4,065	4,115	4,165	4,215	4,265	4,320	4,370	4,420	4,470	4,520	4,570	4,625	4,675	4,725	4,775
22		3,830	3,880	3,930	3,980	4,030	4,080	4,135	4,185	4,235	4,285	4,335	4,385	4,435	4,485	4,535	4,585	4,635	4,685	4,735
24		3,785	3,835	3,885	3,935	3,985	4,035	4,085	4,135	4,185	4,235	4,285	4,330	4,380	4,430	4,480	4,530	4,580	4,630	4,680
26		3,755	3,805	3,855	3,905	3,955	4,000	4,050	4,100	4,150	4,200	4,250	4,300	4,350	4,395	4,445	4,495	4,545	4,595	4,645
28		3,725	3,775	3,820	3,870	3,920	3,970	4,020	4,070	4,115	4,165	4,215	4,265	4,310	4,360	4,410	4,460	4,510	4,555	4,605
30		3,695	3,740	3,790	3,840	3,890	3,935	3,985	4,035	4,080	4,130	4,180	4,230	4,275	4,325	4,375	4,425	4,470	4,520	4,570
32		3,665	3,710	3,760	3,810	3,855	3,905	3,950	4,000	4,050	4,095	4,145	4,195	4,240	4,290	4,340	4,385	4,435	4,485	4,530
34		3,620	3,665	3,715	3,760	3,810	3,855	3,905	3,950	4,000	4,045	4,095	4,140	4,190	4,225	4,285	4,330	4,380	4,425	4,475
36		3,585	3,635	3,680	3,730	3,775	3,825	3,870	3,920	3,965	4,010	4,060	4,105	4,155	4,200	4,250	4,295	4,340	4,390	4,435
38		3,555	3,605	3,650	3,695	3,745	3,790	3,840	3,885	3,930	3,980	4,025	4,070	4,120	4,165	4,210	4,260	4,305	4,350	4,400
40		3,525	3,575	3,620	3,665	3,710	3,760	3,805	3,850	3,900	3,945	3,990	4,035	4,085	4,130	4,175	4,220	4,270	4,315	4,360
42		3,495	3,540	3,590	3,635	3,680	3,725	3,770	3,820	3,865	3,910	3,955	4,000	4,050	4,095	4,140	4,185	4,230	4,280	4,325
44		3,450	3,495	3,540	3,585	3,630	3,675	3,725	3,770	3,815	3,860	3,905	3,950	3,995	4,040	4,085	4,130	4,175	4,220	4,270
46		3,420	3,465	3,510	3,555	3,600	3,645	3,690	3,735	3,780	3,825	3,870	3,915	3,960	4,005	4,050	4,095	4,140	4,185	4,230
48		3,390	3,435	3,480	3,525	3,570	3,615	3,655	3,700	3,745	3,790	3,835	3,880	3,925	3,970	4,015	4,060	4,105	4,150	4,190
50		3,345	3,390	3,430	3,475	3,520	3,565	3,610	3,650	3,695	3,740	3,785	3,830	3,870	3,915	3,960	4,005	4,050	4,090	4,135
52		3,315	3,353	3,400	3,445	3,490	3,530	3,575	3,620	3,660	3,705	3,750	3,795	3,835	3,880	3,925	3,970	4,010	4,055	4,100
54		3,285	3,325	3,370	3,415	3,455	3,500	3,540	3,585	3,630	3,670	3,715	3,760	3,800	3,845	3,890	3,930	3,975	4,020	4,060
56		3,255	3,295	3,340	3,380	3,425	3,465	3,510	3,550	3,595	3,640	3,680	3,725	3,765	3,810	3,850	3,895	3,940	3,980	4,025
58		3,210	3,250	3,290	3,335	3,375	3,420	3,460	3,500	3,545	3,585	3,630	3,670	3,715	3,755	3,800	3,840	3,880	3,925	3,965
60		3,175	3,220	3,260	3,300	3,345	3,385	3,430	3,470	3,500	3,555	3,595	3,635	3,680	3,720	3,760	3,805	3,845	3,885	3,930
62		3,150	3,190	3,230	3,270	3,310	3,350	3,390	3,440	3,480	3,520	3,560	3,600	3,640	3,680	3,730	3,770	3,810	3,850	3,890
64		3,120	3,160	3,200	3,240	3,280	3,320	3,360	3,400	3,440	3,490	3,530	3,570	3,610	3,650	3,690	3,730	3,770	3,810	3,850
66		3,070	3,110	3,150	3,190	3,230	3,270	3,310	3,350	3,390	3,430	3,470	3,510	3,550	3,600	3,640	3,680	3,720	3,760	3,800
68		3,040	3,080	3,120	3,160	3,200	3,240	3,280	3,320	3,360	3,400	3,440	3,480	3,520	3,560	3,600	3,640	3,680	3,720	3,760
70		3,010	3,050	3,090	3,130	3,170	3,210	3,250	3,290	3,330	3,370	3,410	3,450	3,480	3,520	3,560	3,600	3,640	3,680	3,720
72		2,980	3,020	3,060	3,100	3,140	3,180	3,210	3,250	3,290	3,330	3,370	3,410	3,450	3,490	3,530	3,570	3,610	3,650	3,680
74		2,930	2,970	3,010	3,050	3,090	3,130	3,170	3,200	3,240	3,280	3,320	3,360	3,400	3,440	3,470	3,510	3,550	3,590	3,630

From: *Archives of Environmental Health,* February, 1966, Vol. 12, pp. 146–189, E. A. Gaensler, MD and G. W. Wright, MD

FIGURE 21.17 Collins respirometer. This type is commonly used in college biology laboratories.

PROCEDURE

1. *When using the Collins respirometer, the disposable mouthpiece is discarded after use by each subject, and the hose is then detached and rinsed with 70% alcohol.* CAUTION! *This procedure must be repeated with every student using the equipment.*

2. Students should work in pairs, with one student operating the respirometer while the other student is tested.

3. Before starting the recording, a little practice may be necessary to learn to inhale and exhale only through your mouth and into the hose. A nose clip may be used to prevent leakage from the nose.

4. Make sure that the respirometer is functioning. Your instructor may have already prepared the instrument for use.

   a. The leveling screws must be raised/lowered so that the bell does not rub on the metal body.

   b. Paper must be fed from a roll onto the kymograph OR taped onto the kymograph in single sheets.

   c. The soda-lime which absorbs carbon dioxide must be pink (fresh) and not purple (saturated with CO_2).

   d. Pen reservoirs must be full and primed.

5. Raise and lower the drum several times to flush the respirometer of stale air and position the ventilometer pen (the upper pen, usually with black ink) so that it will began writing in the center of the spirogram paper.

6. Set the free-breathing valve so that the opening is seen through the side of the valve.

7. Place a *sterile or disposable mouthpiece in your mouth* and close your nose with a clamp or with the fingers.

8. Breathe several times to become accustomed to the use of this instrument.

9. Have your lab partner close the free-breathing valve (turn it completely to the oppo-

TABLE 21.3
TEMPERATURE VARIATION CONVERSION FACTORS

Temperature °C (°F)	Conversion factor
20 (68.0)	1.102
21 (69.8)	1.096
22 (71.6)	1.091
23 (73.4)	1.085
24 (75.2)	1.080
25 (77.0)	1.075
26 (78.8)	1.068
27 (80.6)	1.063
28 (82.4)	1.057
29 (84.2)	1.051
30 (86.0)	1.045
31 (87.8)	1.039
32 (89.6)	1.032
33 (91.4)	1.026
34 (93.2)	1.020
35 (95.0)	1.014
36 (96.8)	1.007
37 (98.6)	1.000

site direction so that no opening is seen) and turn on the respirometer to the slow speed. At this setting, the kymograph moves 32 mm (the distance between 2 vertical lines) each minute.

10. Perform the following exercises and record your values in Section D.4 of the LABORATORY REPORT RESULTS at the end of the exercise. In each case, * indicates the placement of the mouth on the mouthpiece.

11. Inspire normally, then * exhale normally three times. This volume is your tidal volume. Repeat two more times and average and record the values.

12. Expire normally, then * exhale as much air as possible three times, recording this volume. This value is your expiratory reserve volume. Repeat two more times and record the values.

13. After taking a deep breath, * exhale as much air as possible three times. This volume is your vital capacity. Repeat two more times and record the values. Compare your vital capacity to the normal value shown in Tables 21.1 and 21.2.

14. Because your vital capacity consists of tidal volume, inspiratory reserve volume, and expiratory reserve volume, and because you have already measured tidal volume in step 1 and expiratory reserve volume in step 2, you can calculate your inspiratory reserve volume by subtracting tidal volume and expiratory reserve volume from vital capacity. Inspiratory reserve volume = (vital capacity) − (tidal volume + expiratory reserve volume). Record this value.

In some cases, an individual with a pulmonary disorder has a nearly normal vital capacity. If the rate of expiration is timed, however, the extent of the pulmonary disorder becomes apparent. In order to do this, an individual expels air into a Collins respirometer as fast as possible and the expired volume is measured per unit of time. Such a test is called ***Forced Expiratory Volume (FEV_T).*** The T indicates that the volume of air is timed. FEV_1 is the volume of air forcefully expired in 1 second, FEV_2 is the volume expired in 2 seconds, and so on. A normal individual should be able to expel 83% of the total capacity during the first second, 94% in 2 seconds, and 97% in 3 seconds. For individuals with disorders such as emphysema and asthma, the percentage can be considerably lower, depending on the extent of the problem.

FEV_1 is determined according to the following procedure:

PROCEDURE

1. Apply a noseclip to prevent leakage of air through the nose.
2. Turn on the kymograph.
3. Before placing the mouthpiece in your mouth, inhale as deeply as possible.
4. Expel all the air you can into the mouthpiece.
5. Turn off the kymograph.
6. Draw a vertical line on the spirogram at the starting point of exhalation. Mark this A.
7. Using the Collins VC timed interval ruler, draw a vertical line to the left of A and label it line B. The time between lines A and B is 1 second.
8. The FEV_1 is the point where the spirogram tracing crosses line B.
9. Record your value here ____________
10. Now read your vital capacity from the spirogram and record the value here ____________
11. In order to adjust for differences in temperature in the respirometer, use Table 21.3

as a guide. Determine the temperature in the respirometer, find the appropriate conversion factor and multiply the conversion factor by your vital capacity.

12. If, for example, the temperature of the respirometer is 75.2°F (24°C), the conversion factor is 1.080. And, if your vital capacity is 5600 ml, then

$$1.080 \times 5600 \text{ ml} = 6048 \text{ ml}$$

13. Use the same conversion factor and multiply it by your FEV_1. If your FEV_1 is 4000 ml, then

$$1.080 \times 4000 \text{ ml} = 4320 \text{ ml}$$

14. To calculate FEV_1, divide 4320 by 6048.

$$FEV_1 = \frac{4320}{6048} = 71\%$$

15. Repeat the procedure three times and record your FEV_1 in Section D.4 of the LABORATORY REPORT RESULTS at the end of the exercise.

Compare your results using the Collins respirometer with those using the handheld respirometer.

7. Chemical Regulation of Respiration

The size of the thorax is affected by the action of the respiratory muscles. These muscles contract and relax as a result of nerve impulses transmitted to them from centers in the brain. The area from which nerve impulses are sent to respiratory muscles is located bilaterally in the reticular formation of the brain stem; it is referred to as the ***respiratory center***. The respiratory center consists of a widely dispersed group of neurons that is functionally divided into three areas (1) the medullary rhythmicity area, in the medulla that controls the basic rhythm of respiration; (2) the pneumotaxic (noo-mō-TAK-sik) area, in the pons that facilitates expiration; and (3) the apneustic (ap-NOO-stik) area, in the pons that inhibits expiration.

Although the basic rhythm of respiration is set and coordinated by the respiratory center, the rhythm can be modified in response to the demands of the body by neural input to the center. Among the factors that can alter the rate of respiration are levels of CO_2 (H^+ ions) and O_2 in blood.

Under normal circumstances, arterial blood PCO_2 is 40 mm Hg. If there is even a slight increase in PCO_2—a condition called ***hypercapnia***—the chemosensitive area in the medulla and chemoreceptors in the carotid and aortic bodies are stimulated. Stimulation of the chemosensitive area and chemoreceptors causes the inspiratory area to become highly active, and the rate of respiration increases. This increased rate, ***hyperventilation***, allows the body to expel more CO_2 until the PCO_2 is lowered to normal. If arterial PCO_2 is lower than 40 mm Hg, the chemosensitive area and chemoreceptors are not stimulated and stimulatory impulses are not sent to the inspiratory area. Consequently, the area sets its own moderate pace until CO_2 accumulates and the PCO_2 rises to 40 mm Hg. A slow rate of respiration is called ***hypoventilation***.

The oxygen chemoreceptors are sensitive only to large decreases in the PO_2 because hemoglobin remains about 85 percent or more saturated at PO_2 values all the way down to 50 mm Hg. If arterial PO_2 falls from a normal of 105 mm Hg to about 50 mm Hg, the cells of the inspiratory area suffer oxygen starvation and do not respond well to any chemical receptors. They send fewer impulses to the inspiratory muscles, and the respiration rate decreases or breathing ceases altogether.

We can measure the effect of various factors on respiration by measuring

1. the CO_2 content of exhaled air, or
2. the rate of respiration

In order to measure CO_2 content, a gas is bubbled through lime water and the time is measured until the liquid becomes cloudy. Lime water is a solution of calcium hydroxide [$Ca(OH)_2$]. The cloudiness occurs because carbon dioxide combines with calcium hydroxide to form calcium carbonate, a white precipitate, in the following reaction.

$$CO_2 + CaOH_2 \rightarrow CaCO_2 + H_2O$$

$$\underset{\text{dioxide}}{\text{Carbon}} + \underset{\text{hydroxide}}{\text{Calcium}} \rightarrow \underset{\text{carbonate}}{\text{Calcium}} + \text{Water}$$

PROCEDURE

1. Label 3 test tubes N, H, and E (Normal, Hyperventilation, Exercise). Fill each about half full of calcium hydroxide [$Ca(OH)_2$] solution.

2. Start the stop watch and *exhale* normal expirations through the straw into tube N. Note the time elapsed until the solution becomes turbid.

3. Hyperventilate about 30 seconds. CAUTION! *Stop hyperventilation at the first sign of dizziness or lightheadedness.* Repeat step 2, *exhaling* normally into tube H.

4. After a 3 minute rest, exercise briskly for 2 minutes.

5. Repeat step 2, *exhaling* into tube E.

6. Record all time in Section D.5 of the LABORATORY REPORT RESULTS at the end of the exercise.

In which situation was the CO_2 content of exhaled air the greatest? ________________

Why? ________________________________

____

E. LABORATORY TESTS COMBINING RESPIRATORY AND CARDIOVASCULAR INTERACTIONS

1. ***Experimental setup***—Review earlier explanations on recording the following:

Respiratory movements with a pneumograph (Section D.5 in Exercise 21).

Blood pressure with sphygmomanometer and stethoscope (Section F in Exercise 19).

Radial pulse (Section E.1 in Exercise 19).

CAUTION! *Assuming that your partner has no know or apparent cardiac or other health problems and is capable of such an activity, ask her/him to perform the following activities after attaching the various pieces of apparatus in order to record respiratory movements and blood pressure.*

2. ***Experimental procedure***—Some form of regulated exercise is necessary for this experiment. Choose one of the following forms of exercise to have your subject participate in:

a. *Riding exercise cycle*—If this form is chosen, set the resistance to be felt while riding the cycle, but *not so high* that the subject cannot complete the 4-minute exercise period without difficulty.

b. *Harvard Step Test*—In this form of exercise the subject is to step up onto a 20-in. platform (a chair will substitute quite well) with one foot at a time. The subject is to bring *both* feet up onto the platform before stepping back down to the floor. The subject is also to remain erect at all times, and to do 30 complete cycles (up onto the platform and back down) per minute.

3. ***Experimental protocol***—Record respiratory movements, blood pressure, and pulse during each of the procedures listed. Record your data in the table provided in Section E of the LABORATORY REPORT RESULTS at the end of the exercise. All data should be recorded *simultaneously*, thereby requiring participation of all members of the experimental group.

a. *Basal readings*—Have the subject sit erect and quiet for 3 minutes. Obtain readings for:

Respiratory rate and depth.

Systolic and diastolic blood pressure.

Pulse rate.

When the basal readings have been recorded, obtain additional readings after (1) sitting quietly on the exercise cycle for 3 mintues, if this form of exercise is to be utilized, or (2) standing quietly for 3 minutes in front of the platform that is to be utilize for the Harvard Step Test.

b. *Readings after 1 minute of exercise*—After the subject has exercised for 1 minute, obtain additional readings.

c. *Readings after 2 and 3 minutes of exercise*—Again, obtain additional readings after the subject has completed 2 and 3 minutes of exercise.

d. *Readings upon completion of exercise*—When the subject has completed 4 minutes of exercise, obtain readings for respiratory rate, respiratory depth, heart rate, and systolic and diastolic blood pressure immediately upon completion *while the subject remains seated on the exercise cycle or stands erect on the floor*, depending upon the type of exercise utilized.

e. *Readings 1, 2, 3, and 5 minutes after completion of exercise*—With the subject *still sitting on the exercise cycle or still standing* erect on the floor obtain additional readings at the above time intervals after completion of the exercise period.

ANSWER THE LABORATORY REPORT QUESTIONS AT THE END OF THE EXERCISE

LABORATORY REPORT RESULTS

RESPIRATORY SYSTEM 21

Student ______________________ Date ______________________

Laboratory Section ______________________ Score/Grade ______________________

SECTION D. LABORATORY TEST ON RESPIRATION

1. Measurement of Chest and Abdomen in Respiration

Record the results of your chest measurements in inches, in the following table.

Size after a	Chest	Abdomen
Normal inspiration		
Normal expiration		
Maximal inspiration		
Maximal expiration		

2. Use of Pneumograph

Attach a sample of any one of the following activities.

Reading
Swallowing water
Laughing
Yawning
Coughing
Sniffing

3. Use of Handheld Respirometer

Tidal volume ______________________

Expiratory reserve volume ______________________

Vital capacity ______________________

Inspiratory reserve volume ______________________

Minute volume of respiration (MVR) ______________________

4. Use of Collins Respiromenter

Record the results of your exercises using the Collins respirometer in the following table.

	Tidal volume (1)	Expiratory reserve (2)	Vital capacity (3)	Inspiratory reserve (4)	FEV_1 (5)
First time	ml	ml	ml	ml	ml
Second time	ml	ml	ml	ml	ml
Third time	ml	ml	ml	ml	ml
Your average	ml	ml	ml	ml	ml
Normal value	ml	ml	ml	ml	ml

5. Chemical Regulation of Respiration

	Time until turbidity
Tidal respirations	
After hyperventilation	
After exercise	

SECTION E. COMBINED RESPIRATORY AND CARDIOVASCULAR INTERACTIONS

Record the results of your exercises in the following table.

	Respiratory rate and depth	Systolic and Diastolic pressure	Pulse rate
Basal readings while sitting erect and quiet for 3 minutes			
Basal readings while standing on cycle or in front of platform for 3 minutes			
Readings after 1 minute of exercise			
Readings after 2 minutes of exercise			
Readings after 3 minutes of exercise			
Readings after 4 minutes of exercise			
Readings after 1 minute following completion of exercise			
Readings after 2 minutes following completion of exercise			
Readings after 3 minutes following completion of exercise			
Readings after 5 minutes following completion of exercise			

LABORATORY REPORT QUESTIONS

RESPIRATORY SYSTEM 21

Student ______________________ Date ______________________

Laboratory Section ______________________ Score/Grade ______________________

PART I. MULTIPLE CHOICE

_____ 1. The overall exchange of gases between the atmospherc, blood, and cells is called (a) inspiration (b) respiration (c) expiration (d) none of these

_____ 2. The internal nose is formed by the ethmoid bone, maxillae, inferior conchae, and the (a) hyoid bone (b) nasal bone (c) palatine bone (d) none of these

_____ 3. The portion of the pharynx that contains the pharyngeal tonsils is the (a) oropharynx (b) laryngopharynx (c) nasopharynx (d) pharyngeal orifice

_____ 4. The Adam's apple is a common term for the (a) thyroid cartilage (b) cricoid cattilage (c) epiglottis (d) none of these

_____ 5. The C-shaped rings of cartilage of the trachea not only prevent the trachea from collapsing but also aid in the process of (a) lubrication (b) removing foreign particles (c) gas exchange (d) swallowing

_____ 6. Of the following structures, the smallest in diameter is the (a) left primary bronchus (b) bronchioles (c) secondary bronchi (d) alveolar ducts

_____ 7. The structures of the lung that actually contain the alveoli are the (a) respiratory bronchioles (b) fissures (c) lobules (d) terminal bronchioles

_____ 8. From superficial to deep, the structure(s) that your would encounter first among the following is (are) the (a) bronchi (b) parietal pleura (c) pleural cavity (d) secondary bronchi

PART II. COMPLETION

9. An advantage of nasal breathing is that the air is warmed, moistened, and ______________________ .

10. Improper fusion of the palatine and maxillary bones results in a condition called ______________________ .

11. The protective lid of cartilage that prevents food from entering the trachea is the ______________________ .

12. After removal of the ______________________ an individual would be unable to speak.

13. The upper respiratory tract is able to trap and remove dust because of its lining of ______________________ .

14. The passage of a tube into the mouth and down through the larynx and trachea to bypass an obstruction is called ______________________ .

15. An x-ray of the bronchial tree after administration of an iodinated media is called a ____________________ .

16. The sequence of respiratory tubes from largest to smallest is trachea, primary bronchi, secondary bronchi, bronchioles, ____________________ , respiratory bronchioles, and alveolar ducts.

17. Both the external and internal nose are divided internally by a vertical partition called the ____________________ .

18. The undersurface of the external nose contains two openings called the nostrils or ____________________ .

19. The functions of the pharynx are to serve as a passageway for air and food and to provide a resonating chamber for ____________________ .

20. An inflammation of the membrane that encloses and protects the lungs is called ____________________ .

21. The anterior portion of the nasal cavity just inside the nostrils is called the ____________________ .

22. Groovelike passageways in the nasal cavity formed by the conchae are called ____________________ .

23. The portion of the pharynx that contains the palatine and lingual tonsils is the ____________________ .

24. Each lobe of a lung is subdivided into compartments called ____________________ .

25. A(n) ____________________ is an outpouching lined by squamous epithelium and supported by a thin elastic membrane.

26. The ____________________ cartilage attaches the larynx to the trachea.

27. The portion of a lung that rests on the diaphragm is the ____________________ .

28. Phagocytic cells in the alveolar wall are called ____________________ .

29. The surface of a lung lying against the ribs is called the ____________________ surface.

30. The ____________________ is a structure in the medial surface of a lung through which bronchi, blood vessels, lymphatics, and nerves pass.

PART III. MATCHING

_____ 31. Tidal volume	A. 1200 ml of air
_____ 32. Inspiratory reserve volume	B. 3600 ml of air
_____ 33. Inspiratory capacity	C. 4800 ml of air
_____ 34. Expiratory reserve volume	D. 500 ml of air
_____ 35. Vital capacity	E. 3100 ml of air
_____ 36. Total lung capacity	F. 6000 ml of air

DIGESTIVE SYSTEM 22

Digestion occurs basically as two events—mechanical digestion and chemical digestion. ***Mechanical digestion*** consists of various movements that help chemical digestion. These movements include physical breakdown of food by the teeth and complete churning and mixing of this food with enzymes by the smooth muscles of the stomach and small intestine. ***Chemical digestion*** consists of a series of catabolic reactions that break down the large nutrient molecules that we eat, such as carbohydrates, lipids, and proteins, into much smaller molecules that are usable by body cells.

A. GENERAL ORGANIZATION OF DIGESTIVE SYSTEM

Digestive organs are usually divided into two main groups. The first is the ***gastrointestinal (GI) tract***, or ***alimentary canal***, a continuous tube running from the mouth to the anus, and measuring about 9 m (30 ft) in length. This tract is composed of the mouth, pharynx, esophagus, stomach, small intestine, and large intestine. The small intestine has three regions: duodenum, jejunum, and ileum. The large intestine has four regions: cecum, colon, rectum, and anal canal. The colon is divided into ascending colon, transverse colon, descending colon, and sigmoid colon. The second group of organs composing the digestive system consists of the ***accessory structures*** such as the teeth, tongue, salivary glands, liver, gallbladder, and pancreas (see Figure 22.1).

Using your textbook, charts, or models for reference, label Figure 22.1.

The wall of the gastrointestinal tract, especially from the esphogus to the anal canal, has the same basic arrangement of tissues. The four coats (tunics) of the tract, from the inside to the outside, are the ***mucosa, submucosa, muscularis***, and ***serosa***, or ***adventitia*** (see Figures 22.6 and 22.8).

The outermost layer of the gastrointestinal tract is a serous membrane composed of connective tissue and epithelium, and is called the ***serosa*** or ***peritoneum*** (per′-i-tō-NĒ-um). The ***parietal peritoneum*** lines the wall of the abdominal cavity, and the ***visceral peritoneum*** covers some of the organs. The space between the parietal and visceral protions of the peritoneum is called the ***peritoneal cavity***. Unlike the two other serous membranes of the body, the pericardium and the pleura, the peritoneum contains large folds that weave in between the viscera. The important extensions of the peritoneum are the ***mesentery*** (MEZ-en-ter′-ē) ***mesocolon, falciform*** (FAL-si-form) ***ligament, lesser omentum*** (ō-MENT-um), and ***greater omentum***.

Inflammation of the peritoneum, called ***peritonitis***, is a serious condition because of the peritoneal membranes are continuous with each other, enabling the infection to spread to all the organs in the cavity.

B. ORGANS OF DIGESTIVE SYSTEM

1. Mouth (Oral Cavity)

The ***mouth***, also called the ***oral*** or ***buccal*** (BUK-al) ***cavity***, is formed by the cheeks, hard and soft palates, and tongue. The ***hard palate*** forms the anterior portion of the roof of the mouth and the ***soft palate*** forms the posterior portion. The ***tongue*** forms the floor of the oral cavity and is composed of skeletal muscle covered by mucous membrane. Partial digestion

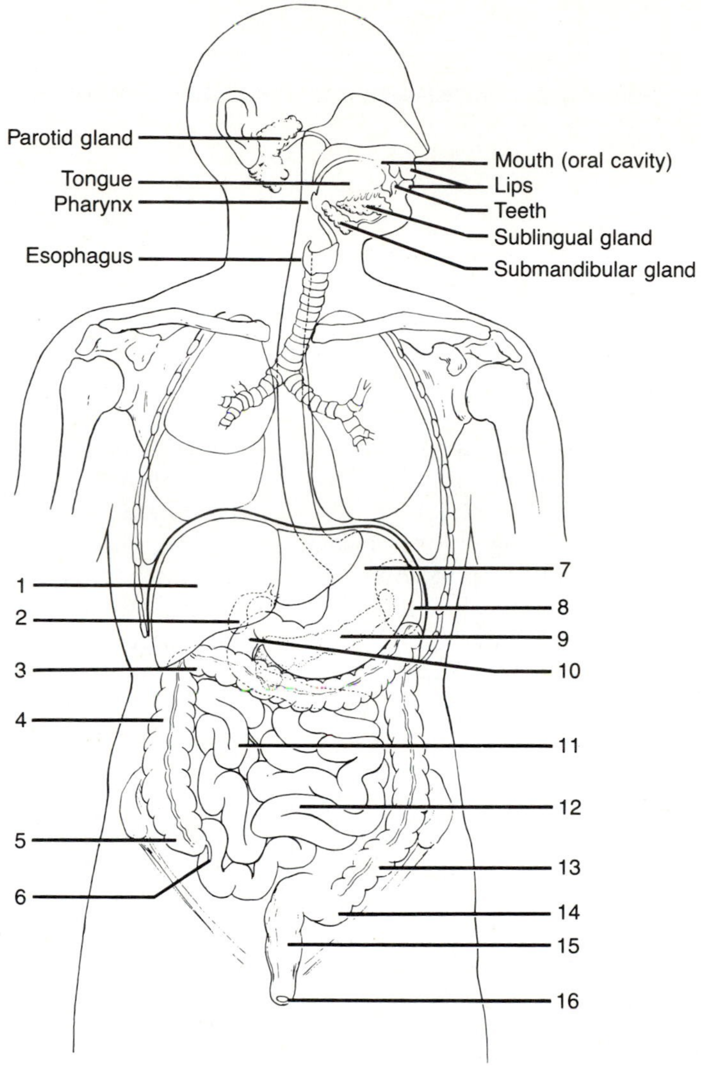

___ Anus
___ Appendix
___ Ascending colon
___ Cecum
___ Descending colon
___ Duodenum
___ Gallbladder
___ Ileum
___ Jejunum
___ Liver
___ Pancreas
___ Rectum
___ Sigmoid colon
___ Spleen
___ Stomach
___ Transverse colon

FIGURE 22.1 Organs of the digestive system and related structures.

of carbohydrates is the only chemical digestion that occurs in the mouth.

a. ***Cheeks***—Lateral walls of oral cavity. Muscular structures covered by skin and lined by stratified squamous epithelium; anterior protions terminate in the ***superior*** and ***inferior labia*** (lips).

b. ***Vermilion*** (ver-MIL-yon)—Transition zone of lips where outer skin and inner mucous membranes meet.

c. ***Labial frenulum*** (LĀ-bē-al FREN-yoo-lum)—Midline fold of mucous membrane that attaches the inner surface of each lip to its corresponding gum.

d. ***Vestibule***—Space bounded externally be cheeks and lips and internally by gums and teeth.

e. ***Oral cavity proper***—Space extending from the vestibule to the ***fauces*** (FAW-sēz), opening of oral cavity proper with pharynx. Area is enclosed by the dental arches.

f. ***Hard palate***—Formed by maxillae and palatine bones and covered by mucous membrane.

g. ***Soft palate***—Arch-shaped muscular partition between oropharynx and nasopharynx. Hanging from free border of soft palate is a conical muscular process, the ***uvula*** (YOU-vyoo-la).

h. ***Palatoglossal arch (anterior pillar)***—Muscular fold that extends inferiorly, laterally, and anteriorly to the side of the base of tongue.

i. ***Palatopharyngeal*** (PAL-a-tō-fa-rin′-jē-al) ***arch (posterior pillar)***—Muscular fold that extends inferiorly, laterally, and posteriorly to the side of pharynx. ***Palatine tonsils*** are between arches and ***lingual tonsil*** is at base of tongue.

j. ***Tongue***—Movable, muscular organ on floor of mouth. ***Extrinsic muscles*** originate outside tongue, insert into it, and move tongue from side to side and in and out to maneuver food for chewing and swallowing; ***intrinsic muscles*** originate and insert within the tongue and alter shape and size of tongue for speech and swallowing.

k. ***Lingual frenulum***—Midline fold of mucous membrane on undersurface of tongue that helps restrict its movement posteriorly.

l. ***Papillae*** (pa-PIL-ē)—Projections of lamina propria on surface of tongue covered with epithelium; ***filiform papillae*** are conical projections in parallel rows over anterior two-thirds of tongue; ***fungiform papillae*** are mushroomlike elevations distributed among filiform papillae and more numerous near tip of tongue (appear as red dots and most contain taste buds); ***circumvallate papillae*** are arranged in the form of an inverted V on the posterior surface of tongue (all contain taste buds).

Using a mirror and tongue depressor, examine your mouth or your partner's and locate as many of the structures (a through l) as you can.

2. Salivary Glands

Most saliva is secreted by the ***salivary glands***, which lie outside the mouth and pour their contents into ducts that empty into the oral cavity.

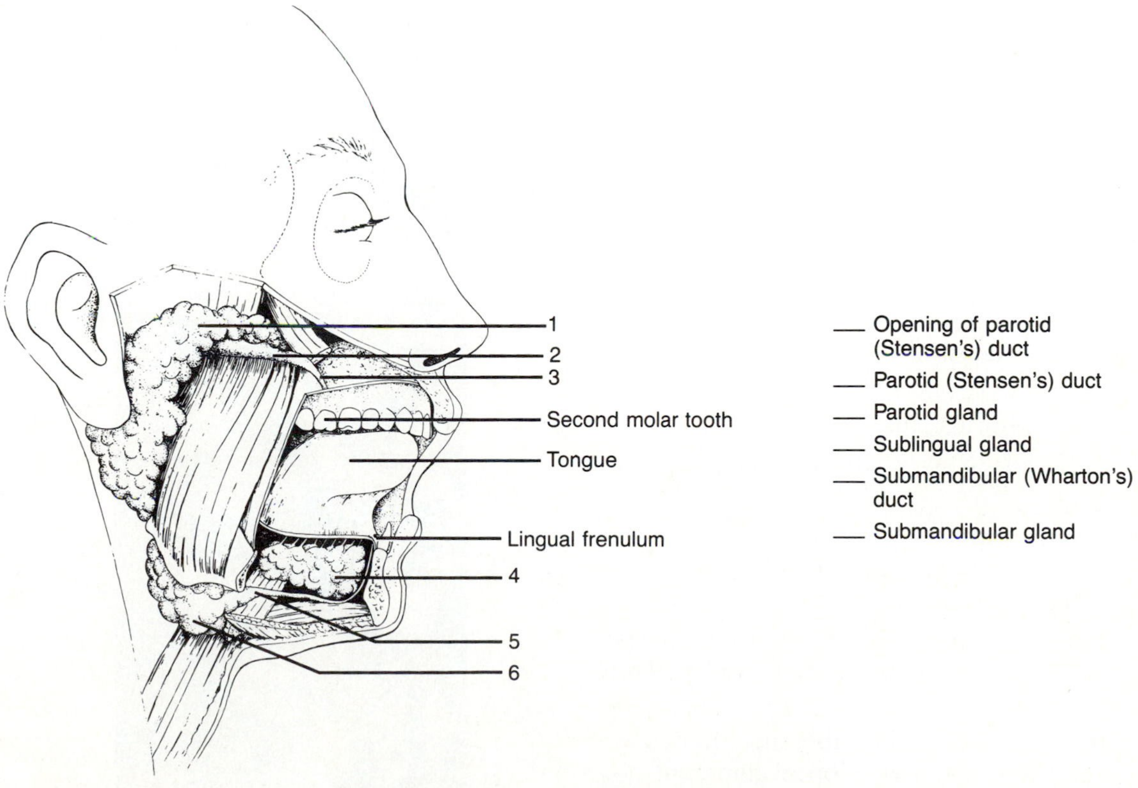

FIGURE 22.2 Location of salivary glands.

The carbohydrate-digesting enzyme in saliva is salivary amylase. The three pairs of salivary glands are: the ***parotid glands*** (in front of and under the ears), ***submandibular glands*** (deep to the base of the tongue in the posterior part of the floor of the mouth), and ***sublingual glands*** (anterior to the submandibular glands).

Label Figure 22.2.

The parotid glands are compound tubuloacinar glands, whereas the submandibulars and sublinguals are compound acinar glands (see Figure 22.3).

Examine prepared slides of the three different types of salivary glands and compare your observations to Figure 22.3.

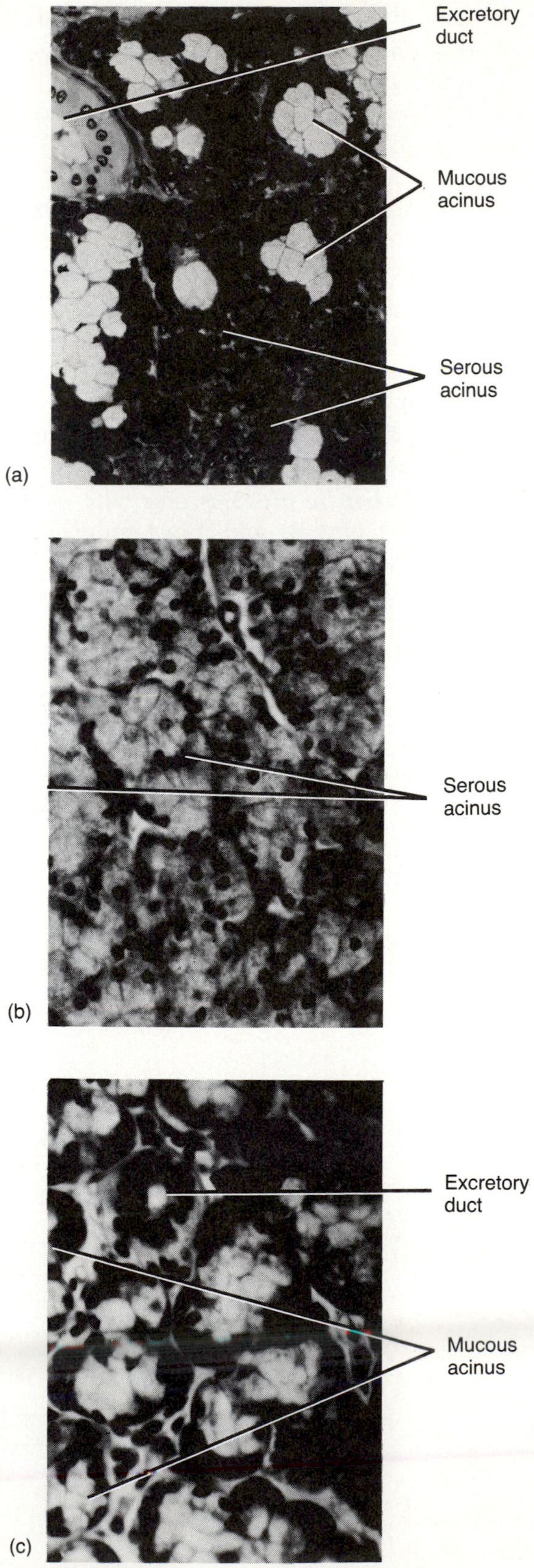

FIGURE 22.3 Histology of salivary glands. (a) Submandibular. (b) Parotid. (c) Sublingual.

3. Teeth

Teeth (dentes) are located in the sockets of the alveolar processes of the mandible and maxillae. The alveolar processes are covered by ***gingivae*** (jin-JĪ-vē) or gums, which extend slightly into each socket, forming a ***gingival sulcus***. The sockets are lined by a dense fibrous connective tissue called a ***periodontal ligament***, which anchors the teeth in position and helps dissipate the force of chewing.

a. ***Crown***—Exposed portion above level of gums.

b. ***Root***—One to three projections embedded in socket.

c. ***Neck***—Constricted junction line of the crown and root near the gum line.

d. ***Dentin***—Calcified connective tissue that gives teeth their basic shape.

e. ***Pulp cavity***—Enlarged part of cavity in crown within dentin.

f. ***Pulp***—Connective tissue containing blood vessels, lymphatics, and nerves.

g. ***Root canal***—Narrow extension of pulp cavity in root.

h. ***Apical foramen***—Opening in base of root canal through which blood vessels, lymphatics, and nerves enter tooth.

i. ***Enamel***—Covering of crown that consists primarily of calcium phosphate and calcium carbonate.

j. ***Cementum***—Bonelike substance that covers and attaches root to periodontal ligament.

With the aid of your textbook, label the parts of a tooth shown in Figure 22.4.

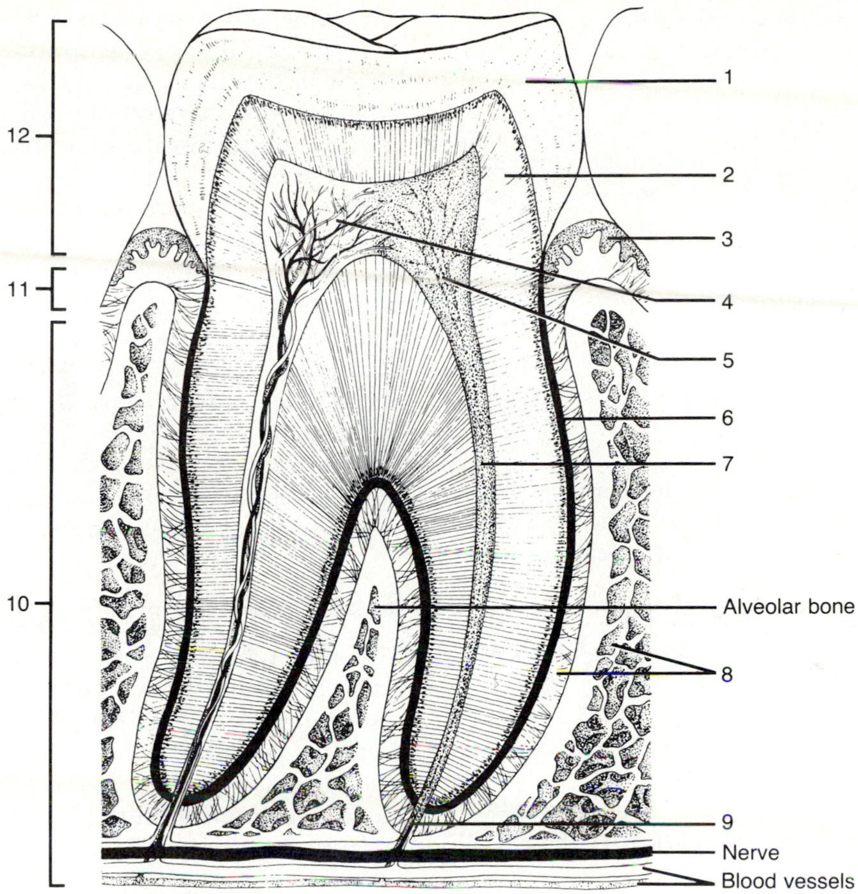

___ Apical foramen	___ Neck
___ Cementum	___ Peridontal ligament
___ Crown	___ Pulp
___ Dentin	___ Pulp cavity
___ Enamel	___ Root
___ Gingiva	___ Root canal

FIGURE 22.4 Parts of a tooth.

4. Dentitions

Dentitions (sets of teeth) are of two types: ***deciduous*** (baby) and ***permanent***. Deciduous teeth begin to erupt at about 6 months of age, and one pair appears at about each month thereafter until all 20 are present. The deciduous teeth are:

a. ***Incisors***—Central incisor closest to midline, with lateral incisor on either side. Incisors are chisel-shaped, adapted for cutting into food, have only one root.

b. ***Cuspids (canines)***—Posterior to incisors. Cuspids have pointed surfaces (cusps) for tearing and shredding food, have only one root.

c. ***Molars***—First and second molars posterior to canines. Molars crush and grind food. Upper molars have four cusps and three roots, lower molars have four cusps and two roots.

All deciduous teeth are usually lost between 6 and 12 years of age and replaced by permanent dentition consisting of 32 teeth that appear between age 6 and adulthood. The permanent teeth are:

a. ***Incisors***—Central incisor and lateral incisor replace those of deciduous dentition.

b. ***Cuspids (canines)***—These replace those of deciduous dentition.

c. ***Premolars (bicuspids)***—First and second premolars replace deciduous molars. Premolars crush and grind food, have two cusps and one root (upper first premolars have two roots).

d. ***Molars***—These erupt behind premolars as jaw grows to accommodate them and do not replace any deciduous teeth. First molars erupt at age 6, second at age 12, and third (wisdom teeth) after age 18.

Using a mirror and tongue depressor, examine your mouth or your partner's and locate as many teeth of the permanent dentition as you can.

With the aid of your textbook, label the deciduous and permanent dentitions in Figure 22.5.

5. Esophagus

The ***esophagus*** (e-SOF-a-gus) is a muscular, collapsible tube posterior to the trachea (see Figure 22.11). The structure is 23 to 25 cm (10 in.) long and extends from the laryngopharynx through the mediastinum and esophageal hiatus in the diaphragm and terminates in the superior portion of the stomach. The esophagus conveys food from the pharynx to the stomach by peristalsis.

Histologically, the esophagus consists of a ***mucosa*** (stratified squamous epithelium, lamina propria, muscularis mucosae), ***submucosa*** (connective tissue, blood vessels, mucous glands), ***muscularis*** (upper third striated, middle third striated and smooth, lower third smooth, and ***adventitia*** (esophagus not covered by peritoneum).

Examine a prepared slide of a cross section of the esophagus that shows its various coats. With the aid of your textbook, label Figure 22.6.

6. Stomach

The ***stomach*** is a J-shaped enlargement of the gastrointestinal tract under the diaphragm in the epigastric, umbilical, and left hypochondriac regions of the abdomen (see Figure 22.1). The superior part is connected to the esophagus; the inferior part empties into the duodenum of the small intestine. The stomach is

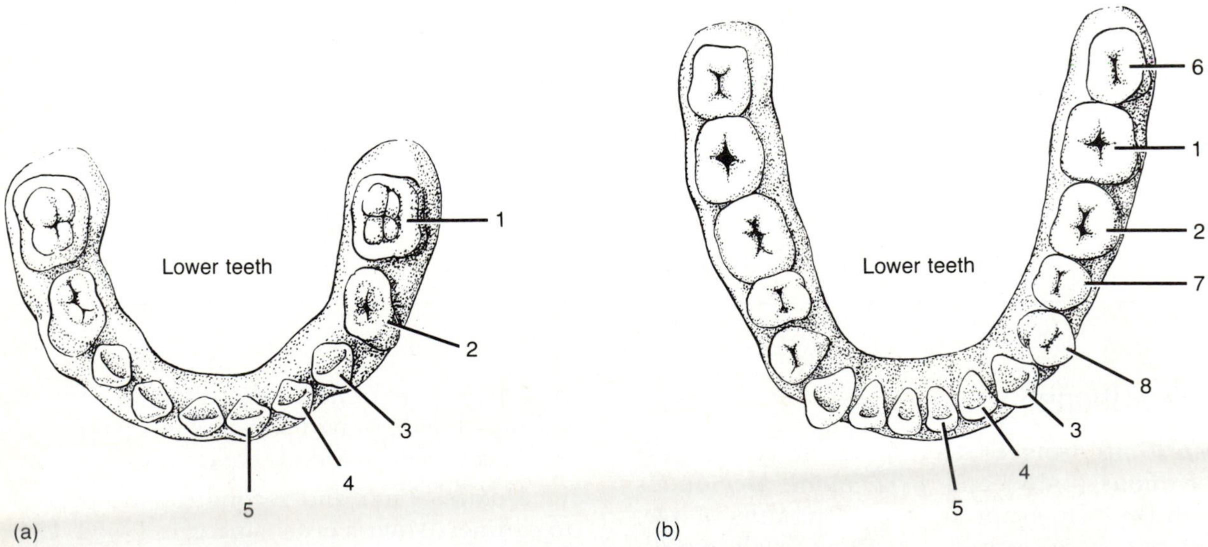

___ Central incisor
___ Cuspid (canine)
___ First molar
___ First premolar (bicuspid)

___ Lateral incisor
___ Second molar
___ Second premolar (bicuspid)
___ Third molar (wisdom tooth)

FIGURE 22.5 Dentitions. (a) Deciduous. (b) Permanent.

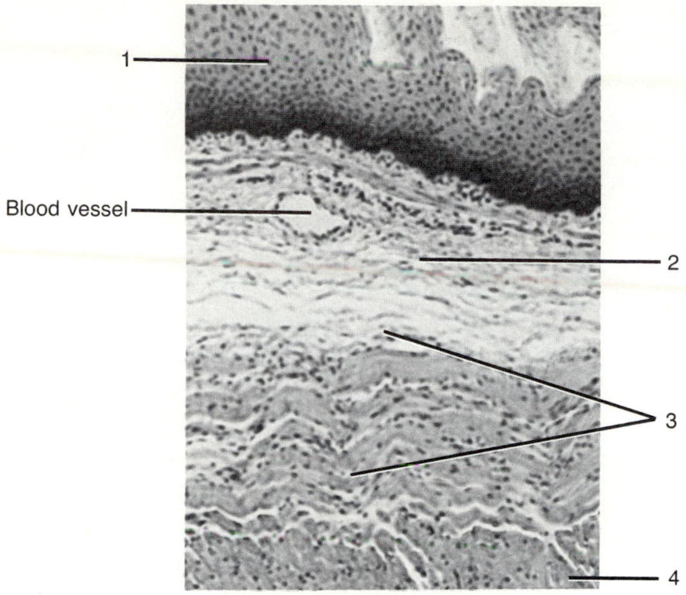

___ Adventitia
___ Mucosa
___ Muscularis
___ Submucosa

FIGURE 22.6 Histology of esophagus.

divided into four areas: cardia, fundus, body, and pylorus. The ***cardia*** surrounds the lower esophageal sphincter, a physiological sphincter in the esophagus just above the diaphragm. The rounded portion above and to the left of the cardia is the ***fundus***. Below the fundus, the large central portion of the stomach is called the ***body***. The narrow, inferior region is the ***pylorus***. The concave medial border of the stomach is called the ***lesser curvature***, and the convex lateral border is ***greater curvature***. The pylorus communicates with the duodenum of the small intestine via a sphincter called the ***pyloric sphincter (valve)***. The main chemical activity of the stomach is to begin the digestion of proteins.

Label Figure 22.7.

The ***mucosa*** of the stomach contains large folds called ***rugae*** (ROO-jē). The columnar epithelium of the mucosa contains many narrow openings that extend down into the lamina propria. These openings are called ***gastric glands (pits)*** and are lined with three kinds of cells: (1) ***zymogenic (peptic) cells*** that secrete inactive pepsinogen, which is converted to active pepsin, a protein-digesting enzyme; (2) ***parietal (oxyntic) cells*** that secrete hydrochloric acid; and (3) ***mucous cells*** that secrete mucus and possibly the intrinsic factor (see Figure 22.8). The ***submucosa*** consists of loose (areolar) connective tissue. The ***muscularis*** has three layers of smooth muscle—outer longitudinal, middle circular, inner oblique. The oblique layer is limited chiefly to the body of the stomach. The ***serosa*** is part of the parietal peritoneum.

Examine a prepared slide of a section of the stomach that shows its varous layers. With the aid of your textbook, label Figure 22.8.

7. Pancreas

The ***pancreas*** is a soft, oblong, tubuloacinar gland posterior to the greater curvature of the stomach (see Figure 22.1). The gland consists of a ***head*** (expanded portion near duodenum), ***body*** (central portion), and ***tail*** (terminal tapering protion).

Histologically, the pancreas consists of ***pancreatic islets (islets of Langerhans)*** that contain (1) glucagon-producing ***alpha cells***, (2) insulin-producing ***beta cells***, and (3) somatostatin-producing ***delta cells*** (see Figure 15.5). The pancreas also consists of ***acini*** that produce pancreatic juice (see Figure 15.5). Pancreatic juice contains enzymes that assist in the chemical breakdown of carbohydrates, proteins, and lipids and nucleic acids.

Pancreatic juice is delivered from the pancreas to the duodenum by a large main tube, the ***pancreatic duct*** (***duct of Wirsung***). This duct unites with the common bile duct from the

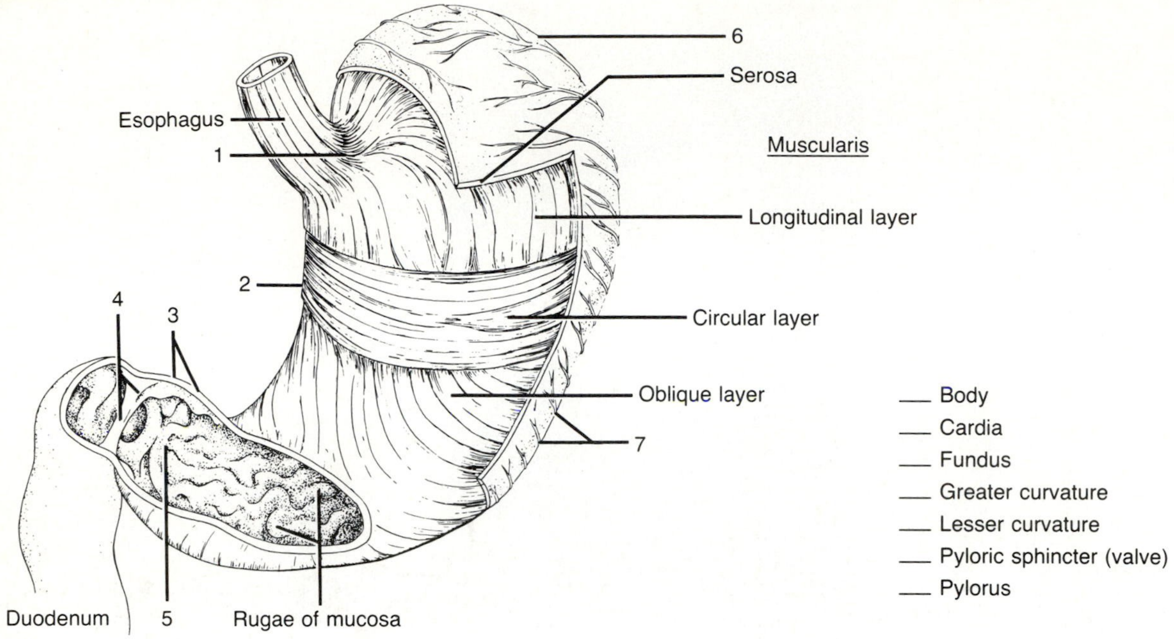

FIGURE 22.7 Stomach. External and internal anatomy.

liver and pancreas and enters the duodenum in a common duct called the ***hepatopancreatic ampulla*** (***ampulla of Vater***). The ampulla opens on an elevation of the duodenal mucosa, the ***duodenal papilla***. An ***accessory pancreatic duct*** (***duct of Santorini***) may also lead from the pancreas and empty into the duodenum about 2.5 cm (1 in.) above the hepatopancreatic ampulla.

With the aid of your textbook, label the structures associated with the pancreas in Figure 22.9.

8. Liver

The ***liver*** is located inferior to the diaphragm, occupying most of the right hypochondriac and part of the epigastric regions of the abdomen (see Figure 22.1). The gland is divided into two principal lobes, the ***right lobe*** and ***left lobe***, separated by the ***falciform ligament***. The falciform ligament attaches the liver to the anterior abdominal wall and diaphragm. The right lobe also has associated with it an inferior ***quadrate lobe*** and a posterior ***caudate lobe***.

Each lobe is composed of microscopic functional units called ***lobules***. Among the structures in a lobule are cords of ***hepatic (liver) cells*** arranged in a radial pattern around a ***central vein***; ***sinusoids***, endothelial lined spaces between hepatic cells through which blood flows; and ***stellate reticuloendothelial*** (***Kupffer***) ***cells*** that destroy bacteria and worn-out blood cells by phagocytosis.

Examine a prepared slide of several liver lobules. Compare your observations with Figure 22.10.

Bile is manufactured by hepatic cells and functions in the emulsification of fats in the small intestine. The liquid is passed to the small intestine as follows: Hepatic cells secrete bile into ***bile capillaries*** that empty into small ducts. The small ducts merge into larger ***right*** and ***left hepatic ducts***, one in each principal lobe of the liver. The right and left hepatic ducts unite outside the liver to form a single ***common hepatic duct***. This duct joins the ***cystic duct*** from the gallbladder to become the ***common bile duct***, which empties into the duodenum at the hepatopancreatic ampulla (ampulla of Vater). When fats are not being digested, a valve around the hepatopancreatic ampulla (ampulla of Vater), the ***sphincter of the hepatopancreatic ampulla (sphincter of Oddi)***, closes, and bile backs up into the gall bladder via the cystic duct. In the gallbladder, bile is stored and concentrated.

With the aid of your textbook, label the structures associated with the liver in Figure 22.9.

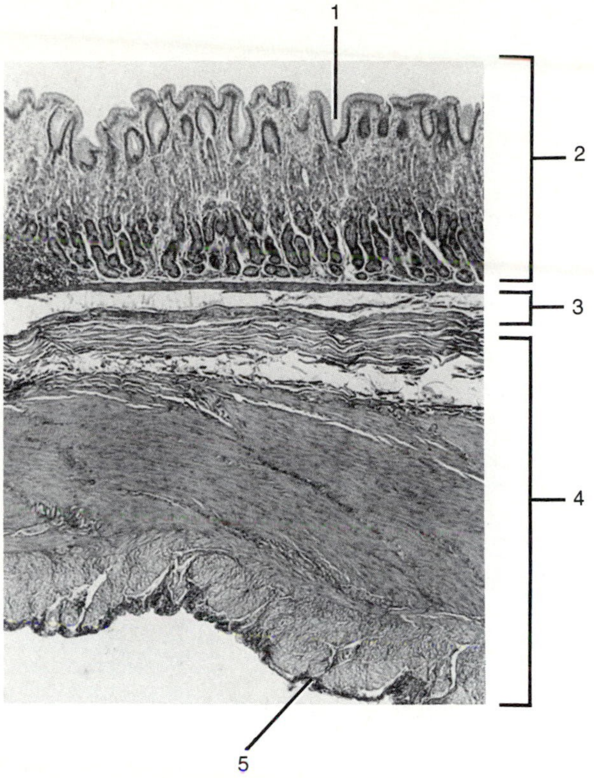

___ Gastric glands (pits)
___ Mucosa
___ Muscularis
___ Serosa
___ Submucosa

FIGURE 22.8 Histology of stomach.

9. Gallbladder (GB)

The ***gallbladder (GB),*** is a pear-shaped sac in a fossa along the undersurface of the liver (see Figure 22.1). The gallbladder stores and concentrates bile. The cystic duct of the gallbladder and common hepatic duct of the liver merge to form the common bile duct. The ***mucosa*** of the gallbladder contains rugae. The ***muscularis*** consists of smooth muscle and the outer coat consists of ***visceral peritoneum.***

With the aid of your textbook, label the structures associated with the gallbladder in Figure 22.9.

10. Small Intestine

The bulk of digestion and absorption occurs in the ***small intestine***, which begins at the pyloric sphincter (valve) of the stomach, coils through the central and lower part of the abdomen, and joins the large intestine at the ileocecal sphincter (see Figure 22.1). The mesentery attaches the small intestine to the posterior abdominal wall. The small intestine is about 6.35 m (21 ft) long and is divided into three segments: ***duodenum*** (doo′-ō-DĒ-num), which begins at the stomach; ***jejunum*** (jē-JOO-num), the middle segment, and ***ileum*** (IL-ē-um), which terminates at the large intestine.

Histologically, the ***mucosa*** contains many pits lined with glandular epithelium called ***intestinal glands (crypts of Lieberkühn)***; they secrete enzymes that digest carbohydrates, proteins, and nucleic acids (see Figure 22.11). Some of the simple columnar cells of the mucosa are ***goblet cells*** that secrete mucus (see Figure 22.11); others contain ***microvilli*** to increase the surface area for absorption (see Figure 3.1). The mucosa contains a series of fingerlike projections, the ***villi*** (see Figure 22.11). Each villus contains a blood capillary and lymphatic vessel called a ***lacteal***; these absorb digested nutrients. The 4 million to 5 million villi in the small intestine greatly increase the surface area for absorption. The mucosa and ***submucosa*** also contain deep permanent folds, the ***circular folds*** or ***plicae circulares*** (PLĪ-kē SER-kyoo-lar-es), which also help to increase the area for absorption. In the submucosa of the duodenum are ***duodenal (Brunner's) glands***, which secrete an alkaline mucus to protect the mucosa from excess acid and the action of digestive enzymes. The ***muscularis*** of the small intestine consists of an outer logitudinal layer and an inner circular layer of smooth muscle. Except for a major portion of the duodenum, the ***serosa*** (visceral peritoneum) completely covers the small intestine.

Obtain prepared slides of the small intestine (section through its tunics and villi) and identify as many structures as you can, using Figure 22.11 and your textbook as references.

11. Large Intestine

The ***large intestine*** functions in the completion of absorption of water that leads to the formation of feces, and in the expulsion of feces from the body. Bacteria residing in the large intestine manufacture certain vitamins. The large intestine is about 1½ m (5 ft) long and extends from the ileum to the anus (see Figure 22.1).

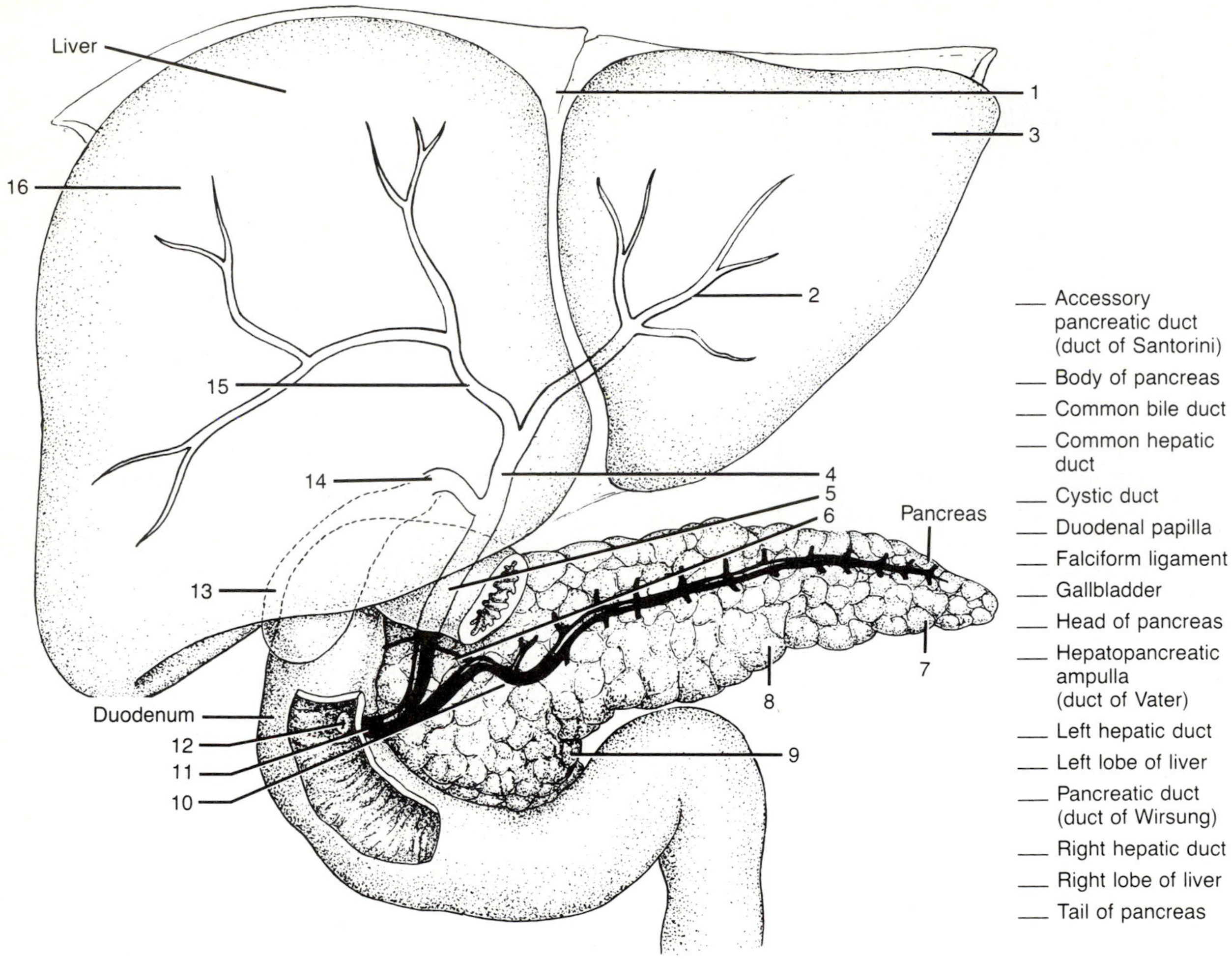

FIGURE 22.9 Relations of the liver, gallbladder, duodenum, and pancreas.

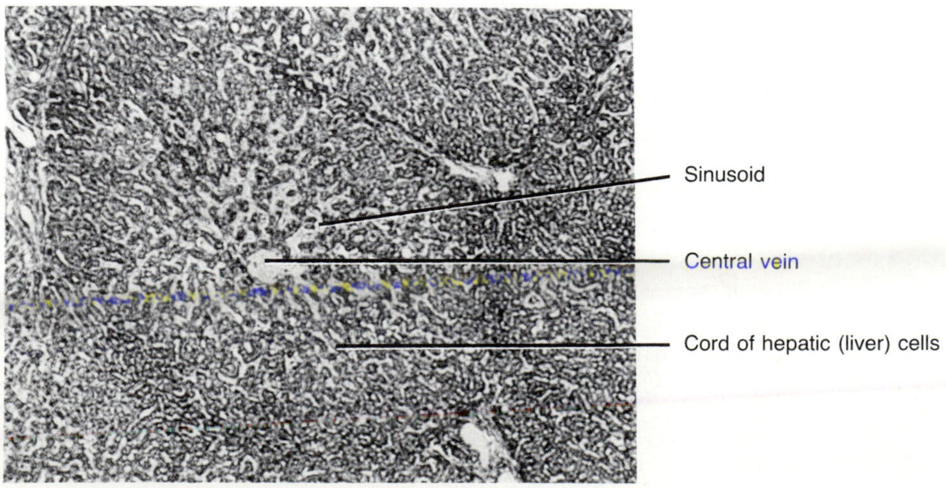

FIGURE 22.10 Photomicrograph of liver lobule.

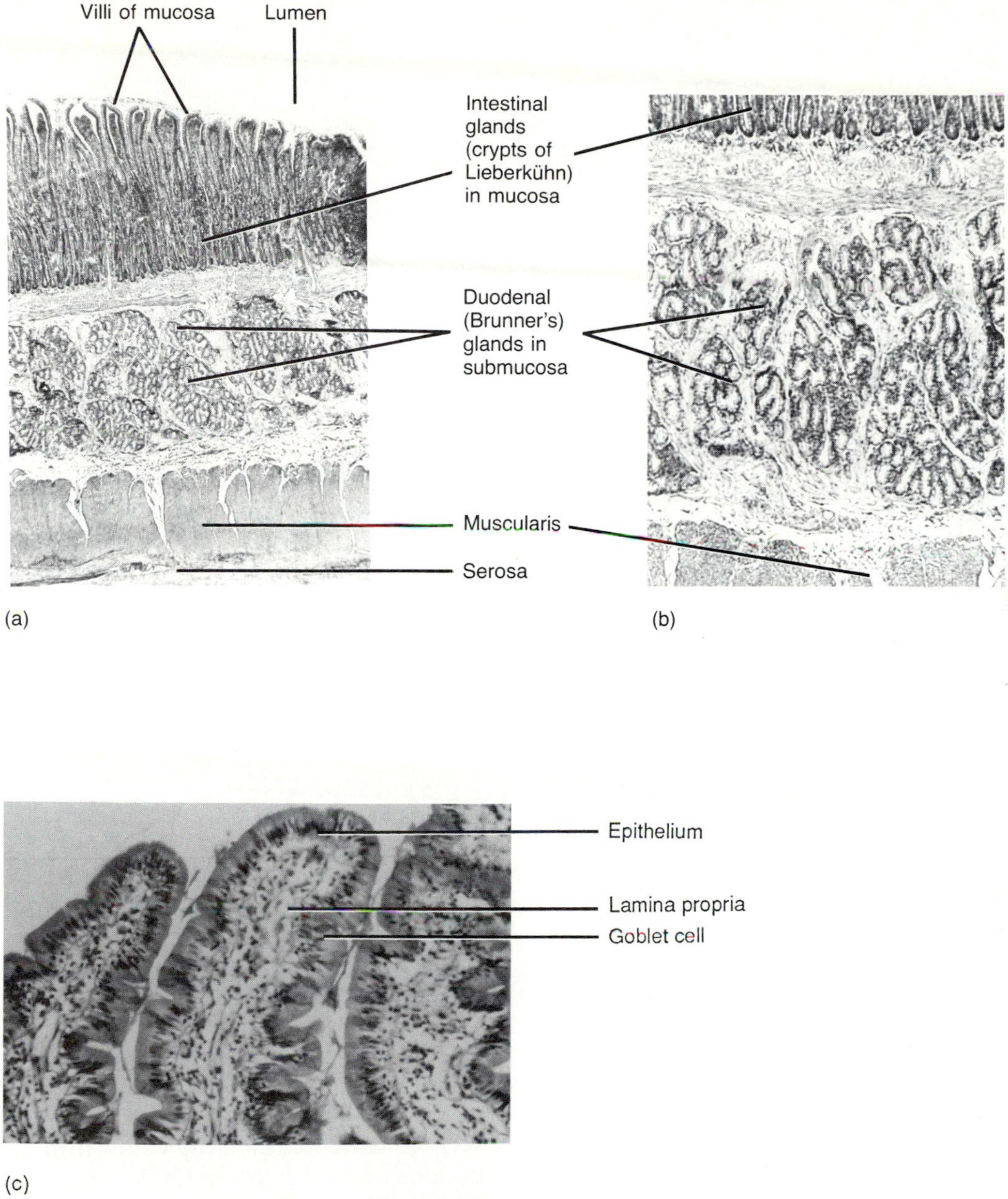

FIGURE 22.11 Histology of small intestine. (a) Photomicrograph of portion of wall of duodenum. (b) Photomicrograph of enlarged aspect of intestinal glands. (c) Photomicrograph of enlarged aspect of several villi.

It is attached to the posterior abdominal wall by an extension of visceral peritoneum called mesocolon. The large intestine is divided into four principal regions: cecum, colon, rectum, and anal canal.

The opening from the ileum into the large intestine is guarded by a fold of mucous membrane, the ***ileocecal sphincter*** (***valve***). Hanging below the valve is a blind pouch, the ***cecum***, to which is attached the ***vermiform appendix*** by an extension of visceral peritoneum called the ***mesoappendix***. Inflammation of the vermiform appendix is called ***appendicitis***. The open end of the cecum merges with the ***colon***. The first division of the colon is the ***ascending colon***, which ascends on the right side of the abdomen and turns abruptly to the left at the undersurface of the liver ***(right colic [hepatic]***

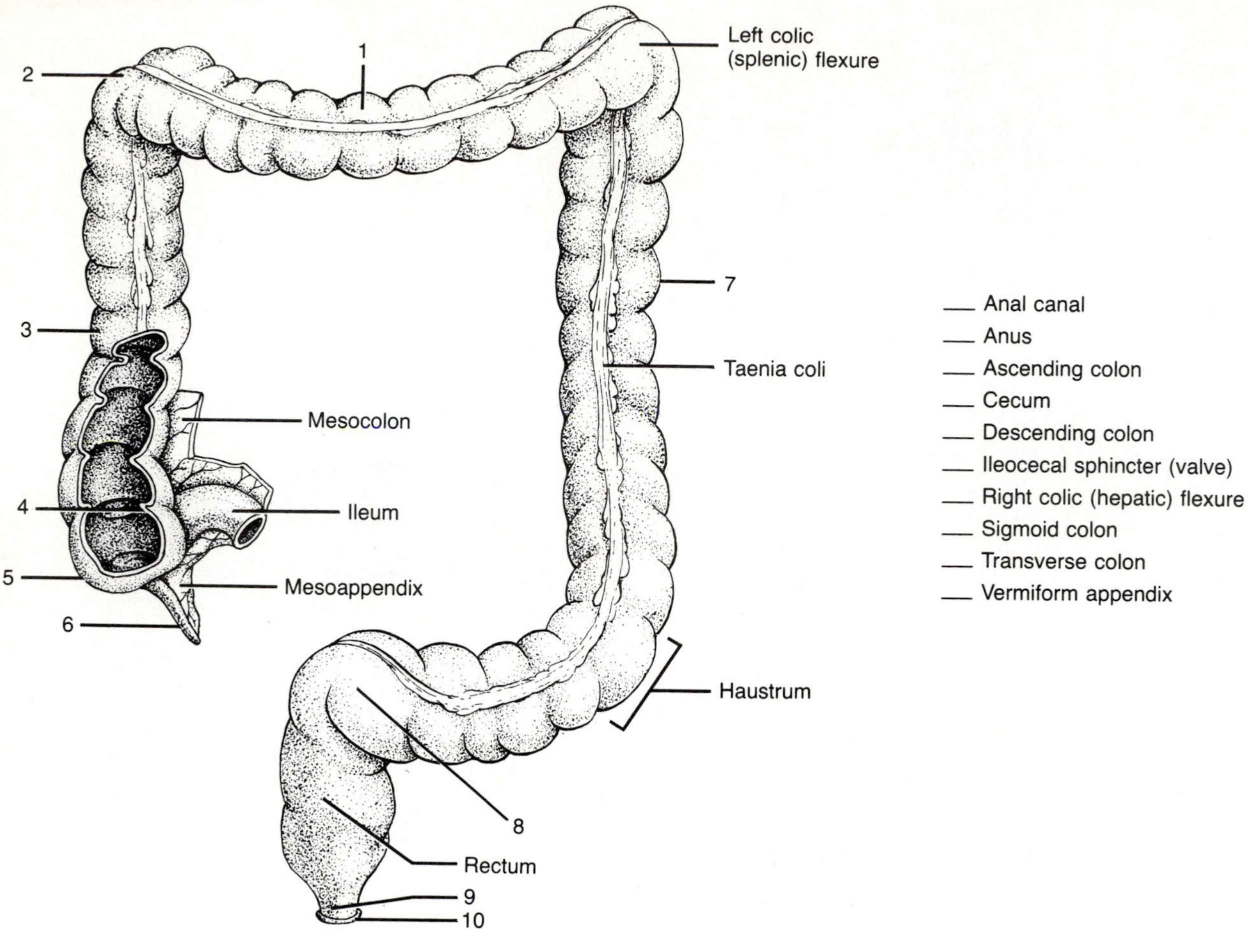

FIGURE 22.12 Large intestine.

flexure). The ***transverse colon*** continues across the abdomen, curves beneath the spleen ***(left colic [splenic] flexure),*** and passes down the left side of the abdomen as the ***descending colon***. The ***sigmoid colon*** begins near the iliac crest, projects inward toward the midline, and terminates at the rectum as S3. The ***rectum*** is the last 20 cm (7 to 8 in.) of the gastrointestinal tract. Its terminal 2 to 3 cm (1 in.) is known as the ***anal canal***. The opening of the anal canal to the exterior is the ***anus***.

With the aid of your textbook, label the parts of the large intestine in Figure 22.12.

The ***mucosa*** of the large intestine consists of simple columnar epithelium with numerous ***goblet cells*** (Figure 22.13). The ***submucosa*** is similar to that in the rest of the gastrointestinal tract. The ***muscularis*** consists of an outer longitudinal layer and an inner circular layer of smooth muscle. However, the longitudinal layer is not continuous; it is broken up into three flat bands, the ***taeniae coli*** (TĒ-nē-a KŌ-lī), which gather the colon into a series of pouches called ***haustra*** (HAWS-tra) (see Figure 22.12). The ***serosa*** of the large intestine is visceral peritoneum.

Examine a prepared slide of the large intestine showing its tunics. Compare your observations to Figure 22.13.

C. DISSECTION OF CAT DIGESTIVE SYSTEM

CAUTION! *Please reread Section D. "Precautions Related to Dissection" at the beginning of the laboratory manual on page xii before you begin your dissection.*

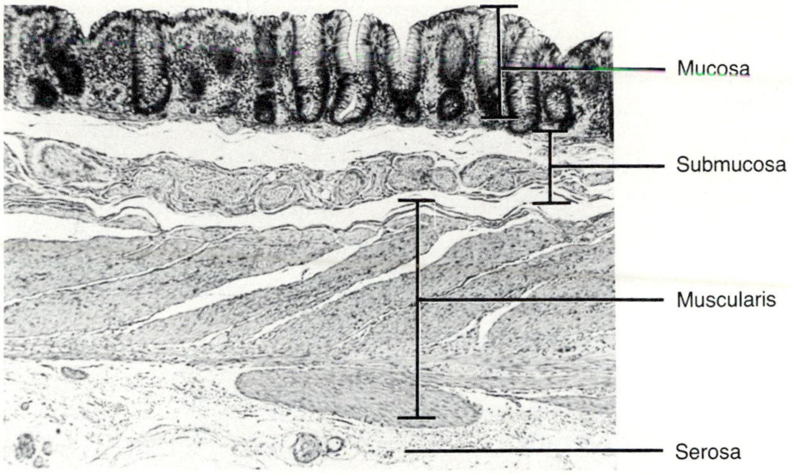

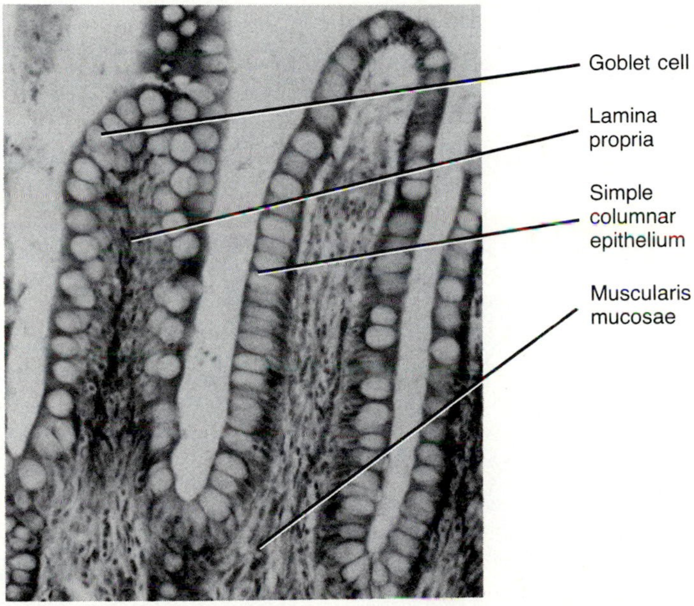

FIGURE 22.13 Histology of large intestine. (a) Photomicrograph of portion of wall. (b) Photomicrograph of enlarged aspect of mucosa.

PROCEDURE

1. Salivary Glands

1. The head of the cat contains five pairs of ***salivary glands***.

2. Working on the right side of the head of your specimen, locate the following glands (Figure 22.14).

a. PAROTID GLAND

1. The largest of the salivary glands, the parotid is found in front of the ear over the masseter muscle.

2. The ***parotid duct*** passes over the masseter muscle and pierces the cheek opposite the last upper premolar tooth.

3. Look inside the cheek to identify the opening of the duct. Do not confuse the parotid duct

with the branches of the anterior facial nerve. One branch is dorsal and one is ventral to the parotid duct.

b. SUBMANDIBULAR GLAND

1. This gland is below the parotid at the angle of the jaw. The posterior facial vein passes over it.

2. Below the submandibular gland is a lymph node that you should remove. Lymph nodes have smooth surfaces, whereas salivary glands have lobulated surfaces.

3. Lift the anterior margin of the submandibular gland to find the ***submandibular duct***.

4. To trace the duct forward, dissect, transect, and reflect the digastric and mylohyoid muscles. The submandibular duct is joined by the duct from the sublingual gland, which is usually difficult to identify.

5. The ducts of the submandibular and the sublingual glands open into the floor of the mouth near the lower incisors.

c. SUBLINGUAL GLAND

This gland is cranial to the submandibular gland. Its duct joins that of the submandibular gland.

d. MOLAR GLAND

This is a very small mass located near the angle of the mouth between the masseter muscle and mandible. Dissection is not necessary.

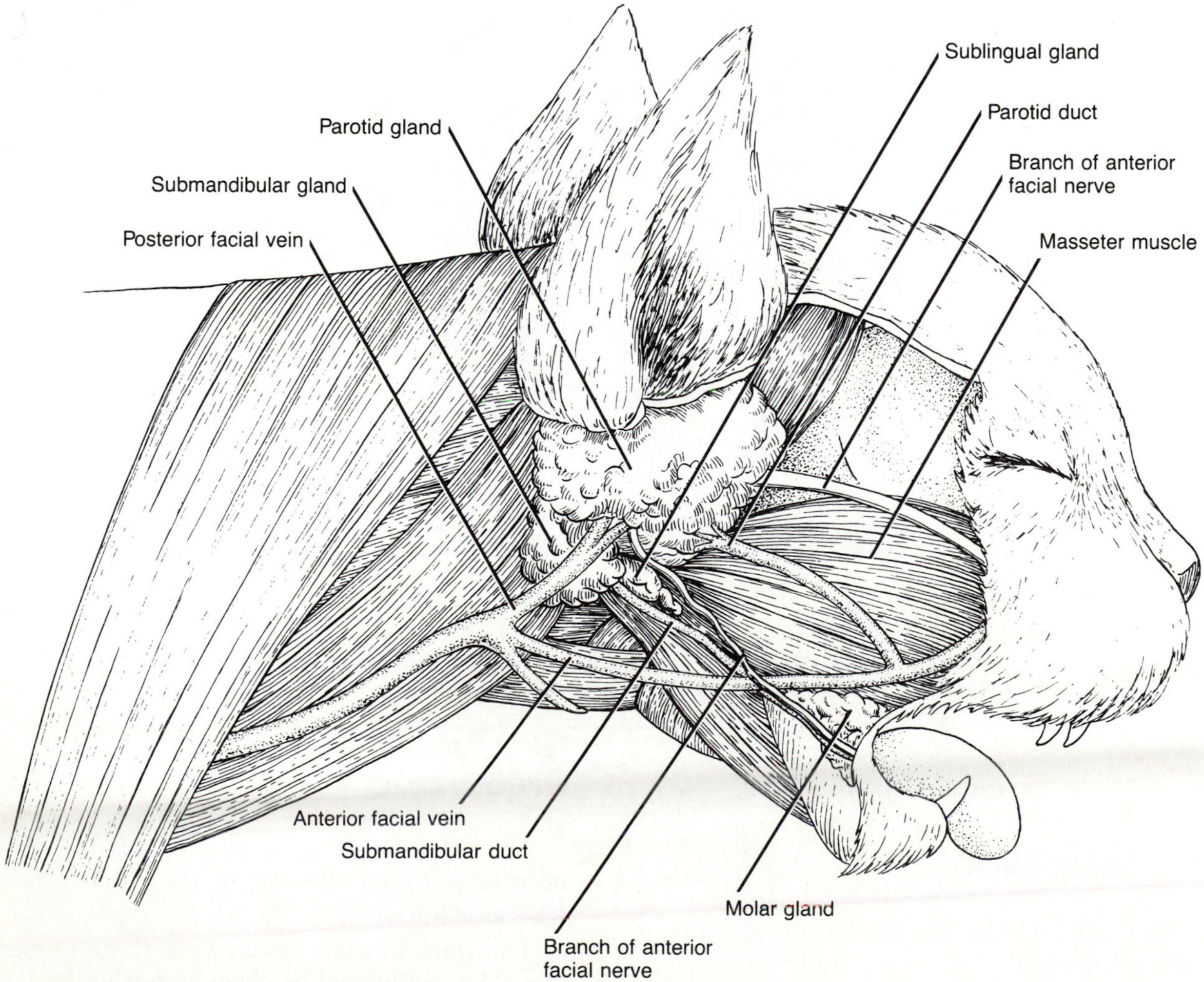

FIGURE 22.14 Salivary glands of cat.

e. INFRAORBITAL GLAND

This gland is within the floor of the orbit (it is not shown in Figure 22.14).

2. Mouth

1. Open the ***mouth*** by cutting through the muscles at each angle of the jaw with bone shears.
2. Press down on both sides of the mandible to expose the mouth.
3. Identify the following parts (Figure 22.15).

a. LIPS

Folds of skin around the orifice of the mouth, lined internally by mucous membrane. They are attached to the mandible by a fold of mucous membrane at the midline.

b. CHEEKS

The lateral borders of the oral cavity.

c. VESTIBULE

The space between lips and teeth.

d. ORAL CAVITY

The part of the mouth behind the teeth.

e. TEETH

1. The ***teeth*** of a cat may be represented by the dental formula $\frac{3-1-3-1}{3-1-2-1}$. The numbers represent, from left to right, the number of incisors, canines, premolars, and molars on each side. The upper row of numbers represents teeth in

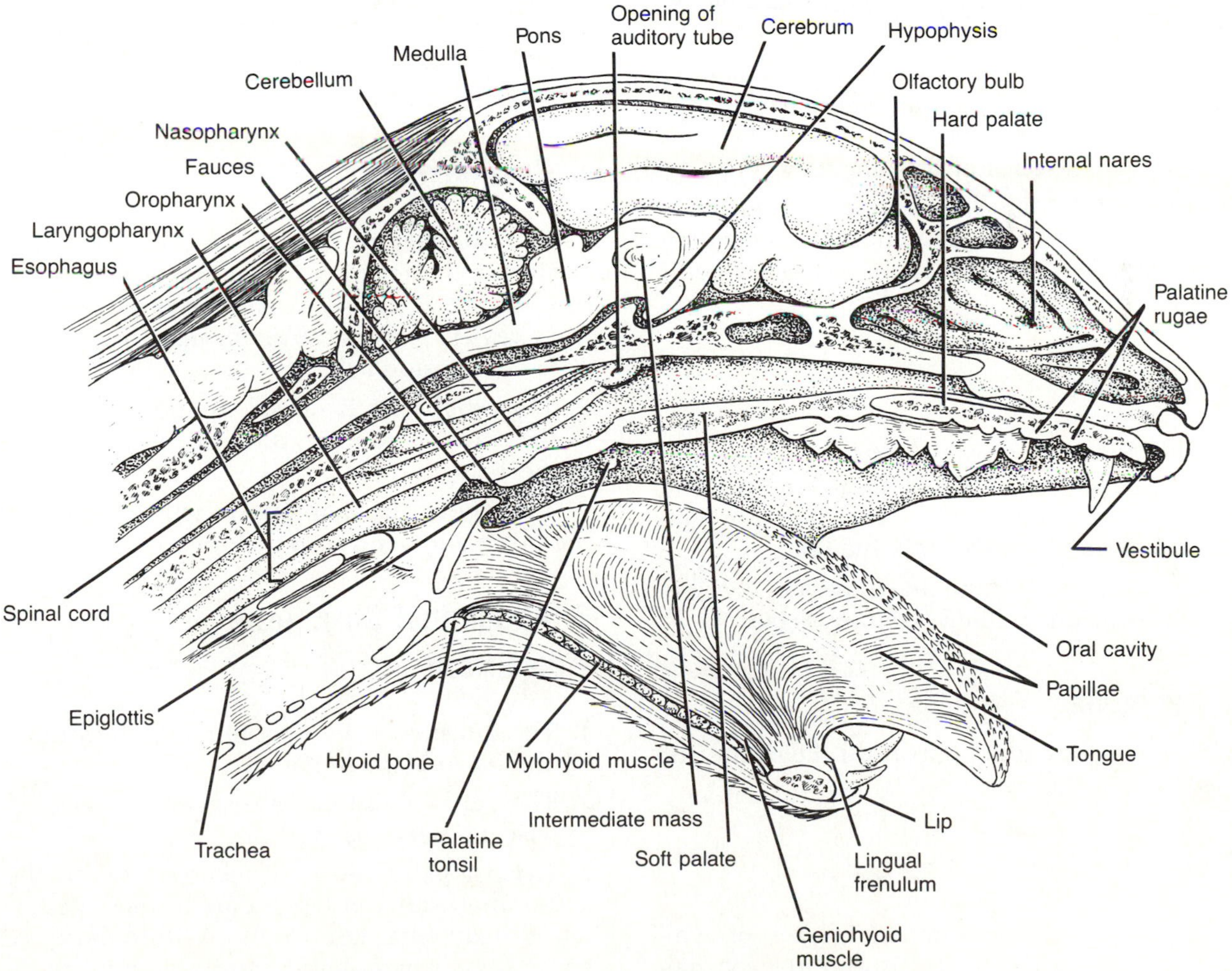

FIGURE 22.15 Midsagittal section through head and neck of cat.

the upper jaw and the lower row represents teeth in the lower jaw.

2. Observe the teeth. How many teeth are in a complete permanent set?

f. TONGUE

1. The ***tongue*** forms the floor of the oral cavity.

2. Observe the ***lingual frenulum***, a fold of mucous membrane that anchors the tongue to the floor of the mouth.

3. Now examine the dorsum (free surface) of the tongue and, using a hand lens, find the papillae.

4. The ***filiform papillae*** are located mostly in the front and middle of the tongue. They are pointed (spinelike) and are the most numerous of the papillae.

5. The ***fungiform papillae*** are small, mushroom-shaped structures between and behind the filiform papillae.

6. The ***circumvallate (vallate) papillae***, about 12 in number, are large and rounded and are surrounded by a circular groove. They are found near the back of the tongue.

7. The ***foliate papillae***, leaf-shaped and relatively few in number, are lateral to the vallate papillae. Taste buds are located between papillae.

g. HARD PALATE

1. This forms the cranial part of the palate. The bones of the hard palate are the palatine process of the premaxilla, the palatine process of the maxilla, and the palatine bone.

2. The mucosa of the hard palate contains transverse ridges called ***palatine rugae***.

h. SOFT PALATE

This forms the caudal portion of the palate and has no bony support.

i. PALATINE TONSILS

1. These are small rounded masses of lymphatic tissue in the lateral wall of the soft palate near the base of the tongue.

2. The tonsils are between lateral folds of tissue called the ***glossopalatine arches***. The arches are best seen by pulling the tongue ventrally.

3. Pharynx

1. The ***pharynx*** (throat) is a common passageway for the digestive and respiratory systems.

2. Although the pharynx will not be dissected at this time, you can locate some of its important parts in Figure 22.15. These include the ***nasopharynx***, the part of the pharynx above the soft palate; the ***auditory (Eustachian) tubes***, a pair of slitlike openings in the laterodorsal walls of the nasopharynx; the ***internal nares (choanae)***, which enter the cranial portion of the nasopharyx; the ***oropharynx***, the portion of the pharynx between the glossopalatine arches and free posterior margin of the soft palate (the ***fauces*** is the opening between the oral cavity and oropharynx); and the ***laryngopharynx***, the part of the pharynx behind the larynx.

3. The laryngopharynx opens into the esophagus and larynx.

4. Esophagus

1. The ***esophagus*** is a muscular tube that begins at the termination of the laryngopharynx (Figure 22.15). It extends through the thoracic cavity, pierces the diaphragm, and terminates at the stomach.

2. The esophagus is dorsal to and a little left of the trachae, running posteriorly toward the diaphragm.

3. Just above the diaphragm, it lies dorsal to the heart and ventral to the aorta.

5. Abdominal Structures

a. PERITONEUM

1. The interior of the abdominal wall is lined by the ***parietal peritoneum.***

2. The various abdominal organs are covered by ***visceral peritoneum.***

3. Extensions of the peritoneum between the abdominal wall and viscera are termed mesenteries, ligaments, and omenta. Within them are blood vessels, lymphatics, and nerves that supply the viscera.

4. Other peritoneal extensions are the greater omentum, lesser omentum, and mesentery (Figure 22.16).

b. GREATER OMENTUM

1. This is a double sheet of peritoneum that attaches the greater curvature of the stomach to the dorsal body wall.

2. The greater omentum encloses the spleen and part of the pancreas and covers the transverse colon and a large part of the small intestine.

3. The cavity between the layers of the greater omentum is called the ***omental bursa***. It contains considerable fat and is often entwined with the intestine.

4. The portion of the greater omentum between the stomach and spleen is called the ***gastrosplenic ligament***.

5. Remove the greater omentum.

c. LESSER OMENTUM

1. This extends from the left lateral lobe of the liver to the stomach and the duodenum.

2. The portion between the stomach and the liver is called the ***hepatogastric ligament***, and the portion between the duodenum and the liver is called the ***hepatoduodenal ligament***.

3. Within the right lateral border of the lesser omentum are the common bile duct, hepatic artery, and portal vein.

d. MESENTERY

1. The ***mesentery proper*** suspends the small intestine from the dorsal body wall.

2. The ***mesocolon*** suspends the colon from the dorsal body wall.

3. Note the blood vessels and lymph nodes in the mesenteries.

e. LIVER

1. The ***liver*** is the largest abdominal organ. It is reddish brown and is located immediately caudal to the diaphragm (Figure 22.16).

2. If you pull the liver and diaphragm apart, you can see the central portion of the diaphragm is formed by a tendon into which its muscle fibers insert. This is the ***central tendon*** of the diaphragm.

3. The liver is divided into ***right*** and ***left lobes***. Between these two main lobes is the ***falciform ligament.***

4. The falciform ligament is continuous with the ***coronary ligament***, which attaches the liver to the central tendon of the diaphragm.

5. At the free edge of the falciform ligament is a fibrous strand that represents the vestige of the umbilical vein. This strand is known as the ***round ligament***.

6. Each main lobe of the liver is further subdivided into two lobes called the lateral and the medial lobes. Thus, there is a ***left lateral lobe***, ***left median lobe***, ***right lateral lobe***, and ***right median lobe***.

7. Caudal to the left lateral lobe is another small lobe, the ***caudate lobe***.

f. GALLBLADDER

The ***gallbladder*** is a saclike structure located in a depression on the dorsal surface of the right median lobe of the liver (Figure 22.16).

g. STOMACH

1. The ***stomach*** lies to the left of the abdominal cavity (Figure 22.16). Its position can be noted by lifting the lobes of the liver.

2. Cut through the diaphragm to note the point at which the esophagus enters the stomach.

3. Identify the following parts of the stomach.

- □ ***Cardiac region***—Portion of the stomach adjacent to the esophagus.
- □ ***Fundus***—Dome-shaped portion extending cranially to the left of the cardiac region.
- □ ***Body***—Large section between the fundus and the pyloric region.
- □ ***Pyloric region***—Narrow caudal portion that empties into the doudenum. Between the pyloric region and the duodenum is a value called the ***pyloric sphincter (valve)***, which can be seen if the stomach is cut open. To see the pyloric sphincter, make

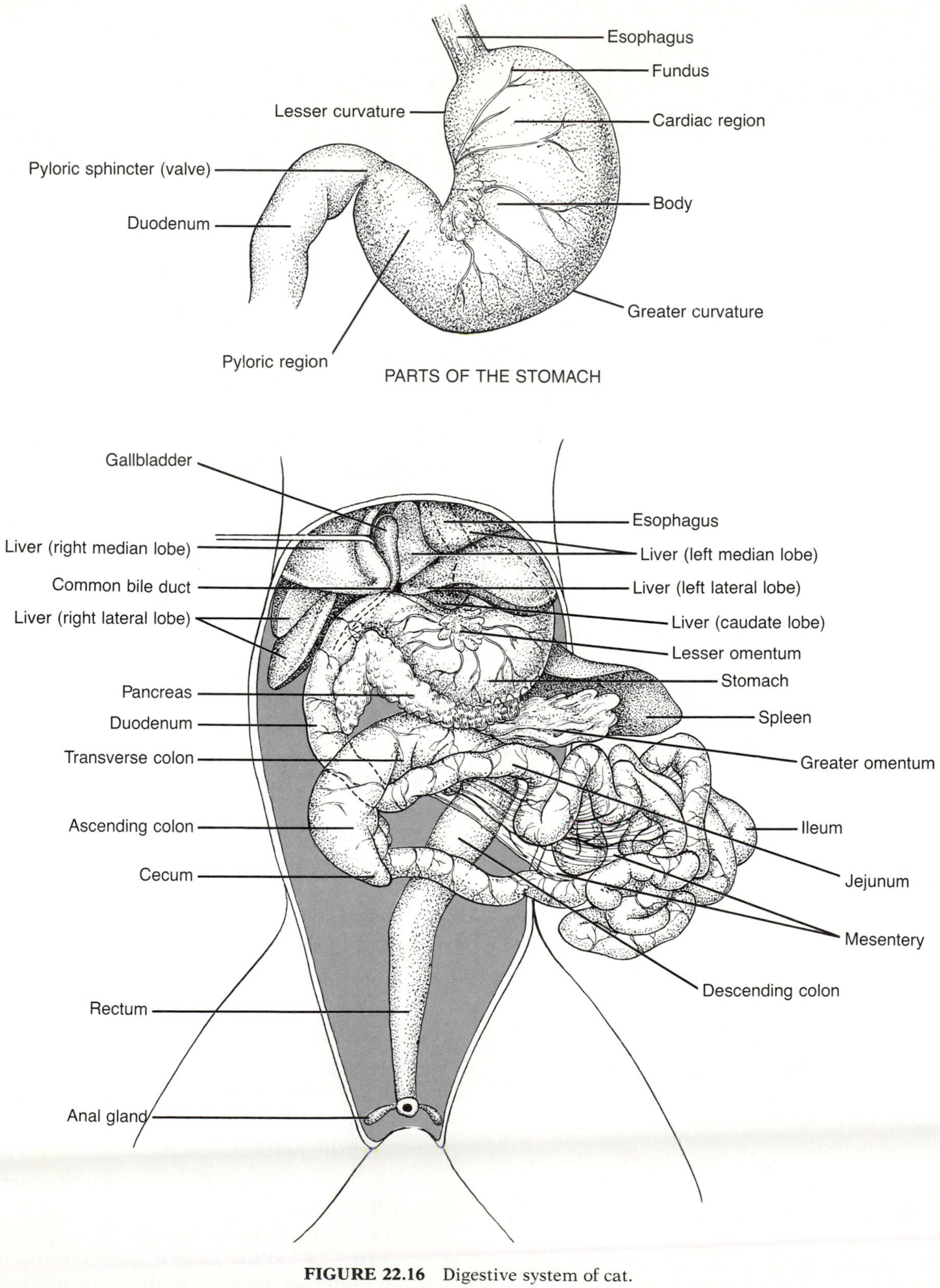

FIGURE 22.16 Digestive system of cat.

a longitudinal incision from the body to the duodenum and then wash the contents out of the stomach. You can also see the longitudinal ***rugae*** if the stomach is cut open.

- □ ***Greater curvature***—The long, left, and caudal margin of the stomach.
- □ ***Lesser curvature***—The short, right, and cranial margin of the stomach.

h. SPLEEN

The ***spleen*** (which is lymphatic tissue rather than part of the digestive system) is a reddish brown strurture on the left side of the abdominal cavity near the greater curvature of the stomach (Figure 22.16).

i. SMALL INTESTINE

1. To examine the small intestine, reflect the greater omentum, which covers the transverse colon and most of the ***small intestine***.

2. The small intestine is divided into three major regions (see Figure 21.16). The first region is the ***duodenum***, which is U-shaped and about 6 in. long. The common bile duct and the pancreatic duct empty into the duodenum near the pyloric sphincter (valve).

3. The enlargment of the duodenum where the ducts unite and enter the duodenum is called the ***hepatopancreatic ampulla*** (Figure 22.17).

4. The ***common bile duct*** is formed by the union of the ***hepatic ducts*** from the lobes of the liver and the ***cystic duct*** from the gallbladder.

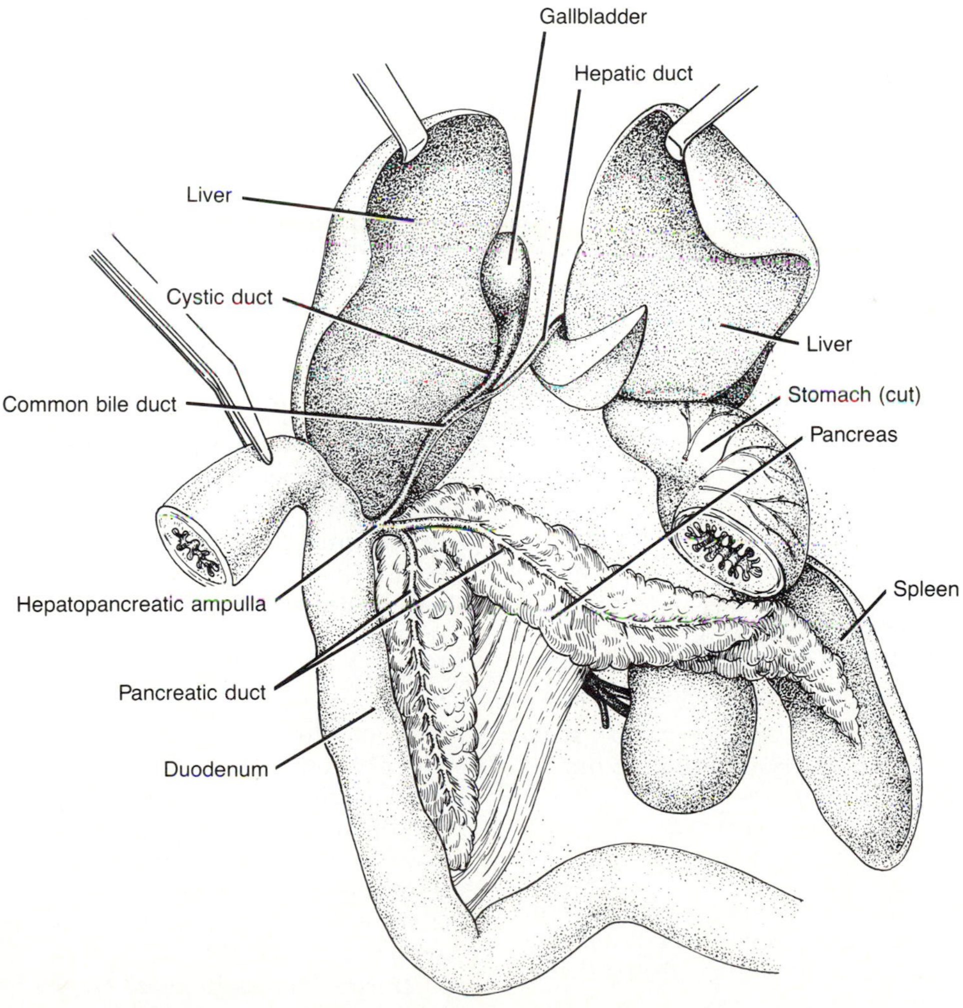

FIGURE 22.17 Relations of the liver, gallbladder, duodenum, and pancreas in cat.

5. The second region of the small intestine is the ***jejunum***, which composes about the first half of its remaining length (see Figure 22.16).

6. The ***ileum***, the third region of the small intestine, composes about the second half of its remaining length. There is no definite demarcation between the jejunum and ileum. Trace the ileum to its juction with the large intestine.

7. If you cut open the jejunum or ileum, you can note the velvety appearance of the mucosa. This appearance is due to villi, microscopic fingerlike projections that carry on absorption.

8. Remove any parasitic roundworms or tapeworms.

j. LARGE INTESTINE

1. The beginning of the ***large intestine*** consists of a short, blind pouch called the ***cecum*** (see Figure 22.16). In humans, the appendix is attached to the cecum. Cats have no appendix.

2. Cut open the wall of the cecum and colon opposite the entrance of the ileum.

3. Identify the ***ileocecal sphincter (valve)*** at the junction of the ileum and colon.

4. Cranial to the cecum is the ***ascending colon*** on the right side of the abdominal cavity. It extends from the cecum to the liver. Here, it passes across the abdominal cavity as the ***transverse colon.***

5. The transverse colon continues as the ***descending colon***, which runs caudally toward the midline. The cat has no sigmoid colon (which is the next part of the large intestine in humans).

6. Following the descending colon in the cat is the ***rectum***, which descends into the pelvic cavity and opens to the exterior as the ***anus***.

7. Lateral to the anus are the ***anal (scent) glands***. Their ducts empty into the rectum.

k. PANCREAS

1. The ***pancreas*** is a lobulated organ lying against the descending portion of the duodenum and across the body of the spleen (see Figure 22.16).

2. If you carefully tease away the pancreatic tissue, you can identify the ***pancreatic duct*** inside the pancreas. This duct is a white, threadlike structure that conveys pancreatic secretions to the duodenum.

3. The pancreatic duct unites with the common bile duct at the hepatopancreatic ampulla.

D. DEGLUTITION

Swallowing, or ***deglutition*** (dē-gloo-TISH-un), is a mechanism that moves food from the mouth through the esophagus to the stomach. It is facilitated by saliva and mucus and involves the mouth, pharynx, and esophagus. Food mixed with saliva is shaped into a rounded mass called a ***bolus***. Swallowing is conveniently divided into three stages: (1) the voluntary stage, in which the bolus is moved into the orophyarynx; (2) the pharyngeal stage, the involuntary passage of the bolus through the pharynx into the esophagus; and (3) the esophageal stage, the involuntary passage of the bolus through the esophagus into the stomach.

Swallowing starts when the bolus is forced to the back of the mouth cavity and into the oropharynx by the movement of the tongue upward and backward against the palate. This represents the ***voluntary stage*** of swallowing.

With the passage of the bolus into the oropharynx, the involuntary ***pharyngeal stage*** of swallowing begins.

The respiratory passageways close and breathing is temporarily interrupted. The bolus stimulates receptors in the oropharynx, which send nerve impulses to the ***deglutition center*** in the medulla and lower pons of the brain stem. The returning nerve impulses cause the soft palate and uvula to move upward to close off the nasopharynx, and the larynx is pulled forward and upward under the tongue. As the larynx rises, it meets the epiglottis, which seals off the glottis. The movement of the larynx also pulls the vocal folds together, further sealing off the respiratory tract, and widens the opening between the laryngopharynx and esophagus. The bolus passes through the laryngopharynx and enters the esophagus in 1 to 2 seconds. The respiratory passageways then reopen and breathing resumes.

During the ***esophageal stage*** of swallowing, food is pushed through the esophagus by involuntary muscular movements called ***peristalsis*** (per′-i-STAL-sis).

Peristalsis is a function of the muscularis and is controlled by the medulla. In the section of the esophagus lying just above and around the top of the bolus, the circular muscle fibers contract. The contraction constricts the esophageal wall and squeezes the bolus downward. Meanwhile, longitudinal fibers lying around the bottom of and just below the bolus also contract. Contraction of the longitudinal fibers shortens this lower section, pushing its walls outward so it can receive the bolus. The contractions are repeated in a wave that moves down the esophagus, pushing the food toward the stomach. Passage of the bolus is further facilitated by glands secreting mucus. The passage of solid or semisolid food from the mouth to the stomach takes 4 to 8 seconds. Very soft foods and liquids pass through in about 1 second.

PROCEDURE

1. Chew a cracker and note the movements of your tongue as you begin to swallow. Chew another cracker, lean over your chair, placing your mouth below the level of your pharynx, and attempt to swallow again.

2. Note the movements of your laboratory partner's larynx during the pharyngeal stage of deglutition.

3. *Clean the earpieces of a stethoscope with alcohol.* Using the stethoscope, listen to the sounds that occur when your partner swallows water. Place the stethoscope just to the left of the midline at the level of the 6th rib. You should hear the sound of water as it makes contact with the lower esophageal sphincter (valve), followed by the sound of water passing through the sphincter after the sphincter relaxes. The sphincter is at the junction of the esophagus and stomach.

4. Record your observations in Section D of the LABORATORY REPORT RESULTS at the end of the exercise. At what point can deglutition be stopped voluntarily? ____________________

____

E. OBSERVATION OF MOVEMENTS OF THE GASTROINTESTINAL TRACT

Contraction of the smooth muscle of the gastrointestinal tract is responsible for various movements that serve to move substances through the tract that result in digestion, absorption, and elimination of wastes. Substances move through the entire gastrointestinal tract by ***peristalsis.*** In addition, several other types of movements occur in various regions of the tract.

Several minutes after food enters the stomach, gentle, rippling, peristaltic movements called ***mixing waves*** pass over the stomach every 15 to 25 seconds. These waves macerate food, mix it with the secretions of the gastric glands, and reduce it to a thin liquid called ***chyme*** (kīm). As digestion proceeds in the stomach, more vigorous mixing waves begin at the body of the stomach and intensify as they reach the pylorus. Each mixing wave forces a small amount of the gastric contents into the duodenum through the pyloric sphincter. Most of the food is forced back into the body of the stomach, where it is subjected to further mixing. The next wave pushes it forward again and forces a little more into the duodenum. The forward and backward movement of the gastric contents are responsible for almost all of the mixing in the stomach.

Pendular movements involve alternate contraction and relaxation of longitudinal muscle, shortening and lengthening the intestine, and thus spilling the contents back and forth.

In the small intestine, ***segmentation*** constricts the intestine into small localized segments. Circular fibers in adjacent segments contract and relax alternately, spilling the contents back and forth. As one set of constrictions relaxes, the next set occurs at a point midway between the previous contractions, thus "chopping" the intestinal contents.

In the large intestine, ***haustral churning,*** a type of segmental movement, churns the contents of one pouch (haustrum) and exposes them to the mucosa where absorption can occur.

In addition, strong peristaltic waves, known as ***mass peristalsis,*** occur along the length of the large intestine about three or four times each day.

CAUTION! *Please reread Section D. "Precautions Related to Dissection" at the beginning of the laboratory manual on page xii before you begin your dissection.*

PROCEDURE

1. Obtain a rat from your laboratory instructor. (The rats will have been sacrificed just prior to the beginning of the laboratory period either by cervical dislocation or by the injection of an air embolism.)

2. Using surgical scissors with a blunt point, quickly make a midline incision through the skin and muscle of the abdomen from just below the xiphoid process to the pubis (see Figure 2.6). Hold your scissors parallel to the abdominal wall so that you do not injure the viscera.

3. Make four lateral incisions, two just beneath the diaphragm and two at the level of the pubis. Do not enter the thoracic cavity. Reflect the abdominal skin to expose the viscera.

4. Note the parietal peritoneum which lines the abdominal wall.

5. Identify the stomach. Describe any movements of the stomach musculature. ________

____

6. Using a blunt forceps, gently squeeze the pyloric sphincter (valve). *Gently* squeeze a loop of the small intestine.

Describe the reaction. ________________

____

7. Identify the greater omentum, a double fold of peritoneum which is suspended from the greater curvature of the stomach.

8. Cover the stomach with a small sponge soaked in Locke's solution or 0.9% sodium chloride (NaCl) solution kept at 37°C. Reflect the greater omentum backward onto the sponge.

9. Examine the mesentery. Note the distribution of blood vessels and the presence of whitish lacteals.

10. Observe the movements of the small intestine. Note pendular, segmental, and peristaltic movements.

Describe the movements: ________________

____

____

ALTERNATE PROCEDURE

If Exercise E is to be done, have your instructor or a laboratory assistant remove a section of jejunum from the rat at this time and immediately immerse it in 37°C Locke's solution.

11. Describe the movements of the intestinal tract as the tissues become ischemic: ________

____

12. Open the stomach along the greater curvature. Test the pH of the gastric contents with pH paper. Slit open the ileum and test the pH of the ileal contents. Record the pH.

Gastric pH ____________

Ileal pH ____________

13. Dispose of the rat according to the procedure given by your laboratory instructor. *Clean your equipment thoroughly.*

F. PHYSIOLOGY OF INTESTINAL SMOOTH MUSCLE

Contraction of smooth muscle in the wall of the intestine is responsible for propelling foods through the tube and mixing the food with digestive enzymes. In these experiments you will have an opportunity to observe some of the factors that effect intestinal smooth muscle contraction. You will also be able to compare the physiology of intestinal smooth muscle to that of skeletal muscle (Exercise 9) and cardiac muscle (Exercise 19).

CAUTION! *Please reread Section A. "General Safety Precautions and Procedures" on page xi and Section C. "Precautions Related to Working with Reagents" on page xii at the beginning of the laboratory manual before you begin any of the following experiments. You should also read the experiments before you perform them to be sure that you understand all the procedures and safety precautions.*

1. Isolation of the Intestinal Segment

PROCEDURE

1. Your instructor or a laboratory assistant will provide you with a two inch segment of intestine from a rat. *Do not stretch it.*

2. Immediately immerse the jejunum in 37°C Locke's solution.

3. Fill a medicine dropper with 37°C Locke's solution and gently flush out the luminal contents of the jejunal segment. Cut off the mesen-

tery, and cut the segment into 2.54 cm (1 in.) pieces.

4. Place the segments in a beaker of clean 37°C Locke's solution.

5. Using a probe, make a small hole in each end of the segment. Tie a 2.54 cm (1 in.) thread loop through one hole; tie an 20 cm (8 in.) length of thread through the other loop.

6. Slip the 2.54 cm (1 in.) loop over the L-tube. Tie the end of the long thread to the hook (Figure 22.18).

7. Adjust the tension so that the string is taut but the tissue is not overly stretched. (Note: Depending on the model of the myograph utilized, either turn up the sensitivity or attach the string to the most sensitive arm of the transducer.)

8. Adjust the air flow so that a continuous, light stream of air passes across the intestinal segment.

9. Adjust the transducer so that the amplitude of the contractions is from 1.5 to 2.5 cm (0.5 to 1 in.). If contractions are not observed after oxygen has bubbled through the solution, raise the heat to 39°C until contractions commence. If contractions are still not observed, use a medicine dropper to add 1 drop of 1:10,000 acetylcholine.

2. Experimental Procedures

a. INHERENT CONTRACTILE ACTIVITY

After allowing an initial 5 to 10 minutes for the intestinal segment to stabilize, observe its

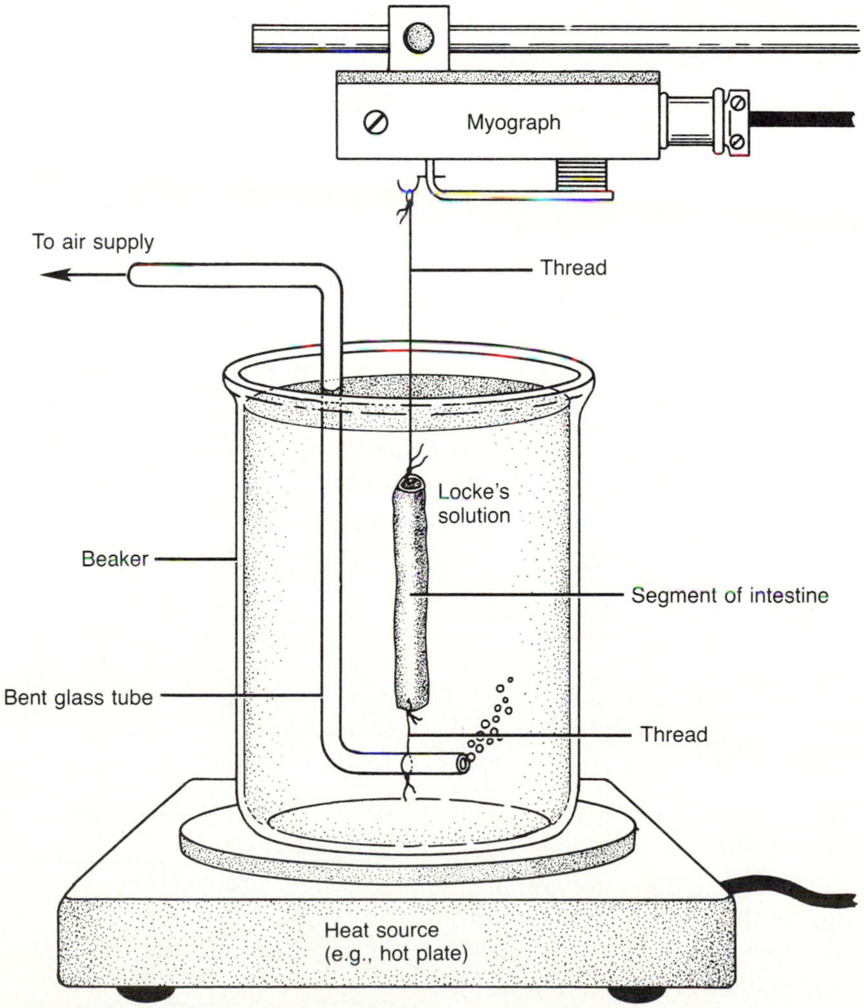

FIGURE 22.18 Experimental setup for the physiology of intestinal smooth muscle.

basal activity while recording with the polygraph.

Is the segment quiescent or does it exhibit intrinsic contractile activity? ______

______

Is the extent of relaxation consistent, or does the preparation exhibit changes? ______

______

Does the preparation resemble skeletal or cardiac muscle with respect to inherent contractile activity? ______

______

Based upon your knowledge of the physiology of skeletal and cardiac muscle, explain your observation. ______

______

______

______

b. EFFECT OF OXYGEN

PROCEDURE

1. Shut off the oxygen supply until you notice a change in contractile activity.

Explain your observations. ______

______

______

2. Resupply oxygen and describe recovery.

______

______

c. EFFECT OF TEMPERATURE

PROCEDURE

1. Lower the temperature of Locke's solution to 20°C by adding ice to it.

2. Record the contractile activity for about 2 minutes by turning on the polygraph. Mark the temperature on the paper.

3. Place a Bunsen burner or alcohol burner under the beaker and warm Locke's solution. *Keep the burner flame low.*

4. Turn on the polygraph and control the heat so that the temperature of the solution rises about 2°C each 2 minutes, up to 40°C.

5. Mark the temperature on the paper each 2 minutes as the temperature rises at 2°C intervals. Turn off the physiograph.

Describe the changes that occurred in the intestinal segment with regard to strength and frequency of contraction as temperature increased. ______

______

______

______

______

At what temperature are the contractions strongest? ______

At what temperature is the frequency of contractions greatest? ______

Explain your observations. ______

______

______

______

______

d. EFFECT OF NOREPINEPHRINE

PROCEDURE

1. Place 5 to 8 drops of a 1:50,000 norepinephrine solution into Locke's solution.

2. Observe any changes in the following: basal tone, strength of contraction, frequency of contraction, and speed of contraction.

3. How do each of these responses compare with those seen when norepinephrine was added to cardiac muscle in Exercise 19? Explain in the spaces provided.

Basal tone ______

______

______

Strength of contraction ______

______

______

Frequency of contraction ______________________

____

____

Speed of contraction ______________________

____

____

e. EFFECT OF ACETYLCHOLINE

PROCEDURE

1. Before performing this experiment, drain Locke's solution containing norepinephrine and add fresh solution. Wait 5 minutes before proceeding.

2. Using a medicine dropper, now place 1 to 5 drops of a 1:1000 acetylcholine solution, one drop at a time, into Locke's solution.

3. Observe any changes in the following: basal tone, strength of contraction, frequency of contraction, and speed of contraction.

4. How do each of these responses compare with those seen when acetylcholine was added to cardiac muscle in Exercise 19? Explain in the spaces provided.

Basal tone ______________________

____

____

Strength of contraction ______________________

____

____

Frequency of contraction ______________________

____

____

Speed of contraction ______________________

____

____

____

Does acetylcholine modify the contractile properties of skeletal muscle? ______________________

____

____

Does the autonomic nervous system play a role in skeletal muscle contractions? ______________________

____

____

f. EFFECT OF ATROPINE

PROCEDURE

1. Before performing this experiment, drain Locke's solution containing acetylcholine and add a fresh solution. Wait 5 minutes before proceeding.

2. Using a medicine dropper, place 5 to 10 drops of a 1:50,000 atropine solution into Locke's solution.

3. Explain the effects of atropine on the following:

Basal tone ______________________

____

____

Strength of contraction ______________________

____

____

Frequency of contraction ______________________

____

____

Speed of contraction ______________________

____

____

g. EFFECT OF ATROPINE AND ACETYLCHOLINE

PROCEDURE

1. Using a medicine dropper, add 1 to 5 drops of a 1:1000 solution of acetylcholine to Locke's solution already containing atropine.

2. Wait 3 to 5 minutes and compare the effect of acetylcholine in the presence of atropine on the following:

Basal tone ______________________

____

____

Strength of contraction ______________________

____

____

Frequency of contraction ______________________

____

____

Speed of contraction ______________________

____

____

Based on your observations, what type of acetylcholine receptors does the smooth muscle segment possess? ______________________

____

____

____

h. EFFECT OF ATROPINE, ACETYLCHOLINE, AND NOREPINEPHRINE

PROCEDURE

1. Using a medicine dropper, add 5 to 8 drops of a 1:50,000 solution of norepinephrine to Locke's solution already containing atropine and acetylcholine.

2. Wait 3 to 5 minutes and compare the effect of norepinephrine in the presence of atropine and acetylcholine on the following:

Basal tone ______________________

____

____

Strength of contraction ______________________

____

____

Frequency of contraction ______________________

____

____

Speed of contraction ______________________

____

____

i. EFFECT OF CALCIUM REMOVAL

PROCEDURE

1. Before performing this experiment, drain Locke's solution containing atropine, acetylcholine, and norepinephrine and fill the beaker with fresh Locke's solution. Allow 5 minutes before proceeding.

2. To determine whether the intestinal segment has completely recovered, record about 3 minutes of activity and then compare your observations to those recorded in Section 3a.

3. Now replace Locke's solution with Locke's solution containing no calcium $O[Ca]_e$.

4. Explain the effects of Locke's solution containing $O[Ca]_e$ on the following:

Basal tone ______________________

____

____

Strength of contraction ______________________

____

____

Frequency of contraction ______________________

____

____

Speed of contraction ______________________

____

____

Would you expect similar results if you were to expose cardiac muscle to Locke's solution containing $O[Ca]_e$? ______________________

____

____

____

Would you expect similar results if you were to expose skeletal muscle to Locke's solution containing $O[Ca]_e$? ______________________

____

____

____

What substance(s) used in this exercise stimulate (depolarize) intestinal smooth muscle?

____

____

____

What substances used in this exercise inhibit (hyperpolarize) intestinal smooth muscle?

____

____

____

G. CHEMISTRY OF DIGESTION

You have learned how food is digested and absorbed. Now you will learn what happens to the food after it reaches the cells of the body. You also will learn what nutrients are needed for survival and why.

Nutrients are chemical substances in food that provide energy, act as building blocks to form new body components, or assist body processes. The six major classes of nutrients are: carbohydrates, lipids, proteins, minerals, vitamins, and water. Cells break down carbohydrates, lipids, and proteins to release energy or use them to build new structures and new regulatory substances, such as hormones and enzymes.

Enzymes are proteins produced by living cells to catalyze or speed many of the reactions in the body. Enzymes act as ***catalysts*** to speed reactions without being permanently altered by the reaction. Digestive enzymes function as catalysts to speed chemical reactions in the digestive tract.

Because digestive enzymes function outside the cells that produce them, they are capable of also reacting within a test tube and therefore provide an excellent means of studying enzyme activity.

CAUTION! *Please reread Section A. "General Safety Precautions and Procedures" on page xi and Section C. "Precautions Related to Working with Reagents" on page xii at the beginning of the laboratory manual before you begin any of the following experiments. You should also read the experiments before you perform them to be sure that you understand all the procedures and safety precautions.*

1. Positive Tests for Sugar and Starch

Salivary amylase (ptyalin) is an enzyme produced by the salivary glands. This enzyme starts starch digestion and ***hydrolyzes*** (splits using water) it into maltose. We measure the among of starch and sugar present before and after enzymatic activity. It is expected that the amount of starch should *decrease* and the sugar level should *increase* as a result of salivary amylase activity.

a. TEST FOR SUGAR

Benedict's test is commonly used to detect sugars. Glucose (monosaccharide), maltose (disaccharide), or any other reducing sugars react with ***Benedict's solution,*** forming insoluble red cuprous oxide. The precipitate of cuprous oxide can usually be seen in the bottom of the tube when standing. Benedict's solution turns green, yellow, orange, or red depending on the amount of reducing sugar present according to the following scale:

blue (−)
green (+)
yellow (+ +)
orange (+ + +)
red (+ + + +)

Test for the presence of sugar (maltose) as follows:

PROCEDURE

1. Using separate medicine droppers, place 2 ml of maltose and 2 ml of Benedict's solution in a Pyrex test tube.

2. *Using a test tube holder,* place the test tube in a boiling water bath and heat for 5 minutes. CAUTION! *Make sure that the mouth of the test tube is pointed away from you and all other persons in the area.* Note the color change.

3. *Using a test tube holder,* remove the test tube from the water bath.

4. Repeat the same procedure using starch solution instead of maltose. Notice that the color does not change, because the solution contains no sugar. Now you have a method for detecting sugar.

b. TEST FOR STARCH

Lugol's solution is a brown colored iodine solution used to test certain polysaccharides, especially starch. Starch, for example, gives a ***deep blue*** to ***black*** color with Lugol's solution (the black is really a concentrated blue color). Cellulose, monosaccharides, and disaccharides do not react. A negative test is indicated by a yellow to brown color of the solution itself, or possibly some other color (other than blue to black) resulting from pigments present in the substance being tested.

PROCEDURE

1. Using separate medicine droppers, place a drop of starch solution on a spot plate and then add a drop of Lugol's solution to it. Notice the black color that forms as the starch-iodine complex develops.
2. Repeat the test using a maltose solution in place of the starch solution. Note that there is no color change, because Lugol's solution and maltose do not combine. Now you have a method for detecting starch.

c. DIGESTION OF STARCH

Salivary amylase from freshly collected saliva is used in this procedure to show digestion of starch.

PROCEDURE

1. Rinse your mouth thoroughly with distilled water two or three times.
2. Chew a piece of paraffin wax or sugarless gum to stimulate saliva flow, collecting 4 ml of saliva in a small container.
3. Add an equal amount of distilled water carefully, so that you have diluted saliva that contains the enzyme to be studied. You may also use commercially prepared amylase.
4. Test the pH of your saliva with pH paper.

pH = ______

5. Using separate medicine droppers, transfer 3 ml of a starch solution to a small beaker, add 3 ml of the saliva solution (or a pinch of amylase powder), and mix thoroughly with a glass rod.
6. Wait 1 minute, record the time, remove one drop of the mixture with a glass rod to the depression of a spot plate, and then test for starch with Lugol's solution.
7. At 1-minute intervals, test one-drop samples of the mixture until you no longer note a positive test for starch. Keep the glass rod in the mixture, stirring it from time to time.
8. After the starch test is seen to be negative, test the remaining mixture for the presence of glucose. Do this by adding 2 ml of the mixture with a medicine dropper to 2 ml of the Benedict's solution in a Pyrex test tube and heat in a boiling water bath as per the procedure outlined in a.2, "Test for Sugar." Answer questions a through c in Section G.1 of the LABORATORY REPORT RESULTS at the end of the exercise.

Why was paraffin or sugarless gum used in this exercise? ______________________________

____

____

2. Effect of Temperature on Starch Digestion

In the following procedure, you wil test starch digestion at five different temperatures to determine how temperature influences enzyme activity. Logol's solution is again used for presence or absence of starch.

A fresh enzyme solution (salivary amylase) should be prepared as before. Five constant temperature water baths should be available. Starting with the lowest temperature, these are: 0°C or cooler, 10°C, 40°C, 60°C, and boiling.

PROCEDURE

1. Prepare 10 test Pyrex tubes, 5 containing 1 ml each of enzyme solution and 5 containing 1 ml each of starch solution.
2. Using rubber bands, pair the tubes (i.e., a tube containing enzyme with one containing starch) and use a test tube holder to place one pair into each water bath. CAUTION! *Make sure that the mouth of the test tube is pointed away from you and all other persons in the area.*
3. Permit the tubes to adapt to the bath temperatures for about 5 minutes, then mix the enzyme and starch solutions of each pair together, and *using a test tube holder* place the single test tube in its respective water bath.

4. After 30 seconds, *using a test tube holder*, remove the test tubes from the water baths and test all five tubes for starch on a spot plate using Lugol's solution.

5. Repeat every 30 seconds until you have determined the time required for the starch to disappear (that is, to be digested).

6. Record and graph your results in Section G.2 of the LABORATORY REPORT RESULTS at the end of the exercise.

3. Effect of pH on Starch Digestion

You can demonstrate the effectiveness of salivary amylase digestion at different pH readings.

PROCEDURE

1. Prepare three buffer solutions as follows:
   Solution A—pH 4.0
   Solution B—pH 7.0
   Solution C—pH 9.0

2. Once again a fresh enzyme solution (salivary amylase) should be prepared as before.

3. Using medicine droppers, mix 4 ml of a starch solution with 2 ml of buffer solution A in a Pyrex test tube.

4. Repeat this procedure with buffer solutions B and C.

5. You now have three test tubes of a starch buffer-solution, each at a different pH (4.0, 7.0, and 9.0).

6. Using separate medicine droppers, place one drop of starch-buffer solution A on a spot plate and immediately add one drop of the saliva.

7. Test for starch disappearance using Lugol's solution, and record the time when starch first disappears completely.

8. Repeat this test for the other two starch buffers (solutions B and C) and record the time when starch is no longer present at each pH.

9. Record and explain your results in Section G.3 of the LABORATORY REPORT RESULTS at the end of the exercise.

4. Action of Bile on Fats

Bile is important in the process of digestion because of its emulsifying effects (breaking down of large globules to smaller, uniformly distributed particles) on fats and oils. *Bile does not contain any enzymes*. Emulsification of lipids by means of bile salts serves to increase the surface area of the lipid that will be exposed to the action of the lipase.

PROCEDURE

1. Place 5 ml of water into one Pyrex test tube and 5 ml of bile solution into a second.

2. Using a medicine dropper, add one drop of vegetable oil that has been colored with a fat-soluble dye, such as Sudan B, into each tube.

3. Place a stopper in both tubes. Shake them *vigorously*, and then let them stand in a test tube rack undisturbed for 10 minutes. Fat or oil that is broken into sufficiently small droplets will remain suspended in water in the form of an ***emulsion.*** If emulsification has not occurred, the fat or oil will lie on the surface of the water.

4. Answer questions in Section G.4 of the LABORATORY REPORT RESULTS at the end of the exercise.

5. Digestion of Fats

You can demonstrate the effect of pancreatic juice on fat by the use of pancreatin, which contains all the enzymes present in pancreatic juice. Because the optimum pH of the pancreatic enzymes ranges from 7.0 to 8.8, the pancreatin is prepared in sodium carbonate. The enzyme activity in this test is ***lipase***, which digests fat to fatty acids and glycerol. The fatty acid produced changes the color of ***blue*** litmus to ***red.***

PROCEDURE

1. Using a medicine dropper, place 5 ml of litmus cream (heavy cream to which powdered litmus has been added to give it a blue color) in a Pyrex test tube, and, *using a test tube holder*, place the test tube in a 40°C water bath.

2. Repeat the procedure with another 5-ml portion in a second tube, but put it in an ice bath.

3. When the tubes have adapted to their respective temperatures (in about 5 minutes), use a medicine dropper to add 5 ml of pancreatin to each tube and, *using a test tube holder*, replace them in their water baths until a color change occurs in one tube.

4. Summarize your results and your explanation in Section G.5 of the LABORATORY REPORT RESULTS at the end of the exercise.

6. Digestion of Protein

Here you demonstrate the effect of pepsin on protein and the factors affecting the rate of action of pepsin. ***Pepsin,*** a proteolytic enzyme, is secreted in inactive form (pepsinogen) by zymogenic (peptic) cells in the lining of the stomach. Pepsin digests proteins (fibrin in this experiment) to peptides. The efficiency of pepsin activity depends on the pH of the solution, the optimum being 1.5 to 2.5. Pepsin is almost completely inactive in neutral or alkaline solutions.

PROCEDURE

1. Prepare and number the following five Pyrex test tubes. For this test, the quantity of each solution must be *measured carefully*.

Tube 1—5 ml of 0.5% pepsin; 5 ml of 0.8% HCl

Tube 2—5 ml of pepsin; 5 ml of water

Tube 3—5 ml of pepsin, boiled for 10 minutes in a water bath; 5 ml of 0.8% HCl

CAUTION! *Make sure that the mouth of the test tube is pointed away from you and all other persons in the area. Using a test tube holder, remove the test tube from the water bath.*

Tube 4—5 ml of pepsin; 5 ml of 0.5% NaOH

Tube 5—5 ml of water; 5 ml of 0.8% HCl

2. First determine the approximate pH of each test tube using Hydrion paper (range 1 to 11) and put the values in the table provided in Section G.6 of the LABORATORY REPORT RESULTS at the end of the exercise.

3. Using forceps, place a small amount of fibrin (the protein) in each test tube. An amount near the size of a pea will be sufficient. *Using a test tube holder,* put the tubes in a 40°C water bath and *carefully* shake occasionally. Maintain 40°C temperature closely. The tubes should remain in the water bath for *at least 1½ hours.*

4. Watch the changes that the fibrin undergoes. The swelling that occurs in some tubes should not be confused with digestion. Digested fibrin becomes transparent and disappears (dissolves) as the protein is digested to soluble peptides.

5. Finish the experiment when the fibrin is digested in one of the five tubes.

6. Record all your observations in the table provided in Section G.6 of the LABORATORY REPORT RESULTS at the end of the exercise.

ANSWER THE LABORATORY REPORT QUESTIONS AT THE END OF THE EXERCISE

LABORATORY REPORT RESULTS

DIGESTIVE SYSTEM 22

Student ______________________ Date ______________________

Laboratory Section ______________________ Score/Grade ______________________

SECTION D. DEGLUTITION

1. Tongue movement at beginning of deglutition: ______________________

____

2. Description of swallowing with head below pharynx: ______________________

____

3. Laryngeal movements during pharyngeal deglutition: ______________________

____

4. Time lapse between two sounds of deglutition: ______________________

____

SECTION G. CHEMISTRY OF DIGESTION

1. Digestion of Starch

a. How long did it take for the starch to be digested? ______________________

b. Did your observation indicate the presence of maltose? ______________________

c. What is the meaning of this result? ______________________

2. Effect of Temperature on Starch Digestion

d. Record the time required for starch to disappear at the temperatures tested.

________ 0°C ________ 40°C ________ Boiling

________ 10°C ________ 60°C

e. What is the optimum temperature for starch digestion? ______________________

f. What happens to amylase when it is boiled? ______________________

g. Is the effect of boiling reversible or irreversible? ______________________

h. Is the effect of freezing reversible or irreversible? ______________________

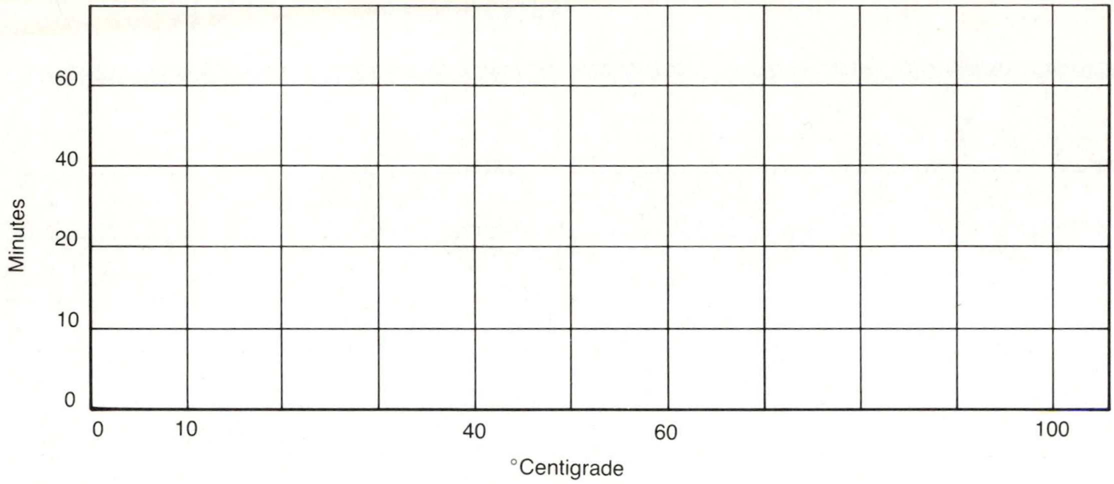

3. Effect of pH on Starch Digestion

i. Record the time required for starch to disappear at the pH readings tested.
solution A: pH 4.0 ____________
solution B: pH 7.0 ____________
solution C: pH 9.0 ____________

j. Explain your results

A: pH 4.0 ____

B: pH 7.0 ____

C: pH 9.0 ____

k. What is the optimum pH for the action of salivary amylase? ____________________

____

4. Action of Bile on Fats

l. What difference can you detect in the appearance of the mixtures in the two test tubes? ____

____

m. How does emulsification of lipids by means of bile aid in the digestion of fat? ____________

____

5. Digestion of Fats

Record the results of the digestion of fats by pancreatic juice in the following table.

Tube no.	Temperature of water bath	Change in pH (color)	Explanation of results
1			
2			

6. Digestion of Protein

Record the results of the digestion of protein by pepsin in the following table.

Tube no.	Tube contents	pH	Digestion observed (yes or no)	Explanation of results
1				
2				
3				
4				
5				

LABORATORY REPORT QUESTIONS

DIGESTIVE SYSTEM 22

Student ______________________ Date ______________________

Laboratory Section ______________________ Score/Grade ______________________

PART I. MULTIPLE CHOICE

_____ 1. If an incision has to be made in the small intestine to remove an obstruction, the first layer of tissue to be cut is the (a) muscularis (b) mucosa (c) serosa (d)submucosa

_____ 2. Mesentery, lesser omentum, and greater omentum are all directly associated with the (a) peritoneum (b) liver (c) esophagus (d) mucosa of the gastrointestinal tract

_____ 3. Chemical digestion of carbohydrates is initiated in the (a) stomach (b) small intestine (c) mouth (d) large intestine

_____ 4. A tumor of the villi and circular folds would interfere most directly with the body's ability to carry on (a) absorption (b) deglutition (c) mastication (d) peristalsis

_____ 5. The main chemical activity of the stomach is to begin the digestion of (a) fats (b) proteins (c) carbohydrates (d) all of the above

_____ 6. Surgical cutting of the lingual frenulum would occur in which part of the body? (a) salivary glands (b) esophagus (c) nasal cavity (d) tongue

_____ 7. To free the small intestine from the posterior abdominal wall, which of the following would have to be cut? (a) mesocolon (b) mesentery (c) lesser omentum (d) falciform ligament

_____ 8. The cells of gastric glands that produce secretions directly involved in chemical digestion are the (a) mucous (b) parietal (c) zymogenic (d) pancreatic islets (islets of Langerhans)

_____ 9. An obstruction in the hepatopancreatic ampulla (ampulla of Vater) would affect the ability to transport (a) bile and pancreatic juice (b) gastric juice (c) salivary amylase (d) succus entericus

_____ 10. The terminal portion of the small intestine is known as the (a) duodenum (b) ileum (c) jejunum (d) pyloric sphincter (valve)

_____ 11. The portion of the large intestine closest to the liver is the (a) right colic flexure (b) rectum (c) sigmoid colon (d) left colic flexure

_____ 12. The lamina propria is found in which coat? (a) serosa (b) muscularis (c) submucosa (d) mucosa

_____ 13. Which structure attaches the liver to the anterior abdominal wall and diaphragm? (a) lesser omentum (b) greater omentum (c) mesocolon (d) falciform ligament

_____ 14. The opening between the oral cavity and pharynx is called (a) vermilion border (b) fauces (c) vestibule (d) lingual frenulum

_____ 15. All of the following are parts of a tooth *except* the (a) crown (b) root (c) cervix (d) papilla

_____ 16. Cells of the liver that destroy worn-out white and red blood cells and bacteria are termed (a) hepatic cells (b) stellate reticuloendothelial (Kupffer) cells (c) alpha cells (d) beta cells

_____ 17. Bile is manufactured by which cells? (a) alpha (b) beta (c) hepatic (d) stellate reticuloendothelial (Kupffer)

_____ 18. Which part of the small intestine secretes the intestinal digestive enzymes? (a) intestinal glands (b) duodenal (Brunner's) glands (c) lacteals (d) microvilli

_____ 19. Structures that give the colon a puckered appearance are called (a) taenia coli (b) villi (c) rugae (d) haustrae

PART II. COMPLETION

20. An acute inflammation of the serous membrane lining the abdominal cavity and covering the abdominal viscera is referred to as _____________________ .

21. The _____________________ is a sphincter (valve) between the ileum and large intestine.

22. The portion of the small intestine that is attached to the stomach is the _____________________ .

23. The portion of the stomach closest to the esophagus is the _____________________ .

24. The _____________________ forms the floor of the oral cavity and is composed of skeletal muscle covered with mucous membrane.

25. The convex lateral border of the stomach is called the _____________________ .

26. The three special structures found in the wall of the small intestine that increase its efficiency in absorbing nutrients are the villi, circular folds, and _____________________ .

27. The three pairs of salivary glands are the parotids, submandibulars, and _____________________ .

28. The small intestine is divided into three segments: duodenum, ileum, and _____________________ .

29. The large intestine is divided into four main regions: the cecum, colon, rectum, and _____________________ .

30. The enzyme that is present in saliva is called _____________________ .

31. The transition of the lips where the outer skin and inner mucous membrane meet is called the _____________________ .

32. The _____________________ papillae are arranged in the form of an inverted V on the posterior surface of the tongue.

33. The portion of a tooth containing blood vessels, lymphatics, and nerves is the _____________________ .

34. The teeth present in permanent dentition, but not in deciduous dentition, that replace the deciduous molars are the _____________________ .

35. The portion of the gastrointestinal tract that conveys food from the pharynx to the stomach is the _____________________ .

36. The inferior region of the stomach connected to the small intestine is the _____________________ .

37. The clusters of cells in the pancreas that secrete digestive enzyme are called _____________________ .

38. The caudate lobe, quadrate lobe, and central vein are all associated with the ________________ .

39. The common bile duct is formed by the union of the common hepatic duct and ________________ duct.

40. The pear-shaped sac that stores bile is the ________________ .

41. ________________ glands of the small intestine secrete an alkaline substance to protect the mucosa from excess acid.

42. The ________________ attaches the large intestine to the posterior abdominal wall.

43. The ________________ is the last 20 cm (7 to 8 in.) of the gastrointestinal tract.

44. A midline fold of mucous membrane that attaches the inner surface of each lip to its corresponding gum is the ________________ .

45. The bonelike substance that gives teeth their basic shape is called ________________ .

46. The ________________ anchors teeth in position and helps to dissipate chewing forces.

47. The portion of the colon that terminates at the rectum is the ________________ colon.

48. The palatine tonsils are between the palatoglossal and ________________ arches.

49. The teeth closest to the midline are the ________________ .

50. The vermiform appendix is attached to the ________________ .

PART III. MATCHING

____ 51. Lipase

____ 52. Benedict's solution

____ 53. Lugol's solution

____ 54. Salivary amylase

____ 55. Pepsin

A. Commonly used solution in the test for starch
B. Capable of digesting starch
C. Capable of digesting protein
D. Commonly used solution for detecting reducing sugars
E. Digests fat to fatty acids and glycerol and changes the color of blue litmus to red

URINARY SYSTEM 23

The ***urinary system*** functions to keep the body in homeostasis by controlling the composition and volume of the blood. The system accomplishes these functions by removing and restoring selected amounts of water and various solutes. The kidneys also excrete selected amounts of various wastes, assume a role in erythropoiesis by forming renal erythropoietic factor, help control blood pH, help regulate blood pressure by secreting renin (which activates the renin-angiotensin pathway), and participate in the activation of vitamin D.

The urinary system consists of two kidneys, two ureters, one urinary bladder, and a single urethra (see Figure 23.1). Other systems that help in waste elimination are the respiratory, integumentary, and digestive systems.

Using your textbook, charts, or models for reference, label Figure 23.1.

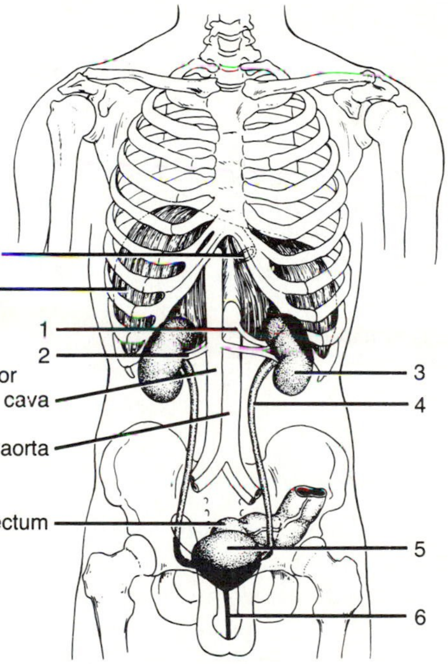

___ Kidney
___ Renal artery
___ Renal vein
___ Ureter
___ Urethra
___ Urinary bladder

FIGURE 23.1 Organs of male urinary system and associated structures.

A. ORGANS OF URINARY SYSTEM

1. Kidneys

The paired ***kidneys***, which resemble kidney beans in shape, are found just above the waist between the parietal peritoneum and the posterior wall of the abdomen. Because they are external to the peritoneal lining of the abdominal cavity, they are referred to as ***retroperitoneal*** (re′-trō-per-i-tō-NĒ-al). The kidneys are positioned between T12 and L3, with the right kidney slightly lower than the left because of the position of the liver. Near the center of the kidney's concave border, which faces the vertebral column, is a notch called the ***hilus***, through which the ureter leaves the kidney and blood vessels and nerves enter and exit the kidney. The hilus is the entrance to a cavity in the kidney called the ***renal sinus***. Three layers of tissue surround each kidney: the ***renal capsule***, ***adipose capsule***, and ***renal fascia***. They function to protect the kidney and hold it firmly in place.

If a coronal (frontal) section is made through a kidney, the following structures can be seen:

a. ***Cortex***—Outer, narrow, reddish area.

b. ***Medulla***—Inner, wide, reddish-brown area.

c. ***Renal (medullary) pyramids***—Striated triangular structures, 5 to 14 in number, in the medulla. The bases of the pyramids face the cortex, and the apices, called ***renal papillae***, are directed toward the center of the kidney.

d. ***Renal columns***—Cortical substance between renal pyramids.

e. ***Renal pelvis***—Large cavity in the renal sinus, enlarged proximal portion of ureter.

f. ***Major calyces*** (KĀ-li-sēz)—Consist of 2 or 3 cuplike extensions of the renal pelvis.

g. ***Minor calyces***—Consist of 8 to 18 cuplike extensions of the major calyces.

Within the cortex and renal pyramids of each kidney are more than 1,000,000 microscopic units called nephrons, the functional units of the kidneys. As a result of their activity in regulating the volume and chemistry of the blood, they produce urine. Urine passes from the nephrons to the minor calyces, major calyces, renal pelvis, ureter, urinary bladder, and urethra.

Examine a specimen, model, or chart of the kidney and with the aid of your textbook, label Figure 23.2.

2. Nephrons

Basically, a ***nephron*** (NEF-ron) is a renal tubule and a tuft (knot) of capillaries called a glomerulus plus its vascular component. A nephron begins as a double-walled epithelial cup, called a ***glomerular (Bowman's) capsule***, lying in the cortex of the kidney. The inner wall of the capsule, the ***visceral layer***, consists of epithelial cells called ***podocytes*** and surrounds a capillary network called the ***glomerulus*** (glō-MER-yoo-lus). A space separates the visceral layer from the outer wall of the capsule, the ***parietal layer***, which is composed of simple squamous epithelium. Together the glomerular capsule and enclosed glomerulus constitute the ***renal corpuscle*** (KOR-pus-sul).

The visceral layer of the glomerular (Bowman's) capsule and endothelium of the glomerulus form an ***endothelial-capsular membrane***, a very effective filter. Electron microscopy has determined that the membrane consists of the following components, given in the order in which substances are filtered (Figure 23.3).

a. ***Endothelium of glomerulus***—Single layer of endothelium with pores averaging 50 to 100 nm in diameter. It restricts the passage of blood cells.

b. ***Basement membrane of glomerulus***—Extracellular membrane that consists of fibrils in glycoprotein matrix. It restricts the passage of large sized proteins.

c. ***Epithelium of visceral layer of the glomerular (Bowman's) capsule***—Epithelial cells are called ***podocytes*** and consist of footlike structures called ***pedicels*** (PED-i-sels) arranged parallel to the circumference of the glomerulus. Pedicels cover the basement membrane of the glomerulus, except for spaces between them called ***filtration slits (slit pores)***. Pedicels contain numerous, thin contractile filaments that are believed to regulate the passage of substances through the filtration slits. In addition, another factor that helps to regulate the passage of substances through the filtration slits is a thin membrane, the ***slit membrane***, that extends between filtration slits. This membrane restricts the passage of intermediate size proteins.

The endothelial-capsular membrane filters blood passing through the kidney. Blood cells and large molecules, such as proteins, are retained by the filter and eventually are recycled into the blood. The filtered substances pass through the membrane and into the space between the parietal and visceral layers of the glomerular (Bowman's) capsule, and then enter the next component of a nephron, the proximal convoluted tubule.

As the filtered fluid (filtrate) passes through the remaining parts of a nephron, substances are selectively added and removed. The end product of these activities is urine. Nephrons are frequently classified into two kinds. A ***cortical nephron*** usually has its glomerulus in the outer cortical zone, and the remainder of the nephron rarely penetrates the medulla. A ***juxtamedullary nephron*** usually has its glomerulus close to the corticomedullary junction, and other parts of the nephron penetrate deeply into the medulla (see Figure 23.4). The following description of the remaining compo-

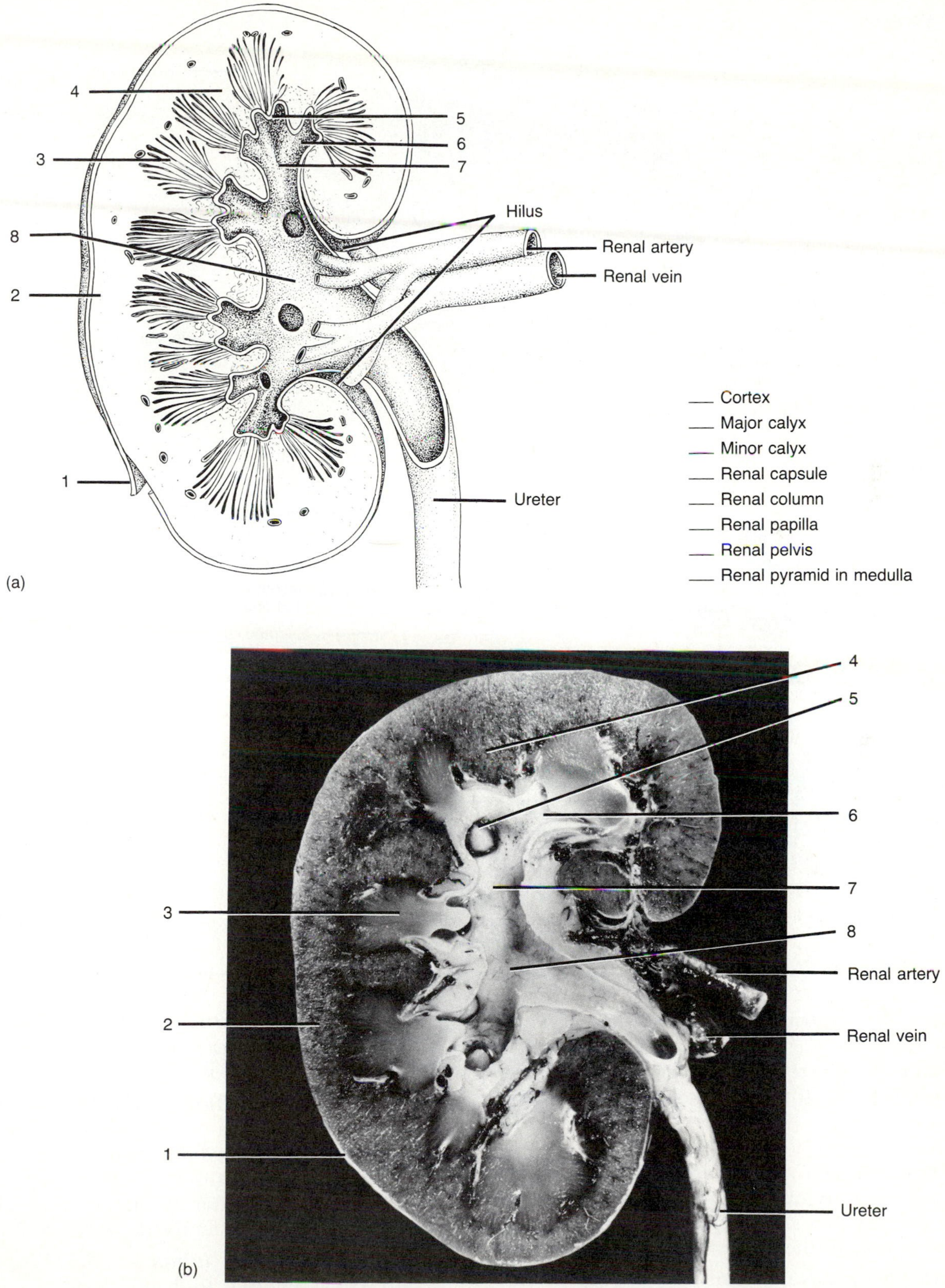

FIGURE 23.2 Coronal section through the right kidney. (a) Diagram. (b) Photograph.

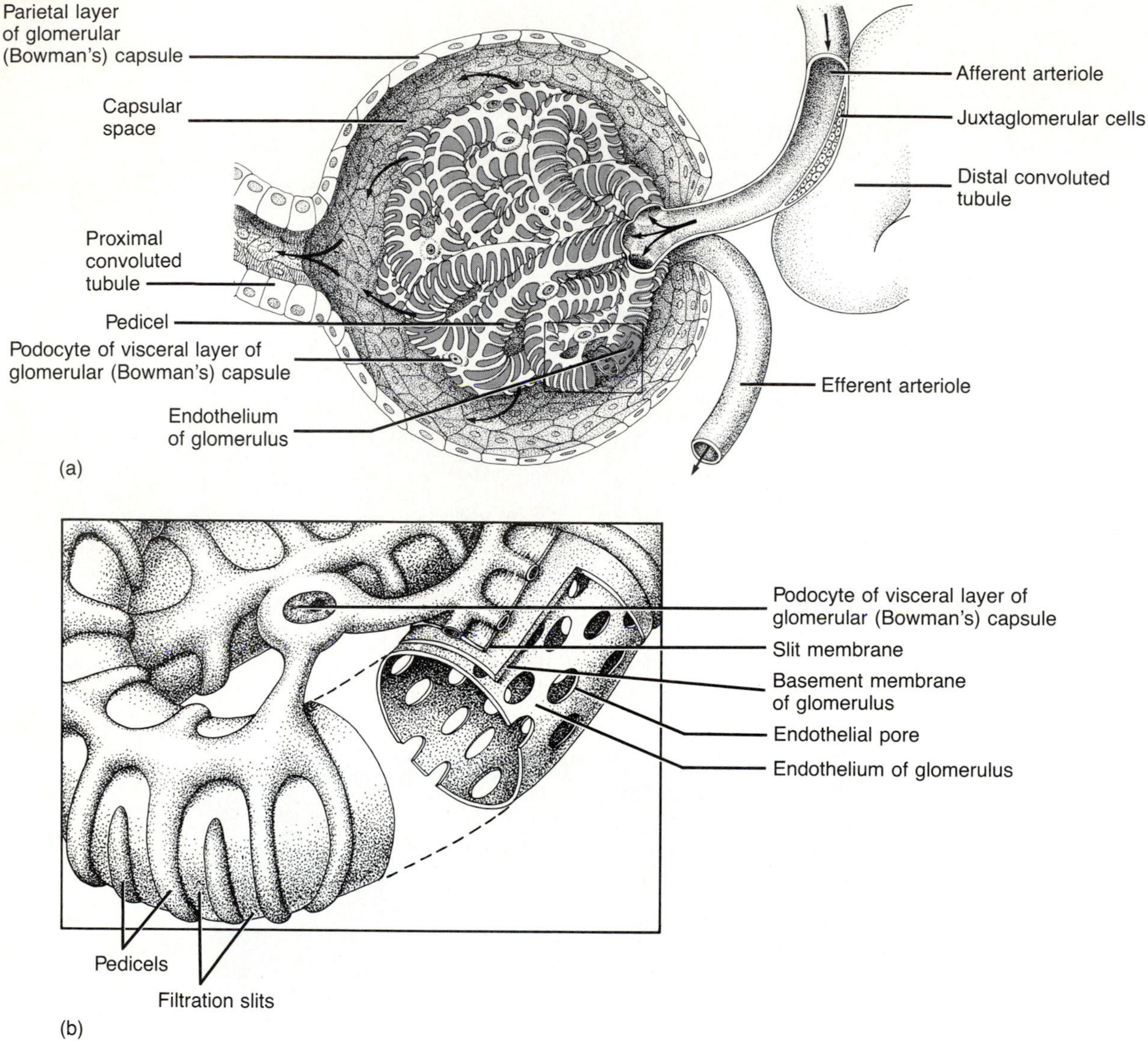

FIGURE 23.3 Endothelial-capsular membrane. (a) Components of renal corpuscle. (b) Details of endothelial-capsular membrane.

nents of a nephron applies to juxtamedullary nephron.

After the filtrate leaves the glomerular (Bowman's) capsule, it passes through the following structures:

a. ***Proximal convoluted tubule***—Coiled tubule in the cortex that originates at the glomerular (Bowman's) capsule; consists of simple cuboidal epithelium with microvilli.

b. ***Descending (thin) limb of the loop of the nephron***—Extension of the proximal convoluted tubule that dips down into medulla; consists of simple squamous epithelium.

c. ***Loop of the nephron (loop of Henle)***—U-shaped tubule in medulla connecting descending and ascending limbs.

d. ***Ascending (thick) limb of the loop of the nephron***—Extension of loop of the nephron that ascends in medulla and approaches cortex; consists of simple cuboidal epithelium, simple columnar epithelium, or both; wider in diameter than the descending limb.

e. ***Distal convoluted tubule***—Coiled extension of ascending limb in cortex; consists of simple cuboidal epithelium with fewer microvilli than in proximal convoluted tubule.

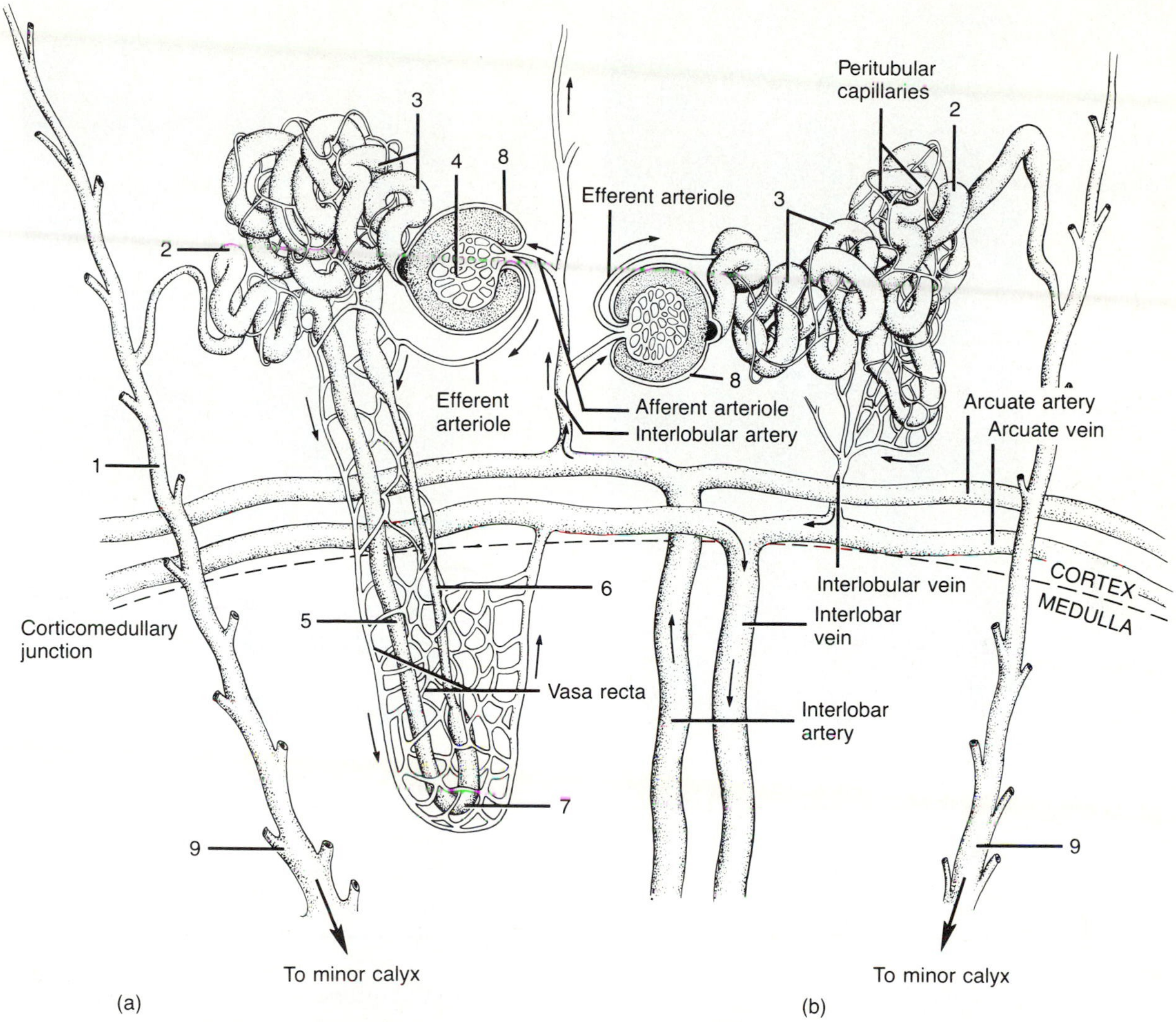

___ Ascending (thick) limb of the loop of the nephron
___ Collecting duct
___ Descending (thin) limb of the loop of the nephron
___ Distal convoluted tubule
___ Glomerular (Bowman's) capsule
___ Glomerulus
___ Loop of the nephron (loop of Henle)
___ Papillary duct
___ Proximal convoluted tubule

FIGURE 23.4 Nephrons. (a) Juxtamedullary nephron. (b) Cortical nephron.

Distal convoluted tubules terminate by merging with straight ***collecting ducts***. In the medulla, collecting ducts receive distal convoluted tubules from several nephrons, pass through the renal pyramids, and open at the renal papillae into minor calyces through about 30 large ***papillary ducts***. The processed filtrate, called urine, passes from the collecting ducts to papillary ducts, minor calyces, major calyces, renal pelvis, ureter, urinary bladder, and urethra.

With the aid of your textbook, label the parts of a nephron and associated structures in Figure 23.4.

Examine prepared slides of various components of nephrons and compare your observations to Figure 23.5.

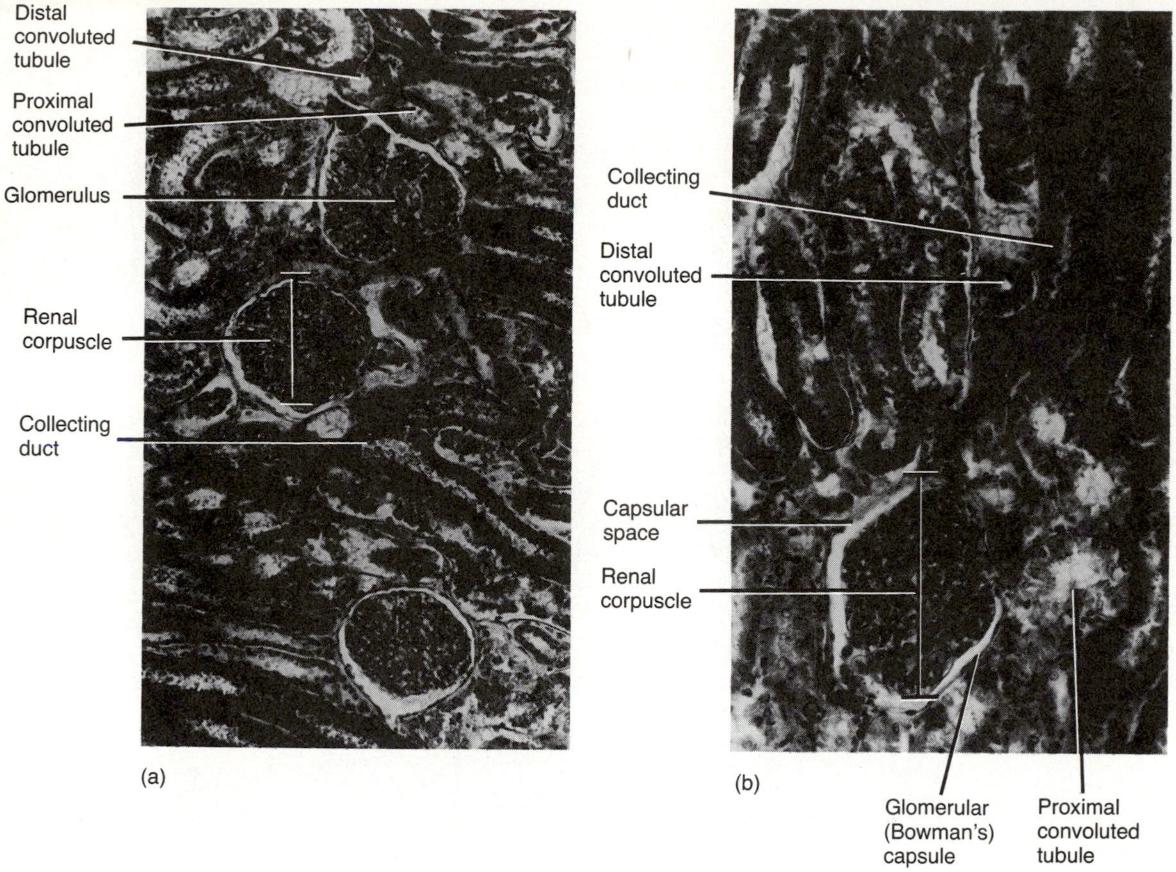

FIGURE 23.5 Nephrons. (a) and (b) Microscopic appearance of components.

3. Blood and Nerve Supply

Nephrons are abundantly supplied with blood vessels, and the kidneys actually receive around one-fourth the total cardiac output, or approximately 1200 ml, every minute. The blood supply originates in each kidney with the ***renal artery***, which divides into many branches, eventually supplying the nephron and its complete tubule.

Before or immediately after entering the hilus, the renal artery divides into several branches, the ***interlobar arteries***, which pass between the renal pyramids in the renal columns. At the bases of the pyramids, the interlobar arteries arch between the medulla and cortex and here are known as ***arcuate arteries***. Branches of the arcuate arteries, called ***interlobular arteries***, enter the cortex. ***Afferent arterioles***, branches of the interlobular arteries, are distributed to the ***glomeruli***. Blood leaves the glomeruli via ***efferent arterioles***.

The next sequence of blood vessels depends on the type of nephron. Around convoluted tubules, efferent arterioles of cortical nephrons divide to form capillary networks called ***peritubular capillaries***. Efferent arterioles of juxtamedullary nephrons also form peritubular capillaries and, in addition, form long loops of blood vessels around medullary structures called ***vasa recta***. Peritubular capillaries eventually reunite to form ***interlobular veins***. Blood then drains into ***arcuate veins*** and ***interlobar veins***. Blood leaves the kidney through the ***renal vein*** that exits at the hilus. (The vasa recta pass blood into the interlobular veins, arcuate veins, interlobar veins, and renal veins.)

As the afferent arteriole approaches the renal corpuscle, the smooth muscle fibers (cells) of the tunica media become more rounded and granular and are known as ***juxtaglomerular cells***. The cells of the distal convoluted tubule adjacent to the afferent and efferent arterioles become narrower and taller, and are known as the ***macula densa***. Together the juxtaglomerular cells and macula densa constitute the ***juxtaglomerular apparatus (JGA)***. The juxtaglomerular apparatus responds to low renal blood pressure by secreting a substance (renin) that begins a sequence of responses that raises renal pressure back to normal.

Using Figures 23.4 and 23.6 as guides, trace a drop of blood from its entrance into the renal artery to its exit through the renal vein. As you do so, name in sequence each blood vessel through which blood passes for both cortical and juxtaglomerular nephrons.

Label Figure 23.7.

The nerve supply to the kidneys comes from the ***renal plexus*** of the autonomic system. The nerves are vasomotor because they regulate the circulation of blood in the kidney by regulating the diameters of the small blood vessels.

4. Ureters

The body has two retroperitoneal ***ureters*** (YOO-re-ters), one for each kidney; each ureter is a continuation of the renal pelvis and runs to the urinary bladder (see Figure 23.1). Urine is carried through the ureters mostly by peristaltic contractions of the muscular layer of the ureters. Each ureter extends 25 to 30 cm (10 to 12 in.) and enters the urinary bladder at the superior lateral angle of its base.

Histologically, the ureters consist of an inner ***mucosa*** of transitional epithelium and connective tissue, a middle ***muscularis*** (inner longitudinal and outer circular smooth muscle), and an outer ***fibrous coat***.

Examine a prepared slide of the wall of the ureter showing its various layers. With the aid of your textbook, label Figure 23.8.

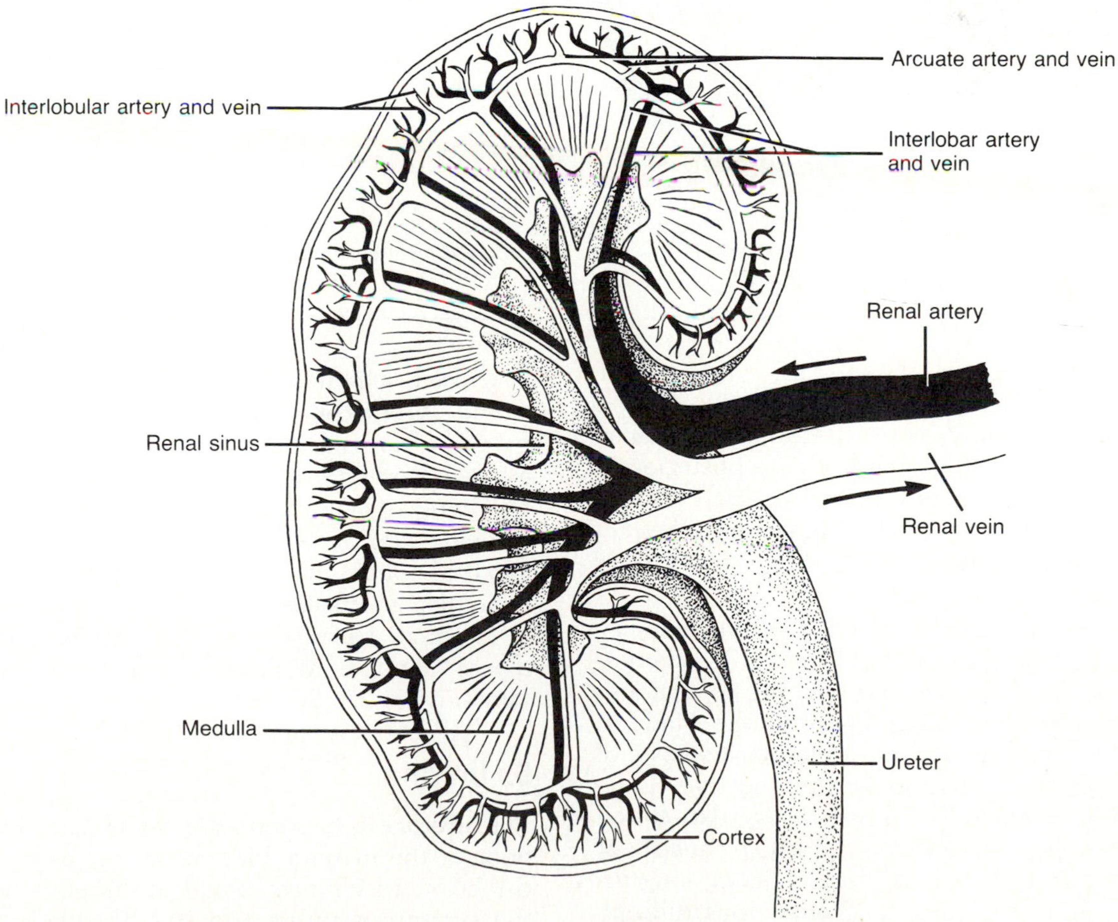

FIGURE 23.6 Macroscopic blood vessels of the kidney. Macroscopic and microscopic blood vessels are shown in Figure 23.4.

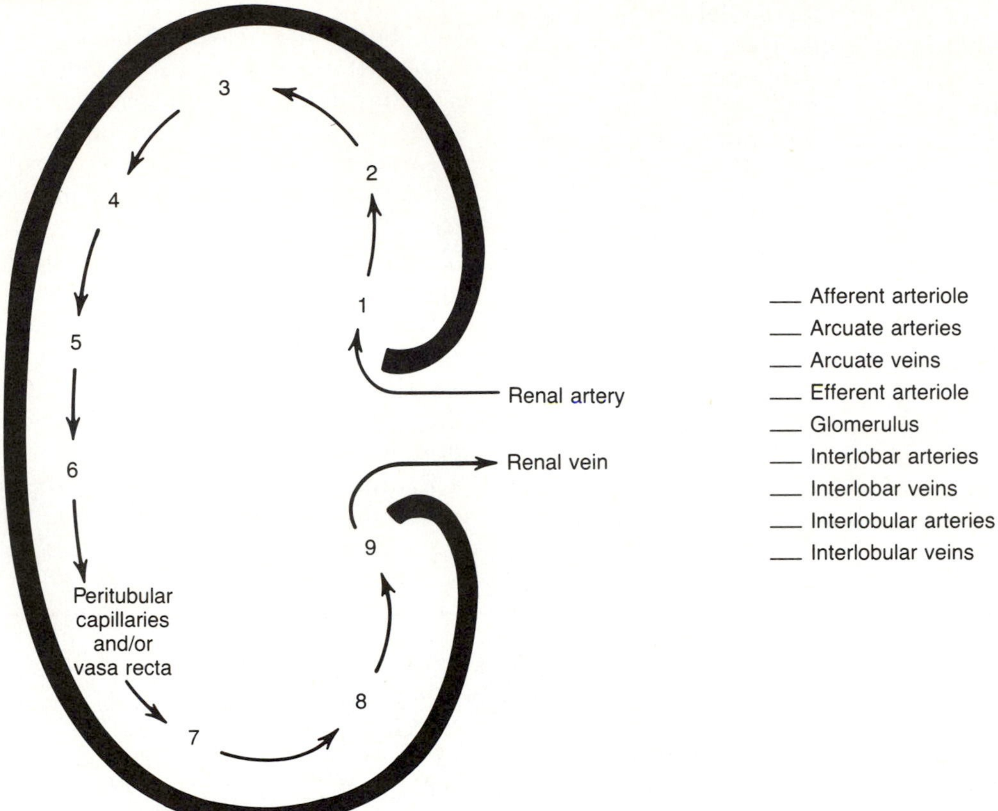

FIGURE 23.7 Blood supply of the right kidney. This view is designed to show the *sequence* of blood flow, not the anatomical location of blood vessels, which is shown in Figure 23.6.

5. Urinary Bladder

The ***urinary bladder*** is a hollow muscular organ located in the pelvic cavity posterior to the symphysis pubis (see Figure 23.1). In the male, the bladder is directly anterior to the rectum; in the female, it is anterior to the vagina and interior to the uterus.

At the base of the interior of the urinary bladder is the ***trigone*** (TRĪ-gōn), a triangular area bounded by the opening to the urethra and ureteral openings into the bladder. The ***mucosa*** of the urinary bladder consists of transitional epithelium and connective tissue that form ***rugae***. The ***muscularis***, also called the ***detrusor*** (de-TROO-ser) ***muscle***, consists of three layers of smooth muscle: inner longitudinal, middle circular, and outer longitudinal. In the region around the opening to the urethra, the circular muscle fibers form an ***internal sphincter***. Below this is the ***external sphincter*** composed of skeletal muscle. The ***serosa*** is formed by visceral peritoneum and covers the superior surface of the urinary bladder.

Urine is expelled from the bladder by an act called ***micturition*** (mik'-too-RISH-un), commonly known as urination or voiding. The average capacity of the urinary bladder is 700 to 800 ml.

Using your textbook as a guide, label the external sphincter, internal sphincter, ureteral openings, and ureters in Figure 23.9.

Examine a prepared slide of the wall of the urinary bladder. With the aid of your textbook, label Figure 23.10.

6. Urethra

The ***urethra*** is a small tube leading from the floor of the urinary bladder to the exterior of the body. In females, this tube is posterior to the symphysis pubis and is embedded in the anterior wall of the vagina; its length is approximately 3.8 cm ($1\frac{1}{2}$ in.). The opening of the urethra to the exterior, the ***urethral orifice***, is between the clitoris and vaginal orifice. In males, its length is around 20 cm (8 in.), and

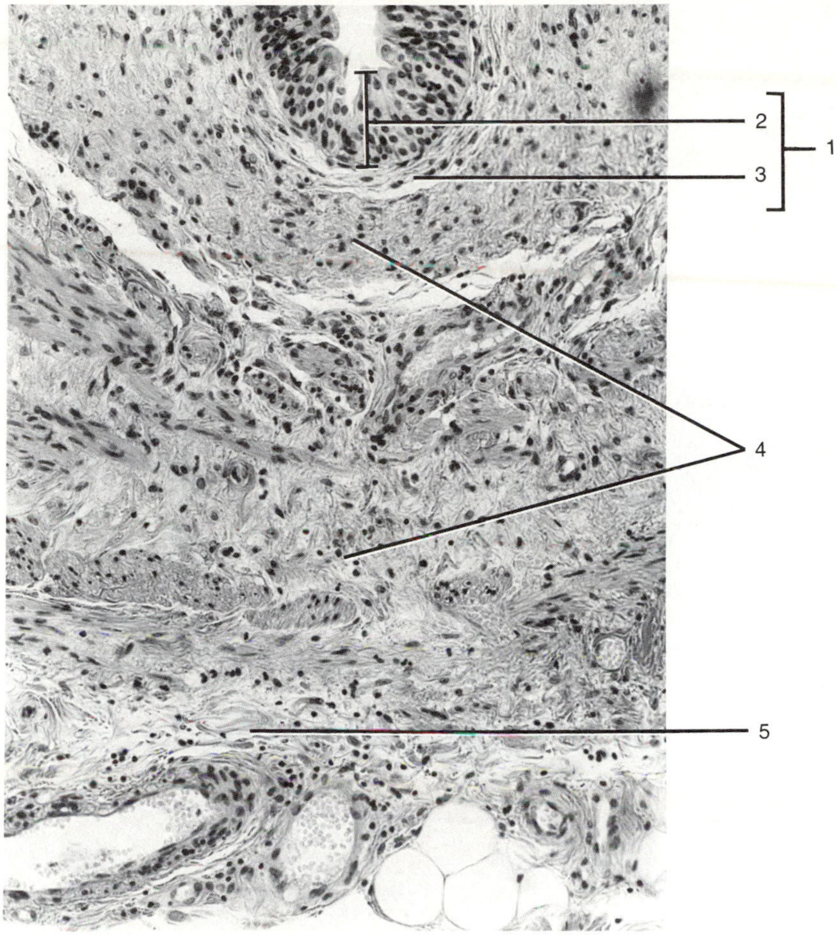

___ Connective tissue of mucosa
___ Fibrous coat
___ Mucosa
___ Muscularis
___ Transitional epithelium of mucosa

FIGURE 23.8 Histology of ureter.

it follows a different route than that of the female. Immediately below the urinary bladder, the urethra passes through the prostate gland, pierces the urogenital diaphragm, and traverses the penis. The urethra is the terminal portion of the urinary system, and serves as the passageway for discharging urine from the body. In addition, in the male, the urethra serves as the duct through which reproductive fluid (semen) is discharged from the body.

Label the urethra and urethral orifice in Figure 23.9.

B. DISSECTION OF CAT URINARY SYSTEM

PROCEDURE

CAUTION! *Please reread Section D. "Precautions Related to Dissection" at the beginning of the laboratory manual on page xii before you begin your dissection.*

1. Kidneys

1. The ***kidneys*** are situated on either side of the vertebral column at about the level of the third to fifth lumbar vertebrae. The right kidney is slightly higher than the left (Figure 23.11).

2. Each is surrounded by a mass of fat called the ***adipose capsule***, which should be removed.

3. Unlike other abdominal organs that are suspended by an extension of the peritoneum, the kidneys are covered by parietal peritoneum only on their ventral surfaces. Because of this, the kidneys are said to be ***retroperitoneal***.

4. Identify the ***adrenal (suprarenal) glands*** located cranial and medial to the kidneys. They are part of the endocrine system.

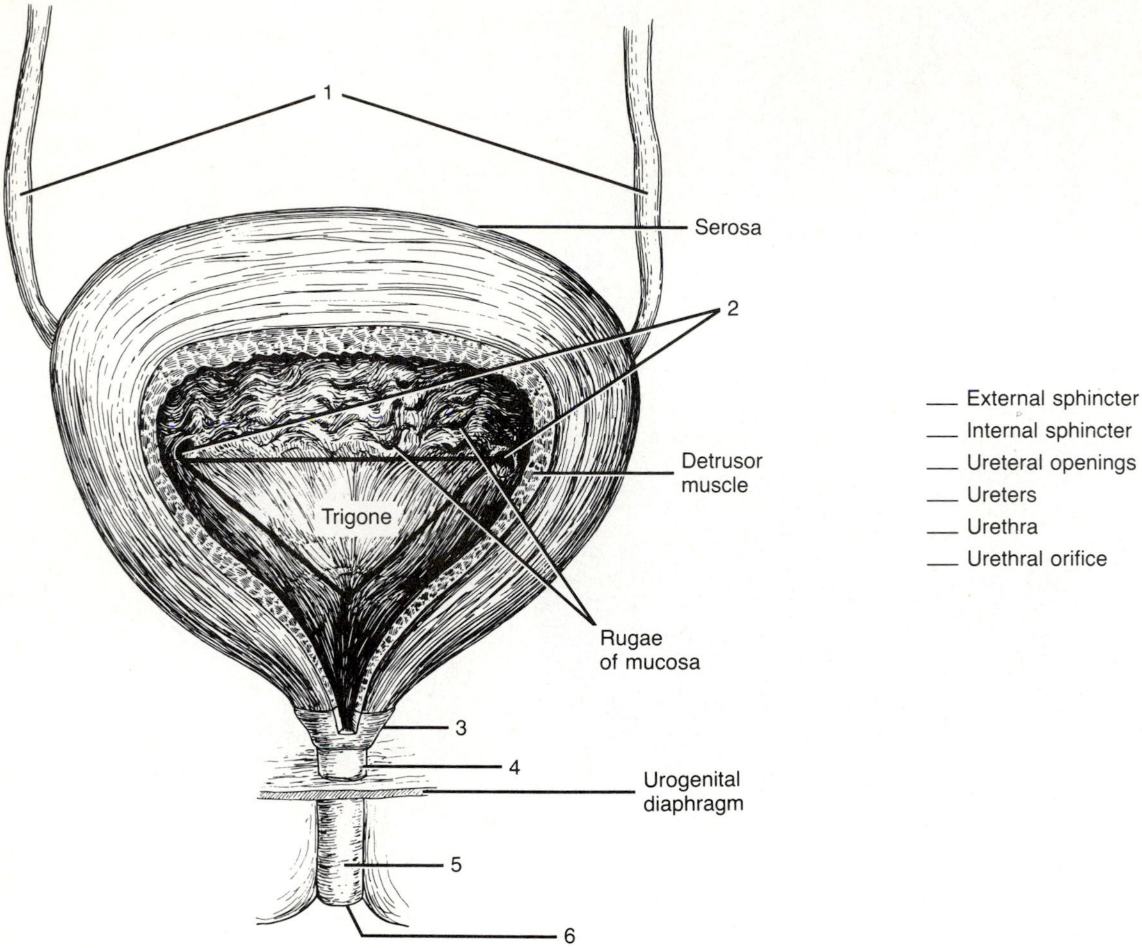

FIGURE 23.9 Urinary bladder and female urethra.

5. Note that the medial surface of each kidney contains a concave opening, the ***hilus***, through which blood vessels and the ureter enter or leave the kidney.

6. Identify the ***renal artery*** branching off the aorta and entering the hilus and the ***renal vein*** leaving the hilus and joining the inferior vena cava. Also, identify the ureter caudal to the renal vein.

7. If a detailed dissection of the cat kidney is to be done, follow the directions of the sheep kidney dissection and identify the following structures of the cat that correlate with the sheep kidney.

   a. Hilus
   b. Renal pelvis
   c. Calyces
   d. Renal cortex
   e. Renal (medullary) pyramids
   f. Renal column
   g. Renal papillae

2. Ureters

1. The ***ureters***, like the kidneys, are retroperitoneal.

2. Each begins as the renal pelvis in the renal sinus and passes caudally to the urinary bladder.

3. The ureters pass behind the urinary bladder and open into its floor.

3. Urinary Bladder

1. The ***urinary bladder***, a pear-shaped musculomembranous sac located just cranial to

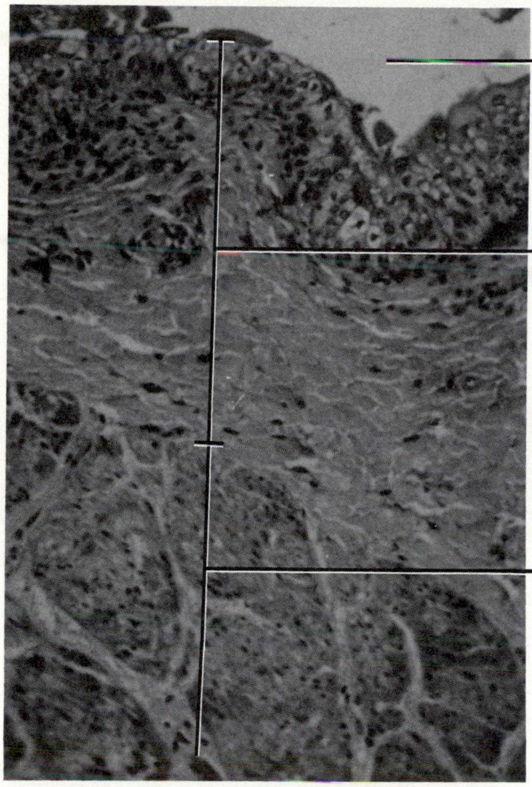

FIGURE 23.10 Histology of urinary bladder.

the symphysis pubis (Figure 23.11), also is retroperitoneal.

2. The urinary bladder is attached to the abdominal wall by peritoneal folds termed ***ligaments***. The ***ventral suspensory ligament*** extends from the ventral side of the bladder to the ***linea alba***. The ***lateral ligaments***, one on either side, connect the sides of the bladder to the dorsal body wall. They contain considerable fat.

3. The broad, rounded, cranial portion of the bladder is called the ***fundus***.

4. The narrow, caudal, attached portion is referred to as the ***neck***.

5. If you have a male cat, identify the ***rectovesical pouch***, the space between the bladder and the rectum.

6. If you have a female cat, identify the ***vesicouterine pouch***, the space between the bladder and the uterus.

7. If you cut open the bladder, you might be able to see the openings of the ureters into the bladder.

4. Urethra

1. The ***urethra*** is a duct which conducts urine from the neck of the bladder to the exterior.

2. Dissection of the urethra will be delayed until the reproductive system is studied.

C. DISSECTION OF SHEEP (OR PORK) KIDNEY

The sheep kidney is very similar to both human and cat kidney. You may use Figure 23.2 as a reference for this dissection.

CAUTION! *Please reread Section D. "Precautions Related to Dissection" at the beginning of the laboratory manual on page xii before you begin your dissection.*

PROCEDURE

1. Examine the intact kidney and notice the hilus and fatty tissue that normally surrounds the kidney. Strip away the fat.

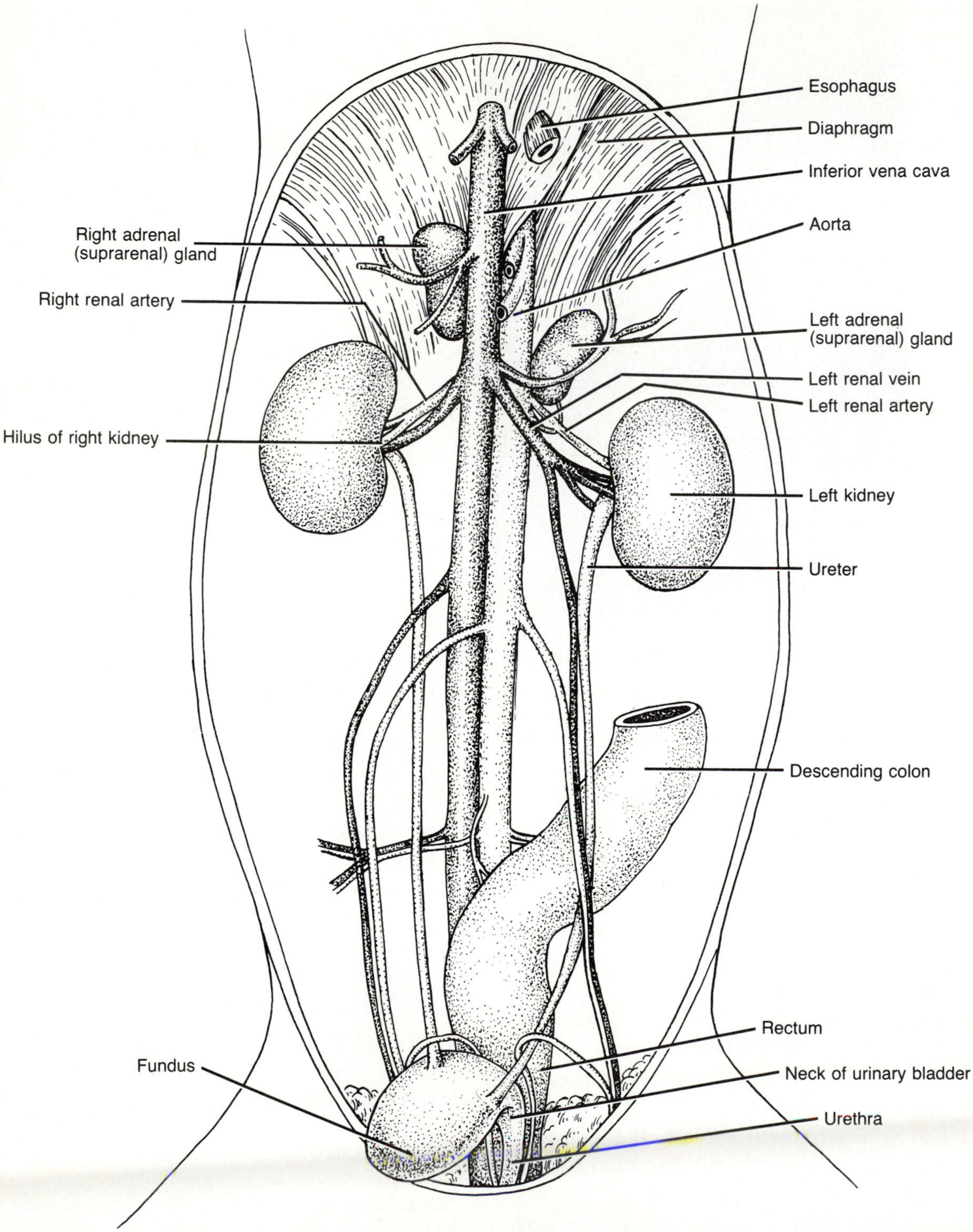

FIGURE 23.11 Urinary system of male cat.

2. As you peel the fat off, look carefully for the ***adrenal suprarenal gland***. This glad is usually found attached to the superior surface of the kidney, as it is in the human. Most preserved kidneys do not have this gland. If it is present, remove it, cut it in half, and note its distinct outer ***cortex*** and inner ***medulla***.

3. Look at the ***hilus***, which is the concave area of the kidney. From here the ***ureter***, ***renal artery***, and ***renal vein*** enter and exit.

4. Differentiate these blood vessels by examining the thickness of their walls. Which vessel has the thicker wall?

5. With a sharp scalpel *carefully* make a longitudinal (coronal) section through the kidney.

6. Identify the ***renal capsule*** as a thin, tough layer of connective tissue completely surrounding the kidney.

7. Immediately beneath this capsule is an outer light-colored area called the ***renal cortex***. The inner dark-colored area is the ***renal medulla***.

8. The ***renal pelvis*** is the large chamber formed by the expansion of the ureter inside the kidney. This renal pelvis divides into many smaller areas called ***renal calyces***, each of which has a dark tuft of kidney tissue called a ***renal pyramid***.

9. The bases of these pyramids face the cortical area. Their apices, called ***renal papillae***, are directed toward the center of the kidney.

10. The calyces collect urine from collecting ducts and drain it into the renal pelvis and out through the ureter.

11. The renal artery divides into several branches that pass between the renal pyramids. These vessels are small and delicate and may be too difficult to dissect and trace through the renal medulla.

D. TUBULAR SECRETION

The glomeruli of the kidneys form a filtrate through a simple filtration process. The material thus formed will pass through the remainder of the nephron and will be modified by several processes involving both active transport and passive diffusion. Movement of a substance from the tubular lumen into the peritubular capillaries is termed ***tubular reabsorption***. This process can occur via either passive diffusion or active transport. Movement of a substance from the peritubular capillaries through the tubular cells and into the tubular lumen is termed ***tubular secretion***. Like reabsorption, this process can occur via either passive diffusion or active transport. Typically materials secreted by the nephron are compounds that are incompletely metabolized or not metabolized at all, weak acids or bases, or a substance foreign to the body. Secretion may occur in both the proximal and distal convoluted tubules.

A very simple technique to demonstrate tubular secretion utilizes isolated goldfish renal tubules. The fish kidney is ideally suited because the lack of supportive connective tissue between the tubules allows easy and rapid dissection of single tubules. Tubules suspended in Ringer's solution remain viable for several hours and readily demonstrate the secretion of several substances.

1. Isolation of Renal Tubules

CAUTION! *Please reread Section C. "Precautions Related to Working with Reagents" and Section D. "Precautions Related to Dissection" on page xii at the beginning of the laboratory manual before you begin any of the following experiments. You should also read the experiments before you perform them to be sure that you understand all the procedures and safety precautions.*

PROCEDURE

1. After decapitating the fish, make a midline ventral incision and expose the kidneys. Identification of the kidneys is aided by their location lateral to the vertebral column and by their reddish-brown color.

2. Gently remove the kidneys and place them in a dish containing Ringer's solution.

3. Utilizing dissection needles, break the kidneys into progressively smaller pieces until the renal tubules are separated. If additional separation is necessary, gently pass some of the tubules into and out of a wide bore medicine dropper.

2. Experimental Procedure

a. SECRETION OF PHENOL RED

PROCEDURE

1. Place several drops of Ringer's solution onto a depression slide and place one or two renal tubules into the depression via the dropper.
2. Observe the tubules under a microscope and become familiar with their anatomy. Be able to locate the lumen of the tubules.
3. Now place several drops of Ringer's solution containing phenol red (0.05 M) into the depression of another slide, and add one or two tubules to this solution.
4. Immediately place the slide under the microscope and determine (a) time required for the *first* appearance of phenol red within the lumen of the tubule and (b) time required for the *maximum concentration* of dye within the lumen of the tubule.
5. Record your results in Section D.1 of the LABORATORY REPORT RESULTS at the end of the exercise.

b. INHIBITION OF SECRETION

Several substances are secreted by similar or identical tubular carrier mechanisms. Therefore, if two substances are competing for the same carrier, the substance that has a higher concentration of greater affinity for the carrier will inhibit the secretion of the second substance. Such is the case with phenol red and penicillin G.

PROCEDURE

1. Place several tubules in a depression slide containing 0.05 M phenol red and 1 M penicillin G.
2. Determine the values for the time required for phenol red to first appear in the lumen of the tubule, and the time required for maximum concentration of the dye within the tubular lumen.
3. Record your results in Section D.2 of the LABORATORY REPORT RESULTS at the end of the exercise.
4. What type of inhibitory mechanism does this experiment demonstrate? ____________

____

____

____

c. METABOLIC REQUIREMENTS FOR SECRETION

Phenol red is secreted via active transport, that is, movement of a substance against a concentration gradient, utilizing a carrier and involving expenditure of energy. The energy in this process is obtained by splitting ATP into ADP and P.

PROCEDURE

1. To demonstrate that the secretion of phenol red requires energy, suspend several tubules in a solution containing 0.05 M phenol red and either 1 mM trichlorophenol or 0.2 mM 2,4 dinitrophenol.
2. Record the times required for dye transport in Section D.3 of the LABORATORY REPORT RESULTS at the end of the exercise.
3. Look up the actions of the chemical you utilized, and discuss how this would affect the transport process in the secretion of phenol red. ________________________

____

____

E. URINE

The kidneys perform their homeostatic functions of controlling the concentration and volume of blood and adjusting pH. The by-product of these functions is the fluid called ***urine***. Urine contains a high concentration of solutes. In a healthy person its volume, pH, and solute concentration vary with the needs of the internal environment. In certain pathological conditions, the characteristics of urine may change drastically. An analysis of the volume and physical and chemical properties of urine tells us much about the state of the body.

1. Physical Characteristics

Normal urine is usually a straw yellow to amber transparent liquid with a characteristic odor. Urine color varies considerably according to the ratio of solutes to water and according to the diet.

Cloudy urine sometimes reflects the secretion of mucin from the urinary tract lining and is not necessarily pathological. The pH of urine is slightly acid, ranging between 5.0 and 7.8. The pH of urine is also strongly affected by diet with a high protein diet increasing acidity and a mostly vegetable diet increasing alkalinity.

Specific gravity is the ratio of the weight of a volume of a substance to the weight of an equal volume of distilled water. Water has a specific gravity of 1.000. The specific gravity of urine depends on the amount of solids in solution and normally ranges from 1.008 to 1.030. The greater the concentration of solutes, the higher the specific gravity.

2. Abnormal Constituents

When the body's metabolism is not working efficiently, many substances not normally found in urine may appear in trace amounts. In addition, normal constituents may appear in abnormal amounts. ***Urinalysis*** is the analysis of the physical and chemical properties of urine, and is a vital tool in diagnosing pathological conditions.

a. ***Albumin—Albumin*** is normally a part of plasma, but it usually does not appear in urine because the particles are too large to pass through the pores in the capillaries. When albumin is found in the urine, the condition is called ***albuminuria***.

b. ***Glucose***—Urine normally contains such small amounts of ***glucose*** that clinically glucose is considered absent. Its presence in significant amounts is called ***glycosuria***, and the most common cause is a high blood sugar level.

c. ***Erythrocytes—Hematuria*** is the condition that occurs when red blood cells are found in the urine in appreciable amounts, and usually indicates a pathological condition.

d. ***Leucocytes—Pyuria*** is the condition that occurs when white blood cells and other components of pus are found in the urine, and usually indicates a pathological condition.

e. ***Ketone bodies***—Normal urine contains small amounts of ***ketone (acetone)*** bodies. Their appearance in large quantities in urine produces the conditions called ***ketosis (acetonuria)*** and may indicate abnormalities.

f. ***Casts—Casts*** are tiny masses of various substances that have hardened and assumed the shape of the lumens of the nephron tubules. They are microscopic and are composed of many different substances.

g. ***Calculi***—Insoluble ***calculi*** (stones) are various salts that have solidified in the urinary tract. They are found anywhere from the kidney tubules to the external opening.

F. URINALYSIS

In this exercise you will determine some of the characteristics of urine and perform tests for some abnormal constituents that may be present in urine. Some of these tests may be used in determining unknowns in urine specimens.

CAUTION! *Please reread Section A. "General Safety Precautions and Procedures" on page xi and Section C. "Precautions Related to Working with Reagents" on page xii at the beginning of the laboratory manual before you begin any of the following experiments. You should also read the experiments before you perform them to be sure that you understand all the procedures and safety precautions.*

1. Urine Collection

PROCEDURE

1. A specimen of urine may be collected at any time for routine tests; urine voided within 3 hours after meals, however, may contain abnormal constituents. For this reason the first voiding in the morning is preferred.

2. Both males and females should collect a midstream sample of urine in a sterile container. A midstream sample is essential to avoid contamination from the external genitalia, and to avoid the presence of pus cells and bacteria that are normally found in the urethra.

3. If not examined immediately, the specimen should be refrigerated to prevent unnecessary bacterial growth.

4. Before testing *always* mix urine by swirling, inverting the container, or stirring with a wooden swab stick.

5. *Keep all containers clean!*

6. Wrap all papers and sticks and put them in the garbage pail.

7. Rinse all test tubes and glass containers carefully with *cold water* after they have cooled.

8. Flush sinks well with cold water.

9. Obtain either your own freshly voided urine sample or a provided sample if one is available.

2. Chemstrip® Testing

Alternate methods can be employed for several of the tests you are about to perform. One alternate method is the use of plastic strips to which are attached paper squares impregnated with various reagents. These strips display a color reaction when dipped into urine with any abnormal constituents.

PROCEDURE

1. At this point, you should take a Chemstrip® and test your urine sample for the following: pH, protein, glucose, ketones, bilirubin, and blood (hemoglobin). Record your results below.

2. Determine which component(s) of urine are measured by the Chemstrip® you are using. If you are using a strip that tests for multiple substances, determine which squares measure which substances. Locate the color chart used to read the results. Note the appropriate time for reading each test.

3. Remove a test strip from the vial and *replace the cap*. Dip the test strip into your urine sample for no longer than 1 second, being sure that all reagents on the strip are immersed.

4. Remove any excess urine from the strip by drawing the edge of the strip along the rim of the container holding your urine sample.

5. After the appropriate time, as indicated on the Chemstrip® vial, hold the strip close to the color blocks on the vial.

6. Make sure that the strip blocks are properly lined up with the color chart on the vial.

Test	Chemstrip result
pH	______
Protein	______
Glucose	______
Ketones	______
Bilirubin	______
Blood	______

3. Physical Analysis

a. COLOR

Normal urine varies in color from straw yellow to amber because of the pigment ***urochrome***, a by-product of hemoglobin destruction. Observe the color of your urine sample. Some abnormal colors are as follows:

Color	Possible Cause
Silvery, milky	Pus, bacteria, epithelial cells
Smoky brown, rust	Blood
Orange, green, blue, red	Medications or liver disease

Record the color of your urine in Section F.3.a of the LABORATORY REPORT RESULTS at the end of the exercise.

How would sickle-cell anemia affect urine color? ______________________________

____

b. TRANSPARENCY

A fresh urine sample should be clear. Cloudy urine may be due to substances such as mucin, phosphates, urates, fat, pus, mucus, microbes, crystals, and epithelial cells.

PROCEDURE

1. To determine transparency, cover the container and shake your urine sample and observe the degree of cloudiness.

2. Record your observations in Section F.3.b of the LABORATORY REPORT RESULTS at the end of the exercise.

c. pH

The pH of urine varies with several factors already indicated. You can test the pH of your urine by using either a Chemstrip® or pH paper. Because you have already determined the pH of your urine by using a Chemstrip®, you might want to verify the results using pH paper.

PROCEDURE

1. Place a strip of pH paper into your urine sample three consecutive times.

2. Shake off any excess urine.

3. Let the pH paper sit for 1 minute and then compare it to the color chart provided.

4. Record your observations in Section F.3.c. of the LABORATORY REPORT RESULTS at the end of the exercise.

How would the consumption of antacid (sodium bicarbonate) affect urine pH? ________

____

____

d. SPECIFIC GRAVITY

The specific gravity is easily determined using a urinometer (hydrometer). The urinometer is a float with a numbered scale near the top that indicates specific gravity directly.

PROCEDURE

1. Familiarize yourself with the scale on the urinometer neck. Determine the change in specific gravity represented by each calibration.

2. Allow your urine to reach room temperature. Urinometers are calibrated to read the specific gravity at 15°C (69°F). If the temperature differs, add or subtract 0.001 for each 3°C above or below 15°C.

3. Fill the cylinder 3/4 full of urine and insert the urinometer. Make sure that it is free floating; if not, spin the neck gently.

4. Read the scale at the bottom of the meniscus when the urinometer is at rest.

5. Record the specific gravity in Section F.3.d of the LABORATORY REPORT RESULTS at the end of the exercise.

6. Rinse the urinometer and cylinder. *Follow your instructor's directions for cleaning them.*

How would dehydration affect the specific gravity of urine? ________

____

____

4. Chemical Analysis

a. CHLORIDE AND SODIUM CHLORIDE

Most of the sodium chloride (NaCl) present in the renal filtrate is reabsorbed; a small amount remains as a normal component of urine. The normal value for chloride (Cl^-) is 476 mg/100 ml; for sodium (Na^+) it is 294 mg/100 ml.

PROCEDURE

1. Using a medicine dropper, place 10 drops of urine into a Pyrex test tube.

2. Using a medicine dropper, add 1 drop of 20% potassium chromate and *gently* agitate the tube. It should be a yellow color.

3. Using a medicine dropper, add 2.9% silver nitrate solution *one drop at a time*, counting the drops and *gently* agitating the test tube during the time the solution is being added.

4. Count the number of drops needed to change the color of the solution from bright yellow to brown.

5. Determine the chloride and sodium chloride concentrations of the sample and record your results in Section F.4.a of the LABORATORY REPORT RESULTS at the end of the exercise. Each drop of silver nitrate added in step 4 is equivalent to: (1) 61 mg of Cl^- per 100 ml of urine, and (2) 100 mg of NaCl per 100 ml of urine.

Thus, to determine the chloride concentration of the sample, multiply the number of drops times 61 to obtain the number of mg of Cl^- per 100 ml of urine.

To determine the sodium chloride concentration of the sample, multiply the number of drops times 100 to obtain the number of mg of NaCl per 100 ml of urine.

What would a high NaCl content in the urine indicate? ________

____

____

b. GLUCOSE

Glycosuria (the presence of glucose in the urine) occurs in patients with diabetes mellitus or other disorders. Traces of glucose may occur in normal urine, but detection of these small amounts requires special tests.

(1) Benedict's Test

Benedict's solution is commonly used to detect reducing sugars in urine and is not specific for just glucose.

PROCEDURE

1. In a Pyrex test tube, combine 10 drops of urine with 5 ml of Benedict's solution. Mix the solution.

2. Using a test tube holder, place the test tube in a boiling water bath for 5 minutes. *CAUTION! Make sure that the mouth of the test tube is pointed away from you and all other persons in the area.*

3. *Using a test tube holder,* remove the test tube from the water bath and read the results according to the following chart.

4. Record your results in Section F.4.b of the LABORATORY REPORT RESULTS at the end of the exercise.

Color	Results
Blue	Negative
Greenish yellow	1 + (0.5 g/100 ml)
Olive green	2 + (1 g/100 ml)
Orange-yellow	3 + (1.5 g/100 ml)
Brick red (with precipitate)	4 + (more than 2 g/ 100 ml)

(2) Clinitest® Reagent Method

An alternative method is to use Clinitest® reagent tablets.

PROCEDURE

1. Using medicine droppers, place 10 drops of water and 5 drops of urine in a Pyrex test tube.

CAUTION! *Place the test tube in a test tube rack because it will become too hot to handle.*

2. With forceps, add one clinitest tablet.

CAUTION! *The concentrated sodium hydroxide in the tablet generates enough heat to make the liquid in the test tube boil. Make sure that the mouth of the test tube is pointed away from you and all other persons in the area.*

3. The color of the solution is graded as in Benedict's test.

4. Fifteen seconds after boiling has stopped, shake the test tube *gently* and evaluate the color according to the Benedict's test table above.

CAUTION! *Make sure the mouth of the test tube is pointed away from you and others.*

Disregard any color change that occurs after 15 seconds.

5. Record your results in Section F.4.b of the LABORATORY REPORT RESULTS at the end of the exercise.

How would a high sugar diet affect the urine?

Explain. ______________________________

____

____

(3) Chemstrip® Method

Record your results in Section F.4.b of the LABORATORY REPORT RESULTS at the end of the exercise.

c. PROTEIN

Normal urine contains traces of proteins that are hard to detect through regular laboratory procedures. Albumin is the most abundant serum protein and is the one usually detected. Because tests for albumin are determined by precipitating the protein either by heat (coagulation) or by adding a reagent, the urine sample should either be filtered or centrifuged (see Figure 23.12). The test for protein will be done by either the Exton's reagent method or the Albutest® reagent method and the Chemstrip® method.

(1) Exton's® Reagent Method

PROCEDURE

1. Using medicine droppers, place equal volumes (about 3–5 ml) of urine and Exton's reagent in a test tube.

2. Roll the tube between your palms to mix the two fluids.

3. If a precipitate is formed, the liquid will appear cloudy because of the presence of albumin.

4. Record your results in Section F.4.c of the LABORATORY REPORT RESULTS at the end of the exercise.

(2) Albutest® Reagent Method

An alternative method is to use Albutest® reagent tablets containing bromphenol blue.

PROCEDURE

1. Using forceps, place an Albutest® tablet on a clean, dry paper towel and add one drop of urine.

2. After the drop has been absorbed, add two drops of water and allow these to penetrate before reading.

3. Compare the color (in daylight or fluorescent light) on top of the tablet with the color chart provided in lab.

4. If albumin is present in the urine, a *blue-green* spot will remain on the surface of the tablet after the water is added. The amount of protein is indicated by the intensity of the blue-green color.

5. If the test is negative, the original color of the tablet will not be changed at the completion of the test.

6. Record your results in Section F.4.c of the LABORATORY REPORT RESULTS at the end of the exercise.

Why is protein not normally found in urine?

____

____

(3) Chemstrip® Method

Record your results in Section F.4.c of the LABORATORY REPORT RESULTS at the end of the exercise.

d. KETONE (ACETONE) BODIES

The presence of ketone (acetone) bodies in urine is a result of abnormal fat catabolism. Reagents such as sodium nitroprusside, ammonium sulfate, and ammonium hydroxide are available in the form of tablets. Ketones turn purple when added to these chemicals.

(1) Acetest® Tablet Method

PROCEDURE

1. Using forceps, place an Acetest® tablet on a clean, dry paper towel, and, using a medicine dropper, place one drop of urine on the tablet.

2. If acetone or ketone is present, a *lavender-purple* color develops within 30 seconds. If the tablet becomes cream-colored from wetting, the results are negative. Compare results with the color chart that comes with the reagent.

3. Record your results in Section F.4.d of the LABORATORY REPORT RESULTS at the end of the exercise.

Why would starvation cause ketones? ________

____

____

(2) Chemstrip® Method

Record your results in Section F.4.d of the LABORATORY REPORT RESULTS at the end of the exercise.

e. BILE PIGMENTS

Bile pigments, biliverdin and bilirubin, are not normally present in urine. The presence of large quantities of bilirubin in the extracellular fluids produces jaundice, a yellowish tint to the body tissues, including yellowness of the skin and deep tissues.

(1) Shaken Tube Test For Bile Pigments

PROCEDURE

1. Fill a Pyrex test tube halfway with urine, stopper the tube, and shake it vigorously, being careful to keep the stopper in the tube.

2. A yellow color of foam indicates the presence of bile pigments.

3. Record your results in Section F.2.e of the LABORATORY REPORT RESULTS at the end of the exercise.

(2) Ictotest® for Bilirubin

PROCEDURE

1. Using a medicine dropper, place a drop of urine on one square of the special mat provided in the Ictotest® kit.

2. Using forceps, place one Ictotest® reagent tablet in the center of the moistened area.

CAUTION! *Do not touch the tablets with your fingers. Recap the bottle.*

3. Add one drop of water directly to the tablet; after 5 seconds, add another drop of water to the tablet so that the water runs off onto the mat. Observe the color of the mat around the tablet at 60 seconds. The presence of bilirubin will turn the mat *blue* or *purple*. A slight *pink* or *red* color is negative for bilirubin.

4. Record your results in Section F.2.e of the LABORATORY REPORT RESULTS at the end of the exercise.

(3) Chemstrip® Method

Record your results in Section F.2.e of the LABORATORY REPORT RESULTS at the end of the exercise.

f. HEMOGLOBIN

Hemoglobin is not normally found in urine.

(1) Chemstrip® Method

Record your results in Section F.4.f of the LABORATORY REPORT RESULTS at the end of the exercise.

5. Microscopic Analysis

If you allow a urine specimen to stand undisturbed for a few hours, many suspended materials will settle to the bottom. A much faster method is to centrifuge a urine sample.

PROCEDURE

1. Place 5 ml fresh urine in a centrifuge tube (Figure 23.12). Follow the directions of your instructor to centrifuge the tube for 5 minutes at a slow speed (1500 rpm).

2. Dispose of the supernatant (the clear urine) and mix the sediment by shaking the test tube.

3. Using a long medicine dropper or a Pasteur pipette with a bulb, place a small drop of sediment on a clean glass slide, add one drop of Sedi-stain® or methylene blue, and place a cover glass over the specimen.

4. Using low power and reduced light, examine the sediment for any of the microscopic elements pictured in Figure 23.13.

5. Increase the light and use high power or oil immersion to look for crystals.

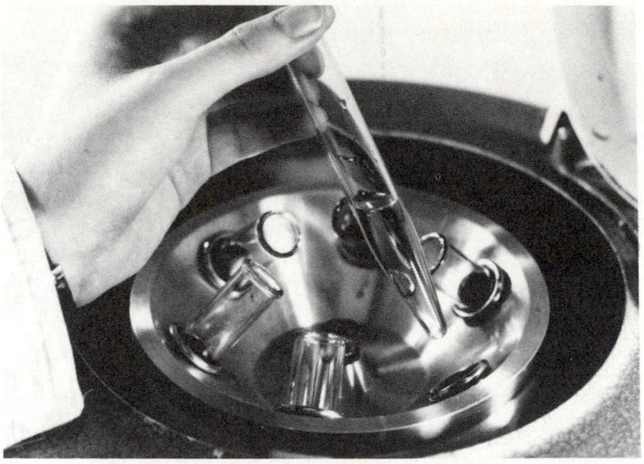

FIGURE 23.12 Tabletop centrifuge used for spinning urine samples to obtain sediment for microscopic analysis.

6. Crystals can be identified as follows (Figure 23.13):

Calcium oxalate—Dumbbell and octahedral shapes.

Calcium phosphate—Very pointed, wedge-shaped formations that may occur as individual crystals or grouped together to form rosettes.

Cholesterol—Spherical crystals that have a crosslike configuration inside.

Hippuric acid—Long, needlelike crystals.

Triple phosphates—Prisms or feathery forms.

Uric acid—Rhombic prisms, wedges, dumbbells, rosettes, irregular crystals. These are pigmented in sediment, and the color varies from yellow to dark reddish brown.

Draw the results of your observation in Section F.5 of the LABORATORY REPORT RESULTS at the end of the exercise.

Why might some red blood cells in urinary sediment be crenated? ______________

____

Why might yeast be seen in the urine of a person with diabetic mellitus? ______________

____

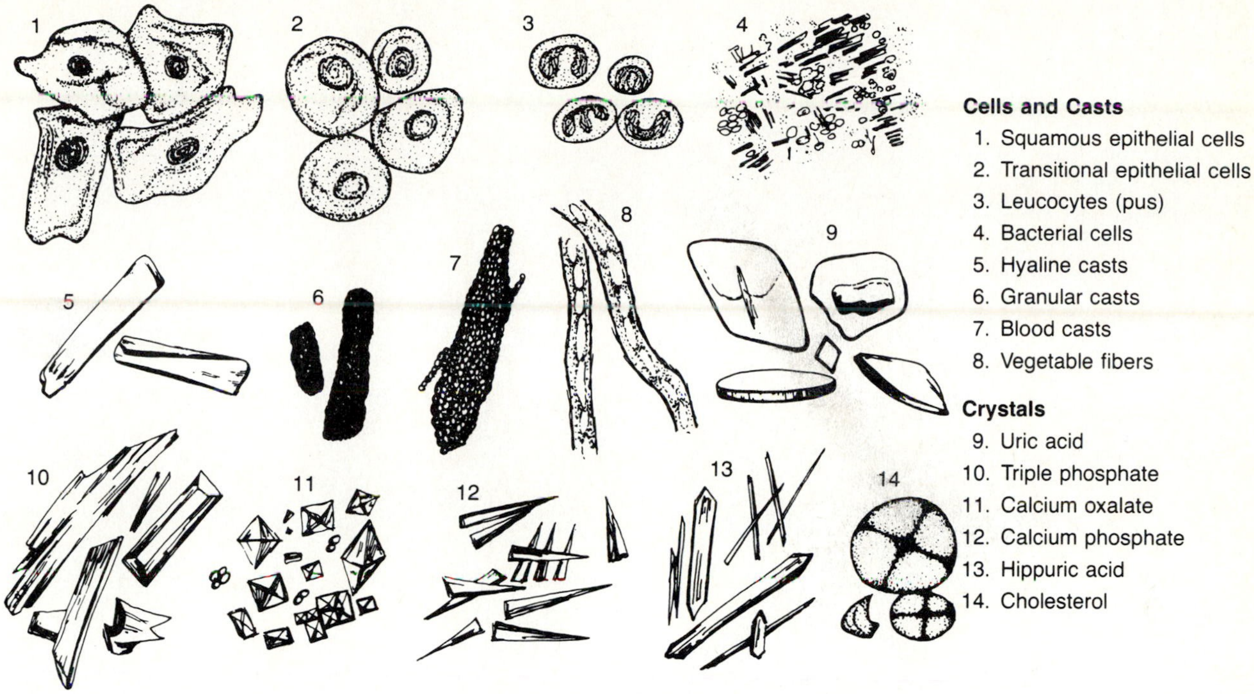

FIGURE 23.13 Diagram of microscopic elements in urine.

6. Unknown Specimens

a. UNKNOWNS PREPARED BY INSTRUCTOR

When the composition of a substance has not been defined, it is called an ***unknown***. In this exercise, the unknowns will be urine specimens to which the instructor has added glucose, albumin, or any other detectable substance. Each unknown contains only one added substance. Perform the previously outlined tests until you identify the substance in your unknown.

Record your results in Section F.6.a of the LABORATORY REPORT RESULTS at the end of the exercise.

b. UNKNOWNS PREPARED BY CLASS (OPTIONAL)

The class should be divided into two groups. Each group adds certain substances, such as glucose, protein, starch, or fat, to normal, freshly voided urine, keeping accurate records as to what was added to each sample. The two groups then exchange samples, and each group does the basic chemical tests on urine to detect which substances were added. Each student should add one substance to one urine sample and see if another student can detect what was added. Record your results in Section F.6.b of the LABORATORY REPORT RESULTS at the end of the exercise.

ANSWER THE LABORATORY REPORT QUESTIONS AT THE END OF THE EXERCISE

LABORATORY REPORT RESULTS

URINARY SYSTEM 23

Student ______________________ Date ______________________

Laboratory Section ______________________ Score/Grade ______________________

SECTION D. TUBULAR SECRETION

1. Secretion of Phenol Red

Time required for first appearance ______________________

Time required for maximum concentration ______________________

2. Inhibition of Secretion

Time required for first appearance ______________________

Time required for maximum concentration ______________________

3. Metabolic Requirements for Secretion

Time required for first appearance ______________________

Time required for maximum concentration ______________________

SECTION F. URINALYSIS

3. Physical Analysis

Characteristic	Normal	Your sample
a. Color	Straw yellow to amber	______________
b. Sediment	None	______________
c. pH	5.0–7.8	______________
d. Specific gravity	1.008–1.030	______________

4. Chemical Analysis

Record your results in the spaces provided.

a. CHLORIDE AND SODIUM CHLORIDE

Chloride concentration ____________ mg/100ml

Sodium chloride concentration ____________ mg/100ml

b. GLUCOSE

Benedict's test ____________________

Clinitest® tablet ____________________

Chemstrip® ____________________

c. PROTEIN

Exton's® reagent method ____________________

Albutest® reagent tablets ____________________

Chemstrip® ____________________

d. KETONE (ACETONE) BODIES

Acetest® tablet ____________________

Chemstrip® ____________________

e. BILE PIGMENTS

Shaken tube (bile pigments) ____________________

Ictotest® (bilirubin) ____________________

Chemstrip® ____________________

f. HEMOGLOBIN

Chemstrip® ____________________

5. Microscopic Analysis

Draw some of the substances (types of cells, types of crystals, or other elements) that you found in the microscopic examination of urinary sediment.

6. Unknown Specimens

a. What substance did you find in the unknown specimen that was prepared by your instructor?

____

b. What substance did you find in the unknown specimen that was prepared by other students?

____

LABORATORY REPORT QUESTIONS

URINARY SYSTEM 23

Student ______________________ Date ______________________

Laboratory Section ______________________ Score/Grade ______________________

PART I. MULTIPLE CHOICE

_____ 1. Beginning at the innermost layer and moving toward the outermost layer, identify the order of tissue layers surrounding the kidney. (a) renal capsule, renal fascia, adipose capsule (b) renal fascia, adipose capsule, renal capsule (c) adipose capsule, renal capsule, renal fascia (d) renal capsule, adipose capsule, renal fascia

_____ 2. The functional unit of the kidney is the (a) nephron (b) ureter (c) urethra (d) hilus

_____ 3. Substances filtered by the kidney must pass through the endothelial-capsular membrane, which is composed of several parts. Which of the following choices lists the correct order of the parts as substances pass through the membrane? (a) epithelium of the visceral layer of glomerular (Bowman's) capsule, endothelium of the glomerulus, basement membrane of the glomerulus (b) endothelium of the glomerulus, basement membrane of the glomerulus, epithelium of the visceral layer of glomerular (Bowman's) capsule (c) basement membrane of the glomerulus, endothelium of the glomerulus, epithelium of the visceral layer of the glomerular (Bowman's) capsule (d) epithelium of the visceral layer of the glomerular (Bowman's) capsule, basement membrane of the glomerulus, endothelium of the glomerulus.

_____ 4. In the glomerular capsule, the afferent arteriole divides into a capillary network called a(n) (a) glomerulus (b) interlobular artery (c) peritubular capillary (d) efferent arteriole

_____ 5. Transport of urine from the renal pelvis into the urinary bladder is the function of the (a) urethra (b) calculi (c) casts (d) ureters

_____ 6. The terminal portion of the urinary system is the (a) urethra (b) urinary bladder (c) ureter (d) nephron

_____ 7. Damage to the renal medulla would interfere first with the functioning of which parts of a juxtaglomerular nephron? (a) glomerular (Bowman's) capsule (b) distal convoluted tubule (c) collecting ducts (d) proximal convoluted tubules

_____ 8. An obstruction in the glomerulus would affect the flow of blood into the (a) renal artery (b) efferent arteriole (c) afferent arteriole (d) intralobular artery

_____ 9. Urine that leaves the distal convoluted tubule passes through the following structures in which sequence? (a) collecting duct, hilus, calyces, ureter (b) collecting duct, calyces, pelvis, ureter (c) calyces, collecting duct, pelvis, ureter (d) calyces, hilus, pelvis, ureter

_____ 10. The position of the kidneys behind the peritoneal lining of the abdominal cavity is described by the term (a) retroperitoneal (b) anteroperitoneal (c) ptosis (d) inferoperitoneal

_____ 11. Of the following structures, the one to receive filtrate *last* as it passes through the nephron is the (a) proximal convoluted tubule (b) ascending (thick) limb of the loop of the nephron (c) glomerulus (d) collecting duct

_____ 12. Peristalsis of the ureter is a function of the (a) serosa (b) mucosa (c) submucosa (d) muscularis

_____ 13. The trigone and the detrusor muscle are associated with the (a) kidney (b) urinary bladder (c) urethra (d) ureters

_____ 14. The notch on the medial surface of the kidney through which blood vessels enter and exit is called the (a) medulla (b) major calyx (c) hilus (d) renal column

_____ 15. Blood is drained from the kidneys by the (a) renal arteries (b) interlobar arteries (c) interlobular veins (d) renal veins

_____ 16. The epithelium of the urinary bladder that permits distension is (a) stratified squamous (b) transitional (c) simple squamous (d) pseudostratified

_____ 17. How many times a day is the entire volume of blood in the body filtered by the kidneys? (a) 100 times (b) 5 times (c) 30 times (d) 60 times

_____ 18. The average urine capacity of the urinary bladder is (a) 1000 to 1200 ml (b) 50 to 100 ml (c) 700 to 800 ml (d) 200 to 300 ml

_____ 19. The normal pH of urine is between (a) 5.0 and 7.8 (b) 2.0 and 4.8 (c) 10.0 and 12.0 (d) none of the above

_____ 20. Normal urine has a specific gravity of approximately (a) 1.008 to 1.030 (b) 1.030 to 1.080 (c) 1.100 to 1.200 (d) none of the above

_____ 21. The special chemical that may be used to detect glucose in the urine is (a) sulfosalicylic acid (b) Benedict's solution (c) Lugol's solution (d) nitric acid

PART II. COMPLETION

22. In addition to the urinary system, other systems that help eliminate wastes are the respiratory, integumentary, and systems ____________________ .

23. The double-walled cup found in a nephron is called a(n) ____________________ .

24. The special capillary network found inside of this double-walled cup is the ____________________ .

25. The major blood vessel that enters each kidney is the ____________________ .

26. The nerve supply to the kidneys comes from the autonomic nervous system and is called the ____________________ .

27. Urine is expelled from the urinary bladder by an act called urination, voiding, or ____________________ .

28. The small tube in the urinary system that leads from the floor of the urinary bladder to the outside is the ____________________ .

29. The apices of renal pyramids are referred to as renal ____________________ .

30. The cortical substance between renal pyramids is called a renal ____________________ .

31. Cuplike extensions of the renal pelvis, usually two or three in number, are referred to as ____________________ .

32. Epithelial cells of the visceral layer of the glomerular (Bowman's) capsule are called ____________________ .

33. Distal convoluted tubules terminate by merging with ____________________ .

34. Long loops of blood vessels around the medullary structures of juxtaglomerular nephrons are called ____________________ .

35. Which blood vessel comes next in this sequence? Interlobar artery, arcuate artery, interlobular artery, ____________________ .

36. The abnormal condition when red blood cells are found in the urine in appreciable amounts is called ____________________ .

37. Various salts that solidify in the urinary tract are called ____________________ .

38. Various substances that have hardened and assumed the shape of the lumens of the nephron tubules are the ____________________ .

39. The pH of urine in individuals on high-protein diets tends to be ____________________ than normal.

40. The greater the concentration of solutes in urine, the greater will be its ____________________ .

REPRODUCTIVE SYSTEMS 24

In this exercise, you will study the structure of the male and female reproductive organs and associated structures.

A. ORGANS OF MALE REPRODUCTIVE SYSTEM

The ***male reproductive system*** includes (1) the testes, or male gonads, which produce sperm, (2) some ducts that either transport or store sperm, (3) accessory glands, whose secretions contribute to semen, and (4) several supporting structures, including the penis.

1. Testes

The ***testes*** or ***testicles*** are paired oval glands that lie in the pelvic cavity for most of fetal life. They usually begin to enter the scrotum by 32 weeks; full descent is not complete until just before birth. If the testes do not descend, the condition is called ***cryptorchidism*** (krip-TOR-ki-dizm). Cryptorchidism results in sterility, because the cells that stimulate the initial development of sperm cells are destroyed by the higher temperature of the pelvic cavity.

Each testis is covered by a dense layer of white fibrous tissue, the ***tunica albuginea*** (al′-byoo-JIN-ē-a), which extends inward and divides the testis into a series of 200 to 300 internal compartments called ***lobules.*** Each lobule contains one to three tightly coiled ***seminiferous tubules*** where sperm production ***(spermatogenesis)*** occurs.

Label the structures associated with the testes in Figure 24.1.

Examine a microscope slide of a testis in cross section. Using oil immersion, note the developing reproductive cells in the seminiferous tubules. From the edge of the tubules toward the lumen, these cells are ***spermatogonia, primary spermatocytes, secondary spermatocytes, spermatids,*** and mature tailed ***spermatozoa (sperm cells).*** Distinguish between these cells and the ***sustentacular (Sertoli) cells*** which nourish the reproductive cells and secrete the hormone inhibin. Locate, in the tissue surrounding the seminiferous tubules, the clusters of ***interstitial endocrinocytes (interstitial cells of Leydig),*** which secrete the male hormone testosterone, the most important androgen. Because they produce both sperm and hormones, the testes are both exocrine and endocrine glands.

Using your textbook, charts, or models as reference, label Figure 24.2.

Spermatozoa (sper′-ma-tō-ZŌ-a) are produced at the rate of about 300 million per day. Once ejaculated, they usually live about 48 hours in the female reproductive tract. The parts of a spermatozoon are as follows:

a. ***Head***—Contains the ***nucleus*** and ***acrosome*** (produces hyaluronic acid and proteinases to effect penetration of egg).

b. ***Midpiece***—Contains numerous mitochondria in which the energy for locomotion is generated.

c. ***Tail***—Typical flagellum used for locomtion.

With the aid of your textbook, label Figure 24.3.

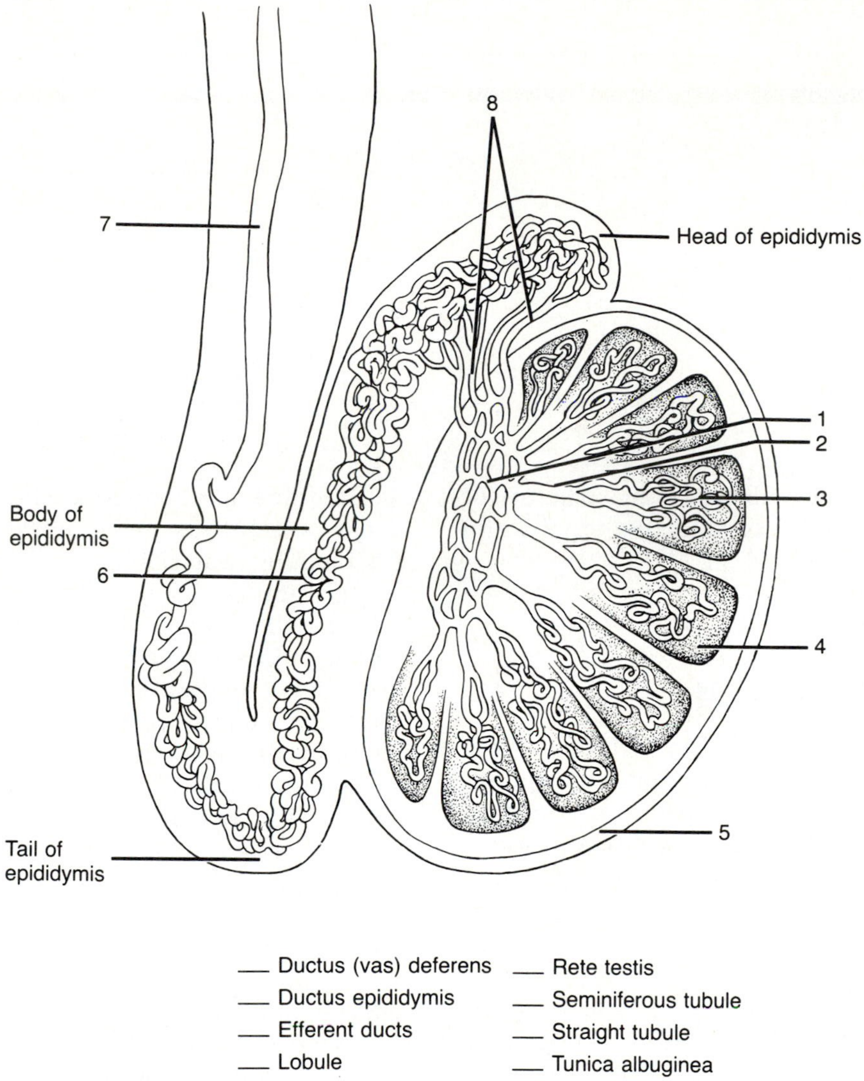

FIGURE 24.1 Sectional view of testis showing its system of ducts.

2. Ducts

As spermatozoa mature, they are moved through seminiferous tubules into tubes called ***straight tubules,*** from which they are transported into a network of ducts, the ***rete*** (RĒ-tē) ***testis.*** The spermatozoa are next transported out of the testes through a series of coiled ***efferent ducts*** that empty into a single ***ductus epididymis*** (ep′-i-DID-i-mis). From there, they are passed into the ***ductus (vas) deferens,*** which ascends along the posterior border of the testis, penetrates the inguinal canal, enters the pelvic cavity, and loops over the side and down the posterior surface of the urinary bladder. The ductus (vas) deferens and duct from the seminal vesicle (gland) together form the ***ejaculatory duct,*** which propels the spermatozoa into the ***urethra,*** the teminal duct of the system. The male urethra is divisible into (1) a *prostatic portion,* which passes through the prostrate gland; (2) a *membranous portion,* which passes through the urogenital diaphragm; and (3) a *spongy (cavernous) portion,* which passes through the corpus spongiosum of the penis (see Figure 24.6). The ***epididymis*** is a comma-shaped organ that is divisible into a head, body, and tail. The head is the superior portion that contains the efferent ducts; the body is the middle portion that contains the ductus

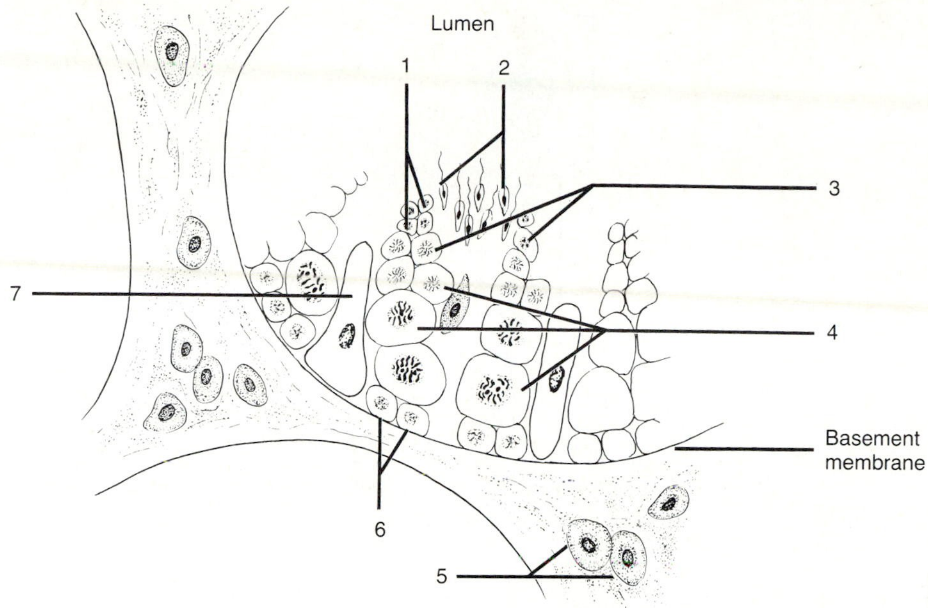

(a)

___ Interstitial endocrinocytes (interstitial cells of Leydig)
___ Primary spermatocytes
___ Secondary spermatocytes
___ Spermatids
___ Spermatogonia
___ Spermatozoa
___ Sustentacular (Sertoli) cell

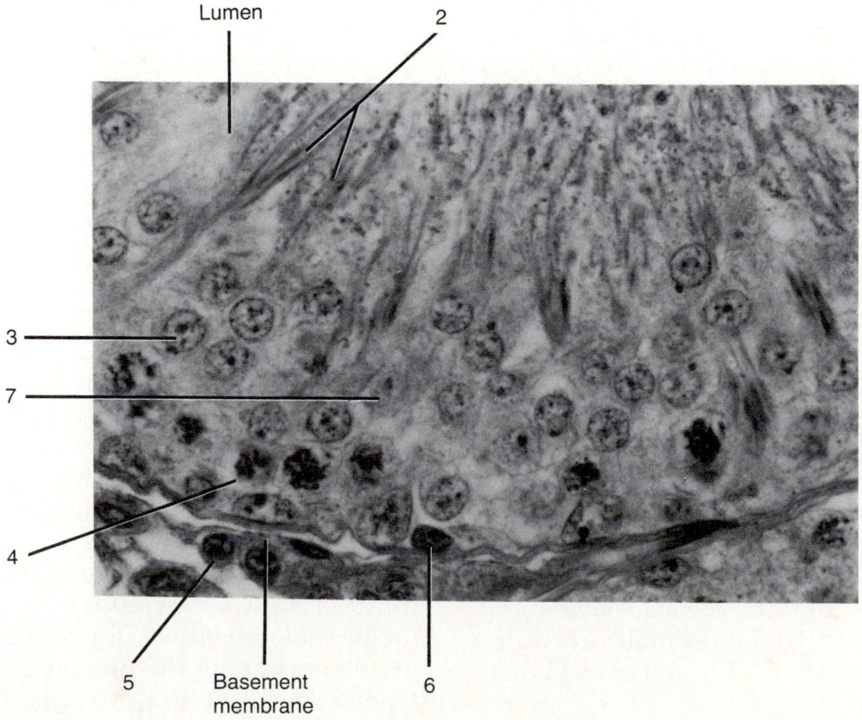

(b)

FIGURE 24.2 Testis. Cross section of seminiferous tubules showing various stages of spermatogenesis. (a) Diagram. (b) Photomicrograph.

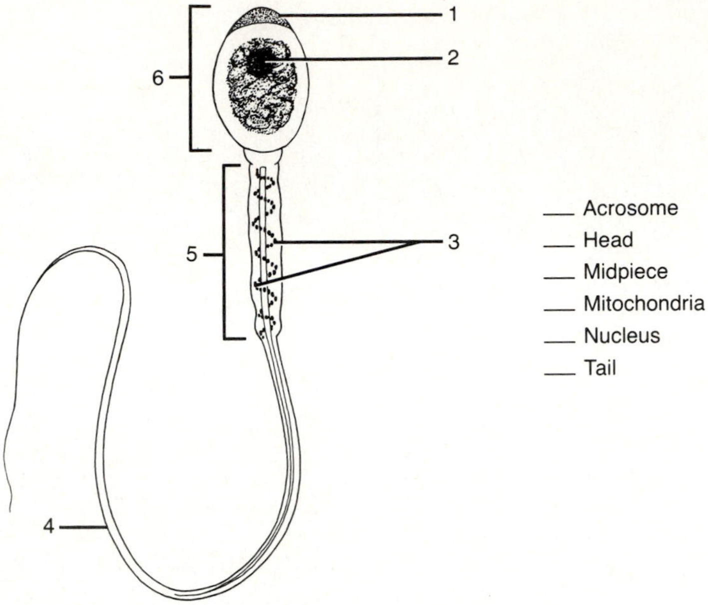

FIGURE 24.3 Parts of a spermatozoon.

epididymis; and the tail is the inferior portion in which the ductus epididymis continues as the ductus (vas) deferens.

Label the various ducts of the male reproductive system in Figure 24.1.

The ductus epididymis is lined with ***pseudostratified epithelium.*** The free surfaces of the cells contain long, branching microvilli called ***stereocilia.*** The muscularis deep to the epithelium consists of smooth muscle. Functionally, the ductus epididymis is the site of sperm maturation. It stores spermatozoa and propels them toward the urethra during ejaculation by peristaltic contractions of the muscularis. After about four weeks of storage, spermatozoa are expelled or reabsorbed.

Obtain a prepared slide of the ductus epididymis showing its mucosa and muscularis. Compare your observations to Figure 24.4.

Histologically, the ***ductus (vas) deferens (seminal duct)*** is also lined with ***pseudostratified epithelium*** and its muscularis consists of three layers of smooth muscle. Peristaltic contractions of the muscularis propel spermatozoa toward the urethra during ejaculation. One method of sterilization in males, ***vasectomy,*** involves removal of a portion of each ductus (vas) deferens.

Obtain a prepared slide of the ductus (vas) deferens showing its mucosa and muscularis. Compare your observations to Figure 24.5.

3. Accessory Sex Glands

Whereas the ducts of the male reproductive system store or transport sperm, a series of ***accessory sex glands*** secrete most of the liquid portion of ***semen (seminal fluid).*** Semen is a mixture of spermatozoa and the secretions of the seminal vesicles, prostate gland, and bulbourethral glands.

The ***seminal vesicles*** (VES-i-kuls) are paired, convoluted, pouchlike structures posterior to and at the base of the urinary bladder in front of the rectum. The glands secrete the alkaline viscous component of semen into the ejaculatory duct. The seminal vesicles contribute about 60% of the volume of semen.

The ***prostate*** (PROS-tāt) ***gland,*** a single doughnut-shaped gland inferior to the urinary bladder, surrounds the prostatic urethra. The prostate secretes a slightly acid fluid into the prostatic urethra. The prostatic secretion constitutes 13% to 33% of the total semen produced.

The paired ***bulbourethral*** (bul′-bō-yoo-RĒ-thral) or ***Cowper's glands,*** located inferior to the prostate on either side of the membranous urethra, are about the size of peas. They secrete an alkaline substance through ducts that open into the spongy (cavernous) urethra.

With the aid of your textbook, label the accessory glands and associated structures in Figure 24.6.

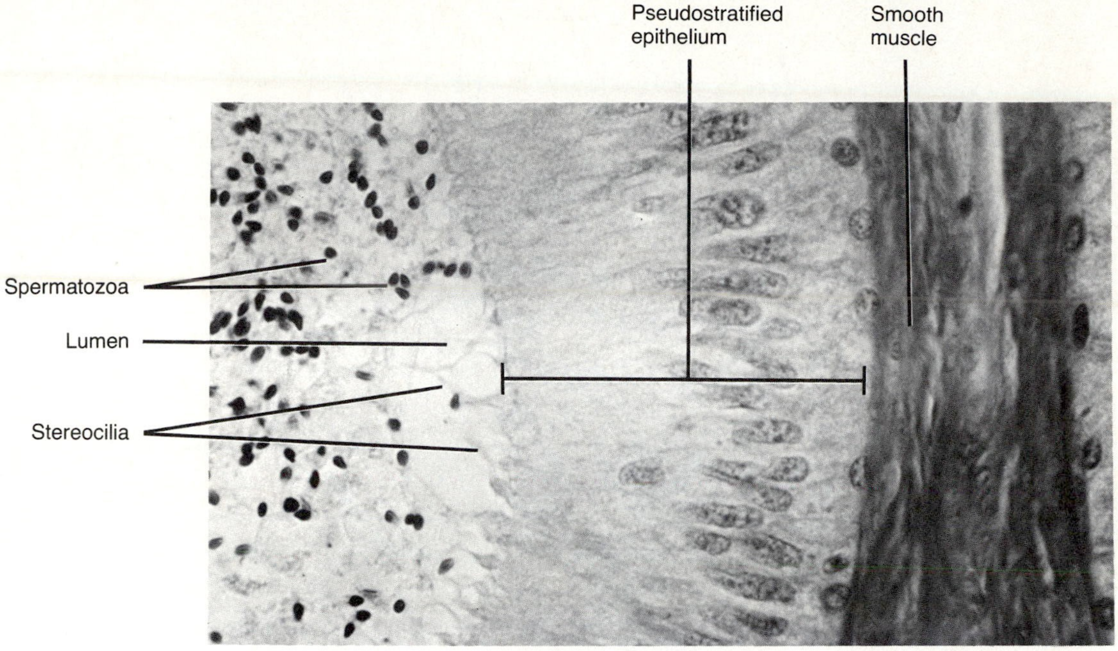

FIGURE 24.4 Histology of ductus epididymis.

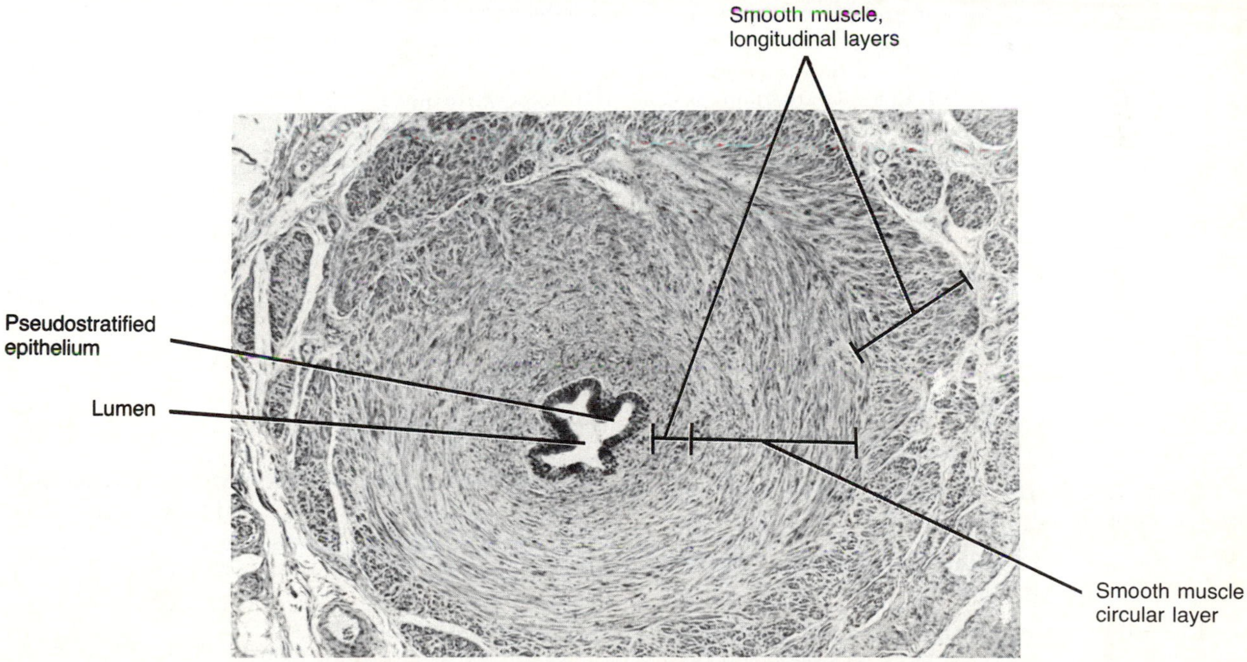

FIGURE 24.5 Histology of ductus (vas) deferens.

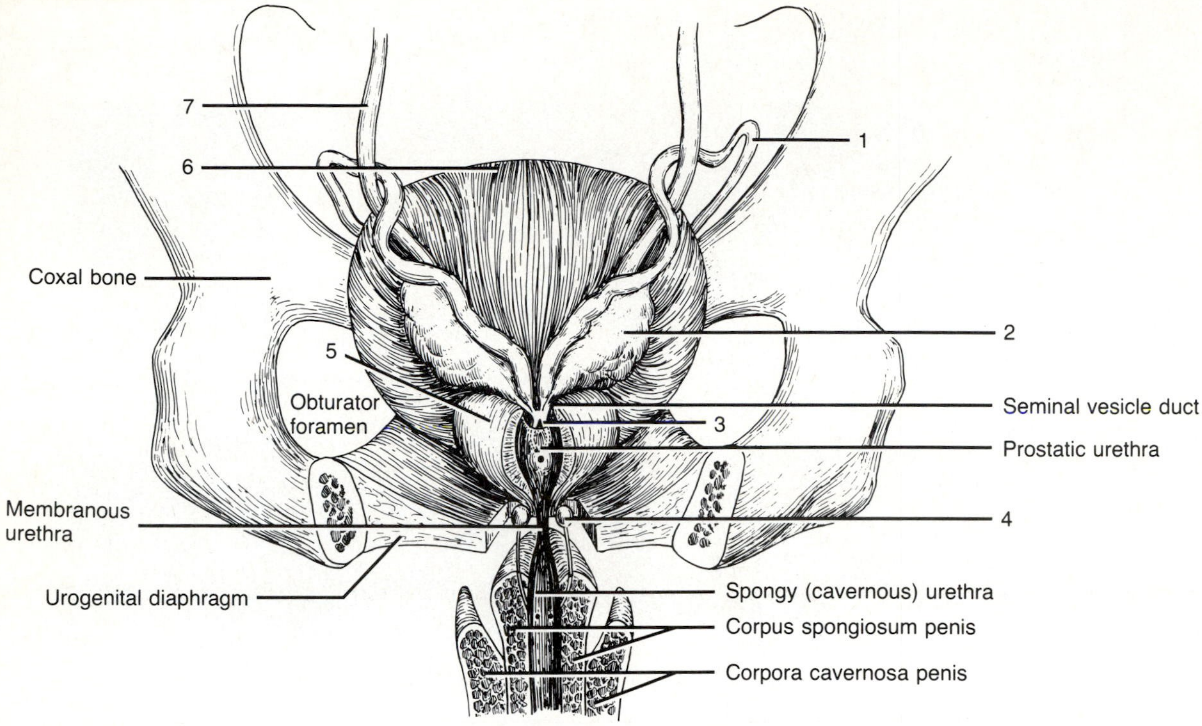

___ Bulbourethral (Cowper's) gland
___ Ductus (vas) deferens
___ Ejaculatory duct
___ Prostate gland
___ Seminal vesicle
___ Ureter
___ Urinary bladder

FIGURE 24.6 Relationships of some male reproductive organs in posterior view.

4. Penis

The ***penis*** conveys urine to the exterior and introduces spermatozoa into the vagina during copulation. Its principal parts are:

a. ***Glans penis***—Slightly enlarged distal end.

b. ***Corona***—Margin of glans penis.

c. ***Prepuce***—Foreskin; loosely fitting skin covering glans penis.

d. ***Corpora cavernosa penis***—Two dorsolateral masses of erectile tissue.

e. ***Corpus spongiosum penis***—Midventral mass of erectile tissue that contains spongy (cavernous) urethra.

f. ***Urethral orifice***—Opening of spongy (cavernous) urethra to exterior.

With the aid fo your textbook, label the parts of the penis in Figure 24.7.

Now that you have completed your study of the organs of the male reproductive system, label Figure 24.8.

B. ORGANS OF FEMALE REPRODUCTIVE SYSTEM

The ***female reproductive system*** includes the female gonads (ovaries), which produce ova; uterine (Fallopian) tubes, or oviducts, which transport ova to the uterus; vagina; external organs that compose the vulva; and the mammary glands.

1. Ovaries

The ***ovaries*** are paired glands that resemble almonds in size in shape. Functionally, the ovaries produce ova (eggs), discharge them about once a month by a process called ovu-

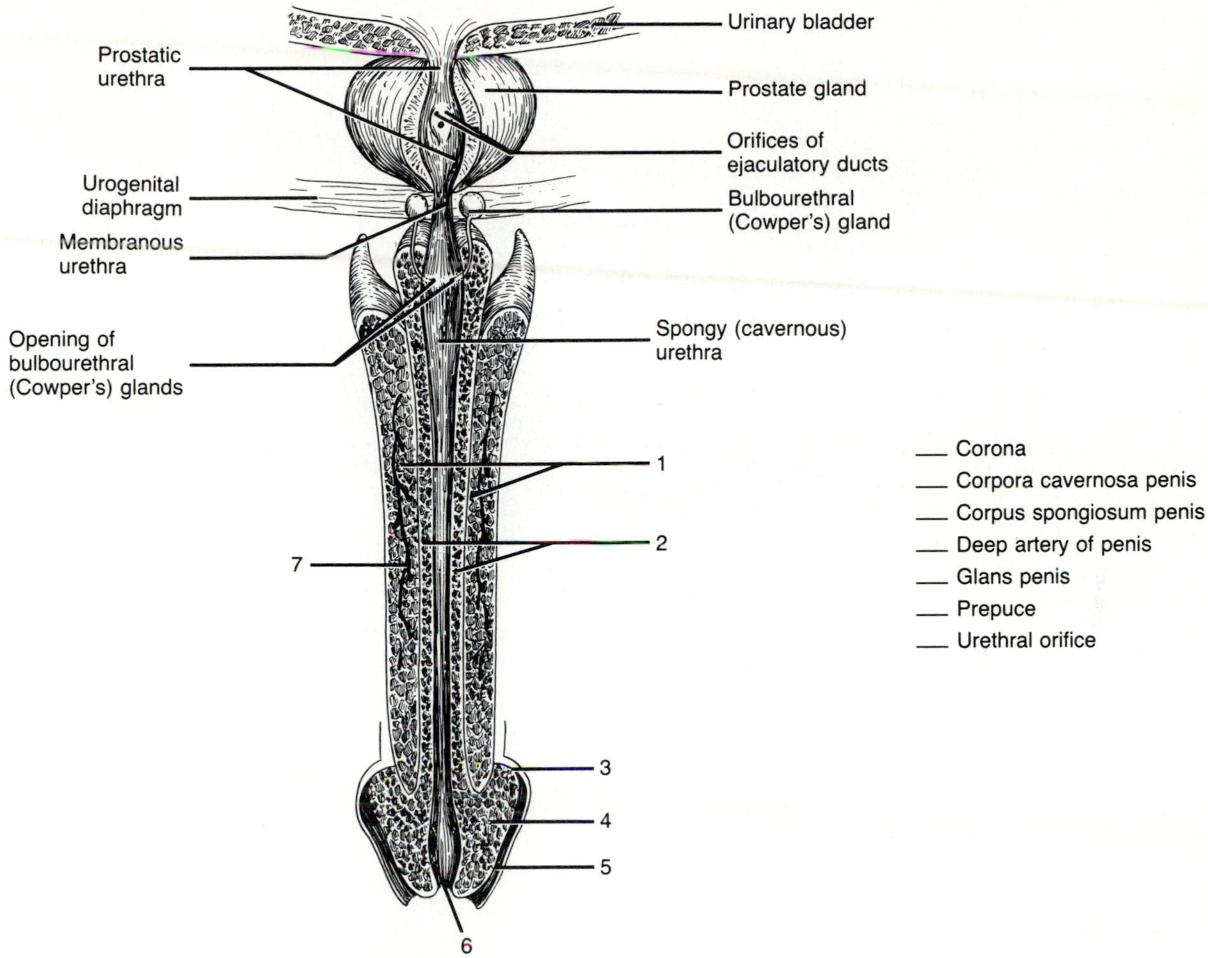

FIGURE 24.7 Internal structure of penis viewed from floor of penis.

lation, and secrete female sex hormones (estrogens, progesterone, and relaxin). The point of entrance for blood vessels and nerves is the ***hilus.*** The ovaries are positioned in the upper pelvic cavity, one on each side of the uterus, by a series of ligaments:

a. ***Mesovarium***—Double-layered fold of peritoneum that attaches ovaries to broad ligament of uterus.

b. ***Ovarian ligament***—Anchors ovary to uterus.

c. ***Suspensory ligament***—Attaches ovary to pelvic wall.

With the aid fo your textbook, label the ovarian ligaments in Figure 24.9.

Histologically, the ovaries consist of the following parts:

a. ***Germinal epithelium***—Layer of simple squamous or cuboidal epithelium covering the free surface of ovary.

b. ***Tunica albuginea***—Connective tissue capsule immediately deep to the germinal epithelium.

c. ***Stroma***—Region of connective tissue deep to tunica albuginea. Outer region, called ***cortex,*** contains ovarian follicles; inner region is ***medulla.***

d. ***Ovarian follicles***—Ova and their surrounding tissue in various stages of development.

e. ***Vesicular ovarian (Graafian) follicle***—Endocrine gland consisting of a potential mature ovum (secondary oocyte) and surrounding tissues that secrete estrogens.

f. ***Corpus luteum***—Glandular body that develops from vesicular ovarian (Graafian) folli-

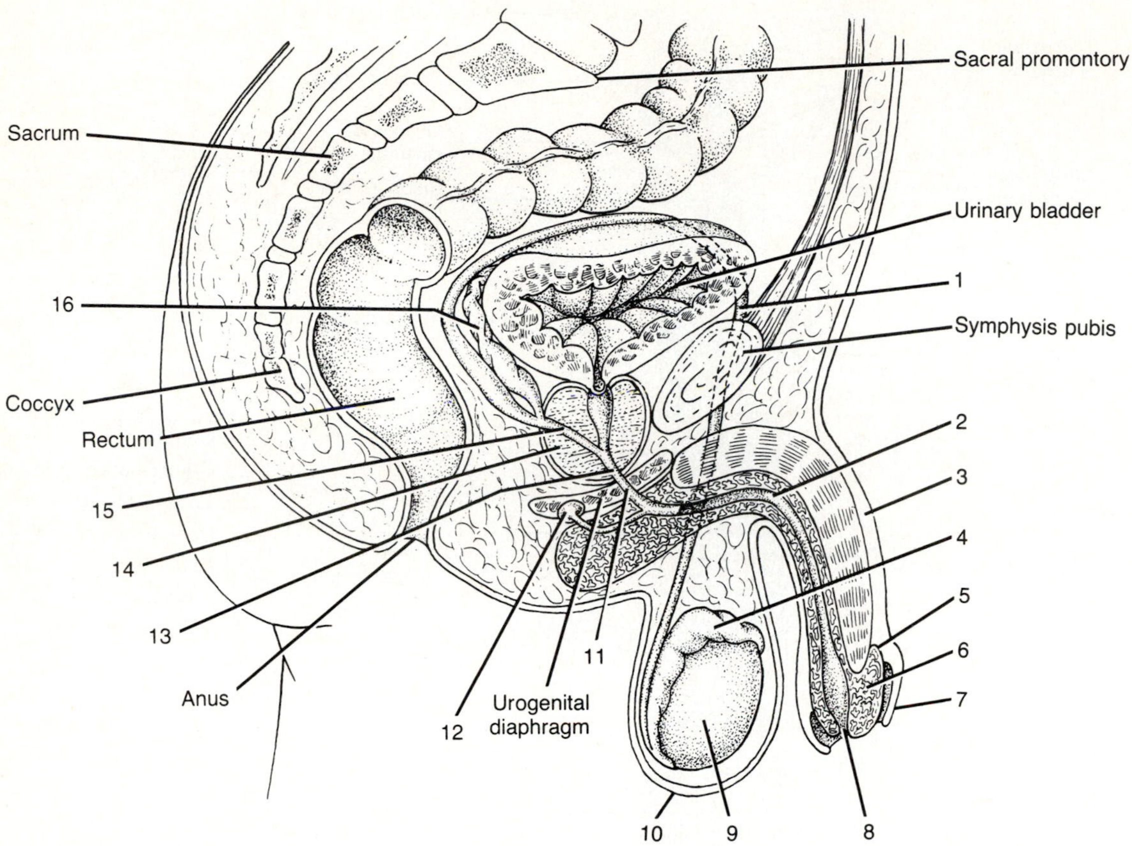

___ Bulbourethral (Cowper's) gland
___ Corona of penis
___ Ductus (vas) deferens
___ Ejaculatory duct
___ Epididymis
___ Glans penis
___ Membranous urethra
___ Penis
___ Prepuce
___ Prostate gland
___ Prostatic urethra
___ Scrotum
___ Seminal vesicle
___ Spongy (cavernous) urethra
___ Testis
___ Urethral orifice

FIGURE 24.8 Male organs of reproduction and surrounding structures seen in sagittal section.

cle after ovulation; secretes estrogens, progesterone, and relaxin.

With the aid of your textbook, label the parts of an ovary in Figure 24.10.

Obtain prepared slides of the ovary, examine them, and compare your observations to Figure 24.11.

2. Uterine (Fallopian) Tubes

The ***uterine (Fallopian) tubes*** or ***oviducts*** extend laterally from the uterus and transport ova from the ovaries to the uterus. Fertilization normally occurs in the uterine tubes. The tubes are positioned between folds of the broad ligaments of the uterus. The funnel-shaped, open distal end of each uterine tube, called the ***infundibulum,*** is surrounded by a fringe of fingerlike projections called ***fimbriae*** (FIM-brē-ē). The ***ampulla*** (am-POOL-la) of the uterine tube is the widest, longest portion, constituting about two-thirds of its length. The ***isthmus*** (IS-mus) is the short, narrow, thick-walled portion that joins the uterus.

With the aid of your textbook, label the parts of the uterine tubes in Figure 24.9.

Histologically, the mucosa of the uterine tubes consists of ciliated columnar cells and secretory cells. The muscularis is composed of inner circular and outer longitudinal layers of smooth muscle. Wavelike contractions of the

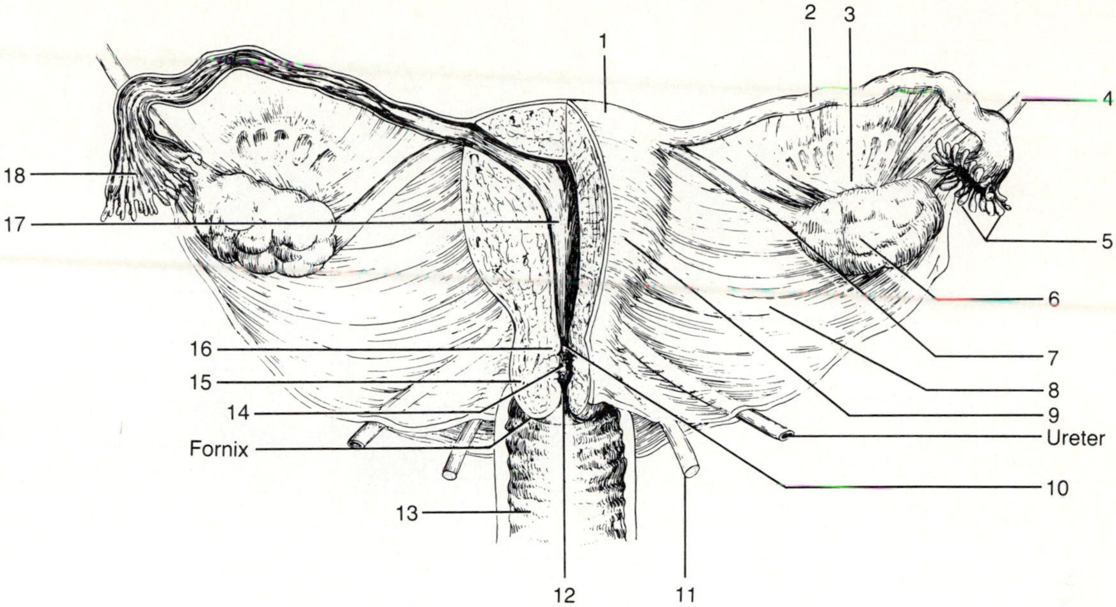

___ Body of uterus
___ Broad ligament
___ Cervical canal
___ Cervix
___ External os
___ Fimbriae
___ Fundus of uterus
___ Infundibulum
___ Internal os
___ Isthmus
___ Mesovarium
___ Ovarian ligament
___ Ovary
___ Suspensory ligament
___ Uterine cavity
___ Uterine (Fallopian) tube
___ Uterosacral ligament
___ Vagina

FIGURE 24.9 Uterus and associated female reproductive structures seen in posterior view. Left side of figure has been sectioned to show internal structures.

muscularis help move the ovum down into the uterus. The serosa is the outer covering.

Examine a prepared slide of the wall of the uterine tube, and compare your observations to Figure 24.12.

3. Uterus

The ***uterus*** is the site of menstruation, implantation of a fertilized ovum, development of the fetus during pregnancy, and labor. Located between the urinary bladder and the rectum, the organ is shaped like an inverted pear. The uterus is subdivided into the following regions:

a. ***Fundus***—Dome-shaped portion above uterine tubes.

b. ***Body***—Major, tapering portion.

c. ***Cervix***—Inferior narrow opening into vagina.

d. ***Isthmus***—Constricted region between body and cervix.

e. ***Uterine cavity***—Interior of the body.

f. ***Cervical canal***—Interior of the cervix.

g. ***Internal os***—Junction of isthmus and cervical canal.

h. ***External os***—Site where cervix opens into vagina.

Label these structures in Figure 24.9.

The uterus is maintained in position by the following ligaments:

a. ***Broad ligaments***—Double folds of parietal peritoneum that anchor the uterus to either side of the pelvic cavity.

b. ***Uterosacral ligaments***—Parietal peritoneal extensions that connect the uterus to the sacrum.

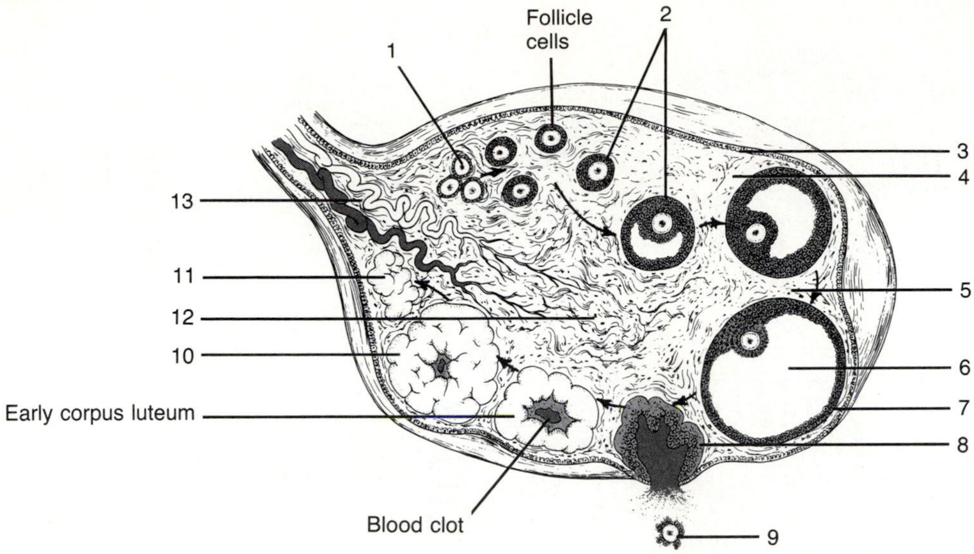

___ Corpus albicans
___ Corpus hemorrhagicum (ruptured follicle)
___ Cortex of stroma
___ Discharged secondary oocyte
___ Follicular fluid
___ Germinal epithelium
___ Hilus
___ Mature corpus luteum
___ Medulla of stroma
___ Primary follicle
___ Secondary follicles
___ Tunical albuginea
___ Vesicular ovarian (Graafian) follicle

FIGURE 24.10 Histology of ovary seen in sectional view. Arrows indicate sequence of developmental stages that occur as part of ovarian cycle.

c. ***Cardinal (lateral cervical) ligaments***—Tissues containing smooth muscle, uterine blood vessels, and nerves. These ligaments extend below the bases of the broad ligaments between the pelvic wall and the cervix and vagina, and are the chief ligaments that maintain position of uterus, helping to keep it from dropping into vagina.

d. ***Round ligaments***—Extend from uterus to external genitals between folds of broad ligaments.

Label the uterine ligaments in Figure 24.9.

Histologically, the uterus consists of three principal layers: endomentrium, myometrium, and perimetrium (serosa). The inner ***endomentrium*** is a mucous membrane with numerous glands and consists of two layers: (1) ***stratum functionalis,*** the layer closer to the uterine cavity that is shed during menstruation, and (2) ***stratum basalis*** (ba-SAL-is), the permanent layer that produces a new functionalis after menstruation. The middle ***myometrium*** forms the bulk of the uterine wall and consists of three layers of smooth muscle. During labor its coordinated contractions help to expel the fetus. The outer layer is the ***perimetrium (serosa),*** part of the visceral peritoneum.

As noted, the uterus is associated with menstruation. The ***menstrual cycle*** is a series of changes in the endometrium of a nonpregnant female that prepares the endometrium to receive a fertilized ovum each month (Figure 24.13). The cycle is controlled by estrogens, progesterone, follicle-stimulating hormone (FSH), luteinizing hormone (LH), and gonadotropin-releasing hormone factor (GnRH). The menstrual cycle is divided into the following stages:

1. ***Menstrual phase (menstruation).*** This is the periodic discharge of blood, fluid, mucus, and epithelial cells due to the sudden drop in estrogens and progesterone. It lasts for about 5 days

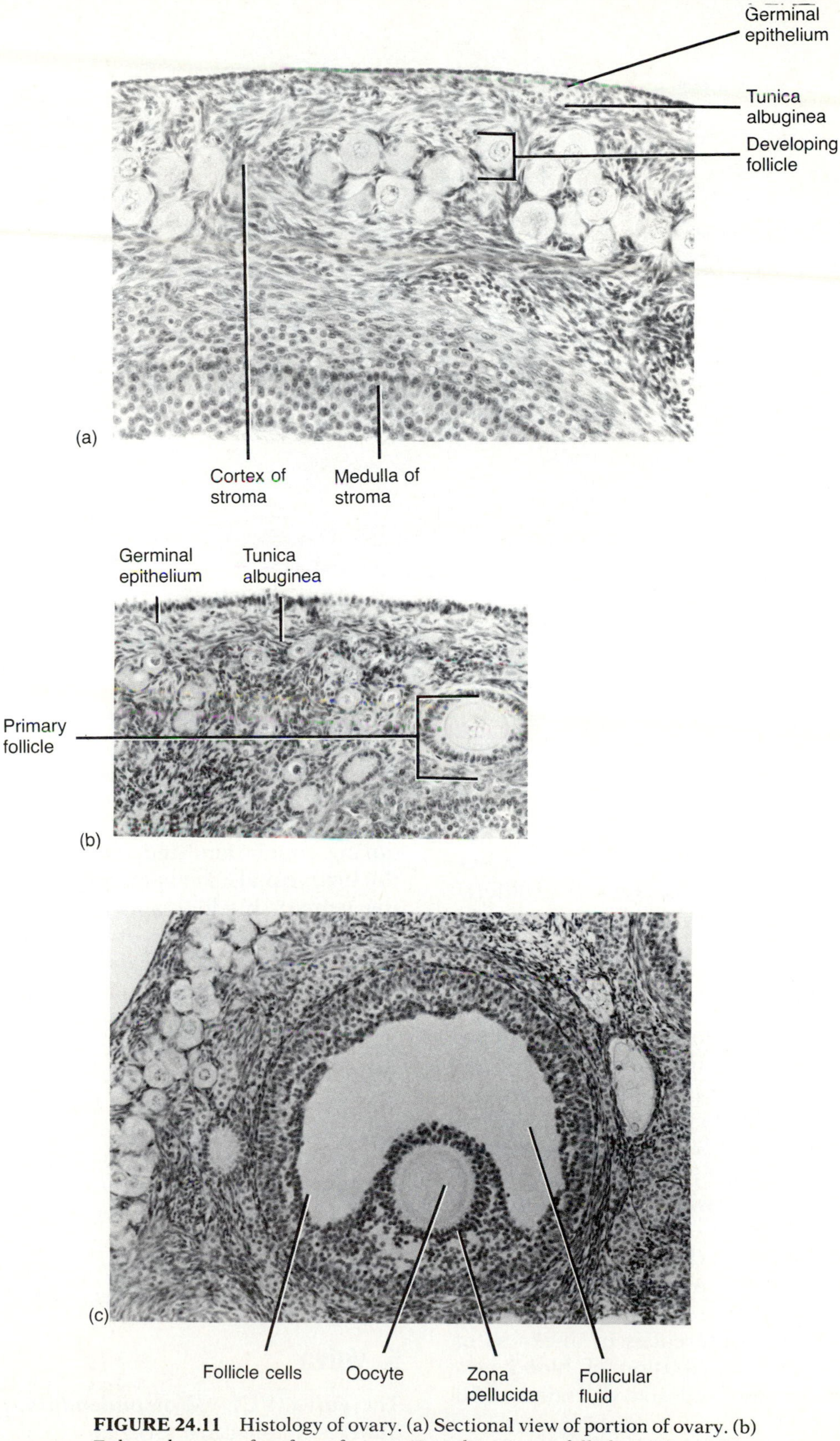

FIGURE 24.11 Histology of ovary. (a) Sectional view of portion of ovary. (b) Enlarged aspect of surface of ovary. Note the primary follicle. (c) Secondary follicle.

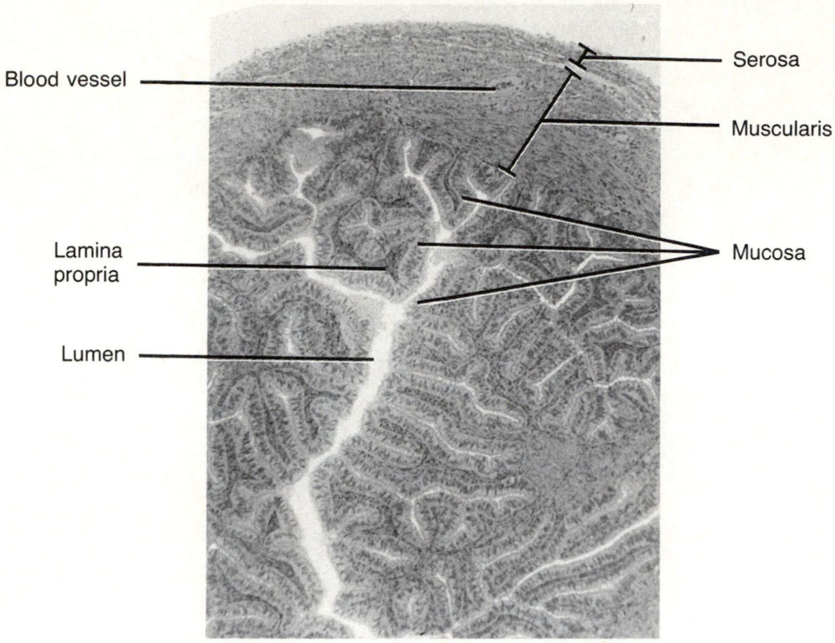

FIGURE 24.12 Histology of uterine (Fallopian) tube.

in an average 28-day cycle. During this phase, the stratum functionalis of the endometrium is sloughed off.

2. ***Preovulatory (proliferative) phase.*** This phase lasts from about day 6 through day 13 in a 28-day cycle. Rises in levels of estrogens stimulate endometrial repair, proliferation of endometrial glands and blood vessels, and result in thickening of the endometrium. Ovulation, the discharge of a secondary oocyte, occurs on the 14th day.

3. ***Postovulatory (secretory) phase.*** This phase lasts from about day 15 through day 28 in a 28-day cycle. Mainly under the influence of progesterone, the vascular supply of the endometrium increases even more. Also, the endometrial glands further increase in size. These changes are greatest about one week after ovulation.

Obtain microscope slides of the endometrium showing the menstrual, preovulatory, and postovulatory phases of the menstrual cycle. See if you can note the differences in thickness of the endometrium, distribution of blood vessels, and distribution and size of endometrial glands.

4. Vagina

A muscular, tubular organ lined with a mucous membrane, the ***vagina*** is the passageway for menstrual flow, the receptacle for the penis during copulation, and the inferior portion of the birth canal. The vagina is situated between the urinary bladder and rectum and extends from the cervix of the uterus to the vestibule of the vulva. Recesses called ***fornices*** (FOR-ni-sēz′) surround the vaginal attachment to the cervix (see Figure 24.9) and make possible the use of contraceptive diaphragms. The opening of the vagina to the exterior, the ***vaginal orifice,*** may be bordered by a thin fold of vascularized membrane, the ***hymen.***

Label the vagina in Figure 24.9.

Histologically, the mucosa of the vagina consists of stratified squamous epithelium and connective tissue that lies in a series of transverse folds, the ***rugae.*** The muscularis is composed of a longitudinal layer of smooth muscle.

5. Vulva

The ***vulva*** (VUL-va) or ***pudendum*** (pyoo-DEN-dum) is a collective term for the external gen-

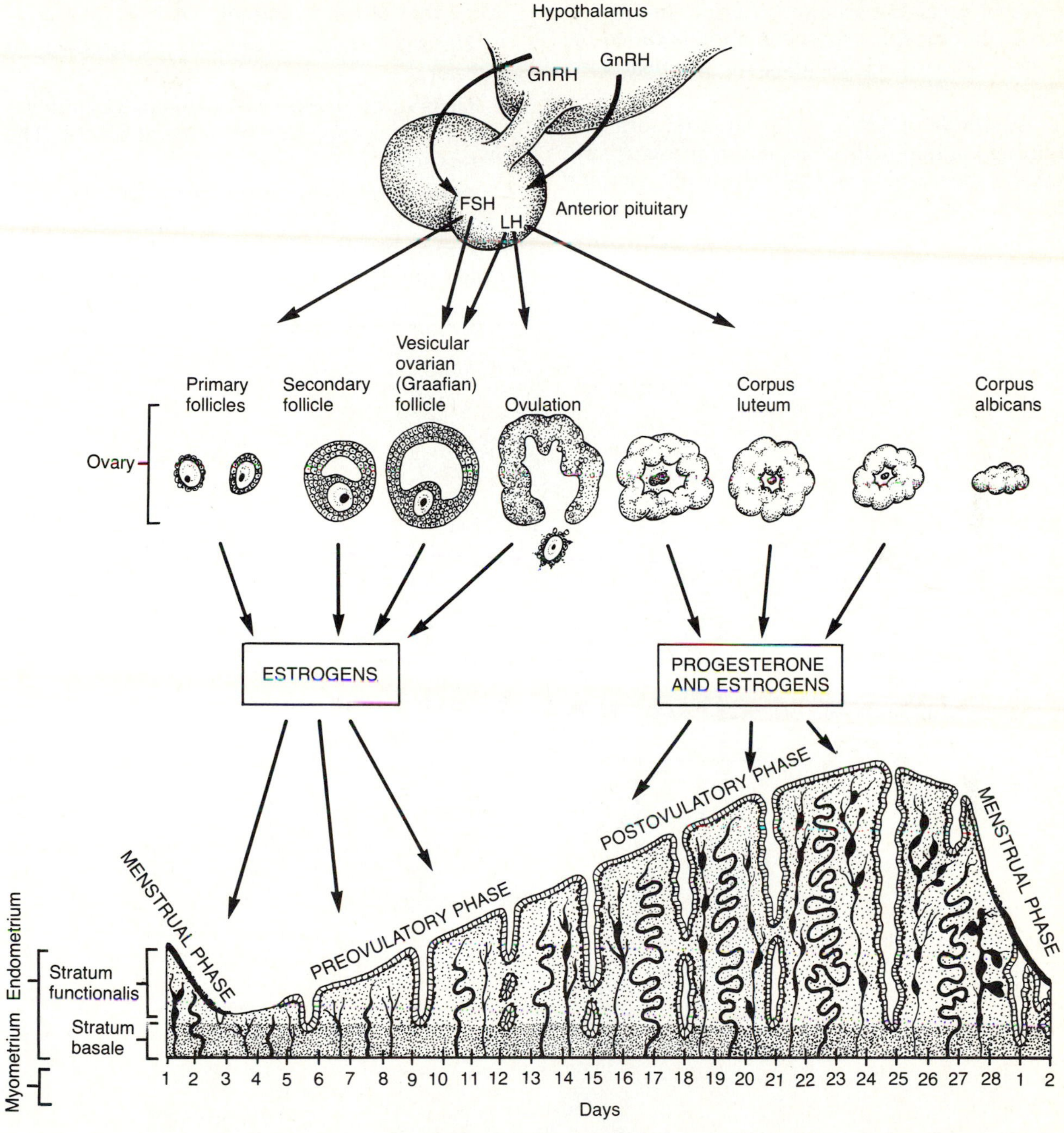

FIGURE 24.13 Menstrual cycle.

itals of the female. It consists of the following parts:

a. ***Mons pubis***—Elevation of adipose tissue over the symphysis pubis covered by skin and pubic hair.

b. ***Labia majora*** (LĀ-bē-a ma-JŌ-ra)—Two longitudinal folds of skin that extend inferiorly and posteriorly from the mons pubis. The folds, covered by pubic hair on their superior lateral surfaces, contain abundant adipose tissue and sebaceous (oil) and sudoriferous (sweat) glands.

c. ***Labia minora*** (MĪ-nō-ra)—Two folds of mucous membrane medial to labia majora. The folds have numerous sebaceous but few sudoriferous glands and no fat or pubic hair.

d. ***Clitoris*** (KLI-to-ris)—Small cylindrical mass of erectile tissue at anterior junction of labia minora. The exposed portion is called the ***glans,*** the covering is called the ***prepuce*** (foreskin).

e. ***Vestibule***—Cleft between labia minora containing vaginal orifice, hymen (if present), urethral orifice, and openings of ducts of lesser and greater vestibular glands.

f. ***Vaginal orifice***—Opening of vaginal to exterior.

g. ***Hymen***—Thin fold of vascularized membrane that borders vaginal orifice.

h. ***Urethral orifice***—Opening of urethra to exterior.

i. ***Orifices of paraurethral (Skene's) glands***—Located on either side of urethral orifice. The glands secrete mucus.

j. ***Orifices of ducts of greater vestibular*** (ves-TIB-yoo-lar) or ***Bartholin's glands***—Located in a groove between hymen and labia minora. These glands produce a mucoid secretion.

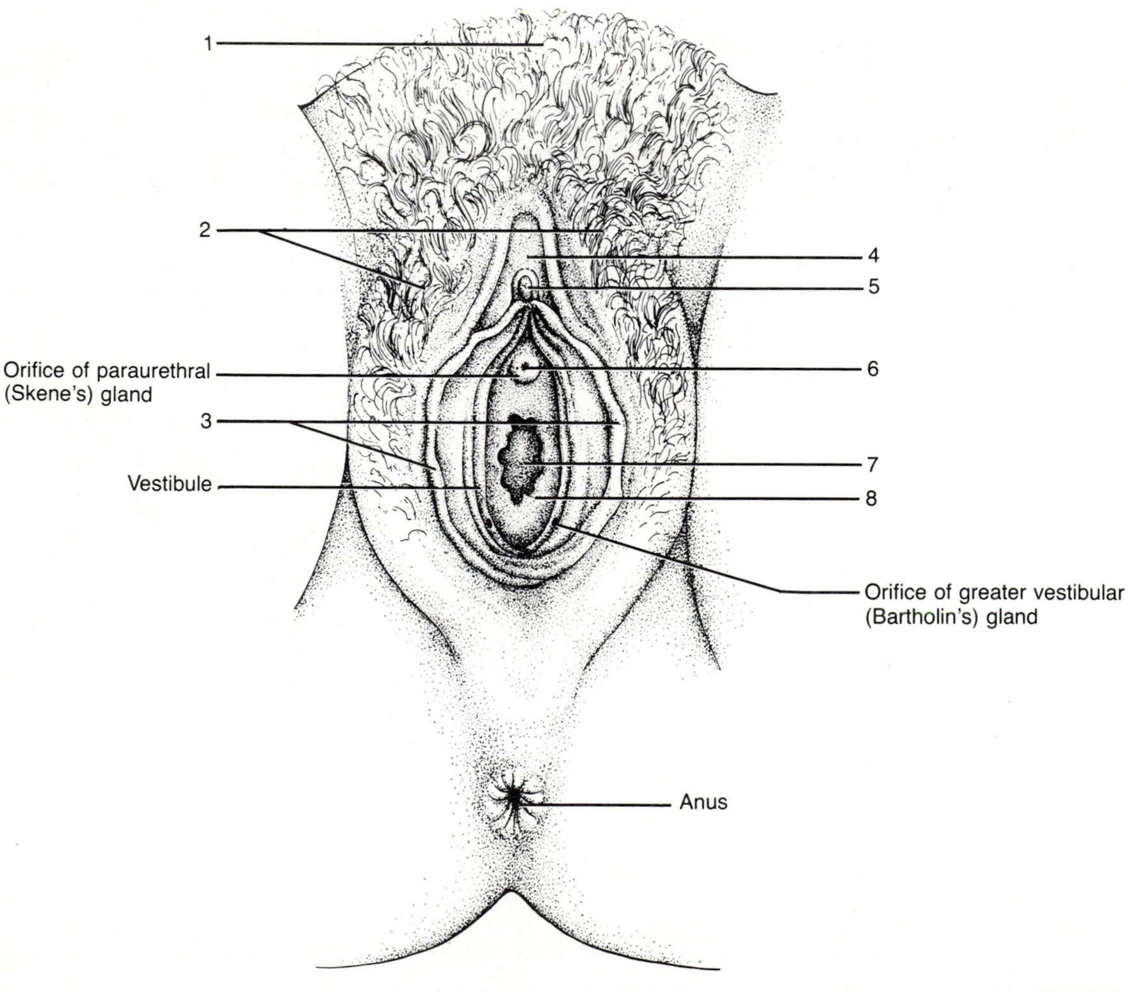

___ Clitoris
___ Hymen
___ Labia majora (spread)
___ Labia minora (spread) exposing vestibule
___ Mons pubis
___ Prepuce
___ Urethral orifice
___ Vaginal orifice (dilated)

FIGURE 24.14 Vulva.

k. ***Orifices of ducts of lesser vestibular glands***—Microscopic orifices opening into vestibule.

Using your textbook as an aid, label the parts of the vulva in Figure 24.14.

6. Mammary Glands

The ***mammary glands*** are modified sweat glands (branched tubuloalveolar glands) that lie over the pectoralis major muscles and are attached to them by a layer of connective tissue. They consist of the following structures:

a. ***Lobes***—Around 15 to 20 separated by adipose tissue.

b. ***Lobules***—Smaller compartments in lobes that contain milk-secreting cells called ***alveoli.***

c. ***Secondary tubules***—Receive milk from alveoli.

d. ***Mammary ducts***—Receive milk from secondary tubules.

e. ***Lactiferous sinuses***—Expanded distal portions of mammary ducts that store milk.

f. ***Lactiferous ducts***—Receive milk from lactiferous sinuses.

g. ***Nipple***—Projection on anterior surface of mammary gland that contains lactiferous ducts.

h. ***Areola*** (a-RĒ-ō-la)—Circular pigmented skin around nipple.

With the aid of your textbook, label the parts of the mammary gland in Figure 24.15.

Examine a prepared slide of alveoli of the mammary gland, and compare your observations to Figure 24.16.

Now that you have completed your study of the organs of the female reproductive system, label Figure 24.17.

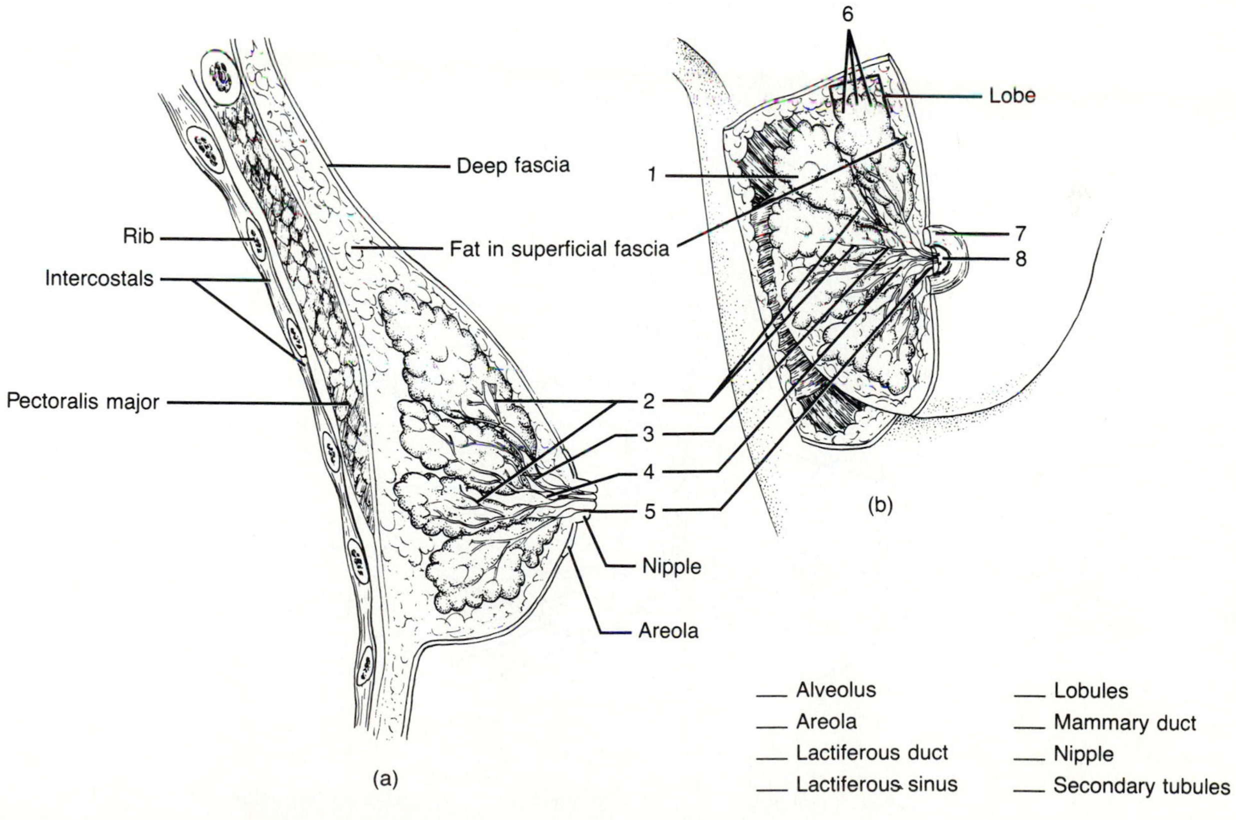

FIGURE 24.15 Mammary glands. (a) Sagittal section. (b) Anterior view, partially sectioned.

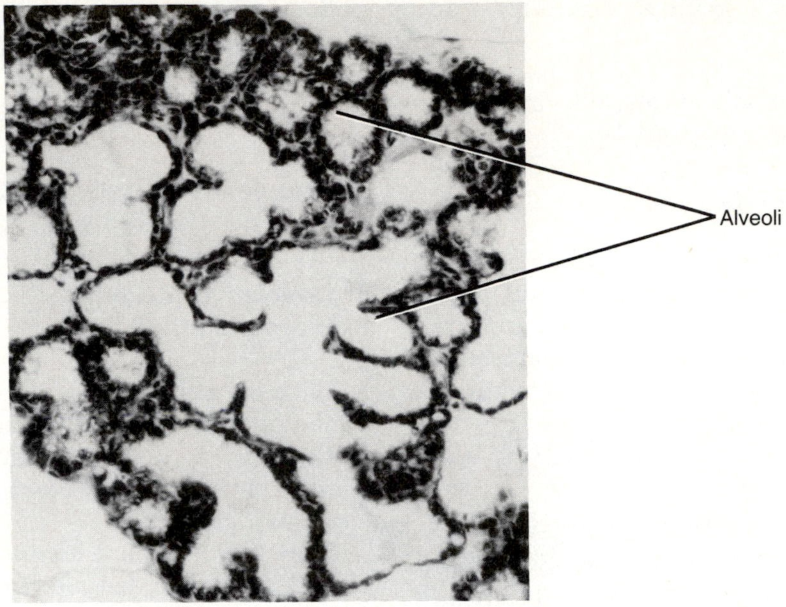

FIGURE 24.16 Histology of mammary gland showing alveoli.

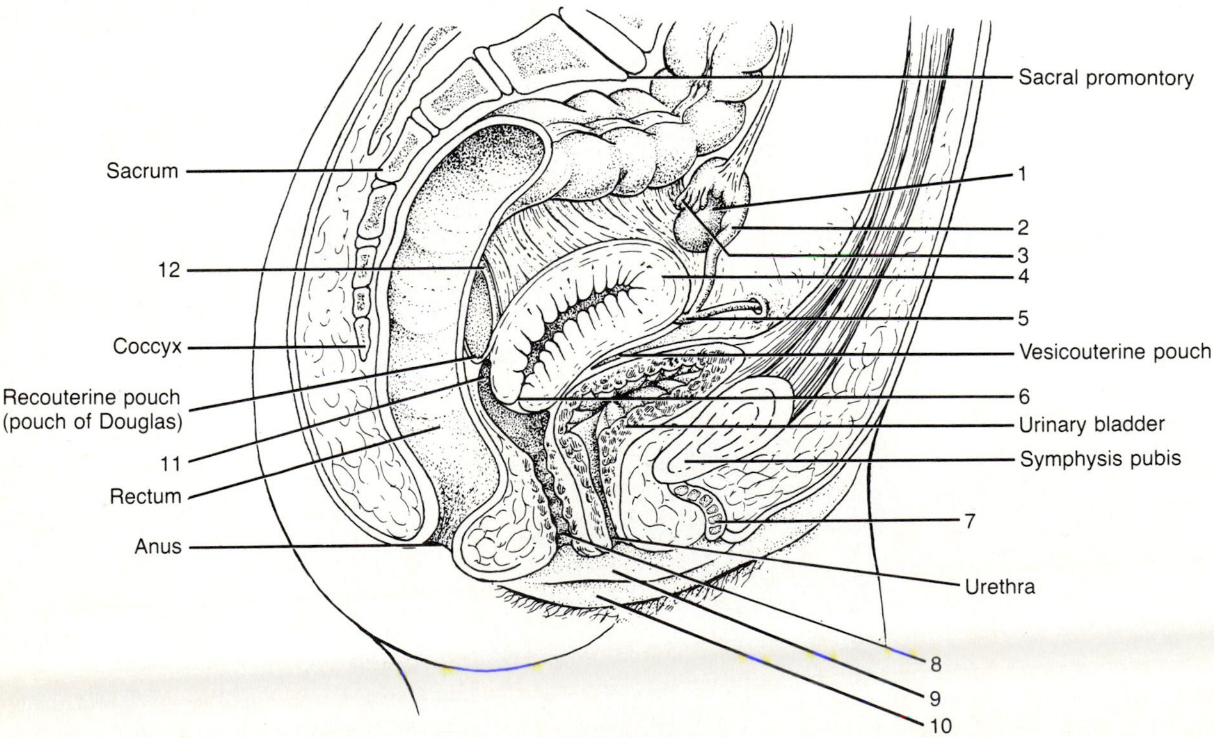

___ Cervix of uterus
___ Clitoris
___ Fimbriae
___ Fornix
___ Labium major
___ Labium minor
___ Ovary
___ Round ligament
___ Uterine (Fallopian) tube
___ Uterosacral ligament
___ Uterus
___ Vagina

FIGURE 24.17 Female organs of reproduction and surrounding structures seen in sagittal section.

C. DISSECTION OF CAT REPRODUCTIVE SYSTEMS

PROCEDURE

CAUTION! *Please reread Section D. "Precautions Related to Dissection" at the begining of the laboratory manual on page xii before you begin your dissection.*

1. Male Reproductive System

The male reproductive system of the cat is shown in Figure 23.18.

a. SCROTUM

1. The ***scrotum,*** a double sac ventral to the anus that contains the testes, consists of skin, muscle, and connective tissue.

2. Internally, the scrotum is divided into two chambers by a ***median septum***. A testis is found in each chamber.

b. TESTES

1. Carefully cut open the scrotum to examine the ***testes.***

2. The glistening covering of peritoneum around the testes is the ***tunica vaginalis.***

c. EPIDIDYMIDES

1. Cut away the tunica vaginalis.

2. Each ***epididymis*** is a long, coiled tube that receives spermatozoa from a testis. It is located anterolateral to its testis.

3. Each epididymis consists of an enlarged cranial end ***(head),*** a narrow middle portion ***(body),*** and an expanded caudal portion ***(tail).***

d. DUCTUS (VAS) DEFERENS

1. Each ***ductus (vas) deferens*** is a continuation from the tail of the epididymis. It carries spermatozoa from the epididymis through the inguinal canal to the urethra.

2. The ***inguinal canal*** is a short passageway through the abdominal musculature.

3. Passing through the inguinal canal with the ductus deferens are the testicular artery, vein, lymphatic, and nerve. Collectively, these structures constitute the ***spermatic cord,*** which is covered by connective tissue.

4. In the abdomen, the ductus deferens separates from the spermatic cord, passes ventral to the ureter, and then medially to the dorsal surface of the urinary bladder. From there it enters the ***prostatic urethra,*** the portion of the urethra that passes through the prostate gland.

5. The pelvic cavity must be exposed by cutting through the symphysis pubis at the midline and opening the pelvic girdle.

e. PROSTATE GLAND

1. At the junction of the ductus (vas) deferens and the neck of the urinary bladder, locate the ***prostate gland.***

2. The ducts of the prostate gland empty into the prostatic urethra.

f. BULBOURETHRAL GLANDS

1. These two glands are located on either side of the membranous urethra, the portion of the urethra between the prostatic urethra and the spongy urethra (the part passing through the penis).

2. The bulbourethral (Cowper's) glands are just dorsal to the penis.

g. PENIS

1. This is the external organ of copulation ventral to the scrotum.

2. The free (distal) end of the penis is enlarged and is referred to as the ***glans penis.*** The loose skin covering the penis is the ***foreskin,*** or ***prepuce.***

3. Cut open the prepuce to examine the glans penis. The opening of the penis to the exterior is the ***urogenital aperture.***

4. Internally the penis consist of two dorsally located cylinders and one ventrally located cylinder of erectile tissue. The dorsal masses are side by side and are known as the ***corpora cavernosa penis.*** Their proximal ends are known as the ***crura.*** Each crus is attached to the ischium on its own side by the ***ischiocavernosus muscle.*** The ventral mass of erectile tissue is called the ***corpus spongiosum penis.*** Through it passes the ***spongy (cavernosus) urethra.***

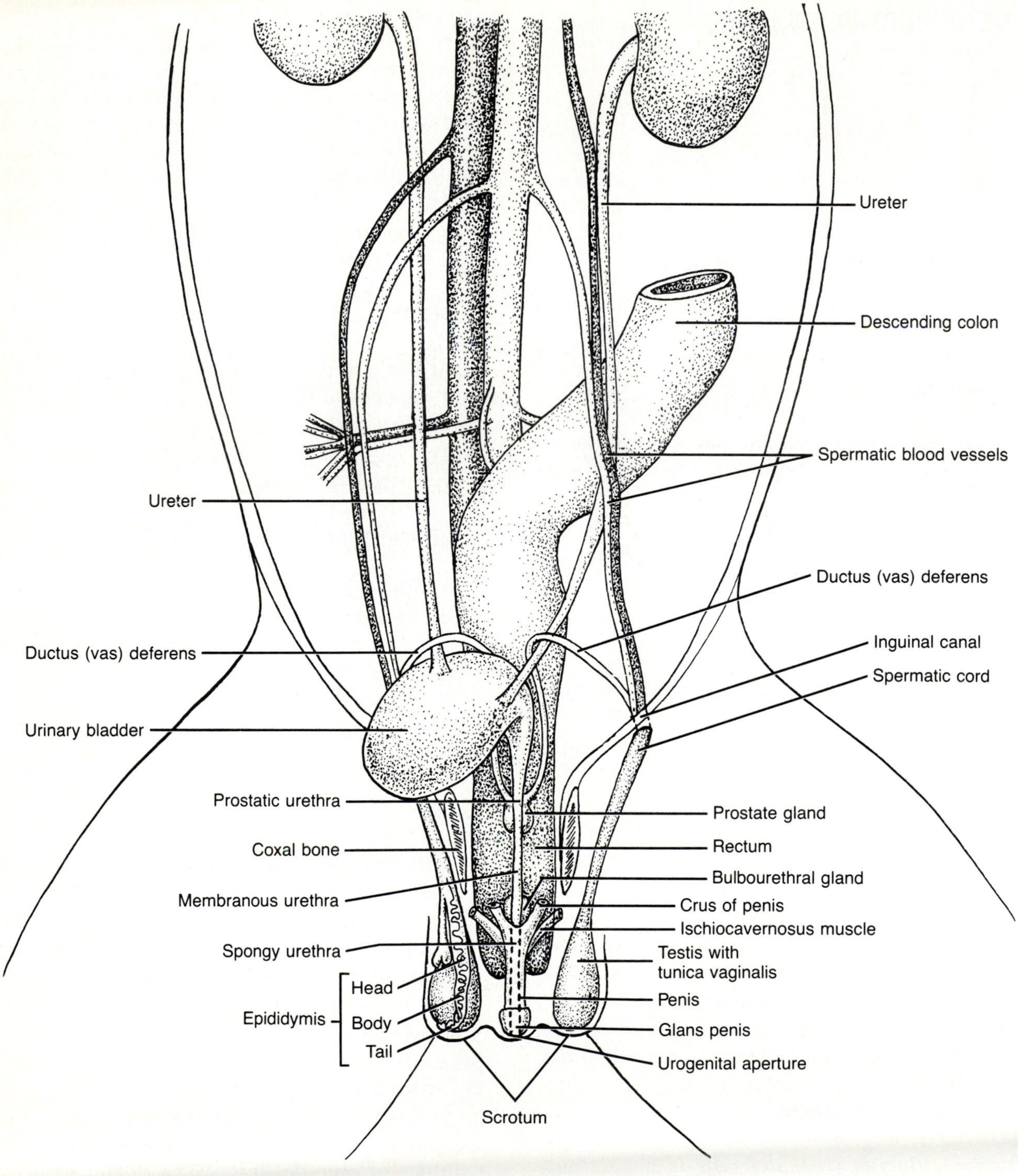

FIGURE 24.18 Male reproductive and urinary systems of cat.

h. OS PENIS

1. This is the small bone embedded on one surface of the urethra in the glans penis.

2. The bone helps to stiffen this part of the penis.

i. ANAL GLANDS

1. These are a pair of round glands beneath the skin on either side of the rectum near the anus.

2. They produce an odoriferous secretion into the anus that is believed to be used for sexual attraction.

2. Female Reproductive System

The female reproductive system of the cat is shown in Figure 23.19.

PROCEDURE

a. OVARIES

1. The ***ovaries*** are a pair of small, oval organs located slightly caudal to the kidneys.

2. The lumpy appearance of the ovaries is due to the presence, within the ovaries, of ***vesicular ovarian (Graafian) follicles***. These are small vesicles that contain ova and protrude to the ovarian surface.

3. Each ovary is suspended by the ***mesovarium,*** a peritoneal fold extending from the dorsal body wall to the ovary.

b. UTERINE (FALLOPIAN) TUBES

1. The paired ***uterine (Fallopian) tubes*** lie on the cranial surfaces of the ovaries.

2. The expanded end of the uterine tube, the ***infundibulum,*** has a fringed border of small fingerlike projections called ***fimbriae*** and also has an opening termed the ***ostium,*** which receives the ovum.

3. The peritoneum of the uterine tubes is called the ***mesosalpinx***. It is continuous with the mesovarium.

c. UTERUS

1. The uterus is a Y-shaped structure consisting of two principal parts—uterine horns and body.

2. The ***uterine horns*** are the enlarged caudal communications of the uterine tubes. The embryos develop in the horns.

3. The ***body*** of the uterus is the caudal, median portion of the uterus. It is located between the urinary bladder and the rectum.

d. BROAD LIGAMENT

1. The uterine horns and body of the uterus are supported by a fold of peritoneum called the ***mesometrium.*** It extends from these structures to the lateral body walls.

2. Together, the mesometrium, mesosalpinx, and mesovarium are referred to as the ***broad ligament.***

e. ROUND LIGAMENT

1. The ***round ligament*** is a thin fibrous band that extends from the dorsal body wall to the middle of each uterine horn.

2. The remainder of the female reproductive system can be seen only if the pelvic cavity is exposed. You will have to cut through the pelvic muscles and symphysis pubis at the midline.

3. Spread the thighs back.

f. VAGINA

1. Before examining the vagina, locate the ***urethra,*** a tube caudal to the urinary bladder.

2. Dorsal to the urethra is the ***vagina,*** a tube leading from the body of the uterus to the ***urogenital sinus (vestibule).*** This is a common passageway formed by the union of the urethra and vagina.

g. VULVA

1. The urogenital sinus opens to the exterior as the ***urogenital aperture,*** just ventral to the anus.

2. On either side of the urogenital aperture are folds of skin called the ***labia majora.***

3. In the ventral wall of the urogenital sinus is the ***clitoris,*** a mass of erectile tissue that is homologous to portions of the penis.

4. The urogenital aperture, labia majora, and clitoris constitute the ***vulva.***

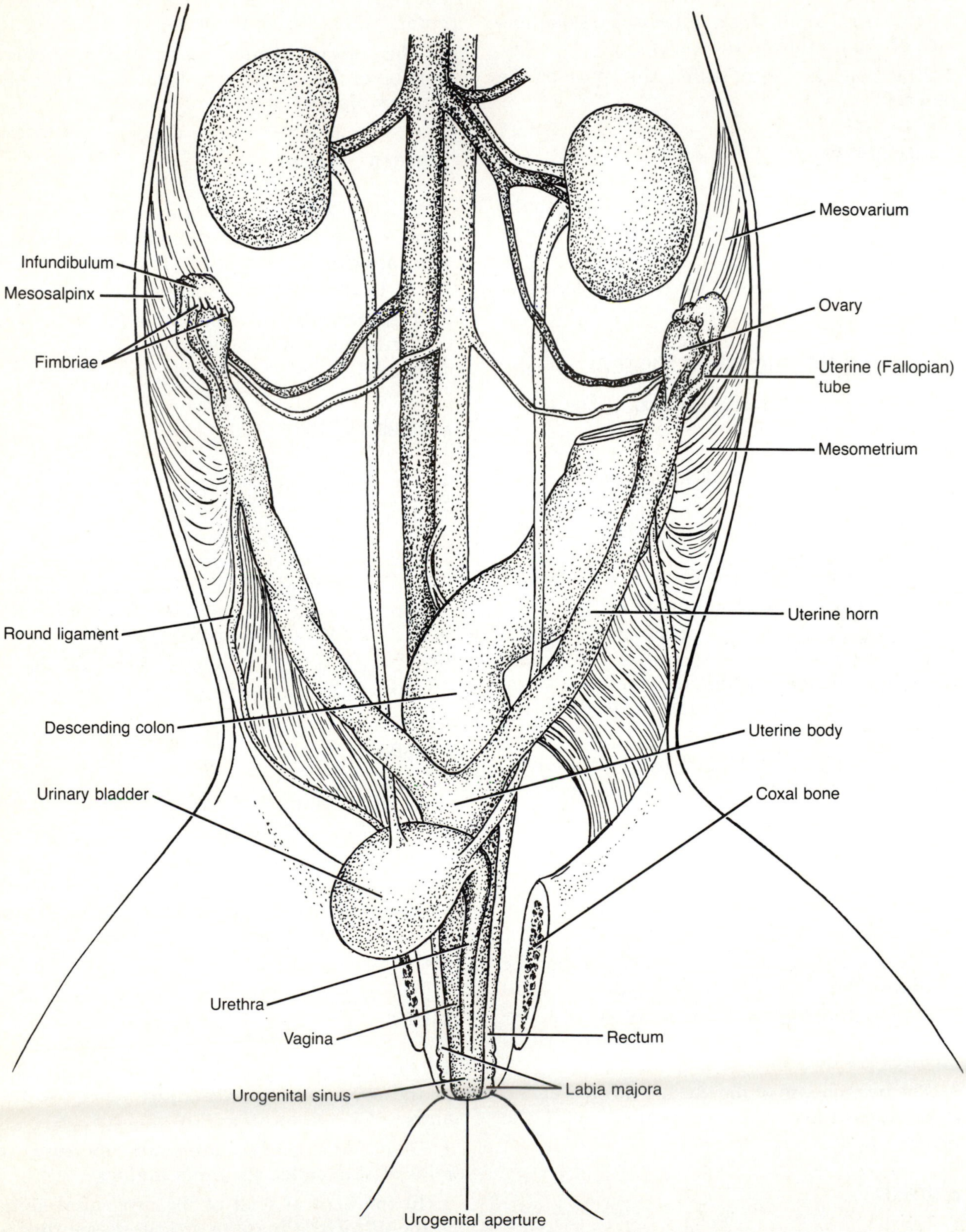

FIGURE 24.19 Female reproductive and urinary systems of cat.

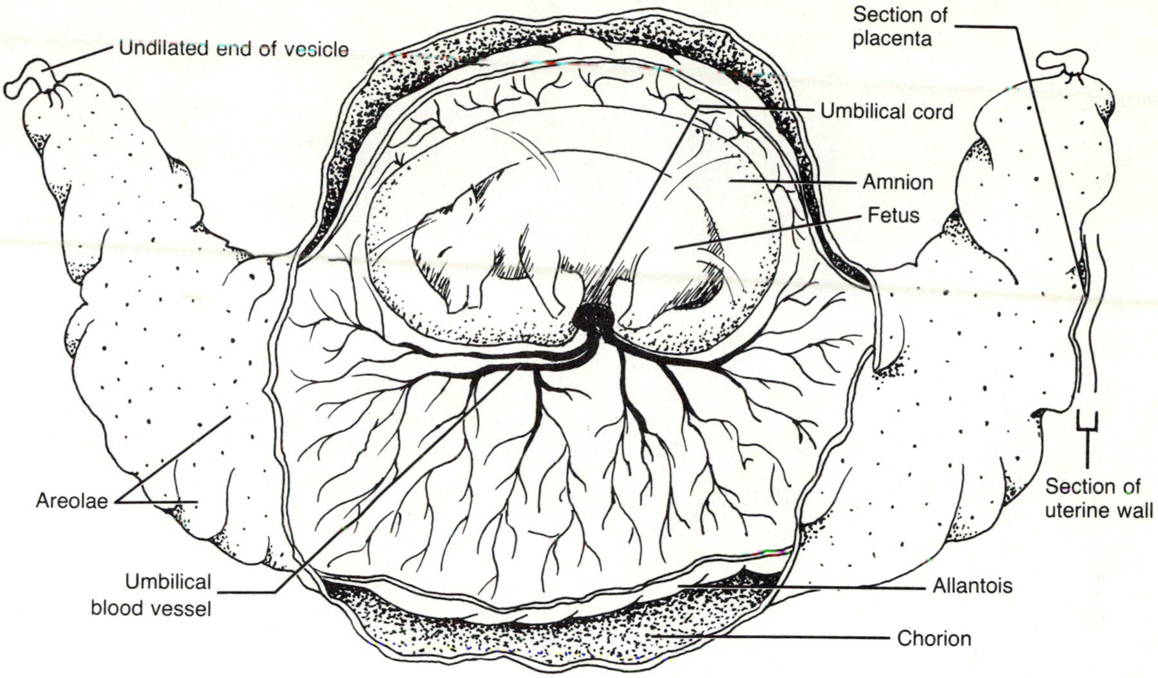

FIGURE 24.20 Fetal pig in opened chorionic vesicle.

D. DISSECTION OF FETUS-CONTAINING PIG UTERUS

Examination of the uterus of a pregnant pig reveals that the fetuses are equally spaced in the two uterine horns. Each fetus produces a local enlargement of the horn. The litter size normally ranges from 6 to 12. Your instructor may have you dissect the fetus-containing uterus of a pregnant pig or have one available as a demonstration (Figure 24.20). If you do a dissection, use the following directions. Also examine a chart or model of a human fetus and pregnant uterus if they are available.

CAUTION! *Please reread Section D. "Precautions Related to Dissection" at the begining of the laboratory manual on page xii before you begin your dissection.*

PROCEDURE

1. Using a sharp scissors, cut open one of the enlargements of the horn and you will see that each fetus is enclosed together with an elongated, sausage-shaped ***chorionic vesicle.***

2. You will also notice many round bumps called ***areolae*** located over the chorionic surface.

3. The lining of the uterus together with the wall of the chorionic vesicle forms the ***placenta.***

4. Carefully cut open the chorionic vesicle, avoiding cutting or breaking the second sac lying within that surrounds the fetus itself.

5. The vesicle wall is the fusion of two extraembryonic membranes, the outer ***chorion*** and the inner ***allantois.*** The allantois is the large sac growing out from the fetus, and the umbilical cord contains its stalk.

6. The ***umbilical blood vessels*** are seen in the allantoic wall spreading out in all directions and are also seen entering the ***umbilical cord.***

7. A thin-walled nonvascular ***amnion*** surrounds the fetus. This membrane is filled with ***amniotic fluid,*** which acts as a protective water cushion and prevents adherence of the fetus and membranes.

ANSWER THE LABORATORY REPORT QUESTIONS AT THE END OF THE EXERCISE

LABORATORY REPORT QUESTIONS

REPRODUCTIVE SYSTEMS 24

Student ______________________ **Date** ______________________

Laboratory Section ______________________ **Score/Grade** ______________________

PART I. MULTIPLE CHOICE

_____ 1. Structures of the male reproductive system responsible for reproduction of sperm are the (a) efferent ducts (b) seminiferous tubules (c) seminal vesicles (d) rete testis

_____ 2. The superior portion of the male urethra is encircled by the (a) epididymis (b) testes (c) prostate gland (d) seminal vesicles

_____ 3. Cryptorchidism is a condition associated with the (a) prostate gland (b) testes (c) seminal vesicles (d) bulbourethral (Cowper's) glands

_____ 4. Weakening of the suspensory ligament would directly affect the position of the (a) mammary glands (b) uterus (c) uterine (Fallopian) tubes (d) ovaries

_____ 5. Organs in the female reproductive system responsible for transporting ova from the ovaries to the uterus are the (a) uterine (Fallopian) tubes (b) seminal vesicles (c) inguinal canals (d) none of the above

_____ 6. The name of the process that is responsible for the actual production of sperm is called (a) cryptorchidism (b) oogenesis (c) spermatogenesis (d) spermatogonia

_____ 7. Fertilization normally occurs in the (a) uterine (Fallopian) tubes (b) vagina (c) uterus (d) ovaries

_____ 8. The portion of the uterus that assumes an active role during labor is the (a) serosa (b) endometrium (c) peritoneum (d) myometrium

_____ 9. Stereocilia are associated with the (a) ductus (vas) deferens (b) oviduct (c) epididymis (d) rete testis

_____ 10. The major portion of the volume of semen is contributed by the (a) bulbourethral (Cowper's) glands (b) testes (c) prostate gland (d) seminal vesicles

_____ 11. The chief ligament supporting the uterus and keeping it from dropping into the vagina is the (a) cardinal ligament (b) round ligament (c) broad ligament (d) ovarian ligament

_____ 12. Which sequence, from inside to outside, best represents the histology of the uterus? (a) stratum basalis, stratum functionalis, myometrium, perimetrium (b) myometrium, perimetrium, stratum functionalis, stratum basalis (c) stratum functionalis, stratum basalis, myometrium, perimetrium (d) stratum basalis, stratum functionalis, perimetrium, myometrium

_____ 13. The white fibrous tissue layer that divides the testis into lobules is called the (a) dartos (b) raphe (c) tunica albuginea (d) germinal epithelium

_____ 14. Which sequence best represents the course taken by spermatozoa from their site of origin to the exterior? (a) seminiferous tubules, efferent ducts, epididymis, ductus (vas) deferens, ejaculatory duct, urethra (b) seminiferous tubules, efferent ducts, epididymis, ductus (vas) deferens, urethra, ejaculatory duct (c) seminiferous tubules, efferent ducts, ductus (vas) deferens, epididymis, ejaculatory duct, urethra (d) seminiferous tubules, epididymis, efferent ducts, ductus (vas) deferens, ejaculatory duct, urethra

_____ 15. The ovaries are anchored to the uterus by the (a) ovarian ligament (b) broad ligament (c) suspensory ligament (d) mesovarium

_____ 16. The terminal duct for the male reproductive system is the (a) urethra (b) ductus (vas) deferens (c) inguinal canal (d) ejaculatory duct

_____ 17. The site of sperm maturation is the (a) ductus (vas) deferens (b) spermatic cord (c) epididymis (d) testes

_____ 18. Which of the following is the site of menstruation, implantation of a fertilized ovum, development of the fetus during pregnancy, and labor? (a) uterus (b) uterine (Fallopian) tubes (c) vagina (d) cervix

_____ 19. Branched tubuloalveolar glands lying over the pectoralis major muscles are (a) lesser vestibular glands (b) adrenal (suprarenal) glands (c) mammary glands (d) greater vestibular (Bartholin's) glands

PART II. COMPLETION

20. Discharge of an immature ovum from the ovary about once each month is a process referred to as _____________ .

21. The inferior, narrow portion of the uterus that opens into the vagina is the _____________ .

22. The milk-secreting cells of the mammary glands are referred to as _____________ .

23. The distal end of the penis is a slightly enlarged region called the _____________ .

24. Covering the slightly enlarged region of the penis is a loosely fitting skin called the _____________ .

25. The circular pigmented area surrounding each nipple of the mammary glands is the _____________ .

26. After a secondary oocyte leaves the ovary, it enters the open, funnel-shaped distal end of the uterine (Fallopian) tube called the _____________ .

27. The portion of a spermatozoon that contains the nucleus and acrosome is the _____________ .

28. Vasectomy refers to removal of a portion of the _____________ .

29. The mass of erectile tissue in the penis that contains the spongy urethra is the _____________ .

30. Both the vesicular ovarian (Graafian) follicle and _____________ of the ovary secrete hormones.

31. The superior dome-shaped portion of the uterus is called the _____________ .

32. The _____________ anchor the uterus to either side of the pelvic cavity.

33. The passageway for menstrual flow and inferior portion of the birth canal is the _____________ .

34. Two longitudinal folds of skin that extend inferiorly and posteriorly from the mons pubis and are covered with pubic hair are the _____________ .

35. The _____________ is a small mass of erectile tissue at the anterior junction of the labia minora.

36. The thin fold of vascularized membrane that borders the vaginal orifice is the ______________________.

37. Complete the following sequence for the passage of milk: alveoli, secondary tubules, lactiferous sinuses, ______________________, lactiferous ducts, nipple.

38. The layer of simple epithelium covering the free surface of the ovary is the ______________________.

39. The phase of the menstrual cycle between days 6 and 13 during which endometrial repair occurs is the ______________________ phase.

40. During menstruation, the stratum ______________________ of the endometrium is sloughed off.

DEVELOPMENT 25

Development refers to the sequence of events starting with fertilization of a secondary oocyte and ending with the formation of a complete organism. Consideration will be given to how reproductive cells are produced and to a few developmental events associated with pregnancy.

A. SPERMATOGENESIS

The process by which the testes produce haploid (n) spermatozoa involves several phases, including meiosis, and is called ***spermatogenesis*** (sper′-ma-tō-JEN-e-sis). In order to understand spermatogenesis, review the following concepts.

1. In sexual reproduction, a new organism is produced by the union and fusion of sex cells called ***gametes.*** Male gametes, produced in the testes, are called sperm cells, and female gametes, produced in the ovaries, are called ova.

2. The cell resulting from the union and fusion of gametes—called a ***zygote***—contains a mixture of chromosomes (DNA) from the two parents. Through repeated mitotic cell divisions, a zygote develops into a new organism.

3. Gametes differ from all other body cells (somatic cells) in that they contain the ***haploid*** (one-half) ***chromosome number,*** symbolized as n. In humans, this number is 23, which comprises a single set of chromosomes. Uninucleated somatic cells contain the ***diploid chromosome number***, symbolized as $2n$. In humans, this number is 46, which composes two sets of chromosomes.

4. In a diploid cell, two chromosomes that belong to a pair are called ***homologous chromosomes (homologues).*** In human diploid cells, the members of 22 of the 23 pairs of chromosomes are morphologically similar and are called ***autosomes.*** The other pair comprises the ***sex chromosomes,*** designated as X and Y. In the female, the homologous pair of sex chromosomes consists of the X chromosomes; in the male, the pair consists of and X and a Y.

5. If gametes were diploid ($2n$), like somatic cells, the zygote would contain twice the diploid number ($4n$), and with every succeeding generation the chromosome number would continue to double.

6. This continual doubling of the chromosome number does not occur because of ***meiosis,*** a process of cell division by which gametes produced in the testes and ovaries receive the haploid chromosome number. Thus, when haploid (n) gametes fuse, the zygote contains the diploid chromosome number ($2n$).

In humans, spermatogenesis takes about 74 days. The seminiferous tubules are lined with immature cells called ***spermatogonia*** (sper′-ma-tō-GŌ-nē-a), or sperm mother cells (see Figure 24.2). Spermatogonia contain the diploid ($2n$) chromosome number and represent a heterogenous group of cells in which three subtypes can be distinguished. These are referred to as pale type A, dark type A, and type B, and are distinguished by the appearance of their nuclear chromatin. Pale type A spermatogonia remain relatively undifferentiated and capable of extensive mitotic division. Following division, some of the daughter cells remain undifferentiated and serve as a reservoir of precursor cells to prevent depletion of the stem cell population. Such cells remain near the basement membrane. The remainder of the daughter cells differentiate into type B spermatogonia. These cells lose contact with the basement membrane of the seminiferous tubule, undergo certain developmental changes, and become known as ***primary spermatocytes*** (SPER-ma-tō-sītz'). Primary spermatocytes, like spermatogonia, are diploid (2*n*), that is, they have 46 chromosomes. Dark type A spermatogonia are believed to represent reserve stem cells, only becoming activated if pale type A cells become critically depleted.

1. Reduction Division (Meiosis I)

Each primary spermatocyte enlarges before dividing. Then two nuclear divisions take place as part of meiosis. In the first, DNA is replicated and 46 chromosomes (each made up of two chromatids) form and move toward the equatorial plane of the nucleus. There they line up by homologous pairs so that there are 23 pairs of duplicated chromosomes in the center of the nucleus. This pairing of homologous chromosomes is called ***synapsis.*** The four chromatids of each homologous pair then become associated with each other to form a ***tetrad.*** In a tetrad, portions of one chromatid may be exchanged with portions of another. This process, called ***crossing-over,*** permits an exchange of genes among chromatids (Figure 25.1) that results in the recombination of genes. Thus, the spermatozoa eventually produced are genetically unlike each other and unlike the cell that produced them—one reason for the great variation among humans. Next the meiotic spindle forms and the chromosomal microtubules produced by the centromeres of the paired chromosomes extend toward the poles of the cell. As the pairs separate, one member of each pair migrates to opposite poles of the dividing nucleus. The random arrangement of chromosome pairs on the spindle is another reason for variation among humans because some of the 23 chromosomes from one parent migrate to each pole. The cells formed by the first nuclear division (reduction division) are called ***secondary spermatocytes.*** Each cell has 23 chromosomes—the haploid number. Each chromosome of the secondary spermatocytes, however, is made up of two chromatids. Moreover, the genes of the chromosomes of secondary spermatocytes may be rearranged as a result of crossing-over.

2. Equatorial Division (Meiosis II)

The second nuclear division of meiosis is equatorial division. There is no replication of DNA. The chromosomes (each composed of two chromatids) line up in single file around the equatorial plane, and the chromatids of each chromosome separate from each other. The cells formed from the equatorial division are called ***spermatids***. Each contains half the original chromosome number, or 23 chromosomes, and is haploid. Each primary spermatocyte therefore produces four spermatids by meiosis (reduction division and equatorial division). Spermatids lie close to the lumen of the seminiferous tubule.

3. Spermiogenesis

The final stage of spermatogensis, called ***spermiogenesis*** (sper'-mē-ō-JEN-e-sis), involves the maturation of spermatids into spermatozoa. Each spermatid embeds in a sustentacular (Sertoli) cell and develops a head with an acrosome and a flagellum (tail). Sustentacular cells extend from the basement membrane to

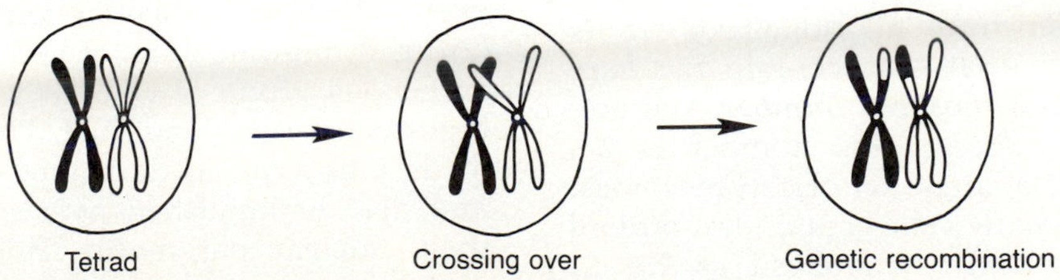

FIGURE 25.1 Crossing over within a tetrad resulting in genetic recombination.

the lumen of the seminiferous tubule where they nourish the developing spermatids. Since there is no cell division in spermiogenesis, each spermatid develops into a single ***spermatozoon (sperm cell).***

Spermatozoa enter the lumen of the seminiferous tubule and migrate to the ductus epididymis, where in 10 to 14 days they complete their maturation and become capable of fertilizing a secondary oocyte. Spermatozoa are also stored in the ductus (vas) deferens. Here, they can retain their fertility for up to several weeks.

With the aid of your textbook, label Figure 25.2.

B. OOGENESIS

The formation of haploid (*n*) secondary oocytes in the ovary involves several phases, including meiosis, and is referred to as oogenesis (ō′-ō-JEN-e-sis). With some important exceptions, oogenesis occurs in essentially the same manner as spermatogenesis.

1. Reduction Division (Meiosis I)

During early fetal development, primordial (primitive) germ cells migrate from the endoderm of the yolk sac to the ovaries. There

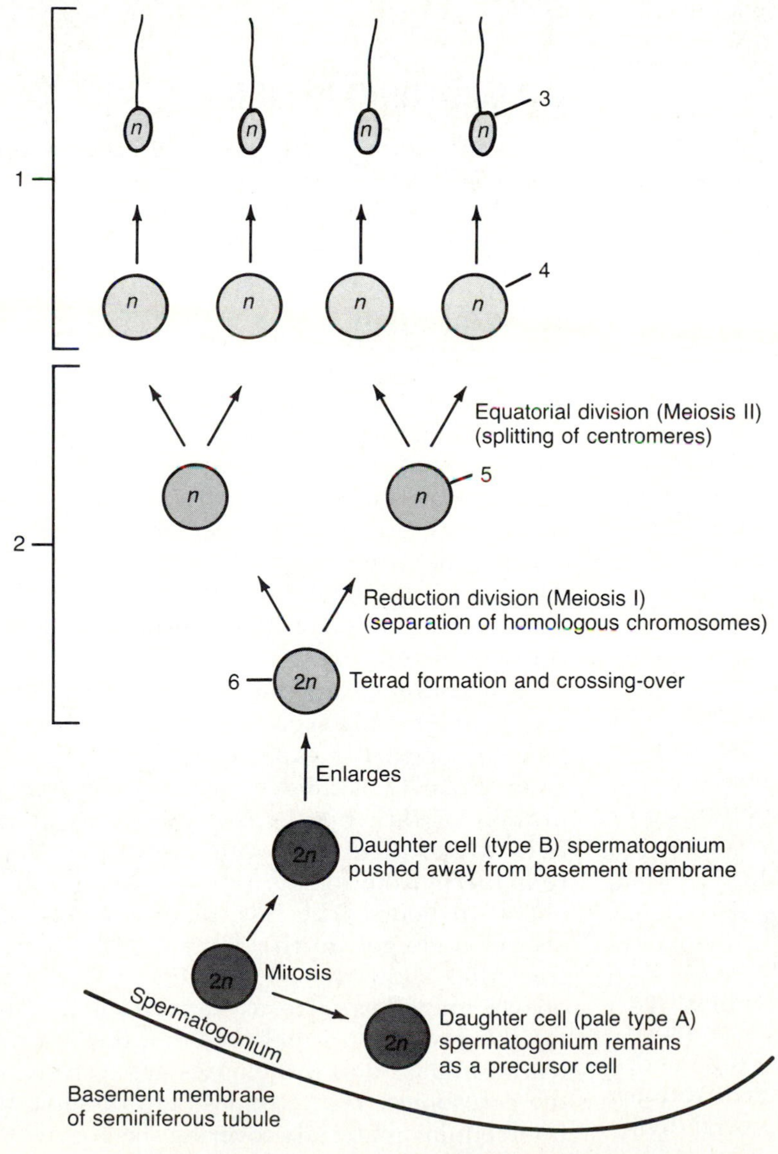

___ Meiosis
___ Primary spermatocyte
___ Secondary spermatocyte
___ Spermatid
___ Spermatozoon
___ Spermiogenesis

FIGURE 25.2 Spermatogenesis.

germ cells differentiate within the ovaries into ***oogonia*** (ō′-o-GŌ-nē-a), diploid ($2n$) cells that divide mitotically to produce a large population of cells. At about the third month of prenatal development, oogonia divide and develop into larger diploid ($2n$) cells called ***primary oocytes*** (Ō′-o-sitz). These cells enter prophase of reduction division (meiosis I) but do not complete it until after the female reaches puberty. Each primary follicle is surrounded by a single layer of flattened epithelial cells (follicular) and the entire structure is called a primary follicle. Primary follicles do not begin further development until they are stimulated by follicle-stimulating hormone (FSH) from the anterior pituitary gland, which, in turn, has responded to gonadotropin releasing hormone (GnRH) from the hypothalamus.

Starting with puberty, several primary follicles respond each month to the rising level of FSH. As the preovulatory phase of the menstrual cycle proceeds, one of the primary follicles reaches a stage in which meiosis resumes and the diploid primary oocyte completes reduction division (meiosis I). Synapsis, tetrad formation, and crossing-over occur, and two cells of unequal size, both with 23 chromosomes (n) of two chromatids each, are produced. The smaller cell, called the ***first polar body,*** is essentially a packet of discarded nuclear material. The larger cell, known as the ***secondary oocyte,*** receives most of the cytoplasm. Each secondary oocyte is surrounded by several layers of cuboidal then columnar epithelial cells and the entire structure is called a secondary (growing) follicle. Once a secondary oocyte is formed, it proceeds to the metaphase of equatorial division (meiosis II) and then stops at this stage. The equatorial division (meiosis II) is completed following ovulation and fertilization.

2. Equatorial Division (Meiosis II)

At ovulation, the secondary oocyte with its polar body and some surrounding supporting cells is discharged. The discharged secondary oocyte enters the uterine (Fallopian) tube and, if spermatozoa are present and fertilization occurs, the second division, the equatorial division (meiosis II) is completed.

3. Maturation

The secondary oocyte produces two cells of unequal size, both of them haploid (n). The larger cell eventually develops into an ***ovum,*** or mature egg; the smaller is the ***second polar body.***

The first polar body may undergo another division to produce two polar bodies. If it does, meiosis of the primary oocyte results in a single haploid (n), secondary oocyte and three haploid (n) polar bodies. In any event, all polar bodies disintegrate. Thus each oogonium produces a single secondary oocyte whereas each spermatocyte produces four spermatozoa.

With the aid of your textbook, label Figure 25.3.

C. EMBRYONIC PERIOD

The ***embryonic period*** is the first 2 months of development, and the developing human is called an ***embryo***. After the second month, it is called a ***fetus.***

1. Fertilization

The term ***fertilization*** refers to the penetration of a secondary oocyte by a spermatozoon and the subsequent union of the sperm nucleus and the nucleus of the oocyte (Figure 25.4a). Fertilization normally occurs in the uterine (Fallopian) tube when the oocyte is about one-third of the way down the tube, usually within 24 hours after ovulation. Peristaltic contractions and the action of cilia transport the oocyte through the uterine tube.

In addition to assisting in the transport of sperm, the female reproductive tract also confers on sperm the capacity to fertilize a secondary oocyte. Although sperm undergo maturation in the epididymis, they are still not able to fertilize an oocyte until they have remained in the female reproductive tract for about 10 hours. The functional changes that sperm undergo in the female reproductive tract that allow them to fertilize a secondary oocyte are referred to as ***capacitation.*** During this process, it is believed that the enzymes hyaluronidase and proteinases are secreted by the acrosomes. The enzymes help dissolve the intercellular materials covering the oocyte, the gelatinous glycoprotein layer called the ***zona***

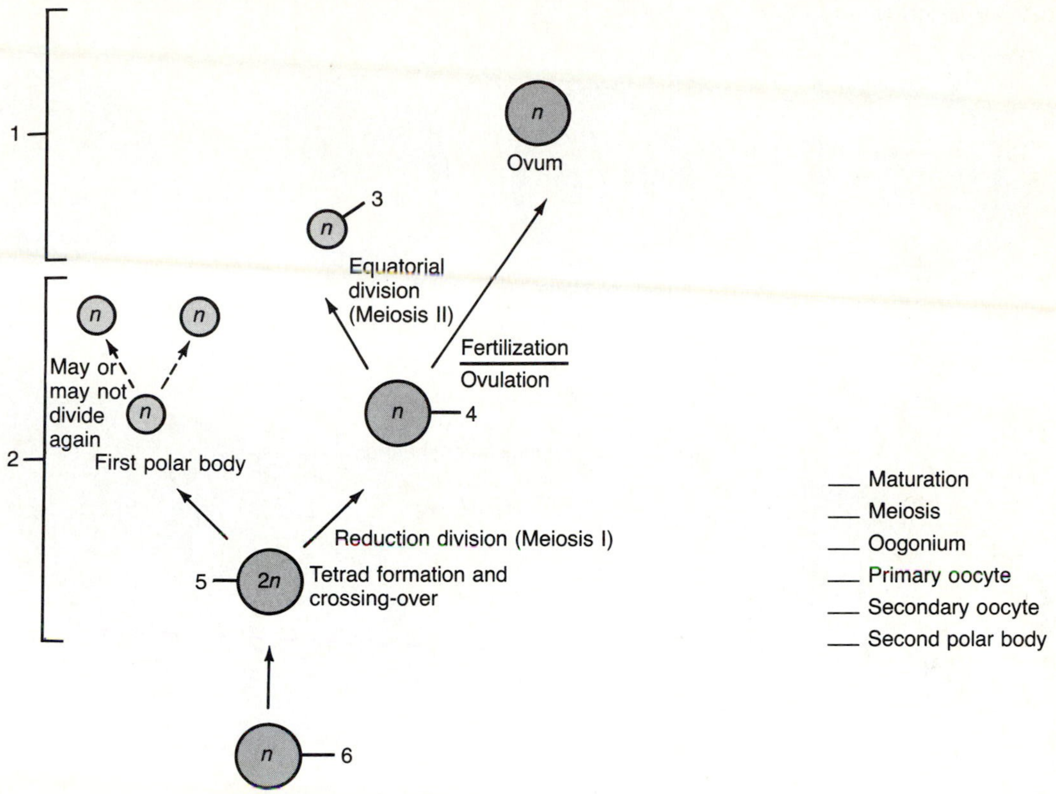

FIGURE 25.3 Oogenesis.

pellucida (pe-LOO-si-da) and several layers of cells, the innermost of which are follicle cells, known as the ***corona radiata*** (Figure 25.4 a and b). Once this is accomplished, normally only one spermatozoon enters and fertilizes and oocyte because once union is achieved, the electrical changes in the surface of the oocyte, block the entry of other sperm, and enzymes produced by the fertilized ovum (egg) alter receptor sites so that sperm already bound are detached and others are prevented from binding. In this way, polyspermy, fertilization by more than one spermatozoon, is prevented.

When a spermatozoon has entered a secondary oocyte, the tail is shed and the nucleus in the head develops into a structure called the ***male pronucleus***. The nucleus of the oocyte develops into a ***female pronucleus***. Once a spermatozoon has entered a secondary oocyte, the oocyte completes equatorial division (meiosis II). The secondary oocyte divides into a larger ovum (mature egg) and a smaller second polar body that fragments and disintegrates. After the pronuclei are formed, they fuse to produce a ***segmentation nucleus***. The segmentation nucleus contains 23 chromosomes (n) from the male pronucleus and 23 chromosomes (n) from the female pronucleus. Thus the fusion of the haploid (n) pronuclei restores the diploid number ($2n$). The fertilized ovum, consisting of a segmentation nucleus, cytoplasm, and enveloping membrane, is called a ***zygote.***

2. Formation of the Morula

Immediately after fertilization, rapid cell division of the zygote takes place. This early division of the zygote is called ***cleavage*** (Figure 25.4a–c). During this time, the dividing cells are contained by the zona pellucida. The progressively smaller cells produced by cleavage are called ***blastomeres*** (BLAS-tō-mērz). A few days after fertilization the successive cleavages have produced a solid mass of cells, the ***morula*** (MOR-yoo-la) or mulberry, which is about the same size as the original zygote (Figure 25.4c).

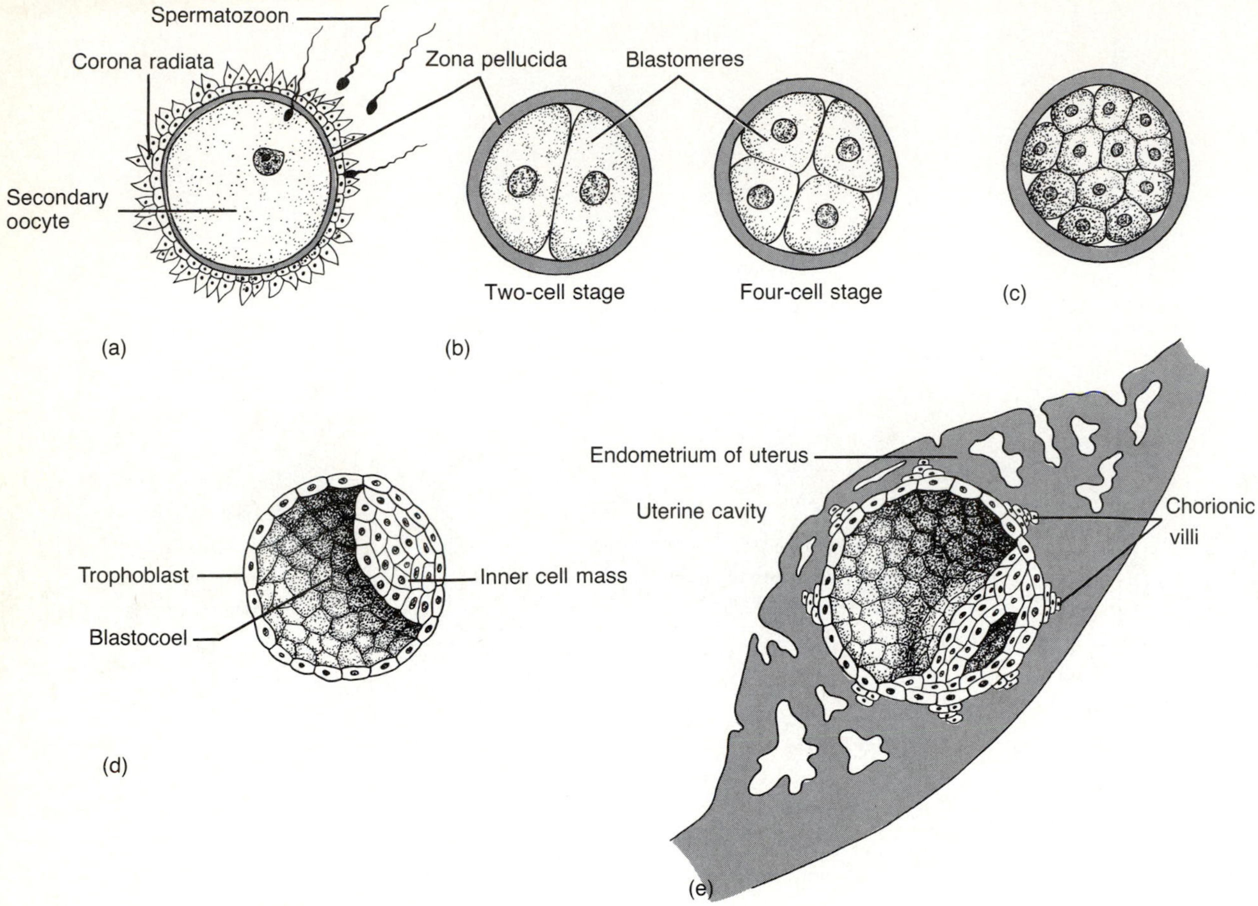

FIGURE 25.4 Fertilization and implantation. (a) Fertilization. (b) Two-cell and four-cell stages of cleavage. (c) Morula. (d) Blastocyst. (e) Implantation.

3. Development of the Blastocyst

As the number of cells in the morula increases, it moves from the original site of fertilization down through the ciliated uterine (Fallopian) tube toward the uterus and enters the uterine cavity. By this time, the dense cluster of cells is altered to form a hollow ball of cells. The mass is now referred to as a ***blastocyst*** (Figure 25.4d). The blastocyst is differentiated into an outer covering of cells called the ***trophoblast*** (TRŌF-ō-blast), an ***inner cell mass (embryoblast)*** and an internal fluid-filled cavity called the ***blastocoel*** (BLAS-tō-sēl). The trophoblast, ultimately forms part of the membranes composing the fetal portion of the placenta; the inner cell mass develops into the embryo.

4. Implantation

The blastocyst remains free within the cavity of the uterus from 2 to 4 days before it actually attaches to the uterine wall. During this time, nourishment is provided by secretions of the endometrium, sometimes called uterine milk. The attachment of the blastocyst to the endometrium occurs 7 to 8 days after fertilization and is called ***implantation*** (Figure 25.4e).

PROCEDURE

1. Obtain prepared slides of the embryonic development of the sea urchin. First try to find a zygote. This will appear as a single cell surrounded by an inner fertilization membrane and an outer, jellylike membrane. Draw a zygote in the spaces provided.

Zygote

Two-cell stage

Four-cell stage

Eight-cell stage

Sixteen-cell stage

Blastula

2. Now find several cleaveage stages. See if you can isolate two-cell, four-cell, eight-cell, and sixteen-cell stages. Draw the various stages in the spaces provided.

3. Try to find a blastula (called a blastocyst in humans), a hollow ball of cells with a lighter center due to the presence of the blastocoel. Draw a blastula in the space provided.

Following implantation, the inner cell mass of the blastocyst differentiates into three ***primary germ layers: ectoderm, endoderm,*** and ***mesoderm***. The primary germ layers are embryonic tissues from which all tissues and organs of the body will develop. Various movements of groups of cells leading to establishment of primary germ layers is called ***gastrulation.*** Cells of the inner cell mass divide by mitosis and form two cavities: the ***amniotic cavity*** and ***extraembryonic coelom*** (Figure 25.5). In gastrulation, the layer of cells nearest the amniotic cavity is the ectoderm; the layer nearest the extraembryonic coelom is the endoderm. Within the ectoderm are cells that form mesoderm between ectoderm and endoderm.

As the embryo develops, the endoderm becomes the epithelium that lines most of the gastrointestinal tract, urinary bladder, gallbladder, liver, pharynx, larynx, trachea, bronchi, lungs, vagina, urethra, and thyroid, parathyroid, and thymus glands, among other structures. The mesoderm develops into all skeletal, all cardiac, and most smooth muscle; cartilage, bone and other connective tissues; bone marrow, lymphoid tissue, endothelium of blood and lymphatic vessels, gonads, dermis of the skin, and other structures. The ectoderm develops into the entire nervous system, epidermis of skin, epidermal derivatives

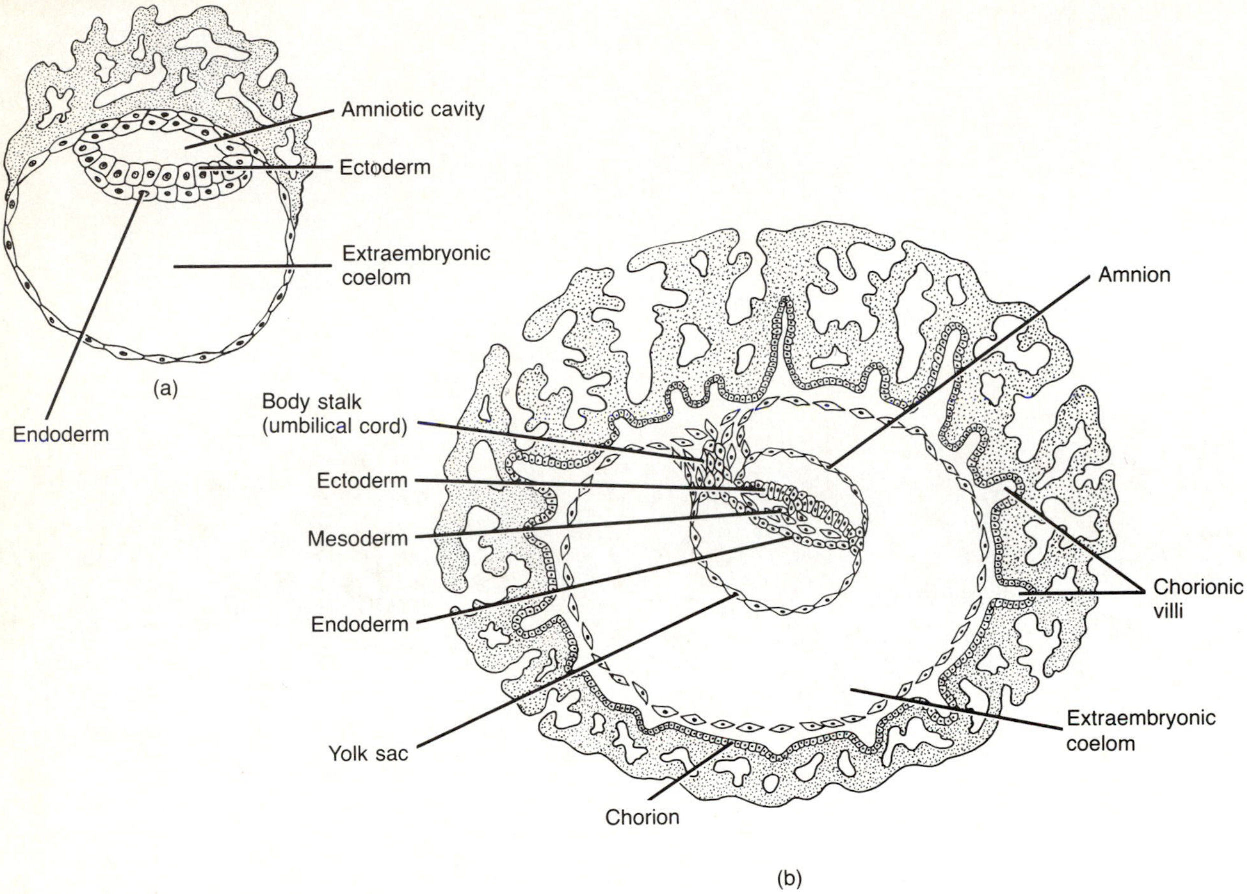

FIGURE 25.5 Formation of the three primary germ layers and embryonic membranes. (a) Early stage. (b) Later stage.

of the skin, and portions of the eye and other sense organs.

During the embryonic period, the ***embryonic membranes*** form, lying outside of the embryo. They function to protect and nourish the embryo and later the fetus. These membranes include the ***yolk sac, amnion, chorion*** (KŌ-rē-on), and ***allantois*** (a-LAN-tō-is) (see Figures 25.5 and 25.6).

The ***placenta*** is a mostly vascular organ formed by the chorion of the embryo and a portion of the endometrium (decidua basalis) of the mother (Figure 25.5). It functions to exchange nutrients and wastes between fetus and mother and secretes the hormones necessary to maintain pregnancy. When the baby is delivered, the placenta detaches from the uterus and is called the afterbirth. The scar that marks the site of the entry of the fetal umbilical cord into the abdomen is called the ***umbilicus.*** The umbilical cord contains blood vessels that (1) deliver fetal blood containing carbon dioxide and wastes to the placenta and (2) return fetal blood containing oxygen and nutrients from the placenta.

D. FETAL PERIOD

During the ***fetal period,*** all the organs of the body grow rapidly from the original primary germ layers, and the organism takes on a human appearance. Some of the principal changes associated with fetal growth are summarized in Table 25.1.

ANSWER THE LABORATORY REPORT QUESTIONS AT THE END OF THE EXERCISE

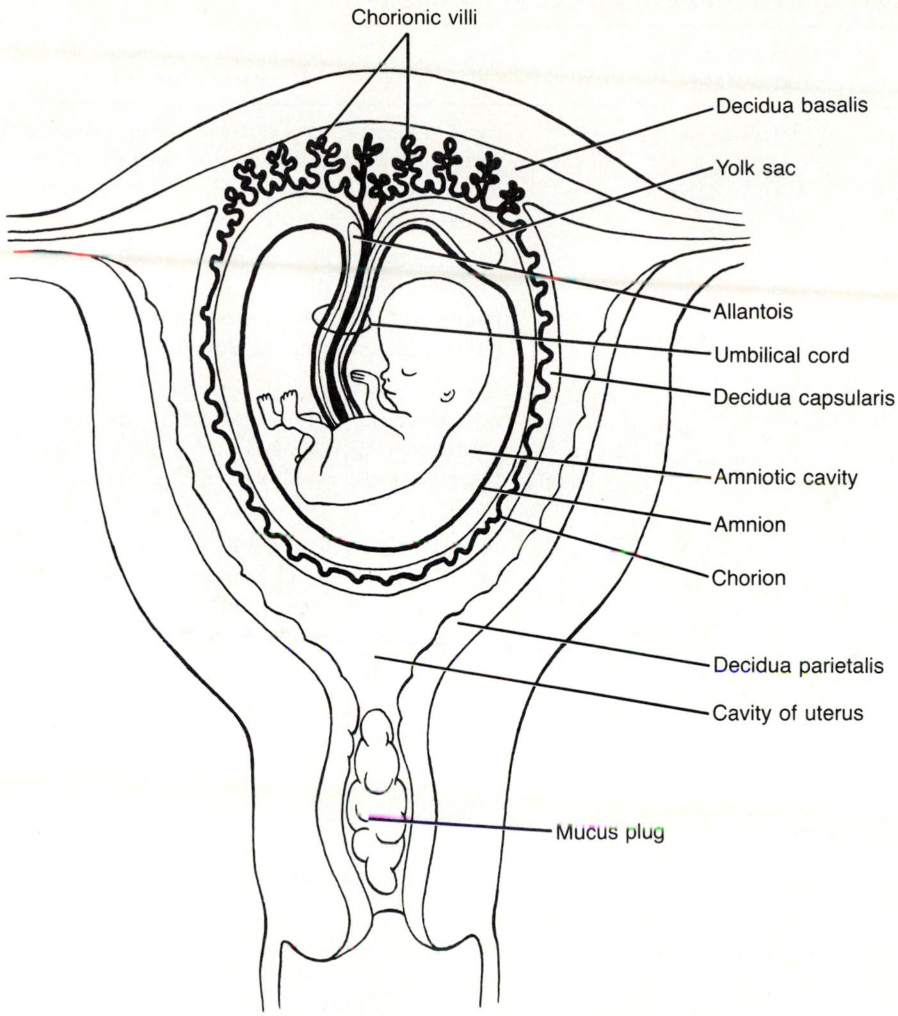

FIGURE 25.6 Embryonic membranes.

TABLE 25.1
CHANGES ASSOCIATED WITH EMBRYONIC AND FETAL GROWTH

End of month	Approximate size and weight	Representative changes
1	0.6 cm (3/16 in.)	Eyes, nose and ears not yet visible. Backbone and vertebral canal form. Small buds that will develop into upper and lower extremities form. Heart forms and starts beating. Body systems begin to form.
2	3 cm (1¼ in.) 1 g (1/30 oz)	Eyes far apart, eyelids fused, nose flat. Ossification begins. Limbs become distinct as upper and lower extremities. Digits well formed. Major blood vessels form. Many internal organs continue to develop.
3	7.5 cm (3 in.) 28 g (1 oz)	Eyes almost fully developed but eyelids still fused, nose develops bridge, external ears present. Ossification continues. Appendages fully formed, nails develop. Heartbeat can be detected. Body systems continue to develop.
4	18 cm (6½–7 in.) 113 g (4 oz)	Head large in proportion to rest of body. Face takes on human features and hair appears on head. Skin bright pink. Many bones ossified, joints begin to form. Continued development of body systems.
5	25–30 cm (10–12 in.) 227–454 g (½–1 lb)	Head less disproportionate to rest of body. Fine hair (lanugo) covers body. Skin still bright pink. Rapid development of body systems.
6	27–35 cm (11–14 in.) 567–681 g (1¼–1½ lb)	Head even less disproportionate to rest of body. Eyelids separate, eyelashes form. Skin wrinkled and pink.
7	32–42 cm (13–17 in.) 1,135–1,362 g (2½–3 lb)	Head and body more proportionate. Skin wrinkled and pink. Seven–month fetus (premature baby) capable of survival.
8	41–45 cm (16½–18 in.) 2,043–2,270 g (4½–5 lb)	Subcutaneous fat deposited. Skin less wrinkled. Testes descend into scrotum. Bones of head soft. Chances of survival much greater at end of eighth month.
9	50 cm (20 in.) 3,178–3,405 g (7–7½ lb)	Additional subcutaneous fat accumulates. Lanugo shed. Nails extend to tips of fingers and maybe even beyond.

LABORATORY REPORT QUESTIONS

DEVELOPMENT 25

Student ______________________ Date ______________________

Laboratory Section ______________________ Score/Grade ______________________

PART I. MULTIPLE CHOICE

_____ 1. The basic difference between spermatogenesis and oogenesis is that (a) two more polar bodies are produced in spermatogenesis (b) the secondary oocyte contains the haploid chromosome number, whereas the mature sperm contains the diploid number (c) in oogenesis, one secondary oocyte is produced, and in spermatogenesis four mature sperm are produced (d) both mitosis and meiosis occur in spermatogenesis, but only meiosis occurs in oogenesis

_____ 2. The union of a sperm nucleus and a secondary oocyte nucleus resulting in formation of a zygote is referred to as (a) implantation (b) fertilization (c) gestation (d) parturition

_____ 3. The most advanced stage of development is the (a) morula (b) zygote (c) ovum (d) blastocyst

_____ 4. Damage to the mesoderm during embryological development would directly effect the formation of (a) muscle tissue (b) the nervous system (c) the epidermis of the skin (d) hair, nails, and skin glands

_____ 5. The placenta, the organ of exchange between mother and fetus, is formed by union of the endometrium with the (a) yolk sac (b) amnion (c) chorion (d) umbilicus

_____ 6. One oogonium produces (a) one ovum and three polar bodies (b) two ova and two polar bodies (c) three ova and one polar body (d) four ova

_____ 7. Implantation is defined as (a) attachment of the blastocyst to the uterine (Fallopian) tube (b) attachment of the blastocyst to the endometrium (c) attachment of the embryo to the endometrium (d) attachment of the morula to the endometrium

_____ 8. Epithelium lining most of the gastrointestinal tract and a number of other organs is derived from (a) ectoderm (b) mesoderm (c) endoderm (d) mesophyll

_____ 9. The nervous system is derived from the (a) ectoderm (b) mesoderm (c) endoderm (d) mesophyll

_____ 10. Which of the following is *not* an embryonic membrane? (a) amnion (b) placenta (c) chorion (d) allantois

PART II. COMPLETION

11. A normal human sperm cell, as a result of meiosis, contains ______________________ chromosomes.

12. The process that permits an exchange of genes resulting in their recombination and a part of the variation among humans is called ______________________ .

13. The result of meiosis in spermatogenesis is that each primary spermatocyte produces four ________________.

14. The stage of spermatogenesis that results in maturation of spermatids into spermatozoa is called ________________.

15. The afterbirth expelled in the final stage of delivery is the ________________.

16. After the second month, the developing human is referred to as a(n) ________________.

17. Embryonic tissues from which all tissues and organs of the body develop are called the ________________.

18. The cells of the inner cell mass divide to form two cavities: amniotic cavity and ________________.

19. Somatic cells that contain two sets of chromosomes are referred to as ________________.

20. At the end of the ________________ month of development, a heartbeat can be detected.

GENETICS 26

Genetics is the branch of biology that studies inheritance. ***Inheritance*** is the passage of hereditary traits from one generation to another. It is through the passage of hereditary traits that you acquired your characteristics from your parents and will transmit your characteristics to your children. If all individuals were brown-eyed, we could learn nothing of the hereditary basis of eye color. However, because some people are blue-eyed and marry brown-eyed people, we can gain some knowledge of how hereditary traits are transmitted. We constantly analyze the genetic bases of the *differences* between individuals. Some of these differences occur normally, such as differences in eye color, blood groups, or ability to taste PTC (phenylthiocarbamide). Other differences are abnormal, such as physical abnormalities and abnormalities in the processes of metabolism.

A. GENOTYPE AND PHENOTYPE

The vast majority of human cells, except gametes, contain 23 pairs of chromosomes (diploid number) in their nuclei. One chromosome from each pair comes from the mother, and the other comes from the father. The two chromosomes that belong to a pair are called ***homologous*** (hō-MOL-ō-gus) ***chromosomes,*** and these homologues contain genes that control the same traits. The homologue of a chromosome that contains a gene for height also contains a gene for height.

The relationship of genes to heredity can be illustrated by the disorder called phenylketonuria, or PKU (see Figure 26.1). People with PKU are unable to manufacture the enzyme phenylalanine hydroxylase. Current belief is that PKU results from the presence of an abnormal gene symbolized as *p*. The normal gene is symbolized as *P*. *P* and *p* are said to be alleles. An ***allele*** is one of many alternative forms of a gene, occupying the same ***locus*** (position of a gene on a chromosome) in homologous chromosomes. The chromosome that has the gene that directs phenylalanine hydroxylase production will either have *p* or *P* on it. Its homologue will also have either *p* or *P*. Thus every individual will have one of the following genetic makeups, or ***genotypes*** (JĒ-nō-tīps): *PP, Pp,* or *pp*. Although people with genotypes of *Pp* have the abnormal gene, only those with genotype *pp* suffer from the disorder because the normal gene masks the abnormal one. A gene that masks the expression of its allele is called the ***dominant gene,*** and the trait expressed is said to be a dominant trait. The homologous gene that is masked is called the ***recessive gene.*** The trait expressed when two recessive genes are present is called the recessive trait. Several dominant and recessive traits inherited in human beings are listed in Table 26.1.

Traditionally, the dominant gene is symbolized with a capital letter and the recessive one with a lowercase letter. When the same genes appear on homologous chromosomes, as in *PP* or *pp,* the person is said to be ***homozygous*** for a trait. When the genes on homologous chromosomes are different, however, as in *Pp,* the person is said to be ***heterozygous*** for the trait. ***Phenotype*** (FĒ-nō-tīp) refers to how the genetic composition is expressed in the body. An individual with *Pp* has a different genotype than one with *PP,* but both have the same phenotype—which in this case is normal production of phenylalanine hydroxylase.

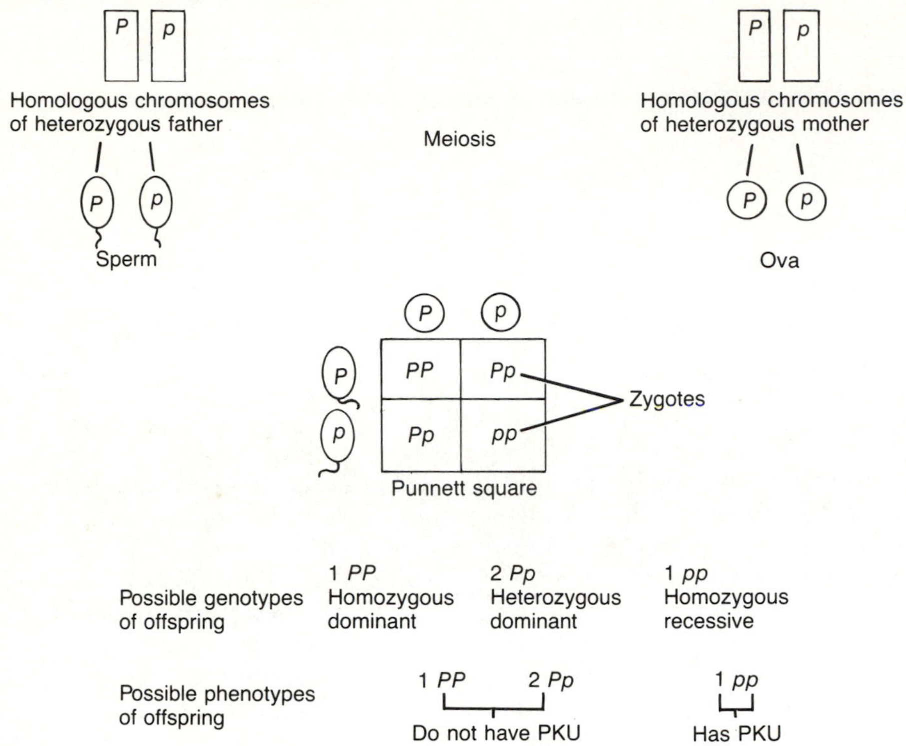

FIGURE 26.1 Inheritance of phenylketonuria (PKU).

TABLE 26.1
HEREDITARY TRAITS IN HUMAN BEINGS

Dominant	Recessive
Curly hair	Straight hair
Dark brown hair	All other colors
Normal skin pigmentation	Albinism
Nearsightedness or farsightedness	Normal vision
Normal color vision	Color blindness
Syndactylism (webbed digits)	Normal digits
Diabetes insipidus	Normal excretion
Huntington's chorea	Normal nervous system
A or B blood factor	O blood factor
Rh blood factor	No Rh blood factor
Normalcy	Phenylketonuria (PKU)

B. PUNNETT SQUARES

To determine how gametes containing haploid chromosomes unite to form diploid fertilized eggs, special charts called ***Punnett squares*** are used. The Punnett square is merely a device that helps one visualize all the possible combinations of male and female gametes, and is invaluable as a learning exercise in genetics. Usually, the male gametes (sperm cells) are placed at the side of the square and the female gametes (ova) at the top (Figure 26.1). The spaces in the chart represent the possible combinations of male and female gametes that could form fertilized eggs. Possible combinations are determined simply by dropping the female gamete on the left into the two boxes below it and dropping the female gamete on the right into the two spaces under it. The upper male gamete is then moved across to the two spaces in line with it, and the lower male gamete is moved across to the two spaces in line with it.

C. SEX INHERITANCE

Lining up human chromosomes in pairs reveals that the last pair (the twenty-third pair) differs in males and in females (see Figure 26.2a). In females, the pair consists of two rod-shaped chromosomes designated as *X* chromosomes. One *X* chromosome is also present in males, but its mate is hook-shaped and

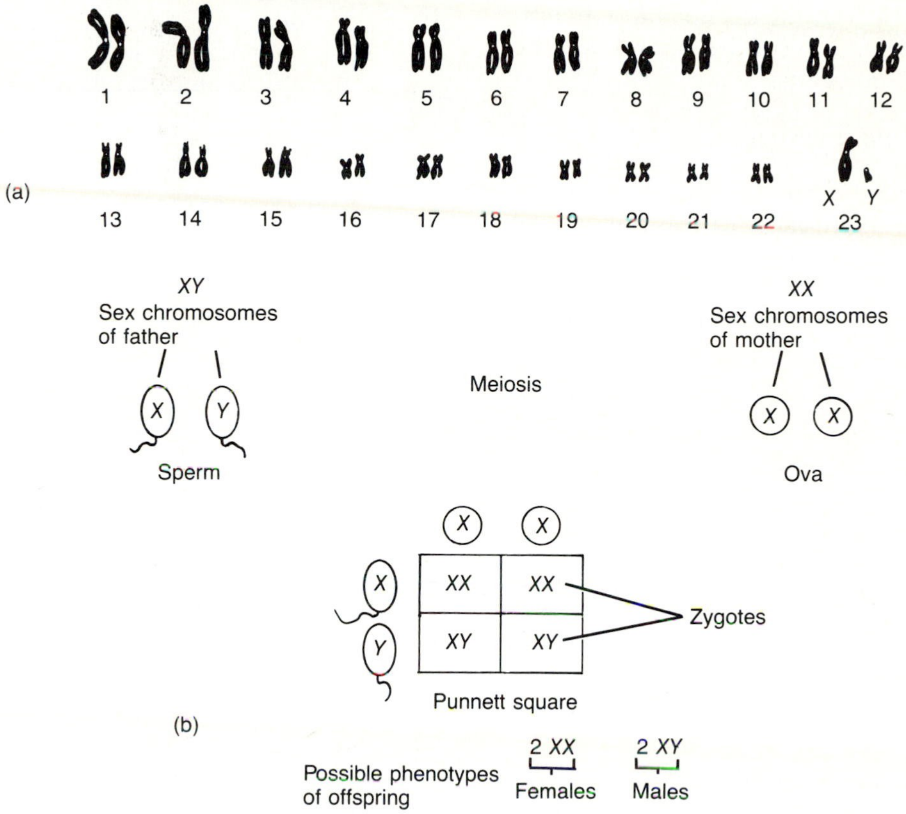

FIGURE 26.2 Inheritance of sex. (a) Human male chromosomes. Note the sex chromosomes, *X* and *Y*. (b) Sex determination.

called a *Y* chromosome. The *XX* pair in the female and the *XY* pair in the male are called the ***sex chromosomes,*** and all other pairs of chromosomes are called ***autosomes.***

The sex of an individual is determined by the sex chromosomes (Figure 26.2b). When a spermatocyte undergoes meiosis to reduce its chromosome number from diploid to haploid, one daughter cell will contain the *X* chromosome and the other will contain the *Y* chromosome. When the secondary oocyte is fertilized by an *X*-bearing sperm, the offspring normally will be a female *(XX)*. Fertilization by a *Y* sperm normally produces a male (*XY*).

Sometimes chromosomes fail to move toward opposite poles of a cell in meiotic anaphase. This is called ***non-disjunction*** and results in one sex cell having 2 members of a chromosome pair while the other receives none. Thus, eggs can contain 2 *X's* or no *X* (symbolized as *0*), and sperm may contain both an *X* and a *Y* chromosome, 2 *X's* or 2 *Y's*, or no sex chromosomes at all.

Because the *X* chromosome contains so many genes unrelated to sex that are necessary for development, a zygote must contain at least one *X* chromosome to survive. Thus, *Y0* and *YY* zygotes do not develop. However, other zygotes with sex chromosome anomalies do develop. Examples are Turner's and Kleinfelter's syndromes.

"Extra" *X* chromosomes (more than 2 in the female and more than 1 in the male) have a surprisingly minor effect on the individual, compared with the significant effect of other additional chromosomes. Studies show that only one *X* chromosome is active in any cell. Any additional *X* chromosomes are randomly inactivated early in development and do not express the genes contained on them.

These inactivated *X* chromosomes remain tightly coiled against the cell membrane and can be seen as what are called ***Barr bodies*** (Figure 26.3). Since *XY* males do not have inactivated *X* chromosomes, no Barr bodies will be seen in normal male cells.

PROCEDURE

1. Make a buccal smear by *gently* scraping the inside of your cheek with the flat end of a toothpick. Discard the toothpick and *gently* scrape the same area again. This will produce more live cells from a deeper layer of the epithelium of the mucous membrane.
2. Spread the material over a clean glass slide.
3. Promptly place the slide in a Coplin jar filled with fixative for one minute.
4. Remove the slide and wash gently under running tap water.
5. Place the slide in a Coplin jar filled with Giemsa stain for 10–20 minutes. Fresh solutions of stain prepared within 2 hours require 10 minutes; older stains require a longer time.
6. Wash the slide *gently* under running water and air dry.
7. Examine the slide under high-power and look for interphase nuclei. Identify Barr bodies, small disc-shaped chromatin bodies lying against the nuclear membrane (see Figure 26.3). Depending on the position of the nuclei and the staining technique, Barr bodies should be seen in 30–70% of the cells of a normal female.

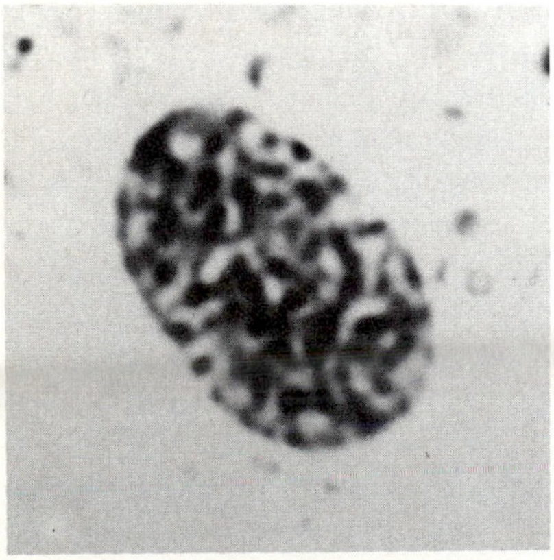

FIGURE 26.3 Barr body in cell from buccal mucosa of human female. Feulgen stain; magnification ×3000.

8. Examine a slide prepared from the buccal epithelium of a class member not of your sex.
9. Draw a cell containing a Barr body in the space provided.

Barr body

D. X-LINKED INHERITANCE

As do the other 22 pairs of chromosomes, the sex chromosomes contain genes that are responsible for the transmission of a number of nonsexual traits. Genes for these traits appear on *X* chromosomes, but many of these genes are absent from *Y* chromosomes. Traits transmitted by genes on the *X* chromosome are called ***X-linked (sex-linked) traits.*** This pattern of heredity is different from the pattern described earlier. About 150 *X*-linked traits are known in humans. Examples of *X*-linked traits are color blindness and hemophilia.

1. Color Blindness

The gene for ***color blindness*** is a recessive one symbolized as *c*. Normal vision, symbolized *C*, dominates. The *C* and *c* genes are located on the *X* chromosome. The *Y* chromosome, however, does not contain the segment of DNA that programs color vision. Thus the ability to see colors depends entirely on the *X* chromosome. The genetic possibilities and the inheritance of color blindness are shown in Figure 26.4. Only females who have two X^c chromosomes are color blind. In X^CX^c females the trait for color blindness is inhibited by the normal dominant gene. Males, however, do not have a second *X* chromosome that would inhibit the trait. Therefore, all males with an X^c chromosome will be color blind. Their pair of sex chromosomes is X^cY.

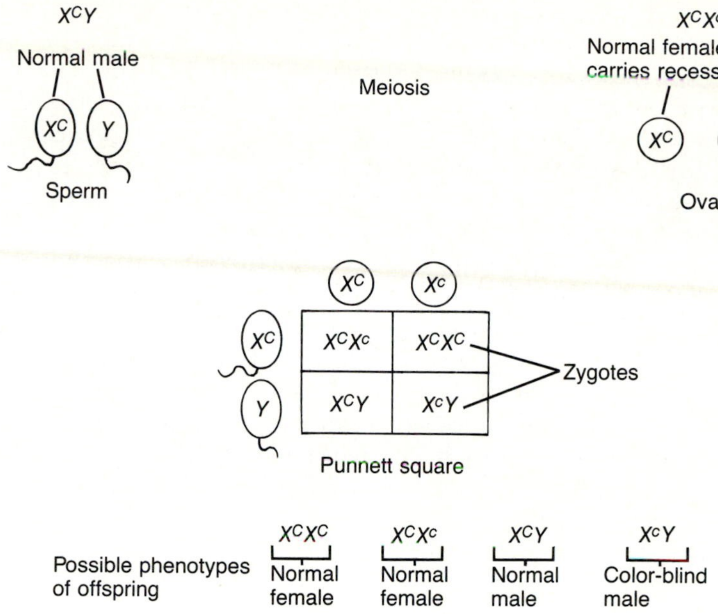

FIGURE 26.4 Inheritance of color blindness.

2. Hemophilia

Hemophilia is a condition in which the blood fails to clot or clots very slowly after an injury. Hemophilia is a much more serious defect than color blindness because individuals with severe hemophilia can bleed to death from even a small cut. Hemophilia is caused by a recessive gene as is color blindness. If H represents normal clotting and h represents abnormal clotting, X^hX^h females will be hemophiliacs. Males with X^HY will be normal and males with X^hY will be hemophiliacs. Other X-linked traits in human beings are certain forms of diabetes, night blindness, juvenile glaucoma, and juvenile muscular dystrophy.

E. MENDELIAN LAWS

In any genetic cross, all the offspring in the first (that is, the parental, or P_1) generation are symbolized as F_1. The F is from the Latin word *filial,* which means progeny. The second generation is symbolized as F_2, the third as F_3, and continues that way. The recognized "father" of genetics is Gregor Mendel, whose basic experiments were performed on garden peas. As a result of his tests, Gregor Mendel postulated what are now called ***Mendelian Laws,*** or ***Mendelian Principles.*** The ***First Mendelian Law,*** or the ***Law of Segregation,*** asserts that, in cells of individuals, genes occur in pairs, and that when those individuals produce germ cells, each germ cell receives only one member of the pair.

This law applies equally to pollen grains (or sperm) and to ova. The genetic cross is represented as follows:

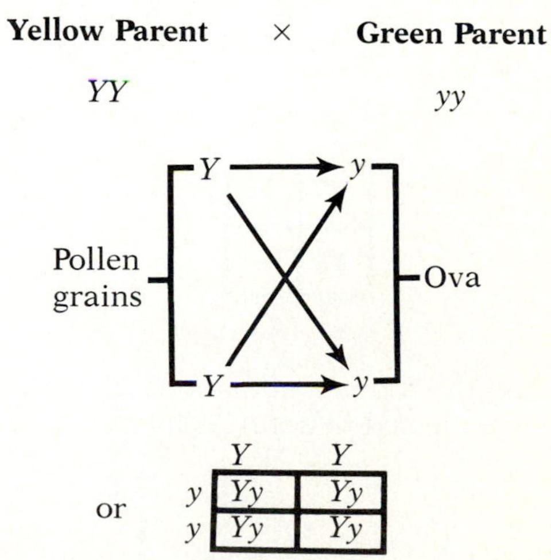

All possible combinations of pollen grains and ova are indicated by the arrows. Notice that all combinations yield the genotype Yy. All these F_1 seeds were yellow, yellow being dominant to green, or, in genetic terms, *Y* being dominant to *y*. These F_1 individuals resembled the yellow parent in phenotype (being yellow) but not in genotype (*Yy* as opposed to *YY*). Both parents were homozygous. Both members of that pair of alleles were the same. The yellow parent was homozygous for *Y* and the green parent for *y*. The F_1 individuals were heterozygous, having one *Y* and one *y*.

When the F_1 plants were self-fertilized, the F_2 seeds appeared in the ratio of 3 yellow/1 green. Mendel found similar 3:1 ratios for the other traits he studied, and this type of result has been reported in many species of animals and plants for a variety of traits. Not only does the recessive trait reappear in the F_2, but also in a definite proportion of the individuals, one-fourth of the total. If the sample is small, the ratio may deviate considerably from 3:1, but as the progeny or sampling numbers get larger, the ratio usually comes closer and closer to an exact 3:1 ratio. The reason is that the ratio depends on the random union of gametes. The result is a 3:1 phenotypic ratio, or a 1:2:1 genotypic ratio.

F_1 **Parents** genotype	**Yellow** *Yy*	×	**Yellow** *Yy*

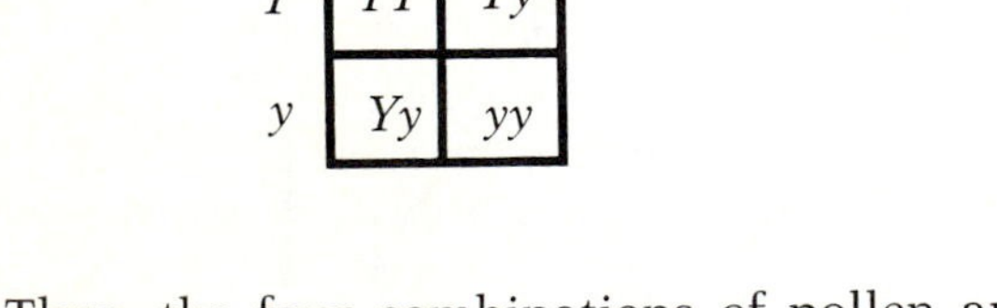

or

	Y	*y*
Y	*YY*	*Yy*
y	*Yy*	*yy*

Thus, the four combinations of pollen and ova are expected to occur as follows:

¼*YY* = yellow ⎤
¼*Yy* = yellow ⎬ ¾
¼*Yy* = yellow ⎦

¼*yy* = green] ¼

It is important to realize that these fractions depend on the operation of the laws of probability. A model using coins will emphasize the point. This model consists of two coins, a nickel and a penny, tossed at the same time. The penny may represent the pollen (male parent). At any given toss, the chances are equal that the penny will come up "heads" or that it will come up "tails." Similarly, at any given fertilization the chances are equal that a *Y*-bearing pollen grain or that a *y*-bearing one will be transmitted. The nickel represents the ovum. Again, the chances are equal for "heads" or "tails," just as the chances are equal that in any fertilization a *Y*-bearing or a *y*-bearing ovum will take part. If we toss the two coins together and do it many times, we will obtain approximately the following:

¼ nickel heads; penny heads	(= *YY*)
¼ nickel heads; penny tails	(= *Yy*)
¼ nickel tails; penny heads	(= *yY*)
¼ nickel tails; penny tails	(= *yy*)

If we assume "heads" as dominant, we find that three-fourths of the time there is at least one "head" and one-fourth of the time no "heads" (both coins "tails"). Hence this gives us a model of the 3:1 ratio dependent on the laws of probability.

The genetic cross just demonstrated considered only one pair of alleles (yellow versus green or "heads" versus "tails") and is therefore called a ***monohybrid cross.*** Mendel's second principle applied to genetic crosses in which two traits or two pairs of alleles were considered. These were ***dihybrid crosses*** and they enabled him to postulate his second principle, the ***Principle of Independent Assortment.*** This principle stated that the segregation of one pair of traits occurred independently of the segregation of a second pair of traits. This is the case only if the traits are caused by genes located on nonhomologous chromosomes.

When Mendel crossed garden peas with round yellow seeds with garden peas with wrinkled green seeds, his F_1 generation showed that yellow and round were dominant. If self-fertilization then occurred, the F_2 generation resulted as follows:

round yellow	¾ × ¾ = 9⁄16
round green	¾ × ¼ = 3⁄16
wrinkled yellow	¼ × ¾ = 3⁄16
wrinkled green	¼ × ¼ = 1⁄16

Therefore, in a dihybrid cross, the expected phenotypic ratio was 9:3:3:1, with 9/16 of the F_2 being doubly dominant, and only 1/16 being doubly recessive.

F. MULTIPLE ALLELES

In the genetics examples we have considered to this point, we have discussed only two alleles of each gene. However, many, and possibly all genes, have ***multiple alleles,*** that is, they exist in more than two allelic forms even though a diploid cell cannot carry more than two alleles.

One example of multiple alleles in humans involves ABO blood groups (Exercise 16). The four basic blood groups of the ABO system are determined by three alleles: I^A, I^B, and i. Alleles I^A and I^B are not dominant over each other (codominant), but are dominant over allele i. These three alleles can give rise to six genotypes as follows:

Genotype	Phenotype (blood type)
I^AI^A or I^AI	A
I^BI^B or I^BI	B
I^AI^B	AB
ii	O

Given this information, is it possible for a child with type O blood to have a mother with type O blood and a father with type AB blood?

Explain ____________________________

____

____

If two children in a family have type O blood, the mother has type B blood and the father has type A blood, what is the genotype of the father? ________________________

What is the genotype of the mother? ________

G. GENETICS EXERCISES

1. Karyotyping

A group of cytogeneticists meeting in Denver, Colorado, in 1960 adopted a system for classifying and identifying human chromosomes. Chromosome ***length*** and ***centromere position*** were the bases for classification. The Denver classification has become a standard for human chromosome studies. By the early 1970s, most human chromosomes could be identified microscopically.

Every chromosome pair could not be identified consistently until chromosome ***banding techniques*** finally distinguished all 46 human chromosomes. Bands are defined as parts of chromosomes that appear lighter or darker than adjacent regions with particular staining methods.

A ***karyotype*** is a chart made from a photograph of the chromosomes in metaphase. The chromosomes are cut out and arranged in matched pairs according to length (see Figure 26.2a). Their comparative size, shape, and morphology are then examined to determine if they are normal or not.

Karyotyping helps scientists to visualize chromosomal abnormalities. For example, individuals with Down's syndrome typically have 47 chromosomes, instead of the usual 46, with chromosome 21 being represented three times rather than only twice. The syndrome is characterized by mental retardation, retarded physical development, and distinctive facial features (round head, broad skull, slanting eyes, and large tongue). With chronic myelogenous leukemia, part of the long arm of a chromosome 22 is missing, resulting in the blood disease. The chromosome is referred to as the Philadelphia chromosome, named for the city where it was first detected.

2. PKU Screening

Phenylketonuria (PKU), an inherited metabolic disorder that occurs in approximately 1 in 16,000 births, is transmitted by an autosomal recessive gene (see Figure 26.1). Individuals with this condition do not have the enzyme phenylalanine hydroxylase, which converts the amino acid phenylalanine to tyrosine. As a result, phenylalanine and phenylpyruvic acid accumulate in the blood and urine. These substances are toxic to the central nervous system and can produce irreversible brain damage. Most states in the United States require routine screening for this disorder at birth. The test is accomplished by a simple color change in treated urine.

The procedure for testing for PKU is as follows:

PROCEDURE

1. A Phenistix® test strip is made specifically for testing urine for phenylpyruvic acid. Dip this test strip in freshly voided urine.

2. Compare the color change with the color chart on the Phenistix® bottle. The test is based on the reaction of ferric ions with phenylpyruvic acid to produce a gray-green color.

3. Record your results in Section G.1 of the LABORATORY REPORT RESULTS at the end of the exercise.

3. PTC Inheritance

The ability to taste the chemical compound known as phenylthiocarbamide, commonly called PTC, is inherited. On the average, 7 out of 10 people, on chewing a small piece of paper treated with PTC, detect a definite bitter or sweet taste. Others do not taste anything.

Individuals who can taste something (bitter or sweet) are called "tasters" and have the dominant allele *T*, either as *TT* or *Tt*. A nontaster is a homozygous recessive and is designated as *tt*.

Determine your phenotype for tasting PTC and record your results in Section G.2 of the LABORATORY REPORT RESULTS at the end of the exercise.

Note: If PTC paper is not available, a 0.5% solution of phenylthiourea (PTT) can be substituted because the capacity to taste PTT is also inherited as a dominant.

4. Corn Genetics

Genetic corn may be purchased and used in this exercise. Each ear of corn represents a family of offspring. Mark a starting row with a pin to avoid repetition. Count the kernels (individuals) for each trait (color, wrinkled, or smooth). Record your results in Section G.3 of the LABORATORY REPORT RESULTS at the end of the exercise.

Develop a ratio by using your lowest number as "1" and dividing it into the others to determine what multiples of it they are. See how close you come to Mendel's ratios. Figure out the probable genotype and phenotype of the parent plants if you can. Monohybrid crosses, test crosses, dihybrid crosses, and trihybrid crosses are available.

5. Color Blindness

Using either Stilling or Ishihara test charts, test the entire class for color blindness. Tests for color blindness depend on the person's ability to distinguish various colors from each other and also on his or her ability to judge correctly the degree of contrast between colors.

Of all men, 2% are color blind to red and 6% to green, so 8% of all men are red-green color blind. Red-green color blindness is rare in the female, occurring in only 1 of every 250 women. Record your results in Section G.4 of the LABORATORY REPORT RESULTS at the end of the exercise.

6. Mendelian Laws of Inheritance

Follow the procedure outlined in the explanation of the Mendelian Law of Segregation, tossing a nickel and a penny simultaneously to prove the law and determine ratios.

PROCEDURE

1. Toss the nickel and the penny together 10 times to get the genotypes of a family of 10. Repeat this procedure for a total of five times to obtain five families of 10 offspring. Record all of the results on the chart in Section G.5 of the LABORATORY REPORT RESULTS at the end of the exercise.

Note: Use the following symbols for the following exercises.

G = gene for yellow
g = gene for green
GG = the genotype of an individual pure (homozygous) for yellow
gg = the genotype of an individual pure (homozygous) for green
Gg = the genotype of the hybrid (heterozygous) individual, phenotypically yellow
♀ = symbol for female
♂ = symbol for male

2. Obeying the Mendelian Law of Segregation and using the Punnett square shown in Section G.6 of the LABORATORY REPORT RESULTS at the end of the exercise, cross yellow garden peas with green garden peas (a monohybrid cross). Show the P_1, F_1, and F_2 generations and all the different phenotypes and genotypes.

3. Obeying the Mendelian Law of Independent Assortment and using the Punnett square shown in Section G.8 of the LABORATORY REPORT RESULTS at the end of the exercise, cross the round yellow seeds with the wrinkled green seeds (a dihybrid cross). Show the P_1, F_1, and F_2 generations and all the different phenotypes and genotypes. The F_2 generation can be generated from a Punnett square comparable to that for the monohybrid cross, but with 16 rather than 4 squares.

7. Observing Phenotypes

The pattern of inheritance of many human traits is complex and involves many genes; the inheritance of other traits is controlled by single genes. You are asked to record your phenotype and your genotype, if it can be determined, for several traits controlled by single genes. For example, if you have the phenotypically dominant trait *A*, your genotype will be *A* –, indicating that the allele symbolized as "–" is not known, since you could be homozygous dominant (*AA*) or heterozygous (*Aa*) for the trait. If you have the recessive trait *a* –, your genotype will be *aa*. Record your phenotype and genotype for the following traits in Section G.10 of the LABORATORY REPORT RESULTS at the end of the exercise.

1. ***Attached ear lobes.*** The dominant gene *E* causes earlobes which develop free from the neck; *ee* results in adherent ear lobes connected to the cervical skin.

2. ***Tongue rolling.*** The dominant gene *R* causes the development of muscles that allow the tongue to be rolled into a U-shape. The *rr* genotype prohibits such rolling.

3. ***Hair whorl direction.*** The dominant gene *W* causes the hair whorl on the cranial surface of the scalp to turn in a clockwise direction; the genotype *ww* determines a counterclockwise whorl.

4. ***Little finger bending.*** The dominant gene *B* causes the distal segment of the little finger to bend laterally. The genotype *bb* results in a straight distal segment.

5. ***Double-jointed thumbs.*** The dominant gene *J* results in loose ligaments which allow the thumb to be bent out of the constricted orientation caused by the recessive genotype *jj*.

6. ***Widow's peak.*** The dominant gene *W* causes the hairline to extend caudally in the midline of the forehead. The recessive genotype *ww* results in a straight hairline.

7. ***Rh factor.*** A dominant gene *Rh* results in the presence of the Rh antigen on red blood cells. This antigen is not present with the recessive genotype *rhrh*. Use the results obtained in Exercise 16.I.4 to determine your phenotype. What additional information would you need to determine your complete genotype if you have the dominant phenotype for these traits? ______________________

____

____

ANSWER THE LABORATORY REPORT QUESTIONS AT THE END OF THE EXERCISE

Student ______________________ Date ______________________

Laboratory Section ______________________ Score/Grade ______________________

SECTION G. GENETICS EXERCISES

PKU Screening

1. ________ negative—cream color
   ________ 15 mg%—light green
   ________ 40 mg%—medium green
   ________ 100 mg%—dark green

PTC Inheritance

2. ________ bitter taste
   ________ sweet taste
   ________ negative (no taste)

Corn Genetics

3. Ratio ________ monohybrid cross
   Ratio ________ test cross
   Ratio ________ dihybrid cross
   Ratio ________ trihybrid cross

Color Blindness

	Male students	Female students
4. Red color blind	________	________
Green color blind	________	________

Mendelian Laws of Inheritance

5. Record the results of tossing a nickel and a penny together 10 times.

	Female (nickel)	**Male (penny)**	**1**	**2**	**3**	**4**	**5**	**Total**	**Class total**
Dominant offspring	A Heads	A Heads							
	A Heads	a Tails							
	a Tails	A Heads							
Recessive offspring	a Tails	a Tails							
Ratio dominant to recessive									

6. Complete the following monohybrid cross. Fill in genotypes (within circles) and phenotypes (under circles).

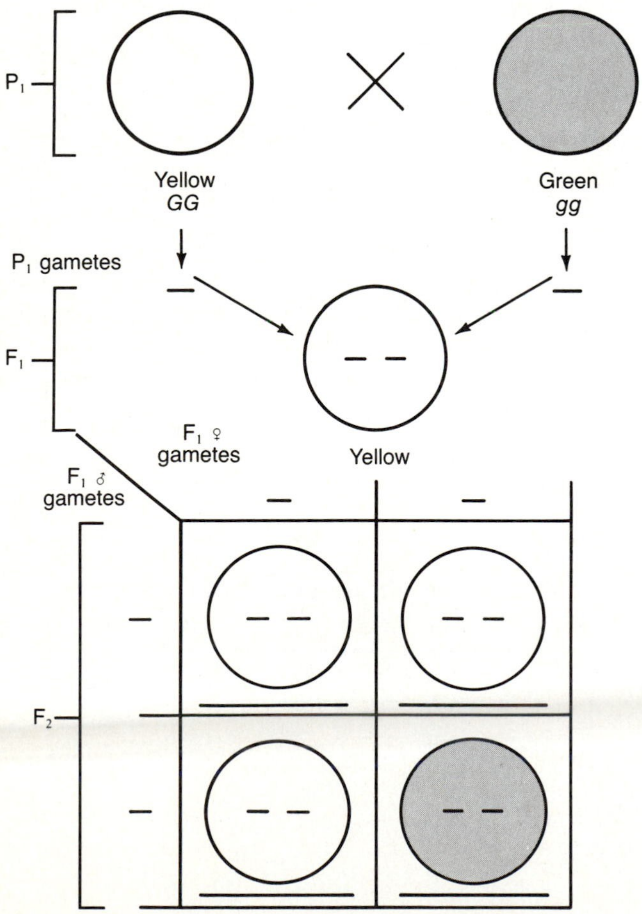

Monohybrid cross in the garden pea (*Pisum sativum*). G = allele for yellow, g = allele for green, P_1 = parental generation, F_1 = first filial generation, F_2 = second filial generation.

7. What is the phenotype ratio of the F_2 generation? ________ yellow/ ________ green.

8. Complete the following dihybrid cross. Fill in genotypes (within circles) and phenotypes (under circles).

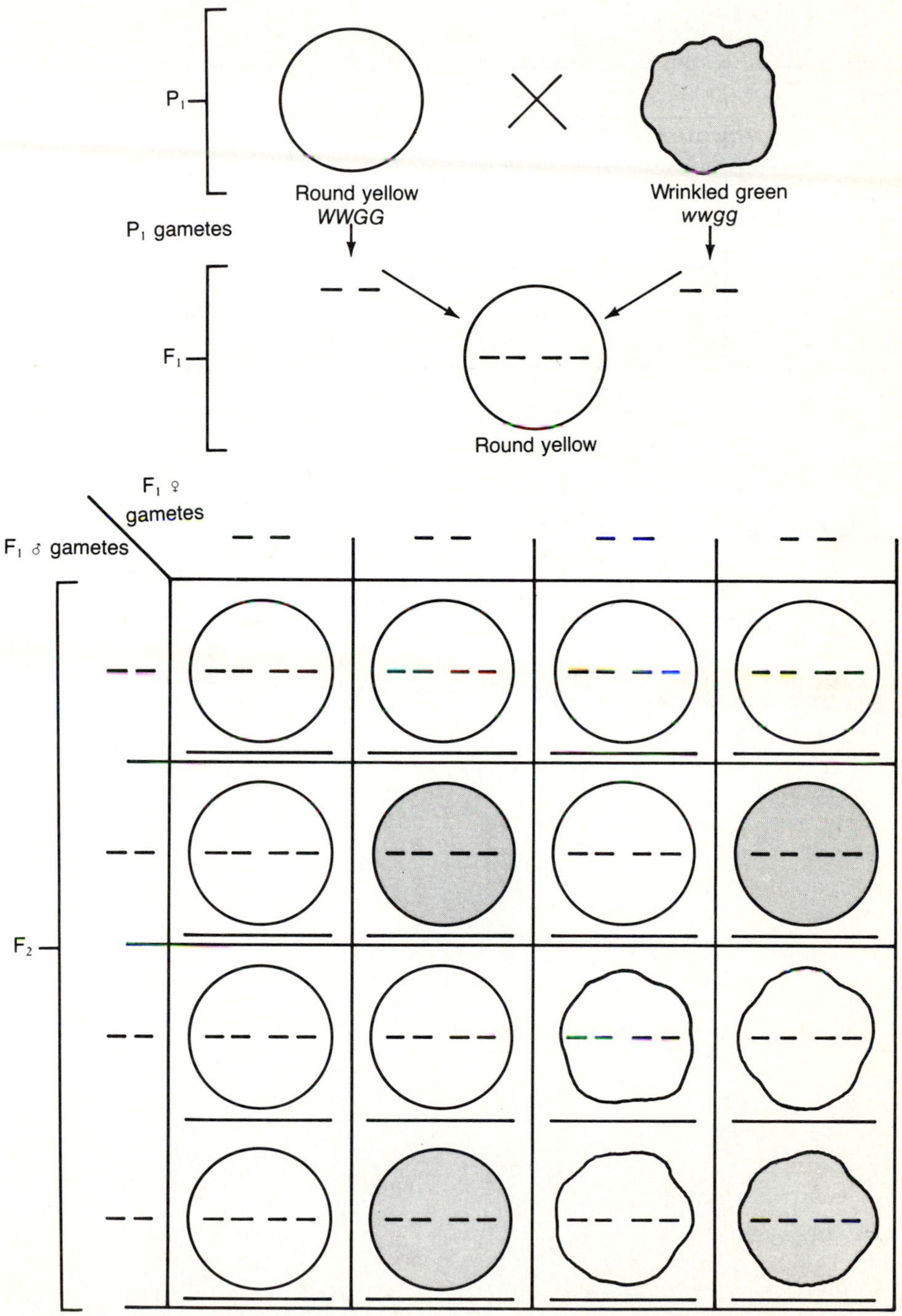

Dihybrid cross in the garden pea *(Pisum sativum)*. *W* = allele for round, *w* = allele for wrinkled, *G* = allele for yellow, *g* = allele for green.

9. What is the phenotype ratio of the F_2 generation?

__________ round yellow/ __________ wrinkled yellow/ __________ round green/

__________ wrinkled green.

10. Observing Phenotypes

Trait	**Phenotype**	**Genotype**
Ear lobes		
Tongue rolling		
Hair whorl		
Double-jointed thumbs		
Widow's peak		
Rh factor		

LABORATORY REPORT QUESTIONS

GENETICS 26

Student ______________________ **Date** ______________________

Laboratory Section ______________________ **Score/Grade** ______________________

PART I. MULTIPLE CHOICE

_____ 1. Using the symbols *Aa* to represent genes, which of the following is true? (a) the trait is homozygous for the dominant characteristic (b) the trait is homozygous for the recessive characteristic (c) the trait is heterozygous (d) *X*-linked inheritance is in operation

_____ 2. Which statement concerning the normal inheritance of sex is correct? (a) all zygotes contain a *Y* chromosome (b) some ova contain a *Y* chromosome, (c) all ova and all sperm contain an *X* chromosome (d) all ova have an *X* chromosome, some sperm have an *X* chromosome, and some sperm have a *Y* chromosome

_____ 3. The genotype that will express characteristics associated with hemophilia (assume that *H* represents the gene for normal blood) is (a) X^HX^h (b) X^hY (c) X^HX^H (d) X^HY

_____ 4. The exact position of a gene on a chromosome is called the (a) homologue (b) locus (c) triad (d) allele

PART II. COMPLETION

5. When the same genes appear on homologous chromosomes, as in *PP* or *pp*, the individual is said to be ______________________ for the trait.

6. If different genes appear on homologous chromosomes, as in *Pp*, the individual is ______________________ for the trait.

7. Genetic composition expressed in the body or morphologically is called the body's ______________________ .

8. The device that helps one visualize all the possible combinations of male and female gametes is called the ______________________ .

9. The twenty-third pair of human chromosomes are the sex chromosomes. All of the other pairs of chromosomes are called ______________________ .

10. Color blindness is an inherited trait that is specifically called ______________________ trait.

11. The recognized "father" of genetics is ______________________ .

12. A genetic cross that involves only one pair of alleles (or traits) is called a(n) ______________________ cross.

13. Passage of hereditary traits from one generation to another is called ______________________ .

14. The genetic makeup of an individual is called the person's ______________________ .

15. One of the many alternative forms of a gene is called its ______________________ .

APPENDIX A
SOME IMPORTANT UNITS OF MEASUREMENT

ENGLISH UNITS OF MEASUREMENT

Fundamental or derived unit		Units and equivalents
Length	12 inches (in.) =	1 foot (ft) = 0.333 yard (yd)
	3 ft =	1 yd
	1760 yd =	1 mile (mi)
	5280 ft =	1 mi
Mass	1 ounce (oz) =	28.35 grams (g); 1 g = 0.0353 oz
	1 pound (lb) =	453g = 16 oz; 1 kilogram (kg) = 2.205 lb
	1 ton =	2000 lb = 907 kg
Time	1 second (sec) =	1/86 400 of a mean solar day
	1 minute (min) =	60 sec
	1 hour (hr) =	60 min = 3600 sec
	1 day =	24 hr = 1440 min = 86 400 sec
Volume	1 fluid dram (fl dr) =	0.125 fluid ounce (fl oz)
	1 fl oz =	8 fl dr = 0.0625 quart (qt) = 0.008 gallon (gal)
	1 qt =	256 fl dr = 32 fl oz = 2 pints (pt) = 0.25 gal
	1 gal =	4 qt = 128 fl oz = 1024 fl dr

METRIC UNITS OF LENGTH AND SOME ENGLISH EQUIVALENTS

Metric unit	Meaning of prefix	Metric equivalent	English equivalent
1 kilometer (km)	kilo = 1000	1000 m	3280.84 ft or 0.62 mi; 1 mi = 1.61 km
1 hectometer (hm)	hecto = 100	100 m	328 ft
1 dekameter (dam)	deka = 10	10 m	32.8 ft
1 meter (m)	Standard unit of length		39.37 in. or 3.28 ft or 1.09 yd
1 decimeter (dm)	deci = 1/10	0.1 m	3.94 in.
1 centimeter (cm)	centi = 1/100	0.01 m	0.394 in.; 1 in. = 2.54 cm
1 millimeter (mm)	milli = 1/1000	0.001 m = 1/10 cm	0.0394 in.

1 micrometer (μm) [formerly micron (μ)]	micro = 1/1 000 000	0.000 001 m = 1/10 000 cm	3.94×10^{-5} in.
1 nanometer (nm) [Formerly millimicrons (mμ)]	nano = 1/1 000 000 000	0.000 000 001 m = 1/10 000 000 cm	3.94×10^{-8} in.
1 angstrom (Å)		0.000 000 000 1 m = 1/100000000 cm	3.94×10^{-9} in.

TEMPERATURE

Unit	K	°F	°C
1 degree Kelvin (K)	1	9/5(K) − 459.7	K + 273.16*
1 degree Fahrenheit (°F)	5/9(°F) + 255.4	1	5/9(°F − 32)
1 degree Celsius (°C)	°C − 273	9/5(°C) + 32	1

VOLUME

Unit	ml	cm^3	qt	oz
1 milliliter (ml)	1	1	1.06×10^{-3}	3.392×10^{-2}
1 cubic centimeter (cm^3)	1	1	1.06×1^{-3}	3.392×10^{-2}
1 quart (qt)	943	943	1	32
1 fluid ounce (fl oz)	29.5	29.5	3.125×10^{-2}	1

*Absolute zero (K) = −273.16°C.

APPENDIX B
PERIODIC TABLE OF THE ELEMENTS

KEY

6 ← Atomic Number C ← Symbol 12.01 ← Atomic Weight Carbon ← Name

1 H 1.0080 Hydrogen																	2 He 4.003 Helium
3 Li 6.940 Lithium	4 Be 9.013 Berilium											5 B 10.82 Boron	6 C 12.011 Carbon	7 N 14.008 Nitrogen	8 O 16.000 Oxygen	9 F 19.00 Fluorine	10 Ne 20.183 Neon
11 Na 22.991 Sodium	12 Mg 24.32 Magnesium											13 Al 26.98 Aluminum	14 Si 28.09 Silicon	15 P 30.975 Phosphorus	16 S 32.066 Sulfur	17 Cl 35.457 Chlorine	18 Ar 39.944 Argon
19 K 39.100 Potassium	20 Ca 40.08 Calcium	21 Sc 44.96 Scandium	22 Ti 47.90 Titanium	23 V 50.95 Vanadium	24 Cr 52.01 Chromium	25 Mn 54.94 Manganese	26 Fe 55.85 Iron	27 Co 58.94 Cobalt	28 Ni 58.71 Nickel	29 Cu 63.54 Copper	30 Zn 65.38 Zinc	31 Ga 69.72 Gallium	32 Ge 72.60 Germanium	33 As 74.91 Arsenic	34 Se 78.96 Selenium	35 Br 79.916 Bromine	36 Kr 83.80 Krypton
37 Rb 85.48 Rubidium	38 Sr 87.63 Strontium	39 Y 88.92 Yttrium	40 Zr 91.22 Zirconium	41 Nb 92.91 Niobium	42 Mo 95.95 Molybdenum	43 Tc (99) Technetium	44 Ru 101.1 Ruthenium	45 Rh 102.91 Rhodium	46 Pd 106.4 Palladium	47 Ag 107.880 Silver	48 Cd 112.41 Cadmium	49 In 114.82 Indium	50 Sn 118.70 Tin	51 Sb 121.76 Antimony	52 Te 127.61 Tellurium	53 I 126.91 Iodine	54 Xe 131.30 Xenon
55 Cs 132.91 Cesium	56 Ba 137.36 Barium	57 La 138.92 Lanthanum	72 Hf 178.50 Hafnium	73 Ta 180.95 Tantalum	74 W 183.86 Wolfram	75 Re 186.22 Rhenium	76 Os 190.2 Osmium	77 Ir 192.2 Iridium	78 Pt 195.09 Platinum	79 Au 197.0 Gold	80 Hg 200.61 Mercury	81 Tl 204.39 Thallium	82 Pb 207.21 Lead	83 Bi 209.00 Bismuth	84 Po (210) Polonium	85 At (210) Astatine	86 Rn (222) Radon
87 Fr (223) Francium	88 Ra (226) Radium	89 Ac (227) Actinium	104 (Russian Proposal Unofficial)														
				58 Ce 140.13 Cerium	59 Pr 140.92 Praseodymium	60 Nd 144.27 Neodymium	61 Pm (147) Promethium	62 Sm 150.35 Samarium	63 Eu 152.0 Europium	64 Gd 157.26 Gadolinium	65 Tb 158.93 Terbium	66 Dy 162.51 Dysprosium	67 Ho 164.94 Holmium	68 Er 167.27 Erbium	69 Tm 168.94 Thulium	70 Yb 173.04 Ytterbium	71 Lu 174.99 Lutetium
				90 Th (232) Thorium	91 Pa (231) Protactinium	92 U 238.07 Uranium	93 Np (237) Neptunium	94 Pu (242) Plutonium	95 Am (243) Americium	96 Cm (247) Curium	97 Bk (249) Berkelium	98 Cf (251) Californium	99 Es (254) Einsteinium	100 Fm (253) Fermium	101 Md (256) Mendelevium	102 No (253) Nobelium	103 Lw 257 Lawrencium

APPENDIX C EPONYMS USED IN THIS LABORATORY MANUAL

Eponym	Current terminology
Achilles tendon	calcaneal tendon
Adam's apple	thyroid cartilage
ampulla of Vater (VA-ter)	hepatopancreatic ampulla
Bartholin's (BAR-tō-linz) gland	greater vestibular gland
Billroth's (BIL-rōtz) cord	splenic cord
Bowman's (BŌ-manz) capsule	glomerular capsule
Bowman's (BŌ-manz) gland	olfactory gland
Broca's (BRŌ-kaz) area	motor speech area
Brunner's (BRUN-erz) gland	duodenal gland
bundle of His (HISS)	atrioventricular (AV) bundle
canal of Schlemm (SHLEM)	scleral venous sinus
circle of Willis (WIL-is)	cerebral arterial circle
Cooper's (KOO-perz) ligament	suspensory ligament of the breast
Cowper's (KOW-perz) gland	bulbourethral gland
crypt of Lieberkühn (LĒ-ber-kyoon)	intestinal gland
duct of Rivinus (re-VĒ-nus)	lesser sublingual duct
duct of Santorini (san′-tō-RĒ-nē)	accessory duct
duct of Wirsung (VĒR-sung)	pancreatic duct
end organ of Ruffini (roo-FĒ-nē)	type II cutaneous mechanoreceptor
Eustachian (yoo-STĀ-kē-an) tube	auditory tube
Fallopian (fal-LŌ-pē-an) tube	uterine tube
gland of Zeis (ZĪS)	sebaceous ciliary gland
Golgi (GOL-jē) tendon organ	tendon organ
Graafian (GRAF-ē-an) follicle	vesicular ovarian follicle
Hassall's (HAS-alz) corpuscle	thymic corpuscle
Haversian (ha-VĒR-shun) canal	central canal
Haversian (ha-VĒR-shun) system	osteon
interstitial cell of Leydig (LĪ-dig)	interstitial endocrinocyte
islet of Langerhans (LANG-er-hanz)	pancreatic islet
Kupffer's (KOOP-ferz) cells	stellate reticuloendothelial cell
loop of Henle (HEN-lē)	loop of the nephron
Malpighian (mal-PIG-ē-an) corpuscle	splenic nodule
Meibomian (mī-BŌ-mē-an) gland	tarsal gland
Meissner's (MĪS-nerz) corpuscle	corpuscle of touch
Merkel's (MER-kelz) disc	tactile disc
Müller's (MIL-erz) duct	paramesonephric duct
Nissl (NISS-l) bodies	chromatophilic substance

Eponym	Current terminology
node of Ranvier (ron-VĒ-ā)	neurofibral node
organ of Corti (KOR-tē)	spiral organ
Pacinian (pa-SIN-ē-an) corpuscle	lamellated corpuscle
Peyer's (PĪ-erz) patches	aggregated lymphatic follicles
plexus of Auerbach (OW-er-bak)	myenteric plexus
plexus of Meissner (MĪS-ner)	submucous plexus
pouch of Douglas	rectouterine pouch
Purkinje (pur-KIN-jē) fiber	conduction myofiber
Rathke's (rath-KĒZ) pouch	hypophyseal pouch
Schwann (SCHVON) cell	neurolemmocyte
Sertoli (ser-TŌ-lē) cell	sustentacular cell
Skene's (SKĒNZ) gland	paraurethral gland
sphincter of Oddi (OD-dē)	sphincter of the hepatopancreatic ampulla
Stensen's (STEN-senz) duct	parotid duct
Volkmann's (FŌLK-manz) canal	perforating canal
Wharton's (HWAR-tunz) duct	submandibular duct
Wharton's (HWAR-tunz) jelly	mucous connective tissue
Wormian (WER-mē-an) bone	sutural bone

FIGURE CREDITS

1.1 Courtesy of Bausch & Lomb, Rochester, NY.

3.5 (a)–(f) By James R. Smail, Macalester College.

4.1 (a), (b), and (e) By Matt Iacobino and Gerard J. Tortora. (c), (d), (f), and (g) By James R. Smail and Russell A. Whitehead, Macalester College.

4.3 (a), (d), (f), (g), and (h) By James R. Smail and Russell A. Whitehead, Macalester College. (b) and (c) Courtesy of Ward's Natural Science Establishment, Inc., Rochester, NY. (e) By Matt Iacobino and Gerard J. Tortora.

5.2 From *Tissues and Organs: A Text-Atlas of Scanning Electron Microscopy* by Richard G. Kessel and Randy H. Kardon. Copyright © 1979, W. H. Freeman and Company.

6.1 (b) From *The Musculoskeletal System in Health and Disease* by Cornelius Rosse and D. Kay Clawson. Courtesy of Lippincott Publishers, 1980. (c) and (d) Courtesy of Fisher Scientific, an Allied Company, and S.T.E.M. Laboratories, copyright © 1975.

6.2 (a) Courtesy of Fisher Scientific, an Allied Company, and S.T.E.M. Laboratories, copyright © 1975. (b) From *Tissues and Organs: A Text-Atlas of Scanning Electron Microscopy* by Richard G. Kessel and Randy H. Kardon. Copyright © 1979, W. H. Freeman and Company.

7.1 (a), (b), (d), (e), 7.2, 7.3, 7.4, 7.7(g), 7.8(a), 7.9–7.12 Courtesy of Matt Iacobino.

7.8 (b) Courtesy of Mayo Foundation, Rochester, MN.

8.3 (c) By John Eads.

8.4 (a)–(e) Copyright © 1983 by Gerard J. Tortora, courtesy of Matt Iacobino and Lynne Borghesi. (f), (g), and (h) Courtesy of Matt Iacobino. (i) and (j) Biomedical Graphics Department, University of Minnesota Hospitals.

9.1, 9.3, 9.4 By James R. Smail and Russell A. Whitehead, Macalester College.

9.7 Courtesy of Narco BioSystems, Inc., Houston, TX.

11.2–11.8, 11.9(a), 11.10–11.12 Biomedical Graphics Department, University of Minnesota Hospitals.

11.9 (b) Courtesy of O. Richard Johnson.

13.13 Adapted from *Fundamentals of Neurophysiology*, ed. 2, R. F. Schmidt, ed. Copyright © 1978, Springer-Verlag, New York.

13.14 (b) Courtesy of Tektronix, Inc., Beaverton, OR.

14.6 By Andrew Kuntzman, Wright State University.

14.7 (b) By James R. Smail and Russell A. Whitehead, Macalester College.

14.10 Biomedical Graphics Department, University of Minnesota Hospitals.

14.18 By Andrew Kuntzman, Wright State University.

14.19 Copyright © 1983 by Gerard J. Tortora, courtesy of James Borghesi.

15.2 By James R. Smail and Russell A. Whitehead, Macalester College.

15.3 By James R. Smail, Macalester College.

15.4–15.6 By Andrew Kuntzman, Wright State University.

16.3 Courtesy of Becton-Dickinson, Division of Becton, Dickenson and Company.

16.4 Modified from *RBC Determination Methods* and *WBC Determination for Manual Methods*, Becton-Dickinson, Division of Becton, Dickinson and Company.

16.6, 16.8 Courtesy of Fisher Scientific, an Allied Company.

16.7 By James R. Smail, Macalester College.

16.9 Courtesy of Lenni Patti.

16.10, 16.11 Courtesy of Cambridge Instruments, Inc., Optical Systems Division, Buffalo, NY.

17.2 From *Photographic Atlas of the Human Body*, Second Edition, by Chihiro Yokoshi and Johannes Rohen. Copyright © 1978, Igaku-Shoin Ltd., Tokyo.

17.3 (b) By John Eads.

18.1 (a), (b), and (c) By James R. Smail and Russell A. Whitehead, Macalester College. (d) From *Tissues and Organs: A Text-Atlas of Scanning Electron Microscopy* by Richard G. Kessel and Randy H. Kardon. Copyright © 1979, W. H. Freeman and Company.

19.8 Copyright © by Gerard J. Tortora, courtesy of Geraldine C. Tortora and Matt Iacobino.

21.3 Copyright © by Gerard J. Tortora, courtesy of Lynne Tortora.

21.5 From *Photographic Atlas of the Human Body*, Second Edition, by Chihiro Yokoshi and Johannes Rohen. Copyright © 1978, Igaku-Shoin Ltd., Tokyo.

21.7 By Andrew Kuntzman, Wright State University.

21.9 From *Tissues and Organs: A Text-Atlas of Scanning Electron Microscopy* by Richard G. Kessel and Randy H. Kardon. Copyright © 1979, W. H. Freeman and Company.

21.15 Courtesy of Lenni Patti.

21.16 Courtesy of Kinetics, Upper Saddle River, NJ.

21.17 Courtesy of Warren E. Collins, Inc., Braintree, MA.

22.3, 22.6, 22.8, 22.10, 22.11(a), (b), 22.13(b) By James R. Smail and Russell A. Whitehead, Macalester College.

22.11(c), 22.13(b) By Andrew Kuntzman, Wright State University.

23.2 (b) From *Photographic Atlas of the Human Body*, Second Edition, by Chihiro Yokoshi and Johannes Rohen. Copyright © 1978, Igaku-Shoin Ltd., Tokyo.

23.5, 23.8 By James R. Smail and Russell A. Whitehead, Macalester College.

23.10, 24.2(b) By Andrew Kuntzman, Wright State University.

23.12 By James R. Smail, Macalester College.

24.4, 24.5 By James R. Smail and Russell A. Whitehead, Macalester College.

INDEX

H

I

J

K

L

M

S

T

Student Survey

Tortora, Tallitsch, and Anagnostakos
Laboratory Exercises in Anatomy and Physiology with Cat Dissections, Third Edition

Students, send us your ideas!

The author and the publisher want to know how well this book served you and what can be done to improve if for those who will use it in the future. By completing and returning this questionnaire, you can help us develop better textbooks. We value your opinion and want to hear your comments. Thank you.

Your name (optional) ____________________ School ____________

Your mailing address ______________________________________

City ____________________ State ____________ ZIP ____________

Instructor's name (optional) ____________________ Course title ____________

1. How does this manual compare with other manuals you have used? (Check one)

   ☐ Superior ☐ Better than most ☐ Comparable ☐ Not as good as most

2. Circle those chapters you especially liked:

   Chapters: 1 2 3 4 5 6 7 8 9 10 11 12 13 14 15 16 17 18 19 20 21 22 23 24 25 26

   Comments:

3. Circle those chapers you think could be improved:

   Chapters: 1 2 3 4 5 6 7 8 9 10 11 12 13 14 15 16 17 18 19 20 21 22 23 24 25 26

   Comments:

4. Please rate the following (Check one for each)

	Excellent	Good	Average	Poor
Logical organization	( )	( )	( )	( )
Readability of material	( )	( )	( )	( )
General layout and design	( )	( )	( )	( )
Match with instructor's course organization	( )	( )	( )	( )
Illustrations that clarify the manual	( )	( )	( )	( )
Up-to-date treatment of subject	( )	( )	( )	( )
Explanation of difficult concepts	( )	( )	( )	( )
Selection of topics in the manual	( )	( )	( )	( )

5. Were instructions for exercises clear and complete ? ☐ Yes ☐ No

Comments: ____

____

6. How useful were these activities to you for learning the course material?

Labeling drawings	☐ Very helpful	☐ Helpful	☐ Not helpful
Laboratory reports	☐ Very helpful	☐ Helpful	☐ Not helpful

7. Please list any chapters that you instructor did not assign. ______________________

____

8. What additional topics did your instructor discuss that were not covered in the manual?

____

9. Did you buy this manual new or used?	☐ New	☐ Used
Do you plan to keep the book or sell it?	☐ Keep it	☐ Sell it
Do you think you instructor should continue to assign it?	☐ Yes	☐ No

10. Author and title of the textbook assigned for the lecture section of the course:

____

Kindly rate the coordination of this laboratory manual with your textbook:

☐ Excellent ☐ Good ☐ Average ☐ Poor

11. After taking the course, are you interested in taking more courses in this filed? ☐ Yes ☐ No

12. GENERAL COMMENTS

May we quote you in our advertising? ☐ Yes ☐ No

Please remove this page and mail to:

Gerard J. Tortora
Natural Sciences and Mathematics, S229
Bergen Community College
400 Paramus Road
Paramus, NJ 07652

THANK YOU!